7 Steps to Success

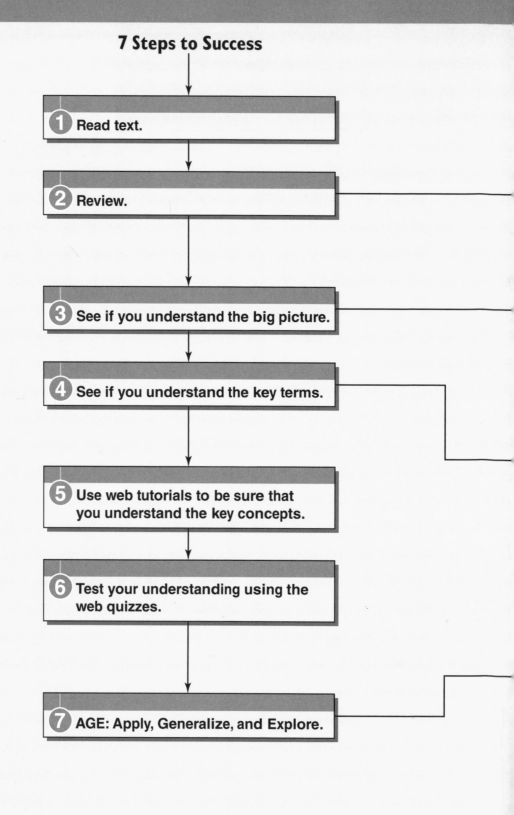

1. Read text.

2. Review.

3. See if you understand the big picture.

4. See if you understand the key terms.

5. Use web tutorials to be sure that you understand the key concepts.

6. Test your understanding using the web quizzes.

7. AGE: Apply, Generalize, and Explore.

Review chapter summary.

Review tables.

See if you can answer the questions in the "key questions" section of the web chapter summary.

Write your own chapter summary. Then compare it to the web chapter summary.

Make your own concept map of the key terms, and compare it to the map on the chapter's website.

Write your own definitions of the key terms, and compare your definitions to the chapter's web glossary.

Do interactive web exercises.

Do end-of-chapter exercises.

Read web appendixes.

Follow links to interesting applications of the chapter material.

Follow links to tips and resources that will help you use the chapter material.

EIGHTH EDITION

RESEARCH DESIGN

Explained

Mark L. Mitchell

Janina M. Jolley

WADSWORTH
CENGAGE Learning·

Australia • Brazil • Japan • Korea • Mexico • Singapore • Spain • United Kingdom • United States

WADSWORTH
CENGAGE Learning·

Research Design Explained, Eighth Edition, International Edition
Mark L. Mitchell and Janina M. Jolley

Senior Publisher: Linda Ganster

Publisher: Jon-David Hague

Acquiring Sponsoring Editor: Timothy Matray

Editorial Assistant: Lauren K. Moody

Marketing Program Manager: Talia Wise

Art Direction, Production Management, and Composition: PreMediaGlobal

Manufacturing Planner: Karen Hunt

Rights Acquisitions Specialist
(Text and Image): Roberta Broyer

Cover Image: © Pichugin Dmitry/
Shutterstock.com

International Edition:

ISBN-13: 978-1-133-49074-6

ISBN-10: 1-133-49074-3

Cengage Learning International Offices

Asia
www.cengageasia.com
tel: (65) 6410 1200

Australia/New Zealand
www.cengage.com.au
tel: (61) 3 9685 4111

Brazil
www.cengage.com.br
tel: (55) 11 3665 9900

India
www.cengage.co.in
tel: (91) 11 4364 1111

Latin America
www.cengage.com.mx
tel: (52) 55 1500 6000

UK/Europe/Middle East/Africa
www.cengage.co.uk
tel: (44) 0 1264 332 424

**Represented in Canada by
Nelson Education, Ltd.**
www.nelson.com
tel: (416) 752 9100/(800)
668 0671

Cengage Learning is a leading provider of customized learning solutions with office locations around the globe, including Singapore, the United Kingdom, Australia, Mexico, Brazil, and Japan. Locate your local office at: **www.cengage.com/global**.

For product information and free companion resources:
www.cengage.com/international.

Visit your local office: **www.cengage.com/global**.

Printed in the United States of America
1 2 3 4 5 6 7 16 15 14 13 12

We dedicate this book to our parents—Anna, Glen, Zoë, and Neal—and to our daughter, Moriah.

BRIEF CONTENTS

CONTENTS

4 Reading, Reviewing, and Replicating Research   111

5 Measuring and Manipulating Variables: Reliability and Validity   143

9 Internal Validity 330

This book focuses on two goals: (1) helping students evaluate the internal, external, and construct validity of studies and (2) helping students write a good research proposal. To accomplish these goals, we use the following methods:

- We use numerous, clear examples—especially for concepts with which students have trouble, such as statistical significance and interactions.
- We focus on important, fundamental concepts; show students why those concepts are important; relate those concepts to what students already know; and directly address common misunderstandings about those concepts.
- We explain the logic behind research design so that students know more than just terminology—they learn how to think like research psychologists.
- We explain statistical concepts (not computations) because (a) students seem to have amnesia for what they learned in statistics class, (b) some understanding of statistics is necessary to understand journal articles, and (c) statistical issues need to be considered before doing research.

FLEXIBLE ORGANIZATION

We know that most professors share our goals of teaching students to be able to read, evaluate, defend, and produce scientific research. We also know that professors differ in how they go about achieving these goals and in the emphasis they place on each of these goals. For example, although about half of all professors of research methods believe that the best way to help students understand design is to cover nonexperimental methods first, about half believe that students must understand experimental methods first. To accommodate professor differences, we have made the chapters relatively self-contained modules. Because each chapter focuses on ethics, construct validity, external validity, and internal validity, it is easy to skip chapters or to cover them in different orders. For example, the first chapter that some professors assign is the last.

CHANGES TO THE EIGHTH EDITION

After the last edition of this book was published, two terrible things happened. First, APA came out with a new style manual. Second, students continued to make mistakes on tests and papers. In response, we took two steps. First, we revised the text to make the chapter on how to write research proposals and reports (Chapter 15), the APA report and proposal checklist (Appendix A), the sample paper (Appendix B), and other sections of the text consistent with APA's new *Publication Manual*. Second, we analyzed the mistakes students made and then revised the text to make it less likely that future students would make those errors.

THE STUDENT WEBSITE

The student website includes many goodies that, together with the book, make it almost impossible for a diligent student to get lower than a "C" in the course. For each chapter, the site contains a concept map, a crossword puzzle, learning objectives, a pretest and a posttest quiz for each chapter based on those learning objectives, and answers to the text's even-numbered exercises.

THE PROFESSOR'S WEBSITE

The professor site has PowerPoint® lectures, interactive PowerPoint chapter quizzes (with explanations of both the right and the wrong answers), chapter summaries, learning objectives, crossword puzzles, demonstrations, and links to videos. In addition, for each chapter, we have a list of articles to assign, a summary of each article, and a "reading guide"—a handout that defines terms, explains concepts, and translates particularly tough passages—so that students can read and understand those articles.

ACKNOWLEDGMENTS

Writing *Research Design Explained* was a monumental task that required commitment, love, effort, and a high tolerance for frustration. If it had not been for the support of our friends, family, publisher, and students, we could not have met this challenge.

Robert Tremblay, a Boston journalist, and Lee Howard, a Connecticut journalist, have our undying gratitude for the many hours they spent critiquing the first six editions of this book. We are also grateful to Darlynn Fink, an English professor and Jamie Phillips, a philosophy professor, for their work on the seventh edition, as well as to the folks at Cengage for sharing and nurturing our vision. In addition to thanking Bob, Lee, Darlynn, Jamie, and Cengage, we need to thank three groups of dedicated reviewers, all of whom were actually coauthors of this text.

First, we would like to thank the competent and conscientious professors who shared their insights with us. We are grateful to the reviewers, whose constructive comments strengthened the book: Jeff Adams, Trent University; Anne DePrince, University of Denver; Karen Fragedakis, Campbell University; Glenn Geher, SUNY–New Paltz; Paula Goolkasian, University of North Carolina–Charlotte; Shelia Kennison, Oklahoma State University; Eugene Packer, William Paterson University; Jodie Royan, University of Victoria; and Donna Stuber-McEwen, Friends University. In addition, we thank the reviewers of past editions: Ruth Ault, Davidson College; Louis Banderet, Quinsigamond Community College; James H. Beaird, Western Oregon State College; John P. Brockway, Davidson College; Tracy L. Brown, University of North Carolina–Asheville; Edward Caropreso, Clarion University; Walter Chromiak, Dickinson College; James R. Council, North Dakota State University; Helen J. Crawford, University of Wyoming; Raymond Ditrichs, Northern Illinois University; Patricia Doerr, Louisiana State University; Linda Enloe, Idaho State University; Mary Ann Foley, Skidmore College; George Goedel, Northern Kentucky University; George L. Hampton III, University of Houston–Downtown; Robert Hoff, Mercyhurst College; Lynn Howerton, Arkansas State University; John C. Jahnke, Miami University; Randy Jones, Utah State University; Sue Kraus, Fort Lewis College; Scott A. Kuehn, Clarion University; R. Eric Landrum, Boise State University; Kenneth L. Leicht, Illinois State University; Charles A. Levin, Baldwin-Wallace College; Joel Lundack, Peru State College; Steven Meier, University of Idaho; Charles Meliska, University of Southern Indiana; Kenneth B. Melvin, University of Alabama; Stephen P. Mewaldt, Marshall University; John Nicoud, Marion College of Fond du Lac; Jamie Phillips, Clarion University; David Pittenger, Marietta College; Carl Ratner, Humboldt State University; Ray Reutzel, Brigham Young University; Anrea Richards, University of California–Los Angeles; Margaret Ruddy, Trenton State College; James J. Ryan, University of Wisconsin–La Crosse; Rick Scheidt, Kansas State University; Gerald Sparkman, University of Rio Grande; Sylvia Stalker, Clarion University; Ann Stearns, Clarion University; Sandra L. Stein, Rider College; Ellen P. Susman, Metropolitan State College of Denver; Russ A. Thompson, University of Nebraska; Benjamin Wallace, Cleveland State University; Paul Wellman, Texas A&M University; and Christine Ziegler, Kennesaw State College.

Second, we would like to thank our student reviewers, especially Susanne Bingham, Mike Blum, Shannon Edmiston, Chris Fenn, Jess Frederick, Kris Glosser, Melissa Gregory, Barbara Olszanski, Shari Poza, Rosalyn Rapsinski, Katelin Speer, and Melissa Ustik.

Third, we would like to thank the English professors who critiqued the previous editions of our book: William Blazek, Patrick McLaughlin, and John Young. In addition to improving the writing style of the book, they also provided a valuable perspective—that of the intelligent, but naïve to psychology, reader.

Finally, we would like to thank our daughter Moriah for allowing us the time to complete this project.

After graduating summa cum laude from Washington and Lee University, Mark L. Mitchell received his MA and PhD degrees in psychology at The Ohio State University. He is currently a professor at Clarion University.

Janina M. Jolley graduated with "Great Distinction" from California State University at Dominguez Hills and earned her MA and PhD in psychology from The Ohio State University. She is currently an executive editor of *The Journal of Genetic Psychology* and *Genetic Psychology Monographs.* Her first book was *How to Write Psychology Papers: A Student's Survival Guide for Psychology and Related Fields,* which she wrote with J. D. Murray and Pete Keller.

In addition to working on this book for more than 100 dog years, Dr. Mitchell and Dr. Jolley coauthored *Developmental Psychology: A Topical Approach.* More recently, they collaborated with Robert O'Shea to write *Writing for Psychology: A Guide for Students* (4th ed.).

Dr. Mitchell and Dr. Jolley are married to research, teaching, and each other—not necessarily in that order. You can write to them at the Department of Psychology, Clarion University, Clarion, PA 16214, or send e-mail to them at either mmitchell@clarion.edu or jjolley@clarion.edu.

Science, Psychology, and You

The whole of science is nothing more than a refinement of everyday thinking.
—Albert Einstein

There are in fact two things, science and opinion; the former produces knowledge, the latter ignorance.
—Hippocrates

CHAPTER OVERVIEW

They lived in a time when scientific giants like Galileo and Newton made tremendous discoveries. Yet, the people of that scientific age executed thousands of women for being witches (Robinson, 1997).

We live in a time just after scientific giants like Einstein and Skinner made tremendous discoveries. Yet, the people of this scientific age spend billions of dollars on magical diets, lotions, potions, and pills (Kolata, 2007; Petersen, 2008); believe in psychics, astrologers, and other modern witches; and cause thousands to die by not believing in science (Specter, 2009). In the United States, thousands die because many doctors do not use the best medical treatments available (Gladwell, 2005; Groopman, 2007), many parents do not immunize their children, and many drivers talk on hands-free cell phones (Specter, 2009). In all societies, much of what is done to attack a wide range of problems—from fighting crime to treating the mentally ill—is based on superstition rather than fact.

As you can see, living in a scientific age does not guarantee that a person will engage in scientific rather than superstitious thinking. In fact, living in a scientific age does not even guarantee that a person will know what science is. Partly because many people do not know what science is, many people do not understand that psychology is a science—and even more people do not understand that it being a science is a good thing for both psychology and for the world.

In this chapter, we will help you understand why psychology is a science. We will begin by showing that science is a set of strategies that you can use to think clearly about real-world problems as well as how to learn from the world's worst teacher: experience. Then, you will see why most psychologists believe that psychology is a science—and that if psychology is to continue to help people, it must continue to be a science. Finally, you will see why it is vital that when you become a professional, you embrace science rather than superstition.

WHY PSYCHOLOGY USES THE SCIENTIFIC APPROACH

For thousands of years, people have asked themselves, "Why are people the way they are?" For most of those thousands of years, asking the question did not lead to accurate answers. What people "knew" was incorrect: Much of common sense was common nonsense, and much traditional "wisdom" was superstition. Fortunately, a little more than 100 years ago, a few individuals tried a new and different approach that led to better answers—the **scientific approach**: a way of using unbiased observation to form and test beliefs. As a result, psychology—the science of behavior—was born.

Today, psychologists and other scientists still embrace the scientific method because it is a useful tool for getting accurate answers to important questions. Crime scene investigators (CSIs) use the scientific approach to solve crimes; biologists use the scientific approach to track down genes responsible for inherited disorders; and behavioral scientists use the scientific approach to unravel the mysteries of human behavior.

Science's Characteristics

What is it about the scientific approach that makes it such a useful tool for people who want answers to important questions? As you will soon see, the eight strengths of the scientific approach are that it

1. finds general rules
2. collects objective evidence
3. makes testable statements
4. adopts a skeptical, questioning attitude about all claims
5. remains open-minded about new claims
6. is creative
7. is public
8. is productive

Seeking Simple Rules: Finding Patterns, Laws, and Order

Just as CSIs assume that criminals have motives, scientists assume that events happen for reasons. Just as CSIs have found certain simple motives (e.g., desire for money, jealousy) that can account for many different crimes, scientists have found general rules that allow them to see connections between seemingly disconnected events (e.g., Newton saw that a general principle—gravity—applies to both an apple falling to the ground and to the moon orbiting the earth).

Scientists hope that by finding the general, underlying reasons for events, they will find simplicity, order, and predictability in what often seems a complex, chaotic, and random universe. Thus, contrary to what happens in some science classes, the goal of science is not to make the world seem more confusing and complicated, but instead to make the world more understandable by finding simple rules that describe, predict, and explain behavior.

Objective: Seeing the Real World for What It Is

Scientists must be careful, however, not to let their desire to see the world as a simple and predictable place blind them to reality. (As Albert Einstein said, "Things should be made as simple as possible, but not any simpler.") We all know people who, in their desire to see the world as simple and predictable, "see" rules and patterns that reflect their prejudices and biases, rather than the facts. For example, some people think they can size up a stranger based on the stranger's race, astrological sign, attractiveness, handwriting, body build, or some other superficial characteristic. However, when we objectively test the ability of these people to predict other people's behaviors, they fail miserably—as more than 100 studies have shown (Dawes, 1994).

Scientists believe in objective facts because human history shows that we can't always accept people's subjective opinions. For example, in the 1940s and 1950s, many physicians believed that destroying parts of patients' brains through a procedure known as a lobotomy made patients better. Some surgeons even issued reports about the number of patients who were "cured" and the number who were "better off." However, "cured" and "better off," rather than reflecting observable improvements in patients (e.g., patients leaving the institution and getting jobs), merely reflected the surgeon's subjective judgments (Shorter, 1997). By not testing their beliefs against objective reality, lobotomists continued to delude themselves about lobotomy's effects.

Most of us tend to forget the twin lessons learned from the practice of lobotomy: (a) that the easiest person to fool is yourself and (b) that to avoid being fooled, you must test your beliefs against observable, objective evidence rather than against your subjective impressions. Because many people ignore the possibility that subjective impressions may be biased,

- many students think their biased views of themselves, their relationships, and other groups are accurate
- many lie detector professionals do not realize that another lie detector professional would disagree with their judgment about whether a suspect is lying
- many therapists do not realize that another therapist would disagree with their interpretation of what a client's response to an inkblot means

To avoid being swept away by either unfounded speculations or biased perceptions, scientists tie their beliefs to concrete, observable, physical evidence that skeptics can double-check. Specifically, to evaluate a claim, scientists look at the **objective evidence**: observable evidence that looks the same to any observer, regardless of the observer's beliefs. Because the scientific laws of gravity, magnetism, and operant conditioning are based on objective observations of physical evidence, these laws are not opinions but facts (Ruchlis & Oddo, 1990).

Test and Testability (Correctability)

Although we have attacked *unsupported* speculation, we are not saying that scientists don't speculate—they do. Like CSIs, scientists are encouraged to make bold, specific predictions that may be proven wrong. In reflecting this "no guts, no glory" approach, psychologist Bob Zajonc says, "You don't do

FIGURE **1.1** Scientists are encouraged to test their beliefs.
To the disappointment of many scientists, this cheer did not catch on.

interesting research if you don't take risks. Period. If you prove the obvious, that's not interesting research."

Similarly, although we would argue that unsupported opinions, like unsupported speculations, are of limited value, we are not saying that scientists don't have opinions. Like everyone else, they do. However, unlike almost everyone else, scientists willingly put their opinions to the test—and willingly throw out opinions that fail the test (see Figure 1.1). In short, scientists agree with Samuel Smiles that, "He who never made a mistake never made a discovery" and with Arthur Guiterman that, "Admitting errors clears the score and proves you wiser than before."

On one level, you probably appreciate scientists' willingness to consider that they might be wrong. You may wish other people—especially certain arrogant ones—asked themselves the "Could I be wrong?" question more often.

On another level, you might ask, "Why should I risk finding out that a cherished opinion is wrong?" Scientists have at least two reasons.

First, a major goal of science is to identify which common sense beliefs are common nonsense. You can only do such "myth busting" by putting beliefs to the test.

Second, one of science's major strengths is that its methods allow scientists to learn from mistakes—and to learn from a mistake, you first must know that you made a mistake. Thus, to be a scientist, you do not need to start with intuitively accurate insights into how the world works. You just need to learn when your initial insights are wrong by making *testable (correctable) statements:* statements that may possibly be shown to be wrong.

The goal in making a testable statement is not to make a statement that *will* be wrong, but rather to put yourself in a position so that if you are wrong, you admit it and learn from your mistake. As Peter Doherty (2002, p. 89) writes, "The scientific method requires that you struggle exhaustively to disprove, or falsify, your best ideas. You actually try, again and again, to find the possible flaw in your hypothesis. We scientists are rather accustomed to falling flat on our faces!"

FIGURE **1.2** Nothing tested, nothing gained.

Elmo claims he has no knowledge of ever missing a free throw. Maintaining this delusion has costs: He is out of touch with reality, and he cannot improve his shooting.

You probably still think it is strange to risk being wrong. However, asking "Am I right or wrong?" is better than asking "Am I right or am I right?" because people who ask "Am I right or am I right?"—that is, people who don't make falsifiable statements—never find out that they are wrong and therefore never allow facts to change their beliefs (See Figure 1.2). For example, one individual predicted that the earth would be destroyed in 1998. Has he admitted his mistake? No, he claims that the earth *was* destroyed in 1998: We are now living on an alternate but identical earth, blissfully unaware of our previous world's destruction. We can't prove him wrong.

We can, however, point out that he has made an untestable (uncorrectable, unfalsifiable) statement—a statement that no observation could possibly disprove. That is, we can point out that his prediction, like the prediction that "Our team will win, lose, or tie its next game," is so flexible that it can fit any result. Because his untestable prediction can fit any result, his belief can never be changed, refined, or corrected by the discovery of new facts.

Because untestable statements do not allow scientists to test beliefs, scientists try to avoid them. Specifically, scientists distrust two types of untestable statements: (a) vague statements and (b) after-the-fact explanations.

Vague Statements May Be Untestable. Vague statements are as useless as they are untestable (see Figure 1.3). For example, a CSI who claims that a murderer was born in this galaxy can neither be proven wrong nor given credit for solving the case.

FIGURE **1.3** Untestable statements.
Vague predictions that are never wrong may be useless.

Vague statements, because they contain loopholes that make them untestable (and therefore unfalsifiable), are often used by politicians, advertisers, pseudoscientists (e.g., astrologists, palmreaders), and con artists. For example, suppose a stranger bets you that Wednesday will be a good day—but he doesn't define what a good day is. No matter what happened on Wednesday, the stranger may come back on Thursday, demanding payment because Wednesday, by the stranger's definition, was a good day. Similarly, one of the authors' horoscopes once read: "Take care today not to get involved with someone who is wrong for you and you for him or her. Trouble could result." This horoscope tells us nothing: No matter what happens, the astrologer could claim to have predicted it.

One reason the astrologer could be so slippery is that she used vaguely defined terms. For example, the astrologer does not give us a clue as to what "wrong for you" means. Thus, if trouble had resulted, the astrologer could say, "The person was wrong for you." If trouble had not resulted, the astrologer could say, "The person was right for you." Note that because the horoscope does not tell us anything, it is both unfalsifiable and useless.

One way that scientists avoid making vague statements is by defining their concepts in specific, concrete, objective terms. Scientists do not describe a concept like stress by saying, "I know stress when I see it" or "You know what I mean." Instead, scientists will tell you how they measured it (e.g., counted the number of stressful events a person claimed to have experienced in the last two years, counted the number of errors the person made on a proofreading task, counted how many times the person's heart beat in one minute, measured levels of cortisol in person's saliva) or scientists will tell you how they produced it (e.g., told participants, "If you do not correctly answer these 60 math problems in an hour, you will be given electric shocks"). That is, when defining a concept like "stress," a psychologist will provide an **operational definition**: a specific, objective set of instructions for either (a) how to measure the concept (i.e., what behaviors to count or record to give participants a score) or (b) how to manipulate the concept (i.e., what to do to participants to increase or decrease their levels of that variable).

As you will see in Chapter 5, these operational definitions may range from measuring brain wave activity to scoring a multiple-choice test. Some of these operational definitions will do a good job of capturing the psychological variable they are supposed to measure, whereas others will not be as good. But no matter what the operational definition is, it is one that other scientists can follow. Because the operational definition is an objective recipe that anyone can follow, there is no disagreement about what each participant's score is.

When researchers state their predictions in such clear, concrete, and objective terms, they can objectively determine whether the evidence supports their predictions, no matter what the researchers' biases. For example, even if two researchers expect to find that happy people have lower IQs, those researchers can objectively establish what the real relationship is between scores on a certain happiness test and scores on a certain IQ test. If, contrary to the researchers' predictions, people with higher scores on the happiness test tend to have higher scores on the IQ test, the researchers would admit that their prediction was wrong. If repeating the study in several different ways obtains the same result, the researchers will change their minds about the relationship between happiness and intelligence. Thus, for scientists, seeing becomes believing.

In contrast, quacks often avoid making testable predictions by avoiding operational definitions. For example, no matter what happens, the quack can continue to insist that everyone the quack views as intelligent is also—in the quack's view—unhappy.

As you have seen, by avoiding operational definitions, quacks avoid making testable predictions and therefore never have to admit being wrong. Another way quacks can avoid making testable predictions that might be proven wrong is to avoid making predictions altogether. Thus, rather than make predictions about what will happen (after all, as physicist Niels Bohr noted, "Prediction is very difficult, especially about the future"), some quacks merely provide explanations about what has already happened: in other words, after the fact.

After-the-Fact Explanations May Be Untestable. After-the-fact explanations (sometimes called *ad hoc explanations*) are difficult to prove wrong. For example, if we say that a person committed a murder because of some event in his childhood, how can we be proven wrong? Most people would accept or reject our claim based on whether it sounded reasonable or not. The problem, however, with accepting "reasonable-sounding" explanations is that after something happens, almost anyone can generate a reasonable-sounding explanation for why it happened.[1] Unfortunately, explanations that sound right can be wrong.

---

[1] According to *Time*, Nancy Reagan's strong trust in an astrologer was cemented by the astrologer showing, *after the fact*, that "her charts could have foretold that period on or around March 30, 1981, would be extremely dangerous for the President." According to *The Skeptical Inquirer*, another individual tried to capitalize on the assassination attempt. That person, a self-proclaimed psychic, realized that predicting the assassination attempt would be persuasive evidence of her psychic powers. Consequently, she faked a videotape to make it look like she had predicted the assassination attempt on Ronald Reagan months in advance. However, analysis showed that the videotape was made the day after the assassination attempt (Nickell, 2005).

To show how explanations that sound right can be wrong, psychologists have asked people to explain numerous "facts," such as why "opposites attract," and why changing your original answer to a multiple-choice test question usually results in changing from a right answer to a wrong answer. Participants were able to generate logical, persuasive reasons for why those "facts" were true—even though all those "facts" were false (Dawes, 1994; Myers, 2009; Slovic & Fischoff, 1977; Stanovich, 2009).

Skeptical: What's the Evidence?

Scientists are not just skeptical about after-the-fact explanations. Scientists, like CSIs, are so skeptical that they want evidence before they believe even the most "obvious" of statements. As Carl Sagan (1993) noted, scientists have the courage to question conventional wisdom. For example, Galileo tested the obvious "fact" that heavier objects fall faster than lighter objects—and found it to be false. Like the skeptical CSI, scientists respond to claims by saying things like "Show me," "Let me see," "Let's take a look," and "Can you verify that?" Neither CSIs nor scientists accept notions merely because an authority says it's true—as Richard Feynman said, "Science is the belief in the ignorance of experts"—or because everyone is sure that it is true. Instead, both CSIs and scientists accept only those beliefs that are supported by objective evidence.

Even after scientists have objective evidence, they continue to be skeptical. They realize that having circumstantial evidence in support of a belief is not the same as having proof that the belief is correct. Consequently, before being convinced by objective evidence in support of a belief, scientists ask themselves two questions.

The first question is, "What is the evidence *against* this belief?" Scientists realize that if they looked only at the cases that support a belief, they could find plenty of evidence for their pet hypotheses. For example, they could find cases—*anecdotal evidence*—to "prove" that going to a physician is bad for your physical health (e.g., by looking only at malpractice cases) and that playing the lottery is good for your financial health (by looking only at lottery winners). Therefore, scientists look for evidence both for and against their beliefs. If there is no known evidence against a belief, a scientist may ask, "What study could I do to find out whether my idea is wrong?"

The second question scientists ask is, "Is there another way to interpret the evidence that has been interpreted as supporting the belief?" For example, people originally believed malaria was caused by breathing the bad-smelling air around swamps (the condition was named after this belief: *Malaria* means "bad air"—"mal" means bad, as in "malpractice," and "aria" means air). As evidence that this belief was correct, people pointed out that malaria cases were more common around swamps—and that swamps contained foul-smelling marsh gas. Scientists asked whether the presence of marsh gas was the only difference between dry areas and swampy areas. As a result of that questioning, we now know that mosquitoes—not marsh gas—infect people with malaria.

Sometimes, as a result of questioning the evidence, we learn that what some considered "convincing proof" may be mere coincidence. A suspect

may be near the victim's house on the night of the murder for perfectly innocent reasons; a patient may suddenly get better even after getting a "quack"
treatment; a correct prediction that a dead body will be found near a body
of water may not be the result of psychic ability; an unexpected noise in a
house may not be due to a ghost; and one cold winter does not disprove
global warming. In short, scientists, as Robert Abelson (1995) says, "give
chance a chance" to explain events.

Open-Minded

Although being skeptical means asking questions, it does not mean being
smug and close-minded. On the contrary, being skeptical means being humble
and open-minded because it means being open to going wherever a careful
consideration of the evidence leads you—even when that consideration leads
you to reject beliefs you once held.

Put another way, the same respect for evidence that allows scientists to be
open to rejecting popular claims that are not supported by evidence allows
them to be open to accepting unpopular claims that are supported by
evidence. Thus, scientists do not deny coincidence, evolution, global warming,
and other inconvenient facts.

But what happens when there isn't evidence for a claim? Just as a good
CSI initially—before the evidence is in—considers everyone a suspect, good
scientists are willing to consider every claim. Scientists have the courage to
be open to the truth and to see the world as it is (Sagan, 1993). Scientists
realize that "cynicism, like gullibility, is a symptom of undeveloped critical
faculties" (Whyte, 2005, p. xi). Consequently, scientists will not automatically
dismiss new ideas as nonsense, not even ideas that seem to run counter to
existing knowledge, such as telepathy. The willingness to consider odd ideas
has led scientists to important discoveries, such as the finding that certain
rainforest plants have medicinal uses.

Creative

To test unconventional ideas and to formulate and test new ideas, scientists
must be creative. Unraveling the mysteries of the universe is not a boring,
unimaginative, or routine task. Scientific giants such as Marie Curie (the
discoverer of radium), Charles Darwin, and Friederich Kekulé (who, while
dreaming, solved the riddle of how carbon-containing benzene-ring molecules
are structured) are called creative geniuses.

We should point out, however, that you don't need to be "naturally creative" to think in a creative way (Rietzschel, De Dreu, & Nijstad, 2007).
Indeed, Darwin, Einstein, and Edison did not attribute their creative success
to natural creative ability, but to persistence. As Einstein said, "The most
important thing to do is to not stop questioning."

Shares Findings

Science is able to capitalize on the work of individual geniuses like Einstein
because scientists, by publishing their work, produce publicly shared knowledge. Most of this published work involves submitting reports of research
studies. These reports, because they make what the scientists did transparent,

allow other scientists to critique, **replicate** (repeat), and build on the original study.

One advantage of scientists publishing their work is that methodological errors can be detected and corrected. Thus, if a researcher publishes a flawed study, critics can see what the researcher did wrong and challenge the researcher's conclusions and research methods.

A second advantage is that a scientist's biases can be detected. For example, if the researcher biased the results of a published study, rival investigators may replicate that study and get different results. Thus, publishing findings and using standard research methods are ways to maintain objectivity by keeping individual scientist's biases in check (Haack, 2004).

A third advantage of scientists publishing their work is that researchers can build on each other's work. You can repeat another scientist's study to see whether the finding currently holds in your location with your population. You can make minor tweaks to the study to see how robust its findings are. Put another way, no scientist has to solve a problem alone. If a scientist doesn't have the time, resources, or energy to solve an entire puzzle, the scientist can work on filling in one of the last remaining pieces of the puzzle that others have almost completed (Haack, 2004). If a scientist does not see how to go about solving the puzzle—or is going about it the wrong way—the scientist can take advantage of the viewpoint of individual geniuses such as Einstein or Piaget, as well as the wide variety of different viewpoints offered by all the other puzzle workers in the scientific community (Haack, 2004). By combining their time, energy, and viewpoints, the community of scientists can accomplish much more than if each had worked alone. As Ernest Rutherford said, "Scientists are not dependent on the ideas of a single person, but on the combined wisdom of thousands."

Without an open sharing of information, science doesn't work, as the debate on cold fusion illustrates. In 1989, two chemists announced at a press conference that they had invented a way of creating nuclear fusion, a potential source of safe electric power, without heating atoms to extreme temperatures. (Before the scientists' announcement, all known ways of producing nuclear fusion used more energy to heat atoms than the fusion reaction produced. Thus, nobody could seriously consider using nuclear fusion to produce electricity commercially.) However, the two scientists did not submit their research to peer-reviewed journals, and they failed to give details of their procedures. Thus, nobody could replicate their work.

All this secrecy worked against science's self-corrective and unbiased nature. By not sharing their work with other scientists before making a public announcement, the two chemists removed the checks and balances that make science a reliable source of evidence. Instead of letting others verify their findings, they expected people to accept claims made at a press conference. Fortunately, most reputable scientists refuse to accept claims, even claims from other scientists, unless those claims are backed up by objective, replicable evidence. Therefore, rather than accepting the chemists' claims, scientists tried to replicate the alleged effect. (Note that, consistent with the view that scientists are open-minded, scientists did not reject the claim outright. Instead, in a fair and open-minded way, they tested the claim.) So far, no one has succeeded in replicating cold fusion.

Thus far, we have skeptically assumed that cold fusion did not really happen. But what if it had? The researchers' lack of openness would still be unfortunate because science progresses only when scientists openly exchange findings.

Productive

Fortunately, scientists usually do share their findings and build on each other's work. As a result, theories are frequently revised, refined, or replaced, and, in some fields of science, knowledge doubles every 5 to 10 years. The evidence that science is a productive tool for making discoveries and advancing knowledge is all around us. The technology created by science has vaulted us a long way from the Dark Ages or even the pre-DVD, pre-personal computer, pre-microwave, pre-cell phone early 1970s.

The progress science has made is remarkable considering that it is a relatively new way of finding out about the world. As recently as the 1400s, people were punished for studying human anatomy and even for trying to get evidence on such basic matters as knowing how many teeth a horse has. As recently as the early 1800s, the scientific approach was not applied to psychology or medicine. Until that time, people were limited to relying on tradition, common sense, intuition, and logic for psychological and medical "knowledge" (and it was not a good idea to go to the doctor).

Once science gained greater acceptance, people used the scientific approach to test and refine notions derived from common sense, intuition, tradition, and logic. By supplementing these other ways of knowing, science helped knowledge progress at an explosive rate.

Almost everyone would agree that science has allowed physics, chemistry, and biology to progress at a rapid rate (see Figure 1.4). Hardly anyone would

"Sometimes I wonder if there's more to life than unlocking the mysteries of the universe."

© Bruce Eric Kaplan/The New Yorker/cartoonbank.com

FIGURE **1.4** Productivity.

Science has made impressive progress.

argue that we could make more progress in understanding our physical world by abandoning science and going back to prescientific beliefs and methods. For example, few would argue for replacing chemistry with its unscientific parent: alchemy (Dawkins, 1998). Indeed, partly because most people respect science, it seems like every field wants to be labeled a science—from creation science, to handwriting science, to library science, to military science, to mortuary science (Haack, 2004). However, not all fields deserve that label.

Psychology's Characteristics

Does psychology deserve the science label? To answer this question, we must ask whether psychologists and psychology can do the eight things that other scientists and sciences do:

1. find general rules
2. collect objective evidence
3. make testable statements
4. be skeptical
5. be open-minded
6. be creative
7. produce publicly shared knowledge that can be replicated
8. be productive

General Rules

Perhaps the most serious question about psychology as a science is, "Can psychologists find general rules that will predict, control, and explain human behavior?" Cynics argue that two unique features of humans make finding general rules to explain the behavior of people impossible.

First, cynics argue that, unlike molecules and other objects that physical scientists study, humans may spontaneously do something for no reason—so rules can't apply. Psychologists counter that most behavior does not just spontaneously appear. Instead, people usually do things for reasons—even if those reasons aren't always obvious. Indeed, if people's behavior did not follow any rules, human interactions (e.g., holding a conversation, driving on a highway, playing a sport) would be extremely difficult, if not impossible.[2]

Second, cynics claim that, unlike the objects that physical scientists study, each person is unique—and thus general rules cannot apply. Psychologists respond by saying that even though people are not all identical, humans are genetically similar. Perhaps because of this genetic similarity, we humans share many common characteristics, from our use of language to our tendency to try to repay those who help us.

Although both cynics and psychologists make good arguments, psychologists win the debate because they have actually found many general rules that govern human behavior (Kimble, 1990). For example, psychologists have discovered laws of operant and classical conditioning, laws of perception,

---

[2] We thank Dr. Jamie Phillips for this example.

laws of memory (Banaji & Crowder, 1989), and even laws of emotion (Frijda, 1988). (If you doubt that emotions follow rules, then ask yourself why people have fairly predictable reactions to certain movies. For example, most people cry or come close to tears the first time they see *Bambi,* whereas most people feel a nervous excitement the first time they see a horror film.)

Like psychology, medicine once faced resistance to the idea that general rules applied to humans. As Burke (1985, p. 197) noted, "Each patient regarded his own suffering as unique and demanded unique remedies." Consequently, one patient's treatment was totally different from another's. Knowledge about cures was not shared, partly because what cured one person was supposed to be totally ineffective for curing anyone else. As a result, medicine did not progress, and many people died unnecessarily. It was only after physicians started to look for general causes of disease that successful cures (such as antibiotics) were found.

Admittedly, general rules do not always work. A treatment that cures one person may not cure another. For example, one person may be cured by penicillin, whereas another may be allergic to it. It would be wrong, however, to say that reactions to drugs do not follow any rules. It's simply that because an individual's response to a drug is affected by many rules, predicting the response would require knowing at least the following: the individual's weight, family history of reactions to drugs, time of last meal, condition of vital organs, other drugs being taken, and level of dehydration.

Because human behavior, like human physiology, is governed by many factors, predicting what one individual will do in a specific situation would be difficult even if you knew all the rules. Consequently, psychologists agree with cynics who claim that predicting an individual's behavior is difficult. However, psychologists disagree with the cynic's assumption that there are no rules underlying behavior. Instead, psychologists know of rules that are useful for predicting the behavior of most people much of the time. As Sherlock Holmes said, "You can never foretell what any man will do, but you can say with precision what an average number will be up to. Individuals may vary, but percentages remain constant."

But can we—and should we—apply general rules to individuals? Yes: general rules can help us help individuals—as long as we don't let the exceptions to those rules fool us. Specifically, we need to avoid two mistakes that some nonscientists make.

First, some nonscientists decide that, because the general rule has exceptions, the rule is wrong. Those people fail to realize that although citing an exception—"That rule is wrong because I know someone who doesn't follow it"—is a way to disprove an absolute rule, it is not a way to disprove a *general* rule. For example, finding that a certain object (e.g., a helium balloon) did not fall to the ground doesn't disprove the general principle of gravity, and finding that a certain patient got worse after being given antibiotics does not disprove the idea that antibiotics generally help people who have bacterial infections.

Second, some nonscientists decide that, because a general rule has exceptions, they shouldn't apply the rule. They realize the obvious: If they consistently apply a general rule, they will make some mistakes. However, they fail to realize the less obvious: If they do not use a valid general rule, they will make even *more* mistakes.

To appreciate why using exceptions to dismiss or ignore a general rule is foolish, imagine that you know that three-quarters of one surgeon's operations are successful, whereas only one quarter of another surgeon's operations are successful. As Stanovich (2009) points out, it would be foolish to say either (a) "It doesn't matter which surgeon I go to because they both succeed sometimes and they both fail sometimes" or (b) "If four of my family members need surgery, I won't send all four to the good surgeon because one might have a bad outcome: Instead, based on my gut feelings, I'll send three to the good surgeon and one to the bad surgeon." After considering this surgery example, you are probably not surprised to learn that psychologists who apply a general rule to all individuals are right more often than (a) psychologists who never use the general rule or (b) psychologists who use intuition to determine when to apply the rule (Stanovich, 2009).

As you have seen, if psychologists have a general rule that allows us to predict behavior, we will make better predictions if we use that rule than if we just guess. But what if psychologists do not have a general rule that will allow us to predict a particular behavior? Does that mean the behavior does not follow general rules? No—if we knew the rules and could precisely measure the relevant variables, the behavior might be perfectly predictable.[3]

To understand how a behavior that we cannot predict may follow simple rules, think about trying to predict whether a coin will come up "heads" or "tails." We can't predict such an outcome with any accuracy. Why not? Is it because the outcome does not follow any rules? No, it follows very simple rules: The outcome depends on what side was up when the coin was flipped and how many times the coin turned over. The problem is that because we do not know how many times the coin will turn over, we can't know whether the coin will land "heads" or "tails." Similarly, most people would agree that the weather is determined by specific events. However, because there are so many events and because we do not have all relevant data on all those events, we can't predict the weather with perfect accuracy.

Objective Evidence

A second question people raise about psychology's ability to be a science is, "Can psychologists collect objective evidence?" There are two reasons for this concern.

The first concern is that psychologists won't be able to keep researcher biases in check. The fear is that the typical researcher will notice information that is consistent with the hypothesis while ignoring information that is inconsistent with the hypothesis; interpret ambiguous evidence as being consistent with the hypothesis; induce participants to behave in a way that

---

[3] In physics, for example, researchers working on chaos theory (also known as complexity theory) have shown that simple processes can produce a complex and hard-to-predict pattern of behavior.

fits with the hypothesis; and, if the study still doesn't support the hypothesis, manipulate statistics to "prove" the hypothesis.

If psychologists engaged in such practices, psychological researchers would "prove" whatever they wanted to prove, and research would simply confirm researchers' pet theories and beliefs. Fortunately, psychologists have found ways to keep their biases in check. Psychologists can separate "the facts of life from wishful thinking" (Myers, 2002b, p. 28). Specifically, by using the techniques you will soon learn (e.g., by using objective measures so that we can't see merely what we want to see and by using statistics correctly so that we can't conclude whatever we want to conclude), the results of psychological research are at least as likely to be as replicable—and thus be at least as objective—as those of physics research (Hedges, 1987; Stanovich, 2009). The high replicability of psychological findings is testimony to the ability of psychologists to make unbiased observations—especially because the reason for repeating a study is often to question its results (Stanovich, 2009).

The second concern is that psychologists won't be able to collect objective evidence about abstract mental concepts because psychologists can't see the mind. Although psychologists can't directly measure abstract concepts such as attitudes, love, and memory, they have developed operational definitions of these concepts (e.g., rating scales to measure attitudes). These operational definitions allow psychology to be objective: indeed, according to one historian of science, psychology's reliance on operational definitions has made psychology more objective than physics (Porter, 1997).

In objectively measuring the unobservable, psychology is following in the footsteps of physics and other older sciences, which have a long history of studying the unseen. Nobody has seen gravity, time, temperature, pressure, magnetism, or electrons, yet these unobservable events can be inferred from observable events. For example, gravity can be inferred from observing objects fall and electrons can be inferred from the tracks in a cloud chamber. Similarly, psychological variables such as love can be assessed by observable indicators such as how long a couple gazes into each other's eyes, pupil dilation at the sight of the partner, physiological arousal at the sight of the partner, passing up the opportunity to date attractive others, and responses on a love scale (Rubin, 1970).

Testable

A third question people have about psychology is, "Can it make testable statements?" If it couldn't, it would share the weaknesses of pseudosciences such as astrology. Fortunately, most published research articles in psychology make testable predictions. Indeed, our journals are full of articles in which predictions made by the investigators were disconfirmed. For example, to his surprise, Kiesler (1982; Kiesler & Sibulkin, 1987) found that many mentally ill individuals are hurt, rather than helped, by being put into mental institutions. In summary, the fact that research frequently disproves researchers' predictions is proof that psychologists make testable statements—and test those statements in an objective manner.

Skeptical

A fourth question about psychology is, "Can psychologists be as skeptical as other scientists?" Some people worry that psychologists will accept, rather than test, existing beliefs about human behavior. After all, many of these beliefs seem logical and have the weight of either popular or expert opinion behind them. Consequently, some fear that, rather than seeing what the evidence says, psychologists will base their decisions about what is true on whether the claim is consistent with either popular or expert opinion.

Although some therapists have ignored the objective evidence, scientific psychologists have been diligent about testing even the most "obviously true" of ideas. For example, Greenberger and Steinberg (1986) performed a series of studies testing the "obviously true" idea that teenagers who have jobs better understand the value of hard work and found that—contrary to conventional wisdom—teenagers who work are more cynical about the value of hard work than nonworking teens. Likewise, Coles (1993) found that "cocaine babies" were not as troubled as many people originally believed. More recently, psychologists have found that women do not talk substantially more than men (Mehl, Vazire, Ramirez-Esparza, Slatcher, & Pennebaker, 2007). In short, psychologists regularly put "obviously true" ideas to the test—and those ideas often fail that test.

In addition to questioning "obviously true facts," psychologists question "obviously true interpretations" of facts. What some see as proof, psychologists see as circumstantial evidence. To illustrate, let's consider three types of claims that psychologists question: (a) claims that knowing or controlling what people are *experiencing* is easy, (b) claims that knowing what *caused* something is easy, and (c) claims that *generalizing* results from a study done with one group in one situation to other groups in other situations is easy.

Psychologists challenge claims that involve assuming that the researchers knew what participants felt and thought because such claims cannot be made based on seeing participants' minds. Instead, such claims are based on seeing some behavior—and then assuming that the behavior is an accurate indicator of what is going on in the participant's mind. Put another way, scientific mind-reading claims are only accurate to the degree to which the *operational definition* of a concept (e.g., scores on the love scale) really represents the concept (e.g., actually being in love).

Psychologists question the degree to which operational definitions accurately capture the concepts those operational definitions are supposed to reflect. Thus, psychologists question the degree to which mental tests truly capture the psychological characteristics that those instruments claim to assess. For example, psychologists are not easily convinced that a set of questions labeled as a "love scale" actually measures love or that an "intelligence test" measures intelligence. If the test uses self-reports (i.e., asking participants questions such as, "How happy are you?"), psychologists are skeptical because they realize that people do not always know their own minds (and, as we shall discuss in Chapter 8, even when people do know, they may not tell). If the test does not use self-report, psychologists are still skeptical because they realize that observing behavior only allows educated guesses about what is going on in the person's mind—and those guesses may be wrong. In short, psychologists do not assume that they know what others are thinking or feeling.

Psychologists are also very skeptical about cause–effect conclusions. They realize that it is hard to isolate the one factor that is causing a certain behavior. Therefore, if psychologists find that better students have personal computers, psychologists do not leap to the conclusion that computers cause academic success. Instead, psychologists would consider at least two alternative explanations. First, psychologists realize that if students were given computers as a reward for doing well in school, computers would be the effect—rather than the cause—of academic success. Second, psychologists realize that the computer-owning students may be doing better than other students because the computer-owning students went to better preschools, had better nutrition, or had better relationships with their parents. Until these and other explanations were eliminated, psychologists would not assume that computers cause academic success.

Finally, many psychologists are skeptical about the extent to which results from a study can be generalized to the real world. Psychologists do not assume that a study done in a particular setting with a particular group of people can be generalized to other kinds of participants in a different setting. For instance, they would not automatically assume that a study originally done with 10-year-olds at a birthday party would obtain the same results if it was repeated with adult participants at a business meeting.

In short, psychologists are extremely skeptical of conventional wisdom and are especially skeptical of conventional wisdom's claim that "the facts speak for themselves." Therefore, psychologists not only question "obvious facts" but also "facts" that involve assumptions about (a) measuring the mind, (b) knowing the causes of behavior, and (c) generalizing the results of a study.

Open-Minded

Paralleling the concern that psychologists do not test "obvious facts" is the concern that psychologists are not open to ideas that run counter to common sense. These concerns are groundless because psychologists are open-minded for the same reason they are skeptical: To psychological scientists, observable facts count more than personal opinions.

Because psychological scientists are skeptical, but not closed-minded, they have tested all sorts of counterintuitive ideas, such as the idea that subliminal, backward messages (back-masking) on records can lead teens to Satanism (Custer, 1985); the idea that people can learn in their sleep; and the idea that ESP can be reliably used to send messages (Bem, 2011; Swets & Bjork, 1990). Although psychologists found little evidence for those particular ideas, psychologists' willingness to test virtually anything has led to tentative acceptance of the ideas that acupuncture may be effective in relieving pain, that meditating helps people to live longer (Alexander, Langer, Newman, Chandler, & Davies, 1989), that more choices can make people less happy (Iyengar & Lepper, 2000), that people can accurately judge another person just by seeing a picture of the other person's room

(Gosling, Ko, Mannerelli, & Morris, 2002), and that unconscious stimuli can affect our behavior (Custers & Aarts, 2010).

Creative

Whereas psychologists' open-mindedness has been questioned, few have questioned psychologists' creativity. Most people realize that it takes creativity to come up with ideas for psychological research. Fortunately, with a little help, most people have the necessary creativity. Indeed, if you follow the tips on idea generation in Chapter 3, you will be amazed at how creative your ideas can be.

Creativity is needed not only to generate a research idea but also to test it. For example, creativity is needed to develop accurate measures of the concepts the researcher plans to study. Imagine the challenge of developing measures of such concepts as love, social intelligence, and prejudice. Fortunately, to measure key variables, you do not always need to rely on your own creativity. Instead, you can often rely on the creativity of others. After all, why reinvent the wheel when creative psychologists have already developed ways of measuring all kinds of concepts—from emotional intelligence (Mayer, Salovey, & Caruso, 2002), to unconscious prejudice (Sriram & Greenwald, 2009), to creativity (Rietzschel, De Dredu, & Nijstad, 2007)?

Even after finding ways of measuring the concepts they wish to study, researchers may still need to use their creativity to develop a situation that will permit them to test their research idea. Like the inventors of the wind tunnel, psychological scientists may need to create a scaled-down model of a real-life situation that is simpler and more controllable than real life, yet still captures the key aspects of the real-life situation. For example, to study real-life competition, social psychologists have developed competitive games for participants to play. Similarly, to model the situation in which nothing you do seems to matter, Seligman (1990) had people try to solve unsolvable puzzles.

Shares Findings

In addition to being creative, psychologists have been very good at sharing their ideas and findings with others in the field, as shown by the hundreds of journals in which psychologists publish their work. Indeed, psychologists may enjoy more candor and cooperation than scientists in other fields because psychologists usually gain little by keeping results secret. For example, if you wanted to be the first to patent a new technology, it would pay you to keep secrets from competitors. In such a race, if you were first, you might make millions. If you were second (like the poor guy who tried to patent the telephone 2 hours after Alexander Graham Bell did), you would make nothing. Although such races for dollars are common in some sciences, they are rare in psychology.

Productive

Partly because psychologists have shared their findings, they have made tremendous progress. One hundred years ago, a person could know everything about every area of psychology. Today, thanks to research, no psychologist could know everything about even one area of psychology—and there are many more areas than there were. We now have textbooks full of research-generated facts in applied areas such as consumer psychology, counseling psychology, political psychology, sports psychology, and organizational psychology. Furthermore, recent research in cognitive, social, and developmental psychology is revealing new insights about almost every aspect of ourselves—from the unconscious mind, to evil, to happiness.

Much recent research has implications for real life. For example, accurate answers to questions like, "How do I tell whether the couple I'm counseling will get divorced?" and "How do I tell when a suspect is lying?" have come from the research lab. Indeed, almost anyone dealing with an applied problem would be well-served to consult the psychological research. For example, suppose you want to cook a meal that will impress your date. In addition to consulting a recipe book, you should consult the psychological research (to learn, for example, that you can affect your date's liking of the food before the meal is served by describing how difficult it was to get some of your "fresh, organic" ingredients; by describing the food using words like "tender," "gourmet," and "traditional"; by letting your date see that there are fancy cooking gadgets and exotic herbs in your kitchen; and by serving the food on your best plates [Wansink, Ittersum, & Painter; 2005]). Because psychological research has been so productive in generating answers to applied questions, professionals in applied areas—such as education, communication, sales, marketing, economics,[4] and medicine—are enthusiastically embracing our research methods.

The Importance of Science to Psychology: The Scientific Method Compared to Other Ways of Knowing

The scientific method is responsible for the tremendous progress in psychology and is also largely responsible for psychology's uniqueness. Whereas many other fields—from astrology to philosophy—are concerned with the thoughts and behaviors of individuals, only psychologists study individuals scientifically (Stanovich, 2009). Thus, as Stanovich points out, it is no accident that every definition of psychology starts out "the *science* of...."

What if psychology were not a science? Psychology would not be useful for helping and understanding people. Without science, psychology might merely be a branch of the popular pseudoscience of astrology (Stanovich, 2009). Without science, psychology might merely be common sense, even though common sense contradicts itself (see Table 1.1). Without science, psychologists might just do what tradition and logic tell us, even when tradition and logic tell us to do things that are actually harmful. For instance, until psychological research showed that premature infants benefit from being

---

[4] For an entertaining and elementary introduction to how one economist uses psychological research methods, read the best-selling book *Freakonomics* (Levitt & Dubner, 2005).

TABLE **1.1** The Inconsistency of Common Sense

1. Absence makes the heart fonder.	BUT	Absence makes the heart wander.
2. Birds of a feather flock together.	BUT	Opposites attract.
3. Look before you leap.	BUT	He who hesitates is lost.
4. Too many cooks spoil the broth.	BUT	Two heads are better than one.
5. To know you is to love you.	BUT	Familiarity breeds contempt.

© Cengage Learning 2013

held, physicians asserted that both logic and tradition dictated that premature infants should not be held (Field, 1993).

Psychology is not the only science that has had to free itself from quackery, tradition, common sense, philosophy, and from the belief that its subject matter follows no rules. Since the beginning of recorded history, some people have argued that finding rules that govern nature is impossible. For centuries, most people believed the stars followed no pattern. Not that long ago, it was believed that diseases followed no patterns. Even today, some people believe that human behavior follows no patterns. Yet, each of these assumptions has been disproven. The stars, the planets, diseases, and humans behave for reasons that we can understand. Admittedly, the rules determining human behavior may be complex and numerous—and it may even be that some behaviors do not follow rules. However, to this point, searching for rules of behavior has been fruitful (see Table 1.2).

Although science is only one way of knowing, it is our most objective way of knowing, and it can work in concert with other ways of knowing (see Table 1.3). Psychologists can use scientific methods to verify knowledge passed down by tradition or from an authoritative expert or to test knowledge obtained by intuition or common sense. By anchoring speculation in reality, psychologists can create, refine, or verify common sense and eliminate superstitions (Kohn, 1988). For example, consider the following 10 findings from research:

1. Punishment is not very effective in changing behavior.
2. Having teens work in low-wage jobs does not instill the "work ethic."
3. Absence makes the heart fonder only for couples who are already very much in love.
4. Multitasking is inefficient.
5. Money does not buy happiness.
6. If you want to make yourself feel better, do charity work.
7. IQ tests predict life success better than tests of emotional intelligence.
8. Married couples' understanding of each other declines over the course of a marriage.

TABLE **1.2** Psychology as a Science

Characteristic	Example
Finds general rules	Helps us understand human behavior through rules such as the laws of operant and classical conditioning, laws of memory (e.g., meaningless information is hard to remember; memorizing similar information, such as Spanish and Italian, leads to memory errors; studying over a period of time—rather than cramming—leads to better recall); predictable reactions to stress (general adaptation syndrome); and a wide range of theories, from social learning theory to cognitive dissonance theory.
Collects objective evidence	Tests whether beliefs and theories are consistent with objective evidence. Obtains objective evidence by recording participants' behaviors: number of words written down on memory test, ratings made on an attitude scale, responses on personality test, reaction times, etc. One index of how effective we are at being objective is that our research findings are as replicable as the research findings in physics.
Makes verifiable statements	Makes specific testable predictions that are sometimes found to be wrong (the hypothesis that rewarding someone for doing a task will always increase their enjoyment of that task has been disproven). That is, we use evidence to correct wrong beliefs.
Skeptical	Demands evidence. Challenges common sense and traditional notions. Does not take evidence (participants' statements or ratings) at face value. Considers alternative explanations for evidence (the group given a memory pill may do better than the other group on a memory task because its members had naturally better memories, because they were tested later in the day, or because they believed the pill would work).
Open-minded	Entertains virtually any hypothesis, from acupuncture relieving pain to meditation prolonging life.
Creative	Measures the mind, generates hypotheses, and devises studies that rule out alternative explanations for findings.
Public	Allows scientists to check and build on each other's work through research published in journals.
Productive	Increases psychological knowledge at a dramatic rate.

© Cengage Learning 2013

9. Nonverbal communication is not very helpful in letting people know what other people are thinking.
10. Psychotherapy, especially grief counseling, can be harmful.

All of these findings are refinements of the common sense of a few years ago. All of these findings are, or will soon become, part of the common sense of this century.

In short, science is a powerful tool that can be used to test our beliefs, refine our knowledge, improve our ability to help people, and stop us from unintentionally hurting people though well-meaning but harmful

TABLE **1.3** Why You Need to Understand Science Rather Than Relying Only on Other Ways of Knowing

Way of Knowing	Problems	Quote/Example	How Science Can Improve, Test, or Work with This Way of Knowing
Expert authorities	• "Experts" are not always knowledgeable in the areas they are discussing (the media needs "experts"—and will create them if necessary). • Experts are not always unbiased—many are influenced by the groups paying them (Kolata, 2007; Peterson, 2008; Tavris & Aronson, 2007). • Experts often give conflicting advice. • The more confident the expert, the less accurate the expert's predictions are (Tetlock, 2005).	• Einstein put too much faith in Freud's theory; Linus Pauling (a two-time Nobel Prize winner) overstated the benefits of Vitamin C. • "It is absurd, as well as arrogant, to pretend that acquiring a PhD some-how immunizes me from the errors of sampling, perception, recording, retention, retrieval, and inference to which the human mind is subject" (Paul Meehl, as cited in Stanovich, 2007, p. 194.) • "If you consult enough experts, you can confirm any opinion."—Arthur Bloch	A true expert is one who has looked at the evidence with both a skeptical and an open mind. Thus, a person could be an expert in one area, but not in another. Simple scientific formulas do a better job of predicting than experts do (Dawes, 1994; Myers, 2002b). Knowing about research will help you to become an expert in whatever field you choose.
Common sense, tradition	• Common sense contradicts itself. • Common sense is not necessarily accurate; it just means that a group has some common beliefs (Duffield, 2007). These common beliefs may be traditions, myths, superstitions, or prejudices (Duffield, 2007; Whyte, 2005). • Some groups have different common sense than others (Duffield, 2007).	• See Box 1.1: "Inconsistencies of Common Sense." • "Common sense is a fable agreed upon." Contrary to common sense, sugar does not make kids hyperactive, you do not need to drink 8 glasses of water a day, and most people do use more than 10% of their brains.	Science can test existing common sense. Science can contribute to common sense: As the saying goes, "today's science is tomorrow's common sense."

(Continued)

Logic and reason	• Logic is not an effective way of obtaining facts that can be obtained through observation. • The conclusion of a logical argument is limited by the "facts" making up that argument—as the saying goes, "garbage in, garbage out."	• Aristotle, father of modern logic, concluded that women have fewer teeth than men. • Delusional systems can be logical. A mental patient's behavior may be completely logical if you accept the patient's premise that she really is from Mars.	After a scientist uses observation to obtain facts, the scientist will use logic to draw reasonable conclusions from those facts.
Experience	• Experience is a very tricky and very bad teacher. We experience more illusions than we think. • Furthermore, people do not learn from their mistakes (Ariely, 2008; Tavris & Aronson, 2007; Tetlock, 2005). • "A man who is so dull that he can learn only by personal experience is too dull to learn anything important by experience." —Don Marquis	• Experience "taught" Aristotle that bees catch honey as it falls from the sky, and it teaches us that the earth is flat and that the sun revolves around it. As Stanovich (2007) points out, thousands of years of observations did not teach people the laws of gravity (e.g., they thought gravity made heavier objects fall faster than lighter objects). • People who work in hospitals believe that there are more admissions during a full moon, but the evidence says otherwise.	Learning to think scientifically can, however, help you to learn better from this bad and tricky teacher: Science helps us to be "wise by other people's experience." — Samuel Richardson Thousands of lives have been saved by neurosurgeons switching from what experience had "taught" them to using scientifically based treatments (Gladwell, 1996).
Intuition and Introspection	• Our own insights about ourselves are often wrong. • There is a difference between what sounds right and what is right (Rosenzweig, 2007). • Much of what people call intuition is prejudice (Whyte, 2005).	• Wilson (2002) showed that we are "strangers to ourselves."	"Science is merely an extremely powerful method of winnowing what's true from what feels good."—Carl Sagan
All positions are somewhat true, but the "real" truth lies in the middle.	• There are facts and there are nonfacts.	• The sun either exists, or it does not exist. The truth is not in the middle.	Science should not compromise about facts.

The truth is just a matter of opinion, and my opinion is as good as anyone else's.	• "Facts are not a matter of opinion" (Whyte, p. 25). It's not really all about you: There is a reality beyond you.	• "Bacteria and planets do not come into or go out of existence depending on what people believe" (Whyte, p. 154). • There is "a distinction between what we have reason to believe and what we have no reason to believe" (Whyte, p. 41).	If you want to know about reality, science is the best tool devised for that purpose.
"Everybody is talking about it," 'I've heard a lot about it lately,' People believe strongly in it."	• Unfortunately, just because "everybody is talking about it" does not mean that "there must be something to it." Where there is smoke, sometimes there is no fire—just smoke that has been manufactured by marketing and political machines. (Corporations know how to produce hype and buzz.) • "Educational" books and seminars are not always what they seem. • Believing does not necessarily translate into accuracy; belief may even lead to bias.	• Because facts are not put to a vote, "information" you get from the web or Wikipedia may be misinformation. As the saying goes, "conventional wisdom is to wisdom what junk food is to food." • Best-selling business books have been full of bunk (Rosenzweig, 2007) and seminars— even continuing education courses for professional counselors—have dealt with nonscientific topics such as improving counselors' psychic abilities (Arkes, 2003; McDougal, 2007).	You need to think scientifically; going with the sheep may get you slaughtered. Eyewitness testimony, for example, shows almost no relationship between confidence and accuracy.

therapies. If this tool will help psychologists solve important problems, shouldn't psychologist use it—especially when it does not rule out the use of other tools?

WHY YOU SHOULD UNDERSTAND RESEARCH DESIGN

Thus far, we have explained why psychologists are interested in scientific research: They see research as a useful tool to obtain answers to their questions. But why should you know how to use this tool? After all, if you don't need to understand the science of agriculture to enjoy its fruits, why do you need to understand the science of psychology to take advantage of its products?

To Understand Psychology

The classic answer is that you can't have an in-depth understanding of psychology—the science of behavior—unless you understand its methods. Without understanding psychology's scientific aspects, you may know some psychological facts and theories, but you will not understand the basis for those facts and theories. Like buying a house without inspecting its foundation, you would not know whether what you had invested in was built on a solid foundation or whether it was, like a house built on sand, ready to collapse.

Even if you trust that the psychological knowledge you gained is based on a solid foundation, you will have trouble selling that knowledge if you can't explain why your advice is sound. If, on the other hand, you understand the foundation on which psychological facts are based, you can be a good ambassador for psychology. You can defend psychological facts from those who claim that such "facts" are just opinions.

Beyond increasing your own credibility and that of your field, explaining the basis for psychological facts will help you to help well-meaning people do good things rather than wasteful things. For example, suppose your organization or community is facing a problem (e.g., increased tensions between two groups). You know certain facts (e.g., that increasing intergroup contact alone will not improve relations; the groups must work as equals toward a common goal) that could be applied to the problem. You want others to use these facts to improve the situation—you do not want them to dismiss these facts as "unsupported opinions" or "merely psychological theory." To convince people that your advice is sound, you must understand the foundation supporting those facts—the research process—so well that you can make others understand it.

We have just discussed the scenario in which psychology has a pre-packaged, fact-based solution to a problem, and your task is to sell that solution. But what if, as is usually the case, psychological science has not yet unearthed a solution that has been applied to your group's particular problem? That's when you really need to understand research methods, because you will need to use research to find a solution. Specifically, you will need to (a) synthesize the relevant scientific literature to find principles that might apply to the problem, (b) develop a solution based on those principles, and (c) monitor your tentative solution to see whether it really works. In short, to be an effective psychologist, you must use psychological research *findings* (its technology) to propose a solution that will probably work and use psychology's research *methods* (its science) to find out whether your solution actually works (Levy-Leboyer, 1988).

To Read Research

When research addresses problems that interest you, you will want to take advantage of that research. To do so, you must be able to read and interpret scientific research reports. For instance, you may want to know something about the latest treatment for depression, the causes of shyness, factors that lead to better relationships, or new tips for improving workplace morale. If you need the most up-to-date information, if you want to draw your own conclusions, or if you want to look at everything that is known about a particular problem, you need to be able to read the research yourself.

You can't rely on reading about research in textbooks, magazines, newspapers, or blogs. Textbooks will give you only sketchy summaries of the few, often out-of-date, studies that the textbook's authors chose to discuss. Magazine articles, newspaper articles, and blogs, on the other hand, often cover up-to-date research, but these reports frequently do not accurately represent what happened (Anderson & Bushman, 2002; Brescoll & LaFrance, 2004; Bushman & Anderson, 2001) and often report the results of poorly done studies (Begley, 2007; Kolata, 2007). Understanding research methods allows you to bypass secondhand accounts of research, thus allowing you to read the original source and come to your own conclusions.

With psychology progressing at a rapid rate, you will need to keep up with the field. As an employee, you will want to make more money every year. How can you justify receiving raises if, with every year, your knowledge is more obsolete? If you see clients, you should be giving them treatments that work, and you should be acting on the best information available. After all, you would not be pleased to go to a physician whose knowledge about your illness was 10 years out of date.

To Evaluate Research

If you understand research, you will not only be able to get recent, first-hand information, but you will also be in a position to evaluate that information. You may even be able to evaluate many secondhand reports of research in magazines, newspapers, and blogs.[5] As a result, you will be able to take full advantage of the knowledge that psychologists are giving away, knowledge that is available to you in newspapers, on the Internet, and on television—without being fooled by the misinformation that is also freely available. You will find that, although scientists can design a study to get results that conform to the scientist's wishes, well-designed studies usually get outcomes that conform to reality (Begley, 2007). You will also find that you need your critical abilities because (a) even studies published in good journals may be flawed, (b) poorly designed studies are, in many areas, much more common than well-designed ones (Begley, 2007; Kolata, 2003), and (c) you will often encounter conflicting research findings.

---

[5] If the secondhand reports provide you with enough information about the study's procedures, you can evaluate the study. If they do not, you will have to read the original scientific publication.

To Protect Yourself From "Quacks"

As common as conflicting research findings are, they may be less common than phony experts. Free speech protects quacks, just as the free market protected "snake oil" salespeople in the days before the U.S. government created the Food and Drug Administration (FDA).[6] Back then, people selling so-called medicines could sell the public almost anything, even pills that contained tapeworm segments (Park, 2000).

Today, "experts" are free to go on talk shows and the Internet to push "psychological tapeworms." Common psychological tapeworms include unproven and sometimes dangerous tips on how to lose weight, quit smoking, discipline children, and solve relationship problems.

We do not mean that all experts are giving bad advice. We mean that it's hard to tell what is good advice and what is not. You cannot assume that something is true just because a well-known expert said it was true: in most fields studied, the more well-known and the more quoted an expert was, the less accurate the expert's predictions were (Tetlock, 2005).

Unfortunately, there is no way you can avoid exposure to bad information. We live in the information age, but we also live in the *mis*information age. Thus, although the truth is often out there, so are lots of lies. Science, nonscience, pseudoscience, and common nonsense exist side by side on the shelves of the psychology section in bookstores, on talk shows, on the Internet, and on televised newsmagazines. Without some training in research design, it is hard to distinguish which information is helpful and which is potentially harmful.

To Be a Better Psychologist

In psychology, as in many fields, professionals who believe that science applies to their field do their job differently than colleagues who don't. Scientifically oriented detectives use fingerprints and DNA analysis, whereas other detectives rely on psychics and lie detectors. Scientifically oriented physicians treat patients based on what research has established as the most effective cure, whereas other physicians rely on their instincts, drug ads, or alternative medicines—and end up causing thousands of their patients to die (Gladwell, 1996). Similarly, scientifically oriented counselors treat patients based on what research has established as the most effective treatment, whereas other counselors sometimes use techniques shown to be ineffective and harmful.

Note that counselors who abandon the scientific approach are not just engaging in "psychoquackery": They are also needlessly harming clients (Begley, 2007; Groopman, 2004; Lilienfeld, 2007). Thus, as Carol Tavris says, "... a cautious, skeptical attitude is the hallmark of good science and caring practice."

---

[6] Although we now do have an FDA, "that agency's 40 analysts can't be counted on to evaluate the accuracy of the more than 30,000 ads and promotions that are made each year" (Schmit, 2005, p. 1A). Furthermore, some treatments (e.g., homeopathic remedies) are exempt from FDA review (Park, 2000).

In short, those who rely on treatments that science has shown to be effective are true professionals, whereas those who do not rely on science are quacks. We do not want you to be a quack.

To Be a Better Thinker

In addition to preventing you from acting like a quack, this course may prevent you from thinking like one. We do not mean that you are a poor thinker now. Indeed, the way you think now will be the foundation for thinking scientifically: As Einstein said, "The whole of science is nothing more than the refinement of everyday thinking." We do mean that the same critical thinking skills, decision-making skills, and problem-solving skills you will learn in this book are taught in books that claim to raise your practical intelligence (e.g., Lewis & Greene, 1982), are measured by some tests of "practical intelligence" (N. Frederikson, 1986), and have increased students' practical intelligence (Lehman, Lempert, & Nisbett, 1988). In short, this course will help you develop the "mindware"—software for the mind—that will help you think better (Stanovich, 2010).

Put another way, people like to be able to separate fact from fiction. They like to believe that they will not accept statements without adequate proof (Beins, 1993; Forer, 1949). However, without understanding science, how can they know what adequate proof is?

To Be Scientifically Literate

Another reason to study psychological research methods is to learn how science works. Intelligent people are supposed to be able to profit from experience, and in today's world many of our experiences are shaped by scientific and technological changes. Yet, many people do not know how science works.

Some argue that this scientific illiteracy threatens our democracy—and they have a point. How can we make intelligent decisions about what to do about global warming if we can't properly interpret data about it? How can juries correctly decide whether to award damages to people who have allegedly been harmed by a product if the jurors can't understand the scientific evidence (Kosko, 2002)? We would like to rely on experts, but experts may contradict each other. Furthermore, some "experts" may be quacks (Kosko, 2002), and others may be unduly influenced by the company or group that is sponsoring them (Peterson, 2008; Tavris & Aronson, 2007). Therefore, if we are going to make informed decisions about how to address global warming, crime, or a host of other problems, we need to know how to interpret scientific research.

Regrettably, it appears that many people are scientifically illiterate. Many high school students (and some high-ranking politicians) believe in astrology. Furthermore, many of astrology's skeptics can easily become believers (Glick, Gottesman, & Jolton, 1989). In addition to astrology, other scientifically invalid procedures such as foot reflexology, numerology, and assessing personality via handwriting analysis also enjoy surprising popularity (Lardner, 1994).

Given this low level of scientific literacy, perhaps it is not surprising that hype often seems to carry more weight than objective facts. Politicians, for example, often say, "We don't need to do research on the problem; we know what we need to do," or "I don't care what the research says." Similarly, many consumers buy products that include "secret, ancient remedies" rather than products that have been proven to be effective through open, public, scientific testing. As a result, Americans spend billions of dollars each year on treatments and products that have been shown to be ineffective.

Even when Americans see evidence for a claim, many don't question how well the evidence supports the claim. As a result, questionable evidence is often used to fool American voters and consumers. Leaders take credit for random or cyclical changes in the economy. Advertisers encourage consumers to believe that some professional models became attractive by using certain products. Talk-show hosts periodically parade a few people who claim "success" as a result of some dieting or parenting technique. Advertisers still successfully hawk products using testimonials from a few satisfied users. Political leaders persistently "prove" what our country needs by telling us stories about one or two individuals rather than "boring" us with facts (Kincher, 1992). Unfortunately, research shows that, to the person not trained in research methods, these nonscientific and often misleading techniques are extremely persuasive (Nisbett & Ross, 1980).

Fortunately, after studying psychological research methods, you will know how to question evidence (Lawson, 1999). You will be more skeptical of misleading information and more receptive to scientific information. Consequently, you will be a better-informed citizen and consumer.

To Increase Your Marketability

Besides making you a more informed citizen and consumer, knowing about research makes you more employable. In today's job market, your being hired will probably not depend on what job-relevant information you have memorized. After all, such information is quickly obsolete and is often instantly accessible through Google or some computer database. Instead, you will be hired because you can find, create, and judge information that your company needs. Like most workers in this century, you will probably be a "knowledge engineer," hired for your ability to evaluate and create information. To illustrate the importance of your ability to think scientifically to your marketability, consider that even marketing majors are told that, at least for their first few years, their scientific skills, not their marketing intuition, are what will pave the way to career success (Edwards, 1990).

In short, if you can distinguish good information from bad and can turn data into information, companies will want you. These same analytical skills will, of course, also be helpful if you plan to go to graduate school in business, law, medicine, or psychology.

To Do Your Own Research

To increase your chances of getting into graduate school or getting a good job, you can conduct your own research. Completing a research project shows that you are organized, persistent, and capable of getting things done—and organizations want people who can get things done.

Increasingly, one thing organizations want to have done is research. Some of our former students have been surprised that they ended up doing research to get government grants or to get more staff for their social services agencies.

Many private organizations—from Walmart to museums—do research to find out whether what they are doing works (Ralof, 1998; Rosenzweig, 2007). Other organizations do research to find out whether what they are planning to do will work. For example, movie moguls do research to determine if a movie's ending is effective—and how to change the ending if it is not.

Beyond the employment angle, you may find that doing research is its own reward. Some students like research because it allows them to *do* psychology rather than simply read about it. Some enjoy the teamwork aspect of working with professors or other students. Some enjoy the creativity involved in designing a study, seeing it as similar to writing a script for a play. Some like the acting that is sometimes involved in studies (some researchers claim that a valuable research skill is the ability to say "oops" convincingly). Some enjoy the challenges of solving the practical problems that go along with completing any project. Others enjoy the excitement of trying to discover the answers to questions about human behavior. They realize that there are so many interesting and important things about human behavior that we don't know—and that they can find out (Ariely, 2008). Reflecting how much they enjoy searching for answers to questions that matter to them, some even refer to psychological research as "me-search" (Field, 1993).

Certainly, many students have found the passion for discovery much more exciting than learning terms and definitions. Thus, not surprisingly, such poor-to-average students as John Watson (the father of behaviorism) and Charles Darwin enjoyed exploring the mysteries of human behavior. Once you start trying to answer one of the many unanswered questions about human behavior, we think you will understand why even the devout humanist Carl Rogers said, "We need to sharpen our vision of what is possible ... to that most fascinating of all enterprises: the unearthing, the discovery, the pursuit of significant new knowledge" (1985, p. 1).

CONCLUDING REMARKS

Understanding research design will help you distinguish between science, pseudoscience, nonscience, and nonsense: a skill that will help you be a better citizen and consumer. If you lived in a world in which companies, governments, and the media were extremely competent and put the truth and your welfare above their own agendas, you might not need that skill. But you don't live in such a world. Instead, you live in a world in which ads and best-selling books push diets that don't work; physicians push newer drugs that are, in some cases, more expensive, more dangerous, and less effective than older ones; government officials lie; the Internet revives ideas that science killed long ago; and well-known television "experts" are repeatedly wrong. In such a world, William Hazlitt's words ring true: "Ignorance of the world leaves one at the mercy of its malice" (see Figure 1.5). On the bright side, refining your research skills will allow you to take better advantage of two of culture's greatest achievements—two of the things that separate us from other animals and two of the things that make it worthwhile for humans to have conscious thoughts: science and writing (Baumeister, 2011).

© Cengage Learning 2013

FIGURE **1.5** Objective reality is important—Although sometimes inconvenient.

People can't always afford the luxury of an anti-scientific attitude.

BOX **1.1** **Nine Reasons to Understand Psychological Research Methods**

1. To understand psychology better
2. To keep up with recent discoveries by reading research
3. To evaluate research claims
4. To protect yourself from quacks and frauds
5. To be a better psychologist
6. To be a better thinker
7. To be scientifically literate and thus a better-educated citizen and consumer
8. To improve your marketability in our information age
9. To do your own research

© Cengage Learning 2013

In addition to helping you in your personal life, understanding research design will also help you in your professional life. If you become a counseling psychologist, being able to read and evaluate research articles will help you find the best, most up-to-date diagnostic tests and treatments for your clients (largely for this reason, licensing exams for counseling psychologists include questions about research design). If you become a manager, being able to evaluate research articles will help you find what techniques were best for motivating your employees. Regardless of what career you choose, understanding research design can help you be an effective and reflective professional (see Box 1.1).

Finally, understanding research design will give you the tools you need to get answers to your own questions. By studying this book, you will learn how to generate research ideas, manipulate and measure variables, collect valid data, choose the right design for your particular research question, treat participants ethically, interpret your results, and communicate your findings. We hope you will use this knowledge to join the most fascinating quest of our time—exploring the human mind.

SUMMARY

1. Science involves making testable predictions—predictions that observable evidence could disprove—and unearthing unbiased, observable, objective evidence that may refute those predictions. Thus, contrary to myth, scientists don't do research to prove their theories.

2. Because scientists make their evidence public, they can check each other's work, as well as build on each other's work.

Because of the public, group-oriented nature of science, scientific progress can be rapid.

3. Because scientists make their evidence public, you can—if you are scientifically literate—take advantage of their new discoveries.

4. Science is both open-minded and skeptical. It is skeptical of any idea that is not supported by objective evidence; it is

open-minded about any idea that is supported by objective evidence.

5. To become a science that dealt with invisible, abstract concepts such as love, psychology defined those concepts in terms of *operational definitions*: objective "recipes" that show what was done to produce the concept (i.e., what events, stimuli, or instructions the participant was exposed to) or to measure the concept (i.e., how behaviors from the participant were collected and turned into scores). Because psychologists can all follow these "recipes" (operational definitions), psychologists can both replicate (repeat) and build on other psychologists' work.

6. Because operational definitions are objective, researchers can test their pet theories objectively.

7. Psychologists realize that an operational definition of a concept may not accurately capture that concept. Therefore, psychologists question the labels researchers give to measures and manipulations.

8. Because chance plays a big role in human behavior, psychologists realize that almost any single event could be due to a coincidence. That is, for any treatment, there will be cases of people getting better after that treatment and people getting worse after that treatment. Because coincidences happen, you can use examples to prove anything. Consequently, you need to use statistics to find out how consistently people's behavior changes after receiving a treatment and whether this improvement rate is more than a coincidence.

9. Psychologists realize that it is difficult to isolate the one factor that causes an effect.

Because psychologists know how difficult it is to rule out all other possible causes of an effect, psychologists usually question cause–effect claims.

10. Psychologists realize that what happens with one group of participants in one setting may not generalize to another type of participant in a different setting. For example, they realize that a study done with one group of students in a lab setting may not apply to a group of people working in a factory. Therefore, psychologists are appropriately cautious about generalizing the results of a study to real-world situations.

11. There is no psychology without science (Stanovich, 2009). Without science, psychology would have fewer facts than it does now and would be little better than pseudosciences such as palmistry, astrology, and graphology (Stanovich, 2009). More specifically, using the scientific approach in psychology has allowed psychologists to (a) improve common sense, (b) disprove certain superstitions, and (c) make enormous progress in understanding how to help people.

12. Science is the best tool we have for obtaining objective and accurate information about the real world. Furthermore, science is a useful tool for testing the accuracy of common sense and intuition.

13. Scientific research is a logical and proven way to obtain important information about human behavior.

14. The skills you learn in this course can help you in the real world.

KEY TERMS

objective evidence *(p. 4)*

operational definition *(p. 7)*

replicate *(p. 11)*

scientific approach *(p. 3)*

EXERCISES

1. Match the following to the qualities of science.

   ___ testable   a. allows science to learn from mistakes

   ___ skeptical   b. provides observable, unbiased evidence

   ___ objective   c. publishing research reports

   ___ public   d. questions authority

   ___ productive   e. makes progress

2. Scientists are skeptical of case studies, examples, and stories. For example, consider the following case: A friend remembers that, during his early teen years, giving up chocolate cleared up his complexion. Why would a skeptical scientist not accept this statement as proof that giving up chocolate reduces acne? Hints: In addition to considering how objective your friend's evidence is, consider main points 8, 9, and 10.

3. Give one example of an untestable statement. Then, change that statement into a testable statement by either making it a prediction rather than an after-the-fact explanation or by making it more specific. Finally, state at least one advantage of scientists making testable, rather than untestable, statements.

4. Provide a dictionary definition of a psychological concept, such as love. Then, provide an operational definition of that concept. (Be sure that your operational definition is an objective recipe that others can follow. The recipe should either the stimuli you will present to manipulate the concept or the behavior you will observe to measure the concept.) Next, explain how your operational definition differs from a dictionary definition of that concept. Finally, explain how operational definitions help psychology to
   a. be objective
   b. make testable statements
   c. be public
   d. be productive

5. How does the ability of psychologists to replicate each other's work help psychology to be
   a. skeptical?
   b. open-minded?
   c. productive?

6. Some early psychologists studied and reported on their own thoughts. For example, a person would solve a mathematical problem and then report on everything that went on in his mind during the time that he worked on the problem. What quality of science was missing in these studies?

7. From what you know about astrology, grade it as "pass" or "fail" on the following scientific characteristics:
   a. makes testable statements
   b. is productive (knowledge is refined, new discoveries are made)
   c. seeks objective, unbiased evidence to test the accuracy of beliefs

8. According to some, iridology is the "science" of determining people's health by looking at their eyes. Practitioners tend not to publish research, they don't try to verify their diagnoses through other means, and different practitioners will diagnose the same patient very differently. From this brief description of iridology, what characteristics of science does iridology lack?

9. Some claim that psychoanalysis is not a science. They attack it by claiming that it lacks certain characteristics of science. Following are three such attacks. For each attack, name the characteristic of science that psychoanalysis is accused of lacking.
   a. "Psychoanalytic explanations for a person's behavior often fit with the facts but are generally made after the fact."
   b. "The unconscious is impossible to observe."
   c. "The effectiveness of psychoanalysis does not appear to have improved in the last 20 years."

WEB RESOURCES

1. Go to the Chapter 1 section of the book's student website and:
   a. Look over the concept map of the key terms.
   b. Test yourself on the key terms.
   c. Take the Chapter 1 Practice Quiz.
   d. Do the interactive end-of-chapter exercises.

2. To learn more about how to market the skills you will develop in this course, read "Web Appendix: Marketing Your Research Design Skills."

3. To learn more about science, read "Web Appendix: Criticisms of Science."

INFOTRAC® COLLEGE EDITION EXPLORATIONS

 To see that pseudoscience is not dead, do an InfoTrac College Edition keyword search on "pseudoscience."

CHAPTER 2

Validity and Ethics:
Can We Know, Should We Know, and Can We Afford Not to Know?

Questions About Applying Techniques From Older Sciences to Psychology

Internal Validity Questions: Did the Treatment Cause a Change in Behavior?

Construct Validity Questions: Are the Variable Names Accurate?

External Validity Questions: Can the Results Be Generalized?

Ethical Questions: Should the Study Be Conducted?

Concluding Remarks

SUMMARY
KEY TERMS
EXERCISES
WEB RESOURCES

Science is not physics, biology, or chemistry ... but a moral imperative ... whose purpose is to give perspective, balance, and humility to learning.
—Neil Postman

Science is a long history of learning how not to fool ourselves.
—Richard Feynman

CHAPTER OVERVIEW

When there is a medical emergency, a natural disaster, or some other problem, most of us believe that trained professionals should use their knowledge to try to help. We would be outraged if the emergency medical technician (EMT) at a scene ignored an accident victim's bleeding or if a clinical psychologist ignored a disaster victim's sobbing. Similarly, we would be shocked if a biologist had an idea for a cure for cancer but did not pursue it, and we are disgusted when we hear that physicians have not bothered to determine whether the standard treatment for a serious disease is effective.

Should research psychologists, like other professionals, try to use their knowledge to help society and individuals? For example, do psychologists owe it to society to try to solve problems that plague us such as prejudice, depression, and violence—as well as to test the effectiveness of existing approaches to such problems?

Before you answer, realize that attempts to help one person may end up hurting that person—or someone else. The EMT's or clinical psychologist's intervention may end up harming an individual who does not need or want treatment. The biologist's potential cure for cancer may not work—and, even if it does, some people may be hurt—even killed—during the early trials. The physician who tests a standard treatment by administering it to some patients and not to others will probably either harm patients in the treatment group (if the treatment is ineffective or actually harmful) or harm patients in the no-treatment group (if the treatment does work, the researcher has withheld a cure). Similarly, research psychologists who try to help society may end up harming individuals.

As you can see, determining whether it is ethical to do any study involves weighing the study's potential for good—its ability to provide a valid answer to an important question—against its potential for harm. Weighing a study's potential benefits and potential risks is especially difficult when studying living beings for at least two reasons. First, as part of evaluating the benefits, psychological researchers must not only ask whether the research question is important but also must ask whether the study will provide a valid answer to that question. Second, evaluating the risks involves trying to predict the reactions of varied, variable, and valuable individuals.

In this chapter, you will learn how researchers determine whether a valid study can and should be done. As we show you some obstacles to getting valid data and some ways to overcome those obstacles, you will begin to learn how to evaluate other people's research as well as how to design your own research. After we discuss how to maximize a study's chances of producing valid data, we will show you how to minimize a study's chances of harming participants and how to conduct a study that is consistent with the American Psychological Association's ethical code. Thus, by the end of the chapter, you will know some basic principles that will help you propose ethical research: research in which risks are minimized, benefits are maximized, and potential risks are outweighed by potential benefits.

QUESTIONS ABOUT APPLYING TECHNIQUES FROM OLDER SCIENCES TO PSYCHOLOGY

To design ethical and valid studies, psychologists use the same tool other sciences use—the scientific method. However, because psychologists study human and animal behavior rather than the behavior of objects, plants, or microbes, psychologists face additional scientific and moral obstacles. To appreciate how sensitive psychologists are to the unique scientific challenges and ethical obligations involved in studying the behavior of living things, let's see how a psychologist would react if someone ignored those additional challenges and responsibilities. For instance, suppose that a novice investigator tried to model his psychological research after the following chemistry experiment:

> A chemist fills two test tubes with hydrogen and oxygen molecules. She leaves the first test tube alone. She heats the second over a flame. She observes that water forms only in the second test tube. She concludes that heating hydrogen and oxygen molecules causes them to react.

The novice investigator then conducts the following study:

> A novice investigator fills two rooms with people. One room is at a normal temperature; the second room is one that he has heated to 80 degrees Fahrenheit (26.7 degrees Celsius). He observes that the people in the second room behave differently from the people in the first room. He concludes that "feeling too hot" makes people "feel aggressive."

Because of the vast differences between humans and molecules, an experienced research psychologist would have four sets of questions about the

TABLE **2.1** Common Threats to the Three Kinds of Validity

Types of Validity	Major Sources of Problems	Mistakes to Avoid	Examples of Problems in Real Life
Internal:			
Establishing a *cause–effect* relationship between the manipulation and a behavior *in a* given study; establishing that a certain observable event caused (was responsible for, influenced) a change in behavior.	Allowing factors other than the manipulation to vary. For example, if the treatment and the no-treatment group differ before the study begins, we can't conclusively establish that the treatment caused the difference between the groups.	Failing to ask, "Is there something other than the treatment that could cause the difference in behavior?" or "Would it (the difference) have happened anyway?"	Misidentifying the causes of a problem. Giving a new president credit or blame for changes in the economy, blaming a new dentist for your existing dental problems, claiming that vaccines are responsible for a child's autism.
Construct:			
Accurately naming our measures and manipulations; making accurate inferences about both (a) what our participants' behaviors mean and (b) what psychological states our manipulations produce.	Faulty measures, resulting in mislabeling or misinterpreting behavior. Poor manipulations can also harm construct validity, as can participants figuring out and playing along with (or against) the hypothesis.	Accepting at face value that a test measures what its title claims it does. Anybody can type up some questions and call it an intelligence test—but that doesn't mean the test really measures intelligence.	Mislabeling a behavior. Thinking that a shy person is a snob, believing that what people *say* they think and feel is exactly what they *do* think and feel, having complete confidence in lie detectors, "knowing" that a cat loves you because it sits in your chair after you get up.
External:			
Generalizing the study's results *outside* the study to other situations and participants.	Studying only a few participants, studying only one subgroup (e.g., White, middle-class men) or studying them in a very controlled environment.	Believing that any survey, regardless of how small or biased the sample, has external validity.	Stereotyping. For example, based on a limited sample, concluding that, "They are all like that; seen one, seen them all."

novice investigator's study. The first three sets (summarized in Table 2.1) deal with the validity of the novice investigator's conclusions. First, did the investigator, by putting participants in different rooms, really *cause* participants to behave differently? Second, did the investigator really manipulate and measure the *variables* ("feeling hot" and "being aggressive") that he claimed he did? Third, would the results *generalize* to other participants and places? The fourth set of concerns is the most serious: Should the study have been done?

Internal Validity Questions: Did the Treatment Cause a Change in Behavior?

The first set of questions deals with the study's **internal validity**: the degree to which the study demonstrates that the treatment *caused* a change in behavior. If the study establishes that putting the participants into different rooms caused participants in one room to behave differently from participants in the other room, the study has internal validity. To establish that one particular factor causes a specific effect, an internally valid study must establish the following three things:

1. Because causes have effects, the research must show that changes in the variable labeled the "cause" were *related* to changes in the variable labeled the "effect": If a study can't demonstrate that two variables are in any way related, it can't possibly establish that two variables are *causally* related. For example, before showing that vaccines cause autism, research must first show that people getting vaccinated are more likely to become autistic than people who are not vaccinated. Because research shows no relationship between vaccines and autism, we know that vaccines do not cause autism. (Similarly, because there is no relationship between full moons and arrests, we know that full moons don't cause crime.)

2. Because causes must come before effects (e.g., parents can't inherit insanity from their kids), the research must show that the alleged "cause" changed *before* the alleged "effect" occurred. Otherwise, what the researcher calls the cause may really be the effect (e.g., frequent doctor visits may be the result—rather than the cause—of poor health).

3. Because effects have many potential causes, the research must show that the alleged cause is the only viable suspect for the effect. Ruling out all other potential causes—whether it be the cause of death in a crime show or the cause of a psychological effect—can be difficult.

How difficult is it to meet these three criteria for internal validity? That depends on what you are studying and what type of research design you are using. For example, consider how our chemist would answer the three questions that need to be answered to establish that a factor caused an effect:

1. Were changes in the effect *related to* changes in the alleged cause? Yes—there was water in the test tube that was exposed to the flame, but no water in the test tube that was not exposed to the flame.

2. Did changes in the effect come *after* changes in the cause? Yes—the difference (water in one test tube, not in the other) occurred *after* the flame manipulation.

3. Could the changes in the effect be due to something other than the alleged cause? No—the difference between the test tubes could not be due to anything other than the flame manipulation

Thus, for the chemist, establishing internal validity is easy.

Psychologists, on the other hand, don't have it that easy. For starters, whereas the chemist is always free to impose a treatment that would be harmful to humans (e.g., burning) on molecules to see what effect that treatment has, psychologists may not even be allowed to expose people to potentially helpful therapies.

For both ethical and practical reasons, psychologists, instead of exposing people to treatments, are often limited to merely observing what people naturally do. For instance, rather than manipulating one variable (living in hot climates) to see whether that first variable has an effect on a second variable (aggression), psychologists are often limited to watching to see whether there is a statistical relationship between those two variables (e.g., are people who live in the southern United States more aggressive than people who live in the northern United States?).

When a researcher cannot manipulate variables, two obstacles prevent the researcher from making cause–effect statements. First, because such researchers often can't determine which variable changed first, they can't say which variable is the cause and which is the effect. For example, if the novice had merely observed that people who were fighting were sweatier than people who were not fighting, the novice would not know whether people (a) were sweaty and then fought or (b) fought and then became sweaty. Put another way, observers who see a relationship between two variables, like basketball referees who see two players fighting, want to know "Which one started it?" Unfortunately, it's quite possible that neither the referees nor the observers can know which one started it.

Even if observers believe they know which suspect started it, they may be wrong. Like the basketball referee who mistakenly believes that the player who fought back is the player who started the fight, observers may be pointing at the effect and calling it the cause. Because knowing which variable changed first is often difficult, some people *may* have cause and effect reversed when they conclude that diet drinks make one fat, antidepressants cause depression, job success causes happiness (Boehm & Lyubomirsky, 2008), or that a company's change in strategy is the cause—rather than a consequence—of its decline (as Rosenzweig, 2007, points out, doing poorly tends to force companies to change strategies).

Fortunately, the novice in our experiment was able to avoid reversing cause and consequence because, instead of merely observing real life events, he put participants in different rooms first, and had them respond later. To appreciate how manipulating the treatment prevents him from confusing cause and effect, ask yourself, "Would he ever think that the hot room became the hot room because the participants fought there?" The answer is "no," because he knows that he made the hot room hot before participants entered it.

Unfortunately, mistaking an effect for a cause is not the only error people can make when trying to figure out what caused an effect. To understand the other mistake, think back to the basketball referee trying to answer the question, "Which one of these two started the fight?" The right answer might be "Neither," because a third party (e.g., a coach, another player), may have provoked the two players to fight.

Just as a third player may really be responsible for a fight between two players, a third factor may be responsible for a statistical relationship between two factors. That is, rather than the two observed variables affecting each other, the two observed variables may both be influenced by some unobserved third variable. For example, if we observe more violent crime in the southern United States than in the northern United States, that relationship could be due to something other than temperature, such as Southerners being more likely than Northerners to view insults as an attack on one's honor that demands a violent response (Nisbett, 1993) or Southerners experiencing more poverty than Northerners.

Similarly, in the novice's temperature–aggression study, some third variable—some difference other than the room manipulation—might have caused the difference in how the two groups behaved. For example, in the novice's study, at least three factors other than the treatment might account for the differences between the two groups.

1.   The people assigned to the hot room may have been naturally more hotheaded and aggressive than the people assigned to the other room. Because the novice could not create (even with cloning) two groups that had identical personalities and backgrounds (the novice can't match the groups on every single variable on which people differ), the novice cannot eliminate the possibility that personality differences between the groups account for the differences in their behavior.
2.   Even if the two groups had identical personalities, they might not have been tested at the same time. For example, if the normal-room group had been tested in the morning and the hot-room group was tested at night, we would be concerned because people tend to be more aggressive at night.

To rule out time-of-day effects, the novice might try to keep the time of day constant. For example, he might test all of the participants at 9:00 a.m.

Alternatively, the novice might make sure that just as many hot-room as normal-room sessions were held at each time of day. One way he could have guaranteed such equality would be to alternate between running hot-room sessions and running normal-room sessions (see table below). For example, during the first week, on Mondays and Wednesdays hot-room participants could be tested in the morning (and the normal-room participants would be tested in the afternoon), and on Tuesdays and Thursdays, hot-room participants would be tested in the afternoon (and the normal room participants would be tested in the morning). In the second week, the novice could reverse the first week's schedule to control for the possibility that some particular day and time combination (e.g., Thursday evening) was a particularly aggressive time. This strategy of systematically balancing out a factor is called **counterbalancing**.

	Week 1				Week 2			
	Monday	Tuesday	Wednesday	Thursday	M	T	W	TH
9:00 a.m.	Hot	Normal	Hot	Normal	Normal	Hot	Normal	Hot
9:00 p.m.	Normal	Hot	Normal	Hot	Hot	Normal	Hot	Normal

3. Even if the novice controlled for time of testing, there would still be many outside events that the novice could neither control nor keep away from participants. Outside events that could make people act more aggressively and could penetrate the lab include the voices of people cursing in the hallway; the noises from jackhammers, lawnmowers, and alarms; and the rumbles, light flashes, air pressure changes, and negative ions of a thunderstorm.

The novice tried to prevent events occurring outside the lab from contaminating the study, but keeping everything constant was not possible. If the novice's lab was soundproof, he would have been able to block out noises from the hallway, but he would not be able to block out all outside influences. A storm, for example, could cause the lights to flicker, the building to shake, the air pressure to drop, the humidity to rise, and the concentration of negative ions to soar. Furthermore, even if he could stop all these outside events from penetrating the lab, they would still affect his participants: The participants who run through a thunderstorm to get to the lab would arrive in a different mood (and in wetter clothes) than those who strolled in on a nice, sunny afternoon.

Researchers who find that two variables are related are in a similar position to crime show detectives who find two dead men next to the gun that killed them. In both cases, such a discovery raises at least three questions. For the detectives, the three questions are (1) did Person A kill Person B and then commit suicide? (2) did Person B kill Person A and then commit suicide? or (3) were Persons A and B killed by some third party? For researchers, the three questions are (1) did Variable A influence Variable B? (2) did Variable B influence Variable A? or (3) were Variables A and B influenced by some third variable?

Whereas crime show detectives avoid confusing the victim for the murderer by relying on the coroner to tell them who died first, the novice dealt with the "what came first" problem by using a technique many research psychologists use: manipulating the treatment. Because the novice manipulated the treatment, he knows that the difference in how the groups were treated (heating the room or not) came *before* the difference in how the groups acted.

However, just as knowing that Person B did not kill Person A fails to establish that Person A—rather than some third party—killed Person B, knowing that Variable B did not influence Variable A fails to establish that A—rather than some third variable—influenced B. Thus, to know that A influenced B, you not only have to establish that changes in A were followed by changes in B but you would also have to show that no third variable caused the changes in B.

In the novice's study, there were many potential third variables—differences between the hot and normal room groups that had nothing to do with the rooms per se. Consequently, it would be irresponsible to say that the difference in the rooms—rather than one of those other differences—was responsible for the differences in how the two groups behaved.

The novice tried to eliminate the possibility that something other than the treatment caused the hot-room participants to behave differently from the normal-room participants by trying to make it so that the only difference between participants was that some participants were in the hot room whereas others were in the normal room. However, he wasn't able to keep everything the same. No matter how hard he tried, he would never be able get two groups of participants who were identical, and he would never be able create two testing situations that were identical.

Given that the novice could not test identical groups of participants under identical conditions, what should the novice have done to establish internal validity? The experienced investigator's answer may surprise you: The novice should have used some *random* (chance) process to *assign* (allocate) participants to either the hot room or the normal room. In this case, **random assignment** would be similar to assigning people to condition based on flipping a coin: "Heads," the participant is put into the hot room; "tails," the participant is put into the normal room.

Note random assignment's two aspects. First, it is assignment, meaning participants are given something. Just as students agreeing to go to a school are assigned advisors, student identification numbers, and class assignments, participants agreeing to be in a study are assigned a treatment. Note that because you cannot assign participants to gender, race, or age, you cannot randomly assign participants to those characteristics.

Second, it is *random*, meaning each participant has *an equal chance* of being assigned to any of the treatments. Note that for researchers, "random" means something very different from "arbitrarily," "unsystematically," "haphazardly," "accidentally," or any other ways that random is used in everyday life. Indeed, researchers usually are very systematic in their efforts to ensure that every participant has an equal chance of being in each group. For example, most researchers use carefully constructed random numbers tables or sophisticated computer programs to randomly assign participants.

Although most researchers use a much more elaborate strategy than assigning people based on the flip of a coin, it may help you appreciate the basic logic of random assignment by imagining that our novice was assigning participants to condition by flipping a coin—or having a computer program simulate flipping a coin—in a way that made the chances of getting a "heads" on any given flip exactly the same as the chances of getting a "tails." If the coin came up "heads," the participant would be assigned to the normal-room group; if the coin came up "tails," the participant would be assigned to the hot-room group.

Note that such a coin would not be systematically biased toward or against any group. For example, it would not have a greater tendency to come up "heads" for aggressive individuals than for nonaggressive individuals. Instead, if the coin did put a greater number of aggressive individuals in the hot-room group than it puts in the normal-room group, it did so only by chance.

Although chance would not necessarily make the groups exactly equal, chance would tend to make the groups approximately equal. If given enough chances, chance will almost balance things out; as a result, almost as many aggressive people would be in the hot-room group as in the normal-room group.

You know that chance tends to balance out: If you flipped a fair coin 100 times, you would get approximately 50 heads and approximately 50 tails. Similarly, if, among your participants, you had 100 who were naturally violent, as you flipped the coin for each of those 100 aggressive participants, you would get about 50 heads and about 50 tails. Thus, if you assigned "heads" participants to the hot-room group and "tails" to the normal-room group, you would have about as many aggressive individuals in the hot-room group as you did in the normal-room group. Note that for random assignment to work, you don't have to know which of your participants are aggressive: After all, the coin is doing the work—and it doesn't know.

Note also that what the coin is doing for aggressive individuals (roughly balancing them between the groups), it is doing for every personal characteristic. For example, if you had 120 women and 80 men, the coin would come up heads approximately 60 times during the 120 flips involving women and come up heads approximately 40 times during the 80 flips involving men. Consequently, you would end up with almost the same number of women (60) in each group and the same number of men (40) in each group. Even participant characteristics that you don't know about (e.g., which participants have obnoxious little brothers) are being distributed *fairly* equally between groups.

When given many chances, chance will tend to distribute random differences fairly equally. However, when given few chances, it may distribute random differences very unequally. Thus, if you assigned each individual to group by flipping a coin and you had many participants, chance would do a good job of making your groups equivalent. Conversely, if you had few participants, chance would probably do a poor job of balancing the effects of individual differences between groups. Indeed, with too few participants, chance has no chance. For example, if you had four people in your study and only one of those was violent, flipping a coin could not give you equal groups. Even if you had eight participants, four of whom were violent, flipping a coin might result in all four violent individuals ending up in the hot-room group. Why? Because, in the short run, chance can be fickle. For instance, it is not that unusual to get four "heads" in a row. To appreciate that chance can be fickle in the short run but dependable in the long run, realize that although a casino may lose several bets in a row, the casino always wins in the end.

Admittedly, even with large samples, random assignment does not guarantee that you will get two identical groups—no method can guarantee you that. At least with random assignment, you know that (a) other than receiving different treatments, chance is the only factor that will cause your groups to differ and that (b) you can use statistics—the science of chance—to estimate the degree to which chance alone would have made your groups different.

To illustrate that you can use statistics to determine the degree to which chance might fail to balance out any non-treatment differences between your conditions, imagine that you do the following study (similar to one done by Batson, Kobrynowicz, Dinnerstein, Kampf, & Wilson, 1997). You have 100 individuals in Condition 1 and 100 in Condition 2. You tell the participants in Condition 1 to flip a coin and tell you the outcome because if it comes up "heads," you will give them a raffle ticket; you tell the people in Condition 2 to flip the coin and tell you the outcome, but you do not tell them that they will win anything. Suppose we obtained the following results: Condition 1 gets—according to their reports—heads 90 times out of 100 flips, and Condition 2 reports heads 50 times out of 100 flips. We can use statistics to determine that this difference is almost certainly not due to chance. Therefore, we would conclude that this difference was probably due to the raffle ticket manipulation changing the behavior of Condition 1 participants.

We can apply statistics to more than just coin-reporting behavior—we can apply statistics to any behavior. As a result, we can use statistics to estimate how much two randomly assigned groups should differ by chance alone, and, if the groups differ by much more than that amount, we can be confident that some variable we did not randomize—ideally, the treatment—is at least partly responsible for the difference in how the two groups behave.

Notice that, if you were to redo the novice's study, random assignment could balance out and account for not only individual differences between groups but also differences between testing sessions—*if you have enough testing sessions*. If you only ran two sessions—one for hot-room participants and one for normal-room participants—the sessions would clearly occur at different times (you can't be in two rooms at the same time), and random assignment could not make the times of those two sessions equivalent. However, if you had many sessions, random assignment could make the times of your sessions equivalent. To illustrate how random assignment deals with differences in the times of testing sessions, suppose the novice had been running participants at the following times:

		Week 1			Week 2			
	Monday	Tuesday	Wednesday	Thursday	M	T	W	TH
9:00 a.m.								
9:00 p.m.								

© Cengage Learning 2013

If the novice randomly assigned each participant to either the hot room or the normal room, chances are that we would somewhat balance out the time of testing. It would be rare for all the hot-room participants to be tested in the morning or for all of them to be tested at night. Instead, it would be much more likely that about half of the hot-room participants would be tested in the morning and about half would be tested in the evening. Admittedly, because random assignment would probably not balance out time of day effects perfectly, the hot-room and normal-room groups' testing times could still differ by chance. However, as we discussed earlier, statistics can be used to factor out the effects of chance. Furthermore, notice that the effects of chance refer not only to chance differences in the time of day of the testing sessions but also to all other chance differences between the testing sessions—even those caused by the experimenter stumbling or lightning striking.

If the novice had tested participants individually and had randomly assigned participants to condition, the novice would have been able to rule out not only the effects of random, outside events but also of other non-treatment factors, such as personality factors. In technical terminology, the novice would have conducted an **experiment**: a particular type of study that allows researchers to make cause–effect statements because it (a) manipulates a treatment to establish that the treatment comes before the difference in behavior and (b) uses some technique—usually random assignment—to establish that the difference in behavior is probably not due to anything other than the treatment. Unfortunately, the novice's study, like most studies, was not an experiment (although he manipulated the treatment, he did not use random assignment)—and because his study was not an experiment, it did not have internal validity.

In short, establishing internal validity is difficult (see Figure 2.1). About the only way to establish internal validity is to do a special type of

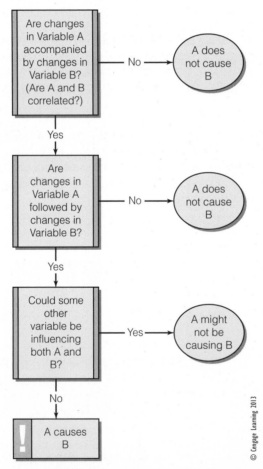

© Cengage Learning 2013

FIGURE **2.1** Establishing internal validity means answering some tough questions.

study—an experiment—that sets up the special conditions that allow a treatment's effect to be isolated (Stanovich, 2009). Experiments typically set up those special conditions by using random assignment. Because random assignment is so important to internal validity, the following table is a useful—although slightly oversimplified—way for you to think about internal validity and designs:

	Treatment Assigned?	Treatment Randomly Assigned?	Internal Validity
Experiment	Yes	Yes	Yes
Quasi-experiment	Yes	No	Questionable
Non-experiments (Surveys, case studies, observational studies, and other studies in which the researcher does not administer a treatment)	No	No	No

© Cengage Learning 2013

Because only experiments have internal validity and because most studies are not experiments, the experienced investigator is skeptical whenever people claim to establish that a treatment causes, increases, decreases, affects, influences, impacts, produces, brings about, triggers, or makes any change in a behavior (Stanovich, 2009).

As we have seen, the novice investigator naively assumed that his study had internal validity. Specifically, he went from (a) *observing* that the participants in the room where he turned up the thermostat behaved differently from those in the other room to (b) *inferring* that "turning up the thermostat *caused* the two groups to behave differently." However, that was not the only questionable inference he made.

Construct Validity Questions: Are the Variable Names Accurate?

The novice investigator also went from (a) *observing* that the participants in the room where he turned up the thermostat behaved differently from those in the other room to (b) *inferring* that "participants who felt hot also felt more like harming others." In making the leap from observable events—what the novice did and saw—to talking about the world inside participants' heads, the novice investigator presumed that his manipulation made hot-room participants feel hot and that what he measured was aggression. In other words, the novice investigator assumed that he accurately manipulated and measured psychological **constructs**: characteristics of individuals that can't be directly observed, such as mental states (e.g., love, hunger, feeling hot), traits (e.g., agreeableness), abilities (e.g., intelligence), and intentions (e.g., aggression: the intent to harm another).

As critics point out, talking about constructs is risky because constructs are not things that are observed. Instead, constructs are concepts that are *constructed*: created, or made up (Levy, 2010). People have invented constructs like "ghosts," "ego strength," "codependency," "motivation," "aggression," and "sense of humor"—but these inventions may be fictions. Thus, Skinner

cathy®   **by Cathy Guisewite**

FIGURE **2.2** The problem with constructs.

Because constructs can't be observed, knowing what participants do is not the same as knowing what they are thinking.

and other radical behaviorists have argued that psychology should stick to talking about observable behavior rather than about invisible constructs.

Even psychological scientists who talk about constructs admit that trying to infer what people are thinking from observing people's behavior can lead to leaping to wrong conclusions (see Figure 2.2). Most non-behaviorists would also admit that using questionable constructs (e.g., "anal retentive," "possessed") to label behavior may lead to leaping to questionable conclusions. Finally, most non-behaviorists would admit that even if the construct is fairly well established (e.g., lying), the person who assumes that a certain response (e.g., avoiding eye contact, fidgeting) is a valid indicator of that construct is leaping to questionable conclusions.

Yet, despite all these reasons not to use constructs, people commonly use constructs to label behavior. For instance, some people are quick to infer that a person who works slowly is unintelligent when the truth may be that the individual is cautious, ill, lazy, or unfamiliar with the task.

Our novice researcher might believe that he was appropriately cautious. He may admit that he was stating his conclusions in terms of the invisible construct—"feeling hot"—but he may also ask, "Why should I be limited to stating my conclusions in terms of the narrow operational definitions (e.g., putting participants in a room that exceeded 80 degrees Fahrenheit)? After all, the chemist's conclusions would deal not with the effects of her operational definition of heat—putting a test tube over a lit Bunsen burner—but rather with the effects of the underlying construct that operational definition was meant to manipulate: heat."

The experienced researcher would admit that the chemist, like the novice, made inferences about unseen constructs. However, the experienced researcher would point out three differences between the chemist's and the novice's studies.

First, the leap from seeing a chemist put a test tube over a lit Bunsen burner to concluding that the chemist is manipulating the heat of molecules in that test tube is a small hop. The flame definitely heats the molecules, and it is unlikely that the burner has any other effects: The molecules do not notice the flame's color, do not hear the gas coming into the burner, and do not smell the gas. The leap from putting participants in a hot room to concluding that participants feel hot, on the other hand, is a gigantic jump. The hot room may not make participants feel hot (instead, they may feel warm, comfortable, or drowsy), and participants may be affected by some of your manipulation's unintended effects (e.g., they may be annoyed by the noise or the odor coming from the heater). Thus, manipulating the temperature of molecules is simpler than manipulating how people feel.

Second, the amount of water produced by a reaction is both a clear indicator of the reaction and easy to measure. Indicators of how aggressive someone feels tend to be either not closely tied to physical aggression (e.g., increased heart rate, scores on a multiple-choice test) or hard to measure (e.g., how insulting the participant's comments were, how physical the participant was with the experimenter). In short, counting drops of water to assess the size of a chemical reaction is simpler than the "mind reading" needed to measure aggression.

Third, molecules don't know when they are in a study, don't try to guess what results the researcher wants, and don't try to act in a way that will give the researcher the desired results. Human participants, on the other hand, often know when they are in a study, try to guess what results the researcher wants, and try to act in a way that will help the researcher get those results.

Because knowing what participants are thinking is so difficult, the research psychologist would question the temperature study's **construct validity**: the degree to which the study measures and manipulates the variables that the researcher claims to be measuring and manipulating (see Figure 2.3). Specifically, the research psychologist would look for potential cracks in the study's construct validity by asking three questions:

1. Was the manipulation poor? If so, the construct "feeling hot" was not manipulated appropriately.
2. Was the measure was poor? If so, the construct "feeling aggression" was not measured accurately.
3. Did participants figure out the hypothesis and play along? If so, participants in the hot room might try to support the researcher's hypothesis by acting in a way that makes them look more aggressive—even though they don't really feel aggressive.

Construct Validity Problems Caused by the Manipulation: What Does the Treatment Really Manipulate?

The experienced researcher would probably begin questioning the construct validity of the novice investigator's study by questioning whether the novice's operational definitions really reflected the variables the novice wanted to manipulate and measure. She might start by questioning the temperature manipulation's construct validity. She would ask herself, "Is it right to call this 'raising-the-thermostat' manipulation a 'feeling-hot' manipulation?"

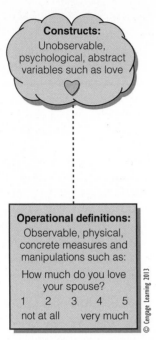

Constructs:
Unobservable, psychological, abstract variables such as love

Operational definitions:
Observable, physical, concrete measures and manipulations such as:

How much do you love your spouse?

1   2   3   4   5
not at all   very much

© Cengage Learning 2013

FIGURE **2.3** Linking the invisible to the visible: The challenge of construct validity.

To begin answering that question, she would ask two other questions. First, "Did the manipulation make the hot-room group really feel hot?" Second, "Did the manipulation have any effect besides making the hot-room group feel hot?"

The answer to the first question is not as simple as you might think. You can't directly get inside participants' minds to change how they feel. Instead, the only possible way for any researcher to manipulate participants' mental states is indirectly—by changing the physical environment and then hoping that participants react to the manipulation in the way the researcher expects. Unfortunately, participants may react to the manipulation differently from the way the researcher intended. Thus, putting one group of participants in a room that is, at the physical level, 10 degrees hotter than another is not the same as making one group, at the psychological level, feel hot. For example, participants may take off jackets and sweaters to cool off, they may find the room's temperature "comfortable," or they might not even notice the difference in temperature.

If the researcher decides that the manipulation does indeed make participants feel hot, she still has to answer the question, "Did the manipulation do anything besides make the hot-room group feel hotter than the other group?" Usually, manipulations are not so pure that their only effect is to change the one thing you intended to change. Instead, manipulations often contain extra ingredients or produce unwanted psychological reactions.

The research psychologist would start her search for the manipulation's extra ingredients by asking, "What did turning up the thermostat do to the participants' environment besides make the room hotter?" She may find that turning up the thermostat also made the room noisier (because the heater was noisy) and decreased the room's air quality (because the heater's filter was dirty). If turning up the thermostat is a temperature manipulation, a noise manipulation, and an air-quality manipulation, how can the novice investigator justify labeling it as a temperature manipulation? It would be more accurate to call it a temperature, noise, and air-quality manipulation.

To figure out the manipulation's impurities, start by asking—and answering—the question: "What differences are there between the study's conditions?" For example, if one room is always the "hot room" and one room is always the "normal room," the rooms are different. Any difference between the rooms—in the lighting, the smell, the quality of the carpet, or even the room number—might influence participants. Thus, the "hot room" manipulation might be a "hot room and carpeting and room number" manipulation.

Even if the manipulation is pure at the physical level, it may not be pure at the psychological level. The novice investigator may have made the participants feel frustrated about being unable to open the windows to cool off the room, or he may have made participants feel angry with him for putting them in such an uncomfortable room. Therefore, in addition to being a manipulation of feeling hot, the treatment may have had the additional side effect of making people frustrated or angry. So, how can the novice investigator justify calling the room manipulation a temperature manipulation when it may actually be a frustration manipulation or an anger manipulation?

As you have seen, it is difficult to manipulate variables. Even seemingly straightforward manipulations may not be what they seem. For example, suppose that an "aspirin" manipulation involves giving aspirins to one group, but not to the other. On the surface, the "aspirin" label would seem to describe the manipulation accurately. However, many aspirin tablets also contain caffeine. Therefore, rather than being a pure manipulation of aspirin, the manipulation may be an "aspirin and caffeine" manipulation. What if getting the aspirin makes participants *expect* to feel better? Then, the manipulation is an "aspirin and positive expectations" manipulation.

In conclusion, you should always question the label that a researcher decides to attach to a manipulation. Because of the difficulties of manipulating what one wants to manipulate, the novice investigator should not expect skeptical scientists to take it on faith that he is manipulating the invisible mental state that he claims to be manipulating.

Construct Validity Problems Caused by the Measure: What Does the Measure Really Measure?

Even if the manipulation of "feeling hot" is valid, the measure of aggression may not be. Psychological constructs such as aggression are abstract and invisible and therefore impossible to measure directly. Because we cannot see directly into participants' minds, the best we can do is to set up situations in which what they are thinking will be reflected in their behavior. Unfortunately, participants' behaviors may be mislabeled. For example, the novice investigator may have misinterpreted "kidding around" and attention-getting behaviors as aggression. Or, the novice investigator may have misinterpreted physiological reactions to being hot (sweating, flushed face) as signs of non-verbal aggression. Or, the novice investigator may have labeled assertive behavior as aggressive. Or, scores on the novice investigator's multiple-choice test of aggression may not have any relationship to aggression. In short, it is reckless to assume that a measure will perfectly capture the construct that the researcher is trying to measure.

Construct Validity Problems Caused by Participants: Is Their Behavior Genuine or an Act?

Even if the novice investigator had used a good manipulation and a good measure, the results may be misleading because participants who know they are in a research study may mask their true feelings. Some participants, rather than reacting to the manipulation, may be acting to "help" the researcher "prove" the hypothesis. In the novice investigator's study, hot-room participants may realize that they have been (a) deliberately placed in an abnormally hot room, (b) given an opportunity to express aggression, and then (c) are expected to behave—or at least act—aggressively. If they like the investigator, they will probably play along.

Review: Comparing Internal Validity and Construct Validity

In this case, our novice investigator would like both internal and construct validity. If he can show that his study has both internal and construct validity, he can safely state that, in his study, feeling hot caused aggression. Specifically, showing that the study had *internal validity* allows him to state that his manipulation *caused* a change in scores on his measure; showing that the study had construct validity allows him to state that he manipulated "feeling hot" and that he measured aggression.

If his study had internal—but not construct—validity, he could make cause–effect statements, but he couldn't legitimately say that he had manipulated and measured the variables he intended to manipulate and measure (the constructs "feeling hot" and "aggression"). Therefore, the only thing he could safely conclude would be that something about his manipulation had *caused* a change in participants' behavior. That is, because he couldn't couch his conclusions in terms of constructs, he would be limited to couching his conclusions in terms of his operational definitions of those constructs. For example, he might be limited to concluding that "Turning up the thermostat caused a difference in how participants filled in circles on a multiple-choice answer sheet."

If, on the other hand, his study had construct—but not internal—validity, he could claim that he was dealing with the variables ("feeling hot" and "aggression") he claimed he was, but he couldn't legitimately make cause–effect statements. He could conclude that, "the group that felt hot was more aggressive," but he would not know *why* that group was more aggressive: He could not conclude that feeling hot *caused* aggression.

External Validity Questions: Can the Results Be Generalized?

Suppose that the novice investigator actually manipulated feeling hot and accurately measured aggression (construct validity). Suppose further that he had established that differences between the two groups in this particular

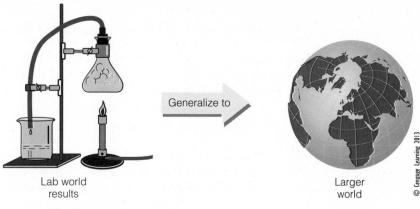

Lab world
results

Generalize to

Larger
world

© Cengage Learning 2013

FIGURE **2.4** External validity.

study were *caused* by the room manipulation *(internal validity)*. In that case, the experienced researcher would still question the study's **external validity**: the degree to which the results could be *generalized* to different participants, places, and time periods (see Figure 2.4). There are at least two reasons to question the aggression study's external validity.

Can the Results Be Generalized to Other Participants?

First, because people differ, a result that holds for one group of people might not hold for a different group of people (see Figure 2.5). For example, the novice investigator might have obtained different results had he studied Russian sixth graders instead of Midwestern U.S. college students, if he had studied people used to working in very hot conditions, or if he had studied less-aggressive individuals.

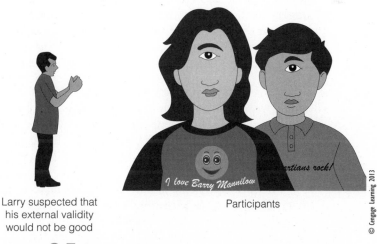

Larry suspected that
his external validity
would not be good

Participants

© Cengage Learning 2013

FIGURE **2.5** Studying a small, unusual sample might harm a study's external validity.

Can the Results Be Generalized to Other Settings?

Second, because people's behavior may change depending on the situation, the results might not hold in another setting. For instance, suppose the novice investigator used a sterile laboratory setting to eliminate the effects of non-treatment factors. By isolating the treatment factor, the novice investigator may have succeeded in establishing internal validity. However, results obtained under such controlled situations may not generalize to more complex situations, such as the workplace or the home, where other factors, such as frustration and pressure, come into play. Therefore, some researchers would advocate that the novice's study be modified to increase its *ecological* validity (also called *mundane realism*): the look and feel of the naturalistic, real-life situation under study. For example, some might urge that the study be repeated in a real-world location (e.g., the dorms), that the lab be made to look more like the family room of a house, or that participants be made to do an everyday activity, such as watching television.

External Validity Questions: Summary and Conclusions

In short, even if temperature did increase aggression in this particular lab, with this particular group of participants, at this particular time, the experienced researcher would not automatically assume that temperature would have the same effect in future studies conducted with different types of participants in different settings. Therefore, to maximize *external* validity (chances that the findings applying *outside* of this particular study), the experienced researcher might repeat the novice's study in different places with different types of participants.

To increase external validity, the experienced investigator might try to select participants who were a representative sample from the most diverse population available to her. Ideally, that diverse population would be the world's population. Realistically, that population would probably be all the students at her school. To get a representative sample from that population, she would probably use **random sampling**: a strategy that makes sure that every person in the population has an equal chance of being selected for the study. In this case, she might get a list of all her school's students and then have a computer randomly select 300 individuals from that list.

Ethical Questions: Should the Study Be Conducted?

Before repeating the study—indeed, before performing it in the first place—the investigator would have to determine whether conducting the study was **ethical**: consistent with the American Psychological Association's principles of right and wrong. If the study could not be conducted ethically, it should not be done (see Figure 2.6).

The idea that some studies should not be conducted is a relatively new one. The first obvious example of such studies came to light after World War II, when some German physicians and administrators were sentenced for "murder, torture, and other atrocities committed in the name of science." Defendants claimed that their experiments were not that different from what U.S. scientists were doing. As part of the verdict, 10 principles of "Permissible Medical Experiments" were produced to (a) prevent scientists from ever again

FIGURE **2.6** The decision to do a study should be based on more than scientific validity.

Not all studies should be performed. Note that the ethical consequences of a study are sometimes hard to predict.

being forced by a government to do such unethical things in the name of science and (b) to illustrate that what the German physicians had done was outside the bounds of acceptable medical research. These 10 principles, which are now called "The Nuremberg Code," are—in weakened form— part of most research ethics codes. Specifically, most research ethics codes incorporate the following three principles:

1. Maximize benefits: The research must have some potential benefits to society, and these benefits should be maximized by having a valid study.
2. Minimize harm: Do not do a study where there are serious risks to the participants. Do what you can to reduce the chances of any harm, including giving participants

   A. Informed consent: Participants should be volunteers and know what the study involves and what risks the study might involve.
   B. The right to withdraw without penalty: Participants should be able to quit the study at any time.

3. Weigh risks and benefits: If risks outweigh benefits, do not do the study.

In deciding whether the study was ethical, the researcher would not rely on the Nuremberg Code. Instead, the researcher would consult the American Psychological Association's *Ethical Principles of Psychologists and Code of Conduct* (American Psychological Association [APA], 2002), often referred to as the *Principles*. A copy of the ethical guidelines from the *Principles* relating to research is included in Appendix D. In addition to the *Principles*, the researcher might also consult the American Psychological Association's *Ethical Principles in the Conduct of Research With Human Participants* (APA, 1982). By consulting both sources, the researcher should be able to make an informed decision about whether the participants' rights had been protected and whether the novice investigator had lived up to his responsibilities.

Has Potential Harm Been Minimized?

As the *Principles* point out, participants have the right to **informed consent:** to understand what will happen in the study and then to agree to participate. Thus, according to the *Principles*, the novice investigator should have told participants that the study would involve sitting in a room that might be hot with a group of people for 30 minutes while filling out a questionnaire about their feelings toward the other participants. Knowing what the study was about, participants should have freely volunteered to be in the study and signed an informed consent form. That consent form, in addition to describing what the study was about and what the participant would do, should

1. explain the potential benefits of the research
2. explain any risks to the participant
3. describe what the researcher will do to protect the participant's privacy
4. explain that participation is voluntary
5. describe any compensation the participant will receive
6. explain that the participant will receive that compensation even if the participant withdraws from the study
7. make it clear to participants that they can quit the study at any point

(To learn more about informed consent and to see a sample consent form, see Appendix D.)

In addition to having the right to refuse to be in the study, participants have the right to confidentiality. Therefore, the experienced investigator would want to know whether the novice investigator took extensive precautions to ensure that no one other than the investigator could find out how each participant behaved during the study. Common precautions include (a) using code numbers (e.g., "Participant 1's response")—rather than participants' actual names—when recording participants' responses, (b) storing data in a locked file cabinet, (c) password-protecting any data files stored on a computer, and (d) signing a pledge to keep all information about participants confidential.

The *Principles* not only address participant rights but also stress investigator responsibilities (see Box 2.1). According to the *Principles*, the investigator's responsibilities begin well before the study begins. As part of the planning phase, the investigator should try to anticipate all possible risks to participants and then protect participants from these risks. In this study, the investigator should consult with physicians to be sure that the temperature was not too hot and to identify types of people who should not participate because they might have a bad physiological reaction to the heat. In addition, the investigator would have to determine how to ensure that the aggression induced by the heat would not get out of hand, leading to someone being harmed either physically or psychologically.

BOX **2.1**   **Selected Ethical Guidelines for Studies Involving Human Participants**

1. Participants must volunteer to be in the study. They should to feel that they can refuse to be in the study. Consequently, bribing people by offering excessive rewards for participation (including awarding extra-credit points that a student could not earn by doing an alternative activity) is forbidden.
2. Participants should have a general idea of what will happen to them if they choose to be in the study. In addition, they should be well-informed about anything that they might perceive as unpleasant. That is, they should know about anything that might cause them to decide not to participate. For example, they should be told about the number and length of sessions and about any foreseeable risks.
3. Participants should be told that they can quit the study at any point, and they should be encouraged to quit the study if, at any point, they find the study upsetting.
4. Investigators should keep each individual participant's responses confidential.
5. Investigators should try to anticipate all possible risks to participants and take steps to prevent these potential problems from occurring.
6. Investigators are responsible for making sure that all people working for them behave ethically.
7. At the end of the study, investigators should probe participants for signs of harm and take steps to undo any harm detected.
8. At the end of the study, investigators should explain the purpose of the study and answer any questions participants may have.
9. Researchers should get approval from appropriate committees (probably their school's institutional review board [IRB]).

© Cengage Learning 2013

While the study is being conducted, the investigator is responsible for behaving in an ethical manner. Furthermore, under some circumstances, the investigator may also be responsible for ensuring that others behave ethically. For example, if the people working with or for the novice investigator on the aggression study had behaved unethically, the novice investigator would have been responsible for their behavior even if he was unaware of what the others were doing.

After each participant has finished taking part in the study, the investigator should **debrief** participants: explain the purpose of the study, answer any questions, address any concerns, and undo any harm that the participant may have experienced. During debriefing, the investigator should actively look for signs of harm because (a) some events that do not bother most people may be traumatic to some participants and (b) some participants may be reluctant to tell the investigator about that harm. If harm is detected, the researcher should try to undo it.

Fortunately, most studies do not harm participants. Thus, the main function of debriefing is usually to explain the study to participants. Educating participants about the study is the least an investigator can do to give something back to those who volunteered to be in the study.

Unfortunately, you can't determine that the novice investigator's study was ethical merely by observing that the novice investigator followed a few simple guidelines. Instead, as the introduction to *Ethical Principles in the Conduct of Research With Human Participants* (APA, 1982) states,

BOX **2.2** **Some Ethically Questionable Studies Conducted in the U.S.**

- 1932–1973: The Tuskegee Study of Untreated Syphilis in the Negro Male (Centers for Disease Control and Prevention, n.d.) studied 399 African American sharecroppers with syphilis to test whether no treatment was better than the dangerous and ineffective treatments of the day. The study also aimed to discover what the most effective treatment was for each stage of syphilis. Participants were told that they had "bad blood," rather than their true diagnosis. Although by the late 1940s, penicillin had proven to be an effective treatment for syphilis, participants were not told that they had syphilis and were denied the new and effective treatment. The study stopped because of a newspaper exposé. In addition to illustrating the need for informed consent and for minimizing harm, the Tuskegee Study emphasized the problem with doing research in which the costs and benefits of the research are not shared fairly. For example, a disadvantaged group may suffer the costs and risks of being experimented on, whereas an advantaged group may reap the benefits of the newer, more expensive treatments resulting from that research.

- 1950s–1960s: Project MK-ULTRA. Senator Ted Kennedy (1977) testified that in 1975, "The Deputy Director of the CIA revealed that over 30 universities and institutions were involved in an 'extensive testing and experimentation' program, which included covert drug tests on unwitting citizens 'at all social levels, high and low, native Americans and foreign.' Several of these tests involved the administration of LSD to 'unwitting subjects in social situations.' At least one death, that of Dr. Olson, resulted from these activities. The Agency itself acknowledged that these tests made little scientific sense."

- 1959–1962: Thalidomide scandal. As part of a "study," pregnant American women were given a drug they assumed was safe and effective. Unfortunately, it wasn't: As a result, some fetuses died, and many more were deformed. Some have implied that the purpose of the study was more about selling physicians on the drug than collecting scientific information (Peterson, 2008). In any event, the "study" violated the Nuremberg Code in that (a) it was started before the animal experiments were completed, (b) the experiment was not conducted by scientifically qualified persons, (c) no preparations were made to protect participants, and (d) the study was not terminated as soon as disabilities and deaths occurred.

- 1960–1964: Studies were conducted in which (a) military personnel were led to believe they were going to die (some were led to believe their plane was about to crash; some were led to believe they would be accidentally killed by artillery fire; and some were led to believe they would die due to an accident involving radioactive fallout); (b) alcoholics volunteering for an experiment that they believed might lead to a cure for alcoholism (but really had nothing to do with alcoholism) got an injection and then found—often to their horror—that they could not breathe; (c) male participants were shown pictures of men and then falsely told that they were homosexually aroused by those pictures; and (d) patients with minor neuroses were given high levels of LSD, electroshock, and sensory deprivation without their permission to see whether erasing their memories could lead to better mental health (Boese, 2007; Lesko, 2009).

- 1963: Medical researchers injected live cancer cells into older adult patients without telling the patients.

- 1966: Henry Beecher published an article in the *New England Journal of Medicine* in which he discusses 22 "examples of unethical or questionable ethical studies." Like the authors of the Nuremberg Code, Beecher suggests that each individual researcher should listen to his or her conscience.

- 1993: *The Albuquerque Tribune* revealed information about a long-term study in which people—some of them mentally retarded children—were exposed to radiation to see its effects.

"The decision to undertake research rests upon a considered judgment by the individual psychologist about *how to best contribute to psychological science and human welfare*" [italics added].

This statement has two important implications. First, it means that even if the novice investigator fulfilled all his responsibilities to the participants, the study might still be unethical if the study was unlikely to contribute to psychological science and human welfare. Second, it means that even if the novice investigator violated certain participant rights (such as not telling participants what the study is trying to find out), the study might still be ethical if the expected benefits of the study would compensate for those violations. Consequently, an important step in determining whether a study is ethical is determining the likelihood that the study will benefit humanity.

Have Potential Benefits Been Maximized?

The experienced researcher would begin to determine the likelihood that the study would benefit humanity by determining the importance of the research question. Unfortunately, determining the value of the research question is highly subjective. One person may find the idea very important, whereas another may find it unimportant. In the aggression study, the novice investigator may believe that determining the relationship between temperature and aggression is extremely valuable, arguing that it might lead to ways of preventing riots. Others, however, may disagree.

To make assessing the potential value of a study even more challenging, no one knows what the researcher will discover. A study that looks promising may discover nothing. On the other hand, many scientific studies designed to answer one question have ended up answering a very important but unrelated question (Burke, 1978; Coile & Miller, 1984). For example, Pavlov set out to discover the role of saliva in digestion, yet ended up discovering classical conditioning. Because it is so hard to judge the value of a research question, the researcher would probably acknowledge that the novice investigator's research question has some merit.

As you have seen, judging the importance of a research question is difficult. Therefore, to estimate the potential value of the novice investigator's study, the research psychologist would put less emphasis on her subjective impression of the importance of the research question and put more emphasis on the more objective judgment of how well the study would answer the research question. That is, she would ask, "Is the study likely to provide valid data?"

By "valid data," the experienced researcher would not necessarily mean that the study must have all three types of validity (i.e., construct, internal, and external). Indeed, few studies even attempt to have all three validities. Rather, her focus would be on determining whether the study has the validity or validities necessary to answer the research question. To illustrate that different research goals require different validities, let's look at three examples.

First, suppose that an investigator wants to describe what most people do on a first date. In that case, because the investigator is not looking for the causes of behavior, the investigator would not strive for internal validity. However, because the investigator is interested in generalizing the results to most people, the investigator would strive for external validity.

Second, suppose that a researcher is trying to develop a test of social intelligence. If the researcher's only goal is to show that the test accurately measures the construct of social intelligence, the researcher needs only construct validity.

Third, suppose that an investigator is trying to explain or control a behavior, such as smoking. In that case, the investigator needs to understand the causes of a behavior and therefore would need internal validity.

After the research psychologist evaluated the extent to which (a) the research question was important and (b) the study would provide a valid answer to that question, the research psychologist would be able to estimate the study's potential benefits. Then, the research psychologist would probably suggest changes that would either maximize the study's potential for benefiting humankind or minimize the study's potential for harming participants. If, after those changes were made, the researcher was satisfied that (a) the research's benefits outweighed the risks and (b) the planned precautions would minimize risks, the researcher would encourage the novice to submit a research proposal to an ethics committee.

Has Permission to Conduct the Research Been Obtained?

Note that even if the researcher believed that the proposed study's risks were minimal, had been minimized, and were outweighed by its potential benefits, the researcher would not grant the novice investigator permission to conduct the study. Indeed, even if the researcher wanted to conduct the study herself, she would not just go out and do it. Instead, she—like most researchers—would consult with others before doing the research.

Consulting with others is vital for at least two reasons. First, when weighing the benefits of one's own research against the costs to participants, it is hard to be fair and impartial. As you can see from Box 2.2, the strategy of trusting individual scientists to follow the Nuremberg Code does not always work. Second, consulting with others may lead to insights about how to maximize the study's benefits and how to protect participants from harm.

Because consulting with others is so important, some researchers will not do a research study until their department's ethics committee has approved the study. At most schools, before conducting a study with human participants, researchers must obtain permission from the school's **institutional review board (IRB)**: a committee of at least five members—one of whom must be a nonscientist—that reviews proposed research and monitors approved research in an effort to protect research participants. As you can see from Figure 2.7, the IRB, when deciding whether to approve research, weighs the potential benefits of the research against the potential risks to participants. In addition to assessing the risks and benefits of the research, the IRB might require additional steps to protect the participants. These steps might include having the investigator

1. make the informed consent form more specific and easier to understand
2. exclude individuals whose ability to give informed consent could be questioned, such as people under 18 or people with mental disabilities
3. exclude individuals who may be more at risk for negative reactions to the treatment, such as pregnant women

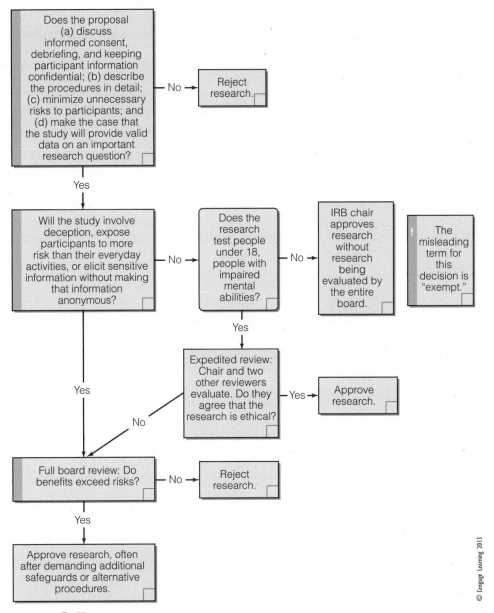

FIGURE **2.7** IRB process: What may happen to a research proposal.

4. eliminate rewards for participation (e.g., extra credit) that might make participants feel obligated to be in the study
5. use alternative procedures that would involve less distress or deception
6. produce a detailed plan for dealing with participants who are upset or harmed by the study
7. take additional steps to protect the participants' privacy

If your school has an IRB, it is a violation of federal law to do research without first submitting that research to the IRB. In any event, a novice investigator should always get approval from a higher authority before conducting a study. *Never conduct a study without first obtaining approval from your professor!*

As you have seen, the psychological researcher's most important concerns about the novice investigator's aggression study are ethical concerns. Indeed, because ethical concerns include concerns about validity and human betterment, one could argue that ethical concerns are the researcher's only concerns (see Box 2.3).

But what if the novice investigator's study had used animals instead of human participants? In that case, some might think that the psychologist would not have been concerned about ethics. As you can see from Box 2.4, nothing could be further from the truth. Indeed, in recent years, animal rights have received more attention from the American Psychological Association than human rights. If the aggression study had used animals as participants, the researcher would have consulted the ethical standards listed in Box 2.4

BOX 2.3   Determining Whether a Research Study Is Ethical

Does It Maximize the Potential Benefits to Psychological Science and Human Welfare?
1. Has the research question already been answered?
2. Is the research question important?
3. Will the research study provide valid answers to the research question? The type of validity needed will depend on the research question.
   - If the research question concerns finding out whether a certain factor *causes* a change in behavior (e.g., "Does a certain type of school environment increase student attendance?"), the study should have *internal validity.* That is, the study should take steps to rule out the possibility that other factors may be responsible for the effect.
   - If answering the research question hinges on accurately measuring abstract psychological concepts, *construct validity* would be vital. In that case, the researchers should be able to make a strong case that the psychological *variables* they are talking about are the variables they actually measured. Construct validity would be the main concern in a research study that was trying to develop a psychological test.
   - If the main purpose of the research is to provide results that can be *generalized* to the real world, *external validity* would be important. In such a case, the researchers might want to show that their participants were representative of a larger group. External validity is important for polls because polls try to determine how most people would respond to certain questions.

Does It Minimize the Potential for Harm to Participants?
1. Does it conform to the ethical principles of the American Psychological Association?
   - Are participants volunteers?
   - Did they know what the study involved before they agreed to participate?
   - Were participants told they could quit the study at any point?
   - Were participants debriefed?
2. If participants will be subjected to stress,
   - have less-stressful alternatives been considered?
   - has the amount of stress been minimized?
   - have procedures for helping distressed participants been established?

BOX **2.4**   **Humane Care and Use of Animals in Research**

The ethical standards that follow are considered enforceable rules of conduct. Violating these rules may result in being expelled from the American Psychological Association and being both sued and arrested.

8.09 Humane Care and Use of Animals in Research
a. Psychologists acquire, care for, use, and dispose of animals in compliance with current federal, state, and local laws and regulations, and with professional standards.
b. Psychologists trained in research methods and experienced in the care of laboratory animals supervise all procedures involving animals and are responsible for ensuring appropriate consideration of their comfort, health, and humane treatment.
c. Psychologists ensure that all individuals under their supervision who are using animals have received instruction in research methods and in the care, maintenance, and handling of the species being used, to the extent appropriate to their role.
d. Psychologists make reasonable efforts to minimize the discomfort, infection, illness, and pain of animal subjects.
e. Psychologists use a procedure subjecting animals to pain, stress, or privation only when an alternative procedure is unavailable and the goal is justified by its prospective scientific, educational, or applied value.
f. Psychologists perform surgical procedures under appropriate anesthesia and follow techniques to avoid infection and minimize pain during and after surgery.
g. When it is appropriate that an animal's life be terminated, psychologists proceed rapidly, with an effort to minimize pain and in accordance with accepted procedures.

Source: From *Ethical Principles of Psychologists and Code of Conduct* (2002). *American Psychologist, 57,* 1060–1073. Reprinted with the kind permission of the American Psychological Association.

(APA, 2002) as well as APA's 1996 booklet *Ethical Principles for the Care and Use of Animals,* a copy of which is included in Appendix D. In addition, the researcher would probably need to have the research approved by the school's institutional animal care and use committee (IACUC). If the study had been done unethically, the investigator would be severely punished.

CONCLUDING REMARKS

In this chapter, you have seen that research psychologists are aware of the challenges and responsibilities of studying human and animal behavior. In the rest of this book, you will see the wide variety of strategies that researchers use to meet these challenges and responsibilities.

SUMMARY

1. Not all data are created equal. Therefore, to determine whether a claim is based on solid data, you should question the internal, external, and construct validity of the research that produced those data.

2. Before accepting a claim that a factor *causes* some effect (that some event produces a certain response), question whether the evidence for that claim comes from a study that has internal validity.

3. To question a study's internal validity—the degree to which it established that a factor caused an effect—ask three questions:
   (I-1) Did the study establish that changes in the alleged cause are *associated* with changes in the alleged effect?
   (I-2) Did the study establish that changes in the cause occurred *before* the changes in the effect?
   (I-3) Did the study establish that it was unlikely that *other factors* besides the alleged cause produced the effect?

   Unless you can answer "yes," to those three questions, the study does not have internal validity. You will probably not be able to answer "yes" to those three questions unless the conclusion is based on a randomized experiment—a special kind of study in which participants are randomly assigned to receive different levels of a treatment.

4. Most studies are not experiments. Thus, most studies do not have internal validity.

5. If a researcher manipulated a treatment but did not use random assignment, you will probably have to answer "no" to question I-3 (Did the study rule out other suspects as causes of the effect?). For example, if a study finds that ice cream consumption is correlated with violence, that study should not be used to claim that ice cream is a cause of violence until other third factors (e.g., hot days) have been ruled out as possible causes.

6. If the study did not manipulate a treatment, you have to answer "no" to question I-3 (Did the study rule out other suspects as causes of the effect?)—and you might also have to answer "no" to question I-2 (Did the "cause" come *before* the effect?). That is, in addition to worrying about whether something other than the so-called "cause" produced the effect, you also have to worry that the so-called "cause" may actually be the effect (e.g., instead of low self-esteem being a cause of teens becoming pregnant, becoming pregnant may be a cause of low self-esteem).

7. If the cause–effect conclusion is not based on a study (or is based on a badly done study), you might even have to answer "no" to question I-1 (Did changes in the cause coincide with changes in the effect?). For example, although people believe that a full moon causes crime, studies find no relationship between full moons and criminal arrests.

8. Question the study's construct validity, that is, the degree to which the labels the researchers give their variables are accurate.

9. A first step to questioning whether researchers are studying the variables the researchers claim to be studying is to question whether the study's measures and manipulations are valid. Do not assume that the researchers did a perfect job of controlling and reading their participants' minds. By questioning a manipulation, you may find that a manipulation does not manipulate what it claims to manipulate (e.g., a guilt manipulation doesn't make participants feel guilty) or that it has some impurities so that it manipulates more than what it claims to manipulate (e.g., the guilt manipulation also makes people feel embarrassed and anxious). Similarly, by questioning a measure, you may find that it does not measuring what it claims to measure (e.g., scores on a so-called measure of social intelligence are not influenced by the participant's actual social intelligence) or that it also measures something besides what it claims to measure (e.g., scores on a measure of social intelligence are influenced by social intelligence and the desire to be the center of attention).

10. A second step to questioning a study's construct validity is to ask whether participants may have figured out the purpose of the research and then played along by

trying to give the researcher the "right" results.

11. If someone claims that a study's findings can be generalized to other people, places, and times, you should question the study's external validity.

12. As a first step to questioning a study's external validity, look at what groups of people were not adequately represented in the study and ask whether there is a reason to expect that the results would not apply to those people.

13. As a second step to questioning a study's external validity, ask under what conditions the results might not apply.

14. As a third step to questioning a study's external validity, ask whether the research has been replicated in different places, at different times, and with different populations.

15. To make the best case that the results apply to a certain population, the researchers should have tested a representative sample from that population. The best way to get such a representative sample is to get a large random sample from that population.

16. Random sampling is different from random assignment. Researchers may use random sampling to obtain a group that represents the larger population. In that case, their goal is to have external validity. Researchers may use random assignment to get groups that, before the treatment is introduced, differ from each other only by chance. In that case, their goal is to have internal validity. It is possible for a study to use both random sampling and random assignment.

17. Human participants in research studies have many rights, including the right to decide whether they want to be in the study, the right to privacy, and the right to learn the study's purpose.

18. *Do not conduct a study without the approval of your professor.* In addition, obtain approval from the appropriate ethics committees. For example, if you are doing animal research, you may need approval from your school's Institutional Animal Care and Use Committee (IACUC). If you are doing research with human participants, you may need approval from your school's institutional review board (in the U.S.) or research ethics board (in Canada).

19. If you are involved with a study that harms a participant, you cannot avoid responsibility by arguing that you did not know the rules; that you did not mean to harm the person; that you were just doing what the lead investigator told you to do; or that your assistant, rather than you, caused the harm.

20. According to APA's ethical principles, the study's potential benefits should outweigh the study's potential for harm. Thus, there are two ways to increase the chances that your study is ethical: reduce the potential for harm and maximize the potential gains of your research.

21. To maximize the gains of your research, you should make sure that your study has the kind of validity that your research question requires. Your research question will determine which type—or types—of validity you need.

22. If your research question is about whether something causes a certain effect, your study should have internal validity.

23. If your research question concerns what percentage of people engage in some behavior, you need a study that has external validity.

24. If your research question involves measuring or manipulating some state of mind (hunger, stress, learning, fear, motivation, love, etc.), you need construct validity.

25. APA's ethical code applies not only to research done on human animals but also to research done on non-human animals. For example, researchers who fail to minimize the risks to the animals they study may be severely punished.

KEY TERMS

construct *(p. 49)*   experiment *(p. 48)*   internal validity *(p. 41)*
construct validity *(p. 51)*   external validity *(p. 55)*   random assignment *(p. 45)*
counterbalancing *(p. 43)*   informed consent *(p. 58)*   random sampling *(p. 56)*
debriefing *(p. 59)*   institutional review board
ethical *(p. 56)*   (IRB) *(p. 61)*

EXERCISES

1. Match the concept to the type of validity.

   ___ construct validity   **a.** generalize
   ___ external validity   **b.** cause–effect
   ___ internal validity   **c.** mental states

2. The professor asks a student, "Do you have any questions?" The student says, "No." Consider the following conclusions that the professor might make from the student's response.
   a. If the professor concludes that the student understood the lecture perfectly, which validity (construct, internal, or external) should be questioned?
   b. If the professor concludes that none of the students would have a question, which validity (construct, internal, or external) should be questioned?
   c. If the professor concludes that the student is saying "no" because of the new way the professor explained a concept, which validity (construct, internal, or external) should be questioned?

3. Match the threat to the type of validity.

   ___ construct   **a.** measure was poor
       validity
   ___ external   **b.** treatment and no-
       validity   treatment groups were
                         unequal before the
                         study began
   ___ internal   **c.** sample of participants
       validity   was not representative

4. Match the threat to the type of validity.

   ___ construct   **a.** no random assignment
       validity
   ___ external   **b.** no random sampling
       validity

   ___ internal   **c.** participants figured out
       validity   the hypothesis

5. An author tells of a case in which a person got much better after receiving a new treatment. The author then concludes that the treatment would work for everyone. How good is the author's evidence for this conclusion in terms of
   (a) internal validity?
   (b) external validity?

6. Is it ethical to treat a patient with a method that has not been scientifically tested? Why or why not? Is it ethical to withhold a treatment that is believed to work in order to find out if it does indeed work? Why or why not?

7. Imagine you were doing a study to see whether people, when frustrated, would be more aggressive toward another person, especially if that person was of a different ethnic group.
   a. How might informed consent hurt the construct validity of your study?
   b. How might a full debriefing of your participants lead to harm?

8. For one of the following television shows—*Survivor, Candid Camera,* or *America's Funniest Home Videos*—state which of the nine APA ethical principles listed in Box 2.1 are violated and explain—or provide an example of—how those principles are violated.

9. Two of the most ethically questionable studies in the history of psychology are Milgram's obedience study (in which participants were told to deliver dangerous shocks to an accomplice of the experimenter)

and Zimbardo's prison study (in which well-adjusted students pretended to be either prisoners or guards). In both of these studies, there would have been no ethical problems at all if participants had behaved the way common sense told us they would; that is, no one would have obeyed the order to shock the accomplice, and none of the "guards" would have mistreated the prisoners.

a. Does the inability to know how participants will react to a research project mean that research should not be done?

b. Does people's inability to know how they and others will react in many situations mean that certain kinds of research should be performed so we can find out the answers to these important questions?

c. What ethical principles, if any, were violated in Milgram's shock experiment? (See Box 2.1.)

d. What ethical principles, if any, were violated in Zimbardo's prison study? (See Box 2.1.)

10. Assume that a participant in a study in which you were involved suffered intense distress. According to the APA ethical guidelines, none of the following is a legitimate excuse that would relieve you of responsibility. For each "excuse," state the principle that is violated (see Box 2.1) and explain how it applies.

a. "I was just following orders."

b. "My assistant conducted the session and behaved inappropriately, not me."

c. "I didn't notice that the participant was upset."

d. "I just didn't think that we had to tell participants that they would get mild electrical shocks."

e. "I didn't think that asking questions about suicide would be upsetting—and for most of my participants it wasn't."

f. "When the participant got upset, it surprised me. I just didn't know what to do and so I didn't do anything."

g. "Our subjects were mice. We can cause mice whatever distress we want." (See Box 2.4.)

 ## WEB RESOURCES

1. Go to the Chapter 2 section of the book's student website and
   a. Look over the concept map of the key terms.
   b. Test yourself on the key terms.
   c. Take the Chapter 2 Practice Quiz.
   d. Do the interactive end-of-chapter exercises.

2. To learn more about IRBs, getting IRB approval for research, and the ethical issues in conducting research, use the "Ethics" link.

Generating and Refining Research Hypotheses

In good science, questions come first. Science is just a tool for answering those questions.
—**John Bargh**

The scientist is not the one who gives the right answers, but the one who asks the right questions.
—**Claude Lévi-Strauss**

CHAPTER OVERVIEW

Research does not begin with variables, equipment, or participants. It begins with questions. Questions often begin when researchers notice that people are behaving in surprising ways. Then, researchers ask the same kinds of questions you ask:

"Did they really do that?"
"Who does that?"
"What were they thinking?"
"When and where did that happen?"
"How did they do that?" and
"Why do they feel that way?"

There are, however, three important differences between the questions that you typically ask and the questions that researchers typically ask. First, whereas you may ask questions that are specific to a particular person (e.g., "Is John rude today because he is in a bad mood?"), researchers look for more general questions that are relevant to a broader range of people (e.g., "Does thinking about money cause people to be ruder?"). Second, whereas your questions may tend to involve the relationship between only two variables (e.g., mood and rudeness), researchers' questions tend to involve more than two variables (e.g., "Is the effect of thinking about money on rudeness increased when more people are around?"). Third, whereas your questions usually come from only two sources (personal experience and common sense), researchers' questions come from a broad range of sources. For example, you probably tend to ask questions only when you are surprised by people who act differently from what your past experience or common sense suggest—and when common sense predictions seem to conflict. Researchers, on the other hand, not only ask questions at those times but also when people act differently from what past research or current theory suggests—and when research findings or theories' predictions conflict.

Perhaps the biggest difference between you and a researcher comes after a question comes to mind. At that point, the researcher knows what steps to take to turn a question into a testable prediction.

In this chapter, you will learn how to close the gap between you and the researcher. You will start by learning what questions to ask of your own experiences and common sense (an informal "theory"). Next, you will learn to ask those same basic questions of research studies (reports of other people's experiences) and of theory (the common sense of knowledgeable people). Finally, you will learn how to convert your questions into workable research **hypotheses**: testable predictions about the relationship between two or more variables.

GENERATING RESEARCH IDEAS FROM COMMON SENSE

Although most of us have many questions about why people behave as they do, some students find developing a research hypothesis intimidating. They ask, "How can I find out something that people don't already know?" One solution is to adopt the skeptical attitude that characterizes science by asking whether what people already "know" is supported by objective evidence. As Abelard said, "The beginning of wisdom is found in doubting; by doubting, we come to question, and by seeking, we may come upon the truth." You can begin your questioning by doubting the effectiveness of "time-tested" treatments or of new treatments. These treatments may range from self-help books to online lectures.

Another avenue for your skepticism is to test common sense. Galileo is famous for his experiments testing the common sense assumption that heavier objects fall faster than lighter objects and for his skepticism about the common sense belief that the sun revolved around the unmoving earth. Likewise, several psychologists have won Nobel Prizes for testing the assumption that humans are rational decision makers. Thus, testing common sense assumptions ("myth busting") is valuable. Indeed, one scientist (Stern, 1993) believes that a major goal of psychology should be to "separate common sense from common nonsense."

Psychologists have a long history of testing common sense by testing old sayings. For example, Schachter (1959) tested the saying that "misery loves company." Zajonc (1968) found the saying "familiarity breeds contempt" to be false in many situations. Berscheid and her colleagues (1971) discovered that birds of a feather do flock together. Byrne (1971) found that opposites don't attract. Latane, Williams, and Harkins (1979) found evidence for the idea that "too many cooks spoil the broth." Pennebaker et al. (1979) learned that "the girls do get prettier at closing time." Wilson and Schooler (1991) discovered evidence against the saying "look before you leap."

More recently, researchers have been testing sayings related to happiness. For example:

- Seligman (2002) found support for the saying "happiness is like a butterfly: If you chase it, you won't catch it" (or, if you prefer Eleanor Roosevelt's version, "happiness isn't a goal, it's a byproduct").
- Emmons and McCullough (2003) discovered evidence to support the saying "focus on what's right in your world instead of what's wrong."

If we mentioned a saying that you were planning to test, do not automatically abandon plans to test that saying. Sayings are usually broad enough that all aspects of them can't be completely tested in a single study. For example, consider Wohlford's (1970) finding that, as would be expected from the saying "like father, like son," fathers who smoked were more likely to have sons who smoked. Researchers still—more than 40 years later—do not have definitive answers about the extent to which a son's behaviors (other than smoking) are modeled after his father's. Similarly, although Vohs, Mead, and Goode (2008) have done many experiments on the saying "money changes people," they admit that there is still much to be known. Thus, you can test a saying that has already been partially tested. However, if you want to test completely untested sayings, there are many from which to choose.

How do you find a common sense assumption that you could dispute? Read anything: fortune cookies, packages of Salada tea (Dillon, 1990), books of quotations, self-help books, song lyrics, bumper stickers, T-shirts, newspaper ads, editorial columns, and headlines. For example, much of what we know about helping someone in trouble comes from two students' efforts to explain an event that shocked New Yorkers: a newspaper article reporting that at least 38 people stood by and did nothing while a young woman was murdered. Darley and Latané (1968) questioned the popular view that the reason the bystanders did not help was because New Yorkers were cold and alienated. Their research suggests that, if the article's report was accurate, rather than saying nobody helped *despite* the presence of 38 witnesses, it would be more accurate to say that nobody helped *because* there were 38 witnesses.

Another way to find an assumption you want to test is to talk to a person who always seems to disagree with you. If both of you make sensible arguments, but neither one of you can prove that the other is wrong, it is time to get objective evidence to show that your acquaintance is wrong.

Yet another way to find questionable assumptions is to attack a real-life, practical problem (cheating, prejudice, rudeness, apathy, too many false fire alarms in the dorms, etc.). Usually, you will find that different people have different "solutions" to almost any practical problem. You could collect objective evidence to find out which of these proposed solutions works best.

If you decide to attack a practical problem, you may find that you have two research projects. The first is to determine the size of the problem; the second is to compare the effectiveness of different approaches to solving the problem. For example, you might first conduct a study to find out how serious the problem of cheating (or prejudice, apathy, superstitious thinking, etc.) is on your campus. Then, your second study might see which approaches to solving the problem are most effective. For instance, you might see if any of the following six methods designed to stop students from cheating on exams are more effective than what teachers normally do:

1. Have a lecturer emphasize the ways that cheating harms the cheater.
2. Have students discuss—in groups that you have set up so that most of the members in the group oppose cheating—whether cheating is unfair to other students.
3. Have students write an essay about why it is wrong to cheat, and then have them read that essay aloud to students in a freshman English class.
4. Have more serious penalties for cheaters.
5. Have observers walk around during the exam.
6. Have students sign a statement at the bottom of their test that says, "I agree that I have abided by the honor system." (Ariely [2008] found that a similar manipulation reduced cheating considerably.)

In summary, questioning common sense is a time-tested way to generate research ideas. In the distant past, famous discoveries—such as that the earth revolves around the sun and that light objects fall just as fast as heavy objects—came from researchers who were willing to question common sense. More recently, a fourth grader made national news by doing research that questioned whether a "healing technique" adopted by over 100 nursing schools was effective (Rosa, Rosa, Sarner, & Barrett, 1998). Even more recently, Cialdini (2005) questioned the persuasiveness of the cards used by thousands of hotels that urge guests to reuse towels, and Strayer and Drews (2008) questioned claims that driving with hands-free cell phones was safer than driving with hand-held cell phones.

As you can see from these examples, just by being skeptical, people have been able to generate important research ideas. If you are naturally somewhat skeptical and use Box 3.1, you too can generate an important research idea. However, testing your own insights is not the only—or even the most preferred—way to generate research ideas.

GENERATING RESEARCH IDEAS FROM PREVIOUS RESEARCH

A more preferred way to generate research ideas is to react to previous research. Most advances in science come from scientists building on each other's work. For a beginning researcher, basing a hypothesis on previous research has at least three major advantages.

First, a hypothesis based on previous research is more than a guess—it is an educated guess. Because your prediction is consistent with both a previous study's results and the logic used to explain those results, your hypothesis is likely to be correct.

Second, regardless of whether your hypothesis is correct, your study will be relevant to what other scientists have done. Consequently, your research will not produce an isolated, trivial fact.

BOX **3.1**   Five Ways to Tap Your Intuition

1. Ask whether an "old saying," a claim made in a song, an assumption made in something you read, or a statement made by an expert is true.
   a. If nobody knows whether it is true, propose a study to find out whether it is true.
   b. If it seems to be true for most of the people most of the time, ask when might it not be true.
      1. For what people is it less likely to be true? Most likely to be true?
      2. In what situations is it not true?
      3. Is it true only in moderation—or does it hold in the extremes as well?
   c. If it is not true (e.g., opposites attract), ask why people believe it is true.
2. Collect data on your own behavior, try to find rules that govern your behavior, and then see if those rules apply to other people.
3. Transform an argument into a research idea—find facts to settle a battle between two opinions.
4. Ask six key questions about any interesting phenomenon (e.g., pick-up lines, being lucky):
   a. Who does it? (How do people who do it frequently or well differ from those who do it rarely or poorly?)
   b. What precisely is it?
   c. When and where is it done? (Under what circumstances and in what situations is it done? In what situations is it effective?)
   d. Why does it happen? (What causes it to happen?)
   e. What are its short- and long-term effects? (What does it cause to happen?)
   f. How is it done?
5. Attack a practical problem (ecology, illiteracy, prejudice, apathy, alcoholism, violence).
   a. Document how common or serious the problem is.
   b. Evaluate the effectiveness of potential cures for the problem.

Third, doing research based on other people's work is easier than starting from scratch, especially when you are a beginning researcher. Just as a beginning cook might find it intimidating to make a pizza from scratch without a recipe, some beginning researchers find it intimidating to design a study from scratch. However, just as the beginning cook would feel comfortable adding a few toppings to a store-bought pizza, a beginning researcher might feel comfortable building on someone else's study.

Specific Strategies

As we just discussed, if you can develop an idea from previous research, you may be able to justify your hypothesis by saying that it is consistent with findings from previous research, you may be able to test your hypothesis using the methods from previous research, and you should be able to show that your results are relevant to previous research. But how can you develop an idea from previous research?

Repeat Studies

The simplest way to take advantage of other people's work is to repeat (**replicate**) someone else's study. Because science relies on skepticism, you should repeat studies when you find the study's results difficult to believe—especially when those results conflict with results from other studies, seem inconsistent with established theory, or have not been replicated. For example, thanks to failures to replicate the "Mozart effect," we know that listening to Mozart does not increase intelligence.

Do a Study Suggested by a Journal Article's Author(s)

Almost as simple as replicating a study is doing a study suggested by an article's authors. At the end of many research articles, the authors suggest additional studies that should be done. Often, they point out that the research should be repeated either using a different sample of participants or using a different setting.

Improve the Study's External Validity by Seeing Where, When, and for Whom the Results Apply

Even if the researchers do not suggest it, you may decide to test the external validity (generality) of their findings by asking whether the results would hold in the real world. To determine where, when, and for whom the results apply, you might ask:

1. Should I redo the study, but include types of participants that were not adequately represented in the original sample? If you have specific reasons to believe that the original study's results would not apply to a certain group—for example, to most women, or to most members of some other group—you probably should redo the study. Thus, even if a study had found that listening to Mozart increased the intelligence of college students, you would probably still want to replicate the study with babies before generalizing the results to babies. Similarly, on reading about Pennebaker et al.'s (1979) finding that bar patrons perceive people of the other gender as being prettier at closing time, you might think of people for whom this result would not apply. For example, like Madey et al. (1996), you might suspect that the effect would not hold for perceivers who are in committed relationships (If you did, you were right.).

2. Should I repeat the study using a more representative sample of the target group? If the study tried to establish what most people in a certain group (e.g., all U.S. citizens) think, feel, or do, the study should have used a random sample of that group. If they did not use random sampling, you could improve the study's external validity by replicating the study using a random sample of that group.

3. Should I repeat the study using stimulus materials more like stimuli that people are exposed to in real life? Sometimes, stimuli in research are highly artificial. For example, many studies have asked participants to form impressions of a person based on reading a list of traits that supposedly describe that person, and many have asked participants to memorize lists of nonsense syllables. You might replace the list of words, nonsense syllables, or traits that participants read in the original study with a videotape of a real-life event.

4. Should I see whether the effect is long lasting? Researchers often only look to see whether a treatment has a short-term effect. You may wish to look for *long-term* effects because (a) long-term effects are important and (b) an action that produces a positive short-term reaction may not produce a positive long-term reaction. Thus, even if playing Mozart had boosted intelligence for 15 minutes, the effects might have worn off after an hour.

5. Can I think of any other *situations* in which the relationship between the variables observed in the original study may not hold? For example, consider the relationship between group size and helping: The *more* people available to help a victim, the *less* likely the victim is to get help. Can you think of some situation—such as an event that increases group identity—that would weaken or reverse the relationship between group size and helping? By asking the "where wouldn't it hold?" question, Nida and Koon (1983) found that Pennebaker's "they get prettier at closing time" occurred at a country western bar but not at a college bar.

Improve the Study's Internal Validity

Instead of improving a study's external validity, you may choose to improve its internal validity. As you learned in Chapter 2, establishing that one particular factor caused an effect is very difficult, partly because it is hard to control all other factors. For example, Gladue and Delaney (1990) argued that Pennebaker's finding that "girls get prettier at closing time" at bars (Pennebaker et al., 1979) left unanswered the question of whether time or alcohol consumption was responsible for increased perceptions of attractiveness. Therefore, they modified the original study to control for the effects of alcohol consumption.

Similarly, although Frank and Gilovich (1988) found that NFL and NHL teams switching to black uniforms were called for more penalties, that finding did not prove that wearing black causes a team to get more penalties. After all, it could be that aggressive coaches like to have their teams wear black. Therefore, Frank and Gilovich devised an *experiment* in which they *randomly assigned* participants to white or dark uniforms. This random assignment allowed them to make sure that uniform color was the only systematic difference between their teams, and allowed them to isolate black as the *cause* of the increased aggression (You can find a write-up of that experiment in Appendix B). The lesson here is simple but important: *If you find a study that did not use random assignment and you can redo that study as an experiment in which you randomly assign participants to a condition, your experiment will have more internal validity than the original study had.*

Improve the Study's Construct Validity

Rather than improving a study's external or internal validity, you may choose to improve a study's construct validity. As you learned in Chapter 2, researchers who try to guess what is going on inside participants' minds may guess wrong. Usually, the problem is either that the researchers used poor operational definitions of their constructs (the manipulation or measure was poor) or the participants figured out the hypothesis and that discovery affected participants' responses. So, when thinking about improving a study's construct validity, start by asking whether you can use a better manipulation or measure than the original researchers used. Thus, if a study claimed to manipulate stress, ask whether there is a better "stress" manipulation than the one the original researchers used.

As you question a study's manipulations and measures, ask whether the researcher's manipulations may have manipulated something other than what the researcher intended and whether the measure measured something besides what the researcher intended. For example, in some studies of the Mozart effect, it may be that (a) the music manipulation, rather than making students think more, put students in a happier, more energetic, and more alert mood; and that (b) the paper-and-pencil maze task, rather than measuring what is typically thought of as intelligence, measured motor skills or speed (Gazzaniga & Heatherton, 2006; Lilienfeld, Lynn, Namy, & Woolf, 2009).

Even if the researchers used a relatively good measure of the construct they wanted to measure, it is unlikely that any single measure of a broad construct will fully capture the entire construct. Therefore, if the original study finds a relationship using one set of operational definitions, it may pay to replicate the study using different operational definitions of the construct(s). For example, when early research suggested that men have greater "spatial ability" than women, critics questioned whether the tasks used to measure spatial ability fully captured the construct of spatial ability. This questioning led to further research. That research has given us a better picture of how men and women differ on spatial ability. (On the average, men are much faster than women at mentally rotating objects, are slightly better at picking out figures that are hidden in a complex background, and are not nearly as good at remembering where objects are.)

Even when the measures and manipulations are good, a study's construct validity will be poor if participants figure out what the hypothesis is. Therefore, when reading a study, you should ask, "If I were a participant, would I know what results the researcher expected?" If your answer to this question is "yes," you may decide to repeat the study but improve it by reducing the biasing effects of participants' expectations.

One way to avoid the biasing effects of participants' expectations is to use the **double-blind technique**: the tactic of keeping both the participants and the research assistants who interact with the participants unaware of which treatment the participants are getting. You are probably most familiar with the double-blind technique from studies examining the effects of new drugs. In such studies, an investigator has physicians give all participants pills, but only some participants get pills that contain the drug being tested. Because neither the physicians nor the participants are told who has received the drug, differences between the medicated group and placebo group will be due to the drug itself rather than to the patients' or physicians' beliefs that the drug will work.

Look for Practical Implications of the Research

Even if you are satisfied with the original study's validity, the study will still leave many questions unanswered. If the study involves basic (nonapplied) research, do the findings apply to a practical situation? For example, can a technique that helps participants remember more words on a list in a laboratory experiment be used to help students on academic probation?

Similarly, if a study finds that a treatment affects the way people think, you could do a study to see whether the same treatment also affects what people actually do. It is one thing to show that a treatment helps participants remember certain information, feel more sympathy for a crime victim, or produce more of a certain kind of chemical; it is something else to show that the treatment changes the way participants actually act in real-life situations.

Try to Reconcile Studies That Produce Conflicting Results

When you find studies that produce conflicting results, try to reconcile the apparent contradictions. One strategy for resolving the contradictions is to look for subtle differences in how the studies were conducted. What do the studies that find one pattern of results have in common? What do the studies that find a different pattern have in common? Asking these questions may alert you to a possible **moderator variable**: a variable that intensifies, weakens, or reverses the relationship between two other variables.

To appreciate the general value of finding a moderator variable, think about children who only know the spelling rule "i before e." They are frustrated by the exceptions; they may even doubt that there is a rule at all. However, when they learn that "c" is the moderator variable—when they are told "i before e, except after c"—they will be happy to know that some aspects of spelling follow rules.

Just as those children had started to doubt whether there was a spelling rule about "i and e," psychologists had started to doubt whether there was rule that predicted the effect an audience had on performance. Many studies found a "social facilitation" effect—the presence of an audience improved performance. Many other studies, however, found a "social inhibition" effect—the presence of others decreased performance.

By comparing the studies that found a social facilitation effect with the studies that found a social inhibition effect, Zajonc (Zajonc, 1965; Zajonc & Sales, 1966) discovered how the two sets of studies differed: Studies finding social facilitation involved tasks that were easy for participants to do, whereas studies finding social inhibition involved tasks that were hard for participants to do. Zajonc then designed several experiments in which he varied both the presence of others and task difficulty. His results supported the hypothesis that the effect of the presence of others depended on—*was moderated by*—how difficult the task was. That is, just as the spelling rule about the relationship between "i" and "e" depends on a moderator variable—"after c" or "not after c"—the relationship between the presence of others and performance depends on a moderator variable—task difficulty (easy task or difficult task).

Conclusions About Generating Research Ideas From Previous Research

Although research studies are designed to answer some questions, no study ends up giving definitive answers to those questions—and most studies raise new questions. Therefore, existing research is a rich source of research ideas.

At the very least, you can always just repeat the original study. If you wish to modify the existing study, you have numerous options. You could improve its internal, construct, or external validity. Or, you may decide to pursue the practical applications of the study. Or, try to find situations where the findings would not hold. Or, try reconciling a study's findings with a study that obtained different findings.

Existing research may give you not only a research idea but also a way to test that idea. If you decide to repeat a study, reading the original study will tell you almost everything you need to know. Even if you want to improve or build on a study, reading the original article will still help you determine how to measure your variables, what to say to your participants, and so forth.

CONVERTING AN IDEA INTO A RESEARCH HYPOTHESIS

If you used any of the strategies we have discussed thus far, you should have some research ideas. However, you still may not have a research **hypothesis**: a testable prediction about the relationship between two or more variables. Note that a research hypothesis is not a general research topic, and it is not a vague question. Instead, a hypothesis is a specific prediction about how two or more variables will be related. Although converting an idea into a workable research hypothesis can be difficult, it is essential because the goal of all research is to test hypotheses.

To be more specific, all researchers test one of two kinds of hypotheses. If researchers want to know *why* something happens, they do experimental research. With *experiments,* researchers can test hypotheses about whether a treatment *causes* a certain effect. If researchers want to test hypotheses about *what* is happening, they do *nonexperimental* research. For example, they might use survey research to test hypotheses about what factors are correlated (associated) with a certain behavior. Thus, rather than finding out why a certain behavior (e.g., laughing) occurs, survey research (like any nonexperimental research) would focus on finding out who is most likely to do that certain behavior, what other behaviors they tend to do, or when, where, and how individuals do that behavior. Because experimental research and nonexperimental research are different, researchers using experiments will have different hypotheses than researchers not using experiments—even when studying the same general topic. For example, an experimenter's hypothesis may involve seeing whether a given intervention stops people from arguing, whereas a survey researcher's hypothesis may deal with finding out whether men are more likely to argue than women are. But although researchers using experiments will have different hypotheses than researchers using other methods, all researchers will have hypotheses.

Not only should all researchers have hypotheses, they should have hypotheses *before* they conduct their research. If they started their research without having a hypothesis to help them know what to look for, it is unlikely that they would find anything. Consequently, before ethics committees will even consider allowing you to do research, they will require you to state the hypothesis you plan to test. Because having a hypothesis is so important, the rest of this chapter is devoted to helping you generate a workable research hypothesis.

Make It Testable: Have Specific Predictions About Observable Events

When converting an idea into a hypothesis, you must be sure that your hypothesis is testable. In general, a testable hypothesis has the same basic characteristics as a fair bet.

As with any bet, you must be able to define your terms. For example, if you bet that "Pat will be in a bad mood today," you need some publicly observable way of determining what a bad mood is. Similarly, if you hypothesize a relationship between two variables, you must be able to obtain operational definitions of your key variables. Thus, if you plan to measure the effects of physical attractiveness on how much a person is liked, you must be able to define both attractiveness and liking according to specific, objective criteria.

Also, as with any bet, your prediction should be specific so that it is clear what patterns of results would indicate that your hypothesis "won" and what results would indicate that your hypothesis "lost." You do not want to conduct a study and then have to debate whether the results supported or refuted your hypothesis. Usually, the easiest way to avoid such disputes is to make your hypothesis as specific as possible. Therefore, when stating your hypothesis, don't just say that increases in one variable will be related to changes in another variable: Specify whether those changes will be increases or decreases. That is, rather than saying aggression will vary with temperature, it would be better to say increases in aggression will correspond to increases in temperature. Ideally, you would be even more specific. For example, you might predict that increasing the temperature from 80 to 90 degrees Fahrenheit will increase aggression more than increasing the temperature from 70 to 80 degrees. To make sure that your prediction is precise enough, ask yourself, "What kind of result would disconfirm my prediction?" and "What kind of result would support my prediction?" Then, graph both of these patterns of results.

By being specific, you can avoid making predictions that are so vague that no pattern of results will disprove them. Unlike some fortune-tellers and unscrupulous gamblers, you want to be fair by giving yourself the opportunity to be proven wrong.

Make It Supportable: Predict Differences Rather Than No Differences

Besides giving your hypothesis a chance to be refuted, you also want to give your hypothesis the opportunity to be supported. That is, you not only have to beware of making bets you can't lose, but also bets you can't win.

You must be especially wary of one kind of bet you can never win—trying to prove the **null hypothesis**: a prediction that there is no relationship between your variables. Even if your treatment group scores exactly the same as the no-treatment group, you will not have proven the null hypothesis.

To understand why you can't prove the null hypothesis, suppose you hypothesize no relationship between honesty and success. Even if you find no relationship, you can't say that there isn't a relationship. You can only say that you *failed to find a relationship*. Failing to find something—whether it be your keys, a murder weapon, a planet, or a relationship between variables—is hardly proof that the thing doesn't exist.[1,2]

The fact that you can't prove the null hypothesis has two important implications. First, you can't do a study to prove that a treatment has *no effect*. If you find no difference between your treatment group and no-treatment group, you can't say that your treatment has no effect: You can say only that you failed to find a treatment effect. Second, you can't do a study to prove that two treatments have the *same effect*. That is, if you find no difference between your two treatment groups, you can't say that the treatments have the same effect: You can say only that you failed to find a difference between them.

Be Sure to Have a Rationale: How Theory Can Help

In addition to making sure that your hypothesis is testable, make sure that you have a solid rationale for your hypothesis. If you can't think of a good reason why your hypothesis should be correct, your hypothesis is probably a "bad bet": It will be a long shot that probably won't pay off. For example, if you hypothesized, without giving any rationale, that people would be more creative after eating three Brussels sprouts, it is doubtful that your prediction would pan out. Instead, it would appear that you were simply going on a hopeless fishing expedition. Therefore, always write out the reasons for making your prediction. Your rationale can come from previous research that has found results similar to what you are hypothesizing, from common sense, or from **theory**: a set of principles that explain existing research findings and that can be used to make new predictions that can lead to new research findings.

Because theories are hypothesis-generating tools, you can—and probably should—use them to generate hypotheses. Even when you don't use theory to generate your hypothesis, you can use theory to provide a rationale for your hypothesis. To illustrate, suppose you had a hunch that taking care of a plant would cause older adults to be more mentally alert and healthy. You might use theory to justify that prediction. For example, according to learned helplessness theory, a lack of control over outcomes may cause depression. Therefore, you could use learned helplessness theory to predict that taking care of a plant may give older adults more of a sense of control and thus make them less vulnerable to helplessness (see Langer & Rodin, 1976).

Demonstrate Its Relevance: Theory Versus Trivia

To have a hypothesis worth testing, not only must you provide at least one reason to expect that your hypothesis will be supported but you must also

---

[1] But why would you fail to find an effect? We'll answer than question in detail in Chapter 10. For now, just realize that failing to find an effect is different from proving there is no effect.

[2] We thank Dr. Jamie Phillips for rewriting much of this section.

provide at least one reason for people to care whether it is supported. Usually, you must explain how testing it will fill a gap in previous research, test a theory, or solve a practical problem. Thus, the hypothesis about what would happen if you gave people three Brussels sprouts—as well as any other hypothesis in which the only rationale for the hypothesis is "let's expose people to an unusual circumstance and see what happens"—is not only a bad bet but also a silly and trivial bet. Scientists frown on doing research just to find isolated bits of trivia. For example, without any other rationale, doing a study to show that alcohol decreases Ping-Pong performance is meaningless—and psychological research should not be a trivial pursuit.

To see how theory can transform your hypothesis from trivial to relevant, consider the following two examples. First, consider this hypothesis: Around age 7, children stop believing in Santa Claus. In its own right, this is a relatively trivial hypothesis. However, when put in the context of Piaget's theory, which states that around age 7, children are able to think logically about concrete events (and thus realize that Santa Claus can't be everywhere at once and can't carry that many toys), the finding has deeper significance. Now, rather than being just an isolated fact about children's thinking, it is evidence supporting Piaget's explanation of how children think.

Second, suppose you were to make a hypothesis about a gender difference. For example, suppose that, like Haselton, Buss, Oubaid, and Angleitner (2005), you predicted that women would be more upset than men to discover that their dating partner was not as wealthy as they were led to believe. At first, such a hypothesis might seem trivial. If you were, however, able to tie your hypothesis to the theory of evolution, your hypothesis would be more interesting.

How could you tie a hypothesis about gender differences to the theory of evolution? The key is to assume that although both genders have the evolutionary goal of bringing as many offspring into adulthood as possible, the strategies that men will use to achieve this aim are different from the strategies that women will use (Buss, 1994). Consequently, it is consistent with the theory of evolution that men would be more

- promiscuous (They have virtually no limit on the number of offspring they can have.)
- jealous (Before DNA testing, men could not be sure that they were the biological parent.)
- impressed by youth (Younger women are more fertile than older women.)
- influenced by physical attractiveness (Attractiveness is a rough indicator of health, and the woman's health is vital because the potential offspring will live inside the woman's body for 9 months.).

You are not limited to using a single theory to make your hypothesis seem relevant. Often, you can show that more than one theory is consistent with your hypothesis. Sometimes, you can show that one theory is consistent with your hypothesis and one is inconsistent with it. Such a hypothesis is very interesting because it puts two theories—two ways of explaining events—in competition. To see the value of putting two theories in competition, imagine how your feelings would change toward someone if you yelled at that person.

According to psychoanalytic theory, if you express hostility toward a person, you'll release pent-up anger and consequently feel better about the person. According to cognitive dissonance theory, on the other hand, if you are mad at a person and then hurt that person, you will justify your aggression by viewing that person in a negative way. Experiments support the dissonance prediction that expressing aggression toward a person leads to feeling more hostility toward that person—and refute the psychoanalytic explanation (Aronson, 1990).

Refine It: 10 Time-Tested Tips

One reason you may have trouble demonstrating your hypothesis's relevance is that you are not used to using past research and theory to justify testing an idea (if so, read Chapter 15). However, another reason you may have trouble selling people on the value of testing your idea is that you need a better hypothesis to sell. The following 10 tips have helped students improve their hypotheses.

1. Don't Be Afraid to Be Wrong: "No Guts, No Glory"

Some beginning researchers mistakenly believe that a good hypothesis is one that is guaranteed to be right (e.g., alcohol will slow down reaction time). However, if we already know your hypothesis is true before you test it, testing your hypothesis won't tell us anything new. Remember, research is supposed to produce new knowledge. To get new knowledge, you, as a researcher–explorer, need to leave the safety of the shore (established facts) and venture into uncharted waters (as Einstein said, "If we knew what we were doing, it would not be called research, would it?"). If your predictions about what will happen in these uncharted waters are wrong, that's okay: Scientists are allowed to make mistakes (as Bates said, "Research is the process of going up alleys to see if they are blind."). Indeed, scientists often learn more from predictions that do not turn out to be true than from those that do.

2. Don't Be Afraid to Deal With Constructs

Some beginning researchers are so concerned about making sure they can measure their hypothesis's key variables that they design their hypothesis around what they can easily observe (e.g., alcohol consumption and reaction time) rather than around the constructs that interest them (e.g., mood and creativity). If you avoid constructs, you may propose a hypothesis that is easy to test but hard to find interesting, such as alcohol slows down reaction time. To avoid proposing hypotheses that lack both constructs and excitement, realize that there are valid ways to manipulate and measure constructs (as you will see in Chapter 5). In short, scientists study what they are interested in—and so should you.

3. Don't Avoid Theory

Good theory explains existing findings and leads to testable new insights. Theory can help you make the leap from just having a general topic to having a specific prediction, especially if your topic is an applied problem. As Kurt Lewin said, "There is nothing so practical as a good theory."

BOX **3.2** **Basic Propositions of Cognitive Dissonance Theory**

1. If an individual has two thoughts that the individual considers inconsistent, then that individual will experience dissonance.
2. Dissonance is an unpleasant state, like anxiety or hunger.
3. An individual will try to reduce dissonance.
4. Changing one of the thoughts to make it consistent with the other is one way of reducing dissonance.

© Cengage Learning 2013

Before seeing how theory can help you attack a practical problem, let's first look at cognitive dissonance theory (see Box 3.2). According to this theory, if a person holds two thoughts that the person considers contradictory, the person will experience an unpleasant state called dissonance (see Figure 3.1). Because dissonance is unpleasant, the person will try to reduce it, much as the person would try to reduce hunger, thirst, or anxiety (Aronson, 1990).

To better understand cognitive dissonance theory, suppose a woman thinks she is generous but also knows she doesn't give money to charity. If she notices and perceives that these two thoughts are inconsistent, this inconsistency will bother her. In other words, she will feel dissonance. To reduce this dissonance, she may change her thoughts or her actions. For example, she may decide that she is not generous, that the charities she refused were not worthwhile, or that she will give some money to charity.

Now that you have some understanding of cognitive dissonance theory, let's see how two sets of researchers used dissonance theory to go from a general practical problem to a specific prediction. The first set of researchers was

FIGURE **3.1** Cognitive dissonance.

concerned with the general problem of how to get people to buy condoms. According to cognitive dissonance theory, people will buy condoms if doing so will reduce their feelings of cognitive dissonance. That is, if John is aware that he has two contradictory thoughts ("I just told people they should use condoms when having sex" and "I have had sex without condoms"), John will feel dissonance. To reduce that dissonance, he can do something that will be consistent with what he has just preached—buy condoms. Thus, cognitive dissonance theory led to this hypothesis: Participants will be motivated to buy condoms if researchers (a) have participants publicly advocate the importance of safe sex and then (b) remind each participant about times when that participant had failed to use condoms. As predicted, participants who were made to see that their past behavior was inconsistent with their publicly stated position were more likely to buy condoms (Stone, Aronson, Crain, Winslow, & Fried, 1994).

The second set of dissonance researchers was concerned about another general problem: Getting introductory psychology students to believe that learning about research methods is important. Although many psychology professors have worried about that problem, Miller, Wozniak, Rust, Miller, and Slezak (1996) seem to be the first ones to use cognitive dissonance theory to find a solution. The hypothesis suggested by cognitive dissonance theory was that having students write essays about why it was important to know about research methods would be more effective than lecturing to students about the value of research. This hypothesis was supported: Students were more likely to believe in the value of research methods when they had convinced *themselves* that it is important.

You have seen that theory can be useful. A theory, however, can help you only if you know about it and understand it. How can you get to know a theory?

Your first step to getting introduced to a theory might be to read textbook summaries of theories. These summaries will allow you to select a theory that can help you. Once you have selected a theory, however, you must go beyond textbook summaries because such summaries may oversimplify the theory. The researcher who relies on textbook summaries may be accused of ignoring key propositions of the theory or of using an exaggerated, oversimplified cartoon version of the theory. Therefore, in addition to reading textbook summaries, you should also consult journal articles that describe studies based on the theory (e.g., "Elation and depression: A test of opponent process theory") so that you can see how other researchers have summarized the theory. The beginnings of these articles usually include a brief description of the theory that the study tests.

Once you have selected a theory, read the original statement of the theory (the citation will be in the texts or articles that you read). Then, to keep up to date about changes in the theory, use *PsycINFO*, *Psychological Abstracts* or *Social Sciences Citation Index* to find books and review articles devoted to the theory. (For more information on how to conduct a literature review, see Web Appendix B.)

4. Act to Make Participants React: Use True Independent Variables

Rather than trying to describe what happens, try to *change* what happens. That is, manipulate variables. We have found that two roadblocks stop students from hypotheses that involve manipulating variables. First, some students believe that people are as unchangeable as stones—a remarkable belief given that some of those students will go into counseling and other applied areas in which their job will be to change people. Second, students are used to watching and listening to people, but are not used to having the ability to manipulate variables. Consequently, students often don't think about how they can manipulate variables. If you can break through these two barriers (and we will soon show you how), you can do some neat things. For example, the authors of the sample article in Appendix B, like many other people, observed that wearing dark uniforms was associated with aggression. They went beyond what others had done, however, by manipulating whether participants wore black or white uniforms and *observing its effect* on aggressiveness.

To see another example of how and why experimenters manipulate variables, consider a study in which experimenters *manipulated* participants' expectations of how likely it was that George W. Bush would be elected president in 2000 (Kay, Jimenez, & Jost, 2002). To manipulate participants' expectations, the experimenters first created five reports that were allegedly based on valid scientific analyses of polling data:

1. a report concluding that Bush would win in a landslide
2. a report concluding Bush would win a narrow victory
3. a report concluding there would be a tie
4. a report concluding Al Gore would win a narrow victory
5. a report concluding Gore would win in a landslide

The experimenters then manipulated (using random assignment) which report each participant read. They found that expecting Bush to win caused people to like Bush more. Note what the researchers would have lost if, instead of manipulating expectations, the researchers had merely asked participant (a) who they expected to win and (b) how much they liked Bush. In that case, if the researchers found that participants who expected Bush to win liked Bush more than those who expected Bush to lose, the researchers could not conclude that expecting Bush to win caused participants to like Bush. After all, it might be the reverse: Liking Bush might cause people to expect Bush to win.

Similarly, Kunstman and Maner (2011) wanted to see whether having power over a woman makes a man more likely to think that the woman is sexually interested in him. Kunstman and Maner could not easily manipulate power in a business setting. However, they could manipulate perceived power in a lab experiment. They assigned half the male participants to the role of leader of a two-person team (composed of the male participant and a female "participant" who was actually a research assistant). As leader, the participant expected to decide both how to accomplish the team's task and how to divide the rewards for completing the task. The other male participants were told that the two team members would be equal partners and would get equal rewards. Compared to the men assigned the "equal partners role," men assigned the "boss" role were more likely to think that the woman was sexually interested in them.

The big lesson here is that if you want to have internal validity (and more interesting research ideas), rather than looking for participants who already differ in some way (e.g., comparing students who sit in the front of the class with students who sit in the back of the class), start with participants who do not differ in that way and then manipulate a factor that will make them differ in that way (e.g., assign them to seats). In technical terminology, you need to use an **independent variable**: a treatment (an intervention or some other stimulus) that you administer to some individuals but not to others.

Using independent variables allows you to find out why people do things. Consequently, many professors will require your hypothesis to be in the form, "_____ affects scores on _____," with the first blank containing the name of your **independent variable** (the factor you are manipulating) and the second blank containing the name of your **dependent variable** (dependent measure): the participant's response that the researcher is measuring.

Despite the terminology, the basic idea for doing an experiment is simple. As Boese (2007, p. x) writes, "An experiment starts when a researcher looks at a situation and thinks, 'What would happen if I changed one part of this?' He or she performs an experimental manipulation and observes the result." You do that same kind of reasoning whenever you tell a story and then note that things probably would have been even worse if _____ had also happened.

If you are having trouble thinking of hypotheses that involve independent variables, remember that one reason people do things is because of the immediate environment—and you can manipulate the immediate environment. Because people are constantly adapting and reacting to their environment, you can tell people things, show people things, and set up situations that will change how they feel, think, and act. Thus, although you may tend to think that individuals do certain things because of the type of person they are, research clearly shows that, in many cases, the immediate environment is much more important in determining individuals' moods, mental states, and behavior than what type of person the individual is. For example,

- People studying to be ministers are much less likely to stop and help a person if they are led to believe that they might be late to give their speech, even when that speech is on the Good Samaritan parable—a story about the virtue of the person who stopped and helped a person in need (Darley & Batson, 1973).
- When playing a game called "The Community Game," students judged by their dorm counselors to be highly competitive were not any more competitive than students judged to be highly cooperative. Similarly, when playing a game called "Wall Street," students considered highly cooperative were just as competitive as student considered highly competitive. What makes this finding more remarkable is that all students were playing the same game—only the name had been changed (Liberman, Samuels, & Ross, 2004).

- In his study that involved a simulated prison, Zimbardo and colleagues (Zimbardo, Haney, Banks, & Jaffe, 1975) found that when well-adjusted undergraduates were assigned to play the role of prison inmate, they became submissive, but the students who were assigned to play the role of prison guard became sadistic.

If you are still failing to come up with a hypothesis because you think that people only do things because of their traits, realize that, for almost any stable *trait* (e.g., how anxious one typically is), there is a corresponding unstable *state* (e.g., how anxious one feels at the moment) that can be manipulated (Revelle, Condon, & Wilt, in press). For example, you could increase participants' momentary levels of

- anxiety by giving them difficult tasks
- arousal by giving participants caffeine or exposing them to noise
- positive mood by showing them humorous movies
- outgoingness by having them act in an outgoing way
- self-esteem by having others praise them
- power by having them supervise others
- materialism by putting them near pictures of money

Manipulating these states allows you to test cause–effect hypotheses. For example, although it was known for years that outgoing people tended to be happier, it was not known whether outgoingness caused happiness (The relationship could have been due to happiness causing outgoingness.). Now, we have direct evidence that outgoingness causes happiness because Fleeson and McNeil (2006) found that happiness increased in the participants they had act in an outgoing way. Similarly, we had known for a long time that, compared to less materialistic people, more materialistic people tended to be less interested in other people (Kasser & Ryan, 1993). However, no one knew whether materialism caused a loss of interest in other people (it could be that people who are not interested in others become interested in money) until Vohs, Meade, and Goode (2008) manipulated how materialistic participants felt and found that making people think about money (e.g., by putting participants near a poster that had a picture of money) made people less social and less helpful.

5. Look for Other Effects: Use Other Dependent Variables

Thinking about money has more than one effect. Not only does it make people less social and less helpful but also more persistent (Vohs, Mead, & Goode, 2008). Almost any treatment will have more than one effect. Effects can be short term, long term, behavioral, physiological, emotional, cognitive, good, and bad. So, if people are looking at the good effects of pursuing the American Dream, you could look for the bad effects (as Kasser & Ryan, 1993 did). Similarly, if others look for the good effects of attractiveness, you could look at the bad effects. The key is to realize that a treatment has many more effects (on beliefs, feelings, thoughts, actions, and bodily reactions) and a predictor may predict many more events than you would first think.

For example, Dabbs and Dabbs (2000) found that high levels of testosterone in one's saliva correlated with

- phony-looking smiles
- rough tactics in domestic disputes
- premeditated murder
- greater apparent confidence when meeting strangers
- crudeness
- higher rates of marriage
- higher rates of divorce
- being a trial lawyer rather than a non-trial lawyer
- being in prison
- lower levels of career achievement

Although being able to think of many different measures is a useful skill for any researcher, it is perhaps the most important skill that an evaluation researcher—a person who looks at the effect of a program (e.g., a training program, an advertising campaign, a new law)—can have. If you have a strong, practical side, you may be able to generate some research ideas by thinking like an evaluation researcher. You could start by looking at flyers posted on bulletin boards for different activities and programs at your school (e.g., tutoring programs, depression screening, art club), picking one, and then listing five possible positive and five possible negative effects that might result from the activity.

6. Reverse Cause and Effect: Switch Independent and Dependent Variables

Rather than adding dependent measures or replacing one dependent variable with another one, you might convert your dependent variable into an independent variable. For example, suppose you have a rather ordinary hypothesis such as if a person is attractive, participants will be more likely to help that person than if the person is not attractive. In other words, your hypothesis is that being attractive (independent variable) causes one to be helped (dependent variable). Your idea is to make a friend look either moderately attractive or very attractive and see if the friend is helped more when she looks very attractive. You could make this hypothesis more interesting by changing which variable is the cause and which is the effect. That is, you could hypothesize that being helped leads to being perceived as attractive. Thus, you might give some participants a chance to do your friend a favor and see if those participants rate your friend as being more attractive than those who are not given that opportunity. Many psychologists have made important contributions by doing studies that reversed conventional cause–effect hypotheses. For example:

- Rather than looking at attitude change leading to behavior change (we do what we believe), Festinger and Carlsmith (1959) found that behavior change led to attitude change (we believe in what we did).
- Rather than looking at the finger pulling the trigger (aggressive people use guns), Berkowitz (1981) found that the trigger can pull the finger (guns can make us aggressive).

- Rather than looking at how seeing leads to believing, some perception researchers have found, as Myers (2002a) puts it, that "believing leads to seeing."
- Rather than looking at how stress on the body affects mental health, many researchers have looked at how mental stress affects physical health.
- Rather than looking at whether increased income leads to happiness, Diener and Seligman (2004) have looked at whether happiness leads to increased income.

7. Ask "How?" to Look for the Missing Links in the Causal Chain: Search for Mediating Variables

What can you do if your hypothesis about the relationship between two variables is so well established that it is a fact rather than a hypothesis? You can do what 3-year-olds do when they ask a question that is too easy (e.g., "Why did the doorbell ring?): Follow it up with another "why" question (e.g., "Why did the doorbell ring when you pushed it?"). By continuing to ask "why," a child may force the parents to realize that although the parents know that $\underline{A} \rightarrow \underline{C}$ (activating the button leads to a consequence: the doorbell to ringing), the parents do not know what happens between A and C. That is, the parents do not know *how* A causes C.

Similarly, even when psychologists know that A → C (e.g., guns increase aggression), they may not know B: the other *intermediate*, in *between* steps in the causal chain. They may suspect that "A," an activating event (e.g., a gun), has an effect by first changing "B" (beliefs or biological reactions), which, in turn, triggers "C," a change in the participant's behavior (e.g., more aggression)—but they may not know what the mediating mechanism is. In technical terminology, they do not know the **mediating variable:** the mechanism—the biological, mental, physical, emotional, or behavioral process—that comes between a cause (e.g., a stimulus) and its effect (e.g., a response). (To help yourself remember what mediator variables are, remember that just as professional mediators *come between* union and management, psychological mediators *come between* stimulus and response.)

Admittedly, you can deal with some things without looking inside to see their mediating mechanisms. For example, you may know that turning the key opens the lock, but not how the key activates the locking mechanism. Similarly, for a while, many psychologists focused on how outside events influenced outward behavior but ignored what happened *inside* people's minds and brains.

The problem with knowing that A (an outside event) causes C (an observable response), but not knowing the hard-to-see, mediating mechanisms (the "B" that comes between "A" and "C") is that you do not know *how* A has its effect. When the mediating mechanism is not clear (i.e., when how something works is unknown), people question whether the relationship actually exists—or even whether one could exist. Consequently, some have questioned whether women's menstrual cycles synchronize, whether antidepressants are effective, and whether being a target of prejudice can harm one's health—and have questioned whether ESP and therapeutic touch are even possible.

To illustrate how not knowing what the mediator is can make a finding seem suspect, consider the early research that found that participants' expectations about whether another person would be shy or outgoing determined whether that other person actually behaved in a shy or outgoing way (Myers, 2004). This finding seemed like magic until researchers found the mediating mechanism: the type of questions participants asked (Myers, 2004). Specifically, participants expecting the other person to act shy asked questions that made the other person act shy (e.g., "Do you ever want to be alone?") whereas participants expecting their partner to be outgoing asked questions that made the other person act outgoing (e.g., "What would you do to liven things up at a party?").

Finding what comes between a stimulus and a response, like finding out what comes between pressing the doorbell and hearing a ring, usually (a) makes the original relationship seem less magical and (b) requires digging below the surface. For example, if "B" (the mediating between variable) is a biological process or a mental process, it may be hard to observe. Because a "B" variable can be so hard to observe, it may be part of a model or theory long before it is observed. Thus, only after researchers had done cognitive dissonance experiments for years did they attempt to measure cognitive dissonance. Similarly, people had noticed that people are more likely to yawn after seeing someone else yawn long before scientists had discovered mirror neurons (and, indeed, it has not yet been established that mirror neurons account for yawning being contagious).

Now that you understand what a mediating variable is, let's see how you could convert even the most mundane hypothesis into a fascinating one by adding a mediating variable to it. For example, the hypothesis that male football fans feel bad when their team loses is not that interesting. It is interesting, however, to hypothesize about the bodily or psychological mechanisms that create this effect. What inside the person causes him to feel bad? Is the mediating variable a physiological process, such as a decrease in testosterone (Bernhardt, Dabbs, Fielden, & Lutter, 1998)? Is the mediating variable a cognitive psychological process, such as a change in self-concept?

Similarly, the hypothesis that increasing the room temperature will cause people to be more aggressive is not that interesting. It is more interesting to find out how warm temperatures increase aggression. What mediates that relationship? Once you have the answer to that question, you have a hypothesis about a mediating variable. For example, if your answer to that question was "liking or disliking others in the room," your hypothesis might be that increasing the room temperature makes people like each other less, and this decreased liking, in turn, causes increased aggression.

To understand mediation better and to see how you might test a hypothesis about mediation, imagine that we have a room containing a massive tangle of electrical cords and power strips. We plug one of those cords (A) into the wall outlet, and a lava lamp (C) lights up. Is the cord directly hooked up to the lava lamp (an A–C connection)—or does it go through a power strip? If so, which power strip does it go through? If we wanted to test whether it goes through the power strip that we labeled "B," we could go one of two routes (see Figure 3.2).

Clearly established observation

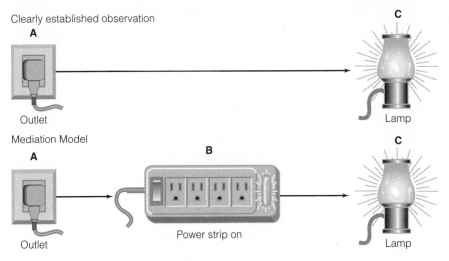

Mediation Model

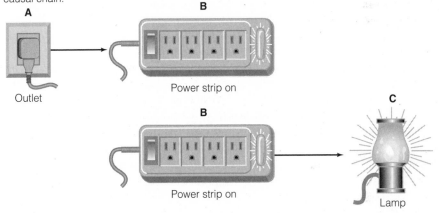

Strategy 1 for testing mediation model. Conduct a two-part test by testing each link in the causal chain.

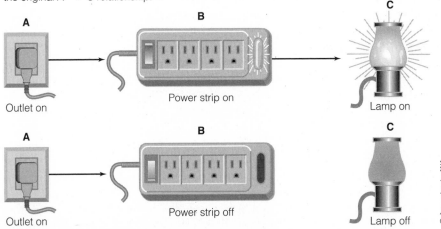

Strategy 2 for testing a mediation model: See if manipulating the mediator moderates (modifies) the original A → C relationship.

© Cengage Learning 2013

FIGURE **3.2** Mediation can be as easy as A B C.

One approach would be to test the A (outlet) → B (power strip) and B (power strip) → C (lamp) connections separately. That is, we would test (1) whether the cord plugged into the wall outlet turned on the power strip and (2) whether the cord plugged into the power strip's outlet turned on the lamp. If plugging and unplugging the cord from the wall turned the power strip on and off and if plugging and unplugging the cord that had been plugged into the power strip's outlet turned the lamp on and off, we could conclude that the wall outlet was powering the power strip, which, in turn, was powering the lamp.

A second, related approach would be to see whether we could mess with the A (plugging the cord into the wall outlet) → C (lamp turns on) by messing with B (the power strip). Specifically, after plugging the cord into the wall outlet and establishing the A–C relationship (plugging the cord into the wall outlet → lamp turns on), we would turn the power strip's switch on and off. If we found that plugging the cord into the wall outlet turned the lamp on only when the power strip was also on, we would conclude that the cord fed into the power strip, which, in turn, went to the lamp.

Let's now see how the same logic we used to figure out whether a power strip mediated an electrical event can be used to figure out whether a variable mediated a psychological event. For our first example, let's look at efforts to try to explain the finding that "after exerting self-control, people are more prone to fail at later efforts at self-control" (Gailliot & Baumeister, 2007, p. 305). If you wanted to see whether blood sugar mediates this relationship (e.g., if A [exerting willpower] → B [lowers blood sugar] → C [decreased willpower]), you could do two experiments. In the first, you would see whether (A) exerting willpower (e.g., persisting on a difficult task) decreased (B), blood sugar levels. In the second, you would see whether decreasing (B), blood sugar levels (e.g., depriving participants of breakfast), would decrease (C), willpower.

For our second example, let's look at trying to explain the **weapons effect:** that the presence of guns increases aggression. If the weapons effect is due to testosterone, we would expect that the presence of (A), guns, increases (B), testosterone levels, and that increases in (B), in turn, lead to increases in (C), aggression. Jennifer Klinesmith, a senior psychology major, did a study in which she found that presenting participants with (A), guns, increased their (B), testosterone levels (Klinesmith, Kasser, & McAndrew, 2006). Although she did not do an experiment to show that increases in testosterone, in turn, lead to increases in aggression, she did point out that some previous studies had shown that increases in testosterone lead to increased aggression and that, in her study, testosterone levels were statistically related to aggression.

Finding a statistical relationship helps make the case that testosterone is a mediator, but a stronger case would have been made if the researcher had been able to manipulate testosterone (for ethical reasons, she could not). Because she could not do an experiment that manipulated testosterone directly, testosterone might not have been the mediating variable: Rather than being the underlying cause of the effect, it might be a side effect of the treatment. To illustrate, suppose you found that the less acupuncture patients sweat during a session, the more effective the acupuncture is at relieving their pain. In such a case, you would not assume that acupuncture leads to less sweating, which, in turn, leads to less pain. Instead, you would suspect that less pain leads to less sweating. Because reduced sweating might be the result—rather than the cause—of reduced pain, you would not conclude that sweating was the mediator of acupuncture's pain-reducing effect.

To establish whether a variable is the treatment's mediator rather than the treatment's by-product, you could do an experiment that manipulated the potential mediating variable (Sigall & Mills, 1998). Such an experiment would use the same logic we used when we turned the power strip off and on to determine whether the current went through the power strip to the lamp. That is, you would see whether blocking the proposed mediating variable blocked the treatment's effect. For example, people suspected that acupuncture (A) reduced pain (C). But how did it reduce pain? Perhaps acupuncture (A) makes the body release its own painkillers called endorphins (B), and those endorphins are what reduces the pain. How did researchers test this A (acupuncture) → B (body releasing endorphins) → C (reduced pain) model? Researchers showed that acupuncture was effective in relieving pain—unless the researchers gave participants a drug that blocked the effect of endorphins. So, A → C, except when B was not allowed to have an effect. Because blocking the effects of endorphins blocked the effect of acupuncture, researchers concluded that endorphins mediated the effect of acupuncture.

In short, one way to expand a simple, general "A" (a treatment) causes "C" (a consequence) hypothesis is to add a mediating variable that may explain *how* the cause has its effect. By adding a hypothesis about the mechanism for the cause, you would expand your "A" causes "C" hypothesis to an "*A causes C due to B*" hypothesis. For example, if your original hypothesis was that small classes cause improved test scores, you might expand your hypothesis to small classes cause increased test scores *due to* more interaction with the professor.

8. Look for Moderator Variables by Asking "Except for When?" "Except for Where?" and "Except for Whom?"

Rather than adding a *mediating* variable to track down *how* the cause has its effect, you could add a *moderating* variable to determine *when*—in what situations and for whom—the cause has its effect. Thus, you could expand your "A causes C" hypothesis to an "*A causes C when D^1 but not when D^2*" hypothesis. For instance, you could say that

1. small classes cause increased test scores when classes are science classes, but not when classes are art classes
2. small classes cause increased test scores when test questions involve applying information, but not when questions involve memorizing information
3. small classes cause increased test scores when students are nontraditional students, but not when students are traditional students
4. small classes cause increased test scores when professors know students

To shorten your hypothesis and to remind yourself that you are looking for exceptions, you could phrase your hypothesis in the form: "*A causes C except when D*^{2}." The problem with phrasing your hypothesis that way is that it encourages you to think only about those moderator variables that weaken or eliminate a relationship—and ignore those moderators that strengthen a relationship as well as those moderators that reverse a relationship. Because a moderator variable is one that modifies (strengthens, weakens, or reverses) the relationship between two other variables (see Figure 3.3), a more complete way to phrase a hypothesis for a moderator variable is "*the more A, the _____ C, when D is absent or low, but the more A, the _____ C when D is present or present in high amounts.*" For now, you may find it easier to use the following format to state a moderator hypothesis: "*The relationship between A and C is generally that the more A, the _____ C, but this relationship depends on D.*"

To illustrate that the above format is a good way to phrase a hypothesis involving a moderator variable, consider the original social loafing hypothesis: The <u>larger the group</u>, the more individuals will <u>loaf on the task</u>. Researchers later found four types of moderators:

1. The larger the group, the more loafing on the task, but the effect depends on type of group: Groups in which members like each other and groups that are competing with another group, loaf less than other groups.
2. The larger the group, the more loafing on the task, but the effect depends on type of task: If individuals like the task, they loaf less.
3. The larger the group, the more individuals will loaf on the task, but the effect depends on the type of individual: For women and people who are group-oriented, increasing group size reduces effort only slightly; for men and people who are individualistic, increasing group size reduces effort greatly.
4. The larger the group, the more loafing on the task, but the effect depends on perceived identifiability: Manipulations that make participants think they are individually identifiable (e.g., making participants aware that an observer is recording each individual's performance, making each participant responsible for one section of a report) reduce loafing.

One way to generate hypotheses involving these four types of moderator variables is to realize that a simple two-variable hypothesis—that changing an aspect of "A" causes a change in "C"—is general in four ways. For example, consider the general hypothesis that increasing a group's size will increase loafing. To see that this hypothesis is general in four ways, let's rewrite it: For any type of group, increasing its size will increase loafing on any type of task for any type of person in any type of situation. As you can see, the hypothesis makes four generalizations:

1. It implies that changing one aspect of the group (its size) will have an effect for *all types of groups*.
2. It implies that changing one aspect of the group (its size) will have an effect on *all types of tasks*.

Original relationship

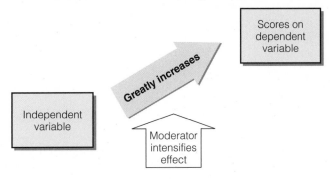

1. Moderator increases treatment's original effect.

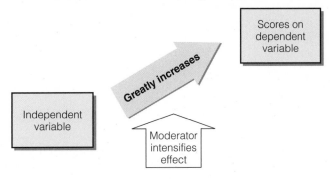

2. Moderator reverses treatment's original effect.

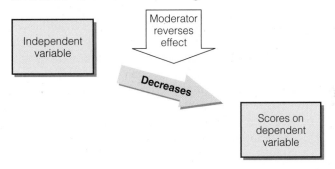

3. Moderator neutralizes treatment's original effect.

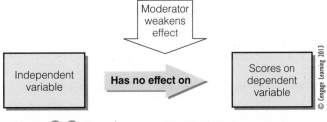

FIGURE **3.3** Three types of moderators.

3. It implies that increasing group size will increase loafing for *all types of individuals*.
4. It implies that increasing group size will increase loafing in *all types of situations*.

Thus, to start your search for moderators, you could ask whether each of these four generalizations is valid. So, if you were looking at the size of the group–loafing relationship, you might first ask whether increasing group size would have less of an effect on some types of groups (e.g., close friends) than on other groups (strangers). Second, you might ask if size would increase loafing to a lesser degree on some types of tasks (e.g., interesting tasks) than on others (e.g., boring tasks). Third, you might ask if the size–loafing effect might hold less true for some types of individuals (e.g., people high in conscientiousness) than for others (e.g., people low in conscientiousness). Fourth, you might ask if the effect of group size would be less in some types of situations (e.g., when competing against another team) than in others (e.g., when not competing against another team).

If you don't want to go through such a structured approach, simply ask yourself external validity questions: "Under what circumstances might this rule (social loafing in large groups) not apply?" and "For what groups might this relationship not hold? To help you answer these questions, think of exceptions to the rule—and then try to see how these exceptions to the rule may suggest an addition to the old rule.

To see how asking these questions helps you generate good research ideas, imagine that Zebrowitz had decided to test the old saying "people from a different race all look alike." This hypothesis would not be interesting because much research has shown support for it. Instead, she and her colleagues thought about exceptions to that rule. They (Zebrowitz, Montepare, & Lee, 1993) found that, under certain conditions, people could do a good job of distinguishing among members of other racial groups. For example, attractiveness moderated the "all look alike" effect: People could distinguish between attractive and unattractive members of a different race.

In addition to looking for moderating variables that allow you to state conditions under which common sense rules are wrong, you could look for moderator variables that allow you to reconcile conflicts among common sense rules. For example, when does "like attract like," and when do "opposites attract"? When does "absence make the heart grow fonder," and when does "absence make the heart wander?" Under what circumstances are "two heads better than one," and under what circumstances is it better to do it yourself (after all, "too many cooks spoil the broth" and "nothing is worse than a committee")?

One type of moderator variable you might look for could be an intervention program. For example, you might predict that people who went through your "how to work better in groups" training program might follow the "two heads are better than one" rule, whereas those who had not would follow the "too many cooks spoil the broth" rule. In other words, untrained people might work better alone whereas trained people might work better in groups.

If you are having trouble thinking of a variable that may moderate an effect, go back and think about variables that may mediate the effect. If you know or suspect the process by which the treatment causes the effect, anything you do that will modify that process may modify (moderate) the treatment's effect. To take a simple example, suppose we had the following mediating variable hypothesis: Having a child in day care causes parents to get sick due to the existence of viruses and bacteria within the day-care center. That mediating hypothesis suggests at least three moderator variable hypotheses.

1. Having a child in day care causes parents to get sick, but this relationship will depend on whether the parents wash their hands with antibacterial soap.
2. Having a child in day care causes parents to get sick, but this relationship will depend on whether the parents have been vaccinated.
3. Having a child in day care causes parents to get sick, but this relationship will depend on how often and thoroughly we kill the viruses and bacteria within the day-care center.

To take a psychological example of how thinking about mediating variables (variables that are often mind or brain activities that come between a cause and an effect) can lead to moderator variables, suppose you suspect that the reason individuals loaf when they are in a group (the "social loafing effect") is that they do not think that their efforts can be identified. In that case, perceived identifiability would be the mediating variable (as you can see from Figure 3.4, large group → low perceived identifiability → loafing). If your hypothesis is correct, variables that affect perceived identifiability should moderate the social loafing effect (Williams, Nida, Baca, & Latané, 1989). For example, manipulations that make individuals think they are more identifiable (e.g., having observers record each individual's performance or making an individual responsible for one section of a report) should moderate the social loafing effect by weakening it.

In conclusion, if you have a hypothesis about the relationship between two variables that is too obvious and too well documented, you may be able to rescue your hypothesis by adding a moderator variable. For example, although the hypothesis, increasing group size will decrease effort, has already been extensively tested, the hypothesis, increasing group size will decrease effort when anxiety is low but not when anxiety is high, has not.

Even if your hypothesis does not need to be rescued, chances are good that your hypothesis can be improved by adding a moderating variable. For instance, Frederick and Hazelton (2007) added a moderator to the hypothesis that women will prefer muscular men to make it "women will prefer muscular men more when considering a short-term partner than when considering a long-term partner."

9. Make Precise Predictions About When More May Be Too Much—Or Not Enough

Besides adding a moderator variable, Frederick and Hazelton (2007) made their hypothesis about muscularity and attraction more interesting by being precise about the relationship between muscularity and attractiveness.

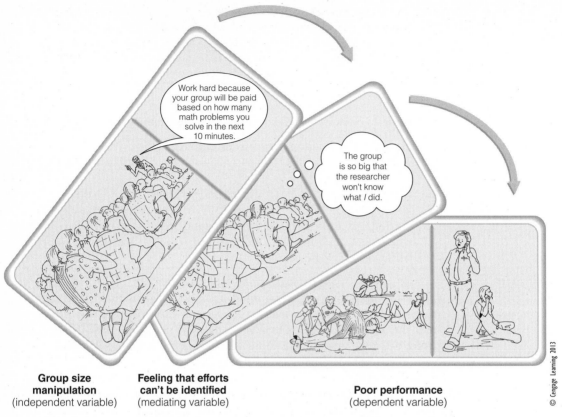

Group size manipulation
(independent variable)

Feeling that efforts can't be identified
(mediating variable)

Poor performance
(dependent variable)

© Cengage Learning 2013

FIGURE **3.4** Perceived identifiability as mediator of social loafing.

Frederick and Hazelton did not just say that the more muscular guys would tend to be seen as more attractive than less muscular guys. Instead, Frederick and Hazelton specified that women would find low levels of muscularity unappealing, moderate and high levels of muscularity appealing, and extremely high levels of muscularity unappealing. Thus, they hypothesized that if you used muscularity as your predictor variable and attractiveness as your outcome variable, the **functional relationship**—the relationship between how much of the predictor variable corresponds to how much of the outcome variable—would be complex. That is, at low levels of muscularity, perceived attractiveness would *increase* as muscularity increased; at high levels of muscularity, perceived attractiveness would *decrease* as muscularity increased. Thus, to know how much a man's perceived attractiveness would change as a *function* of him becoming more muscular, we would have to know how muscular he currently is.

To see that experiments that look at functional relationships can be both useful and easy to do, consider the work of one of the earliest psychologists, Sir Francis Galton, who explored the functional relationship between the temperature of tea and the taste of tea. From his research, he figured out

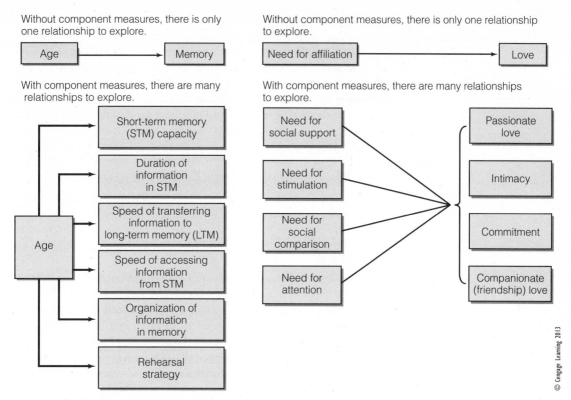

FIGURE **3.6** Two examples of how finding measures of specific components allows you to look for new relationships.

general traits: openness, agreeableness, conscientiousness, extraversion, and neuroticism. That research had found only two differences: Conservatives tend to be more conscientious, and liberals tend to be more open-minded. Thus, many people were under the impression that conservatives and liberals did not differ on any aspect of agreeableness and that conservatives scored higher than liberals on every aspect of conscientiousness. However, by looking beyond the five general traits to the specific traits that make up those five broad traits, Hirsh made two discoveries. First, although conservatives are not more agreeable than liberals, conservatives and liberals do differ on *aspects* of agreeableness: Conservatives are more polite whereas liberals are more compassionate. Second, conservatives score higher than liberals on conscientiousness because conservatives are more orderly than liberals—but conservatives and liberals do not differ on other aspects of conscientiousness such as self-discipline, cautiousness, and striving to achieve goals.

When doing experimental research, looking at the different components of a treatment allows you to see how important—or unimportant—different aspects of a treatment are to producing an effect. By comparing how baby monkeys responded to "wire mothers" who provided milk with how baby monkeys responded to "cloth mothers" who did not provide milk, Harlow (1958) showed that, for facilitating attachment, it is more important for monkey moms to be soft than to provide food. More recently, researchers, by comparing participants who were given either handheld or hands-free cell phones, showed that it is talking on the phone—not holding the phone—that impairs driving performance (Strayer, Drews, & Crouch, 2006). The same researchers were also able to show that it was not listening that impaired driving (people were not impaired when they listened to books on tape or radio broadcasts) and that it was not even having a conversation that impaired driving (people were not impaired when talking with another person who was also in the (simulated) car. Instead, Strayer and Drews (2008) were able to show that impairment occurs only when talking on the cell phone, partly because, unlike a passenger, the person on the other end of the call will keep talking when the driver is in a challenging driving situation.

Make Sure That Testing the Hypothesis Is Both Practical and Ethical

Once your hypothesis is testable, reasonable, and relevant, you must still ask two additional questions. The first question is, "Can I test this hypothesis?" Sometimes, you will lack the equipment, experience, or money to test it. For example, testing some hypotheses in physiological psychology may require equipment or surgical skills that you do not have. The second question is, "*Should* I test this hypothesis?" That is, can the hypothesis be tested in an ethical manner?

You have a serious obligation to make sure that your study is ethical. Clearly, you do not have the right to physically or psychologically harm another. Reading Appendix D can help you decide whether your study can be done in an ethical manner. However, because conducting ethical research is so important, do not make the decision to conduct research without consulting others. Before doing a study, you and your professor will probably need to have your project reviewed by an ethics committee. In any event, *never conduct a study without your professor's approval!* (See Box 3.3.)

CHANGING UNETHICAL AND IMPRACTICAL IDEAS INTO RESEARCH HYPOTHESES

In their present form, some of your ideas may be impractical or unethical. However, with a little ingenuity, many of your ideas can be converted into workable research hypotheses. As you will see, many practical and ethical obstacles can be overcome by making the key variables more abstract, constructing a smaller scale model of the situation, or toning down the manipulation. To understand how these principles can turn even the most impractical and unethical idea into a viable research hypothesis, consider the following hypothesis: Receiving severe beatings causes one to be a murderer. How could we convert this idea into a workable research hypothesis?

Make Variables More General

One possibility is to consider your original variables as specific instances of more general variables and then reformulate your hypothesis using these

---

BOX **3.3**   Questions to Ask About a Potential Hypothesis

1. Can it be proven wrong?
   - Can you obtain operational definitions of the variables?
   - Is the prediction specific?
2. Can it be supported?
   - Are you predicting that you will find an effect or a difference? (Remember, your results can never prove the null hypothesis.)
3. Are there logical reasons for expecting the prediction to be correct?
   - Is it predicted by theory?
   - Is it consistent with past research findings?
   - Does it follow from common sense?
4. Would the results of the test of your prediction be relevant to
   - previous research?
   - existing theory?
   - a practical problem?
5. Is it practical and ethical to test your prediction?
   - Do you have the physical and financial resources to test this idea?
   - Would testing the hypothesis cause physical or psychological harm to the participants? (See Appendix D.)
   - Do you have approval from your professor?
   - If you are planning on doing research with human participants and your school has an institutional research review board (IRB), do you have approval from that board?
   - If you are planning on doing research with nonhuman animals and your school has an institutional animal care and use committee (IACUC), do you have approval from that committee?

© Cengage Learning 2013

---

more abstract, more psychological variables. In our murder example, you could view murder as a more specific instance of aggression and view beating as a specific instance of aggression, pain, or punishment. Thus, you now have three research hypotheses that have been studied in controlled settings: (1) *aggression leads to more aggression*, (2) *pain causes aggression*, and (3) *punishment causes aggression*.

Use Smaller Scale Models of the Situation

Of course, for both ethical and practical reasons, you are not going to measure aggression by having human participants hit each other. Instead, you may give participants an opportunity to destroy something that supposedly belongs to the target of their anger, an opportunity to write a negative evaluation of another person, an opportunity to press a button that supposedly delivers (but doesn't really deliver) a mild shock to another person, an opportunity to decide how much hot sauce to put in a glass of water that the other participants will supposedly have to drink, and so on. As you can imagine, working with a small-scale model of an aggressive situation is more ethical than manipulating real-life aggression.

Smaller scale models of the situation not only have ethical advantages but also have practical advantages. For example, if you are interested in the effects of temperature on aggression, you can't manipulate the temperature outside. However, you can manipulate the temperature in a room. Similarly, you can't manipulate the size of a crowd at a college football game to see the effect of crowd size on performance, but you can manipulate audience size at a dart contest that you sponsor. By using a dart contest, testing your audience-size hypothesis is not only possible but also practical. For instance, if audience size has an effect, you could probably find it by varying the size of the audience from zero (when you are hiding behind a one-way mirror) to three (you and two friends).

Once you have a small-scale model of a phenomenon, you can test all kinds of ideas that previously seemed impossible to test. For example, can you imagine using the dart contest situation to test the effects of audience involvement or size of reward on performance?

Smaller scale models can include simulations (e.g., putting people in a driving simulator to test the effects of cell-phone use on driving), simulated worlds (e.g., having people obey a command to "hurt" an avatar in "Second Life"), or scenarios (having participants imagine what they would do if their partner broke up with them). Because smaller scale models of situations are so valuable, researchers often review research literature to discover if someone else has already made a smaller scale model of the phenomenon they wish to study. That is, just as an airplane designer may use someone else's wind tunnel to test new airplane designs, researchers may use someone else's model of a situation to see if their ideas fly (Myers, 2004).

Carefully Screen Potential Participants

In some research, you might avoid ethical problems by avoiding participants who are likely to be harmed by the manipulation. Therefore, if you were to do a frustration–aggression study, you might only use participants who

- were, according to a recently administered personality profile, well-adjusted,
- were physically healthy, and
- volunteered after knowing about the degree of discomfort they would experience.

Use "Moderate" Manipulations

Another way to prevent people from being harmed by your manipulation is to make your manipulation less harmful. One way to make your manipulation less harmful is to avoid unpleasant stimuli entirely by replacing them with either positive or neutral stimuli. Thus, rather than comparing a positive manipulation (e.g., raising self-esteem) with a negative manipulation (e.g., decreasing self-esteem), you might compare a positive manipulation (e.g., praising a person to increase their self-esteem) with a neutral manipulation (e.g., neutral or no feedback).

If you must use an unpleasant manipulation, consider making it moderately—rather than extremely—unpleasant. Thus, if you were to induce frustration to observe its effect on aggression, you might decide not to use a very high level of frustration. Even though a high level of frustration would be more likely to produce aggression, you might decide to use lower levels of frustration to lower the risks of harming your participants. Similarly, although it would be illegal, immoral, and unethical to cause someone permanent brain damage, researchers have used transcranial magnetic stimulation (TMS) to temporarily decrease activity of neurons in parts of the brain. For example, temporarily deactivating parts of the brain has temporarily improved participants' performance on drawing tasks (Schachter, Gilbert, & Wegner, 2009).

As a Last Resort, Do Not Manipulate Variables

Finally, as a last resort, you may decide not to manipulate the variables at all. Unfortunately, if you choose this last resort, you must change your hypothesis. To understand why, let's return to the original hypothesis: Receiving severe beatings causes one to be a murderer. If you pursue this idea by interviewing murderers and nonmurderers to see whether murderers were more likely to report being beaten as children, your research will probably be ethical. However, your research will not test your hypothesis: Even if you found that murderers were more likely than nonmurderers to have been beaten, your results would *not* necessarily mean that the beatings *caused* the murders. Beatings may have no impact on murders. Instead, murderers may have been beaten more than nonmurderers because, even when they were younger, murderers were more aggressive and more disobedient than nonmurderers. Thus, when you give up manipulating variables, you give up the ability to test cause–effect hypotheses.

Fortunately, however, although giving up on manipulating variables means giving up on testing cause–effect hypotheses, it does not mean you have to give up on testing all hypotheses. For example, although interviewing wouldn't allow you to discover whether beatings cause children to become murderers, it might allow you to address a related research hypothesis: Murderers are more likely to claim to have been beaten by their parents than nonmurderers.[3]

■ CONCLUDING REMARKS

In this chapter, you have learned how to generate research ideas and how to refine those ideas. Consequently, if you spend a little time reviewing this chapter, you should be able to generate several hypotheses about how two or more variables are related.

SUMMARY

1. The purpose of scientific research is to test ideas. One way to get research ideas is to test common sense ideas.

2. **Constructs** (also referred to as hypothetical constructs and theoretical constructs) are abstract variables that can't be directly

---

[3] Unfortunately, you will not know whether murderers actually were beaten more than other people because murders may exaggerate the extent to which they were beaten.

observed (e.g., love, learning, thirst, etc.). Researchers deal with abstract constructs by devising recipes for these variables called **operational definitions**: concrete ways of manipulating stimuli (e.g., stress could be defined as putting participants hands in ice water for 3 minutes) or measuring responses (e.g., stress could be defined as an increase in heart rate).

3. Building on other people's research is an easy way to get good research ideas.

4. Strategies for developing research ideas from previous research include improving the original study's external, internal, or construct validity; repeating the study; seeing if the finding has any practical implications; doing a follow-up study suggested by the study's authors; and trying to determine why two studies produced conflicting findings.

5. You can sometimes improve a study's construct validity by using the double-blind technique.

6. If a study did not use random assignment, it does not have internal validity. Thus, if you can modify the study so that you do randomly assign participants to condition (i.e, you use a random process to determine who gets what treatment), your study will have more internal validity than the original study.

7. If you think that a study's results wouldn't hold in a different situation, in a different place, or a different time, you are questioning the study's external validity. If you can think of a factor that will influence whether the relationship the study found holds, that factor is a potential **moderator variable**: a variable that changes the relationship between two other variables.

8. Just as the moderator of a talk show may inflame, defuse, or even mend a tense relationship between two individuals, a moderator variable may increase, decrease, or reverse the relationship between two variables.

9. How do two individuals have a relationship even when they refuse to speak to each other? How does something out in the world

(e.g., music) cause you to do something else (e.g., exercise more intensely)? The answer to both of these "how" questions is "through a mediator." In psychology, a **mediator** (also called a **mediator variable** or a **mediating variable**)—is what comes between an outside event ("A") and your observable reaction to it ("C"); it is *how* the cause has its effect. Usually, we can look for at least two kinds of mediating mechanisms through which a cause has its effect. First, we can try to see what the event did to the body (e.g., made it release adrenaline) that triggered the response (e.g., more intense exercising). Second, we can try to see what the event did to your mind that changed behavior (e.g., the music makes people feel more energized, which, in turn, makes them exercise more intensely").

10. Never do a study without first obtaining your professor's permission.

11. A null hypothesis states that there is no relationship between two variables. Although the null hypothesis can be disproven, it can't be proven. Therefore, you cannot prove hypotheses (a) predicting that a treatment will have no effect or (b) predicting that there will be no difference between treatments.

12. When possible, use theory and past research to provide a rationale for your prediction and to show that the results of your study may have implications for evaluating theory and for understanding previous research findings.

13. Knowing the functional relationship between two variables enables us to know what amount of the predictor variable will result in what amount of the outcome variable. If we want to know how much benefit we will get by increasing a treatment, if we want to know how much of a treatment is too much, if we want to know how much of a treatment is too little, or if we want to make precise predictions about a variable's effect, we need to know the functional relationship between the treatment and the outcome variable.

14. If your hypothesis involves a prediction that one variable influences a second variable, you can refine that hypothesis by (a) asking "how

much" questions to find the functional relationship between those two variables, (b) asking "how" questions to find the physiological or mental variable mediating that relationship, or (c) asking "when," "where," and "who" questions to find a variable that moderates (alters) that relationship.

15. A research hypothesis must be testable and must be testable in an ethical manner.

16. Even the most impractical and unethical of ideas may be converted into a practical and ethical hypothesis if you carefully screen your participants, use a small-scale model of the phenomenon you wish to study, tone down the intensity of your manipulation, or—as a last resort—do not use manipulations. However, realize that if you do not use a manipulation, you will not be able to test a cause–effect hypothesis.

KEY TERMS

hypothesis (p. 80)
double-blind technique (p. 78)

moderator variable (p. 79)
null hypothesis (p. 81)

mediating variable (p. 91)
functional relationship (p. 100)

EXERCISES

1. Generate the following three types of hypotheses:
   a. A cause–effect hypothesis. You should be able to fit that hypothesis into this format: Increasing_____(increases/decreases) _____.
   b. A cause–effect hypothesis that includes a moderator variable. You should be able to fit that hypothesis into this format: Increasing_____(increases/decreases) _____ depending on _____.
   c. A cause–effect hypothesis includes a mediating variable. You should be able to fit that hypothesis into this format: Increasing _____(increases/decreases) _____ due to _____.

2. Look up a research study that tests a common-sense notion or proverb. (If you are having difficulty finding an article, consult Web Appendix B.) What is the title of the article? What are its main conclusions?

3. Writing an essay that expresses opinions that go against your beliefs may cause you to change your beliefs. According to dissonance theory, what factors would moderate the effect of writing such an essay?

4. According to dissonance theory, what is an important variable that mediates attitude change?

5. Find a research article that tests a hypothesis derived from theory. Give the citation for the article and describe the main findings.

6. Describe the relationship between moderator variables and external validity.

7. Design a study to improve the construct validity of the study reported in Appendix B.

8. Design a study to test the generalizability of the findings of the study reported in Appendix B.

9. The study reported in Appendix B finds a relationship between two variables. Design a study that maps out the functional relationship between those two variables. Alternatively, propose a hypothesis that explores a practical implication of the study's findings.

10. Taking into account the problems with the null hypothesis, discuss what is wrong with the following statements:
    a. "There is no difference in outcome among the different psychological therapies."
    b. "Viewing television violence is not related to aggression."
    c. "There are no gender differences in emotional responsiveness."

WEB RESOURCES

1. Go to the Chapter 3 section of the book's student website and
   a. Look over the concept map of the key terms.
   b. Test yourself on the key terms.
   c. Take the Chapter 3 Practice Quiz.
   d. Do the interactive end-of-chapter exercises.
   e. Download the "Idea Generator," and develop a research idea.
   f. Use the "C3Tester" link to help spell out your predictions.
   g. Practice evaluating hypotheses using the "C3Evaluator" link.

2. Get a better sense of what research is like by using Chapter 3's "Participate in a Study" link.

3. Get more ideas on how to use theory to support your hypothesis by reading "Web Appendix: Using Theory to Generate Ideas."

4. If you have a research hypothesis that you want to test, use Chapter 3's "Getting Started on Writing Your Introduction" link.

Reading, Reviewing, and Replicating Research

It's not what you don't know that's the problem. It's what you know that just ain't so.
—Will Rogers

Science, in the very act of solving problems, creates more of them.
—Abraham Flexner

CHAPTER OVERVIEW

Science produces much more information than you can possibly learn. Furthermore, thanks in part to the disinformation campaigns waged by politicians, advertisers, and the media, some of what you think you know is false. Most of the time, what you don't know won't hurt you. However, there are times—when making a decision about a medical treatment, when deciding on a way to help a child, when deciding what charity to support, or when deciding whether a country has nuclear weapons—when not knowing the facts can have life-changing, and even life-ending, consequences.

When you need the best information, you need to read the research and question that research. In this chapter, we will focus on helping you read and question psychological research presented in journal articles. We chose psychological research articles because they are rich gold mines of information about a wide variety of topics: from how to be happier to how the genders differ to how the mind works. However, if you wish to use the critical thinking skills we discuss in this chapter to help you mine other sources that are relevant to making informed purchasing or political decisions, you will be able to do so.

You will start by learning how to make sense of a research article. After you learn the anatomy of an article, you will learn how to spot flaws and limitations in research. Finally, you will learn how you can get research ideas by reading research: You will see how ideas breed ideas. Thus, the aim of this chapter is to make you an intelligent consumer and producer of research.

READING FOR UNDERSTANDING

You wouldn't find a "how to" manual about how to download ringtones for your cell phone very useful unless you were reading it while you were downloading ringtones. Similarly, you will find this "how to read an article" chapter little more than a review of what you already know unless you read it while you are reading an article. Therefore, before you finish the next section, get an article!

Choosing an Article: Browse Wisely and Widely

But don't just read the first article you find. You are going to be spending a lot of time with whichever article you choose, so shop around before you invest your time reading an article. At the very least, choose an article that uses a type of study you understand (e.g., a survey or a simple experiment) and that deals with an area that you find interesting.

To start your quest for such an article, you could

1. Look at sections of textbooks that you find particularly interesting, and look up the articles they reference. For example, you might want to look up a study referenced in this textbook, such as the study by Iyengar and Lepper (2000) showing that people are happier when they have fewer choices than when they have many choices.
2. Consult Web Appendix B, "Searching the Literature," to learn how to search for articles on a topic that interests you.
3. Browse the table of contents of current journals.

Your first clue to whether an article is interesting is its title. Almost any title will tell you what the general topic or research question was. A good title will also contain the study's key variables. For example, in articles describing an experiment, the main manipulated factor (the main independent variable) and the outcome variable (the dependent measure) may be in the title. In some cases, the title may hint at what the hypothesis was or even what the main findings were.

Once you find an article that has an interesting title, the next step is to read a brief, one-paragraph summary of that article. This one-paragraph overview of the research's purpose, methodology, and results is called the **abstract**.

Even if you don't have the original article, you can read its abstract—provided you have access to one of the resources described in Web Appendix B, such as **Psychological Abstracts** or **PsycINFO** (the computerized version of Psychological Abstracts). If you have the original article, the only problem in finding the abstract is that it is usually not labeled. To find the abstract, turn to the article's first page. The article's first paragraph (it usually starts right below the author's names) is the abstract.

Reading the Abstract

By reading the abstract, you should get a general sense of what the researchers' hypotheses were, how they tried to test those hypotheses, and whether the results supported those hypotheses. But most importantly, you will get an idea about whether you want to read the article. Just as you probably would continue to channel surf if the short, onscreen summary of a TV show did not interest you in that show, you should look for another article if its abstract turns you off. If the abstract seems promising, skim the article and read the first paragraph of the Discussion section. If you can't understand that paragraph, consider looking for another article. Looking at other articles before committing to one pays off: When we have students analyze an article, we find that the students who look at more than five abstracts before choosing an article are the happiest with their choices.

Reading the Introduction

Once you find an article that has an interesting title and abstract, you are ready to start reading the rest of the article. For the beginning student, the best place to start reading an article is at the beginning. Although not labeled as such, the beginning of the article is called the **introduction**. The introduction is the most difficult, most time-consuming, and most important part of the article to understand. You must understand the introduction because it is where the authors tell you

1. how they came up with the hypothesis, including reasons why they think the hypothesis will be supported,
2. why testing the hypothesis is important, and
3. why the authors' way of testing the hypothesis is the best way to test the hypothesis (see Figure 4.1).

One way of thinking of the introduction is as an ad for the article. The authors try to sell you on the importance of their research. They may try to sell you on their study by claiming that, relative to previous research ("our competitor's brands"), their methodology—their way of testing the hypothesis—is clearly superior.

Sometimes they argue that their methodology is superior because their study has better *construct validity*: the degree to which it is studying the

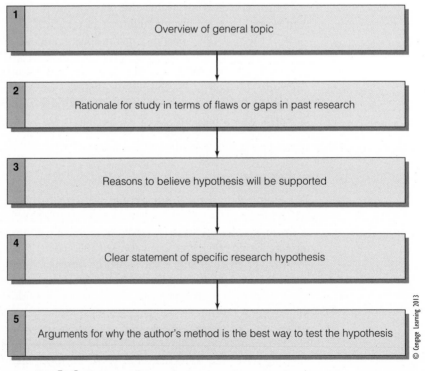

FIGURE **4.1** General flow chart of an introduction.

variables it claims to be studying. They may argue that they use a better measure, a better manipulation, or a better way of preventing participants and experimenters from biasing the results. For example, the authors of the sample article in Appendix B argue that, unlike related, previous research, their study did not set up a situation in which participants could guess and play along with the hypothesis.

Sometimes, authors argue that their methodology is superior because their study has more *internal validity*: the degree to which it can legitimately make *cause–effect* statements. They may point out that, unlike other studies, they compared the participants who received the treatment to a control group—participants who did not receive the treatment—so they could see whether the changes depended on the treatment or whether the changes would have happened anyway. If they used a control group, they should point out that they used random assignment (a random process similar to assigning participants on the flip of a coin) to determine which participants received the treatment and which did not. Authors might not spell out why their study's internal validity is superior to previous researchers'. Instead, they may merely state that they, unlike previous researchers, are able to demonstrate that the treatment, rather than some other factor, causes changes in the participants because they, unlike previous researchers, used an **experimental design**: a design in which (a) a treatment manipulation is administered and (b) that manipulation is the only variable that systematically varies between treatment conditions.[1] They may state this idea in an even more abbreviated form by saying, for example, "In contrast to the case study and correlational designs used in previous research, this study uses an experimental design."

Sometimes, researchers argue that their study is superior to previous research in terms of *external validity*: the degree to which the results can be generalized to different people, settings, and times. One way they may make the case that their study has more *generalizability* than previous research is by arguing that they studied participants in a situation that was more like real life than the situations used in previous research (e.g., participants were tested in a more naturalistic setting or did a task that was more similar to what people do in real life). To emphasize such a difference, they may write that their study has more *ecological validity* than previous research had. A second way to argue that their study has more generalizability than previous research is by arguing that their study's sample is more representative of (is a better mirror of) the entire population than previous research was. In survey research, for example, researchers may emphasize that they used a *random sample* (everyone in their population had an equal chance of

---

[1] As you will see in Chapters 10–13, the only factors that vary systematically in an experimental design are the treatment factors. Therefore, in an experiment, if manipulating a treatment is followed by a significant behavioral change, the experimenter can be confident that the behavioral change is due to the treatment. As you will also see in Chapters 10–13, one key to making sure other variables do not systematically vary along with the treatment is *random assignment*. In its simplest form, random assignment involves using a coin flip or some other random process to determine which of two treatments a participant receives.

being asked to participate) and that almost everyone they sampled answered their questions.

Sometimes researchers argue that their study is superior to previous research because it has more **power:** the ability to detect differences between conditions. Just as a more powerful microscope may reveal differences that a lower-powered microscope missed, a powerful study may find relationships between variables that a less powerful study failed to find. To make the case that their study has more power, authors may stress that they are using a more sensitive design, more sensitive measures, or more participants than the original study.

If the authors don't try to sell you on the methodological superiority of their study, they may try to sell you on their study by telling you that, relative to previous hypotheses, their hypothesis is "new and improved" because it has a special ingredient that other hypotheses don't have. Thus, they will try to get you to say, "It's incredible that people have done all this other, related research but not tested this hypothesis! Why didn't anyone else think of this?" Often, such authors extend existing research by

1. studying a different sample than previous research (e.g., women versus men).
2. looking at a different outcome measure (e.g., manner of walking instead of facial expression, gambling rather than bar-pressing, problem-solving skills rather than aggressiveness, persistence rather than helpfulness, happiness rather than depression, behavior rather than feelings).
3. looking at a different predictor. The different predictor could be a newer form of the original predictor (video games instead of television, online classes instead of traditional classes) or a more specific form of the original (types of video games instead of just video games, types of music instead of just music).
4. looking at when, where, or for whom a previous relationship holds—and when it doesn't hold. They will refer to the variable that moderates (*modifies*) the relationship between two other variables as a *moderator variable.* For example, Zajonc found that an audience's effect on performance was moderated by task difficulty: for easy tasks, an audience helped; for difficult tasks, an audience hurt (Zajonc & Desales, 1966).
5. looking at *how* a variable has its effect. Often, this involves looking for changes that occur in either the mind (e.g., feelings of being overwhelmed) or the body (e.g., decreased blood sugar levels, decreased oxytocin) that are triggered by the stimulus and that then, in turn, trigger the response. For example, Zajonc (1965) found that the presence of others had its effect by increasing arousal. Thus, whereas previous research had established the A (presence of others) → C (change in behavior) causal chain, Zajonc filled in a missing part of the chain to make it: A (presence of others) → B (increased arousal) → C (change in behavior). More recently, researchers have been trying to find the missing in-between steps (the "B") in many (A → C) causal chains. For instance, researchers are looking at how being a victim of prejudice leads to poor health (e.g., perhaps by increasing blood pressure), how—through what processes—religious faith leads to good health (Ai, Park, Huang, Rodgers, & Tice, 2007), and

how wearing red helps athletes win contests (probably because refs are biased toward athletes who wear red [Hagemann, Strauss, & Leising, 2008]). Because these "B" variables *mediate*—come between—the cause and the effect, they are often called either *mediating* variables or *mediators*.

6. testing a competing explanation for a relationship (e.g., one theory's explanation versus a competing theory's explanation, such as biological versus cultural explanations for a gender difference).

7. attempting to reconcile studies that have produced conflicting results.

Rather than look at the introduction as an ad for the study, you could look at the introduction as the opening scene that prepares you for the rest of the article. The authors start by giving you an overview of the research area. Then, they go into more detail about what other researchers have found in exploring a specific research topic. Next, the researchers point to some problem with past research that prevented a hypothesis from being tested. Sometimes, the problem is that no studies have tried to test a certain hypothesis. Sometimes, the problem is that the studies that have tried to test the hypothesis are either flawed or have obtained conflicting results. Whatever the problem with past research, the authors will claim that their study will fix it. Then, after stating their research questions (their hypotheses), the authors may explain why their method for testing the hypothesis is the best way to test the hypothesis (see Figure 4.1 on page 114). For example, they may justify their choice of design, their choice of measures, and their choice of participants. Consequently, if you understand the introduction, you should be able to predict much of what the authors will say in the rest of the article.

Unfortunately, understanding the introduction is not always easy. The main reason the introduction may be hard for you to understand is that the authors are not writing it with you in mind. Instead, they are writing it to other experts in the field. Their belief that the reader is an expert has two important consequences for how they write their article. First, because they assume that the reader is an expert in the field, they do not think that they have to give in-depth descriptions of the published articles they discuss. In fact, authors often assume that just mentioning the authors and the year of work (for instance, Miller & Smudgekins, 2011) will make the reader instantly recall the essentials of that article. Second, because they assume that the reader is an expert, they do not think they have to define the field's concepts and theories.

Because you are not an expert in the field, the authors' failure to describe studies and define concepts may make it difficult to understand what they are trying to say. Fortunately, you can compensate for not having the background the authors think you have by doing two things. First, read the abstract of the main study that the authors are going to refute, repeat, or build on. Second, look up unfamiliar terms or theories in a textbook. If you can't find the term in a textbook, consult the sources listed in Box 4.1.

To encourage yourself to look up all relevant terms and theories, photocopy or print out a copy of that article. On the that copy, use a yellow

BOX 4.1 Deciphering Journal Articles

Even experts may need to read a journal article several times to understand it fully. During your first reading, highlight any terms or concepts that you do not understand on a copy of the article. The highlighting shows you what you don't understand (if you highlight the entire article, maybe you should find another article). Once you identify the terms that you don't understand, decipher those terms by using one of the techniques listed below.

To Decipher Highlighted Terms
- Consult an introductory psychology text.
- Consult an advanced psychology text.
- Consult a psychological dictionary or encyclopedia.
- Consult a professor.
- Consult general sources such as Psychological Science, Psychological Bulletin, Annual Review, and American Psychologist to understand key theories.
- Consult other articles that were referenced in the article.
- Look up the term in a search engine, such as http://www.google.com or http://scholar.google.com.

highlighter to mark any terms or concepts you do not understand (Brewer, 1990). Do some background reading on those highlighted concepts, and then reread the introduction. As you reread it, highlight any terms or concepts you do not understand with a pink marker. Do some more background reading to get a better understanding of those terms. Then, reread the introduction using a green marker to highlight terms you still do not understand (Brewer, 1990).

By the third time you go through the introduction, you should see much less green than yellow, so you can see that you are making progress (Brewer, 1990). However, even if you know all the individual terms, how do you know that you understand the introduction? One test is to try to describe the logic behind the hypothesis in your own words. A more rigorous test is to design a study to test the hypothesis and then describe the impact of those results for current theory and further research. If you can do that, you not only understand the introduction, but you probably also have a good idea of what the authors are going to say in their discussion section.

To reiterate, do not simply skim the introduction and then move on to the method section. The first time through the introduction, ask yourself two questions:

1. What concepts do I need to look up?
2. What references do I need to read?

Then, after doing your background reading, reread the introduction. Do not move on to the method section until you can answer these six questions:

1. What variables are the authors interested in?
2. What is the prediction (hypothesis) involving those variables? (What is being studied?)

3. Why does the prediction (hypothesis) make sense?
4. How do the authors plan to test their prediction? Why does their plan seem to be a reasonable one?
5. Does the study correct a weakness in previous research? If so, what was that weakness? That is, where did others go wrong?
6. Does the study fill a gap in previous research? If so, what was that gap? That is, what did others overlook?

Reading the Method Section

After you are clear about what the authors predicted and why the authors made those predictions, read the method section to find out what the authors did and who they did it with.

In the **method section**, the authors will tell you what was done in terms of

1. who the participants were and how they were recruited and selected,
2. what measures and equipment were used, and
3. what the researchers said and did to the participants (what participants experienced).

An efficient way to tell you about each of these three aspects of the method is to devote a section to each aspect. Thus, many method sections are subdivided into these three subsections: participants, apparatus (or measures), and procedure. However, some method sections have fewer than three sections, and some have more.

Studies with more than three sections may have an overview subsection. You are most likely to see an overview subsection if the article reports several studies, all of which use similar procedures. By using an overview section, the author of a five-experiment paper can describe the aspects of the method that are the same for all five studies once, rather than repeating those details in all five method sections. You may also see a brief overview section for any method section that is so long or so detailed that readers need to get a general idea of the method before they can make sense of the method section.

Some method sections have a separate design subsection. For instance, the design subsection might tell you whether the design was a survey, a between-subjects design (in which one group of participants is compared against another group), or a within-subjects—also called "repeated measures"—design (in which each participant is compared against himself or herself). However, instead of a separate design section, authors may put information about the design in the participants section or in some other section.

Just as authors often do not include a design subsection, authors often do not include either a materials or an apparatus section. Instead, they may incorporate information about the apparatus or materials in the procedure section.

In short, there is no one rule for how many subsections a method section should have. Many will have only two: a participants section and a procedure section. Others may have an overview section, a participants and design section, a procedure section, and a dependent-measures section. Other method sections will use still different formats.

Regardless of its structure, the method section should be easy to understand for two reasons. First, the main purpose of the method section is to tell you what happened in the study—who the participants were, how many participants there were, and how they were treated. The authors should make it easy for you to imagine what it would be like to be a participant in the study. Indeed, some good procedure sections almost make you feel like you are watching a video, shot from the participants' perspective, of what happened in the study.

Second, even though the introduction probably foreshadowed how the authors planned to test the hypothesis, the authors are still going to take you, step-by-step, through the process so that you could repeat their experiment This "How we did it" section will be easy to follow unless (a) the authors give you too many details (details that might be useful for redoing the study but aren't essential for understanding the basics of what the researchers did), (b) the authors avoid giving you too many details by using a shorthand for a procedure (e.g., "we used the same procedure Hannibal & Lector [2012] used"), or (c) the authors use some task (e.g., a Stroop task) or piece of equipment (e.g., a tachistoscope) that you are unfamiliar with.

What should you do if you encounter an unfamiliar procedure, apparatus, or measure? If knowing the details of the procedure is essential, find the article the authors referenced (e.g., Hannibal & Lector, 2012). Look up any unfamiliar apparatus on Google or in the index of an advanced textbook. If all that fails, ask your professor. If you encounter an unfamiliar measure, find a source that describes the measure in detail: Such a source should be referenced in the original article's reference section. If the source is not referenced in the original article, look up the measure in the index of one or more textbooks. If that fails, search for concept the measure is claiming to assess in PsycArticles, or Google Scholar. That search should lead you to an article that will describe the measure.

After reading the method section, take a few moments to think about what it would have been like to be a participant in each of the study's conditions. Would you have been engaged in the study? Would you have figured out the hypothesis? Would you have interpreted the situation the way the researchers expected you to? Then, think about what it would have been like to be the researcher. Would you have been able to avoid biasing the results?

Realize that the method section contains the information you need to evaluate the study's internal, external, and construct validity. Consequently, to critique a study, you will need to reread the method section.

When evaluating the study's internal validity, you will want to know

- whether the study was an experiment (surveys, polls, and other studies in which a treatment is not administered are not experiments. Most studies that do not use random assignment are not experiments. If the study was not an experiment, assume that it does not have internal validity.)
- whether an apparent effect might be due to more people dropping out of one condition than another

When evaluating external validity, you will want to know

- the population from which the sample was drawn
- how participants were recruited
- what criteria were used to exclude people from the study
- whether random sampling was used
- what the dropout rate was
- what the gender, age, racial, and ethnic composition of the sample was

When evaluating the construct validity of a study, you will want to know

- the degree to which the measure is **valid**: measures what it claims to measure (e.g., does the aggression measure really measure aggression?)
- the degree to which the researcher has used techniques to prevent researcher and participant bias, such as (a) having the dependent measure collected and coded by assistants who do not know what condition the participant is in, (b) having the treatment administered by assistants who do not know what the hypothesis is, and (c) having participants in the comparison group believe they are receiving a treatment

Do not leave the method section to go on to the results section until you can answer these five questions:

1. What were the participants like (species, gender, age), and how did they come to be in the study?
2. What was done to the participants?
3. What tasks or actions did the participants perform?
4. What were the key variables in this study, and how did the researcher operationally define those variables? For example, how was the dependent variable measured?
5. What was study's design? (If it was not an experiment, was it a survey or an observational study? If it was an experiment, how many independent variables were there—and did all participants get the same treatments or did some participants get one type of treatment whereas others got a different treatment?)

Reading the Results Section

Now, turn to the results section of the article to find out what happened. Just as a sports box score tells you whether your team won, lost, or tied, the **results section** tells you whether the hypothesis won (was supported), lost (was refuted), or tied (results were inconclusive—like a "no decision" in boxing). Similarly, just as the box score provides statistical details that help you understand what happened, the results section provides statistical details to help you understand what participants did. (Although you might not understand these statistical details, you should still be able to figure out whether the hypotheses were supported. If, however, you feel overwhelmed by the numbers in the results section, skip ahead to the first paragraph of the discussion section [that paragraph will summarize the results without using numbers], and then return to the results section.)

Of course, there are many differences between box scores and results sections. One difference is that authors of box scores do not have to explain what the numbers in the box scores mean. For example, any baseball fan knows that a "1" in the "HR" column means that the batter hit one home run. But, in a study, what does it mean that the participants averaged a "6"? The meaning of a 6 would depend on the study; therefore, at the beginning of the results section (if they did not do so in the method section), the authors will briefly explain how they got the numbers that they later put into the statistical analysis. That is, they will describe how they scored participants' responses. Often, the scoring process is straightforward. For example, researchers may say, "The data were the number of correctly recalled words."

Occasionally, computing a score for each participant involves a little more work. In one study, researchers were looking at whether participants believed a person had a mental illness (Hilton & von Hippel, 1990). To measure these beliefs, the researchers had participants answer two questions. First, participants answered either "yes" or "no" to a question about whether the person had a mental illness. Then, researchers had participants rate, on a 1–9 scale, how confident participants were of their decision. How did the researchers turn these two responses into a single score? To quote the authors,

> In creating this scale, a value of –1 was assigned to "no" responses and a value of +1 was assigned to "yes" responses. The confidence ratings were then multiplied by these numbers. All ratings were then converted to a positive scale by adding 10 to the product. This transformation led to a scale in which 1 indicates a high degree of confidence that the person is normal and 19 represents a high degree of confidence that the person is pathological.
>
> —Source: From Hilton, J. L., & von Hippel, W. (1990). The role of consistency in the judgment of stereotype-relevant behavior. *Personality and Social Psychology Bulletin, 16,* 723–727. Reprinted by permission of Sage Publishing, Inc.

Do not merely glance at the brief section describing the scores to be used. Before leaving that section, be sure you know what a low score indicates and what a high score indicates: If you do not understand what the numbers being analyzed represent, how will you be able to understand the results of analyses based on those numbers?

After the authors explain how they got the scores for each participant, they will explain how those scores were analyzed. For example, they may write, "The prison sentence the participant recommended was divided by the maximum prison sentence that the participant could have recommended to obtain a score on the dependent measure. These scores were then subjected to a 2 (attractiveness) × 3 (type of crime) between-subjects analysis of variance (ANOVA)."[2]

Usually, authors will report more than one analysis. Indeed, for some kinds of studies, authors may report the following four kinds of results.

Basic Descriptive Statistics

The first kind of analysis that may be reported—but often is not—is an analysis focusing on basic, descriptive statistics that summarize the sample's

---

[2]Note that even if you have never heard of an *F* test or ANOVA before, you would still be able to have a basic understanding of such a results section. However, if you want to learn more about ANOVA, see Appendix E.

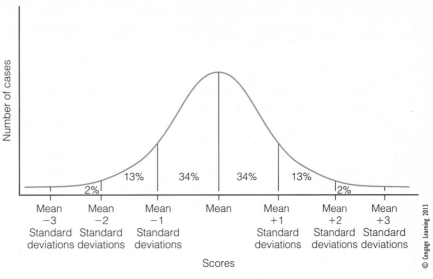

FIGURE **4.2** A normal curve.

scores on one or more measures. Typically, authors will describe the average score using the mean (which they will abbreviate as *M*). In addition, they will describe how closely most scores are to the mean. This measure of variability—of how spread out the scores are—will either be the range (high score minus low score) or, more commonly, the standard deviation (abbreviated *SD*). In addition to stating how much the scores are spread out (distributed), they may report whether the scores are normally distributed.[3] For instance, authors may report, "Overall, recall was fairly good (*M* = 12.89, *SD* = 2.68), and recall scores were normally distributed."

Knowing that the data are normally distributed is useful because many statistical tests, such as the *t* test and the *F* test, assume that data are normally distributed. If the data are not normally distributed, the researcher has three choices. First, the researcher may do the test and hope that violating the normality assumption will not unduly affect the results. Second, the researcher can decide to use a statistical test that does not assume that scores are normally distributed (often, such tests are called "nonparametric tests"). Third, the researcher may be able to perform some mathematical operation (transformation) on the scores to get a set of transformed scores that are normally distributed (hence, the joke that statisticians are not normal, but can be transformed). Occasionally, transforming scores involves only basic math. For instance, rather than analyze how much time it took participants to scan a

---

[3] As you can see from Figure 4.2, normally distributed usually indicates that (a) the most common score is the mean, (b) most scores are near the mean, (c) as many scores are above the mean as below the mean, and (d) plotting the scores on the graph produces a bell-shaped pattern in which the graph's left half is the mirror image of its right half.

word, a researcher may analyze the scanning speed. Thus, a scanning time of half (1/2) a second per word may be transformed into a speed of 2 words per second (2/1).

Results Supporting the Measure's Validity

If a new measure is used, the authors may report evidence that they think will increase your confidence in the measure. Often, they will want to convince you that scores on the measure are reliable, consistent, and stable by showing you that individuals who take the test twice score about the same both times. To show you that an individual's score the first time that person is tested will be consistent with that individual's score when retested, authors may report, "test–retest reliability was .90." (Test–retest reliability can range from 0 to 1; .90 test–retest reliability is excellent.)

If raters are rating participants, the authors will want to convince you that the raters are reliable (consistent) by showing you that raters' judgments consistently agree with each other. That is, if one rater judges a participant to be very outgoing, the other raters should also judge the participant to be very outgoing.

To present evidence that judges' ratings agree, authors will present some measure of inter-rater agreement. Sometimes, it will be obvious that they are reporting interobserver agreement (e.g., "Raters agreed on 96% of the ratings"); sometimes, it will be a little less obvious ("Inter-rater reliability was extremely high [$r = .98$])"; and sometimes, it will be far from obvious (e.g., "Cohen's kappa = .82") Despite the differences between these measures, scores near 1 usually indicate excellent agreement, whereas numbers below .70 (70%) indicate poor agreement.

If the authors have a scale composed of several questions, each question on that scale is, in a sense, like a judge rating participants. To increase your confidence in those "judges," the authors will try to show you that, when judging a participant, the judges are consistent—they agree with each other. For example, if, based on a participant's answer to question 1, the participant is scored as very outgoing, the other questions should also "judge" the participant to be very outgoing. Note that if the judges did not agree (e.g., one question "judged" the participant as extremely outgoing whereas the other judged the participant as extremely shy), we would wonder how good the "judges" were and what they could possibly be "seeing." When talking about how well the questions seem to be agreeing in their judgments, authors will talk about *internal consistency* and refer to some measure of it. Although authors could talk about correlations between the scale's questions, they will be more likely to refer to Cronbach's alpha (abbreviated as α), expecting to show that it is at least respectable (above .70) and hoping to show that it is excellent (above .90). For example, authors might report, "the scale was internally consistent ($\alpha = .91$)."

As you have seen, authors may provide evidence that the test agrees with itself from one day to the next (test–retest reliability), that raters agree with each other (inter-rater reliability), and that the individual items on the test agree with each other (internal consistency). Beyond showing that the measure is consistent, authors may provide evidence that the test is measuring what it claims to be measuring by showing that the measure agrees with other valid indicators of the construct. For instance, authors may report that their emotional intelligence test (a) correlates with other tests of emotional intelligence and (b) predicts a behavior associated with the construct (e.g., people scoring higher on the emotional intelligence test have more positive interactions with friends).

Results Supporting the Manipulation's Validity: Manipulation Check Data

In addition to showing that their measures have construct validity, researchers may try to show that their manipulations have construct validity. Similar to how some researchers show that their measure of a construct correlates with other measures of that construct, some researchers show that their manipulation of a construct affects a measure of that construct. Specifically, researchers will try to see whether different treatments (e.g., an unattractive photo versus an attractive photo) result in different scores on the *manipulation check*: a question or set of questions designed to check whether participants perceived the experiment's manipulation in the way that the researcher intended (e.g., judgments of how attractive the person in the photo is).

Usually, the manipulation check results will be statistically significant (unlikely to be due to chance alone and, thus, probably due to the treatment) and unsurprising. For example, if a study manipulates attractiveness of a defendant in a trial, the researchers might report that "Participants rated the attractive defendant ($M = 6.2$ on a 1 [*unattractive*] to 7 [*attractive*] scale) as significantly more attractive than the unattractive defendant ($M = 1.8$), $F(1,44) = 11.56$, $p < .001$, $d = 1.03$."[4] After the authors have shown you evidence that they manipulated the factor they said they manipulated, they are ready to show you whether that factor had the effect that they had hypothesized.

Results Relating to Hypotheses

Even if the researchers did not have a manipulation check, they will discuss the findings that relate to the hypotheses. After all, the researchers' main goal in the results section should be to let the reader know what the results of testing the hypothesis were. Just as listeners of a radio sports results show should be able to find out whether their team won, lost, or tied, readers of a results section should know whether the hypothesis was supported, was refuted, or was not supported (i.e., results were nonsignificant and thus inconclusive). For example, if the hypothesis was that attractive defendants

---

[4] "$F(1,44) = 11.56$" means that a statistical test called an F test was calculated and the value of that test was 11.56; $p = .001$ means that if there was no effect for the manipulation, the chances of finding a difference between the groups of that size or larger is about 1 in a 1000. Traditionally, if p is less than 5 in 100, the difference is considered reliable and is described as "statistically significant." In some journals, authors use p_{rep} (probability of replication) instead of p. For example, they might say $p_{rep} = .95$, meaning that the chances of a replication of the study getting the same pattern of results (e.g., the one group again having a higher average score than the other) would be 95%. To learn more about why some people want to use p_{rep} rather than p and to learn more about statistical significance, see Appendix E.

would receive lighter sentences than unattractive defendants, the author would report what the data said about that hypothesis:

> The hypothesis that attractive defendants would receive lighter sentences was not supported. Attractive defendants received an average sentence of 6.1 years whereas the average sentence for the unattractive defendants was 6.2 years. This difference was not significant, $F(1,32) = 1.00$, $p = .32$.

Other Significant Results

After reporting results relating to the hypothesis (whether or not those results are statistically significant), authors will dutifully report any other statistically significant (reliable) results. Even if the results are unwanted and make no sense to the investigator, significant results must be reported. Therefore, you may read things like, "There was an unanticipated interaction between attractiveness and type of crime. Unattractive defendants received heavier sentences for violent crimes whereas attractive defendants received heavier sentences for nonviolent crimes, $F(1,32) = 18.62$, $p < .001$." Or, you may read, "There was also a significant four-way interaction between attractiveness of defendant, age of defendant, sex of defendant, and type of crime. This interaction was uninterpretable." Typically, these results will be presented last: Although an author is obligated to report unexpected and unwelcomed findings, an author is not obligated to emphasize them.

Conclusions About Reading the Results Section

In conclusion, reading the results section may be difficult—especially if the authors used a statistical test that you are not familiar with. After reading through the results section, you probably will not understand everything. However, before moving on to the discussion, you should be able to answer these four questions:

1. What are the scores they are putting into the analysis?
2. What are the average scores for the different groups? Which types of participants score higher? Lower?
3. What type of statistical analysis did the authors use?
4. Do the results appear to support the authors' hypothesis?

Reading the Discussion

Finally, read the **discussion**. The relationship between the discussion and the results section is not that different from the relationship between a sports article and the box score. The box score focuses on which team won the game, whereas the article about the game reiterates key points from the box score and focuses on putting the game in a larger context—what the team's performance means for the team's play-off hopes, the team's place in history, or even for the league itself. Similarly, whereas the results section analyzes what the results mean for the hypothesis, the discussion section looks at the bigger picture: What are the implications of the results for the real world, for theory, for interpreting past research, and for doing future research?

The discussion section should hold few surprises. In fact, before reading the discussion, you can probably write a reasonable outline of it if you take the following three steps.

1. Summarize the main findings.
2. Relate these findings to the introduction.
3. Speculate about the reasons for any surprising results.

Because many discussion sections follow this three-step formula, the discussion is mostly a reiteration of the highlights of the introduction and results sections. If the authors get the results they expect, the focus of these highlights will be on the consistency between the introduction and the results. If, on the other hand, the results are unexpected, the discussion section will attempt to reconcile the introduction and results sections.

After discussing the relationship between the introduction and the results, the authors will discuss some of their study's limitations, suggest follow-up research that will overcome those limitations, and conclude by explaining why their study is important. To be sure you understand what they are saying, keep studying the discussion section until you can answer these five questions:

1. How well do the authors think the results matched their predictions?
2. How do the authors explain any discrepancies between their results and their predictions?
3. Do the authors admit that their study was flawed or limited in any way? If so, how?
4. What additional studies, if any, do the authors recommend?
5. What are the authors' main conclusions?

DEVELOPING RESEARCH IDEAS FROM EXISTING RESEARCH

Once you understand the article, you can take advantage of what you know about internal, external, and construct validity to question the article's conclusions. In addition, if the researchers failed to find a relationship, you can question the study's **power**: its ability to detect differences between conditions.

As you can see from Appendix C, there are many questions you can ask of any study. Asking questions of a study pays off in at least two ways.

First, because you become aware of the study's limitations, you avoid the mistake of believing that something has been found to be true when it has not. Consequently, you are less likely to act on misinformation and thus less likely to make poor choices when buying medicines, voting in elections, making business decisions, or treating clients.

Second, because you are aware that no single study answers every question, you realize that additional studies should be done. In other words, a common result of asking questions about research is that you end up designing additional studies that will document, destroy, or build on the previous research. Thus, familiarity with research breeds more research.

The Direct (Exact) Replication: Making Sure That Results Are Reliable

Whenever you read a study, one obvious research idea always comes to mind—repeat the study. In other words, do a **direct replication** (also called an **exact replication**): a copy of the original study.

One reason to do a direct replication is to develop research skills. Many professors—in chemistry, biology, and physics, as well as in psychology—have their students repeat studies to help students develop research skills.

But, from a research standpoint, isn't repeating a study fruitless? Isn't it inevitable that you will get the same results the author reported? Not necessarily—for two main reasons.

First, the results of a study done in a different place, at a different time, and with different participants may not apply to current students at your school. Thus, one reason to do an exact replication is to test the original study's external validity.

Second, the original study's results may be in error. Sometimes, this error may be due to fraud. More often, the error is due to either underestimating or overestimating the effects of chance.

Using Direct Replications to Fight Fraud

Although scientific fraud is rare, it does occur. Some cheat for personal fame. Others cheat out of the misguided notion that, if the evidence doesn't support their hypothesis, there must be something wrong with the evidence. Consequently, they may decide to "fix" the evidence.

Although thousands of researchers want to be published, cheating is unusual because the would-be cheat knows that others may replicate the study. If these replications don't get the results the cheat reported, the credibility of the cheat and of the original study would be questioned. Thus, the threat of direct replication keeps would-be cheats in line. Some scientists, however, worry that science's fraud detectors are becoming ineffective because people are not doing replications as often as they once did (Broad & Wade, 1982). Given that one large-scale study found more than one-third of scientists confessing to unethical behavior (Wadman, 2005) and that researchers funded by a sponsor obtain results that support their sponsor's position much more often than independent researchers do (Tavris & Aronson, 2007), this worry is well founded.

Using Direct Replications to Fight Statistical Errors

Although fraud is one reason that some published "findings" are inaccurate (Broad & Wade, 1982), significance testing is a more common reason for reported findings being in error. Significance testing is a technique that many researchers use to predict whether a finding would replicate. In other words, researchers often use significance testing to see whether a finding is a reliable one rather than the result of a coincidence.

To understand significance testing and some of its problems, imagine you do a study in which you randomly assign participants to two groups. After administering the treatment to one of the groups, you obtain scores for all participants on the outcome measure. Suppose that the treatment group scores higher than the no-treatment group. Is this difference due to the treatment? Not necessarily. You know that chance might make the groups score differently. For example, you know that because random assignment isn't perfect, just by chance, most of the individuals who naturally score high on your measure may have been assigned to the treatment group. Because you know that chance may have affected your study's results, you want to know whether your results reflect a true pattern (your results are statistically reliable) or whether your results are due to a fluke. That is, you want to know whether you would get the same pattern of results if you did the study again. Consequently, you might ask this question: "*Given a difference of this size, what are the chances that the treatment has no effect?*"

Until recently, most psychologists acted like significance tests would accurately answer that question.[5] Specifically, if a significance test yield a probability (p) value of .05 or less, researchers assumed that there was less than a 5% chance that the treatment had no effect—and thus concluded that the treatment probably had an effect. Even if this assumption were accurate, the researcher would still be coming to wrong conclusions some of the time for two reasons.

First, even if the treatment had no effect, the researcher would falsely claim it did 5% of the time. In other words, if the treatment had no effect, the researcher is taking a 5% ($p = .05$) risk of making a **Type 1 error**: concluding that variables are related when they are not. Type 1 errors are statistical significance "false alarms" that involve mistaking a coincidence for a real relationship (i.e., a Type 1 error is like crying "wolf" when there is no wolf). A Type 1 error would occur if a chance difference between a group who took an ineffective drug and a group who took a placebo (a sugar pill) was declared statistically significant, leading the researcher to conclude—incorrectly—that the ineffective drug had an effect.

Second, even if the treatment had an effect, the researcher will fail to detect that effect some of the time. In technical terms, the researcher will make **Type 2 errors**: failing to detect that the variables are related. Type 2 errors are cases in which the statistical significance alarm fails to go off when it should, resulting in us failing to recognize that the pattern we see is more than a coincidence (i.e., a Type 2 error is like not yelling "wolf" when there is a wolf). A Type 2 error would occur if the new drug really had more side effects than the placebo, but the statistical test was unable to determine that the new drug was significantly (reliably) more dangerous than the placebo. Partly because of psychologists' concerns about Type 1 and Type 2

---

[5] Actually, the question significance test answers is different: "What are the chances that I would get these results if the treatment has no effect?" To see that such similar-sounding questions can have very different answers, consider that the answer to the question: "What are the chances I won the lottery given that I bought a ticket?" is very different from the answer to the question: "Given that I won the lottery, what are the chances that I bought a ticket?" (Mlodinow, 2008). To answer the question that the researcher wants answered, some people believe that researchers should use the statistic p_{rep} (probability of the results replicating) instead of p. However, not everyone agrees.

errors, some journals solicit and accept studies that replicate—or fail to replicate—previously published research.

Fighting Type 1 Errors

To understand how an original study's results may have been significant due to a Type 1 error, imagine that you are a crusty journal editor who allows only simple experiments that are significant at the $p = .05$ level to be published in your journal. If you accept an article, you believe that the chances are only about 5 in 100 that the article will contain a Type 1 error (the error of mistakenly declaring that a chance difference is a real difference). Thus, you are appropriately cautious. But what happens once you publish 100 articles? Then, you may have published five articles that have Type 1 errors.

Actually, you may have published many more Type 1 errors than that because researchers do not send you nonsignificant results. They may have done the same experiment eight different times, but they send you the results of the eighth replication—the one that came out significant. Or, if 20 different teams of investigators do basically the same experiment, only the team that gets significant results (the team with the Type 1 error) will submit their study to your journal. The other teams will just keep their nonsignificant results in their filing cabinets. As a result, you have contributed to what is called the **file drawer problem**: a situation in which the research not affected by Type 1 errors languishes in researchers' file cabinets, whereas the Type 1 errors are published.

To see how the file drawer problem might bias published ESP research, imagine you do a study looking for an effect of ESP. If you fail to find an effect, your study would not be published (the data would probably never make it out of your file cabinet). However, if, because of a Type 1 error, you found an effect, your study could be published, thereby contributing to a "file drawer problem."

Knowing about the file drawer problem can prevent you from contributing to it. For example, while serving as editor for the *Journal of Personality and Social Psychology,* Greenwald (1975) received an article that found a significant effect for ESP. Because Greenwald was aware that many other researchers had done ESP experiments that had not obtained significant results, he asked for a replication. The authors could not replicate their results. Thus, the original study was not published because the original results were probably the result of a Type 1 error (Greenwald, 1975).

Fighting Type 2 Errors

Just as studies that find significant effects may be victimized by Type 1 errors, studies that fail to find significant effects may be victimized by Type 2 errors. Indeed, Type 2 errors (the failure to find a statistically significant difference when a reliable difference exists) are probably more common than Type 1 errors. Realize that, in a typical study, the chance of a Type 1 error is usually about 5%. However, in most studies, the chance of a Type 2 error is much higher. To give you an idea of how much higher, Cohen (1990), who has urged psychologists to make their studies much less vulnerable to Type 2 errors than they currently are, wants psychologists to set the chance of a Type 2 error at about 20%. Even if researchers would reach Cohen's relatively high standards, the risk of making a Type 2 error in a study would still be at least four times ($4 \times 5\% = 20\%$) higher than the risk of making a Type 1 error!

Few researchers conduct studies that come close to Cohen's standards. Cohen (1990) reports that, even in some highly esteemed journals, the studies published ran more than a 50% chance of making a Type 2 error. Similarly, when reviewing the literature on the link between attributions and depression, Robins (1988) found that only 8 of 87 published analyses had the level of power that Cohen recommends. No wonder some studies found relationships between attributions and depression whereas others did not! Thus, when a study fails to find a significant effect, do not assume that a direct replication would also fail to find a significant effect. Repeating the study may yield statistically significant results.

Systematic Replications: Minor Tweaks to Improve the Study

Rather than merely repeating the study, you could do a **systematic replication**: a study that copies (replicates) the original study, except that it deliberately (systematically) differs from the original in terms of at least one of the following: participants, equipment, setting, or procedures. Usually, your systematic replication will differ from the original for one of two reasons.

First, you may make a small change that improves the study. For example, your systematic replication may use more participants, more sensitive equipment, a more realistic setting, or more standardized procedures than the original study used.

Second, you may make different trade-offs than the original researcher did. If, for example, the original researcher sacrificed construct validity to get power, you may sacrifice power to get construct validity.

You now know what a systematic replication is, but why should you do one? There are two main reasons.

First, as we suggested earlier, you might do a systematic replication for any of the reasons you would do a direct replication. Because the systematic replication is similar to the original study, the systematic replication, like the direct replication, can alert us that the results reported by the original author may be due to a Type 1 error, a Type 2 error, or to fraud.

Second, you might do a systematic replication to make new discoveries. The systematic replication may uncover new information because it will either do things differently or do things better than the original study. Because you can always make different trade-offs than the original researcher and because most studies can be improved, you can almost always do a useful systematic replication.

In the next few sections, we will show you how to design a useful systematic replication. Specifically, we will show you how to design systematic replications that have more power, more external validity, or more construct validity than the original study. We will begin by showing you how to change the original study to create a systematic replication that has more power than the original.

BOX **4.2**   **How to Devise a Systematic Replication That Will Have More Power Than the Original Study**

1. Reduce random differences that could hide differences caused by the treatment effect by
   a. using participants who are similar to one another,
   b. administering the study in a consistent way, and
   c. training and motivating raters to be more consistent.
2. Balance out the effects of random error by using more participants.
3. Use a more sensitive dependent measure.

© Cengage Learning 2013

Improving Power by Tightening the Design

If the original study fails to find a relationship between variables, that failure could be due to that study not looking hard enough or smart enough for the relationship. Therefore, you might want to repeat the study, but add a few minor refinements to improve its ability to detect relationships (see Box 4.2). Although we will discuss the logic and techniques for improving power in other chapters, you already have an intuitive understanding of what to do.

To show that you do have an intuitive understanding of how to design a study that can find relationships, let's look at an absurdly designed study that needs some help:

> Dr. F. Ehl wants to see whether aspirin improves mood. He enlists two participants: one person who is depressed and one who is not. One participant receives 1/16 of an aspirin tablet, the other receives 1/8 of an aspirin tablet. The assistant running the study is extremely inconsistent in how she treats participants: Sometimes, she is warm, smiling, and professional as she ushers the participant into a nice, air-conditioned room, provides the participant with a nice clean glass of cool water, and carefully administers the dose of aspirin. Other times, she is rude, grumpy, and unprofessional as she dumps the participant into a hot, stinky room that used to be the janitor's closet, provides the participant with a dirty glass of warm water, and drops the aspirin on the floor before administering it. To measure mood, the assistant asks the participant whether he or she is in a good mood. If the participant says "yes," that is to be coded as "1"; if the participant says "no," that is to be coded as "2." Unfortunately, the assistant sometimes codes "yes" as "2" and "no" as "1."

To improve the study's ability to find an effect for the treatment variable (aspirin), you would make six improvements.

First, you would use participants who were more similar to one another. If one participant is depressed and the other is not (as in Dr. F. Ehl's experiment), the "groups" will clearly be different from each other before the treatment is administered. Consequently, even if the treatment has an effect, that effect will probably not be detected.

Second, in addition to trying to reduce differences between participants, you would try to reduce differences between research sessions. That is, you would run the study in a more consistent, standard way. You would try to keep the assistant's behavior, the room, and the glass's cleanliness the same for each participant.

That way, you wouldn't have to worry as much that the difference between the groups' behavior was due to inconsistencies in how participants were treated.

Third, you would try to make your measuring system more reliable. The assistant must be consistent in how she codes "yes" and "no" responses so you know that differences between the groups aren't due to unreliable coding.

Fourth, you would use more participants. You know that with only one participant in each condition, it would be impossible to say that the treatment—rather than differences between participants—caused the "groups" to be different. You know that as you add participants to the two groups, it becomes easier to say that a difference between the groups is at least partly due to the treatment rather than entirely due to chance. For example, with two participants, it is quite likely that random assignment will produce one "group" that is naturally much happier than the other; with 100 participants, it is unlikely that random assignment will produce one group that is naturally much happier than the other. Similarly, with more participants, it becomes less likely that a random measurement error (e.g., miscoding a "1" as a "2") or random differences in how the experiment was conducted (e.g., the assistant being in a grumpy mood) will affect participants in one condition much more than participants who were in the other condition.

Fifth, you would use a measure that was more sensitive to small differences. For example, rather than asking whether participants were in a good mood, you might ask participants to rate their mood on a 1 (poor) to 9 (excellent) scale.

Sixth, you might improve your study's power to find a difference by making your study create a difference that would be too big for you to overlook. To create bigger differences in how your participants behave, you would try to create bigger differences in the treatment amounts you give each group. In this case, you might give one group a pill containing no aspirin and the other group an entire aspirin. Another way to have a more powerful manipulation is to use a manipulation that is so dramatic that participants can't ignore it. For example, when Frederickson et al. (1998) had men sit in a room with a mirror while wearing a bathing suit, the men did not do worse on a math test than men wearing a sweater. Thus, there was no evidence that wearing swimsuits made men self-conscious about their bodies. However, Hebl, King, and Lin (2004) noted that the men wore swim trunks rather than very brief Speedo swimsuits. When Hebl, King, and Lin replicated the study by having men wear Speedo swimsuits, the men did do more poorly on the math test.

Improving External Validity

If the original study had adequate power, this power to obtain a statistically significant relationship may have come at the expense of other valued characteristics (see Table 4.1), such as external validity. To illustrate, let's look at two cases in which attempts to help power hurt the generalizability of the results.

In the first case, a researcher realizes that if all the individuals in the study were alike, it would be easier to find out whether the treatment has an effect. If, on the other hand, individuals in the study were all quite different, it would be difficult to distinguish differences caused by the treatment from those individual differences. To be more specific, if the individuals in the study are very different from each other, random assignment of those

TABLE **4.1** Trade-Offs Involving Power

Conflict	Steps to Improve Power
Power versus construct validity	Using empty control group (a group that is just left alone), despite its failure to control for *placebo effects:* effects due to expecting the treatment to work. For example, to look at the effects of caffeine, a power-hungry researcher might compare an experimental group that drank caffeinated colas with an empty control group that drank nothing. However, a less powerful, but more valid, manipulation would be to use a control group that drank decaffeinated colas.
	Using a sensitive self-report measure (e.g., "on a 1–10 scale, how alert do you feel?"), despite its vulnerability to self-report biases.
	Testing the same participants under all experimental conditions (a within-subjects design) even though using that design alerted participants to the hypothesis (e.g., if participants played a peaceful video game, filled out an aggression scale, then played a violent video game and filled out an aggression scale, participants would assume they were supposed to act more aggressively after playing the violent game).
Power versus external validity	Using a restricted sample of participants so that differences between participants won't hide the treatment effect.
	Using a simple, controlled environment to reduce random error due to uncontrolled situational variables, but losing the ability to generalize to more realistic environments.
	Maximizing the number of participants per group by decreasing the number of groups. Suppose you can study 120 participants, and you choose to do a two-group experiment that compares no treatment with a medium level of the treatment instead of a four-group experiment that compares no, low, medium, and high levels of the treatment. With 60 participants per group, you have good power but, with only two treatment amounts, the degree to which you can generalize results to different amounts of treatment is limited.
	Using a within-subjects design in which participants got both treatments (e.g., psychoanalysis and behavioral therapy) even though—in real life—individuals receive either psychoanalysis or behavior therapy (Greenwald, 1976).
Power versus internal validity	Using the more powerful within-participants design (in which you compare each participant with herself or himself) rather than between-participants design (in which you compare participants getting one treatment with participants getting another treatment) even though the between-participants design may have better internal validity. As you will see in Chapter 13, when you compare participants with themselves, they may change for factors unrelated to getting the current treatment (e.g., getting better at or bored with the task, getting tired, having a delayed reaction to an earlier treatment).
Power versus statistical conclusion validity	Increasing the chances of declaring a real difference significant by being more willing to risk declaring a chance difference significant. For example, if a researcher increases the false alarm (Type 1 error) rate to $p < .20$ (rather than the conventional $p < .05$ rate), the study will have more power to find real effects but will also be more likely to mistake chance effects for real ones.

individuals into two groups may produce two groups that are substantially different from each other before the study starts. Consequently, if the treatment has a small positive effect, that effect will probably not be detected for two reasons. First, if, before the treatment is introduced, the treatment group is much lower on the key variable than the no-treatment group, the treatment's small effect may not be enough to make the treatment group score higher than the no-treatment group. Thus, random differences between the groups would have overwhelmed the treatment's effect. Second, even if the treatment group does score slightly higher than the no-treatment group, the researcher might conclude that this modest difference was due to random error: After all, if the groups could have been substantially different even before the treatment was administered, finding that the groups are somewhat different after the treatment was administered is hardly proof of a treatment effect.

To make it easier to prevent individual differences from overwhelming or masking treatment effects, the researcher tries to reduce the impact of individual differences on the study by choosing a group of individuals who are homogenous (all similar). The researcher knows that studying participants who are alike will tend to boost power: The researcher is more likely to obtain a statistically significant result. However, choosing homogeneous participants will decrease the extent to which the results can be generalized to other kinds of participants. What applies to a particular group of 18-year-old White, male, middle-class, first-year college participants may not apply to other groups of people, such as retirees, people who did not go to college, and members of minority groups.

In our second case of trading power for external validity, a researcher is worried that doing a study in a realworld setting would allow uncontrolled, nontreatment factors to have effects that might hide the treatment's effect. To prevent nontreatment effects from masking treatment effects, the researcher performs the study in a lab rather than in the field. The problem is that we do not know whether the results would generalize outside of this artificial environment.

Suppose you find a study that you believe lacks external validity. For example, suppose that some students performed a lab experiment at their college to examine the effects of defendant attractiveness. There are at least four things you can do to improve the study's generalizability (see Box 4.3).

BOX **4.3** How to Devise a Systematic Replication That Will Have More External Validity Than the Original Study

1. Use a more diverse group of participants or use a participant group (for instance, women) that was not represented in the original study.
2. Repeat a lab experiment as a field experiment (to see how, go to the field experiment section of our website).
3. Use more levels (amounts) of the independent or predictor variable.
4. Delay measurement of the dependent variable to see if the treatment effect persists over time.

First, you can use a sample that differs from the original study's sample. Your study might include a more representative sample of participants (perhaps by using random sampling from a broad population) than the original or it might include a group that was left out of the original study. For example, if their study tested only men, you might test only women.

Second, you can change a lab experiment into a field experiment. For example, suppose that the defendant study used college students as participants. By moving the defendant study to the field, you might be able to use real jurors as participants rather than college students.

Third, you can use different amounts of the independent (treatment) variable to see whether the effects will generalize to different levels of the independent variable. In the defendant study, researchers may have only compared attractive versus unattractive defendants. Therefore, you might replicate the study to see whether extremely attractive defendants have an advantage over moderately attractive defendants.

Fourth, you can wait a while before collecting the dependent measure to see whether the effect lasts. Fearing the effect will wear off, researchers often measure the dependent variable almost immediately after the participant gets the treatment to maximize their chances of finding a significant effect. However, in real life, there may be a gap between getting the treatment and having the opportunity to act.

Improving Construct Validity

We have discussed doing a systematic replication to improve a study's power and external validity. You can also do a systematic replication to improve a study's construct validity, especially if you think the original study's results could be due to participants guessing the hypothesis and then deciding to "give the researcher results that will 'prove' the hypothesis" (see Box 4.4).

To illustrate how a systematic replication could prevent participants from essentially telling the researcher what the researcher wants to hear, imagine a two-group lab experiment in which one group gets caffeine (in a cola), whereas the other group gets nothing (an empty control group). You could design a study that had more construct validity by

- replacing the no-cola empty control group with a placebo treatment (a caffeine-free cola) and making the study a double-blind experiment

---

BOX 4.4 How to Devise a Systematic Replication That Will Have More Construct Validity Than the Original Study

1. Replace an empty control group (a no-treatment group) with a placebo treatment group (a fake treatment group).
2. Use more than two levels of the independent variable.
3. Alter the study so that it is a double-blind study.
4. Add a cover story or improve the existing cover story.
5. Replicate it as a field study.

(an experiment in which neither the participant nor the assistant interacting with the participant knows which treatment the participant received)

- misleading the participants about the purpose of the study by giving them a clever cover story (e.g., they are doing a taste test).
- not letting them know they were in a study by doing your study in the real world: if participants do not know they are in a study, they will probably not guess the hypothesis

In short, the systematic replication accomplishes everything a direct replication does and more. By making some slight modifications in the study, you can improve the original study's power, external validity, or construct validity.

Conceptual Replications: Major Changes to Boost Construct Validity

Suppose you believe there were problems with the original study's construct validity—problems that cannot be solved by making minor procedural changes. Then, you should perform a **conceptual replication**: a study that is based on the concept (idea) of the original study but that is very different from the original study, usually because it improves on the original study's construct validity by using different measures or manipulations.

Because there is no such thing as a perfect measure or manipulation, virtually every study's construct validity can be questioned. Because confidence in the validity of a finding is increased when the same basic result is found using other measures or manipulations, virtually any study can benefit from conceptual replication. Therefore, you should have little trouble finding a study you wish to redo as a conceptual replication.

There are multiple ways to design a conceptual replication (see Box 4.5). For example, you could use a different way of manipulating the treatment variable. The more manipulations of a construct that find the same effect, the more confident we can be that the construct actually has that effect. Indeed, you might use two or three manipulations of your treatment variable and use the type of manipulation as a factor in your design.

For instance, suppose a study used photos of a particular woman dressed in either a "masculine" or "feminine" manner to manipulate the variable

---

BOX **4.5** **How to Devise a Conceptual Replication That Will Have More Construct Validity Than the Original Study**

1. Use a different manipulation of the treatment variable and add a manipulation check.
2. Use a different dependent measure, such as one that
   a. is closer to accepted definitions of the construct; or
   b. is less vulnerable to social desirability biases and demand characteristics, such as
      i. a measure of overt behavior (actual helping rather than reports of willingness to help), or
      ii. a measure that is unobtrusive (how far people sit from each other, rather than reports of how much they like each other).

"masculine versus feminine style." You might use the original experiment's photos for one set of conditions, but also add two other conditions that use your own photos. Then, your statistical analysis would tell you whether your manipulation had a different impact from the original study's manipulation.

Realize that you are not limited to using the same type of manipulation as the original study. Thus, instead of manipulating masculine versus feminine by dress, you might manipulate masculine versus feminine by voice (masculine-sounding versus feminine-sounding voices).

Although varying the treatment variable for variety's sake is worthwhile, changing the manipulation to make it better is even more worthwhile. One way of improving a treatment manipulation is to make it more consistent with the definition of the construct. Thus, in our previous example, you might feel that the original picture manipulated "fashion sense" rather than masculine versus feminine style. Consequently, your manipulation might involve two photos: one photo of a woman fashionably dressed in a feminine way, one of a woman fashionably dressed in a masculine manner. To see whether you really were manipulating masculinity–femininity instead of fashion sense or attractiveness, you might add a manipulation check. Specifically, you might ask participants to rate the masculine and feminine photos in terms of attractiveness, fashion sense, and masculinity–femininity.

When trying to see whether you can improve a manipulation, don't just look at the treatment condition. Look at the no-treatment or comparison condition because that's often where the real problem lies. For example, the treatment group might do one hour of meditation three times a week, whereas the comparison group does half an hour of meditation once a week. The problem is not so much with the meditation group but with the comparison group: That group is not only getting less meditation than the other group—they are also getting less attention from the researchers, going to fewer research sessions, and spending less time doing a novel activity.

Because no manipulation is perfect, replicating a study using a different treatment manipulation is valuable. Similarly, because no measure is perfect, replicating a study using a different measure is valuable. Often, you can increase the construct validity of a study by replacing a self-report measure that asked people what they would do with a behavioral measure that allows you to see what participants actually do. By replacing a self-report measure with a behavioral measure, you don't have to worry as much about participants lying or misremembering.

The Value of Replications

Replications are important for advancing science. Direct replications are essential for guaranteeing that the science of psychology is rooted in solid, documented fact. Systematic replications are essential for making psychology a science that applies to all people. Conceptual replications are essential for making psychology a science that can make accurate statements about constructs. Conceptual replications help us go beyond talking about relationships between specific procedures and test scores to knowing about the relationships between broad constructs such as stress and mental health.

In addition to replicating previous research, systematic and conceptual replications extend previous research. Consider, for a moment, the conceptual replication that uses a better measure of the dependent variable or the systematic replication that shows the finding occurs in real-world settings. Such conceptual and systematic replications can transcend the original research.

Extending Research

Systematic and conceptual replications are not the only ways to extend published research. Of the many other ways to extend published research (see Box 4.6), let's briefly discuss the two easiest.

First, you could both replicate and extend research by repeating the original study while adding a variable that you think might moderate the observed effect. For instance, if you think that being attractive would hurt a defendant if the defendant had already been convicted of another crime, you might add the factor of whether or not the defendant had been previously convicted of a crime.

BOX **4.6** **Extending Research**

1. Conduct studies suggested by authors in their discussion section.
2. If the study describes a correlational relationship between two variables, do an experiment to determine whether one variable causes the other. For example, after finding out that teams wearing black were more likely to be penalized, the authors of this textbook's sample paper (Appendix B) did an experiment to find out whether wearing black causes one to be more violent.
3. Look for related treatments that might have similar effects For example, if additional time to rehearse is assumed to improve memory by promoting the use of more effective rehearsal strategies, consider other variables that should promote the use of more effective rehearsal strategies, such as training in the use of effective rehearsal strategies.
4. See if the effects last. For example, many persuasion and memory studies look only at short-term effects.
5. See what other effects the treatment has.
6. Replicate the research, but add a factor (participant or situational variable) that may moderate the effect. That is, pin down under what situations and for whom the effect is most powerful.
7. Instead of using a measure of a general construct, use a measure that will tap a specific aspect of that construct. This focused measure will allow you to pinpoint exactly what the treatment's effect is. For example, if the original study used a general measure of memory, replicating the study with a measure that could pinpoint what aspect of memory (encoding, storage, or retrieval) was being affected would allow a more precise understanding of what happened.
8. If the study involves basic (nonapplied) research, see if the finding can be applied to a practical situation. For example, given divers who either learned words on land or under water recalled more words when they were tested where they learned the words, should medical students be taught material in the hospital rather than in the classroom (Koens, Cate, & Custers, 2003)?
9. Do a study to test a competing explanation for the study's results. For example, if the researchers argue that people wearing black are more likely to be violent, you might argue that there is an alternative explanation: People wearing black are more likely to be perceived as violent.

Second, you could extend the research by doing the follow-up studies that the authors suggest in their discussion section. Sometimes, authors will describe follow-up studies in a subsection of the discussion titled "Directions for Future Research." At other times, authors will hint at follow-up studies in the part of the discussion section in which they talk about their study's limitations. Thus, if the authors say that a limitation of their study was that it looked only at a treatment's short-term effects, they are suggesting that someone do a replication that looks at the treatment's longer-term effects. If they say that a limitation was that they used self-report measures, they are suggesting a replication using other types of measures. If they say their study was correlational and so cause–effect statements cannot be made, they may be suggesting that someone should do an experiment based on their study.

CONCLUDING REMARKS

As you have seen, much of the work done by scientists is a reaction to reading other scientists' work. Sometimes, researchers get excited because they think the author is onto something special, so they follow up on that work. Other times, the researchers think that the author is wrong, so they design a study to prove the author wrong. Regardless of the reaction, the outcome is the same: The publication of an article not only communicates information but also creates new questions. As a result of scientists reacting to each other's work, science progresses.

After reading this chapter, you can be one of the scientists who reacts to another's work and helps science progress. You know how to criticize research as well as how to improve it. Thus, every time you read an article, you should get at least one research idea.

SUMMARY

1. Not all articles are equally easy and interesting to read. Therefore, if you are given an assignment to read any article, you should look at several articles before committing to one.

2. Reading the title and the abstract can help you choose an article that you will want to read.

3. The abstract is a short, one-paragraph summary of the article. In journals, the abstract is the paragraph immediately following the authors' names and affiliations.

4. In the article's introduction, the authors tell you what the hypothesis is, why it is important, and justify their method of testing it.

5. To understand the introduction, you may need to refer to theory and previous research.

6. The method section tells you who the participants were, how many participants there were, and how they were treated.

7. In the results section, authors should report any results relating to their hypotheses and any statistically significant results.

8. The discussion section either reiterates the introduction and results sections or tries to reconcile the introduction and results sections.

9. When you critique the introduction, question whether (a) testing the hypothesis is vital, (b) the hypothesis follows logically from theory or past research, and (c) the authors have found the best way to test the hypothesis.

10. When you critique the method section, question the construct validity of the measures and manipulations and ask how easy it would have been for participants to have played along with the hypothesis.

11. When you look at the results section, question any null (nonsignificant) results. The failure to find a significant result may be due to the study failing to have enough power.

12. In the discussion section, question the authors' interpretation of the results, try to explain results that the authors have failed to explain, find a way to test your explanation, and note any weaknesses that the authors concede.

13. The possibility of Type 1 error, Type 2 error, or fraud may justify doing a direct replication.

14. You can do a systematic replication to improve power, external validity, or construct validity.

15. If minor changes can't fix problems with a study's construct validity, you should do a conceptual replication.

16. Replications are vital for the advancement of psychology as a science.

17. Reading research should stimulate research ideas.

KEY TERMS

abstract *(p. 113)*
conceptual replication *(p. 137)*
direct or exact replication
   (p. 128)
discussion *(p. 126)*
experimental design *(p. 115)*

file drawer problem *(p. 130)*
introduction *(p. 114)*
method section *(p. 119)*
power *(p. 116)*
Psychological Abstracts
   (p. 113)

PsycINFO *(p. 113)*
results section *(p. 121)*
systematic replication
   (p. 131)
Type 1 error *(p. 129)*
Type 2 error *(p. 129)*

EXERCISES

1. Find an article to critique. If you are having trouble finding an article, consult Web Appendix B (Searching the Literature) or critique the article in Appendix B. To critique the article, question its internal, external, and construct validity. If you want more specific help about what questions to ask of a study, consult Appendix C.

2. What are the main strengths and weaknesses of the study you critiqued?

3. Design a direct replication of the study you critiqued. Do you think your replication would yield the same results as the original? Why or why not?

4. Design a systematic replication based on the study you critiqued. Describe your study. Why is your systematic replication an improvement over the original study?

5. Design a conceptual replication based on the study you critiqued. Describe your study.

Why is your conceptual replication an improvement over the original study?

6. Evaluate the conclusions of these studies. Then, recommend changes to the study.

   a. A study asked teens whether they had taken a virginity pledge and found that those who claimed to have taken a pledge were more likely to abstain from sex than those who claimed not to have taken that pledge. The researchers conclude that abstinence pledges cause students to abstain from sex.

   b. A study finds that teens, after completing a three-year, voluntary, after-school abstinence education program, are better informed about the diseases that may result from sex. The researchers conclude that abstinence pledges cause students to abstain from sex.

WEB RESOURCES

1. Go to the Chapter 4 section of the book's student website and
   a. Look over the concept map of the key terms.
   b. Test yourself on the key terms.
   c. Take the Chapter 4 Practice Quiz.
2. Get a better idea of the steps involved in actually conducting a study by reading "Appendix D: Practical Tips for Conducting an Ethical and Valid Study."
3. To learn more about how to use PsycINFO and other databases to find articles, go to Chapter 4's "Computerized Searches" link.

4. To learn more about the value of reading the original source, click on Chapter 4's "Misinformation From Textbooks, Newspaper Articles, and Other Secondhand Sources" link.
5. If you want to read articles that are available on the web (including articles written by students), click on Chapter 4's "Web Articles" link.
6. If you want to start writing the introduction to either your research proposal or your research report, use Chapter 4's "Getting Started on Writing Your Introduction" link.

CHAPTER 5

Measuring and Manipulating Variables:

Reliability and Validity

Science begins with measurement.
—**Lord Kelvin**

An experiment is a question which science poses to Nature, and a measurement is the recording of Nature's answer.
—**Max Planck**

CHAPTER OVERVIEW

To state a hypothesis, you usually propose a relationship between two or more variables. For example, you might propose that "bliss causes ignorance." To test this hypothesis, you must define the fuzzy, general, and abstract concepts "ignorance" and "bliss" in terms of **operational definitions**: clear, specific, and concrete recipes for manipulating or measuring variables.

Because operational definitions are objective recipes for variables, they allow you to talk about your variables in objective, rather than subjective, terms. Thus, rather than saying, "My opinion is that they are happy," you can say, "They scored 94 on the happiness scale." By letting you talk about objective procedures rather than subjective opinions, operational definitions enable you to test your hypothesis objectively. In addition, because they are specific recipes that others can follow, operational definitions make it possible for others to repeat (replicate) your study.

Most people recognize that the ability of psychology to test hypotheses objectively and to produce publicly observable facts—in short, its ability to be a science—depends on psychologists' ability to develop publicly observable ways to measure psychological variables objectively and accurately. Unfortunately, most people also believe one of two myths about measuring psychological variables.

At one extreme are cynics who believe the myth that psychological variables cannot be measured. For example, they believe shyness is a subjective concept that can't be measured with a multiple-choice test or any other objective measure and that arousal can't be measured by increases in heart rate or changes in brain waves. These people think that psychology is not—and cannot be—a science.

At the other extreme are trusting, gullible innocents who believe the myth that psychological variables are easy to measure and that anyone who claims to be measuring a psychological variable is doing so. For example, they believe that the polygraph (lie detector) test accurately measures lying and that tests in popular magazines accurately measure personality. Because these naïve individuals can't distinguish between accurate (valid) measures and inaccurate (invalid) measures, they can't distinguish between scientific and pseudoscientific claims.

The truth is that developing measures and manipulations of psychological variables is not easy. However, developing objective measures of abstract constructs such as love, motivation, shyness, religious devotion, or attention span is not impossible. By the end of this chapter, you will know not only how to develop operational definitions of such abstract concepts but also how to determine whether such operational definitions have a high degree of construct validity. Put another way, by the end of this chapter, you will have completed a short course in psychological testing.

CHOOSING A BEHAVIOR TO MEASURE

If your hypothesis is about a behavior, such as smoking, yawning, jaywalking, typing, exercising, or nose-picking, your hypothesis (e.g., people will be more likely to smoke after being told that they should not be allowed to smoke) tells you what behavior to measure. Indeed, your hypothesis may even spell out whether you should measure the behavior's

- rate (how fast—if you're measuring smoking, rate might be measured by how many cigarettes the participant smoked in 2 hours)
- duration (how long—if you're measuring smoking, duration might be measured by how many minutes the participant spent smoking)
- cumulative frequency (how many—if you're measuring smoking, cumulative frequency might be measured by the *total* number of cigarettes the participant smoked during the observation period)
- intensity (how vigorously—if you're measuring smoking, intensity might be measured by how much smoke the participant inhaled with each puff)
- latency (how quickly the behavior began—if you're measuring smoking, latency [also called either response time or reaction time] might be measured by how much time passed before the participant lit up a cigarette)
- accuracy (how mistake-free—if you're measuring typing, accuracy might be measured by number of typos)

Thus, if you have a hypothesis about a specific behavior, obtaining accurate scores from each participant seems manageable: All you have to do is accurately measure the right aspect of the behavior.

But what if your hypothesis is about an abstract construct? At first, objectively measuring a construct may seem impossible: You cannot see abstract, invisible, psychological states such as love. As much as you might want to see what people are feeling, you can see only what they do. Fortunately, what they do may give you an indication of what they feel. Thus, although you cannot see love, you may be able to see love reflected in one of four types of behavior:

1. verbal behavior—what participants say, write, rate, or report, such as how a participant fills out a "love scale"

2. overt actions—what participants do, such as the extent to which a participant passes up opportunities to date attractive others
3. nonverbal behavior—participants' body language, such as the amount of time a person spends gazing into a partner's eyes
4. physiological responses—participants' bodily functions, such as brain wave activity, heart rate, sweating, pupil dilation, the degree to which a person's blood pressure increases when the partner approaches (Rubin, 1970)

To choose a specific behavior that is a valid indicator of your construct, you should consult theory and research. If you don't, you may choose a behavior that research has shown is not a valid marker of that construct. For example, if you choose self-reported social intelligence as a measure of actual social intelligence, handwriting as a sign of personality, or blood pressure as a gauge of lying, you are in trouble because those behaviors aren't strongly related to those constructs.

ERRORS IN MEASURING BEHAVIOR

If you have chosen a behavior that is a valid indicator of your construct—or if you are interested in measuring a certain behavior (e.g., smoking) rather than a construct—your search for a measure is off to a good start. However, it is only a start—you don't have a measure yet.

To understand why choosing a measure involves more than choosing a relevant behavior, imagine that you want to measure how fast participants run a 40-yard dash. To measure this behavior, (1) you must set the stage for the behavior to occur, (2) participants must perform the behavior, and (3) you must record the behavior. What happens at each of these three stages will affect what time you write down as the participant's 40-yard dash time.

First, by controlling the testing conditions, you, as the person administering the test, will affect how fast each participant runs. For example, variations in the instructions you give, in what participants wear, in the temperature at the time of the test, and in how many people watch the test all affect how fast participants run—and are all factors you may be able to control. In technical terms, any variations in testing conditions will introduce error. Thus, the accuracy of participants' times will depend on you minimizing error by keeping the testing conditions constant.

Second, the participant will determine how fast he or she runs. Specifically, two types of participant characteristics will affect the runner's speed: (a) characteristics that tend not to vary, such as the runner's height and athletic ability, and (b) characteristics that can vary, such as the participant's mood, health, energy level, and desire to run. Variations in these variable factors—other than changes you wanted your manipulation to cause—will introduce error. For example, if the participant becomes ill right before the run, you will not be measuring the runner's typical performance.

Third, you, as observer and recorder, will determine the participant's *recorded* time by what time you write down, and the time you write down will usually be affected by when you start and stop the stopwatch. For instance, if you stop the stopwatch before a participant crosses the finish line, that participant's recorded time will be faster than the participant's actual time.

All three of these factors—testing conditions, participants' psychological and physiological states, and observers—can vary. Testing conditions, such as weather conditions and what instructions the researcher gives, may not be the same from day to day, moment to moment, and participant to participant. Participants and their energy levels, efforts, and expectations may vary. Observers' accuracy may vary.

Overview of Two Types of Measurement Errors: Systematic Bias and Random Error

When testing conditions, participants, or observers change, those changes make participants' scores change. Thus, in an ideal study, these three factors would not change unless the researcher deliberately made them change. In actual research, however, these three factors often do vary independently of researchers' conscious wishes—and the impact of this variation on participants' scores depends on whether the three factors vary (a) systematically or (b) randomly.

Bias: Systematic Error

If these factors vary systematically (in a way that pushes scores in a certain direction), the result is bias. **Bias** may cause a researcher to "find" whatever he or she expects to find. For example, suppose a researcher believes that one group of individuals—the individuals given a special treatment—will run faster than the no-treatment group. The researcher could unintentionally bias the results in at least three ways.

First, the researcher could create biased testing conditions by consistently giving the participants who received the treatment more time to warm up than the other participants. Second, the researcher could bias the participants' expectations by telling participants who received the treatment that the treatment should improve their performance. Third, the researcher could bias observations by clicking off the stopwatch just *before* the participants who received the treatment reach the finish line, but clicking off the stopwatch just *after* the other participants reach the finish line. If the researcher, consciously or unconsciously, does any of these things, the researcher will "find" that the group that was expected to run faster will have faster recorded times than the other group.

Random Error: Unsystematic Error

You have seen that when the testing conditions, the researcher's expectations, or the scoring of the test consistently favor one group, the result is systematic bias. But what if testing conditions, researcher's expectations, and the scoring of the test vary but do not vary in a way that consistently favors any group? Then, the result is unsystematic **random error** of measurement. For instance, suppose the wind at the time participants run the race varies in an unsystematic way. It unpredictably blows at the back of some runners, in the face of other runners, but, on the average, it probably does not aid the runners receiving the treatment substantially more than it aids the other runners.

This unsystematic random measurement error makes individual scores less trustworthy. For some runners, the wind will be at their backs and help them. For others, the wind will blow in their faces and slow them down. Occasionally, if there is a lot of wind, a slower runner may end up with a better time than a faster runner.

Although random measurement error has a strong effect on individual scores, it has little effect on a group's average score. Why? Because random measurement error, like all random error—and unlike bias—does *not* consistently push one group's scores in one direction. To be more specific and technical, *random error tends to average out to zero* (see Figure 5.1).

Because random error tends to balance out to zero, if a group is large enough (e.g., more than 60), the seconds random error added to some members' times will be balanced out by the seconds random error subtracted from other members' times. Consequently, random error's average effect on the group will tend to be near zero. Because random measurement error will probably have little to no effect on either the treatment group's or the no-treatment group's average score, random measurement error will probably

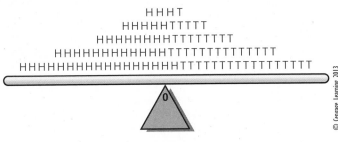

FIGURE **5.1** Random error balancing out to zero.

Notes:
1. These are the results of 90 coin flips. "H" stands for heads; "T" stands for tails. Of the first four flips (see the first row), three were heads. However, of the next 26 flips (see the second and third rows), 13 were heads and 13 were tails. Thus, whereas 75% of our first four flips were heads, only 53% of our first 30 flips were heads. Put another way, when we had only four flips, the difference between the percentage of heads we expected (50%) and what we obtained (75%) was large, but when we had 30 flips, the difference between the percentage we expected (50%) and what we obtained (53%) was small.
2. In our example, after 90 flips, we had 45 heads and 45 tails. This is an unusual result, especially given that we started out with 3 heads and 1 tail. Regardless of what happened in the first 4 flips, we would have expected to obtain approximately 43 heads and 43 tails in our next 86 flips. Thus, given that we started out with two more heads than tails, we would have expected—after 90 flips—to still have two more heads than tails. That is, we don't expect the coin to have a memory and for it (or chance) to correct for past errors. If we had obtained 43 heads and 43 tails in the next 86 flips, our total number of heads would have been 46, which is 51% heads, which is 1% more than the 50% we would get without random error. Thus, it might be best for you to think of random error balancing out toward—rather than evening out to—zero.

not create significant differences between those groups. For example, random gusts of wind, rather than helping the runners who received the treatment much more than runners who did not receive the treatment, will tend to affect both groups equally.

Admittedly, random measurement error will probably not balance out perfectly: The wind may help one group slightly more than the other. Fortunately, however, you can use statistical techniques to estimate the extent to which random measurement error might fail to balance out. To appreciate the value of such statistical analyses, imagine that a statistical analysis told you that all sources of random error combined (variations in wind, variations in instructions, variations in scoring, etc.) would probably not cause the groups to differ by more than 5 seconds.

In that case, if your groups' average times differed by 10 seconds, you could conclude that the difference between the groups is due to something more than random error—the treatment. But what if the groups differed by 4 seconds?

The good news is that if this 4-second difference is due solely to random error, you will not be fooled into claiming that the groups really differ. You know, thanks to the statistical analysis, that the groups could reasonably be expected to differ by as much as 5 seconds by chance alone.

The bad news is that if some of this 4-second difference is due to a real difference between the groups, you will *fail* to recognize that effect. You are not going to claim that the observed difference of 4 seconds represents a treatment effect when you know, thanks to the statistical analysis, that the groups could reasonably be expected to differ by as much as 5 seconds by chance alone. In such a case, unsystematic random measurement error would hide true differences between groups.

To see the benefits of reducing random measurement error, suppose you had reduced random measurement error to the point that it was unlikely that wind and other random error would have caused the groups to differ by more than 2 seconds. Then, because your observed difference (4) was more than the 2 seconds that chance could account for, you would be able to see that the groups really did differ. Thus, by reducing random error, you reduced its ability to overshadow a treatment effect.

The Difference Between Bias and Random Error: Poisoning Validity is Worse Than Watering It Down

Although random measurement error and bias in measurement are both measurement errors, the two errors are different. Bias is systematic (it pushes scores in a certain direction), and statistics cannot account for its effects. Thus, bias can often fool you into thinking that two groups differ when they do not. Random error, on the other hand, is unsystematic, and statistics can partially account for its effects. Thus, if you use statistics, you will rarely mistake the effects of random error for a genuine difference between your groups—no matter how much random error is in your measurements.

We have argued that whereas random error is a nuisance, bias is a serious threat to a study's validity. To appreciate this point, imagine that two people are weighing themselves over a period of days. Although neither is losing weight, the first is content with her weight, whereas the second is trying to lose weight. They might record the following data:

Day	Person 1	Person 2
Day 1	150	151
Day 2	149	150
Day 3	151	149

© Cengage Learning 2013

In the case of Person 1, the errors are random; the errors do not make it look like she is losing weight. Despite the errors, we know that her weight is around 150 pounds. Although the weight of the clothes she is wearing while being weighed, the time of day she weighs herself, and how she reads the needle on the scale are not exactly the same from measurement to measurement, they do not vary in a systematic way. In the case of Person 2, on the other hand, the errors are systematic; they follow a pattern that makes it look like he is losing weight. Maybe the pattern is due to moving the scale to a more sympathetic part of the floor, maybe the pattern is due to weighing himself at a time of day when he tends to weigh less (before meals), or maybe the pattern is due to his optimistic reading of the scale's needle. Regardless of how he is biasing his measurements, the point is that he is seeing what he wants to see: He is not being objective.

Errors Due to the Observer: Bias and Random Error

To help you better understand how the bias from participants or researchers that creates fake differences between conditions contrasts with the inconsistent, random error that hides real differences between conditions, we will show you how these two types of errors can come from the following three sources: (1) the person administering the measure, (2) the participant, and (3) the person scoring the measure. We will begin by discussing how bias and random error come into play when researchers observe, score, and record behavior.

Observer Bias (Scorer Bias)

The first, and by far the most serious, type of observer error occurs when people's subjective biases prevent them from making objective observations. Observers may be more likely to count, remember, or see data that support their original point of view. In other words, a measure of behavior may be victimized by **observer bias**: observers recording what they expect participants will do rather than what participants are actually doing.

To see how serious a problem observer bias can be, suppose that biased observers record the cigarette-smoking behavior of smokers before and after the smokers go through a "stop smoking" seminar. Before a smoker entered the program, if she took one puff from a cigarette, the observer counted that as smoking an entire cigarette. However, after the smoker completed the program, the observer did not count smoking one puff as smoking. In such a case, observer bias would be systematically pushing cigarette-smoking scores in a given direction—down. By decreasing the average smoking score, observer bias may lead us to believe that a smoking prevention program worked—even when it did not.

If you can't control observer bias, you can't do scientific research. There is no point in doing a study if, regardless of what actually happens, you are going to "see" the results you want to see. If, on the other hand, you can control observer bias, you move toward the scientific ideal that the findings will be the same no matter who does the study. Because objective measures are not vulnerable to observer bias, scientists value objective measures.[1]

Random Observer Error

The second type of mistake that observers make in scoring behavior is making unsystematic random errors that will inconsistently increase and decrease individual scores. For example, a participant who should get a score of 3 could get a score of 2 one moment but a 4 the next.

If your observers are that inconsistent, the bad news is that you can't trust individual scores. You can't say "Participant X scored a 3, so Participant X's true score is a 3." Instead, the most you can do is use the observed score to estimate the range of scores in which the participant's true score might fall.[2] For instance, you might say, "Participant X scored a 3, but random observer error may easily have added or subtracted a point from that score. Because random observer error has made that score inaccurate, we shouldn't think of it as a 3, but as a score somewhere between 2 and 4."

The good news is that because random errors are unsystematic, they will probably not substantially affect a *group's* overall average score. The points that random observer errors add to some group members' scores will tend to be balanced out by the points that random observer errors subtract from other group members' scores. Thus, unlike observer bias, random observer error will probably not substantially change a group's average score.

Minimizing Observer Errors

Although we would like to reduce the influence of both observer bias and random observer error, reducing observer bias is more important than reducing random error.

---

[1] Fortunately, science does have a safeguard against subjective measures: replication. If a skeptic with different beliefs replicates the study, the skeptic will obtain different results—results consistent with the skeptic's beliefs. The failure to replicate the original study's results may expose the measure's subjectivity.

[2] To calculate how big the range will be, first use the formula for the standard error of measurement (*SEM*): standard deviation (*SD*) $\times \sqrt{1 - \text{reliability}}$. For example, if the *SD* was 10 and the reliability was .84, the *SEM* would be $10 \times \sqrt{1 - .84} = 10 \times \sqrt{.16} = 10 \times .4 = 4$. Next, determine how confident you want to be that your range includes the true score. You can be 68% confident that the true score is within 1 *SEM* of the observed score, 95% confident that the true score is within 2 *SEMs* of the observed score, and 99% confident that the true score is within 3 *SEMs* of the observed score. Thus, with a SEM of 4, we could be 68% confident that the person's true score was within 4 points of the observed score. Put another way, our range would extend 8 points—from 4 points below the observed score to 4 points above.

Why It Is More Important to Reduce Observer Bias Than Random Error. To understand why observer bias is more of a problem than random error, let's consider two error-prone basketball referees. The first makes many random errors; the other is biased. Which would you want to referee your team's big game?

Your first reaction might be to say, "Neither!" After all, the referee who makes many random errors is aggravating. Who wants an inattentive, inconsistent, and generally incompetent ref? However, in the course of a game, the "random" ref's errors will tend to balance out. Consequently, neither team will be given a substantial advantage. On the other hand, a referee who is biased against your team will consistently give the opposing team a several-point advantage. Thus, if you had to choose between the two error-prone officials, which one would you pick? Most of us would pick the one who made many random errors over the one who was biased against us.

Eliminating Human Observer Errors by Eliminating the Human Observer. Often, we don't have to choose between minimizing random error and minimizing observer bias because the steps that reduce observer bias also tend to reduce random observer error. For example, one way to eliminate observer bias is to replace the human observer with scientific instruments, such as computers or other automated data recorders, and to replace subjectively graded essay tests with computer-scored multiple-choice tests. Eliminating the human observer not only eliminates bias due to the human observer but it also eliminates random error due to the human observer.

Limiting Human Observer Errors by Limiting the Human Observer's Role. If you can't eliminate observer error by eliminating the observer, you may still be able to reduce observer error by reducing the observer's role. For instance, rather than having observers rate *how* aggressive a participant's behavior was, observers could simply decide whether the participant's behavior *was* aggressive. Or, rather than judging how aggressive essays are, observers could count how many aggressive words were used. For more tips on how to make the observer's job easier—and thus reduce both random observer bias and random observer error—see Table 5.1.

Reducing Observer Bias—But Not Random Error—by Making Observers "Blind." Although Table 5.1 includes a wide variety of strategies that will help reduce observer bias, those tactics may not eliminate observer bias. To understand why limiting the observer's role may not eliminate observer bias, suppose you were having observers judge essays to determine whether men or women used more "aggressive" words. Even if you conducted a thorough training program for your raters, the raters might still be biased. For example, if they knew that the writer was a man, they might rate the passage as more aggressive than if they thought the same passage was written by a woman.

To reduce such bias, you should not let your raters know whether an essay was written by a man or a woman. Instead, you should make your raters **blind** (also called **masked**): unaware of the participant's characteristics and situation.

> TABLE **5.1** Techniques That Reduce Both Random Observer Error and Observer Bias
>
> 1. Replace human observers and human recorders with machines (such as computers and automatic counters).
> 2. Simplify the observer's task:
>   a. Use objective measures such as multiple-choice tests rather than essay tests.
>   b. Replace tasks that require observers to judge a behavior's intensity with tasks that merely require observers to count how many times the behavior occurs.
>   c. Reduce the possibility for memory errors by making it very easy to immediately record their observations. For example, give your observers checklists so they can check off a behavior when it occurs, or give observers mechanical counters that observers can click every time a behavior occurs.
> 3. Tell observers that they are to record and judge observable behavior rather than invisible psychological states.
> 4. Photograph, tape record, or videotape each participant's behavior so that observers can recheck their original observations.
> 5. Carefully define your categories so that all observations will be interpreted according to a consistent, uniform set of criteria.
> 6. Train raters, and motivate them to be accurate.
> 7. Use only those raters who were consistent during training.
> 8. Keep observation sessions short so observers don't get tired.

© Cengage Learning 2013

The importance of making observers blind has been illustrated in several studies. In one such study, people rated a baby in a videotape as much more troubled when they were told they were watching a baby whose mother had used cocaine during pregnancy than when they were not told such a story (Woods, Eyler, Conlon, Behnke, & Wobie, 1998).

Conclusions About Reducing Observer Bias. Because eliminating observer bias is vital, scientists often eliminate observer bias by eliminating the observer. Consequently, measures that do not require an observer, such as multiple-choice tests, rating scale measures, and reaction time measures, are popular.

You are probably familiar with the basic techniques that researchers use to reduce observer bias because many of the tactics a researcher would use to reduce observer bias are the same tactics a professor would use to avoid favoritism in grading. Instead of determining students' grades solely by sitting down at the end of the term and trying to recall the quality of each student's work, the favoritism-conscious professor probably give computer-scored multiple-choice tests. If the favoritism-conscious professor were to give an essay exam, she would establish clear-cut criteria for scoring the essays, follow those criteria to the letter, and not look at students' names while grading the essays.

Conclusions About the Relationship Between Reducing Observer Bias and Reducing Random Observer Error. Making observers blind should eliminate observer bias, but it will not eliminate random observer error. Blind observers can still be careless, inattentive, forgetful, or inconsistent about how they interpret behavior. Suppose, for example, that a history professor grades 100 essay exams over the weekend. Even if the professor avoids bias by grading all those exams "blind," he may still fail to grade consistently from test to test. Thus, random error can creep in due to variations in how closely the professor reads each paper, variations in how much partial credit he gives for a certain essay answer, and even in variations (errors) in adding up all the points.

We do not mean to imply that the steps you take to reduce observer bias will never reduce random observer error. On the contrary, except for the blind technique, every step that you take to reduce observer bias will also tend to reduce random observer error.

Errors in Administering the Measure: Bias and Random Error

By using blind procedures and by reducing the observer's role, you can reduce the amount of measurement error that is due to scoring errors. However, not all errors in measurement are due to the scorer. Some errors are made in administering the measure. As was the case with scoring, there are two kinds of errors that researchers can make in administering the measure: bias and random error.

When you administer the measure, you hope to avoid introducing either bias or random error. But to avoid both these errors completely, you would have to keep everything in the testing environment exactly the same from session to session. For example, if you were administering an IQ test, you would have to make sure that noise level, lighting, instructions to participants, your facial expressions, your gestures, and the rate, loudness, and pitch at which you spoke did not vary from session to session.

Keeping all these factors perfectly constant is impossible. However, most researchers—and most people who administer psychological tests—strive for a high level of **standardization**: treating each participant in the same (standard) way. Ideally, you would test all your participants in the same soundproof, temperature-controlled setting. Although you probably won't be able to do that, you can—and should—write out a detailed description of how you are going to test your participants and stick to those procedures. For example, you might write down whatever instructions you were going to give participants and read those instructions to every participant. To standardize your procedures even more, you might present your instructions on videotape, put the instructions and measures in a booklet, or you might even have the instructions and measures administered by a computer program.

Because perfect standardization is usually impossible, there will usually be some measurement error due to imperfect standardization. If you must have such error, you would prefer that this error be random error rather than bias. As was the case with observer error, random error will not push scores in a certain direction whereas bias will. Thus, although it would be annoying if you were randomly inconsistent in how you treated participants, it would be disastrous if you biased the results by being more attentive, enthusiastic, and patient when administering the test to the treatment group than to the no-treatment group.

To prevent bias from creeping in when your measure is administered, you should try to keep the person who administers the measure blind. For example, you might have one researcher administer the treatment and a second researcher—who does not know which treatment the first researcher gave the participant—administer the measure. That way, the second researcher can't bias the study's results (but, if the second researcher administers the measure inconsistently, the second researcher could introduce additional random error into the study).

Errors Due to the Participant: Bias and Random Error

To this point, we have focused on two sources of measurement error: errors made by the person administering the measure and errors made by the person scoring the measure. We will now turn to a third source of measurement error: participants.

Random Participant Error

Participants may produce responses that don't perfectly reflect their true behavior or feelings because they themselves are not perfectly consistent. Their behavior is variable, and some of this variability is random. One moment they may perform well; the next moment they may perform poorly. For example, participants may misread questions, lose their concentration, or make lucky guesses.

One way to overcome this random variability in participants' behavior is to get a large sample of their behavior. For example, if you wanted to know how good a free-throw shooter someone was, you wouldn't have her shoot only two free throws. Instead, you would probably have her shoot at least 20. Similarly, if you wanted to know how outgoing she was, you wouldn't base your conclusion on a two-item test. Instead, you would probably use a test that had at least 20 questions on it so that random participant error would tend to balance out.

Because psychologists want to give random participant error a chance to balance out, they often avoid trying to measure a construct with a single question. Instead, they tend to ask multiple questions on the same topic—even if that means restating the same question in several ways. Thus, if you have filled out a psychological test, you may have wondered, "Why is it so long—and why are they asking me what seems to be the same questions over and over?" Now you know one answer to your question: to balance out random participant error.

Two Types of Subject Biases

When trying to know what participants are like from their behavior, random participant error is a problem because it may cause us to think that a random, uncharacteristic action is characteristic of the participant. However, a more serious obstacle to deducing participants' thoughts from their actions is subject (participant) bias: participants changing their behavior to impress you or to help you (or, sometimes, even to thwart you).

One of the earliest documented examples of the problem of subject bias was the case of Clever Hans, the mathematical horse (Pfungst, 1911). Hans would answer mathematical problems by tapping his hoof the correct number of times. For example, if Hans's owner asked Hans what 3×3 was, Hans would tap his hoof nine times. Hans's secret was that he watched his owner. His owner would stop looking at Hans's feet when Hans had reached the right answer. Although people believed Hans's hoof tapping meant that Hans was performing mathematical calculations, his hoof tapping only meant that Hans was reacting to his owner's gaze. Hans didn't know math, but he did know how to give the "right" answer.

If animals can produce the right answer when they know what you are measuring, so can humans. In fact, for humans, there are two kinds of right (biased) responses: (1) ones designed to make the researcher look good and (2) ones designed to make the participant look good.

Obeying Demand Characteristics. The first kind of right answer is the one that makes you, the researcher, look good by ensuring that your hypothesis is supported. Orne (1962) believed that participants are so eager to give researchers whatever results the researcher wants that participants look for clues as to what responses will support the researcher's hypothesis. According to Orne, if the research setting or the researcher gives the participant a hint about how to support the hypothesis, the participant will follow that hint as surely as if the researcher had demanded that the participant follow it. Consequently, Orne named such hints **demand characteristics**.

To give you some idea of the power of demand characteristics, consider how they operate in everyday life. Imagine you and a friend are at a restaurant. The service is slow, and the food is bad. You and your friend grumble about the food through much of the meal. Then, at the end of the meal, your server asks you, "Was everything all right?" Do you share your complaints, or do you give in to demand characteristics and say that everything was fine?

To see how demand characteristics might affect the results of a study, imagine that you do the following study. First, you have participants rate how much they love their partner. Next, you give them fake feedback, supposedly from their partner, showing that their partner loves them intensely. Finally, you have participants rate how much they love their partner a second time. Participants may realize that they are supposed to rate their love higher the second time. Therefore, if participants reported that they loved their partner more the second time, you would not know whether learning about their partners' devotion changed participants' feelings or whether participants merely obeyed the study's demand characteristics.

Participants might have obeyed the study's demand characteristics because of two problems with your measure. First, your measure tipped them off that you were trying to measure love. Once participants knew that you were trying to measure love, they were able to guess why you showed them their partners' ratings. Your measure gave them all the clues (demand characteristics) they needed to figure out what you would consider a "good" response. Second, you made it easy for them to give that response.

So, to improve your study, you need to choose a measure that doesn't have both the problems of your original measure. At the very least, you

TABLE **5.2** Ways to Avoid Subject Biases When Measuring Love

Technique	Example
Measure participants in nonlaboratory settings	Observe hand-holding in the college cafeteria.
Unobtrusive observation	Observe hand-holding in the lab through a one-way mirror.
Unobtrusive measures (nonverbal)	Observe how much time partners spend gazing into each other's eyes.
Unobtrusive measures (physical traces)	Measure how close together the couple sit by measuring the distance between their chairs.
Unexpected measures	Lead the participant to believe that partner has damaged something by accidentally knocking it over, and then ask the participant to repair the alleged damage
Disguised measures	Ask participants to rate themselves and their partners on several characteristics. Then, infer love from the extent to which they rate their partner as being similar to themselves.
Physiological responses	Measure pupil dilation to see if it increases when their partner comes into the room.
Important behavior	See if the participant passes up the opportunity to date a very attractive person.

© Cengage Learning 2013

should use a measure that either (a) makes it more difficult for participants to figure out what the hypothesis is or (b) makes it more difficult for participants to play along with that hypothesis.

As Table 5.2 shows, there are at least two ways you could make it hard for participants to figure out what you are measuring, thereby making it hard for participants to figure out your hypothesis. Unfortunately, both ways raise ethical questions because they both involve compromising the principle of **informed consent**: Participants should freely decide whether to participate in the study only after being told what is going to happen to them.[3]

The first way to make it hard for participants to know what you are measuring is to make it hard for participants to know that you are observing them: You spy on them. In technical terminology, you use **unobtrusive measurement**: recording a particular behavior without the participants knowing you are measuring that behavior. For example, you might spy on them when they are in the real world or you might spy on them through a one-way mirror when they are in the waiting room.

The second way involves disguising your measure. You might let participants think you were measuring one thing when you were actually measuring something else. For instance, you might take advantage of the fact that people in love tend to overestimate how similar they are to their partner. Therefore, you could have participants rate themselves and their partners on a variety of

---

[3] In the next chapter, we discuss the ethical issues involved in choosing a measure.

BOX **5.1**   **The Logic of a Disguised Prejudice Measure**

Saucier and Miller (2003) had participants rate, for 16 different paragraphs and conclusions, how well each paragraph supported its conclusion. Although participants were asked to rate how logical the argument for a position was (e.g., spending more money on research to find a cure for sickle-cell anemia), the researchers were using participants' ratings as a measure of prejudice. For example, participants scored high on prejudice to the degree that they (a) gave low ratings on the degree to which the paragraph supported a conclusion when those conclusions were favorable toward Blacks and (b) gave high ratings on the degree to which the paragraph supported a conclusion when those conclusions were unfavorable toward Blacks.

After reading this chapter, you may want to read Saucier and Miller's (2003) article to see how they made the case that their measure really did measure prejudice. As you might expect, they found that participants' ratings of how logical the argument was correlated with the degree to which participants agreed with the argument's conclusion. Also, as you might expect, Saucier and Miller correlated their measure with other measures of prejudice to see whether their measure predicted prejudiced behavior. In addition, they tried to show that their measure was not strongly affected by (a) random error or (b) social desirability bias.

© Cengage Learning 2013

characteristics. Participants would probably think you are interested in how accurately or positively they rate their partners. Instead, you'd be seeing the extent to which participants believed that they were similar to their partner—and using perceived similarity as a measure of love. (To see how a disguised measure can be used to measure prejudice, see Box 5.1.)

But what if you can't stop your participants from figuring out the hypothesis? Even if participants figure out the hypothesis, you can still do two things to prevent participants from playing along with it.

First, you could make it almost impossible for participants to play along with the hypothesis. For instance, you might use a physiological measure of love that most people can't voluntarily control, such as brain wave activity, pupil dilation, or contraction of certain facial muscles that are associated with happiness. If you wanted to use a nonphysiological measure, you might use a measure based on reaction time because such a measure is also hard for participants to fake. (To see how reaction time can be used to measure prejudice, see Box 5.2.)

Second, you could make it costly for participants to play along with the hypothesis. For example, if you made it so participants would have to spend more time performing a dull task (watching people fill out questionnaires) to help out their partner, many would not be willing to put themselves through that much aggravation just to play along with your hypothesis.

Social Desirability Bias.   Even if you prevent participants from giving answers that they think will make you look good, you have prevented only one kind of subject bias. You have not prevented a second kind of bias, called the **social desirability bias:** the participant acting in a way that makes the participant look good. On most questionnaires, it is easy for participants to choose the answer that makes them look good. Indeed, research has shown that

BOX **5.2** **The Implicit Attitude Test (IAT): A Reaction Time Measure That Is Hard to Fake**

Imagine that you are a white person and that you consider yourself unprejudiced. You go to the Implicit Association Test website (https://implicit.harvard.edu/implicit/) and take a test designed to measure whether you are biased toward Blacks. At first, the task is ridiculously easy: When you see a black face, hit a key on the left (e.g., the letter "e" key); when you see a white face, hit a key on the right (e.g., the letter "i" key). Then, you have yet another easy task: When you see a positive word (e.g., "good") hit a key on the left; when you see a negative word (e.g., "bad), hit a right key. In the next phase, the task becomes slightly more challenging. Now, you have to deal with two categories at once. If you see either a white face or a positive word, you have to hit the left key. If you see either a black face or a negative word, you have to hit the right key. You're still doing fine—responding accurately and quickly. In the final phase, the rules are changed. Now, if you see either a black face or a positive word, you have to hit the left key. If you see either a white face or a negative word, you have to hit the right key. If you are like most (88%) Whites, you will find the last task the hardest and perform it the slowest, indicating some degree of bias against Blacks.

How do researchers know that slower reaction times on this last task indicate implicit bias? Researchers know because, among other things, studies have shown that

- Blacks are less likely than Whites to have slower reaction times on this last task.
- Conservatives are more likely than liberals to have slower reaction times on this last task.
- People in favor of racial profiling are more likely than others to have slower reaction times on this last task.
- Whites with slower reaction times on this last task are less likely to choose to work with a black partner.
- Scores on the test do not correlate with hand–eye coordination.

© Cengage Learning 2013

people claim to be much more helpful (Latané & Darley, 1970) and less conforming (Milgram, 1974) than they really are.

To reduce social desirability bias, you could use any of the four main measurement strategies that work for reducing demand characteristics: (a) not letting participants know they are being measured, (b) not letting participants know what is being measured, (c) using physiological and other measures that are impossible to fake, and (d) using behavioral measures that would be costly (in terms of time, money, energy, or fun) to fake. Put another way, participants won't produce fake responses to impress the researcher if (a) they don't know they are being watched, (b) they don't know what the right response is, (c) they can't fake the right response, or (d) they don't care to pay the price of impressing the researcher. For example, they probably won't try to be more generous than they are if they don't know you are watching them or if it costs them time and money to show how generous they are.

Although the techniques to reduce demand characteristics can be used to reduce social desirability bias, the easiest and most commonly used tactic to deal with the social desirability bias is a technique that is *not* used to reduce demand characteristics: having participants *not* put their names on their answer sheets so that their responses are anonymous. If participants cannot

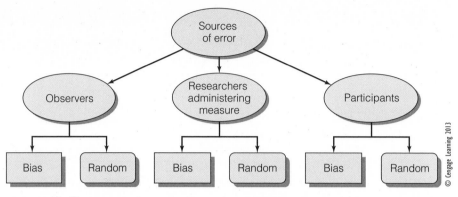

FIGURE **5.2** Sources and types of measurement error.

get credit for their answers, participants will probably not be motivated to make socially desirable, but false, responses.[4]

Although anonymous participants cannot make themselves look good, they can still try to make you look good by producing results that they think will support your hypothesis. Thus, although making responses anonymous eliminates social desirability bias, it doesn't eliminate bias due to obeying demand characteristics.

To prevent participants from following demand characteristics, you would remove demand characteristics by making participants blind (unaware of what condition they are in). Making participants blind, however, would not reduce social desirability bias because it would not stop participants from trying to make themselves look good.

Summary of the Three Sources and Two Types of Measurement Error

We have discussed three major sources of measurement error: errors due to the person scoring the measure, errors due to the person administering the measure, and errors due to the participant. We have also stressed that each of these sources can contribute two types of measurement error: random error and systematic bias (see Figure 5.2). Furthermore, we stressed that bias is a much more serious threat to validity than random error. We showed how observer bias was worse than random observer error, how researcher bias was worse than random errors in administering a measure, and how participant bias was worse than random participant error. To combat bias, we advocated using three strategies that target bias but do not reduce random error—blind (masked) techniques, unobtrusive measurement, and making responses anonymous (see Table 5.2)—and two strategies that reduce both bias and random error—standardizing how the measure is administered and simplifying how the measure is scored (see Table 5.1).

---

[4] Unfortunately, anonymous participants may still give false or misleading information. For example, some adolescents may display a sense of humor or a sense of rebelliousness by putting outrageous answers on an anonymous questionnaire.

RELIABILITY: THE (RELATIVE) ABSENCE OF RANDOM ERROR

As you have seen, bias is a much more serious threat to a measure's validity than random error (see Figure 5.3). Indeed, at this point, you might be saying to yourself, "Bias is bad. It makes me see what I want to see rather than what is really there. I should try to eliminate it. Random error, on the other hand, doesn't seem that serious. Why should we bother to develop a measure that is free of random error?" In the next section, we will answer that question by explaining why you should want a measure that is **reliable**: producing stable, consistent scores that are not strongly influenced by random error.

The Importance of Being Reliable: Reliability as a Prerequisite to Validity

You want scores on your measure to be stable over time when you are measuring a construct that, rather than changing from minute to minute, is stable over time. Thus, if you are accurately measuring a person's intelligence, shyness, or height at two different times, you should get the same results each time. For example, if someone is 5 feet tall (152 cm) and your measure is valid (accurate), you should consistently (reliably) measure that person as 5 feet (152 cm) tall. Thus, if your measure of height or any other stable characteristic is valid, your measurements must be reliable. In short (when talking about a stable characteristic), *validity guarantees reliability*; that is, *valid measures must be reliable.*

Reliability, however, does not guarantee validity; that is, *reliable measures may not be valid*. For example, if we reliably measure someone's

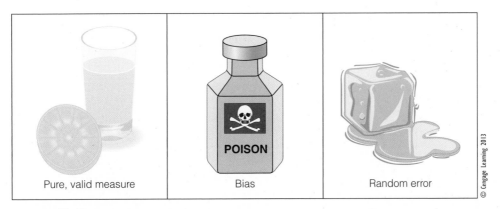

FIGURE **5.3** Bias poisons a measure's validity, whereas random error merely dilutes a measure's validity.

If you loved pure orange juice, you wouldn't want your juice to be poisoned or watered down. However, if you had to choose, you would rather have your drink watered down than poisoned. Similarly, if you have to have error in your measure, you would prefer that the error be random error (which dilutes your measure's validity) rather than bias (which poisons your measure's validity).

height at 5 feet tall (152 cm) but the person is actually 6 feet tall (180 cm), our measurements are reliably wrong (and, therefore, invalid).

Although reliability does not guarantee validity, *reliability is a prerequisite for validity*. To be more specific, *reliability puts a ceiling on how high validity can be*. That is, only to the degree that your measurements of a stable trait are stable can your measurements of that stable trait be accurate. For example, suppose you measure a person's height twice. If you measure the person as 5′ 5″ (165 cm) both times, your measure's reliability is perfect and your measure's validity *could* be perfect. Indeed, if the person is 5′ 5″ tall (165 cm), your measure's validity would be perfect. However, suppose you measure the person as 5′ 6″ (167 cm) one time and 5′ 4″ (162 cm) the next. In that case, your measure is not perfectly reliable and your average error of measurement would be at least 1 inch (2.54 cm). If your measurements were so unreliable that you measured someone's height to be 5′ 10″ (175 cm) one day and 5′ 0″ (152 cm) the next, your average error of measurement would be at least 5 inches (12.70 cm).

You have seen that the *less* reliable your measurements, the *more* random error your measurements contain. The more scores are affected by unstable, unsystematic random factors (random error), the less opportunity scores have to be affected by the stable factor (e.g., height) that you want to measure. Because reliability puts a ceiling on validity (see Figure 5.4), you want a reliable measure: one that produces scores that will not be bounced around by the erratic winds of chance. But how can you know the extent to which your measure is contaminated by random error?

Using Test–Retest Reliability to Assess Overall Reliability: To What Degree Is a Measure "Random Error Free"?

To find out the full extent to which your measurements are contaminated by random error, you should find out the measure's overall reliability. Perhaps the most direct way to find out the degree to which the measure is reliable is to obtain the measure's **test–retest reliability**.

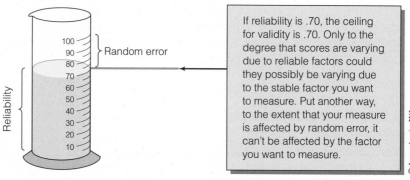

If reliability is .70, the ceiling for validity is .70. Only to the degree that scores are varying due to reliable factors could they possibly be varying due to the stable factor you want to measure. Put another way, to the extent that your measure is affected by random error, it can't be affected by the factor you want to measure.

© Cengage Learning 2013

FIGURE **5.4** Reliability puts a ceiling on validity.

As the name suggests, test–retest reliability requires participants to be tested and then retested. For example, a psychological test developer may test participants on a measure and then test those same participants again on the same measure 3 months later. The developer would then calculate a test–retest coefficient by comparing the scores that participants received the first time they took the test with the scores participants received the second time.

The more participants' scores on the first measurement correspond to their scores on the second measurement, the higher the test–retest coefficient. Although test–retest coefficients can range from 0 (no reliability) to 1 (perfect reliability), most are between .60 and .98.[5]

The test–retest coefficient tells you, in percentage terms, the degree to which scores are stable and consistent from test to retest. Thus, a test–retest coefficient of 1.00 would mean that there was a perfect (100%) correspondence between the first and second time of measurement: Everyone who scored high the first time also scored high the second time. The data that follow reflect a 1.00 test–retest coefficient.

Participant	Score first time	Score second time
Hinto	3	3
Nato	4	4
Misu	5	5

© Cengage Learning 2013

Put another way, the test–retest coefficient tells you the *percentage* of variation in scores that is *not* due to random error. Thus, a test–retest coefficient of 1.00 tells us that 100% of the differences between scores are *not* due to random error. The measure is 100% (completely) free from random error.

What if a measure, rather than being perfectly reliable, was perfectly unreliable? Then, the measure would have a test–retest coefficient of zero, meaning that there was absolutely no (0%) relationship between the scores participants received the first time they took the test and the scores they received when they were retested. In the next table, we have displayed data from a measure with a zero test–retest coefficient. As you can see, there is no connection between a participant's test and retest scores: The only way a participant gets the same score on both the test and retest is by chance. Put another way, in the case of a measure with a zero reliability coefficient, scores are 0% (not at all) free from random error.

---

[5] The test–retest reliability coefficient is usually not a correlation coefficient. If it were, it could range from −1 to +1. Instead, the test–retest reliability coefficient is usually the square of the test–retest reliability correlation (Anastasi, 1982). This squared term represents the percentage of variation that is *not* due to random error. To find out how much of the variation of scores is due to random error, you subtract the test–retest coefficient from 1. Thus, if your test–retest coefficient is 1, none (0%) of the variation in your scores is due to random error (because $1 - 1 = 0$). If, on the other hand, your test–retest coefficient is 0, then all (100%) of the variation in your scores is due to random error (because $1 - 0 = 1 = 100\%$).

Participant	Score first time	Score second time
Hinto	3	4
Nato	4	3
Misu	5	4

© Cengage Learning 2013

Note that because this measure is completely affected by random error, it can't be affected by the stable trait it is supposed to measure. Put another way, because it has zero reliability, it has zero validity: Because it is so unstable that it does not even correlate with itself, it can't correlate with the stable trait it is supposed to measure.

What if, rather than having zero reliability, the measure has .40 reliability? In that case, because only 40% of the variability in scores is not due to random error, only 40% of the variation in scores could possibly be due to the stable trait the measure is supposed to measure. Put another way, because it has .40 reliability, it can't have more than .40 validity. To express the idea that a measure can't correlate with what it is supposed to measure (validity) more than it correlates with itself (reliability), experts often say, "reliability puts a ceiling on validity."

Because reliability puts a ceiling on validity, you would like to know your measure's reliability. If you are examining a previously published measure, somebody may have already calculated its test–retest reliability coefficient for you. To find that coefficient, check the article in which the measure was published. If the measure was not published in an article, check articles that used the measure or check the *Directory of Unpublished Experimental Mental Measures* (Goldman & Mitchell, 2007). If you are using a psychological test and you still can't find the test–retest reliability, check the test's manual or *Test Critiques* (Keyser & Sweetland, 2007).

When interpreting a measure's test–retest reliability coefficients, remember that these coefficients are telling you the extent to which the measure is *not* affected by random error. To find out the extent to which the measure *is* affected by random error, you have to subtract the reliability coefficient from 1. Like the ads that boast that their product is 70% fat-free rather than saying that 30% of the fat remains ($1.00 - .70 = .30 = 30\%$), and like the optimist who sees the glass as 70% full rather than as 30% empty, the test–retest reliability coefficient focuses on the positive by telling us how random error free the measure is.

However, if you are trying to avoid using a poor measure, you may need to focus on the negative. For example, although a test–retest coefficient of .70 does not sound bad, it means two things. First, it means that your validity can't possibly be above .70 (because reliability puts a ceiling on validity). Second, it means that 30% ($1.00 - .70 = .30 = 30\%$) of the differences between participants' scores on the measure are due to random error (see Figure 5.5).

Normally, you would not choose a measure in which more than 30% of the differences between participants' scores was due to random error. In other words, you would probably not choose a measure that had a test–retest reliability coefficient below .70.

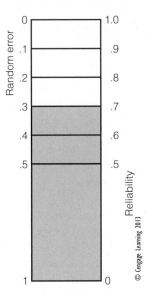

FIGURE **5.5** Reliability and random error are opposites.

In this example, the reliability coefficient (in blue) is .70, and random error (the white) makes up the remaining .30 of the variation in scores. How much variation in scores would be due to random error if the reliability coefficient was (a) 1.0? (b) 0? (c) .5? (d) .8?

Identifying (and Then Dealing with) the Main Source of a Measure's Reliability Problems

What if your measure's test–retest reliability is below .70? Then, more than 30% of the total variation in scores (i.e., more than 30% of the differences between participants' scores) is due to random error. Where is this random error coming from? The three likely sources are

1. random error due to the observer
2. random error due to the participant
3. random error due to the way the measure is administered

If you knew which of these possible sources was the main source of a measure's unreliability, you might be able to fix the measurement's reliability problem at the source. If the main reason your participants' scores were not consistent was because your observers were inconsistent, you would work on making your observers more consistent. On the other hand, if inconsistent scores were due to inconsistencies in how the measure was administered, you would work on administering the measure the same way every time. Unfortunately, test–retest reliability, because it is affected by all three sources of random error, doesn't let you pinpoint which of those three sources is the main source of your measure's unreliability. So how do you determine which of the three likely sources is most to blame for the measure's poor reliability?

Are Observers to Blame for Low Test–Retest Reliability? Assessing Observer Reliability

Most researchers start by seeing whether the observer is to blame for a measure's poor reliability. If the scoring system is relatively objective, researchers can immediately determine that the scorer is *not* a major source of random error. For example, in a multiple-choice test, observers are probably not going to make many errors.

But what about when observers rate behavior? Then, you should determine the extent to which random observer error is the cause of your measure's low test–retest reliability.

You can estimate the extent to which random observer error is a problem by having two or more observers independently (without talking to one another) rate the same behavior. If the observers are unaware of which condition the participant is in and their ratings agree, you don't have an observer related problem. If, however, the two trained raters score the same behavior differently, these differences may be due to random observer error. For instance, two observers may produce different scores because one or both observers guessed about which category to put the behavior in, misread the stopwatch, failed to pay attention, wrote down one number when they meant to write down another, or made any number of other random mistakes.

To determine how well your independent raters agreed, you need to calculate an index of the degree to which observers agree. Two of the most common indexes that researchers use are interobserver agreement and interobserver reliability.

Sometimes, researchers will report the **interobserver (judge) agreement**: the percentage of times the raters agree. For example, the researchers might report that the raters agreed 98% of the time. Interobserver agreement is simple to calculate and understand. If observers are agreeing 100% of the time, there is no random error due to the observer.

Rather than report the percentage of times the raters agree, most researchers will report some index of interobserver reliability. They may use Cohen's kappa or Krippendorf's alpha. However, most of the time, they use the simplest index of interobserver reliability: the **interobserver (scorer) reliability coefficient**. To obtain the interobserver reliability coefficient, researchers calculate a correlation coefficient between the different raters' judgments of the same behaviors and then square that correlation.[6]

Like test–retest reliability coefficients, interobserver reliability coefficients can range from 0 to 1. An interobserver reliability coefficient of 1.00 means there is a 100% correspondence between the raters. Knowing how one observer rated a behavior allows you to know perfectly (with 100% accuracy)

---

[6] To show you the connection between interobserver agreement and interobserver reliability, imagine that we are having observers rate whether a behavior falls into one of two categories. If observers were just flipping a coin to determine what category the behavior belonged, they would agree 50% of the time. Thus, if the observers' judgments are completely affected by chance, interobserver agreement would be 50% and the interobserver reliability coefficient would be zero. If, on the other hand, random error had no effect on judgments, interobserver agreement would be 100% and the interobserver reliability coefficient would be 1.00.

how the other observer rated the behavior. The following data reflect an interobserver reliability coefficient of 1.00.

Participant	Observer 1's rating	Observer 2's rating
Jorge	1	1
Nia	2	2
Dalia	3	3
Malik	4	4
Naval	5	5
Maria	6	6

© Cengage Learning 2013

The 1.00 interobserver reliability coefficient tells you the extent to which your measure is *not* affected by random observer error. Specifically, the 1.00 interobserver reliability coefficient shows that the measure is 100% (completely) free of random observer error.

An interobserver reliability coefficient of 0, on the other hand, indicates that the measure is not at all (0%) free from random observer error. That is, 100% of the differences between scores are due to random observer error. In such a case, there is no relationship (0) between the observers' ratings: Knowing how one observer rated a behavior tells you nothing about how the other observer rated the same behavior. To see a case in which observers' judgments are completely a function of random error, look at the data in the next table.

Participant	Observer 1's rating	Observer 2's rating
Jorge	1	3
Nia	2	5
Dalia	4	3
Malik	5	3
Naval	2	3
Maria	6	4

© Cengage Learning 2013

As you can see, there is no connection between Observer 1's ratings and Observer 2's ratings. Because scores are completely a function of random observer error, the measure has no interobserver reliability.

Because observers usually agree to some extent, and because journal editors will usually publish only articles that have a high degree of interobserver reliability, you will almost never see a published study that includes an interobserver reliability coefficient below .60.[7] Therefore, when reading a

---

[7] Indeed, observer reliability coefficients—or Cohen Kappa's—below .70 are usually considered unacceptable.

study, your question will not be, "Did the observers agree?" but rather, "To what extent did the observers agree?" Generally, you will expect interobserver reliability coefficients of around .90.

You want a measure with a high interobserver reliability coefficient for two reasons. First, you want your measure to be objective—you want trained raters to report the same scores. Second, *interobserver reliability puts a ceiling on overall (test–retest) reliability*. For example, if interobserver reliability is .60, test–retest reliability can't be above .60.

If interobserver reliability is low, you probably need to reduce random observer error. Sometimes, you can reduce random observer error by training or motivating your observers. Often, however, the most effective way to reduce—and prevent—random observer error is to simplify the observer's job.

To illustrate the benefits of simplifying the observer's job, consider Ickes's (2003) research on "everyday mind reading." One way he studies such mind reading is by having two strangers interact. He tapes the interaction and then has each participant view the tape. Participants are to stop the tape at different points and say what they are thinking. Then, participants see the tape again and are to stop it at different points and write down what their interaction partner was thinking at that point. Observers rate the degree to which the participant's guess about what the partner was thinking matches what the partner was actually thinking. Ickes could have had observers make their judgments on a 0 (*not at all*) to 100 (*completely*) scale. However, he had observers use a 3-point scale with "0" being "essentially different content," "1" being "similar, but not the same, content," and "2" being "essentially the same content." By using fewer categories, raters found the job of rating easier and raters were able to make reliable ratings.

Ickes knew the observers' ratings were agreeing with each other because he calculated interobserver reliabilities. He was able to calculate inter-rater reliabilities because he had more than one person rate each participant.

Being able to calculate inter-rater reliabilities is one benefit of having more than one observer rate each participant. A second benefit is that rather than each participant's score being based on a single observer's rating, each participant's score can be based on the average of two or more observers' ratings.

The advantage of average ratings is that average ratings are more reliable (more stable; less influenced by random observer error) than individual ratings. Average ratings are more reliable than individual ratings because observer error, like all random error, tends to average out to 0. That is, random errors made by one observer tend to be cancelled out by random errors made by another. Thus, even if Ickes had found that *individual* ratings had low interobserver reliability, random observer error would not have been a huge problem for his study because he was basing participants' scores on the *average* of five observers' ratings.

If you can't get multiple observers to rate each behavior and you can't get reliable ratings, you may need to find a way of measuring behavior that does not involve observers. One reason multiple-choice tests and rating scale measures are so popular in research is that these measures essentially eliminate human observers, thereby eliminating the random observer error that human observers produce.

Estimating Random Error Due to Participants

So far, we have discussed a situation in which interobserver reliability is low, thus dooming test–retest reliability to be low. But what if interobserver reliability is high, yet test–retest reliability is still low? For example, suppose you have the following data:

	Test		Retest	
Participant	Observer 1	Observer 2	Observer 1	Observer 2
Jordan	3	3	5	5
Asia	4	4	3	3
Deja	5	5	5	5

© Cengage Learning 2013

In such a case, you know your low test–retest reliability is not due to erratic observers but instead must be due to inconsistencies in (a) how the measure was administered and/or (b) the participant.

Ideally, you would figure out how much of the random error was due to poor standardization and how much was due to random changes in the participant. Unfortunately, it is impossible to directly assess how much random error is due to poor standardization. Fortunately, it is sometimes possible to get a rough idea of how much random error is due to the participant—if your measure is a multiple-choice test or some other objectively scored measure.

Internal Consistency: Test Questions Should Agree With Each Other. To estimate how much of the random error in participants' scores is due to the participant, you must make some assumptions. You must start by assuming that each question on the test is measuring the same thing. Thus, if you have a shyness test, you would assume that all the questions on that test are measuring shyness. If that assumption is correct, people who score "shy" on one question should score shy on all the other questions. In technical terminology, the test should be internally consistent: a participant's score on any one question should be a good predictor of how that participant will score on the other questions. That is, if the test questions are "judging" the same thing, they should "agree" with each other.

But what if the "scores" given by the different questions are not consistent? For example, what if a participant is shy according to the participant's answers to some questions but outgoing according to the participant's answers to other questions?

If the assumption that all the questions are measuring the same thing is correct, this inconsistency in how questions are answered is due to random error. Is this random error due to the participant? Not necessarily. After all, as you may recall, random error, ordinarily, could be due to changes in one or more of the following:

1. the observer/scorer
2. the testing environment
3. the participant

Random Error Due to Participants May Cause Low Internal Consistency. To know that this random error is due to the participant, you need to rule out two other possible sources of random error: (1) the observer/scorer and (2) the testing environment. If you use an objective, multiple-choice test, you can rule out the scorer as a source of random error.

If you look only at random error in participants' responses during a single testing session, you may be able to rule out the testing environment as a source of random error. You would argue that the testing environment did not change much during that testing session. For example, you would claim that the testing environment was about the same when participant answered question 2 as it was when the participant answered question 3. Therefore, if a participant answers question 2 differently than question 3, the testing environment is probably not to blame.

To recap, you must do four things before you can estimate the degree to which your measure is affected by random error due to the participant. First, you must assume that inconsistencies in a participant's scores on individual questions are due to random error (rather than to the questions measuring different things). Second, you must use objective measures so that you can safely assume that your measure's random error is probably not due to the scorer. Third, you must look only at random variations in participant responses that occur within a single testing session, so that you can reasonably assume that those variations are not due to variations in the testing environment. Fourth, you need to recognize that if the random error in responses is not due to the scorer, or the testing environment, it must be due to

1. ~~the observer~~
2. ~~the testing environment~~
3. the participant

Specifically, the measure's inconsistency may reflect (a) participants experiencing random, momentary variations in mood or concentration or (b) participants randomly guessing at the answers to some questions.

Two Solutions to Problems Caused by Random Participant Error

If your measure's reliability problems are due to the participant, what can you do? Your plan of attack will depend on whether you can make participants behave more consistently.

Add Questions to Let Random Participant Error Balance Out. If participants fluctuate considerably from moment to moment in how they think or feel, the best you can do is make sure your measure has many questions. By having participants provide many responses, you allow random fluctuations to balance out. Asking numerous questions should also help balance out the effects of guessing. For example, suppose you are unprepared for a physics quiz and the only thing you can do is guess at the answers. If the quiz is composed of one multiple-choice question, you might get 100% just by guessing. However, if the quiz is composed of 100 multiple-choice questions, random guessing is not going to get you a high score.

Ask Better Questions to Reduce Random Participant Error. Asking more questions is not the only way to deal with the problem of guessing. Sometimes, the solution is to ask better questions. Your participants may be guessing at the answers because some questions are so poorly worded that participants are guessing at what the questions mean. In a sense, participants are mentally flipping a coin to answer the question. As a result, their answers to such a question will be so inconsistent that such answers won't consistently correlate with their answers to anything—including their answers to the same question the next day and their answers to other questions on the test. If, however, you reword or eliminate questions that participants misinterpret, you should be left with questions that participants answer reliably. If the remaining questions are reliable and are all measuring the same variable, a participant's answers to one question should now consistently agree with that participant's answers to any of the other questions on the test. In other words, by reducing random error due to participants guessing what your questions mean, you should have boosted your measure's **internal consistency**: the degree to which answers to each question correlate with the overall test score; the degree to which the test agrees with itself.

Measuring Internal Consistency

But how would you know whether you have boosted your measure's internal consistency? How would you know whether your measure's internal consistency was poor in the first place? In other words, how do you tell how well your questions agree with each other?

To estimate your measure's internal consistency, you would find or calculate an index of internal consistency, such as an average inter-item correlation, a split-half reliability, or Cronbach's alpha (often abbreviated as alpha, Cronbach's α, or just α). All of these indexes measure the degree to which answers to one item (question) of the test correspond to answers given to other items on the test. Thus, the following data would produce a high score on any index of internal consistency:

Participant	Question 1	Question 2	Question 3
Miles	1	1	1
Theodora	3	3	3
Becky	5	5	5

© Cengage Learning 2013

Average Inter-Item Correlations as Indexes of Internal Consistency. One index that directly assesses the extent to which answers to one test item (question) correlate with answers to other test items is the average inter-item correlation. As the name suggests, this index involves computing a correlation between the answers to each question (item) and then averaging those correlation coefficients.

There is more than one kind of average. Thus, there is more than one average inter-item correlation. If you use the mean as your average (you add up all the correlation coefficients and divide by the number of correlation coefficients), you have the *mean inter-item correlation.* If your average is the median (you arrange the correlation coefficients from lowest to highest and pick the middle one), you have the *median inter-item correlation.*

Usually, there is little difference between the median inter-item correlation and the mean inter-item correlation. For example, if you had a three-item test, you might find the following:

Correlation of item 1 with item 2	.2
Correlation of item 1 with item 3	.3
Correlation of item 2 with item 3	.4
Mean inter-item correlation:	.3
Median inter-item correlation:	.3

Split-Half Reliability Coefficients as Indexes of Internal Consistency. Other indexes of internal consistency are less direct than the average inter-item correlation. Many rely on essentially splitting the test in half and comparing how participants scored on one half versus how they scored on the other half. For instance, researchers may (a) calculate each participant's score for the first half of the test, (b) calculate each participant's score for the second half of the test, and then (c) correlate scores on the first half of the test with scores on the last half of the test. This correlation between the score for the first half of the test and the score for the second half is a type of split-half reliability. Thus, the following data would yield a perfect (1.00) split-half reliability and suggest that the scale was internally consistent:

Participant	Score on first 10 questions	Score on last 10 questions
Lionel	50	50
Alexi	10	10
Lothar	30	30

© Cengage Learning 2013

Another type of split-half reliability involves splitting the test in half by comparing answers to the odd-numbered questions (1, 3, 5, etc.) with answers to the even-numbered questions (2, 4, 6, etc.). Specifically, researchers may calculate a score based only on the answers to the odd-numbered questions, a score based only on the answers to even-numbered questions, and then correlate each participant's "odds" score with that participant's "evens" score. That correlation would be the measure's "odd–even correlation."

Additional Indexes of Internal Consistency. In addition to the measures of internal consistency that we have described, there are more mathematically sophisticated measures of internal consistency, such as Cronbach's alpha and Kuder-Richardson reliabilities. At this point, we do not want you to know the advantages and disadvantages of each measure of internal consistency. Instead, we want you to realize whenever you see odd–even correlations, average inter-item correlations, Cronbach's alpha, split-half reliabilities, or Kuder-Richardson reliabilities, the researchers are just trying to tell you the extent to which their measure is internally consistent.

In general, you can treat these indexes of internal consistency as all being pretty much alike—except that the score suggesting good internal consistency is lower for the average inter-item correlations than for the other indexes of internal consistency. For the other indexes we have mentioned, you need a score of at least .70 (and preferably above .80) to say that the measure is internally consistent. Therefore, you would probably not use a measure with an odd–even correlation of .60 or a Cronbach's alpha of .50. However, the cutoff for acceptable internal consistency of the average (mean or median) inter-item correlation index is around .30. For example, most experts would say that a measure that has a median inter-item correlation of .35 has an adequate degree of internal consistency.

Conclusions About Internal Consistency's Relationship to Reliability

An adequate degree of internal consistency suggests that your measure's reliability problems are not due to its questions or to minute-to-minute fluctuations in your participants. Instead, your reliability problems, if you have any, are probably due to participants changing over time or to improper standardization.

Low internal consistency, on the other hand, suggests that there are problems with the questions on your test. These problems will tend to hurt your test's overall reliability, especially if your test is relatively short. Therefore, you may want to boost your test's internal consistency by eliminating or refining some of your test's less reliable questions.

Conclusions About Reliability

Up to now, this chapter has focused on reliability. We have shown you why reliability is important, how to determine if a measure has sufficient reliability, and how to determine where a measure's reliability is breaking down (for a review, see Table 5.3, Box 5.3, and Figure 5.6). Specifically, we have stressed that:

1. Reliability is a prerequisite for validity.
2. Test–retest reliability tells you the **total** extent to which random error from observers, from participants, *and* from testing situations is influencing your measure.

Low test–retest reliability tells you that the combined effects of random error (from all sources: observers + testing environment + participants) on your measure's scores is too large—and should encourage you to pinpoint the source of your measure's reliability problems. For example, by calculating the interobserver reliability coefficient, you could determine to what extent inconsistent observers are responsible for your measure's reliability problems.

TABLE **5.3** Reliability Indexes and the Type of Random Error They Detect

	Random error due to the observer	Random error due to random changes in participants	Random error due to random changes in the testing situation
Measures of observer reliability (Cohen's kappa, Krippendorf's alpha, interobserver reliability coefficient)	Yes	No	No
Measures of internal consistency (Cronbach's alpha, split-half reliability, Kuder-Richardson reliability, mean inter-item correlation, median inter-item correlations)	No	Only for changes that occur during a testing session	Only for changes that occur during a testing session
Test–retest reliability	Yes	Yes	Yes

© Cengage Learning 2013

BOX **5.3**   **Key Points to Remember About Reliability**

1. Two major *avoidable* sources of unreliability are
   a. random fluctuations in the measurement environment, and
   b. random fluctuations in how observers interpret and code observations.
2. All reliability coefficients are not the same.
3. Test–retest reliability tells you the total extent to which random error is influencing your measure.
4. Other types of reliability can help you find the source of the measure's unreliability. For example, a low interobserver reliability coefficient tells you that random observer error is seriously reducing your measure's overall reliability.
5. Reliability is necessary for construct validity: Valid measures are reliable.
6. Reliability does not guarantee construct validity: Reliable measures are not always valid.
7. Unreliability weakens a measure's validity but does not introduce systematic bias into the measure.
8. Reliability is an important, but not all-important, consideration in choosing a measure.

© Cengage Learning 2013

However, we have not emphasized the main limitation of reliability: *Reliability does not guarantee validity.* For example, consider the following data:

Participant	Score on physical aggression test	Score on retest	Number of fights
Paco	50	50	3
Lacrishia	60	60	4
Eldora	70	70	1

© Cengage Learning 2013

As you can see from the previous table, individuals score the same when retested as they did when they were first tested. Thus, the test has high

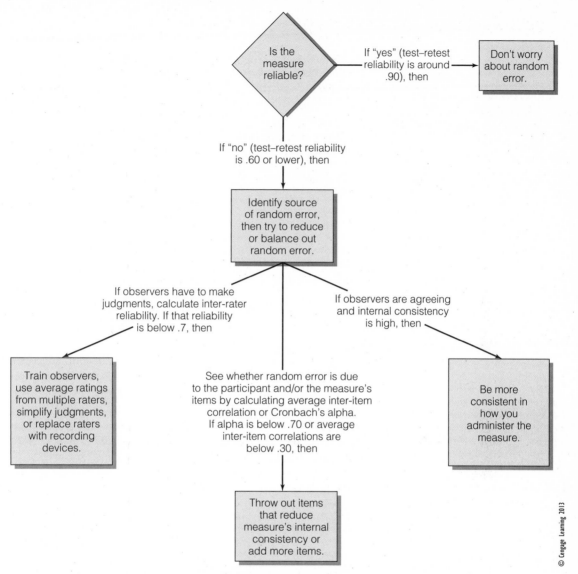

FIGURE **5.6** Determining whether—and how—to improve a measure's reliability.

test–retest reliability. However, individuals' scores on the physical aggression test do not correspond with how many fights they have. Thus, the test is not a valid measure of physical aggression. Consequently, the measure is reliable but not valid.[8]

---

[8] As one reviewer pointed out, a classic example of a measure that would be reliable but not valid is using head circumference as a measure of IQ.

BEYOND RELIABILITY: ESTABLISHING CONSTRUCT VALIDITY

One reason a reliable measure may not be valid is that it is reliably and consistently measuring the wrong thing. How can we know that a measure has **construct validity**: the degree to which the measure is measuring the construct or variable that it claims to measure?

We can never prove that our measure has construct validity. However, we can make a case for a measure's construct validity, especially if it has good

- content validity
- internal consistency
- convergent validity
- discriminant validity

Content Validity: Does Your Test Have the Right Stuff?

Often, the first step in devising a valid test is to establish its **content validity**: the extent to which it represents a balanced and adequate sampling of relevant dimensions, knowledge, and skills. Before writing any test questions, you should consult the established theories and definitions of the concept you wanted to measure. Once you know what you are measuring, use your definition to guide you in writing questions that measure all your concepts' dimensions. For example, if you define love as, "feeling sexual attraction toward a person and a willingness to make sacrifices for that person," you would make sure your measure had questions that measured both sexual attraction and willingness to sacrifice. In fact, to make sure that you had an adequate number of both types of questions, you might break your scale into two subscales: a "sexual attraction" subscale and a "sacrifices" subscale.

As you can see, there are two main points to content validity: (1) having *content* (questions) tapping every important aspect of your construct and (2) having enough questions to provide an adequate *sampling* of questions from each of those aspects. Consequently, content validity is sometimes called *content sampling*.

When evaluating classroom tests and other tests of knowledge or skills, content validity may be extremely important. For instance, a test to assess everything you have learned about psychology should not consist of only one multiple-choice question. Beyond having many questions, such a test should cover all areas of psychology, not just one. For example, if such a test consisted of 500 questions, all of which were about classical conditioning, it would not have content validity.

Internal Consistency Revisited: Evidence That You Are Measuring One Characteristic

Content validity is important. If you can't argue that your measure logically follows from an accepted definition of the concept, few would accept your measure as valid. However, you must do more than claim that you have sets of questions that measure each important aspect of your construct: You need objective, statistical evidence of that claim. For example, if you write some questions that you believe reflect the willingness to sacrifice for one's partner and call it a "personal sacrifices for partner scale," scientists are not going to accept your scale as valid without more evidence.

A first step toward making an evidence-based case that your measure really is measuring the right thing (your construct) is to show that all the questions making up your instrument are measuring the same thing. To do this, you have to show that your scale is internally consistent: that participants who score a certain way on one of the "personal sacrifices for partner" subscale questions score similarly on all the other "personal sacrifices for partner" questions. For example, if a participant's answer to one question suggests the participant is willing to sacrifice for his partner, the participant's answers to the other questions should echo that suggestion.

To understand the value of internal consistency, you may want to think of each question as a judge of whether participants have a certain characteristic. If one judge's scores are very different from everyone else's, you would doubt that judge's competence. You might ask, "Was she watching the same thing everybody else was?" Similarly, if answers to a certain question on the test do not correspond to answers to the other questions, you would have doubts about that question. For example, if the students who did well on a class test got question 13 wrong, whereas the students who did poorly on that test got question 13 right, there would be questions about question 13. If there were many questions like question 13 on the test, there would be questions about that test. Thus, as you can see, low internal consistency raises doubts about your test. Therefore, you want your measure to have high internal consistency.

As we mentioned earlier, there are several ways you can determine whether a measure has high internal consistency. If you found high inter-item correlations (e.g., above .35), split-half reliabilities (e.g., above .85), or Cronbach's alphas (e.g., above .85), you would be more confident that all the questions on the test were measuring the same construct. The following data would support the view that the questions agree with each other:

Participant	Question 1	Question 2	Question 3
Omar	1	1	1
Destiny	3	3	3
Osana	5	5	5

© Cengage Learning 2013

However, what if the data were like this?

Participant	Question 1	Question 2	Question 3
Omar	1	1	5
Destiny	3	3	1
Osana	5	5	3

© Cengage Learning 2013

Then, you would probably want to reword or eliminate question 3.

By eliminating and rewording questions, you should be able to achieve a reasonable level of internal consistency. Thus, if you were to take participants' scores on the first half of a 20-question test and correlate them with their scores on the last half of the test, you should find a high degree of agreement, like this:

Participant	Score on first 10 questions	Score on last 10 questions
Omar	50	50
Destiny	10	10
Osana	30	30

© Cengage Learning 2013

Because scores on the first half of the test are the same as scores on the second half, the test seems to be internally consistent. This internal consistency strongly suggest that the two halves of the test are measuring one thing. Thus, internal consistency seems like a good idea if you are measuring a simple construct that has only one dimension.

But what if you are measuring a complex construct that you think has two separate aspects? For instance, suppose that, consistent with our love example, you assumed that love had two different dimensions (being sexually attracted to one's partner and being willing to sacrifice for one's partner). Furthermore, assume that these dimensions are relatively independent: You believe that people who are above average in terms of being sexually attracted to their partner are not more likely be willing to sacrifice for their partner than people who are below average in terms of being sexually attracted to their partner. In such a case, you don't expect a high level of internal consistency for your scale as a whole because your scale is measuring two different things.

As suggested earlier, one solution would be to make up a love scale that had two different subscales. You would then expect that each of the individual subscales would have a high degree of internal consistency but that the subscales would not correlate highly with each other. That is, all the responses to questions related to sexual attraction should correlate with one another; all the responses to items related to sacrifice should correlate with one another; but the sexual attraction scale should not correlate highly with the sacrifice scale.[9] The following results would support the case that your two subscales were measuring two different constructs:

Split-half reliability for sacrifice subscale	.84
Split-half reliability for sexual attraction subscale	.89
Correlation between the sacrifice and sexual attraction subscales	.10

---

[9] Another commonly used strategy (discussed at the end of Appendix E) is to use factor analysis. In this case, you would want at least three outcomes from that factor analysis. First, you want the factor analysis to extract two independent factors (sacrifice and sexual attraction) that together accounted for much (over 60%) of the variability in scores. Second, you want items in the sacrifice subscale to correlate with (have factor loadings above .4 on) the factor you labeled sacrifice and not load on (not correlate with) the factor you labeled sexual attraction. Third, you want items in the sexual attraction subscale to load heavily on the factor you labeled sexual attraction and not load heavily on the factor you labeled sacrifice.

Internal consistency can help you build the case that you are measuring the number of constructs you claim to be measuring. The data just presented strongly suggest that our love test is measuring two things. However, they do not tell us what those two things are.

The data do not give us objective evidence that we are measuring "sacrifice" and "sexual attraction." Indeed, if one subscale was measuring intelligence and the other scale was measuring political attitudes, we might obtain those same data.

Convergent Validation Strategies: Statistical Evidence That You Are Measuring the Right Construct

To get evidence that your measure is measuring a certain construct, you need to do more than look at your measure. You need to show that high scorers on your measures differ from low scorers in ways other than that the high scorers answer questions on your test differently than the low scorers. You need to show that high scorers differ from low scorers in ways that relate to your construct. To be more precise, you need to obtain evidence for your measure's **convergent validity**: the extent to which your measure correlates with other indicators of the construct. The idea behind convergent validity is that your measure correlates with these other indicators of your construct because your measure and these other indicators are all converging on the same thing—your construct.

Although there are several ways to show that your measure correlates with what it should correlate with, perhaps the most obvious step in convergent validation is to show that your measure correlates with other measures of the same construct. Thus, if you were measuring love, you might correlate your measure with another measure of love, such as Rubin's Love Scale (Rubin, 1970). Because both measures are supposed to be measuring the same thing, the two measures should correlate highly with one another. That is, participants scoring high on your love measure should score high on the other love measure, and participants scoring low on your love measure should score low on the other love measure. Ideally, you would find a convergent validity correlation between your measure and the existing measure of .80 or higher, as in this example:

Participant	Established love measure	Your measure
Basil	100	100
Marisol	20	20
Jaivin	60	60

© Cengage Learning 2013

Another obvious convergent validity tactic is to find two *groups*: one *known* to possess a high degree of the characteristic you want to measure and one known to possess a low degree of that characteristic. You would hope that participants known to have a high level of the construct would score higher on your measure than participants known to have a low level of the construct. This tactic is called the **known-groups technique**. Thus, in validating your love scale, you might give your scale to two groups—one that is known to be in love (dating couples) and one that is known not to be in love (strangers). Scores on the love scale might approximate the following:

Strangers	Dating couples
55	90

© Cengage Learning 2013

In addition to seeing whether different existing groups score differently on your measure, you could see whether individual scores on your measure predict future group membership. For instance, you might see if your measure could predict which dating couples would get engaged and which would soon split up (like Rubin did).

You could also see if your measure distinguishes between two groups exposed to different experimental manipulations. For example, if you had an experimental group that was expecting a shock and a control group that was not, you would expect the experimental group to score higher on your measure of anxiety than the control group.

Finally, you could determine whether your measure correlated with nonverbal or physiological indicators of the concept. Thus, you might correlate scores on the love scale with a behavior that lovers tend to do, such as look into each other's eyes, and find that couples with low scores on the love scale were less likely to gaze at each other than couples with high scores (like Rubin did).

Discriminant Validation Strategies: Showing That You Are Not Measuring the Wrong Construct

Convergent validity builds the case for construct validity by showing that the measure correlates with what it should correlate with. This approach is useful, but when you only look for evidence that supports your measure's validity, you may fool yourself. For example, suppose you guessed people's weights—but your guesses were really based entirely on people's heights. If you correlated your "weights" with other measures of weight, you might obtain a respectable correlation (perhaps as high as .80). Thus, your measure of "weight" would have some convergent validity even though you were really measuring height. Similarly, an "intelligence test," despite correlating with some measures of intelligence, may really measure mathematical knowledge.

To see the limits of convergent validity, suppose you correlate your love scale with another love scale. Imagine you have the following data:

Person	Score on your scale	Score on other love scale
Basil	30	38
Marisol	52	49
Jaivin	70	60

© Cengage Learning 2013

You have some good evidence for convergent validity because people who score one way on your test score about the same way on an established love test. But are you really measuring love? Before you say "yes," let's look at a table that combines data from the previous table with data from a liking scale:

Person	Liking scale	Your scale	Other love scale
Basil	30	30	38
Marisol	52	52	49
Jaivin	70	70	60

© Cengage Learning 2013

As you can see from this new table, your scale seems to be measuring liking rather than love. Because you cannot show that your so-called love measure is different from a liking measure, you cannot say that the "love" measure is valid.

The moral of the previous two tables is that convergent validity alone is not enough to establish construct validity. It is not enough to show, as the first table did, that a measure correlates with measures of the right construct. We must also show that our measure has **discriminant validity**: showing that it is *not* measuring a different construct by showing that it does not correlate with what it should not correlate with. Typically, you show that your measure is not measuring the wrong thing—that is, you establish discriminant validity—by showing that your measure (a) does not correlate with measures of unrelated constructs and (b) does not correlate too highly with measures of related constructs.

Showing Discriminant Validity Relative to Unrelated Constructs

If you need to show discriminant validity relative to a construct that is unrelated, unassociated, and irrelevant to your construct, you would need a near zero (anywhere from −.20 to +.20) correlation between your measure and a measure of that unrelated construct. For example, you, like many test developers, may need to show that your measure is not affected by social desirability bias. That is, you want to avoid a big problem with "tests" that appear in magazines: The more a test taker gives the answer that will put him or her in a good light, the higher that person will score on the desirable "trait" (e.g., "being a good friend") that the magazine claims to measure. How could you show that your measure was immune to that bias?

The first step is to have participants complete your scale and a social desirability scale. Social desirability scales, sometimes called "lie scales," measure the degree to which the participant gives answers that—rather than being truthful—would impress most people. Basically, a social desirability scale consists of questions that ask you to describe yourself by choosing between two responses:

1. a socially desirable response ("Yes, I always help people out") that makes a good impression but is not honest
2. a less impressive response ("No, I do not always help people out") that is honest

People who lie by picking the socially correct responses will score high on the social desirability scale; people who pick the truthful, but less flattering, answers will score low in social desirability.

After administering both your measure and the social desirability scale, you would correlate the two measures. A correlation of around zero (between −.20 and +.20) would suggest that your measure is not strongly influenced by the social desirability bias. That is, a near-zero correlation between your scale and a social desirability scale suggests that you are measuring people's true feelings rather than people's willingness to make a good impression.

If, on the other hand, scores on the social desirability scale correlate highly with responses on your scale, your scale may be strongly affected by social desirability bias: Rather than measuring what people really think, you may just be measuring their willingness to make a good impression. For example, do you think the following data support the idea that the "love test" really measures being in love?

Person	Love score	Social desirability score
Basil	30	30
Marisol	52	52
Jaivin	70	70

© Cengage Learning 2013

These data don't support the idea that the so-called love test is measuring being in love. Instead, they suggest that the so-called love test is really a measure of social desirability. As you can imagine, showing that a new measure has a near-zero correlation with social desirability is often one of the first steps that researchers take in establishing discriminant validity.

Showing Discriminant Validity Relative to Related Constructs

In addition to showing that you are not measuring social desirability, you may also need to show that your love measure has discriminant validity relative to a variety of related—but different—constructs, such as liking, lust, loyalty, and trust. When trying to show discriminant validity relative to a related construct, you do not want a zero correlation between your measure and a measure of that related construct. After all, you hope that scores on your measure would be somewhat related to scores on a measure of that related construct. For instance, you might expect that your love scale would correlate moderately (.60) with the liking scale because people who love each other tend to like each other. However, if your love scale correlated very highly with the liking scale, you would be worried that you were measuring liking instead of love. You would especially be concerned if your scale correlated more highly with a liking scale than it did with other love scales.

Similarly, if you had a measure of practical intelligence, you would not be alarmed if it correlated moderately with a measure of conventional intelligence. However, if the correlation was around .80 (about as high as different IQ tests correlate with each other), you would have a hard time arguing that you had devised a test of a different type of intelligence. Instead, the evidence would suggest that you had devised yet another measure of conventional intelligence.

Conclusions About Discriminant Validity

In short, to make a convincing case that a measure really measures the construct it is supposed to measure (construct validity), you need to show that it really isn't measuring a construct that it shouldn't be measuring (discriminant validity). Therefore, when trying to decide whether to use a certain measure, you should ask two questions:

1. What other constructs might this measure be measuring?
2. Is there evidence that this measure does not measure those other constructs?

If you are trying to show that you are not measuring an unrelated construct, you need to show that there is a near-zero correlation between your measure and a measure of the unrelated construct. If, on the other hand, you are trying to show that you are not measuring a related construct, even a moderately high correlation of .60 might provide evidence of discriminant validity—as long as that was significantly lower than your measure's convergent validity correlations.[10]

Summary of Construct Validity

As you can see from Figure 5.7 and Table 5.4, building a strong case for your measure's construct validity involves several research projects. To assess convergent validity, you may need to consult theory and past research to see what manipulations affect your construct and then see whether those manipulations affect scores on your measure. In addition, you may need to consult theory and past research to see what behaviors correlate with high levels of the construct and then see whether people scoring high on your measure exhibit those behaviors. You will probably also need to correlate your measure with other measures of the same construct. To assess discriminant validity, you would have to correlate your measure with measures of other constructs. To assess internal consistency, you need to administer the measure and calculate some measure of internal consistency.

---

[10] Two common ways of showing that a measure is not measuring a related construct are (1) partial correlations and (2) tests of the differences between correlations. Partial correlations are correlations between two variables that try to control for the possibility that the correlation between those two variables is due to some third variable. For example, suppose a social intelligence test and a conventional intelligence test both correlated .60 with grades. In that case, you might wonder whether the social intelligence test was just another intelligence test. To find out, we could compute a partial correlation between social intelligence and grades that would control for IQ scores. If the partial correlation between social intelligence and grades was now zero, we would conclude that the social intelligence test did *not* have discriminant validity relative to conventional intelligence tests. If, on the other hand, the partial correlation was significantly different from zero, we would conclude that the social intelligence measure did have discriminant validity relative to conventional intelligence. The test between correlations might involve doing a statistical significance test to determine whether the correlation between your social intelligence test and another social intelligence test was significantly higher than the correlation between your social intelligence test and a conventional intelligence test.

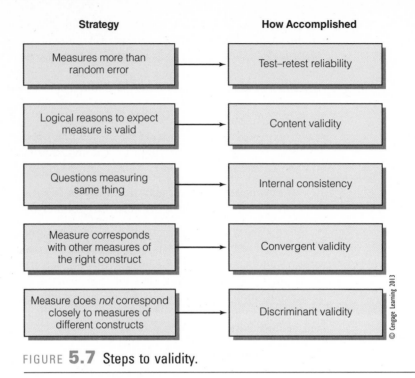

Strategy	How Accomplished
Measures more than random error	Test–retest reliability
Logical reasons to expect measure is valid	Content validity
Questions measuring same thing	Internal consistency
Measure corresponds with other measures of the right construct	Convergent validity
Measure does *not* correspond closely to measures of different constructs	Discriminant validity

© Cengage Learning 2013

FIGURE **5.7** Steps to validity.

TABLE **5.4** How Rubin Validated His Love Scale

Requirements for construct validity	How Rubin met the requirement
Reliability	Showed that the measure was not excessively affected by random error: test–retest reliability of .85.
Content validity	All three dimensions of love are represented (predisposition to help, dependency, and possessiveness).
Convergent validity	Predicts how much two individuals will gaze at each other. Predicts probability that individuals will eventually get married. People who are engaged score higher than people who are casually dating.
Discriminant validity	Love scores correlate only moderately with scores on a liking scale, suggesting that the scale is not a measure of liking.

© Cengage Learning 2013

Because validating a measure takes so much time and because most researchers are interested in finding out new things about a construct rather than finding new ways to measure it, most researchers do not invent their own measures. Instead, researchers use measures that others have already validated. (In fact, as you read about what it would take to validate your own love scale, you may have been saying to yourself, "Let's use Rubin's Love Scale instead.") To find such measures, go to the Chapter 5 section of this text's website.

MANIPULATING VARIABLES

We have devoted most of this chapter to measuring, rather than manipulating, variables for two reasons. First, all research involves measuring variables, whereas not all research involves manipulating variables. For example, if a researcher wants to know whether people who tend to exercise are happier than people who don't, researchers can simply measure both variables. Although a researcher who wanted to know whether exercise *causes* one to be happier, would have to *manipulate* how much participants exercise, that researcher would still have to measure happiness. Second, as you can see from both Table 5.5 and the next section of this chapter, most of the things you should think about when measuring variables are the same things you should think about when manipulating variables.

Common Threats to a Manipulation's Validity

When evaluating manipulations, you have the same four concerns as you have when you measure variables:

1. Can we reduce random error?
2. Can we reduce bias due to the researcher?
3. Can we reduce bias due to the participant?
4. Can we provide evidence that the operational definition we are using is valid?

TABLE **5.5** Similarities Between Measuring and Manipulating Variables

Measure	Manipulation
Reduce random error by standardizing administration of the measure.	Reduce random error by standardizing administration of the manipulation.
Reduce observer bias by training, standardization, instruments, and making researcher "blind" to the participant's condition.	Reduce researcher bias by training, standardization, instruments, and making the researcher "blind" to the participant's condition.
Participants may figure out what the measure is measuring and then act in such a way as to make a good impression or give the researcher the "right" results. Sometimes, the problem of subject biases is dealt with by not letting participants know what the measure is or what the hypothesis is.	Participants may figure out what the manipulation is designed to do and then act in such a way as to make a good impression or to give the researcher the "right" results. Sometimes, the problem of subject biases is dealt with by not letting participants know what the manipulation is or what the hypothesis is.
Show that your operational definition is consistent with the theory's definition of the construct.	Show that your operational definition is consistent with the theory's definition of the construct.
Convergent validity is shown by correlating the measure with other measures of the construct.	Convergent validity is sometimes demonstrated by showing that the manipulation has the same effect that other manipulations of the construct have and that it has an effect on the manipulation check.

1. Reducing Random Error

Just as you want to minimize random error when measuring variables, you want to minimize random error when manipulating variables. Therefore, just as you standardized the administration of your measure, you want to standardize the administration of your treatment. You want to administer the treatment the same way every time.

2. Reducing Experimenter (Researcher) Bias

Just as you were worried about researchers being biased when they score the measure, you will be worried about **experimenter bias**: experimenters being biased when they administer the treatment. For instance, researchers may be friendlier to the participants who are getting the treatment. As was the case with observer bias, you can reduce researcher bias by

- using scientific equipment to administer the manipulations
- using written or computerized instructions
- standardizing procedures
- making the researcher *blind* to what treatment the participant received

3. Reducing Subject (Participant) Biases

Just as you were concerned that your *measure* might tip off participants to how they should behave, you should also be concerned that your *manipulation* might tip off participants as to how they should behave. One of the most frequently cited examples of how a treatment could lead to demand characteristics was a series of studies begun in the 1920s at the Hawthorne Electric Plant. The investigators, Roethlisberger and Dickson (1939), were looking at the effects of lighting on productivity. At first, everything seemed to go as expected: Increasing illumination increased productivity. However, when they reduced illumination, productivity continued to increase. The researchers concluded that the treatment group was reacting to the special attention rather than to the treatment itself. This effect became known as the **Hawthorne effect**.

Although many experts now believe that Roethlisberger and Dickson's results were not due to the Hawthorne effect, no one disputes that participants may act differently simply because they think they are getting a treatment. Therefore, researchers use a wide variety of techniques to avoid the Hawthorne effect. Some of these techniques are similar to the techniques used to make a measure less vulnerable to subject biases. Just as researchers may reduce subject biases by measuring participants in nonresearch settings, experimenters may reduce subject biases by manipulating the treatment in a nonresearch setting.

A more common way of reducing subject biases is to give the "no-treatment" group a **placebo treatment**: a treatment that is known to have no effect. For example, in most studies examining the effect of a drug, some participants get the pill that contains the drug (the treatment), whereas others get a sugar pill (the placebo). If both groups improve equally, researchers would be concerned that the treatment group's improvement might be due to participants expecting to get better. If, however, the treatment group improves more than the placebo group (and neither the researcher nor the participant knew whether the pill the participant got contained the real drug), we know that the improvement was not due to participants' expectations.

4. Making the Case for a Manipulation's Construct Validity

As with measures, you would like to provide evidence that your operational definition is what you claim it is. The difference is that making a case for the validity of a treatment is usually less involved than making a case for the validity of a measure.

Indeed, making the case for a treatment's validity usually involves only two strategies: (1) arguing that your treatment is consistent with a theory's definition of the construct and (2) using a **manipulation check**: a question or set of questions designed to determine whether participants perceived the manipulation in the way that the researcher intended.

Consistency With Theory

To illustrate the value of these two ways of establishing construct validity, suppose that you wanted to manipulate cognitive dissonance: a state of arousal caused when participants are aware of having two inconsistent beliefs. To create dissonance, you ask a group of smokers to write an essay about why people shouldn't smoke. You would want to argue that your manipulation meets three general criteria that dissonance theory says must be met to induce dissonance:

1. Participants must believe they are voluntarily performing an action that is inconsistent with their attitudes (a smoker writing an essay about why people shouldn't smoke).
2. Participants should believe that the action is public and will have consequences (before writing the essay, participants must believe that others will not only read their essay but know that the participant wrote it).
3. Participants must *not* feel that they engaged in the behavior for a reward (that they are being bribed to write an essay that goes against their beliefs).

To make the case that the manipulation is consistent with dissonance theory, you might argue that

1. You told participants that their cooperation was voluntary and that they could refuse.
2. You told them that their essay would be signed and that children who were thinking about smoking would read it.
3. You did not pay participants for writing an antismoking essay.

Manipulation Checks

Your procedures would seem to induce the mental state of dissonance—assuming that participants perceived the manipulation as you intended. To check on that assumption, you might use a manipulation check. For example, you might ask participants whether they felt aroused, uncomfortable, coerced, that their attitudes and behavior were inconsistent, that their behavior was public, whether they foresaw the consequences of their behavior, and so on. Many researchers believe that you should always use a manipulation check when doing research on human participants (Sigall & Mills, 1998).

But what if the manipulation check tips off participants to the study's hypothesis? In that case, manipulation check advocates would say you have two options. First, use the manipulation check—but only after the participant has responded to your measure. Second, conduct a mini-study in which the only thing that happens is that participants receive the manipulation and then respond to the manipulation check.

But what if it's obvious that you are manipulating whatever you think you are manipulating (physical attractiveness, concrete versus abstract words, etc.)? Even then, manipulation check advocates would urge you to go ahead with a manipulation check for two important reasons. First, because you are doing research to test assumptions rather than to make assumptions, you should be willing to test the assumption that you are manipulating what you think you are manipulating.[11] Rather than relying on your opinion that "more attractive" photos were shown to one group and the "less attractive" photos were shown to a second group, ask each group to rate how attractive the photos were. Second, a manipulation check could establish the discriminant validity of your treatment. For example, wouldn't it be nice if you could show that your attractiveness manipulation increased perceptions of attractiveness but did not change perceptions of age or wealth?

Pros and Cons of Three Common Types of Manipulations

Choosing a manipulation usually involves making trade-offs because there is no such thing as the perfect manipulation. Different manipulations have different strengths and weaknesses. In the next sections, we will briefly highlight the strengths and weaknesses of three common ways of giving participants in different conditions different experiences: (1) giving them different instructions (instructional manipulations), (2) changing their physical environment (environmental manipulations), and (3) varying the behavior of their "co-participants"—research assistants pretending to be participants (stooge manipulations).

Instructional Manipulations

Perhaps the most common treatment manipulation is the **instructional manipulation**: manipulating the variable by giving written or oral instructions. One advantage of an instructional manipulation is that you can standardize

---

[11] In addition to using the manipulation check to test the hypothesis that your manipulation (A) had the predicted psychological effect (B), you could also use the manipulation check data to test the hypothesis that the psychological effect (B), in turn, changed (C) behavior (Sigall & Mills, 1998). Thus, without a manipulation check, you are limited to testing the hypothesis that A (the manipulation) → C (the dependent measure). With a manipulation check, on the other hand, you can test the A (the manipulation) → B (variable measured by the manipulation check) → C (the dependent measure) hypothesis.

your manipulation easily. Often, all you have to do is give each participant in the treatment condition one photocopied sheet of instructions and give each participant in the no-treatment condition another sheet of instructions. If you use a computer to compose the different sets of instructions, you can easily ensure that the instructions are identical except for the manipulation. If you have a cover sheet for the instructions, the person handing out the instructions can be blind to which set of instructions the participant receives. Because written instructions are easily standardized and because written instructions allow you to make the researcher blind, written instructions can reduce both random error and experimenter bias.

Unfortunately, instructional manipulations, although easy for you to administer, are easy for participants to misinterpret, ignore, or play along with. Therefore, just because you can consistently present instructions to participants, don't assume that your instructions will be perceived the same way every time.

To get participants to notice, remember, and understand your instructions, "hit participants over the head" with your manipulation by repeating and paraphrasing your most important instructions and, if necessary, by quizzing participants over those instructions. Thus, if you were manipulating anonymity, you would make a big deal of forbidding participants in the "private" condition from writing their name on any of the materials—and you would tell participants that their responses will be anonymous, confidential and private and that no one will know how they responded. Then, you might have them fill out a form in which they had to indicate whether their responses were anonymous or public. In the public condition, you would do just the opposite: You would make a big deal of making participants write their names on their paper, and you would tell them that many people would see their paper. Then, you might have them fill out a form in which they had to indicate whether their responses were anonymous or public.

By making sure that participants notice, remember, and understand the manipulation, you run the risk that they may understand your manipulation too well: They may figure out what you are trying to manipulate and then play along. Fortunately, you can reduce this threat to your study's construct validity by using placebo treatments, counterintuitive hypotheses (hypotheses that make a prediction that is opposite of what most people—and most participants—would predict), and clever ways of measuring your construct. (To see examples of clever measures, see Table 5.2.)

Environmental Manipulations

If you are concerned that participants will play along with an instructional manipulation, you might use an **environmental manipulation**: changing the participants' surroundings. Some environmental manipulations take the form of "accidents." For instance, smoke may fill a room, the participant may be induced to break something, or the participant may "overhear" some remark.

When considering an environmental manipulation, ask two questions. First, will participants notice the manipulation? Even when manipulations have involved dramatic changes in participants' environments (smoke filling a room), a sizable proportion of participants report not noticing the manipulation (Latané & Darley, 1970).

Second, can you present the manipulation the same way every time? Fortunately, many environmental manipulations can be presented in a consistent, standardized way. Most animal research, for example, involves environmental manipulations that can be consistently presented (food deprivation). Likewise, research in perception, sensory processing, cognition, and verbal learning usually involves environmental manipulations (presenting illusions or other stimuli). These manipulations vary from the routine—presentation of visual stimuli by computer, tachistoscope, memory drum, or automated slide projector—to the exotic. For example, Neisser (1984) has done studies in which the manipulation consists of silently moving the walls of the participant's cubicle.

"Stooge" (Confederate) Manipulations: Using Fake Participants to Engage Real Participants

A special kind of environmental manipulation employs **stooges** (**confederates**): people who pretend to be real participants but who are actually the researcher's assistants. By using stooges, social psychologists and others get participants to respond openly, thus avoiding the demand characteristics that accompany instructional manipulations. However, using stooges leads to two problems.

First, using stooges raises ethical questions because by deceiving your participants, you are violating the principle of informed consent. Your attempt to reduce demand characteristics is coming at the cost of participants' rights. The decision to try to deceive participants should be made only after careful consideration of the alternatives. Thus, for ethical reasons, you, your professor, or your school's ethics committee may decide that you shouldn't use stooges (for more on ethics, see Appendix D).

Second, it's hard to standardize the performance of a stooge. At best, inconsistent performances by stooges create unnecessary random error. At worst, stooges may bias the results. Some researchers solve the standardization problem by having participants listen to tapes of actors rather than relying on stooges to give exactly the same performance time after time. For example, both Aronson and Carlsmith (1968) and Latané and Darley (1970) made participants believe they were listening to people talking over an intercom when participants were actually listening to a tape recording.

The Type of Manipulation You Should Use Depends on Your Concerns

As you can see from Table 5.6, choosing manipulations usually means making trade-offs. To choose the right manipulation for your study, you must determine what your study needs most. Is experimenter bias your biggest concern? Then, you might use an instructional manipulation. Is subject bias your biggest concern? Then, you might use an environmental manipulation.

Conclusions About Manipulating Variables

As you have seen, when manipulating variables, you have many of the same concerns you have when measuring variables, such as random error, participant bias, and researcher bias. As a result, when manipulating variables, you want to use some of the same techniques you use when measuring variables, such as standardizing procedures and keeping both participants and researchers blind.

TABLE **5.6** Comparing the Advantages and Disadvantages of Three Different Kinds of Manipulations

Instructional	Environmental	Stooges
Easy to do.	Not as easy to do.	Hardest to do.
Easily standardized.	Not as easily standardized.	Hardest to standardize.
Reduces: 1. Random error. 2. Potential for experimenter biases.	May lead to concerns about: 1. Random error. 2. Potential for experimenter biases.	May lead to concerns about: 1. Random error. 2. Potential for experimenter biases.
Vulnerable to subject biases.	Less vulnerable to subject biases.	Least vulnerable to subject biases.

© Cengage Learning 2013

 ## CONCLUDING REMARKS

In Chapter 3, you developed a research idea: a prediction about how two or more variables were related. In this chapter, you learned how to determine whether you had valid operational definitions of those variables. Now that you have the raw materials to build a research design, you can take advantage of the rest of this book.

SUMMARY

1. *Reliability* refers to whether you are getting consistent, stable measurements. Reliable measures are relatively free of random error.

2. One way to measure the extent to which a measure is free of random error is to compute its test–retest reliability.

3. Three major sources of unreliability are random errors in scoring the behavior, random variations in how the measure is administered, and random fluctuations in the participant's performance.

4. You can assess the degree to which random errors due to the observer are affecting scores by computing an interobserver reliability coefficient. Interobserver reliability puts a ceiling on test–retest reliability.

5. For objective tests, you may get some idea about the degree to which scores are affected by random, moment-to-moment fluctuations in the participant's behavior by using an index of internal consistency. Popular indexes of internal consistency are Cronbach's alpha, split-half reliabilities, and average inter-item correlations.

6. Random error is different from bias. Bias is a more serious threat to validity. In a sense, random error dilutes validity, whereas bias poisons validity.

7. *Validity* of a measure refers to whether you are measuring what you claim you are measuring.

8. Reliability puts a ceiling on validity; therefore, an unreliable measure cannot be valid. However, reliability does not guarantee validity; therefore, a reliable measure may be invalid.

9. A valid measure must (a) have some degree of reliability and (b) be relatively free of both observer and subject biases.

10. Two common subject biases are social desirability (trying to make a good impression) and obeying the study's demand characteristics (trying to make the researcher look good by producing results that support the hypothesis).

11. By not letting participants know what you are measuring (unobtrusive measurement), you may be able to reduce subject biases (see Table 5.2).

12. Masking (also called blinding) can reduce bias due to demand characteristics—but it does not reduce social desirability bias. Making responses anonymous can reduce social desirability bias—but it does not reduce bias due to demand characteristics. Neither making participants blind nor making participants anonymous reduces random error.

13. Establishing internal consistency, discriminant validity, convergent validity, and content validity are all ways of building the case for a measure's construct validity.

14. With convergent validity, you are trying to show that you are measuring the right construct. Therefore, you show that your measure correlates with what it should correlate with—other measures or indicators of the construct you are trying to measure.

15. With discriminant validity, you are trying to show that you aren't measuring the wrong construct. Therefore, you show that your measure does not correlate with what it should not correlate with—measures of different constructs.

16. Choosing a manipulation involves many of the same steps as choosing a measure.

17. Placebo treatments and unobtrusive measurement can reduce subject bias.

18. "Blind" (masked) procedures and standardization can reduce experimenter bias.

19. You can use manipulation checks to make a case for your manipulation's validity.

KEY TERMS

bias (p. 147)
blind masked (p. 152)
construct validity (p. 176)
content validity (p. 176)
convergent validity (p. 179)
demand characteristics (p. 156)
discriminant validity (p. 181)
environmental manipulation
   (p. 189)
experimenter bias (p. 187)
Hawthorne effect (p. 186)

informed consent (p. 157)
instructional manipulation
   (p. 188)
internal consistency (p. 171)
interobserver (judge) agreement
   (p. 166)
interobserver (scorer) reliability
   coefficient (p. 166)
known-groups technique (p. 179)
manipulation check (p. 187)
observer bias (p. 150)

operational definitions (p. 144)
placebo treatment (p. 186)
random error (p. 147)
reliable, reliability (p. 161)
social desirability bias (p. 158)
standardization (p. 154)
stooges (confederates) (p. 190)
subject bias (p. 155)
test–retest reliability (p. 162)
unobtrusive measurement
   (p. 157)

EXERCISES

1. Why is bias considered more serious than random error?

2. What are the two primary types of subject bias? What are the differences between these two types?

3. Suppose a "social intelligence" test in a popular magazine had high internal consistency. What would that mean? Why would you still want to see whether the test had discriminant validity? How would you do a study to determine whether the test had discriminant validity?

4. Given that IQ tests are not perfectly reliable, why would it be irresponsible to tell someone his or her score on an IQ test?

5. What is content validity? How does it differ from internal consistency? For what measures is it most important?

6. Swann and Rentfrow (2001) wanted to develop a test "that measures the extent to which people respond to others quickly and effusively." In their view, high scorers would tend to blurt out their thoughts to others immediately and low scorers would be slow to respond.

   a. How would you use the known-groups technique to get evidence of your measure's construct validity?

b. What measures would you correlate with your scale to make the case for your measure's discriminant validity? Why? In what range would the correlation coefficients between those measures and your measure have to be to provide evidence of discriminant validity? Why?

c. To provide evidence of convergent validity, you could correlate scores on your measure with a behavior typical of people who blurt out their thoughts. What behavior would you choose? Why?

7. A researcher wants to measure "aggressive tendencies" and is trying to decide between a paper-and-pencil test of aggression and observing actual aggression.

a. What problems might there be with observing aggressive behavior?

b. What would probably be the most serious threat to the validity of a paper-and-pencil test of aggression? What information about the test would suggest that the test is a good instrument?

8. Think of a construct you would like to measure.

a. Name that construct.

b. Define that construct.

c. Locate a published measure of that construct (if you are having trouble finding a published example, see Web Appendix B), and write down the reference for that source.

d. Develop a measure of that construct.

e. What could you do to improve or evaluate your measure's reliability?

f. If you had a year to try to validate your measure, how would you go about it? (Hint: Refer to the different kinds of validities discussed in this chapter.)

g. How vulnerable is your measure to subject and observer bias? Why? Can you change your measure to make it more resistant to these threats?

9. What problems do you see with measuring "athletic ability" as 40-yard-dash speed? What steps would you take to improve this measure? (Hint: Think about solving the problems of bias and random error.)

10. Think of a factor that you would like to manipulate.

a. Define this factor as specifically as you can.

b. Find one example of this factor being manipulated in a published study (if you are having trouble finding a published example, see Web Appendix B). Write down the reference citation for that source.

c. Would you use an environmental or instructional manipulation? Why?

d. How would you manipulate that factor? Why?

e. How could you perform a manipulation check on the factor you want to manipulate? Would it be useful to perform a manipulation check? Why or why not?

 WEB RESOURCES

1. Go to the Chapter 5 section of the book's student website and

a. Look over the concept map of the key terms.

b. Test yourself on the key terms.

c. Take the Chapter 5 Practice Quiz.

d. Do the interactive end-of-chapter exercises.

2. If you are ready to draft a method section, click on the "Method Section Tips" link.

CHAPTER 6

Beyond Reliability and Validity:

The Best Measure for Your Study

When possible, make the decisions now, even if action is in the future. A reviewed decision usually is better than one reached at the last moment.
—William B. Given, Jr.

It is the mark of an educated mind to rest satisfied with the degree of precision which the nature of the subject permits—and not to seek exactness where only an approximation is possible.
—Aristotle

CHAPTER OVERVIEW

Some people buy the most highly rated software available, only to find out that it doesn't work well for them. The program may be incompatible with other software they have, may be too difficult to use, or may lack a certain feature they want. As a result, they have a decent program, but one that doesn't work well for what they want to do. Similarly, when selecting a measure, sometimes people choose the most valid measuring instrument available, yet find out that it doesn't work well for what they want to do.

At first, choosing the most valid measure seems perfectly sensible. After all, who wouldn't want to make sure that their instrument is the best at measuring what it is supposed to measure?

After giving it some thought, however, you probably realize that most important decisions involve weighing more than one factor. Every measure, like every computer program, has weaknesses. The key is to find the measure whose weaknesses are least likely to get in the way of what you want to do.

To choose the measure whose weaknesses won't stop you from doing what you want to do, you need to know not only the measure's weaknesses but also your particular study's strengths and weaknesses. For example, imagine that you find three measures that, although they have similar overall levels of validity, are vulnerable to different threats to validity. The first measure is vulnerable only to biased observers. The second measure is a rating scale measure vulnerable only to subject biases. The third measure's only weakness is unreliability. Which measure should you choose?

The answer depends on how the measure's strengths and weaknesses complement your study's strengths and weaknesses. For example, suppose your design made it impossible for observers to bias the results in favor of the treatment group (because you didn't let observers know whether the participant they were rating was in the treatment group or in the no-treatment group). Because your design eliminates observer bias, you would choose the measure that is vulnerable only to observer bias.

If you had a hypothesis that is easy to figure out, you would avoid the measure that is vulnerable to subject bias. Suppose, for example, you hypothesized that participants would like a product more after seeing an ad for the product. Unfortunately, even if the ad was ineffective, participants might still rate the product higher after seeing the ad because they thought you wanted the ad to be effective. In this case, the combination of a hypothesis that is easy to guess and a rating scale measure that is easy to fake would be deadly to your study's construct validity. Therefore, you would not use the rating scale measure. Suppose, however, you had a hypothesis that participants probably wouldn't figure out. For example, suppose your hypothesis was that the ad would actually decrease liking for the product. In that case, you might choose the rating scale measure because its vulnerability to subject biases would not hurt you.

Finally, if your design did not eliminate either subject or observer bias, you would choose the measure that was unbiased but unreliable. Admittedly, the measure's low reliability guarantees low validity—and its unreliability will make it hard for you to find that a treatment has a statistically reliable effect. However, if, despite the fog created by the measure's random error, you were able to find a significant difference between the treatment and no-treatment groups, you could be confident that the difference wasn't due to measurement bias.

As you can see, even if validity were your only concern, you would not always choose the measure that, in general, was most valid. Instead, you would choose the measure that would be most valid for your particular study. But validity should never be your only concern. Instead, validity may need to take a backseat to four other concerns.

First, you must put practical concerns above construct validity. If you can't afford the measure or aren't qualified to administer it, you can't use it. Second, you should place ethical concerns above validity. If the most valid measure would humiliate or endanger participants, you shouldn't use it. Third, you may need to be more concerned about a measure's ability to tell you how and to what degree participants differ than about its validity. For example, if the most valid measure can tell you only that the two groups differ but you need to know which group is best, or how much better one group is than another, or how many times better one group is than another, you shouldn't use that measure. Fourth, you may need to be more concerned about the measure's ability to find small differences than about its validity. For example, if the most valid measure is too insensitive to detect your treatment's small effect, you shouldn't use that measure.

In short, choosing the right measure involves weighing many factors. In this chapter, we will show you how to shop for—and, if necessary, alter—a measure so that you have the right measure for your study.

SENSITIVITY: WILL THE MEASURE BE ABLE TO DETECT THE SMALL DIFFERENCES YOU MAY NEED TO DETECT?

To understand how an insensitive measure could prevent you from answering your research question, imagine a cell biologist's reaction to being told that she could use only a magnifying glass to distinguish the DNA of cancer cells from the DNA of normal cells. Obviously, she would be surprised. Without a microscope, the cell biologist could not detect small differences between cells.

Like cell biologists, psychologists often look for subtle but important differences. Consequently, like cell biologists, psychologists usually want their measure to be **sensitive**: to have the ability to detect differences among participants on a given variable. To illustrate the need for sensitive measures, imagine that you are doing an experiment to see whether we can increase participants' empathy for others. Realistically, you realize that one short treatment is probably not going to have enormous effects on a trait that has been shaped by heredity and a lifetime of experiences. If you have been able to make even a small improvement in this characteristic, you want to know about it.

Achieving the Necessary Level of Sensitivity: Three Tips

How can you find or develop a sensitive measure? Often, you can evaluate or improve a measure's sensitivity by evaluating it on the same three characteristics you would use to evaluate the sensitivity of a system for comparing people's weights: validity, reliability, and ability to provide a wide variety of scores.

First, you would want your measurements to be valid. For example, you wouldn't weigh people with their shoes and coats on because the difference in your recorded weights might be due to differences in the weight of their clothes rather than to differences in their body weights. Similarly, if you were interested in differences in body fat, you would want to use a much more direct and valid measure of body fat than body weight.

Second, you would want your measurements to be reliable. If the scale can't give the same person the same weight from one minute to the next, you know the scale is not reliable and not valid. You also know that such a scale is not sensitive because if it is not accurately measuring people's weights, it cannot be accurately measuring differences between people's weights. Put another way, if measurement error is creating large, random differences between people's measured weights, those errors may prevent researchers from detecting small, real differences between people's weights that do exist.

Third, you want your measurements to be able to range from low to high and to include small amounts in between. For example, you would want the scale to go high enough that you could distinguish people weighing 400 pounds from those weighing 500 pounds, and you would want the marks on the scale to be close enough together that you could tell the difference between someone weighing 150.0 pounds and someone weighing 150.5 pounds. Consequently, if you found a measure in which participants must receive either a score of "1" or "2," you would know that the measure could not be sensitive to subtle differences among participants.

You now have a general idea of what makes a measure sensitive: validity, reliability, and ability to provide a variety of scores. In the next sections, you will learn why these three qualities are important and how you can increase your measure's sensitivity by increasing the degree to which your measure has these three qualities.

Look for High Validity

The desire for sensitivity is a major reason that researchers often insist on having the most valid measure available. Even though they have several valid measures to choose from, they want the most valid one because it will tend to be the most sensitive.

Why does the most valid measure tend to be the most sensitive? To answer this question, keep two related facts in mind. First, the most valid measure is the one in which scores are least affected by factors that are irrelevant to what you are trying to measure. Second, the less a measure's scores are affected by irrelevant factors, the more it will be sensitive to changes in the relevant factor. For example, if you were weighing people to determine whether a diet had an effect, you would be more likely to find the diet's effect if participants were weighed unclothed rather than clothed. Although both the clothed and unclothed measures would be valid, the unclothed measure, because it is not assessing the weight of the clothes, is more valid and more sensitive to actual weight changes.

For similar reasons, measures that involve fewer inferences tend to be both more valid and more sensitive. Consequently, scientists would prefer to measure a person's weight on a scale rather than by the depth of the impression the person's footprint left in the sand. Likewise, they would prefer to measure body fat using calipers rather than by estimating it from overall body weight.

You now know that valid measures tend to be more sensitive and that valid measures are usually more pure and more direct than less valid measures. To make your measure more pure, more direct, more valid, and more sensitive, take two steps.

First, spend the time to figure out precisely what it is you want to measure. When you initially come up with a research idea, you may have a general sense of what you want to measure. For example, you may start out thinking that you want to measure attraction. Upon reflection, however, you may decide that you really want to measure lust. One way of helping you focus on what you want to measure is to look up the term you think you want to measure in the *Psychological Thesaurus* (see Web Appendix B).

The *Thesaurus* will help you clarify what you want to measure by alerting you to more specific terms, as well as to related terms.

Second, ask if there is a more direct way of measuring your construct. For instance, if you are interested in measuring aggression in football, do not simply measure how many penalties a team gets. Instead, measure how many penalties they get for unsportsmanlike conduct. Similarly, rather than assuming that fear will lead children to sit closer to each other and then measuring fear by how closely children sit to each other, take the more direct approach of asking children how afraid they are. By thinking about the simplest, most direct way to measure what you want to measure, you can often reduce the extent to which your measure is affected by things you don't want to measure.

Look for High Reliability

One thing you don't want to measure is random error. When you are measuring random error, you aren't measuring the quality you wish to measure. The less reliable the measure, the more scores on that measure are being influenced by random error. Therefore, all other things being equal, **the less reliable the measure, the less sensitive it will be** at detecting different amounts of the quality you wish to measure.[1]

To illustrate that unreliability hurts sensitivity, imagine that you have two scales: a reliable one and an unreliable one. Suppose that you go for a while without gaining weight and then you gain 2 pounds one weekend. If you always weighed yourself on the reliable scale, the scale would probably register the same weight day after day until that weekend. Consequently, you would probably notice your weight gain. If, on the other hand, you weighed yourself every day on the highly unreliable scale, the scale would make you think your weight was bouncing around even when it wasn't. In that case, if you were to gain 2 pounds one weekend, it would be hard to determine whether you had really gained 2 pounds for two reasons.

First, the scale wouldn't consistently register you as 2 pounds heavier. Indeed, if the unreliability of the scale made it fluctuate by 8 pounds, the scale, rather than indicating that you had gained 2 pounds, might indicate that you had lost 6 pounds.

Second, if the scale did register 2 pounds heavier, you wouldn't know whether that was due to you gaining 2 pounds or whether it was due to random fluctuations you were used to seeing. Thus, instead of realizing that you'd gained weight, you might think that the different readings on the scale were due entirely to random error.

As this example suggests, unreliability in your data is like static interfering with your ability to hear a radio news bulletin. With a little static, you

---

[1] The less reliable the measure, the more it is affected by random error—provided that you are measuring a stable characteristic. If you are measuring intelligence—and if intelligence is stable—any unreliability in your measure reflects random error. If, however, you are measuring something that changes (knowledge about research methods), unreliability might not reflect random error.

can still hear the program. But as the static increases, you will find it increasingly difficult to make out what is being said. Similarly, with a lot of random error in your measurements, it becomes hard to pick up the news your data are sending you. Even when the signal isn't completely drowned out, you cannot be sure you really heard it. That is, even if you notice a large difference between your groups, you may not know whether those large differences are due to random measurement error or whether they represent actual differences.

To visualize how random error makes it hard to see the message in your data, imagine you were measuring the time it took two different rats to run a maze. Suppose that Rat A and Rat B ran the maze four times each. Below are their actual times.

	Trial 1	Trial 2	Trial 3	Trial 4
Rat A	6 seconds	6 seconds	6 seconds	6 seconds
Rat B	5 seconds	5 seconds	5 seconds	5 seconds

© Cengage Learning 2013

If you had used a perfectly reliable and valid measure, you could have clearly seen that Rat B was the faster rat. However, suppose your measuring system was unreliable. For example, suppose you were having some problems with the stopwatch or you weren't always paying close attention. Then, you might record the rats' times as follows:

	Trial 1	Trial 2	Trial 3	Trial 4
Rat A	7 seconds	6 seconds	5 seconds	6 seconds
Rat B	8 seconds	4 seconds	6 seconds	2 seconds

© Cengage Learning 2013

Despite the random error in your measurements, you correctly calculated that Rat A averages 6 seconds to run the maze and that Rat B averages 5 seconds to run the maze. Thus, random error does not bias your observations. However, because of the unreliable, erratic nature of your measuring system, it is hard to determine whether Rat B really is the faster rat. The unreliability of the measuring system causes static that makes it harder to get a clear picture of the message your data should be sending you.

You have seen that too much random error in your measuring system can prevent you from detecting true differences between participants. In other words, all other things being equal, the more reliable the measure is, the more sensitive it is. Therefore, if you want to have a sensitive measure, you should probably choose a measure that has a high (above .80) test–retest reliability coefficient.

Find Measures That Provide a Variety of Scores

Thus far, we have discussed cases in which you could increase sensitivity by increasing both reliability and validity. The more scores on the measure are affected by the characteristic you want to measure (rather than by bias, a

FIGURE **6.1** An insensitive but reliable measure.

Having only two scale points (light and heavy) makes this scale insensitive. Adding scale points (e.g., marks for every pound or every half a kilogram) would make the scale more sensitive.

different trait, or random error), the more the measure is likely to be sensitive to differences between individuals on that characteristic. A reliable and valid measure might still be insensitive, though, because—like a scale that will measure you only to the nearest 100 pounds—it fails to allow participants who differ slightly on a trait to receive different scores (see Figure 6.1).

If a measure is to be sensitive to subtle differences between participants, participants who differ on the characteristic must get different scores. Thus, if you measured participants who varied widely on the characteristic, you should get a wide range of scores. Some participants should get extremely low scores, and others should get extremely high scores. Few participants should get the same score.

What could prevent a valid measure from producing the wide variety of scores necessary to reflect the full extent of the variation among your participants? What should you do to increase a measure's sensitivity? What should you avoid doing? To answer these questions, let's imagine that you are trying to detect small changes in how much a man loves a woman. What could prevent you from detecting changes in the man's love?

Avoid Behaviors That Are Resistant to Change. One reason you might be unable to detect small changes in love is that you chose to measure a behavior that is resistant to change. As a result, when the man's love changed, his behavior did not change along with his love. Thus, you should not choose to measure a behavior that is resistant to change. But what behaviors are resistant to change?

Important behaviors, such as proposing marriage or buying a car, and well-ingrained habits, such as smoking or cursing, are resistant to change. Such behaviors are especially insensitive to subtle changes in a construct. For example, suppose your measure of love was whether a man asked a woman to marry him. Because a man would ask a woman to marry him only after his love had reached an extremely high level, this measure would be insensitive to many subtle changes. It would not be able to detect a man's love changing from a near-zero level of love to a moderate level.

So, if you are interested in sensitivity, stay away from measures that cannot detect low levels of a construct. Don't use death as a measure of stress, tile erosion in front of a painting as a measure of the painting's popularity, quitting smoking as a measure of willpower, or any other measure that stays at zero until a high level of the variable is present. Instead, if sensitivity is a big concern, base the measure of your construct on a behavior that will change as quickly and easily as participants change on that construct.

Avoid "All or Nothing" Measures.   A second thing that could prevent you from distinguishing between the subtly different levels of the man's love is if your measure did not represent all these different levels. Consequently, a second reason that a marriage proposal is an insensitive measure is that there are only two scores the man could receive (asked or didn't ask). You are trying to distinguish between numerous subtly differing degrees of love, but you are letting your participant respond in only two different ways. If a measure is going to discriminate between many different degrees of love, participants must be able to give many different responses.

Ask "How ___" Rather Than "Whether."   One way to allow participants to get a variety of scores is to avoid asking whether the participant did the behavior. Instead, ask *how much* of the behavior the person did, *how quickly* the participant did the behavior, or *how intensely* the person did the behavior. Thus, if you are measuring generosity, don't just record whether someone gave to charity. Instead, record how much she gave or how long you had to talk to her before she was willing to give. Similarly, if you are using maze running to measure motivation, don't simply record whether the rat ran the maze. Instead, record how fast the rat ran the maze.

Asking "how much?" instead of whether is an especially good tactic when your original measure involved asking people about themselves or others. For example, rather than measuring love by asking the question: "Are you in love? (1—no, 2—yes)," ask, "How much in love are you? (1—not at all, 2—slightly, 3—moderately, 4—extremely)." Similarly, rather than having an observer judge whether a child was aggressive, you could have the observer rate how aggressive the child was.

Add Scale Points.   If your measure already asks how much, you may still be able to improve its sensitivity by having it ask *precisely* how much. That is, just as adding 1/8-inch marks to a yardstick makes the yardstick more useful for detecting subtle differences in the lengths of boards, adding scale points to your measure may increase its sensitivity.

Using scientific equipment may help you add scale points to your measure. For instance, with the proper instruments, you can measure reaction time to the nearest thousandth of a second. Similarly, by using a sound meter to measure how loudly a person is speaking, you can specify exactly how many decibels the person produced.

Adding scale points to a rating scale measure is simple. You can change a 3-point scale to a 5-point scale, a 5-point scale to a 7-point scale, or a 7-point scale to a 100-point scale.

There comes a point, however, where adding scale points to a measure will not enhance sensitivity. Asking people to report their weight to the nearest thousandth of a pound or asking them to report their love on a 1,000-point scale will probably not boost sensitivity. After a certain point, any apparent gains in accuracy are wiped out by the fact that responses are unreliable guesses. In addition, such questions may cause participants to be frustrated or to doubt your competence. Therefore, to boost sensitivity without frustrating your participants, you should not add scale points beyond a certain point. But what is that point?

According to conventional wisdom, that point could be after you reach 3 points or after you reach 11 points, depending on the kind of question you are asking. If you are asking about something that your participants think about a lot, you might be able to use an 11-point scale. If, however, you are asking about an issue that your participants are relatively ignorant of (or uninterested in), you may be fine with a 3-point scale. When in doubt, use either a 5- or 7-point scale.

Pilot Test Your Measure. If you have followed our advice, you now have a measure that potentially provides a range of scores. But, just because there are many possible scores a participant *could* get on your measure, that does not mean there are many different scores that your participants *will* get.

To determine whether scores will actually vary, *pilot test* your measure: Try out your study and your measure on a few participants before conducting a full-blown study. If you do a pilot test, you will often find that participants' scores on the measure do not vary as much as you expected.

If you do not conduct a pilot test, you will discover the problem with your measure only after you have completed the study. For example, one investigator performed an experiment to see whether participants who read a story while also watching the story on video (the reading plus video group) would remember more about the story than participants who only read the story (the reading-only group).

To measure memory, she asked the children 24 questions about the story. She thought participants' scores might range from almost 0 (none correct) to 24 (all correct). Unfortunately, the questions were so hard that all of the children got all of the questions wrong. Put another way, the measure's *floor*— the lowest score participants could get—was too high. Because of the high floor, all of the children got the same score (0), even though the children probably did differ in terms of how well they knew the story. Consequently, the investigator didn't know whether the video had no effect or whether the measure's high floor prevented her from detecting the effect.

The previous example points out why you might pilot test a measure that you devised. But should you pilot test a published measure? To answer this question, suppose our student researcher had, instead of devising her own measure, found a published measure that appeared to be sensitive and that consisted of 24 questions about a story.

If her participants were less skilled readers than the participants in the published study, all her participants might still have scored near the bottom (the floor) of the measure. In that case, if participants' memories really were worse in the videotape condition (because the videotape distracted participants), this decrease in memory wouldn't be detected because the videotape group couldn't score lower than the no-videotape group. In technical terminology, if adding the videotape had a negative effect on memory, this harmful effect would probably be hidden by a **floor effect**: the effect of a treatment or combination of treatments being underestimated because the measure is not sensitive to values below a certain level. In such a case, if the investigator had pilot tested the measure, she would have known that she needed to either abandon the measure or to modify it by making the questions easier.

What if her participants had been better readers than the participants in the published study? In that case, pilot testing would still have been useful. Although she would not need pilot testing to avoid floor effects, she might need pilot testing to avoid the opposite problem: All of her participants might have scored close to 24—the measure's highest score, its *ceiling*.

To see why there are problems when most of the participants score near or at the ceiling, suppose that all the participants in the reading-only group are scoring at the ceiling. Even if the reading plus videotape group remembered the story better than the other group, the reading plus videotape group can't show it on this measure because the reading plus videotape group can't get better than the perfect score the other group is getting. In technical terminology, the reading plus videotape group's superiority would have been hidden by a **ceiling effect**: the effects of the treatment or combination of treatments being underestimated because the measure places too low a ceiling on what the highest response can be (for more on floor and ceiling effects, see Figure 6.2).

If the researcher had pilot tested the measure, she would have known that using it would lead to ceiling effects that would hide any effects the treatment might have. Therefore, she would have modified the measure by making the questions more difficult, or she would have used a different measure.

If you pilot test a measure and find that participants' scores vary widely on the pilot test, you probably will not have to worry about your measure's sensitivity being ruined by floor effects, ceiling effects, or some other factor that restricts the range of scores. However, you may still need to worry about your measure's sensitivity being ruined by random error. Even if you are using a measure that was highly reliable in one study, that measure may not be so reliable when you administer it to your participants (Wilkinson & the Task Force on Statistical Inference, 1999). Therefore, as part of pilot testing the measure, you may wish to collect some data to determine how reliable the measure is with your participants.

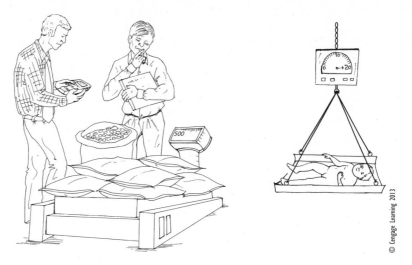

FIGURE **6.2** Measures may lead to ceiling or floor effects.

If we were looking at the effects of exercise on how much people weigh, the scale on the left would lead to a floor effect (because everyone under 500 pounds would get the lowest reading), whereas the one on the right would lead to ceiling effects (because everyone would weigh more than the scale's highest reading of 20 pounds).

Conclusions About Sensitivity

You have seen that if a measure is to be sensitive to differences between participants, two things must happen. First, different participants must get different scores. Second, different participants must get different scores because they differ on what the measure is supposed to be measuring. If participants are getting different scores due to random error, the measure will not be sensitive. For example, if you have people respond, on a 100-point scale, to a question they don't understand, you will get a wide range of scores, but your measure will be insensitive. In general, to the extent that participants' scores vary because of factors unrelated to your construct, the measure is not sensitive.

To boost sensitivity, you should minimize the extent to which participants' scores vary because of factors unrelated to your construct. Thus, if you use simple, direct, anonymous behaviors as measures (such as responses to self-rating scale questions like "How much do you like your partner?"), you may be more likely to detect differences between participants than if you observe complex, public behaviors (such as time spent with a partner) that are influenced by many factors other than your construct.

As you can imagine, the goal of sensitivity sometimes conflicts with the goal of validity. For example, to avoid subject biases, you might want to use a complex, public behavior (sacrificing for one's partner) as your measure of love. However, to have a sensitive measure, you might want to use a simple rating scale. Do you choose the complex behavior that might be insensitive?

Or, do you use the rating scale, even though it would be invalid if participants simply give you the ratings they think you want?

In certain situations, some researchers would choose the more sensitive rating scale. To understand why, realize that a sensitive measure can help you find small differences so that you can make discoveries. An insensitive measure, on the other hand, may stop you from making discoveries. Consequently, some scientists might select a more sensitive measure and worry about construct validity only after they have found differences. They would prefer debating what a difference meant to not finding any differences at all.

In short, even though validity is important, it is not the only factor to consider when selecting a measure. Depending on the circumstances, having the ability to detect subtle differences may be equally important. After all, an insensitive measure may—by preventing you from finding anything—prevent you from being able to answer your research question.

SCALES OF MEASUREMENT: WILL THE MEASURE ALLOW YOU TO MAKE THE KINDS OF COMPARISONS YOU NEED TO MAKE?

Whereas an insensitive measure may prevent you from answering your research question because it fails to detect that there is a difference between conditions, other measures—no matter how sensitive—may prevent you from answering your research question because they fail to detect *what kind of* difference there is between your conditions. To be more specific, some measures won't allow you to make the kind of comparison you need to make to answer your research question. To see that different research questions make different kinds of comparisons, consider these four questions:

1. Do the two groups *differ* on the quality?
2. Does one group have *more* of the quality than the other?
3. *How much more* of the quality does one group have than the other group?
4. *How many times more* of a quality does one group have than the other group?

All four of these questions could be answered by using numbers—and all measures can provide numbers. However, very few measures could help you answer the fourth question. Why?

The short answer is that different measures produce different kinds of numbers. Before you can understand how different measures produce different kinds of numbers, you first need to understand that, although all measures assign numbers to observations, *not all those numbers are alike.* Just as some descriptive phrases are more informative and specific than others ("Dion doesn't look the same as Elon" versus "Dion is twice as attractive as Elon"), some numbers are more informative than others.

The Four Different Scales (Levels) of Measurement

Rather than saying that some numbers provide more specific information than other numbers, researchers say that some numbers represent a *higher scale (level) of measurement* than others. To be more specific, social scientists have identified four different kinds of numbers: nominal, ordinal, interval, and ratio. In the next few sections, we will show you

1. how numbers representing different scales of measurement differ,
2. why some measures provide more informative numbers than others, and
3. how to determine what kind of numbers you need.

Nominal Scale Numbers: When 3 Is Different From But Maybe Not More Than 2

The least informative numbers do not have many of the qualities that you associate with numbers. With these numbers, bigger numbers don't represent bigger amounts, and higher scorers don't necessarily have more of a quality than lower scorers. Consequently, you shouldn't try to add, subtract, multiply, or divide using these numbers. However, these numbers do have one quality in common with numbers you are familiar with: individuals assigned one number (e.g., the 1s) are different from individuals assigned a different number (e.g., the 2s).

These low level numbers are called **nominal scale numbers (data)**: numbers that do not represent amounts but instead represent *kinds*. With nominal numbers, different numbers represent different qualities, types, or categories.

Nominal numbers substitute for names. Like names, nominal numbers can be used to identify, label, and categorize things. Things having the same number are alike (they belong in the same category); things having different numbers are different (they belong to different categories).

Like names, nominal numbers cannot be ordered from lowest to highest in a way that makes sense. Just as we do not say that Xavier is a bigger name than Sofia, we do not say that someone having the uniform number 36 is necessarily better than someone wearing number 35.

In everyday life, we often see these name-like, order-less, nominal numbers. For example, social security numbers, student ID numbers, charge card numbers, license plate numbers, and serial code numbers are all nominal numbers.

In psychological research, the best use of nominal numbers is when the participants can be clearly classified as either having a certain quality (e.g., married) or not. In those cases, we can use numbers to substitute for category names. For example, we may put people into categories such as male/female or student/faculty. Note, however, that the number we give to category names is completely arbitrary: We could code male as 1, female as 2; male as 2, female as 1; male as 0, female as 5,000. (If men are coded as 2, then, to paraphrase Shakespeare, a "2" by another nominal number would smell just as sweet.) This most basic way of using numbers is ideal for when you aren't interested in measuring different *amounts*, but are interested in different *kinds* or *types*. For instance, if you were measuring types of love, someone scoring a "1" might think of love as an addiction, a person scoring a "2" might think of love as a business partnership, a person scoring a "3" might think of love as a game, and a person scoring a "4" might think of love as "lust" (Sternberg, 1994).

Unfortunately, sometimes we would like to measure different amounts of a construct, but we can't because our measuring system is too primitive. Consequently, we can get only nominal numbers. That is, in the early stages of developing a measure, we may have such a poor idea of what scores on the measure mean that we can't even say that a high score means we have more of a construct than a low score. Suppose that when participants see their partner, some participants produce one pattern of brain waves, whereas others produce a different pattern. Labeling the first brain wave pattern "1" and the other pattern "2" is arbitrary. We could have just as easily labeled the first pattern "2" and the other pattern "1." Consequently, we have nominal scale measurement because we do not know whether "2" indicates a greater reaction than "1." We only know that "2" is a different pattern than "1."

Once we find out that one pattern indicates more love than another, it would be meaningful to give that pattern the higher number. At that point, we would have moved beyond nominal scale measurement.

Ordinal Scale Numbers: When 3 Is Bigger Than 2 But We Don't Know How Much Bigger

As you shall see, we often want to move beyond nominal scale measurement. Rather than always being limited to saying only that participants getting different numbers differ, we often want to say that participants receiving higher scores have more of a given quality. Rather than being limited to saying only that people scoring "3" are similar to each other and are different from people scoring "1," we also want to say that people scoring "3" have *more* of a certain quality than those scoring "1." In other words, we may want to be able to meaningfully *order* scores from lowest to highest, with higher scores indicating more of the quality. For example, we would often like to say that people scoring a "5" feel more love than people scoring "4," who feel more love than people scoring a "3," and so on.

If you can assume that higher numbers indicate more love than lower numbers, your measure is producing at least **ordinal scale numbers (data)**: numbers that can be meaningfully *ordered,* in terms of the characteristic they represent, from lowest to highest. When you assume that you have ordinal data, you are making a very simple assumption: The numbers have a meaningful order.

One way to get ordinal numbers is to *rank* participants. For example, if you record runners' times according to what place (first, second, third, etc.) they finished, you have an ordinal measure of time. Runners getting the higher numbers (e.g., "3") took more time to finish the race than runners getting lower numbers (e.g., "1"). Note, however, that you are not assuming that the difference in time between the runner who finished first and the runner who finished second is the same as the difference between the runner who finished second and the runner who finished third. For example, there might be a tenth of a second difference between the first and second runners but a 2-second difference between the second and third runners.

To illustrate what ordinal scaling does and does not assume, suppose you successfully ranked 10 couples in terms of how much they loved each other. Because the numbers can be ordered meaningfully from highest to lowest,

© Cengage Learning 2013

FIGURE **6.3** An ordinal scale.

This ordinal scale lets us know which weighed more—but not how much more.

these are definitely ordinal data. Yet, because they are only ordinal data, the actual difference between the couple scoring a "1" and the couple getting a "2"—on the characteristic being measured—may be very different from the actual difference between the couple getting a "9" and the couple getting a "10." For example, there might be little difference between how much in love the couple getting rank 1 and the couple getting rank 2 are, but there might be an enormous difference between how much in love the couple getting rank 9 and the couple getting rank 10 are. In short, ranked data, like all ordinal data, can tell you whether one participant has more of a quality than another but are limited in that they can't tell you how much more of a quality one participant has than another (see Figure 6.3).

Interval Scale Numbers: When We Know How Much Bigger 3 is Than 2 But Not How Many Times Bigger

Because of the limitations of ordinal numbers, you may decide you want a higher scale of measurement. For example, you may want numbers that will let you know how much of a quality an individual has or how much more of a quality one group has than another group. You want to make the same kind of statements about your measurements as you make about temperature: With temperature, you can say that the difference between 10 degrees and 20 degrees is the same as the difference between 30 degrees and 40 degrees. You can make these statements because you can assume that (a) the numbers follow an order (higher numbers always mean hotter temperatures) and (b) the distance (the *interval*) between any two consecutive numbers (e.g., 10 degrees and 11 degrees; 102 and 103 degrees) is, in terms of temperature, the same.

Similarly, if you are going to talk about how much of a psychological quality an individual or group has, you need to assume that (a) the numbers follow an order (higher scores mean more of the characteristic) and (b) the difference (the *interval*), between any two consecutive numbers (e.g., between 1 and 2 and between 2 and 3) is, in terms of the characteristic being measured, exactly the same. In technical terminology, you must be able to assume that you have **interval scale numbers (data)**: numbers can be ordered from lowest to highest in terms of the characteristic being measured and for which equal differences in scores represent equal differences in the characteristic being measured.

Although you may want to assume that you have an interval scale measure, be aware that you are making a big assumption. You are assuming that you have equal intervals—that a 1-unit difference (*interval*) in scores (e.g., between a 1 and a 2, between a 4.5 and a 5.5, and between a 6 and a 7) always represents that same (*equal*) amount of the characteristic being measured.

Some measures clearly do not allow you to make the assumption that you have equal intervals. For example, ranked data are typically assumed to be only ordinal—not interval.

No matter what measure you use, the assumption of equal intervals is difficult to defend. Thus, even if you use a measure of nonverbal behavior, you could still fail to meet the assumption of equal intervals.

Suppose, for example, that during the 10 minutes you had couples wait in a small room, you recorded the total amount of time the couple stared into each other's eyes. It would be risky to assume that the difference in the *amount of love* between a couple who looks for 360 seconds and a couple who looks for 300 seconds is the same as the difference between a couple who looks for a total of 60 seconds and a couple who does not look at all.

Likewise, if you use a physiological measure, it is hard to justify the assumption that equal changes in bodily responses correspond to equal changes in psychological states. It seems unlikely that changes in the body correspond perfectly and directly to changes in the mind. For example, if, on seeing their partners, one participant's blood pressure increases from 200 to 210 and another's goes from 90 to 100, would you say that both were equally in love?

How could you get interval scale data? One possibility is to ask participants to do the scaling for you. That is, ask participants to rate their feelings on a scale, trusting that participants will view the distances between each scale point as equal psychological distances.

Although many psychologists assume that rating scales produce interval scale data, this assumption of equal intervals is controversial. To see why this assumption is hard to justify, suppose you had people rate how they felt about their spouse on a −30 (hate intensely) to a +30 (love intensely) scale. Would you be sure that someone who changed from −1 to +1 had changed to the same degree as someone who had changed from +12 to +14?

Ratio Scale Numbers: Zeroing in on Perfection So That 4 is 2 × 2

If you are extremely demanding, it may not be enough for you to assume that your measure's numbers can be meaningfully ordered from lowest to highest and that equal intervals between numbers represent equal psychological distances. You may want to make one last, additional assumption—that your measure has an absolute zero. In other words, you might assume that someone scoring a zero on your measure feels absolutely no love. If a score of zero on your love measure represented absolutely no love, and you had equal intervals, then you could make ratio statements such as: "The couple who scored a '1' on the love measure was 1/2 (a ratio of 1 to 2) as much in love as the couple scoring a '2.'" In technical terminology, you can make ratio statements because your measure provides **ratio scale numbers (data)**: numbers that have both (a) an absolute zero and (b) equal intervals.

Meeting one key assumption of ratio scale measurement—the assumption of having an absolute zero—is not easy. Indeed, even when measuring physical reality, you may not have an absolute zero. To illustrate, 0 degrees Fahrenheit doesn't mean the absence of (zero) temperature. If 0 degrees Fahrenheit meant no temperature, we could make ratio statements such as saying that 50 degrees is half as hot as 100 degrees. Similarly, if you were timing a runner using a handheld stopwatch, even if the runner somehow took zero seconds to cross the finish line, you would time the runner at about .1 seconds because it takes you time (a) to react to the runner crossing the finish line and (b) to push the stop button. Thus, your measure of time would provide interval rather than ratio scale data. To get near ratio scale data, you would need to use automated electronic timers.

Note that if our measure of zero is off by even a little bit, it may prevent us from making accurate ratio statements. To illustrate, suppose you have a 2-pound weight and a 1-pound weight. If your scale's zero point is off by one quarter of a pound, instead of seeing a 2:1 ratio between the weights, you will "find" a 2.25 to 1.25 (1.8 to 1) ratio.

Even if the scale's zero point is perfectly accurate, any inaccuracy in our measurements may prevent us from making accurate ratio statements. Thus, if we weigh the 1-pound weight with perfect accuracy but weigh the 2-pound weight as 1.9 pounds, our statement about the ratio between the weights would be inaccurate. In other words, to make perfectly accurate ratio statements, we need perfectly accurate measurements.

Although meeting the assumptions of ratio scale measurement is difficult when measuring physical reality, it is even more difficult to meet those requirements when measuring a psychological characteristic. It is difficult to say that a zero score means a complete absence of a psychological characteristic or that the numbers generated by a measure correspond *perfectly* to psychological reality. It's tough enough to have some degree of correspondence between scores on a measure and psychological reality, much less to achieve perfection.

Because of the difficulty of achieving ratio scale measurements, most researchers do not ask participants to try to make ratio scale judgments. They usually do not ask participants to think of "zero" as the absence of the quality. Indeed, they often do not even let participants have a zero point. Instead, participants are more likely to be asked to make their ratings on a 1-to-5 scale than on a 0-to-4 scale. Furthermore, even when participants rate on a 0-to-4 scale, they are rarely asked to think of "2" as having twice as much of the quality as "1," "3" as three times "1," and "4" as four times as much as "1."

Occasionally, however, in a process called *magnitude estimation*, participants are asked to make ratio scale judgments. For example, participants might be told that the average amount of liking that people feel for a room-mate is a "50." If they feel one-fifth as much liking toward their roommate as that, they should estimate the magnitude of their liking toward their room-mate as 10. If they like their best friend twice as much as they think most people like their roommates, they should estimate the magnitude of their lik-ing for their best friend as 100. Participants would then be asked to rate the magnitude of their liking for a variety of people.

Magnitude estimation doesn't always involve using numbers. Instead, participants might draw lines. For example, they may be shown a line and told to imagine that the length of that line represents the average extent to which people like their roommates. Then, they may be asked to draw lines of different lengths to express how much they like various people. If a partici-pant likes Person A three times as much as Person B, the participant's line representing his or her liking for Person A should be three times longer than the line for Person B.

Advocates of magnitude estimation believe that the numbers or line lengths that participants produce provide ratio scale data. However, as you might imagine, critics have doubts. They point out that, "Just because you asked participant to make ratio scale judgments, that doesn't mean that parti-cipants could or did accurately make such judgments."

Why Our Numbers Do Not Always Measure Up

You can see why participants' subjective ratings on scales and on estimates of magnitude might not provide ratio scale numbers. But why don't you get ratio scale numbers from your behavioral measures of love? For example, why isn't time staring into each other's eyes a ratio scale measure of love?

If you were interested only in gazing behavior, time spent gazing would be a ratio scale measure: Zero would be the complete absence of gazing, and 3 seconds of gazing would be three times as much gazing as 1 second. However, suppose you were not interested in gazing for gazing's sake. Instead, you were interested in love. You were using gazing behavior (an observable behavior) as an indicator of love (an unobservable psychological state). As an indirect, imperfect reflection of love, time of gaze does not allow you to estimate amount of love experienced with ratio scale precision (see Box 6.1). Similarly, although we can all agree that a heart rate of 100 beats per minute is twice as fast as a heart rate of 50 beats per minute, we can't all agree that a person with a heart rate of 100 beats per minute is twice as excited as a person with a heart rate of 50 beats per minute.

Likewise, we can all agree that a person who donates $2.00 to our cause has given twice as much as a person who gives us $1.00. We have a ratio scale mea-sure of how much money has been given. If, however, we are using dollars given as a measure of a construct such as "generosity," "kindness," "empathy," or "gullibility," we do not have a ratio scale measure. The person who gives $2.00 is not necessarily twice as kind as the person who gives $1.00.

To reiterate, you cannot measure generosity, excitement, love, or any other construct directly. You can measure constructs only indirectly, capturing their reflections in behavior. It is unlikely that your indirect

BOX **6.1**   Numbers and the Toll Ticket

The toll ticket shows us many kinds of numbers in action. For example, the numbers representing vehicle class (1–4) at the top of the ticket (under toll by vehicle class) are nominal numbers. The only reason the toll people used numbers instead of names is that numbers take up less room. So, instead of writing "car," "16-wheeled truck," "small truck," and so on, they wrote 1, 2, 3, and 4. There's no particular order to these numbers as shown by the fact that a "3" is charged more than any other number.

On this toll ticket, the exit numbers refer to the order in which the exits appear. Thus, a "1" on the toll ticket refers to the first exit and "7" refers to the seventh exit. The exits, when used as an index of distance, represent ordinal data. You know that if you have to get off at exit 5, you will have to go farther than if you get off at exit 4, but—without looking at the miles column—you don't know how much farther. When you do check the miles column, you realize that missing exit 4 isn't too bad— the next exit is only 4 miles away. Missing exit 6, on the other hand, is terrible—the next exit is 66 miles farther down the road!

Money, as a measure of miles, is also an ordinal measure. Although you know that the more money you spend on tolls, the farther you have gone, you can't figure out how much farther you have gone merely by looking at how much money the toll was. For example, if you are vehicle class number 1, it costs you 25 cents to go 3 miles, 15 cents more to go 7 additional miles, and only 10 more cents gets you 30 additional miles.

As you have seen, both the amount of money spent and the number of exits passed are just ordinal measures when they are used to try to estimate the amount of another variable (distance). Similarly, some behavioral and physiological measures (eye-gazing or blood pressure increases) may merely be ordinal measures when used to estimate the amount of another variable, such as the invisible psychological state of love.

Toll (in Dollars) by Vehicle Class

Exit	No. of Miles	Vehicle Class			
		1	**2**	**3**	**4**
1	3	0.25	0.35	0.60	0.35
2	10	0.40	0.45	1.00	0.60
3	40	0.50	0.60	1.35	0.80
4	45	0.80	0.90	2.15	1.30
5	49	0.90	1.10	2.65	1.55
6	51	1.45	1.65	3.65	2.15
7	117	3.60	4.15	9.95	5.85

© Cengage Learning 2013

measure of a construct will reflect that construct with the perfect accuracy that ratio scale measurement requires.

Which Level of Measurement Do You Need?

You have seen that there are four different levels of measurement: nominal scale, ordinal scale, interval scale, and ratio scale. As you go up the scale from nominal to ordinal to interval to ratio scale measurement, the numbers become increasingly more informative (for a review, see Table 6.1 and Figure 6.4).

TABLE **6.1** The Meaning and Limitations of Different Scales of Measurement

Scale	What Different Scores Represent	What We Can Say	What We Can't Say
Nominal	Different scores indicate *different* amounts, kinds, or types.	People scoring a "3" experience a different kind (or amount) of love than people scoring a "1."	Because there is no order to nominal numbers, we can't say that "3" indicates *more* love than "1."
Ordinal	Different scores indicate different amounts *and* *higher scores represent greater amounts of the measured variable.*	People scoring a "3" are more in love than people scoring a "1."	Because the distances between numbers do not correspond to psychological reality, we can't say *how much* more of a quality one participant has than another.
Interval	Different scores indicate different amounts and higher scores represent greater amounts of the measured variable. *and* equal distances between numbers represent equal psychological differences.	We can say *how much* more love one participant feels than another. For example, people scoring "3" are more in love than people scoring "1" to the same extent that people scoring "5" are more in love than people scoring "3."	Because we do not have an absolute zero, we cannot say *how many more times* in love one participant is than another.
Ratio	Different scores indicate different amounts and higher scores represent greater amounts of the measured variable. and equal distances between numbers represent equal psychological differences. *and* *zero means a complete absence of the measured variable.*	The mathematical ratio between two scores perfectly corresponds to reality. People scoring "3" are three *times as much* in love as people scoring "1."	None.

You have also seen that as you go up the scale, it becomes harder to find a measure that provides the required level of measurement. For instance, if you need ordinal data, you can use almost any measuring system—from ranked data to magnitude estimation. However, if you need ratio scale data, magnitude estimation might be your only option; you cannot use a measure that involves ranking participants from lowest to highest—no matter how valid that ranking system is. Thus, if you need ratio scale measurement, the scale of measurement you need, rather than validity, will determine what measure you should use.

(a) Ratio Scale Ruler: Absolute zero and equal intervals. We have perfect measurement and we can make ratio statements such as, "The object we measured as '4' is two times as long as the object we measured as '2.'"

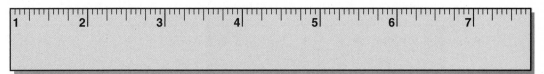

(b) Interval Scale Ruler: No absolute zero but equal intervals. Because we do not have an absolute zero, we shouldn't say, "The object we measured as '4' is twice as long as the object we measured as '2.'" Because we have equal intervals, we can say, "The difference between the object we measured as '7' and the object we measured as '6' is the same as the difference between the object we measured as '4' and the object we measured as '3.'"

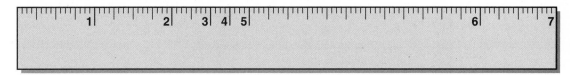

(c) Ordinal Scale Ruler: Order but not equal intervals. Because we do not have equal intervals, we shouldn't say, "The difference between the object we measured as '7' and the object we measured as '6' is the same as the difference between the object we measured as '4' and the object we measured as '3.'" Because we do have order, we can say, "The object we measured as '6' is longer than the object we measured as '5.'"

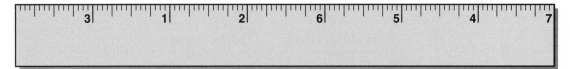

(d) Nominal Scale Ruler: No order—higher numbers (e.g., "3") represent a different length than lower numbers (e.g., "1") but not more length. Because we do not have order, we can't say, "The object we measured as '6' is longer than the object we measured as '5.'" However, we can say, "The object we measured as '6' has a different length than the object we measured as '5.'".

FIGURE **6.4** Different rulers, different scales of measurement.

The scale of measurement you need to test your hypothesis should always influence what measure you use. Therefore, when choosing a measure for a study, you should ask two questions:

1. What scale of measurement do I need to answer the research question?
2. Which of the measures that I am considering will give me this level of measurement?

The next sections and Tables 6.2 and 6.3 will help you answer these two key questions.

TABLE **6.2** Different Research Questions Require Different Levels of Measurement

Research Question	Scale of Measurement Required
Can more members of Group A be categorized as _____ (in love, neurotic, etc.) than members of Group B?	At least nominal
Is Group A more _____ than Group B?	At least ordinal
Did Group A change more than Group B?	At least interval
Is the difference between Group 1 and Group 2 more than the difference between Group 3 and Group 4?	At least interval
Is Group A three times more _____ than Group B?	Ratio

© Cengage Learning 2013

TABLE **6.3** Measuring Instruments and the Kind of Data They Produce

Scale of Measurement	Measuring Tactics Assumed to Produce Those Kinds of Numbers
Ratio	Magnitude estimation
Interval	Rating scales (Magnitude estimation)
Ordinal	Ranks (e.g., first, second, etc.)
	Nonverbal measures
	Physiological measures
	(Rating scales)
	(Magnitude estimation)
Nominal	Any valid measure
	(All of the above)

Note: Any measurement technique that provides data that meet a certain level of measurement also provides data that meet the less stringent requirements of lower levels of measurement. Thus, a measure that provides data that meet the requirements of interval scale measurement also provides data that meet the requirements of ordinal and nominal scale measurement.

© Cengage Learning 2013

When You Need Ratio Scale Data

Suppose you want to find out whether engaged couples are twice as much in love as dating couples who are not engaged. Because you are hypothesizing a 2-to-1 ratio, you need a measure that gives you ratio scale numbers. Similarly, suppose you had love scores from the following three groups:

Didn't go to counseling at all	3.0
Went to counseling for 1 week	6.0
Went to counseling for 8 weeks	8.0
(The higher the score, the more in love. Scores could range from 1 to 9.)	

© Cengage Learning 2013

If you wanted to say that people who went to counseling for 1 week were twice as much in love as those who did not go to counseling, you would need ratio scale data. Unfortunately, as Table 6.3 indicates, there are few measures that you can use if you need ratio scale numbers. If you want a ratio scale measure of a construct (e.g., happiness), you probably need to use magnitude estimation. If you want a ratio scale measure of behavior (e.g., number of cigarettes smoked), you need to measure that behavior with perfect accuracy. Fortunately, you need ratio scale level of measurement only if you are trying to make ratio statements like "Married women are two times as happy as widows."

When You Need at Least Interval Scale Data

Because you would rarely have a hypothesis that would specify a ratio (e.g., Group A will be 1/3 as anxious as Group B), you will rarely need to assume that your measure has ratio properties. However, because you will often be concerned about how much more of a quality one group has than another, you will often need to assume that your measure has interval properties. To illustrate, consider the data from our previous example:

Didn't go to counseling at all	3.0
Went to counseling for 1 week	6.0
Went to counseling for 8 weeks	8.0
(The higher the score, the more in love. Scores could range from 1 to 9.)	

© Cengage Learning 2013

We might want to be able to say that the first week of counseling does more good than the next seven weeks. In that case, we would need interval data (because we want to be able to say that the difference between 3 and 6 is greater—psychologically—than the difference between 6 and 8).

To see a more common case in which we would need interval data, let's look at another study you might do to estimate the effects of therapy on relationships. Before relationship counseling is offered, you measure the degree to which couples are in love. Next, you observe who goes to counseling and who doesn't. Finally, at the end of the term, you measure the couples' love again. Let's say that you got the following pattern of results:

	Beginning of term	End of term
Didn't go to counseling	3.0	4.0
Went to counseling	5.0	7.0
(The higher the score, the more in love. Scores could range from 1 to 9.)		

© Cengage Learning 2013

Did the couples who went for counseling change more than those who didn't? At first glance, the answer seems obvious. The no-counseling group changed 1 unit and the counseling group changed 2 units, so isn't 2 units more than 1 unit? At the mathematical and score levels, the answer is "yes" However, at the psychological level, the answer is "not necessarily."

If we have interval or ratio scale data, we can assume that each unit of change represents the same psychological distance. Therefore, we can say that—at the mental, psychological, emotional level (as well as at the score level)—2 units of change is more than 1 unit of change. Thus, if we have interval or ratio scale data, we can say that the counseling group changed more than the no-counseling group.

But what if we had nominal or ordinal data? With nominal or ordinal data, we can't safely assume that each unit of change represents the same psychological distance. If our data were nominal or ordinal, the psychological distance between 3 and 4 could be much more than the psychological distance between 5 and 7. Thus, if we had nominal or ordinal data, we would *not* be able to answer the question, "Did couples who went for counseling change more than those who didn't? The lesson from this example is that if your research question involves asking whether one group changed more than another group, you must use a measure, such as a rating scale, that has at least interval properties.

When Ordinal Data Are Sufficient

Suppose you don't care how much more in love one group is than the other. All you want to know is which group is most in love. For example, suppose you want to be able to order these three groups in terms of amount of love:

Didn't go to counseling at all	3.0
Went to counseling for 1 week	6.0
Went to counseling for 8 weeks	8.0
(The higher the score, the more in love. Scores could range from 1 to 9.)	

© Cengage Learning 2013

If you had ordinal data, you could conclude that participants who went to counseling for 8 weeks were most in love, those who went for 1 week were less in love, and those who didn't go to counseling were least in love. So, if you simply want to know which group is higher on a variable and which group is lower, all you need is ordinal data. If that's the case, you are in luck. As you can see from Table 6.3, most measures produce data that meet or exceed the requirements of ordinal level measurement.

When You Need Only Nominal Data:

It's conceivable that you aren't interested in discovering which group is more in love. Instead, you might have the less ambitious goal of trying to find out whether the different groups differ in terms of their love for each other. If that's the case, nominal data are all you need. Because you need to make only the least demanding and safest assumption about your numbers (that different numbers represent different things), any valid measure you choose will measure up.

Conclusions About Scales of Measurement

As you have seen, different research questions require different scales of measurement. If you are asking only whether two groups *differ,* any scale of measurement, even *nominal,* will do. If, however, you are asking whether one group has *more* of a quality than another, you need at least *ordinal* level data. If your research question involves asking *how much* more of a quality one group has than another, you need to use a measure that provides at least *interval* data. If you need to find out *how many times more* of a quality one group has than another, you need *ratio* level data.

If your research question requires a given level of measurement, you must use a measure that provides at least that level of measurement. Consequently, you may find that the type of data you need will dictate the measure you choose—and that the only measure that will give you the type of data you need is not as sensitive or as free from biases as another measure.

To illustrate that the type of data you need may dictate what measure you use, suppose you want to know whether a treatment is more effective for couples who are less in love than it is for couples who are more in love. You are reluctant to use a rating scale measure because rating scale measures are extremely vulnerable to subject biases, such as participants lying to impress the researcher or participants providing the answers that they think will support the researcher's hypothesis.

Although validity questions make you hesitant to use rating scales, rating scale measures are commonly assumed to produce interval data—and your research question requires at least interval scale data. To see why your research question requires interval data, imagine that the average love score for the unhappy couples increases from a "1" to a "3," whereas the average love score for the happy couples increases from an "8" to a "9." To say that the unhappy couples experienced more improvement, you must assume that the difference between "1" and "3" is greater than the difference between "8" and "9." This is *not* an assumption you can make if you have either nominal or ordinal data. It is, however, an assumption you can make if you have interval data (because with interval data, the psychological distance between "1" and "2" is the same as the distance between "8" and "9").

Because your research question requires interval data, you must use a measure that provides at least interval data. Consequently, if the rating scale is the only measure that gives you interval scale data, you will have to use it—despite its vulnerability to subject bias—because it is the only measure that will allow you to answer your research question.

ETHICAL AND PRACTICAL CONSIDERATIONS

Clearly, you want to use a measure that will allow you to answer your research question. However, there may be times when you decide not to use a certain measure even though that measure is the one best able to answer your research question. For example, suppose you have a measure that gives you the right scale of measurement and is more valid than other measures because it is not vulnerable to subject biases. However, it avoids subject bias by surprising or tricking participants. In such a case, you may decide against using that measure because you believe participants should be fully informed about the study before they agree to participate. Similarly, you may reject field observation because you feel those tactics violate participants' rights to privacy.

When choosing a measure, you should always be concerned about ethical issues (see Appendix D). In addition, at times, you may also have to be concerned about practical issues. You may have to reject a measure because it is simply too time consuming or expensive to use. Practical concerns may even force you to either reject or use a measure based on its **face validity**: the extent to which a measure looks, on the face of it, to be valid. Note that face validity has nothing to do with actual, scientific validity. Therefore, you would usually not choose a measure for its face validity any more than you would judge a book by its cover. Typically, you would choose a measure based on a careful evaluation of scientific evidence, rather than on participants' opinions.

Indeed, under many circumstances, high face validity could harm real validity. To illustrate, if people think that questions from a "test" in a popular magazine are measuring a certain trait, people can "fake" the test to get the results they want. Conversely, a measure with no face validity may be valid precisely because participants don't see what it is measuring.

As we have pointed out, when evaluating the validity of a measure, you will usually have little use for face validity. However, face validity may be important to the consumer (or the sponsor) of your research. For example, imagine you are doing research on factors that affect effort. How loud a person yells and how many widgets a person produces may be equally valid measures of effort. But if you were going to get a factory manager to take your research seriously, which measure would you use?

CONCLUDING REMARKS

In this chapter, you have seen that choosing a measure is a complex decision. It is not enough to pick the most valid measure. Instead, you need to pick the measure that will be most likely to answer your research question. For your particular study, you must decide what threats to validity are most serious, decide how much sensitivity you need, decide what level of measurement your research question requires, and carefully weigh both ethical and practical considerations.

In designing a research project, choosing a measure is an important decision. However, it is only one of several design decisions you will make.

For example, you must decide whether to do a survey, an experiment, or some other type of research. Then, if you decide to do an experiment, you must decide on what type of experiment to do. In the next few chapters, you will learn how to make those key design decisions.

SUMMARY

1. Because no measure is perfect, choosing a measure involves making trade-offs. Thus, there is more to choosing a measure than choosing the one with the highest validity.

2. Sensitivity, reliability, and validity are highly valued in a measure.

3. Sensitivity is a measure's ability to detect small differences.

4. Because reliability is a prerequisite for sensitivity, an unreliable measure cannot be sensitive. However, because reliability doesn't guarantee sensitivity, a reliable measure could be insensitive.

5. You may be able to increase a measure's sensitivity by asking "how much" rather than "whether," by knowing what you want to measure, by avoiding unnecessary inferences, and by using common sense.

6. Whereas a measure's sensitivity affects whether you can find a difference, the measure's scale of measurement affects what you can say about the differences you find.

7. Different kinds of measures produce different kinds of numbers. These numbers range from the least informative (nominal) to the most informative (ratio).

8. Nominal numbers let you say only that participants differ on a characteristic, but they do not let you say that one participant has more of a characteristic than another.

With nominal measurement, higher numbers don't mean more of a quality because the number you assign to a category is arbitrary. For example, if you coded men as "1" and women as "2," your coding is entirely arbitrary. You could even go back and recode women as "1" and men as "2."

9. Ordinal numbers let you say that one participant has more of a quality than another. However, ordinal numbers do not allow you to talk about specific amounts of a quality. They let you talk only about having more of it or less of it, but not about how much more or less.

10. Interval and ratio numbers let you say how much more of a quality one participant has than another. So, if you want to know about the *amount* of a difference, you need interval or ratio data.

11. Ratio scale numbers let you say how many times more of a quality one participant has relative to another.

12. A measure's sensitivity is different from its scale of measurement. So, a sensitive measure could have a low scale of measurement.

13. Depending on the research question, a measure's sensitivity and its level of measurement may be almost as important as validity.

14. You must always consider ethical and practical issues when choosing a measure.

KEY TERMS

ceiling effect *(p. 204)*
face validity *(p. 220)*
floor effect *(p. 204)*
interval scale numbers (data)
   (p. 210)

nominal scale numbers (data)
   (p. 207)
ordinal scale numbers (data)
   (p. 208)

ratio scale numbers (data)
   (p. 211)
sensitive, sensitivity
   (p. 197)

EXERCISES

1. Suppose that in a study involving only 40 participants, researchers look at self-esteem differences between two groups. They find a small, but statistically significant, difference between the self-esteem of the two groups. Based on this information, would you infer that the measure's reliability was low or high? Why?
2. List the scales of measurement in order from least to most accurate and informative.
3. Becky wants to know how much students drink.
   a. What level of measurement could Becky get? Why?
   b. Becky asks participants: How much do you drink?
      1. 0–1 drinks
      2. 1–3 drinks
      3. 3–4 drinks
      4. more than 4 drinks
   What scale of measurement does she have?
   c. Becky ranks participants according to how much they drink. What scale of measurement does she have?
   d. Becky assigns participants a "0" if they do not drink, a "1" if they primarily drink wine, and a "2" if they primarily drink beer. What scale of measurement is this?
   e. Becky asks participants: How much do you drink?
      1. 0–1 drinks
      2. 1–3 drinks

3. 3–4 drinks
4. more than 4 drinks
5. don't know
   What scale of measurement does she have? Why?
4. Assume that facial tension is a measure of thinking.
   a. How would you measure facial tension?
   b. What scale of measurement is it on? Why?
   c. How sensitive do you think this measure would be? Why?
5. Suppose a researcher is investigating the effectiveness of drug awareness programs.
   a. What scale of measurement would the investigator need if she were trying to discover whether one drug awareness program was more effective than another?
   b. What scale of measurement would the investigator need if she were trying to discover whether one program is better for informing the relatively ignorant than it is for informing the relatively well informed?
6. In an ideal world, car gas gauges would have which scale of measurement? Why? In practice, what is the scale of measurement for most gas gauges? Why do you say that?
7. Find or invent a measure.
   a. Describe the measure.
   b. Discuss how you could improve its sensitivity.
   c. What kind of data (nominal, ordinal, interval, or ratio) do you think that measure would produce? Why?

WEB RESOURCES

1. Go to the Chapter 6 section of the book's student website and
   a. Look over the concept map of the key terms.
   b. Test yourself on the key terms.
   c. Take the Chapter 6 Practice Quiz.
2. Go to the "Measure Chooser" link to practice choosing the right measure for the situation.
3. Try one of the "Scales of Measurement" tutorials.

CHAPTER 7

Introduction to Descriptive Methods and Correlational Research

The invalid assumption that correlation implies cause is probably among the two or three most serious and common errors of human reasoning.
—**Stephen Jay Gould**

Remember, correlation does not equal causality.
—**Unknown**

CHAPTER OVERVIEW

In this chapter, you will learn how to refine techniques you use every day—watching people, asking them questions, and paying attention to records of their behavior (e.g., diaries, police reports, news reports)—into descriptive research: methods that will provide objective, reliable, and scientifically valid descriptions of what people think, say, and do. Perhaps more importantly, you will learn when to use descriptive research—and when not to.

Like all research, the goal of descriptive research is to test hypotheses and answer questions. However, unlike experimental research, descriptive research is not equipped to test cause–effect hypotheses and therefore can't answer questions about the "whys" (causes) of behavior. Instead, it can help us answer "what," "who," "when," and "where" questions.

Descriptive researchers often start by trying to answer "what" questions about a single variable, such as "What is the behavior?" and "What percentage of people have that characteristic?" For example, the earliest research on flirting focused on what people did when flirting; early work on laughter dealt with describing laughter; the earliest work on unconscious prejudice focused on seeing what percentage of people held these prejudices; and the earliest work on happiness counted how many people were happy.

Usually, descriptive researchers quickly expand their focus from "what" questions describing a single variable to "who," "when," and "where" questions describing that variable's relationship to other variables. Thus, soon after researchers had described laughter, they were finding factors that related to it, such as who laughs (women more than men), where people laugh (in public), and when (women laugh when listening to a man they like; Provine, 2004). Similarly, happiness researchers quickly went from finding out that most people were happy (Diener & Diener, 1996) to finding (a) factors that predict happiness, such as exercise, extroversion, marriage, and religious faith, as well as (b) factors that happiness seems to predict, such as a longer life (Danner, Snowden, & Friesen, 2001). Because researchers using descriptive methods almost always look at relationships between two or more variables to see whether those variables covary (*correlate*), most research using descriptive methods is called *correlational research*.

USES AND LIMITATIONS OF DESCRIPTIVE METHODS

When you use descriptive methods, you gain the ability to test hypotheses about virtually any variable in virtually any situation. For example, you can use them even when you can't—for either practical or ethical reasons—manipulate variables. You can use descriptive methods even when you can neither control irrelevant variables nor account for their effects. In short, if your hypothesis is that two or more measurable variables are statistically related (e.g., soccer playing and IQ scores, being spanked and aggressive behavior, parental warmth and autism, mother's skill at reading her child's mind and child's self-esteem, smoking and hyperactivity, church attendance and happiness), descriptive methods give you the flexibility to test that hypothesis.

Descriptive Research and Causality

But this flexibility comes at a cost. Without being able to manipulate variables and account for the effects of irrelevant variables, you cannot legitimately make cause–effect statements. In other words, you can find out that two variables are related, but you cannot find out why they are related. For example, if you find a relationship between church attendance and happiness, you cannot say *why* church attendance and happiness are related. Certainly, you cannot say that church attendance *causes* (produces, results in, affects, creates, brings about, influences, changes, increases, triggers, has an effect on) happiness. That is, you cannot say that people who go to church are happy be*cause* they go to church.

Why Descriptive Methods Cannot Test Causal Hypotheses

Why not? There are two reasons.

First, rather than church attendance being a *cause* of happiness, church attendance may be an *effect* of happiness. Because you did not manipulate variables, you don't know which variable came first—happiness or church attendance—and thus, you may be wrong about which variable caused which. You may believe church attendance causes happiness, but you may have it backward: Maybe happiness causes church attendance. For example, happy people may be more likely to be out of bed and ready to face the outside world in time to go to church than people who are depressed.

Second, rather than church attendance being a cause of happiness, both church attendance and happiness may be effects of some other variable. Because you have neither controlled for nor accounted for any other factors that might lead both to being happy and to going to church, many factors might be responsible for the relationship between happiness and church attendance. Just a few of those possible factors are listed here:

- Disciplined people may be happier and more likely to go to church.
- People who like structured social activities may be more likely to be happy and more likely to go to church.
- Having genes that predispose one to be conventional and outgoing may cause one to be happy and go to church.
- Having church-going friends may cause one to be happy and to go to church.
- Married people may be more likely to be happy and more likely to go to church than unmarried people.
- Unhealthy people may be both less likely to make it to church and less likely to be happy.
- Optimistic people may be more likely to go to church (perhaps because they are more likely to believe in an afterlife) and may be more likely to be happy.

To repeat a key point, correlational methods do not have internal validity, so they do not allow you to make cause–effect statements. When you use a correlational method to find a relationship between two variables, you do not know whether the relationship is due to changes (1) in the "first" variable causing changes in the "second" variable, (2) in the "second" variable[1] causing changes in the "first" variable, or (3) in a third variable causing the changes in both variables (see Figure 7.1). If you think you know what causes the relationship, you are like a game show contestant who thinks he knows behind which of the three doors is the prize: You don't know; you're probably wrong; and, if you're right, it's by luck.

How Descriptive Methods Can Stimulate Cause–Effect Hypotheses

Although data from descriptive research cannot allow you to make cause–effect assertions, as you can see from Box 7.1, such data may help you think of cause–effect questions. Two strategies can help you use findings from correlational studies to think of experimental (cause–effect) hypotheses.

First, if you find a relationship between two variables, try to think an experiment to determine whether the relationship is a cause–effect relationship. This strategy let scientists go from knowing that there was a correlation between smoking and lung cancer to doing experiments that tested whether smoking caused lung cancer.

Second, even if you believe that the two factors do not directly influence each other, try to find out what causes them to be statistically related. In other words, try to identify the third factor that accounts for their

---

[1] Often, you don't know which variable came first, so labeling one variable the "first" variable and the other the "second" is arbitrary. Given that you don't know which variable is really the first, you can't say which caused which (the situation is like the "Which came first—the chicken or the egg?" question).

1. The "first" factor causes a change in the "second" factor.

2. The "second" factor causes a change in the "first" factor.

3. Some "third" factor could cause a change in both the "first" and "second" factors.

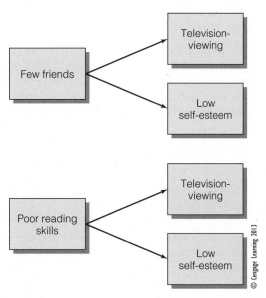

FIGURE **7.1** Three basic possibilities for an observed relationship between low self-esteem and television-viewing.
Note that there are many "third" factors that might account for the relationship. We have listed only two (friends and reading skills).

relationship. For example, suppose you find that, in general, students who study more have lower grade-point averages. This finding may suggest the following idea: Perhaps some students are using study strategies that are both time-consuming and ineffective. This idea may lead you to do experiments to test whether some study strategies are more effective than others. Alternatively, if you think the relationship is due to students with lower

---

BOX **7.1**   **Generating Causal Hypotheses From Correlational Data**

For each correlational finding listed here, develop an experimental hypothesis.

1. Listeners of country music tend to be more depressed than people who listen to other types of music.
2. There is a correlation between school attendance and good grades.
3. Attractive people earn higher salaries than less attractive people.
4. In restaurants, large groups tend to leave lower tips (in terms of percentage of bill) than individuals.
5. Teams that wear black uniforms are penalized more often than other teams.
6. People report being more often persuaded by newspaper editorials than by television editorials.
7. Students report that they would be less likely to cheat if professors walked around the classroom more during exams.
8. Students who take notes in outline form get better grades than those who don't.

© Cengage Learning 2013

---

grades being less effective readers, you might design an experiment to see whether training in reading skills improves grades.

In summary, descriptive research does *not* allow you to infer causality (see Box 7.2). However, descriptive research may stimulate experimental research that will allow you to infer causality: Once you use a *descriptive* design to find out *what* happens, you can use an *experimental* design to try to find out *why* it happens.

Description for Description's Sake

By hinting at possible causal relationships, descriptive research can indirectly help psychologists achieve two goals of psychology—explaining behavior and controlling behavior. But the main purpose of descriptive research is to achieve another important goal of psychology—describing behavior.

Is description really an important scientific goal? Yes—in fact, description is a major goal of every science. What is chemistry's famed periodic table but a description of the elements? What is biology's system of classifying plants and animals into kingdom, phylum, genus, and species but a way of describing living organisms? What is astronomy's mapping of the stars but a description of outer space? What is science but systematic observation and measurement? Thus, one reason psychologists value descriptive methods is that description is the cornerstone of science. Besides, psychologists, like everyone else, want to be able to describe what people think, feel, and do.

Description for Prediction's Sake

Psychologists also like descriptive methods because knowing what *is* happening helps us predict what *will* happen. In the case of suicide, for example, psychologists discovered that certain signals (giving away precious possessions, abrupt changes in personality) were associated with suicide. Consequently, psychologists now realize that people sending out those signals are more likely to attempt suicide than people not behaving that way.

| BOX **7.2** | **Three Sets of Questions to Answer Before Claiming That One Variable Affects (Influences, Controls, Changes, Causes) Another** |

1. **Is there a relationship between the two variables in the sample?**
   * Did the researchers accurately measure the two variables?
   * Did the researchers accurately record the two variables?
   * Did the researchers accurately perceive the degree to which the variables were related?
2. **If the variables are related in the sample, are the variables related in the population?**
   * Is the sample a random sample of the population?
   * Even if the sample is a random sample of the population, is the sample large enough—and the relationship strong enough—that we can be confident that the relationship really holds in the population?
3. **If the variables are related, did the predictor variable cause changes in the variable it was designed to predict?**
   * Is it possible that the "criterion" (outcome) variable caused changes in the predictor variable? In other words, is our "cause" really the effect?
   * Do we know which variable came first?
   * Do we have data that suggest to us which came first? For example, if we are looking at self-esteem and delinquency, high school records might provide information about people's self-esteem before they became criminals. If there is no difference between delinquents' and nondelinquents' self-esteem prior to committing crimes, we would be more confident that self-esteem was not a cause of delinquency.
   * Can we logically rule out the possibility that one variable preceded the other? For example, if height and being a delinquent were correlated, we can make a good case that the person's height was established before he or she became a delinquent.
   * Is it possible that a third variable could be responsible for the relationship? That is, neither variable may directly influence (cause) the other. Instead, the two variables might be statistically related because they are both effects of some other variable. For example, increases in assaults and ice cream consumption may both be consequences of temperature.
   * Were all other variables randomized or held constant? (This control over other variables happens only in experimental designs. For more on experimental designs, see Chapters 10–13.)
   * Does the researcher know what the potential third variables are? If so, the researcher may be able to statistically control for those variables. However, it is virtually impossible to know and measure every potential third variable.

© Cengage Learning 2013

WHY WE NEED SCIENCE TO DESCRIBE BEHAVIOR

Certainly, describing behavior is an important goal of psychology. But do we need to use scientific methods to describe what's all around us? Yes! Intuition alone cannot achieve all four steps necessary to describe behavior accurately:

1. Get objective measurements of key variables.
2. Keep track of these measurements.
3. Use these measurements to determine the degree to which the variables are related in the sample of behavior that was observed.
4. Accurately infer that the observed pattern of results reflects what typically happens (see Figure 7.2).

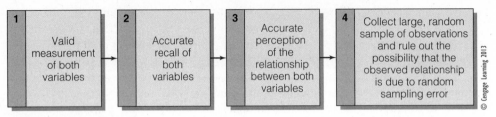

FIGURE **7.2** Four steps involved in determining whether there is a relationship between two variables.

People who draw conclusions based on their own personal experience could be making mistakes at every single one of these steps. Do you know anyone who executes all four steps correctly? Probably not—even some scientific studies fail to execute all four steps correctly.

Indeed, intuition alone fails at each of these four steps. Thus, as you will soon see, we need science to measure variables, keep track of those measurements, determine the degree to which variables are related, and to make accurate inferences about the degree to which the observed pattern of results reflects what typically happens.

We Need Scientific Measurement

We need scientific methods to accurately measure the variables we want to measure. As you saw in Chapter 5, reliable and valid measurement of psychological variables is not automatic. If you are to observe psychological variables in a systematic, objective, and unbiased way, you must use scientific methods. Imagine using intuition to measure a person's level of motivation, intelligence, or some other psychological variable!

We Need Systematic, Scientific Record-Keeping

Even if you could intuitively get accurate measurements of psychological variables, you could not rely on your memory to keep track of your observations. Your memory can fool you, especially when it comes to estimating how often things occur. For example, our memories may fool us into believing that more people die from plane crashes than actually do, more people die from shark attacks than actually do, and that more words start with *r* than have *r* as their third letter (Myers, 2002b). Therefore, if you are to describe behavior accurately, you need to record your observations systematically so that your conclusions are not biased by memory's selectivity.

We Need Objective Ways to Determine Whether Variables Are Related

Obviously, if you're poor at keeping track of observations of one variable, you are going to be even worse at keeping track of two variables—plus the relationship between them. Therefore, you cannot rely on your judgment to determine whether two things are related.

People are so eager to see relationships that they sometimes "see" variables as being related, even when those variables are not related. In several experiments on illusory correlation (Chapman & Chapman, 1967; Ward & Jenkins, 1965), researchers showed participants some data that did not follow any pattern and that did not indicate any relationship among variables. Remarkably, participants usually "found" patterns in these patternless data and found relationships (**illusory correlations**) between the unrelated variables.

Out of the lab, we know that people see systematic patterns in the stock market, even though the stock market behaves in an essentially random fashion (Shefrin & Statman, 1986). Similarly, many people believe that the interview is an invaluable selection device, even though research shows that interviews have virtually no validity (Dawes, 1994; Schultz & Schultz, 2006).

Even when there is a relationship between two variables, the relationship people perceive between those two variables may be exactly opposite of the relationship that exists. To illustrate, basketball coaches swear that if a player makes a shot, that player will be *more* likely to make the next shot. However, as Gilovich, Vallone, and Tversky (1985) discovered, a shooter is *less* likely to make the next shot if he has made the previous shot.

Coaches are not the only ones to misperceive relationships. Many bosses and parents swear that rewarding people doesn't work whereas punishing them does, even though research shows that rewards are more effective than punishments. Many students swear that cramming for exams is more effective than studying consistently, even though research contradicts this claim. Similarly, psychiatric hospital nurses believe that more patients are admitted during a full moon, even though admissions records dispute that. You probably know a biased person whose prejudices cause him or her to see relationships that don't exist. In short, because people may misperceive the relationship between variables, we need to do research to determine the real relationship between variables.

We Need Scientific Methods to Generalize From Experience

Even if you accurately describe your own experience, how can you generalize the results of that experience? After all, your experience is based on a limited and small sample of behavior.

One problem with small samples is that they may cause you to overlook a real relationship. Thus, it's not surprising that one man wrote to "Dear Abby" to inform her that lung cancer and smoking were not related: He knew many smokers, and none had lung cancer.

Another problem with small samples is that the relationship that exists in the sample may not reflect what typically happens in the population. Thus, our experiences may represent the exception rather than the rule. In other words, the relationship you observe may be due simply to a coincidence. For example, if we go by some people's experiences, playing the lottery is a great investment.

As you have seen, even if you accurately observe a pattern in your experiences, you must take one additional step—determining whether that pattern is simply a coincidence. But how can you determine the likelihood that a pattern of results is due to a coincidence?

To discount the role of coincidence, you need to do two things. First, you need to have a reasonably large and random sample of behavior. Second, you need to use probability theory to determine the likelihood that your results are due to random error. Thus, even if you were an intuitive statistician, you would still face one big question: What's to say that your experience is a large, random sample of behavior? Your experience may be a small and biased sample of behavior. Results from such biased samples are apt to be wrong. For example, in 1988, George H. W. Bush defeated Michael Dukakis by one of the biggest margins in the history of U.S. presidential elections. However, right up to election eve, some ardent Dukakis supporters thought that Dukakis would beat Bush. Why? Because everybody they knew was voting for Dukakis.

In summary, generalizations based on personal experience are often wrong. These informal generalizations are error prone because they (1) are based on small or biased samples and (2) are based on the assumption that what happens in one's experience happens in all cases. To avoid these problems, researchers who make generalizations about how people typically act or think (1) study a large, random sample and then (2) use statistics to determine how likely it is that the pattern observed in the sample holds in the population.

Conclusions About Why We Need Descriptive Research

As you can see, we need descriptive research if we are to accurately describe and predict what people think, feel, or do. Fortunately, descriptive research is relatively easy to do. To describe how two variables are related, you need to get a representative sample of behavior, accurately measure both variables, and then objectively assess the association between those variables. The bottom line in doing descriptive research is getting accurate measurements from a representative sample.

The key to getting a representative sample is to get a large and random sample. But how can you get accurate measurements from such a sample?

SOURCES OF DATA

In the next few sections, we'll look at several ways to get measurements. We'll start by examining ways of making use of data that have already been collected; then we'll move to going out and collecting new data.

Ex Post Facto Data: Data You Previously Collected

One possible source of data for descriptive research is data that you have already collected. For example, you may have done an experiment looking at the effects of time pressure on performance on a verbal task. At the time you did the study, you may not have cared about the age, gender, personality type, or other personal characteristics of your participants. For testing your experimental hypothesis (that the treatment had an effect), these individual difference variables were irrelevant. Indeed, experimenters often call such individual difference variables "nuisance variables." However, you collected this individual difference information anyway.

After the experiment is over, you might want to go back and look for relationships between these nuisance variables and task performance. This kind of research is called **ex post facto research**: research done after the fact.

External Validity

Suppose your ex post facto research revealed that women did better than men on the verbal task. Although this finding is interesting, you should be careful about generalizing your results. Unless the men and women in your study are a random sample drawn from the entire population of men and women, you cannot say that women do better at this verbal task than men do. Your effect may simply be due to sampling men of average intelligence and women of above-average intelligence. This sampling bias could easily occur, especially if your school was one that had higher admissions standards for women than for men. (Some schools had higher admissions standards for women than for men when they switched from being all-women colleges to coeducational institutions.)

You could have a bit more confidence that your results were not due to sampling error if you had also included a mathematical task and found that although women did better than men on the verbal task, men did better on the mathematical task. If, in this case, your results are due to sampling error, they aren't due to simply having sampled women who are above average in intelligence. Instead, your sampling error would have to be due to something rather strange, such as sampling women who were better than the average woman in verbal ability but who were worse than the average woman in mathematical ability. Although such a sampling bias is possible, it is not as likely as having merely sampled women who are above average in intelligence. Therefore, with this pattern of results, you would be a little more confident that your results were not due to sampling bias.

Construct Validity

Even if you could show that your results are not due to sampling error, you could not automatically conclude that women had greater verbal ability than men. To make this claim, you would have to show that your measure was a valid measure of verbal ability and that the measure was just as valid for men as it was for women. For example, if your verbal ability measure used vocabulary terms relating to different colors, women's fashions, and ballet, critics would argue that your measure was biased against men.

Internal Validity

If you had carefully chosen a valid measure and, if, by randomly sampling from a representative sample, you had carefully selected a representative sample, you might be able to claim that women had better verbal ability than men. However, you could not say why women had superior verbal ability. As you'll recall, correlational methods are not useful for inferring causality. Therefore, you could not say whether the difference in men's and women's verbal ability was due to inborn differences between men and women or due to differences in how men and women are socialized.

Conclusions About Ex Post Facto Research

In summary, ex post facto research takes advantage of data you have already collected. Therefore, the quality of ex post facto research depends on the quantity and quality of data you collect during the original study. The more information you collect about your participants' personal characteristics, the more ex post facto hypotheses you can examine. The more valid your measures, the more construct validity your conclusions will have. The more representative your sample of participants, the more external validity your results will have. Therefore, if you are doing a study, and there's any possibility that you will do ex post facto research, you should prepare for that possibility by using a random sample of participants and collecting a lot of data about each participant's personal characteristics.

Archival Data

Rather than use data that you have collected, you can use **archival data**: data that someone else has already collected. Basically, there are two kinds of archival data—coded data and uncoded data.

Collected and Coded Data

As the name suggests, coded data are not mere records of behavior (e.g., diaries, videotapes, pictures) but rather are data that have been scored (coded) so that numbers have been assigned to the recorded behaviors. Market researchers, news organizations, behavioral scientists, and government researchers are all collecting and tabulating data daily. How much data? To give you some idea, more than 5,000 Americans are surveyed every day—and surveys are just one way that these researchers collect data. Not only can you get access to some of these survey results but you can also get access to many statistics relating to people's behaviors—including statistics on accidents, attendance (church, school, sporting events, etc.), bankruptcy, baseball, chess, crime, mental health, income, IQ, literacy, mortality, movie viewing, obesity, voting, and all kinds of sales and spending (Trzesniewski, Donnellan, & Lucas, 2011).

Sometimes, to test a hypothesis, you need to comb through records and collate (pull together, assemble, compile) the data yourself. For example, researchers found support for the idea that hotter temperatures are associated with aggression by looking through the baseball and weather sections of newspapers and finding there was a relationship between game time temperature and how many batters were hit by pitches (Larrick, Timmerman, & Carton, & Abrevaya, 2011; Reifman, Larrick, & Fein, 1991).

Many times, the data have already been collated (brought together and organized) for you. You simply need to make the connection between the data and your hypothesis—and that often involves making the connection between two different sources of data. For example, using existing measures of wealth (inflation-adjusted U.S. gross national product) and life satisfaction, Diener and Seligman (2004) found that although U.S. citizens became wealthier from 1950 to 1998, U.S. citizens did not become happier during that time.

Sometimes, collated data can help you test hypotheses derived from theory. To test the limits of modeling theory (a theory that describes when and why people imitate others), David Phillips (1979) used suicide and traffic accident statistics to find that both suicides and one-car accidents increased after a well-publicized suicide—but only among people who were similar in age to the person who committed suicide. To test a hypothesis derived from social loafing theory—that songwriters would not work as hard on group-authored songs as they would on their own solo efforts—two music-loving graduate students looked at the equivalent of Top 40 charts. Specifically, the students found that songs written by members of the 1960s rock band The Beatles were better (measured by popularity in terms of chart rankings from *Billboard Magazine*) when written alone than when jointly written (Jackson & Padgett, 1982). Using a hypothesis derived from the theory of evolution—that left-handedness survives because it is helpful in hand-to-hand combat—two researchers found that, in eight societies that use knives rather than guns as weapons, the societies that had the most killings had the most left-handed people (Faurie & Raymond, 2005).

Sometimes, the data have been collected and collated but not published. In that case, you just need to ask for the information. For example, a journalist wanted to test the hypothesis that many heads are better than one by seeing whether the opinion expressed by the majority of the studio audience on the game show *Who Wants to Be a Millionaire?* is more accurate than the opinion given by the "expert" friend the contestant has selected. The journalist did not have to look at tapes of all the shows to compile the relevant data. Instead, because the show had already compiled that data, just interviewing the show's spokesperson gave him the information he needed: The answers provided by the friend-selected experts were right 65% of the time, whereas the answers provided by the studio audience were right 91% of the time (Surowiecki, 2004).

There is good news and bad news about archival data that have been coded for you. The good news is that if you can get access to archival data, you can often look at data that you would never have collected yourself because you wouldn't have had the time or resources to do so—and it has already been coded for you. The bad news is that most archival data are data that you would never have collected yourself because they were collected in a way that is inappropriate for answering your research question—and even those data that you would possibly have collected, you would never have coded that way.

Collected but Uncoded Data

If you are willing to code the data yourself, you can avoid the problem of inappropriately coded data. You will also gain access to a vast amount and variety of preserved records of behavior, including

- letters to the editor
- transcripts of congressional hearings
- videotapes of television shows
- yearbook photos (which you can code for smiling and type of smile)
- diaries and autobiographies
- *Playboy* centerfolds (an indicator of the physical characteristics that *Playboy* readers consider ideal)
- comments made in Internet chat rooms and discussion groups
- personal ads for a dating partner

The main advantage of using records of behavior is that the basic data have already been collected for you. All you have to do is code them—and you can code them to suit your needs. If you want to study happiness, for example, you can code a wide range of data, such as

- Videotapes. By having students rate the happiness of Olympic athletes on the podium during medal ceremonies, Medvec, Madey, and Gilovich (1995) coded videotapes and found that Bronze (third place) winners were happier than Silver (second place) winners.
- College yearbook photos. Using a coding strategy that involved looking at the position of two facial muscles, Harker and Keltner (2001) judged the happiness expressed by people in their college yearbook photos and found that the students who were rated as showing more positive emotion in their yearbooks were, 30 years later, more likely to be happily married.
- Essays. By coding the happiness expressed in short essays that nuns wrote when they were first accepted into the sisterhood, Danner, Snowden, and Friesen (2001) found that the nuns expressing the most happiness lived the longest.

Content Analysis: Objectively Coding the Uncoded. The challenge of using such data is that you must convert the photos, videotapes, transcripts, or other records into a form that you can meaningfully and objectively analyze. To succeed at this task, use **content analysis**.

Content analysis has been used to categorize a wide range of free responses—from determining whether a threatening letter is from a terrorist to determining whether someone's response to an ambiguous picture shows that they have a high need for achievement. In content analysis, you code behavior according to whether it belongs to a certain category (aggressive, sexist, superstitious, etc.).

To use content analysis successfully, you must first carefully define your coding categories. To do so, you should review the research to find out how others have coded those categories. If you can't borrow or adapt someone else's coding scheme, do a mini-study (often called a pilot study) to get an idea of the types of behavior you will be coding, and to help you choose and define the categories you will use to code the data.

After you have defined your categories, you should provide examples of behavior that would fit into each of your categories. Then, train your raters to use these categories.

The primary aim in content analysis is to define your categories as objectively as possible. Some researchers define their categories so objectively that all the coder has to do is count the number of times certain words come up. For example, to get an indication of America's mood, a researcher might count the number of times words like *war, fight,* and so on appear in *The New York Times*. These word-counting schemes are so easy to use that even a computer can do them. In fact, one set of researchers invented a computer program that can tell genuine suicide notes from fake ones (Stone, Smith,

Dunphy, & Ogilvie, 1966), and another set invented a computer program that can, with a fair degree of accuracy, tell poetry written by poets who committed suicide from poetry written by poets who did not (Stirman & Pennebaker, 2001). Thus, objective coding can be simple and have construct validity.

Is Objective Coding Valid?   Unfortunately, objective criteria are not always so valid. To get totally objective criteria, you often have to ignore the context—yet the meaning of behavior often depends on the context. For example, you might use the number of times the word *war* appears in major newspapers as a measure of how eager people are for war. This method would be objective, but what if the newspaper was merely reporting wars in other countries? Or, what if the newspaper was full of editorials urging us to avoid war or urging us to expand the war on poverty? In that case, our measure would be objective, but invalid.

Indeed, context is so important that completely objective scoring criteria of certain variables is virtually impossible. Whether a remark is sarcastic, humorous, or sexist may depend more on when, where, and how the statement is said than on the content of what is said. However, despite the difficulties of objectively and accurately coding archival data, researchers often have successfully developed highly objective ways of coding archival data.

An Example of Archival Research

To get a clearer picture of both the advantages and disadvantages of archival research, suppose you wanted to know whether people were more superstitious when they were worried about the economy. As your measure of concern about the economy, you use government statistics on unemployment. As your measure of how superstitious people are, you have the computer count the number of key words such as *magic, superstition,* and *voodoo* that appear in local newspapers and then divide this number by the total number of words in the newspaper. This would give you the percentage of superstitious words in local newspapers.[2]

Internal Validity

Once you had your measures of both economic concern and of superstitiousness, you would correlate the two. Suppose you found that the higher the unemployment rate is, the more superstitious words were used in the newspaper. Because you have done a correlational study, you cannot say why the two variables are related. That is, you do not know whether

1.   the economy caused people to become superstitious;
2.   superstitious beliefs caused the downfall of the economy; or
3.   some other factor (bad weather ruining crops, a poor educational system) are responsible for both an increase in superstitious beliefs and a decline in the economy.

---

[2] Padgett and Jorgenson (1982) did a study similar to this one.

Construct Validity

In addition to the internal validity problems that you have anytime you use correlational data, you have several construct validity problems specific to archival data. You are using measures of a construct, not because they are the best, but because they are the only measures that someone else bothered to collect. Although you are using unemployment records as an index of how insecure people felt about the economy, you would have preferred to ask people how they felt about the economy. To the degree that the relationship between how many people are unemployed and how people feel about the economy is questionable, your measure's construct validity is questionable.

Even if there is a strong relationship between actual unemployment and feelings about the economy, your measure may not be valid because it may not accurately assess unemployment. The unemployment records may give an inaccurate index of unemployment because of **instrumentation bias**: scores on the measure changing due to (1) changes in the measure itself, (2) changes in how the measure is scored, or (3) changes in who is being measured and recorded.

In measuring unemployment, we would be most concerned about two sources of instrumentation bias: changes in the definition (scoring) of unemployment and changes in whose data is included when calculating the unemployment statistics. A change in scoring, such as the government changing the definition of unemployment from "being unemployed" to "being unemployed for 6 weeks and showing documentation that he or she looks for three jobs every week," would reduce the number of people recorded as unemployed.

Any change in how thoroughly data are collected and collated could affect unemployment statistics. For example, because of the introduction of unemployment compensation in 1935 and more recent computerization of national statistics, current unemployment statistics are more complete than they were in the early 1900s. Thus, better record-keeping may increase the number of people currently recorded as unemployed. Another change might be found in data collection: Because the people who would collect unemployment statistics—social workers and other government workers—are sometimes laid off during hard economic times, unemployment statistics might be less complete during periods of high unemployment. Still another change that would reduce the number of people labeled as unemployed would be if some politicians distorted unemployment data to make things seem better than they were.[3] (Similarly, a commonly used index of the U.S. economy and of U.S. stock prices—the Dow Jones Industrial—fell sharply during the global financial crisis that started in 2008. By early 2011, that index had rebounded to close to its pre-crisis levels. However, part of this rebound was due to removing stocks of companies that were doing extremely poorly from that index. For example, AIG—which, at one point was responsible for about a 3,000 point loss to the index—has been removed from the index.)

To illustrate how the three sources of instrumentation bias (changes in the instrument, changes in scoring, and changes in sampling and tabulating)

---

[3] According to Levitt and Dubner (2005, p. 92), violent crime statistics in Atlanta were altered as part of that city's attempt to host the 1996 Olympics.

can make it difficult to do a meaningful study, imagine you wanted to compare average SAT scores from 2011 with average SAT scores of 1981. Your first problem is that the SAT instrument has changed. The types of questions the current SAT asks are different from the ones used in the past. In 2005, a section on analogies was dropped from the test, an essay section was added, and the math segment went through a major change. Your second problem is that the way the instrument is scored has changed. For example, in 1985, the SAT was based on a perfect score of 1,600. However, starting in 2005, the new perfect score is 2,400. Your third problem is that the sample of students that are measured has changed. Specifically, a greater percentage of high school students took the SAT in 2005 (at one time, only the very best high school students took the SAT; in 2005, about half of all high school students took it).

Thus, even if you developed a formula to compare 1981 scores with 2011 scores, you would have difficulty making meaningful comparisons. For instance, if scores on the SAT went down, you could not conclude that students were learning less in high school because (a) rather than testing what it once did, the test is testing different knowledge and (b) rather than testing the top 10% of students, the test is testing the top 50%. If, on the other hand, scores on the SAT went up, you couldn't conclude that students were learning more in high school because (a) the test is testing different knowledge and (b) the scoring system has been changed.

Instrumentation is also a problem in understanding the correlates and incidence of autism. The definition of autism has recently expanded, and so the numbers of people diagnosed with autism has recently exploded (Radford, 2007). As a result, anything else that has recently become more popular (e.g., cell phones, ultrasounds, vaccines) will correlate with autism (Gernsbacher, 2007). In addition, because physicians will usually not diagnose autism at age 1, parents whose child is diagnosed as autistic at age 2 may assume that some event between age 1 and 2 caused their child to become autistic (Novella, 2007).

Fortunately, in the economy–superstition study, you only have to worry about instrumentation bias ruining the measure of unemployment; you do not have to worry about instrumentation bias ruining the superstition measure. However, you still have to worry about your superstition measure's construct validity. Is the number of times superstitious terms are mentioned in newspapers a good index of superstition? Perhaps these articles sell papers, and major newspapers stoop to using these articles only when sales are low. Rather than relying on the number of times "superstition" appears in papers, you would prefer to have results of some nationwide survey that questioned people directly about their superstitious beliefs. However, your measure has one advantage over the poll—it is a **nonreactive measure**: Collecting it does not change participants' behavior.

External Validity

Because you can collect so much data so easily, your results should have good external validity. In some cases, your results may apply to millions of people because you have data from millions of people. Specifically, you can easily get unemployment statistics for the entire United States. Furthermore, because you can collect data for a period of years rather than for just the immediate present, you should be able to generalize your results across time.

The Limits of Aggregate Data

Gaining access to group data (for instance, the unemployment rate for the entire United States for 1931) is convenient and may aid external validity. However, as psychologists, we are interested in what individuals do; therefore, we want data on individuals. Consequently, even if we find that there is a correlation between unemployment for the nation as a whole and superstition for the nation as a whole, we are still troubled because we do not know which individuals are superstitious. Are the individuals who are unemployed the ones who are superstitious? Or, are the superstitious ones the people whose friends have been laid off? Or, are the superstitious ones the people who are doing quite well? With aggregate (group) data, we can't say.

Conclusions About Archival Research

By using archival data, you can gain access to a great deal of data that you did not have to collect, which may allow you to test hypotheses you would otherwise be unable or unwilling to test. For example, you can test hypotheses about relationships between type of prison and violence in prisons (Briggs, 2001), color of uniform and violence in professional sports (Frank & Gilovich, 1988), economic conditions and what men want women to look like (Pettijohn & Jungeberg, 2004), and competitiveness and violence (Wilson & Daly, 1985). Because archival data often summarize the behavior of thousands of people across a period of years, your results may have impressive external validity.

But relying on others to collect data has its drawbacks. You may find that others used measures that have less construct validity than the measures you would have used. You may find that others did not collect the data as carefully and as consistently as you would have. You may find that you have data about groups but no data about individuals. Because the data that others collected will usually not be ideal for answering the question you want to answer, you may decide to collect your own data.

Observation

One way to collect your own data is through observation. As the name implies, observation involves watching (observing) behavior.

You may be familiar with using observational techniques in an experiment to measure participants' responses to a treatment. For example, an experimenter administering different levels of a drug to rats might observe and categorize each rat's behavior. Similarly, an experimenter manipulating levels of televised violence might observe and categorize each participant's behavior.

In this chapter, we will focus on observation as a method for describing behavior that we are interested in for its own sake. Describing behavior is a vital concern of every field of psychology. Developmental psychologists use observation to describe child–parent interactions, social psychologists to describe cults, clinical psychologists to describe abnormal behavior, counseling psychologists to describe human sexual behavior, and comparative psychologists to describe animal behavior.

Types of Observational Research

There are three basic types of observation: laboratory observation, naturalistic observation, and participant observation. In both naturalistic and participant observation, you study real behavior in the real world. In contrast, laboratory observation, as the name suggests, occurs in a laboratory (under controlled conditions).

Laboratory observation, however, is not always as artificial as you might think. The lab experience is often very real to participants—and participants' behavior may strongly relate to real-world behavior. For example, consider Mary Ainsworth's "strange situation." To oversimplify, a mother and her 1-year-old child enter the lab. The child has a chance to explore the room. Next, a stranger enters. Then, the mother leaves. Later, the mother reunites with the child (Ainsworth & Bell, 1970). This situation is very real to the child. Many children were extremely upset when their mother left and very happy when she returned. How children behave in the strange situation also seems to relate to how the child behaves in real life and even relates to the child's social skills and self-confidence 10 years later (Elicker, Englund, & Sroufe, 1992).

The lab experience can be real to adults as well. For example, Ickes and some of his students (Ickes, Robertson, Tooke, & Teng, 1986) brought pairs of opposite-sex strangers to the lab. The strangers sat down next to each other on a couch, supposedly to view slides that they were to judge. As the slide projector warmed up, the projector bulb appeared to pop. As the researcher left to find a bulb, the students began to talk—and their talking was what Ickes observed. The situation was quite real to the students—and not that different from real-life situations in which two students who arrive early to class find themselves talking to each other.

Even when people know they are being videotaped and have sensors clipped to their ears and fingers, they may behave naturally. For example, under these conditions, married couples will still argue with each other freely—and their behavior during conflict predicts with greater than 94% accuracy whether they will be married 5 years later (Carrere & Gottman, 1999). Yet, despite the impressive generalizability of lab observation, many researchers want to observe behavior in a more realistic setting. Such researchers use either naturalistic observation or participant observation.

In **naturalistic observation**, you try to observe the participants in a natural setting *unobtrusively*: without letting them know you are observing them. Often, naturalistic observation involves keeping your distance—both physically and psychologically.

In **participant observation**, on the other hand, you actively interact with your participants. In a sense, you become "one of them."

Both types of observation can lead to ethical problems because both may involve collecting data without participants' informed consent.[4] Naturalistic observation may involve spying on your participants from a distance; participant observation may involve spying on a group that you have infiltrated.

---

[4] According to most ethical guidelines, people in studies should be volunteers who know what they have volunteered for. For more on ethics, see Chapter 2 and Appendix D.

Because the participant observer is more likely to have a direct effect on participants, most people consider participant observation to be more controversial than naturalistic observation is.

But which method provides more valid data? Not everyone agrees on the answer to this question. Supporters of participant observation claim that you get more "inside" information by using participant observation. Fans of naturalistic observation counter that the information you get through participant observation may be tainted. As a participant, you are in a position to influence (bias) what your participants do. For example, in a classic study of cults (Festinger, Riecken, & Schachter, 1956), five researchers infiltrated a cult, thus increasing the size of the cult from 9 to 14. Critics also point out that, as an active participant, you may be unable to sit back and record behavior as it occurs. Instead, you may have to rely on your (faulty) memory of what happened.

Problems With Observation

Whether you use participant or naturalistic observation, you face two major problems. First, if participants know they are being watched, they may not behave in their normal, characteristic way. That is, observational research can be *reactive:* participants may *react* to being watched. Second, even if participants act "naturally," you may fail to record their behavior *objectively.* That is, your personality and motives may affect what things you ignore and how you interpret those things you do pay attention to.

Dealing With Effects of the Observer on the Observed: Making Observation Less Reactive. To deal with the reactivity problem, the problem of changing behavior by observing it, you might observe participants unobtrusively (without their knowledge). For example, you might want to observe participants through a one-way mirror.

If you can't be unobtrusive, try becoming less noticeable. One way to do this is to observe participants from a distance, hoping that they will ignore you. Another way is to let participants become familiar with you, hoping that they will eventually get used to you. Once participants are used to you, they may forget that you are there and revert back to normal behavior.

Dealing With Difficulties in Objectively Coding Behavior. Unfortunately, steps you might take to make observers less reactive, such as observing participants from a distance, may make observers less accurate. For example, if observers can't easily see or hear participants, they may record what they expected the participant to do rather than reporting what the participant actually did. However, even when observers can observe behavior at close range, observations may lack objectivity. That is, as with archival research (which you could consider indirect observation), one problem with observation is that different observers may code the same behavior differently.

As with archival data, the way to see whether different observers are coding the same behavior differently is to have more than one observer rate the same behavior and then obtain some index (percent of times they agree, correlation between raters, Cohen's kappa) of interjudge reliability. As was the case with archival data, the way to maximize interjudge agreement is to use a clear coding scheme. You need to

1. define your categories in terms of specific target behaviors;
2. develop a check sheet to mark off each time a target behavior is exhibited; and
3. train and motivate raters to use your check sheet.

Training and motivating your raters are even more important in observational research than in archival research because, in observational research, there are often no permanent records of the behavior. Thus, unmotivated or disorganized raters do not get a second chance to rate a behavior they missed: There is no instant replay. Furthermore, without permanent records, you cannot check or correct a rater's work.

We have shown you why training is so important. But how do you train observers to categorize behavior? Training should involve at least three steps. First, you should spell out what each category means, giving both a definition of each category and some examples of behaviors that belong and do not belong in each category. Second, you should have your observers rate several videotaped examples of behavior, and you should tell them why their ratings are right or wrong. Third, you should continue the training until each rater is at least 90% accurate in his or her categorizations.

Conclusions About Observation

In conclusion, observation can be a powerful technique for finding out what people do. However, the observer may influence—rather than merely record—the individuals being observed, and the observer's biases may influence what the observer "sees," "remembers," and records.

Tests

If you do not want to rely on observers, you may decide to use tests. Tests are especially useful if you want to measure ability, knowledge, or personality variables. For instance, you might correlate scores on an extroversion test with scores on a happiness test.

External Validity

As was the case with ex post facto research, the external validity of a study that uses tests depends on the representativeness of the sample. You cannot generalize your results to a population unless you have a random sample of that population. Therefore, you cannot say that women score more extroverted on an extroversion test than men unless you have a random sample of all men and women. Similarly, you cannot say that extroverts are happier than introverts unless you have a random sample of all introverts and extroverts.

Internal Validity

As is the case with all correlational research, if you find a relationship between test scores, that relationship is not necessarily a causal relationship. For example, if extroverts are happier than introverts, we don't know whether extroversion causes happiness, happiness causes extroversion, or some other factor (supportive parents, social skills, etc.) causes both extroversion and happiness.

The fact that correlation does not prove causation is important to keep in mind. Without an understanding of this concept, you may mistake circumstantial evidence for proof. For example, certain authors try to show a genetic basis for some characteristics (career preferences, schizophrenia, introversion, etc.) by showing that identical twins score similarly on a test of a particular trait. However, identical twins could be similar on the trait because they share a similar environment or because they have influenced one another.

Conclusions About Using Tests

By using tests, you can take advantage of measures that other people have spent years developing. As a result, construct validity is usually less of a problem than if you had devised your own measures. Furthermore, tests are often easier to use than other measures. Because of these advantages, tests are often used in experimental as well as nonexperimental research. When used in nonexperimental research, however, this research has the same weaknesses as other correlational research: It doesn't allow you to make cause–effect statements, and the generalizability of your results will only be as good as the representativeness of your sample (to compare different descriptive designs, see Table 7.1).

TABLE **7.1** Comparing Different Correlational Methods

Validity	Ex Post Facto	Archival	Observation	Tests
Internal validity	Poor	Poor[a]	Poor	Poor
Construct validity	Fair	Fair to poor	Fair to poor	Fair to good
Objective—Avoids observer bias	Good	May be good	May be poor	Good
Nonreactive—Avoids subject bias	Often a problem	Often good	Can be poor	Reactive—But steps can be taken to control for subject biases
Operational definition is consistent with definition of the construct	Fair to good	Often poor	Fair	Good
External validity				
Ease of getting a large representative sample	Depends on original study	May be easy	Difficult	May be easy

[a]Internal validity will be poor unless you find a situation in which a random process determines which treatment people receive. For example, two anthropologists found that Olympians who were randomly assigned to wear red were more likely to win than those who were randomly assigned to wear blue.

ANALYZING DATA FROM DESCRIPTIVE STUDIES: LOOKING AT INDIVIDUAL VARIABLES

Once you have coded your data, you should compile and summarize them. You want to know what your data "look like."

You may start by describing participants' scores on one or more key variables. Often, summarizing those scores will involve calculating both (a) the average score as well as (b) an index of the degree to which scores vary from either that average or from each other. For example, you might report that the mean (average based on adding up all the scores and then dividing by the number of scores) score on the personality test was 78 and the *range* (the highest score minus the lowest score) was 50. Instead of reporting the range, you will probably report the **standard deviation** (SD): an index of the extent to which individual scores differ (deviate) from the mean, a measure of the degree of scatter in the scores.[5] For example, you might say that the mean was 78 and the standard deviation was 10.

Researchers must mention both the average score and an index (like the range or, better yet, the standard deviation) of the extent to which scores vary. If researchers mentioned only the average score, it would lead to many problems. One problem would be that descriptive research, rather than providing a deeper and richer appreciation of people, might lead to labeling, stereotyping, and other oversimplifications. For example, consider the problems caused by people knowing that the average age when infants begin talking is 12 months. The problem is that half of all infants are going to talk later than that. Many of those infants' parents, not understanding the wide range at which children begin to talk, needlessly worry that their child's development is delayed. Similarly, take the research suggesting that the average teenager today is as stressed as the average teenager in therapy in the 1950s (Twenge, 2002). Without considering the variability, knowing this fact might cause some people to stereotype today's teenagers as all being neurotic kids.

One way to describe the variability of scores is to display the **frequency distribution**—how often each score occurs—in a graph. From left to right, across the bottom of the graph, the possible scores are arranged from lowest to highest. Thus, the lowest score is on the bottom left of the graph and the highest score is on the bottom right of the graph. The frequency of a particular score is indicated by the height of the point above that score. Thus, if there is no point above a score, no one had that score (the score's frequency

---

[5] The lowest the standard deviation can be is zero. You would get a zero only if everyone in the group scored at the mean. In that case, there would be zero (no) deviations from the mean. If you want a rough estimate of the standard deviation, divide the range by 6. If you want a more precise estimate of a population's standard deviation and you have a random sample from that population, (a) get the differences between each score and the mean by subtracting each score from the mean, (b) square each of those differences, (c) get the sum of those squared differences (also called "sum of squares") by adding (summing) up all those squared differences, (d) get the variance by dividing the sum of the squared differences by one less than the number of scores, and (e) get the standard deviation by taking the square root of the variance. For more on calculating and using the standard deviation, see Appendix E.

is zero). Conversely, the highest point will be above the **mode**: the score that occurred most often; the most frequent score. To draw a crude frequency distribution, start near the top left corner of a page and draw a line straight down almost to the bottom of the page. This vertical line is called the y-axis. Because you will use this line to represent how frequently scores occur, label this line "frequency" (see Figure 7.3a).

Your next step is to draw a line that goes from the bottom of the y-axis straight across to the right side of the page. (If you are using lined paper, you may be able to trace over one of the paper's horizontal lines.) This horizontal line is called the x-axis and will represent your scores, so label this x-axis with numbers representing possible scores on your measure (see Figure 7.3b). For example, if the scores could range from 0 to 10, the bottom left-hand part of the graph would be labeled "0" and the bottom right-hand part of the graph would be labeled "10." Then, find the mode, the most common score. For each person who scored at the mode, put an "X" above the mode (see Figure 7.3c). After making a column of "Xs" at the mode (each "X" representing one person who scored at the mode), repeat the process for the rest of the possible scores.

Once you are done plotting the scores, your distribution will probably look like the normal distribution in Figure 7.3d. This bell-shaped distribution shares at least three characteristics with every *normal distribution.*

First, the center of the distribution is at the mean. One indication that the mean is at the middle of the distribution is that the mean is the most common score. In other words, the mean is also the mode in a normal distribution, as indicated by the fact that the tallest row of "Xs" is at the mean. A stronger indication that the mean is the distribution's middle point is that just as many scores are above the mean as below the mean: If you count the "Xs" below the mean, you will know how many are above the mean. In other words, for the normal curve, the mean is the same as the **median**: the middle score, the score at which just as many scores are above as are below (just as the median of the highway is in the middle of the road, the median of a set of scores is in the middle of the scores).

Second, not only is the distribution balanced on the mean but the distribution is symmetrical. That is, if you fold the distribution in half at the mean, the two halves will match.

Third, the distribution extends for about three standard deviations in both directions from the mean, with about 2/3 of the scores being within one standard deviation of the mean. Relatively few of the scores (less than 5%) are more than two standard deviations from the mean.

But what if your frequency distribution does not look like a normal distribution? That is, what if, instead of having a symmetrical normal distribution, your frequency distribution is skewed (tilted) to one side of the mean, like the distribution in Figure 7.3e? Such a skewed distribution is likely if you use a reaction-time measure—and reaction times are used to measure many constructs, from unconscious prejudice to intelligence to personality (Robinson, Vargas, Tamir, & Solberg, 2004).

One problem with skewed distributions is that a few extreme scores (those causing the skew) can distort (skew) the mean. For example, if a

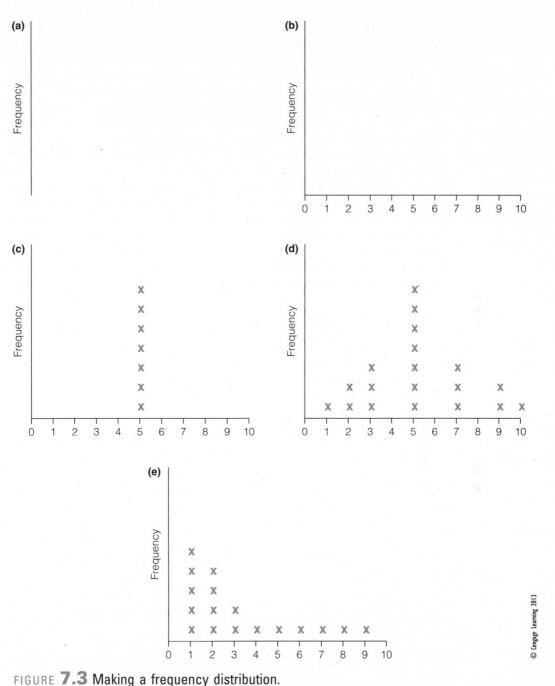

FIGURE **7.3** Making a frequency distribution.

participant's reaction times were 0.1, 0.1, 0.1, 0.2, and 5 seconds, the participant's mean score would be 1.1 seconds. Note that although the one extreme score (the 5-second reaction time) throws off the mean, it does not throw off the median score (indeed, in this example, no matter what the last score is, the median—the middle score—will be 0.1). Therefore, if the researcher has a skewed distribution, the researcher may want to use the median (middle) score rather than the mean.[6]

If you put people into categories such as "helped" or "didn't help," you have nominal (qualitative, categorical) data. Calculating a mean on qualitative data makes no sense. For example, it doesn't make sense to say, "mean helping was 0.4." With categorical (nominal) data, you should use percentages (e.g., "40% of people helped") to summarize your data. If you must use an average, use the mode (e.g., "the most common [modal] behavior was to avoid eye contact with the person needing help").

In short, the most appropriate average score for your data could be a mean, a median, or a mode. However, most of the time, the most appropriate average will be the mean.

If your mean is based on a random sample of a larger group, you may want to estimate the mean of that population.[7] For example, Brescoll and LaFrance (2004), using a random sample of major U.S. newspapers, looked at the degree—on a scale of 1 (extremely opposed) to 5 (extremely in favor of)—to which newspapers opposed or supported women being allowed to enter military academies. The average rating for the newspaper editorials in their sample was 3.48. Thus, the best guess about the average extent to which all editorials in all major U.S. newspapers opposed or supported women being allowed to enter military academies—the population mean—would also be 3.48.

This estimate of the population average, the sample mean, may differ from the actual population average. Therefore, you may want to provide not only your estimate of the population mean, but also an estimate of how good your estimate is. In that case, you would probably establish a range in which the population mean is likely to fall. Often, researchers establish a **95% confidence interval**: a range in which you can be 95% sure that the population mean falls. You can establish 95% confidence intervals for any

---

[6] Some researchers still use the mean with reaction-time data. However, these researchers often (a) throw out reaction times that are abnormally long (or replace those times with a certain value, such as 2 seconds) and (b) use some mathematical transformation of the data to make the data normally distributed.

[7] If you have categorical data (e.g., number of people who helped), you can still use the confidence interval technique we describe in the next section. The only differences are that (a) instead of the mean, you use the proportion of participants who did the behavior (e.g., .40 [40%] of the participants helped) and (b) instead of basing your standard error on the standard deviation, you calculate it by (1) multiplying p by $(1 - p)$, (2) dividing that quantity by n, and then (3) taking the square root. For example, suppose p was .40 and you had 240 observations. Your sample mean equivalent is p, which is .40. Now you need the standard error, which you can calculate in three steps. The first step would be to multiply p by $(1 - p)$. That would be $.4 \times (1 - .4)$, which is $.4 \times .6$, which is .24. The second step would be to divide .24 by 240, which would give you .001. The third step would be to take the square root of .001, which would be .03. Once you have the standard error, setting up the confidence interval for proportion is just like doing the confidence interval for the mean. Thus, in this case, the confidence interval for the population proportion would go from approximately 2 standard errors below the sample proportion to approximately 2 standard errors above the sample proportion (i.e., approximately $.40 \pm (2 \times .03) = .40 \pm .06 = .34$ to .46). For more specifics, see the Chapter 7 website.

population mean from the sample mean if you know the **standard error of the mean**.[8] You establish the lower limit of your confidence interval by subtracting approximately 2 standard errors from the sample mean.[9] Then, you establish the upper limit of your confidence interval by adding approximately 2 standard errors to the sample mean.

In this example, because the sample size was above 60, we can be more than 95% confident that the true population mean is within 2 standard errors of the mean. Thus, because the average was 3.48 and the standard error was .06, we can be more than 95% confident that the true population mean is somewhere between 3.36 (2 standard errors below the mean) and 3.60 (2 standard errors above the mean). Because 3.60, the upper limit of our confidence interval, is below 4, we can be very confident that the true population mean is below 4. The fact that the true population mean is below 4 is of interest because, on the scale, 4 represented "supporting the women's right to go to a military academy—but with reservations." Consequently, the results suggest that there is a conservative bias in newspapers, at least as far as a woman's right to serve in the military is concerned.

In this example, we used confidence intervals to find that the population mean is probably below 4. However, we could have found the same result using the one-sample t test. That is, a one-sample t test on that same data would find that the extent to which newspapers supported women being allowed to enter military academies was rated, on average, as being significantly (reliably) different from 4. The results of such an analysis might be reported as, "The mean rating of 3.48 was significantly less than 4, $t(325) = 8.46$, $p < .001$."[10]

Should you use confidence intervals or one-sample t tests? To answer that question, let's compare the results of the confidence interval with the results of the one-sample t test. The one-sample t test told us one thing that the

---

[8] Many calculators and web pages can calculate the standard error of the mean for you (our website has links to some of those calculators). If you need to calculate the standard error, take the standard deviation and divide it by the square root of the number of observations. If you don't have the standard deviation, you can calculate it by following the steps in Footnote 5 or by using the formula $\sqrt{\Sigma(X - M)^2/(N - 1)}$. In this case, the standard deviation was 1.11 and the number of observations was 326. Thus, the standard error was .06 (or $1.11/\sqrt{326} = 1.11/18.06 = .06$).

[9] The exact number will usually vary from 1.96 to 2.776, depending on how many participants you have. To be more precise, the exact number will depend on your degrees of freedom (df)—and your df will be 1 less than your number of participants. For example, if you have a mean based on 11 participants' scores, your *df* will be 10. Once you have calculated your df, go to the *t* table (Table 1) in Appendix F. In that table, look under the .05 column (it starts with 12.706) and find the entry corresponding to your df. Thus, if you have a *df* of 10, you would multiply your standard error by 2.228; if you had a *df* of 120, you would multiply your standard error by 1.98.

[10] "Significantly" means reliably. The *t* of 8.46 was calculated by subtracting the observed mean (3.48) from 4 and then dividing by the standard error of the mean (.06). If the real mean was 4, the chances of getting an observed mean as low or lower than 3.48 are fewer than 5 in 100 ($p < .05$). The "325" refers to the degrees of freedom (df) for the test, which is the number of observations minus 1 (1 df is lost computing the 1 sample mean).

confidence interval told us: that it was unlikely that the true mean of our sample was 4.0. However, the *t* test failed to tell us two things that the confidence interval did.

First, it didn't tell us how close our population mean could be to 4—it told us only that the population mean is reliably different from 4. Thus, whereas our confidence interval told us where the population mean is likely to be as well as what the population mean isn't, the *t* test only told us what the population mean isn't.

Second, the one-sample *t* test didn't tell us how accurate our estimate of the population mean was. For instance, the *t* test doesn't tell us whether our estimate is highly accurate, as indicated by a narrow confidence interval, such as one between 3.4 and 3.5, or whether the estimate is imprecise, as indicated by a wide confidence interval, such as one between 1.1 and 3.9.

In conclusion, despite the one-sample *t* test's popularity, it really doesn't tell you anything more than a confidence interval does—and it sometimes tells you less (Cumming & Finch, 2005).[11] Therefore, APA now recommends that when you report the *t* test for your data, you also report the confidence interval for those data (APA, 2010; Cooper, 2011).

ANALYZING DATA FROM DESCRIPTIVE STUDIES: LOOKING AT RELATIONSHIPS BETWEEN VARIABLES

Although you can answer an interesting question by describing how one group of participants scored on one measure, you will usually answer more interesting questions if you also look at how participants' scores on one measure relate to their scores on some other measure. For example, rather than just knowing at what ages children begin to talk, you may want to know what relationship age of talking has with future success. Similarly, rather than knowing the average anxiety levels of teenagers, you might want to know whether boys are less anxious than girls. Sometimes, the simplest way to describe relationships between two variables is to look at 2 means. For example, you might compare the means for men and the means for women on your measure. Or, you might compare the happiness of a group of lottery winners against a group of people who, other than winning the lottery, seem similar to those lottery winners.

Comparing Two Means

To begin to compare the 2 means, you could subtract the smaller mean from the larger mean to find the difference between means. Then, you could

---

[11] Both analyses involve comparing the mean and the standard error, so they both are similar. Indeed, you could use some algebra on the formula for confidence intervals to get the *t*.

calculate a 95% confidence interval for the difference between the means.[12] You would be interested in seeing whether that confidence interval included 0 because 0 would indicate no (0) difference between the 2 means. If your confidence interval did not include 0 (e.g., the lower and upper limits were both positive, or the lower and upper limits were both negative), you would be relatively confident that the difference between the means is not zero. In that case, you could say that the means are reliably different. If, on the other hand, your confidence interval included 0 (e.g., the confidence interval includes both a negative number and a positive number), there may be no (0) real difference between your means. In that case, you couldn't say that the means were reliably different.

In addition to the confidence interval, you could compute an independent (between) groups t test.[13] Like confidence intervals that did not include 0, a significant t test would tell you that the groups were reliably different. However, note what the significant t test does not tell you. Whereas a confidence interval might tell you about how big the difference was (e.g., between .1 and .2 points or between 3 and 5 points), a significant t test tells you only that the difference was probably not zero.

The t test is even less informative when it is not significant. In that case, its results are completely inconclusive. In contrast, using confidence intervals on the same data will tell you two things.

First, confidence intervals give you some idea about how big the relationship might be. That is, a confidence interval between $-.1$ and $+.1$ indicates the relationship, even if it exists, is small. A confidence interval between $-.1$ and 10, on the other hand, hints that a sizable relationship may exist.

Second, confidence intervals give you some idea about how big a role random error played in your study. For example, if the confidence interval is from $-.1$ to $+.1$, you have done a good job of dealing with random error. In such a case, you would probably conclude that the difference either doesn't

---

[12] To estimate the 95% confidence interval, you would (a) multiply the standard error of the differences by 2, (b) subtract that number from the mean to get the lower limit of the confidence interval, and (c) add that number to the mean to get the upper limit of the confidence interval. To get the exact confidence interval, rather than multiplying by 2, you would multiply by the number in the .05 column of the t table (Table 1) in Appendix F corresponding to your degrees of freedom (df). Note that your df would be 2 less than your number of participants. Thus, if you had 12 participants, you would have a df of 10 (12–2), and you would multiply your standard error of the differences by 2.228. If you need to calculate the standard error of the differences and you have the same number of participants in each group, you can simply (a) square the standard error of the mean of each group, (b) add those squared terms together, and (c) take the square root. If you need help calculating the standard error of the mean, see footnote 8.

[13] To compute a t, you would subtract your two group means and then divide by the standard error of the differences. To calculate the standard error of the differences by hand, you have three options: (1) use the formula: standard error of the differences $= \sqrt{(s_1^2/N_1) + (s_2^2/N_2)}$, where $s_1 =$ standard deviation of group 1, $s_2 =$ standard deviation of group 2, $N_1 =$ number of participants in group 1, and $N_2 =$ number of participants in group 2; (2) follow the brief instructions at the end of footnote 12; or (3) follow the more detailed set of instructions in Appendix E.

exist or, if it does, is too small to be of interest. Therefore, you would probably not redo the study to see if you could find a difference. If, on the other hand, the confidence interval ranged from –20 to +20, your failure to find a difference may be due to random error causing you to have an extremely imprecise estimate of the difference between the means. Therefore, you might try to redo the study by making changes that would (a) reduce random error, such as using more reliable measures, or (b) balance out random error's effects, such as using more participants.

In short, just as confidence intervals of means provide more information than one-sample t tests, confidence intervals of differences between means provide more information than independent group t tests. However, independent group t tests are useful and popular.

Doing a Median Split to Set Up the t Test

To do an independent group t test, you need two groups that differ on your predictor variable. But what if you don't have two groups? For example, what if you only have participants' self-esteem scores, which you are using to predict their grade-point averages (GPA)? In that case, you could use participants' self-esteem scores to create two groups: participants scoring in the top half on the self-esteem measure ("highs") and participants scoring in the bottom half ("lows"). Then, you could compare the GPA of the highs to the GPA of the lows. Dividing participants into two groups depending on whether they scored above or below the median (the middle score) on a predictor variable is called a **median split**.

Doing a median split and then conducting a t test is a common way of analyzing correlational data. You will frequently encounter such analyses in published articles (MacCallum, Zhang, Preacher, & Rucker, 2002).

The Case Against Doing a Median Split. Although the median split is popular, most statisticians argue that there are many reasons not to do it (MacCallum et al., 2002). The main reason is that using a t test based on median splits reduces your ability to find relationships (Cohen, 1990). This is because you have less information with which to work. Put another way, you have less power—ability to find differences—because you are recoding data in a way that hides differences. Instead of using participants' specific scores, you are using the median split to lump together all the participants who scored above average. Thus, a participant who scores 1 point above average gets the same score—as far as the analysis is concerned—as a participant who scores 50 points above average. Similarly, you are lumping together everyone who scored below average, despite the differences in their scores. In a sense, you are deliberately ignoring participants' actual scores.

Not surprisingly, some experts object to this waste. Cohen (1990), for example, argues that researchers should not lose power and information by "mutilating" variables. Instead of "throwing away" the information regarding a participant's specific score by doing a median split, Cohen believes

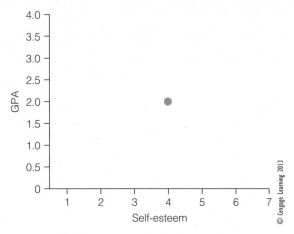

FIGURE **7.4** The beginning of a scatterplot.

that researchers should do correlational analyses that use participants' actual scores.

Graphing Scores

To begin using participants' actual scores, graph your data. Start by labeling the *x*-axis (the line that goes straight across the page) with the name of your predictor variable. More specifically, go a few spaces below the bottom of the graph and then write the name of your predictor variable. Next, label the other axis, the *y*-axis (the vertical line on the left side of the graph), with the name of your criterion (outcome) measure. Then, plot each observation.

For example, suppose we were trying to see whether we could use self-esteem to predict grade-point average (GPA). Figure 7.4 shows the beginning of such a graph. As you can see, we have plotted the score of our first participant, a student who has a score of 4 on the self-esteem scale and a 2.0 GPA. As we *plot* more and more of our data, the points will be *scattered* throughout the graph. Not surprisingly, then, our graph will be called a **scatterplot**. There are four basic relationships that the scatterplot could reveal.

A Positive Relationship

The scatterplot in Figure 7.5 shows a pattern that indicates that the higher one's self-esteem, the higher one's grade-point average is likely to be. Put another way, the lower one's self-esteem, the lower one's grade-point average will be. This kind of relationship indicates a **positive correlation** between the variables. One common example of a positive correlation is the relationship between height and weight: The taller you are, the more you are likely to weigh. Intriguing psychological examples are that smoking is positively correlated with sex drive, coffee drinking, stress, risky behavior, external locus of control (feeling that outside events control your life), negative affect (being in a bad mood), having problems in school, and rebelliousness.

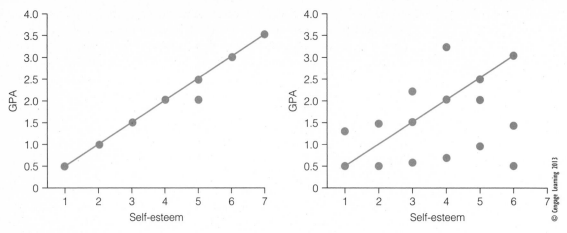

FIGURE **7.5** Scatterplots revealing positive correlations.

If a line through the points slopes upward, you have a positive correlation. The closer the points to that line, the stronger the relationship is. Thus, the graph on the left indicates a strong positive correlation; the graph on the right indicates a weaker positive correlation.

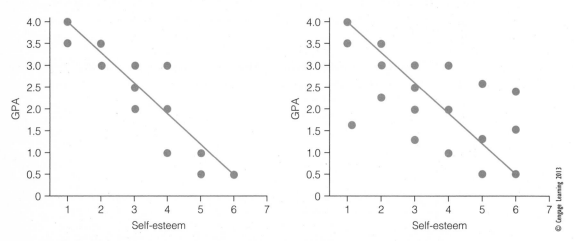

FIGURE **7.6** Scatterplots revealing negative correlations.

If a line through the points slopes downward, you have a negative correlation. The closer the points to that line, the stronger the relationship is. Thus, the graph on the left indicates a strong negative correlation; the graph on the right indicates a weaker negative correlation.

A Negative Relationship

The scatterplot in Figure 7.6 shows a second pattern: The higher one's self-esteem, the lower one's grade-point average tends to be. Put another way, the lower one's self-esteem, the higher one's grade-point average tends to be. This "reverse relationship"—as if the two variables were on a teeter-totter—indicates a **negative correlation** between the variables. Many variables are negatively (inversely) related. One common example of a negative correlation is the relationship between miles run and weight: The more miles you run, the less you tend to weigh. Smoking is negatively correlated with internal locus of control (feeling in control of your life), positive affect (being in a good mood), doing well in school, and conformity.

Note that whether we have a positive or a negative relationship may depend on how we label or measure our variables. For example, suppose we find that people with high self-esteem answer math questions more quickly than people with low self-esteem. In that case, the type of correlation (positive or negative) we obtain will depend on whether we measure quickness in terms of speed (number of questions answered in 1 minute) or in terms of time (average time it takes to answer one question). If we measure speed in terms of questions answered per minute, we would find a positive correlation between self-esteem and speed (higher self-esteem, more questions answered in one minute). If, on the other hand, we measure time taken to answer a question, we would find a negative correlation between self-esteem and time (more self-esteem, less time to answer a question). Similarly, if high self-esteem individuals did better on math tests, we would find a positive correlation between self-esteem and questions correct (higher self-esteem, higher percentage correct) but a negative correlation between self-esteem and questions missed (higher self-esteem, fewer questions missed).

No Relationship

The scatterplot in Figure 7.7 shows a third pattern: no relationship between self-esteem and grade-point average. This pattern reflects a **zero correlation** between the two variables.

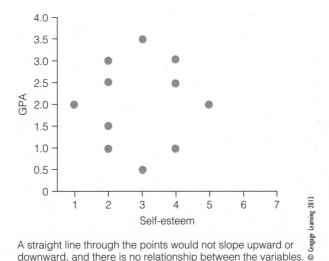

A straight line through the points would not slope upward or downward, and there is no relationship between the variables. © Cengage Learning 2013

FIGURE **7.7** A scatterplot revealing a zero correlation.

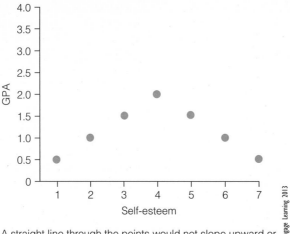

A straight line through the points would not slope upward or downward, but there is a relationship between the variables. © Cengage Learning 2013

FIGURE **7.8** A scatterplot revealing a nonlinear relationship.

A Nonlinear Relationship

The scatterplot in Figure 7.8 shows a fourth pattern: a nonlinear relationship between self-esteem and grade-point average (GPA). As you can see, in a complex, nonlinear relationship, the relationship between self-esteem and GPA may vary, depending on the level of the variables. Thus, in the low ranges of self-esteem, self-esteem may be positively correlated with GPA, but in the high ranges, self-esteem may be negatively correlated with GPA. Such a pattern could emerge in any situation in which a low amount of a variable could be too little, a medium amount of a variable could be just right, and a high level of the variable could be too much. For example, with too little motivation, performance may be poor; with a moderate amount of motivation, performance could be good; and with too much motivation, performance might be poor.

Correlation Coefficients

Although a graph gives a good picture of your data, you may want to summarize your data with a single number that expresses the "go-togetherness" of the two variables: a **correlation coefficient**. The kind of correlation coefficient you use will depend on the nature of your data (see Box 7.3). Probably, you will use the most commonly calculated correlation coefficient—the Pearson *r*.

The Pearson *r*, like most correlation coefficients, ranges from –1 to +1. More importantly, like all correlation coefficients, the Pearson *r* summarizes the relationship described in a scatterplot with a single number.

The Pearson *r* will be consistent with the data in your scatterplot. When your scatterplot indicates a positive correlation between the two variables,

BOX **7.3**   **Different Kinds of Correlation Coefficients**

When reading journal articles, you may come across terms for correlation coefficients other than the Pearson *r*. Most of these terms refer to a different type of correlation coefficient (one notable exception is that the term "zero-order correlation" is usually just another name for a Pearson *r*). In addition to reading about different types of correlations, you may be asked to compute correlations other than the Pearson *r*. This table should help you understand the distinctions among these different coefficients.

   Although this box focuses on the differences between these correlation coefficients, these coefficients share commonalities. For example, all of them yield coefficients between −1 (a perfect negative correlation) and +1 (a perfect positive correlation). Furthermore, as Cohen and Cohen (1983) point out, the Pearson *r*, the point biserial, the phi coefficient, and Spearman's rho can all be computed using the same formula: the formula for the Pearson *r*. The difference in calculating the various correlation coefficients comes from what data are entered into that formula (see fourth column of the table). For example, if you were calculating Spearman's rho, you would not enter participants' actual scores into the formula. Instead, you would convert those scores to ranks and then enter those ranks into the formula.

Name of Coefficient	Level of Measurement Required	Example	Data Entered	Significance Test
Pearson product-moment correlation (r)	Both variables must be at least interval	Height with weight	Actual scores	t test[a]
Point biserial (r_{pb})	One variable is interval, the other nominal or dichotomous (having only two values)	Weight with gender	Actual scores for the interval variable, 0 or 1 for the nominal variable	t test[a]
Spearman's rho (r_s)	Ordinal data	High school rank with military rank	Ranks	Chi-square test
Phi coefficient (Φ)	Nominal data	Gender with learning style	Zeros and ones: zero if the participant is not a member of a category; 1 if the participant is a category member	Chi-square test

[a]To calculate *t*, you could use the following formula (*n* refers to the number of participants). [a]$t = r/\sqrt{(1 - r^2)/(n - 2)}$

your correlation coefficient will also be positive. When the scatterplot indicates that your variables are not related, your correlation coefficient will be close to zero. Finally, when your variables are negatively correlated (inversely related), the correlation coefficient will be negative.

Pearson r and the Definition of Correlation

If you want to compute a Pearson r, you can use a computer, a calculator (links for online calculators are on our website), or a formula.[14] At this point, however, we do not want you to focus on how to compute the Pearson r. Instead, we want you to focus on understanding the logic behind the Pearson r either by relating the Pearson r to (a) the definition of correlation or (b) graphs of correlational data.

As you know, the correlation coefficient is a number that describes the relationship between two variables. If the variables are positively correlated, when one variable is above average, the other is usually above average. In addition, when one variable is below average, the other tends to be below average. If the variables are negatively correlated, the reverse happens: When one is above average, the other is usually below average.

To see how the Pearson r mathematically matches that description, suppose that we have a pair of scores for each of 20 students: (a) one score telling us whether that student scored above or below average on a vocabulary test that was based on words that a teacher had just tried to teach, and (b) a second score telling us whether that student scored above or below average on a test of picking up on nonverbal cues.[15] To see whether student learning is correlated with sensitivity to nonverbal cues, we go through a two-step process.

First, we add one point for each student whose pair of scores matches and subtract a point for each student whose two scores do not match.[16] Given that we have 20 participants, our total could range from –20 (mismatches between the pairs of scores for all 20 participants) to +20 (matches between the pairs of scores for all 20 participants).

---

[14] For example, you could use the following formula:

$$r = \frac{N\Sigma XY - (\Sigma X)(\Sigma Y)}{\sqrt{(N\Sigma X^2 - (\Sigma X)^2)(N\Sigma Y^2 - (\Sigma Y)^2)}}$$

N refers to the number of pairs of scores, X refers to scores on the first variable, and Y refers to scores on the second variable. To use this formula, you need to know that ΣX^2 means you square everyone's score on the first variable and then add up all those squared terms, but $(\Sigma X)^2$ means you add up everyone's scores on the first variable and then square that sum. Similarly, ΣY^2 means you square everyone's score on the second variable and then add up all those squared terms, but $(\Sigma Y)^2$ means you add up everyone's scores on the second variable and then square that sum. Finally, whereas ΣXY refers to multiplying each person's score on the first variable by their score on the Y variable and then adding up all those results, $(\Sigma X)(\Sigma Y)$ means to get the total of all the X scores and multiply that by the total of all the Y scores. If you are not comfortable with formulas and want step-by-step directions for computing Pearson r, see Appendix E.

[15] For a published example of a Pearson r calculated on these two variables, see Bernieri (1990).

[16] Mathematically, we could do this by first giving the students either a "+1" if they were above average on the test of definitions or a "–1" if they were below average. Then, we would give them either a "+1" if they were above average on the reading nonverbal cues test or a "–1" if they were below average. Finally, we would multiply each person's scores together. If the person scored the same on both tests, the result would be +1 (because +1 × +1 = 1 as does –1 × –1). If the person scored differently on the two tests, the result would be –1 (because +1 × –1 = –1).

Second, to give us a number that could range from –1 to +1, we divide our total by 20 (the number of participants). This number would be a crude index of correlation.

If most students' scores on the one test match their scores on the other test, our correlation will be positive. If all students' scores on one test match their scores on the other test, as in the case below, our correlation would equal +1.

	Sensitivity to nonverbal cues	
Words learned	Below average	Above average
Below average	10	0
Above average	0	10

© Cengage Learning 2013

Conversely, if participants who are above average on one variable are usually below average on the other variable, we will end up with a negative correlation. Indeed, if everyone who is high on one variable is also low on the other, as is the case below, our correlation index would equal –1.

	Sensitivity to nonverbal cues	
Words learned	Below average	Above average
Below average	0	10
Above average	10	0

© Cengage Learning 2013

Finally, consider the case below in which there is no relationship between the variables. In that case, the mismatches (–1s) cancel out the matches (+1s), so the sum of the points is 0, and our coefficient will end up being 0.

	Sensitivity to nonverbal cues	
Words learned	Below average	Above average
Below average	5	5
Above average	5	5

© Cengage Learning 2013

Mathematically, the Pearson r is a little more complicated than what we have described. However, if you understand our description, you understand the basic logic behind the Pearson r.

Pearson r and the Scatterplot

We have discussed how the Pearson r produces a number that is consistent with a table of the data. Now, we will show how the Pearson r produces a number that is consistent with a graph of the data.

Pearson r could be estimated by drawing a straight line through the points in your scatterplot. If the line slopes upward, the correlation is positive. If the line slopes upward and every point in your scatterplot fits on that line, you have a perfect positive relationship, reflected by a +1.00 correlation. Usually, however, there are points that are not on the line (if the line represents the rule, "the higher an individual is on one variable, the higher that individual will be on the second variable," the points not on the line represent exceptions to the rule). For each point that is not on the line, the correlation coefficient is made closer to zero by subtracting a value from the coefficient. The farther the point is from the line, the larger the value that is subtracted. Once all the misfit points are accounted for, you end up with the correlation coefficient.

If, on the other hand, the line that fits the points slopes downward, the correlation is negative. If every point fits on that line, you have a perfect negative relationship, reflected by a −1.00 correlation. However, perfect negative relationships are rare. Most of the time, many points will not be on that line. For each point that is not on the line, the correlation coefficient is made closer to zero by adding a value to the coefficient. The farther the point is from the line, the larger the value that is added. After all the misfit points are accounted for, you end up with the correlation coefficient.

As we have just discussed, the correlation coefficient describes how well the points on the scatterplot fit a straight *line*. That is, the correlation coefficient describes the nature of the *line*ar relationship between your variables. But what if the relationship between your variables is not described by a straight line, but by a curved line? For example, suppose the relationship between your variables was *nonlinear*, like the nonlinear relationship depicted in Figure 7.8 on p. 256.

The fact that the correlation coefficient examines only the degree to which variables are linearly related is not as severe a drawback as you may think. Why? First, completely nonlinear relationships among variables are rare. Second, even if you encounter a nonlinear relationship, you would know that you had such a relationship by looking at your scatterplot. That is, you would notice that the points on your scatterplot fit a nonlinear pattern, such as a U-shaped curve.

If there is a linear relationship between your variables, the correlation coefficient can tell you how strong this relationship is—if you know what to look for and what to ignore. Ignore the sign of the coefficient. The *sign* tells you only the *kind* of relationship you have (the direction, either positive or negative). The sign does not tell you how *strong* the relationship is.

To get a general idea of how strong the relationship is, look at how far the correlation coefficient is from zero. The farther the correlation coefficient is from zero (no relationship), the stronger the relationship. Thus, because −.4 is farther from 0 than +.2, a −.4 correlation indicates a stronger relationship than a +.2 correlation.

The Coefficient of Determination

To get a better idea of the strength of relationship between two variables, square the correlation coefficient to get the **coefficient of determination**: an index of the degree to which knowing participants' scores on one variable helps in predicting what their scores will be on the other variable. The coefficient of determination can range from 0 (knowing participants' scores on one variable is no [0] help in predicting what their scores will be on the other variable) to 1 (knowing participants' scores on one variable allows you to know exactly what their scores will be on the other variable). To use more technical terminology, the coefficient of determination can range from 0 (the predictor accounts for 0% of the variation in the other variable) to 1.00 (the predictor explains 100% of the variance in the other variable).

When we square the correlation coefficients from our previous example, we find that the coefficient of determination for the relationship described by a −.4 correlation (.16) is much bigger than the coefficient of determination for the relationship described by a +.2 correlation (.04). In journal articles, researchers might describe the first relationship by saying that "16% ($-.4^2 = .16 = 16\%$) of the variability in scores was explained (accounted for) by the relationship between the variables" and might describe the second relationship by saying that "the relationship explained only 4% ($.2^2 = .04 = 4\%$) of the variance."

Note how small a coefficient of determination of .04 is. It is close to the lowest possible value: 0 (knowing participants' scores on one variable is absolutely no help in predicting what their scores will be on the other variable). It is far away from the highest possible value: +1.00 (knowing participants' scores on one variable allows you to know exactly what their scores will be on the other variable). Realize that correlation coefficients between −.2 and +.2 will produce coefficients of determination of .04 or below. Thus, if the correlation between your predictor and outcome variables is between −.2 and +.2, basing your predictions on your predictor will be only slightly better than simply predicting that everyone will score at the mean.

Determining Whether a Correlation Coefficient Is Statistically Significant

On rare occasions, you may want to describe—but not generalize from—a particular sample. If you just want to describe the relationship between self-esteem and grade-point average in one particular class during one particular term, a scatterplot, the correlation coefficient, and the coefficient of determinations are all you need.

Most of the time, however, you are interested in generalizing the results obtained in a limited sample to a larger population. You know what happened in this sample, but you want to know what would happen in other samples.

To generalize your results to a larger population, you first need a random sample of that population. If you want to generalize results based on observing a few students in your class to all the students in your class, the participants you examine should be a random sample of class members. If you want to generalize the results based on measuring a few people to all Americans, you must have measured a random sample of Americans. If you want to generalize results based on observing two rats for an hour a day to all the times that the rats are awake, the times you observe the rats must be a random sample from the rats' waking hours.[17]

---

[17] Many researchers do not randomly sample from a population, but they still generalize their results. How? They argue that their sample could be considered a random sample of an unknown population. Then, they use statistics to determine whether the results are due to sampling error or whether the results hold in the larger population. If their results are statistically significant, they argue the results hold in this unspecified population. (The "unspecified population" might be "participants I would study at my institution.")

Random samples, however, are not perfect samples. Even with a random sample, you are going to have sampling error. For example, suppose you studied a random sample of sophomores at your school and found a correlation of −.30 between grade-point average and self-esteem. Clearly, you found a negative correlation in your random sample. However, you can't say that if you had studied all sophomores at your school, you would have obtained a negative correlation coefficient.

To convince yourself that what happens in a sample does not necessarily precisely mimic what happens in the population, you could conduct the following study. Find three people. Have each of these three people flip a coin one time. Record each person's height and the number of "heads" (0 or 1) the person flipped. Do this for 10 different samples. Graph each sample individually. Even though there is no reliable relationship between a person's height and how many "heads" he or she will flip, some of your graphs will reveal a positive correlation, whereas others will reveal a negative correlation.

As you have seen, even if the two variables are not related, they will appear to be related in some samples. That is, *a relationship that exists in a particular sample may not exist in the population.* Consequently, if you observe a relationship in your sample, you will want to know if you have observed (a) a real pattern that is characteristic of the population or (b) a mirage caused by random sampling error alone.

Fortunately, there is a way to determine whether what is true of your sample is true of the population: Use inferential statistics. Inferential statistics will allow you to determine (infer) how likely it is that the relationship you saw in your sample could be due to random error. Specifically, inferential statistics allow you to ask the question: "If there is no relationship between these variables in the population, how likely is it that I would get a correlation coefficient this large in this particular random sample?"

If the answer to this question is "not very likely," you can be relatively confident that the correlation coefficient in the population is not zero. Therefore, you would conclude that the variables are related. To use proper terminology, you would conclude that your correlation coefficient is *significantly* (reliably) different from zero (see Figure 7.9).

Exactly which statistical test you use to determine whether a correlation coefficient is statistically different from zero depends on which kind of correlation coefficient you have (if you want to review the different types of correlation coefficients, see Box 7.3). But regardless of which test you use, the test will estimate how unlikely it is that your sample's correlation

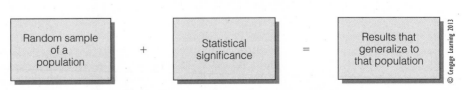

FIGURE **7.9** Necessary conditions for producing generalizable results.

coefficient came from a population in which the coefficient between those variables was zero. To determine whether your correlation coefficient came from such a population, the test will take advantage of two facts about random samples from populations in which the correlation is zero.

First, if the population correlation coefficient is zero, the sample's correlation coefficient will tend to be near zero. Consequently, *the farther the sample's correlation coefficient is from zero, the less likely the population coefficient is zero.* Thus, a correlation coefficient of .8 is more likely to be significantly different from zero than a correlation coefficient of .2.

Second, if the population correlation coefficient is zero, the larger the sample, the more likely that the sample's correlation coefficient will be near zero. Therefore, *the larger the sample, the more likely that a nonzero correlation coefficient indicates that the variables are related in the population.* Consequently, a correlation coefficient of .30 is more likely to be significantly different from zero if it comes from a sample of 100 observations than if it comes from a sample of 10 observations.

Interpreting Significant Correlation Coefficients

If a correlation coefficient is significantly different from zero, it should mean that there is a relationship between your variables. That is, the relationship between your variables, rather than being due to random error, is a reliable relationship (see Table 7.2). Note that we have *not* said that statistically significant results

- have external validity,
- allow you to make cause–effect statements,
- always indicate a reliable relationship, or
- are large.

TABLE **7.2** The Different Meanings of Statistical Significance

Question Asked About a Statistically Significant Result	Answer If You Conduct a Correlational Study	Answer If You Conduct an Experiment
Are the variables related?	Yes	Yes
Do we know whether the predictor variable *caused* changes in the criterion variable?	No	Yes
		1. The experimental design guaranteed that the treatment came before the change in the criterion (dependent) variable.
		2. The experimental design also guaranteed that the treatment was the only systematic difference between treatment conditions. Thus, the relationships between the variables could not be due to some third factor.

Significant Results May Not Have External Validity

A significant correlation indicates that the relationship you observed probably also exists in the population from which you randomly sampled. So, in your random sample of everyone in your country, if you find a significant correlation, you can generalize your results to your population—everyone in your country. If, however, your random sample is of students in your class—or if you didn't use a random sample—significant results do not necessarily generalize to your entire country.

Significant Results Do Not Allow You to Make Cause–Effect Statements

Ideally, statistical significance allows you to say that two variables are really related: The relationship you observed is not merely the result of a coincidence. However, even if you know that two variables are related, you do not know that they are *causally* related. As we said earlier in this chapter, to establish a cause–effect relationship between two variables, you must do much more than establish that your variables are statistically related.

For example, to infer that self-esteem caused low grade-point averages, you would have to show not only that self-esteem and grade-point average are related, but also that

1. The low self-esteem students had low self-esteem *before* they got low grades, and the high self-esteem students had high self-esteem *before* they got high grades.
2. No other differences between your high and low self-esteem individuals could account for this relationship (there were no differences between groups in terms of parental encouragement, IQ, ability to delay gratification, etc.).

Significant Results May Be False Alarms

As we've seen, significant results in correlational research do not mean that changes in one variable *caused* a change in the other. At best, significant results mean only that both variables are correlated. However, all too often, significant results don't even prove that two variables are correlated.

To understand why significant results don't prove correlation, suppose that a researcher is trying to determine whether two variables are correlated. The researcher uses the conventional $p < .05$ significance level, suggesting that a significant result means that if there is no relationship between these variables in the population, the probability (p) of obtaining a correlation coefficient this large or larger in this particular random sample is less than 5 in 100. Suppose further that, in reality, the variables aren't correlated. What are the chances that the researcher will obtain significant results?

You might be tempted to say "about 5%." You would be correct—if the researcher had conducted only one statistical test. However, because correlational data are often easy to obtain, the researcher might correlate hundreds of variables with hundreds of other variables. If the researcher does hundreds of statistical tests, many of these tests will be significant by chance alone.[18]

---

[18] The exception is if they use a sophisticated multivariate statistical test that controls for making multiple comparisons.

FIGURE **7.10** When people do multiple tests, there may be something fishy about significant results.

Put another way, if the researcher uses a $p = .05$ level of significance and does 100 tests, the researcher should expect to obtain 5 significant results, even if none of the variables were related. Thus, if you aren't careful, disciplined, and ethical, you will "find" relationships that are really statistical errors (see Figure 7.10). Therefore, we urge you to resist temptation to have the computer calculate every possible correlation coefficient and then pick out the ones that are significant. Instead, decide which correlation coefficients relate to your hypotheses and test to see whether those correlation coefficients differ from zero.

If you are doing more than one statistical test, there are at least two things you can do to avoid mistaking a coincidence for a correlation. One option is to make your significance level more conservative than the traditional $p < .05$ level. For example, use a $p < .01$ or even a $p < .001$ level. A second option is to repeat your study with another random sample of participants to see whether the correlation coefficients that were significantly different from zero in the first study are still significant in a second study.

Significant Results May Be Tiny and Insignificant: Bigger Than Nothing Isn't Everything

Even if you establish that the relationship between the variables is reliably different from zero, you have not shown that the relationship is large or important. If you had enough observations, a correlation as tiny as .02 could be statistically significant.

In many cases, the issue is not whether there is a relationship, but whether the relationship is large enough. Put another way, the question you should be asking is often not "Are they completely unrelated?" but rather "Are they strongly related?" For example, if you have two people rating the same behavior, the relevant question usually isn't "Do the raters agree at all?" but rather "To what extent do the raters agree?" Thus, experts would not be reassured by a correlation between raters of .10 that was significantly different from zero. Instead, they would usually want a correlation of at least .85.

Similarly, if you correlate your measure of a construct with another measure of the same construct, the question isn't whether the correlation is greater than zero but rather whether the correlation is strong enough to suggest that the two measures are measuring the same thing: You're not trying to show that there is some overlap between what the two tests measure; you are trying to show that there is considerable overlap. A correlation of .20 between two measures does not suggest that the two measures are measuring the same thing; a correlation of .80 does.

Finally, if you find a significant correlation between responses on two questionnaires you handed out, few psychologists will be impressed. This is because if both your measures are affected to any degree by the same response bias, scores on the measures will correlate to some extent because of response bias. For example, if giving socially acceptable answers tends to increase scores on both your scales, people who are more likely to give such answers will tend to score higher than other people on both measures. Likewise, if an individual tends to agree with items, this may make their responses on one questionnaire similar to their responses on another. Thus, if you find a small correlation between two questionnaires, the correlation does not mean that there is a relationship between the two variables that the two questionnaires were designed to measure. Instead, it may mean that both questionnaires are vulnerable to the same response bias.

In short, do not just look at whether a correlation is significantly different from zero. Instead, also look at the correlation's size, especially its coefficient of determination. Also, consider testing whether your correlation coefficient is significantly (reliably) greater than a certain meaningful value (e.g., .60) rather than just whether it is significantly greater than zero.

Interpreting Null (Nonsignificant) Correlation Coefficients

If your results are not statistically significant, it means that you *failed* to show that the correlations you observed were due to anything other than random error. It does not mean your variables are unrelated—it means only that you have failed to establish that they were related.

If there is a relationship between your variables, why would you fail to find it? There are four main reasons.

First, you may not have had enough observations. Just as you cannot determine whether a coin is biased by flipping it a few times, you cannot determine whether two variables are related by studying only a few participants. With few participants, even a strong relationship in your sample could be dismissed as being due to chance (just as getting 3 heads in 3 flips could be dismissed as a common coincidence). With more observations, on the other hand, you could argue that chance would be an unlikely explanation for your results (just as getting 100 heads in 100 flips would be unlikely).

Second, you may have failed to find a significant relationship because of **restriction of range**: You sampled from a population in which everyone is similar on one of the variables. Restriction of range is a problem because to say that both variables vary together, you need both variables to vary. If both variables don't vary, you end up with correlations of zero. To take an absurd example, suppose you were looking at the relationship between IQ and grade-point average (GPA), but everyone in your study had a 4.0 GPA. In that case, there would be no relationship in your study between IQ and GPA: No matter what participants' scored on the IQ test, their GPA would be 4.0. To take a more typical example, suppose that everyone in your sample scored between 125 and 130 on the IQ test. In that case, the correlation between IQ and GPA would be near zero; consequently, the correlation might not be significant. If, on the other hand, your participants' IQs had ranged from 75 to 175, you would probably have a sizable and statistically significant correlation between IQ and GPA.

Third, you may fail to find a significant relationship because you had insensitive measures. By preventing you from seeing how one variable varies, an insensitive measure also prevents you from seeing how that variable co-varies with another variable.

Fourth, your variables may be related in a nonlinear way. This is a problem because most statistical tests are designed to detect straight-line (*line*ar) relationships. Fortunately, you can easily tell whether you have a nonlinear relationship by looking at a scatterplot of your data. If you can draw a straight line through the points of your scatterplot, you don't have a nonlinear relationship. If, on the other hand, a graph of your data revealed a nonlinear relationship, such as a definite U-shaped curve, you would know that a conventional correlation coefficient underestimates the strength of the relationship between your variables.

Nonlinear Relationships Between Two Variables

What if the scatterplot suggests that your variables are related in a nonlinear way? Or, suppose that you hypothesized that there was a nonlinear (curvilinear) relationship between two variables, such as temperature and aggression. That is, suppose you don't believe that with each degree the temperature rises, aggression increases. Instead, you think there is some curvilinear relationship (see Figure 7.11). You might think that temperature only increases aggression after the temperature goes above 80 degrees Fahrenheit (21°C), or you might think that when the temperature goes over 90 degrees (32°C), aggression declines.

Other cases in which you might look for—and be likely to find—a curvilinear relationship include

- Accuracy of married couples in reading each other's minds increases during the first few years of marriage and then decreases (Thomas, Fletcher, & Lange, 1997).

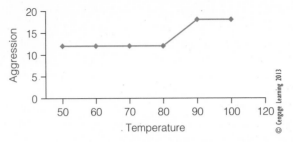

FIGURE **7.11** A potential nonlinear relationship between temperature and aggression.

- Happiness increases rapidly as income increases until income is above the poverty level, above which point there is little to no relationship between income and happiness (Helliwell, 2003).
- As scores on happiness (measured on a 1–10 scale) increase from 1 to 7, income increases, but as happiness increases from 8 to 10, income actually decreases (Oishi, Diener, & Lucas, 2007).

To test these kinds of curvilinear hypotheses, experts will often use a type of correlational analysis that uses each person's actual scores.[19] However, you might test for these relationships by using a less sensitive test that looks for differences between group means. Specifically, you could use a more flexible version of the *t* test: analysis of variance (ANOVA). (To see the similarities between the *t* test, ANOVA, and determining whether a correlation is statistically significant, see Box 7.4.)

To set up your ANOVA, you would divide your participants into three or more groups based on their scores on the predictor variable. For example, if you were studying self-esteem's relationship to GPA, you might divide participants into three groups: (1) a low self-esteem group, (2) a moderate self-esteem group, and (3) a high self-esteem group. Then, you would compare the means of the three groups. If there was a curvilinear relationship, you might find that the group with moderate self-esteem had higher GPAs than the groups with either low or high self-esteem (see Figure 7.12).

To find out whether the curvilinear pattern you observed was reliable (i.e., was not due to random error; applies to the population you randomly sampled from), you would first do an ANOVA. Using a computer program or statistical calculator, you would enter each person's group number (a "1" for the low self-esteem group, a "2" for the moderate self-esteem group, or a "3" for the high self-esteem group) as the predictor

---

[19] The technique is called polynomial regression. Normal regression, called linear regression, enters values on the predictor and looks for the best straight line that can fit the outcome variable. By also adding in the square (the value to the second power) of each value on the predictor to the equation, a researcher can look at the best line with one bend in it that can fit the data. By also adding the cube of each value of the predictor (the value to the third power), a researcher can look at the best line with two bends in it that can fit the data. By taking predictors to even higher powers, researchers can see how well even more complex curves fit the data.

BOX **7.4** **The Similarities Between a *t* Test, an *F* test, and a Test to Determine Whether a Correlation Coefficient Is Significantly Different From Zero**

When you have only two groups, doing a *t* test, an *F* test, and an analysis of correlation are essentially the same. Thus, in the simple experiment, the three procedures are quite similar. In all three cases, you are seeing whether there is a relationship between the treatment and the dependent variable—that is, whether the treatment and the dependent variable co-vary.

The only difference is in how you measure the extent to which the treatment and dependent measure co-vary. In the *t* test, you use the difference between means of the two groups as your measure of covariation; in the *F* test, you use a variance between means of the two groups as the measure of covariation; and in testing the significance of a correlation, you use the correlation between the treatment and the dependent variable as the measure of covariation. Consequently, regardless of which technique you use to analyze the results of a simple experiment, significant results will allow you to make cause–effect statements. Furthermore, regardless of which technique you use to analyze the results of a correlational study, significant results will *not* allow you to make cause–effect statements.

To show you that the three analyses are the same, we have done these three analyses of the same data.

t Test Analysis

Group	N	Mean	*df*	*t*	Probability[a]
Group 1	59	9.429	115	.87	.3859
Group 2	58	8.938			

Standard error of the difference = .565

$$t = \frac{9.429 - 8.938}{.565} = \frac{.491}{.565} = .87$$

[a]Probability (often abbreviated as *p*) refers to the chances of finding a relationship in your sample that is as large as the one you found if the two variables were really unrelated in the population. Thus, the smaller *p* is, the less likely it is that the relationship observed in your sample is just a fluke—and the more likely it is that the variables really are related in the population.

Analysis of Variance

Source	*df*	Sum of Squares	*MS*	*F* Value	Probability
Treatment	1	4.155	4.155	.758	.3859[b]
Error	115	630.768	5.485		

Correlational Analysis

Count	R	Probability
117	.081	.3859[c]

[b]Note that the probability (*p*) value is exactly the same, no matter what the analysis. That is, our *p* is .3859 whether we do a *t* test, an *F* test, or a correlational analysis. [c]The *t* that led to this probability value is 0.87, just as it was when we calculated the *t* between group means. However, because we were testing a correlation, we used a different formula. Applying that formula $t = [r \times \sqrt{(N-2)}]/\sqrt{(1 - [r \times r])}$ to our data led to the following computations:

$$t = \frac{.081 \times \sqrt{115}}{\sqrt{.993}} = \frac{.868}{.996} = 0.87$$

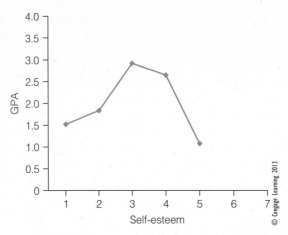

FIGURE **7.12** A curvilinear relationship between self-esteem and GPA.

and each person's GPA as the dependent measure. If your ANOVA was statistically significant, you would know that there was a relationship between self-esteem and GPA in your population. Then, you could do a follow-up test to see whether the curvilinear relationship you observed in your sample applied to the population by following the instructions in Table 4 in Appendix F.

Relationships Involving More Than Two Variables

You can also use ANOVA to look at hypotheses involving two or more predictors. For example, with ANOVA, you could look at how self-esteem and gender together predict grade-point average. Specifically, an ANOVA would allow you to compare the grade-point averages of (1) women with low self-esteem, (2) women with high self-esteem, (3) men with low self-esteem, and (4) men with high self-esteem. By doing this analysis, you might find that high self-esteem is related to high grade-point averages for women (i.e., for women, self-esteem and grades are positively correlated), but that high self-esteem is related to low grade-point averages for men (i.e., for men, self-esteem and grades are negatively correlated). In such a case, gender would be a **moderator variable**: a variable whose presence modifies (strengthens, weakens, reverses) the relationship between two other variables.

If you want to look for moderating variables, you do not have to use ANOVA. Instead, you can use multiple regression. In multiple regression, the computer uses a few predictors (in this case, gender, self-esteem, and a variable that represents the combined effects of both variables) to try to predict participants' scores on the dependent measure.

ANOVA and multiple regression are similar. Indeed, if you have a computer do an ANOVA for you, the devious computer will probably actually do a multiple regression and then just format the output to make it look like it did an ANOVA analysis. Because ANOVA is similar to multiple regression, many people do an ANOVA with correlational data. However, with most correlational data, multiple regression is a more powerful technique than ANOVA.

Multiple regression is more powerful than ANOVA for the same reason a test of the significance of a correlation coefficient is more powerful than a *t* test based on a median split: Multiple regression uses each individual's actual score on the predictor rather than ANOVA's trick of giving everyone in the group the same score. For example, in multiple regression, if someone scores 17 on the self-esteem test, that's the score that is put in the analysis. In ANOVA, on the other hand, you artificially create groups (e.g., "a low self-esteem group" and a "high self-esteem group") and give everyone in a group the same score on the predictor (e.g., all "lows" get a 1 on self-esteem, and all "highs" get a 2). The costs of lumping together participants into arbitrary groups is that you lose information about the extent to which participants differ on your predictors—and that loss of information, in turn, causes you to lose power. (See Table 7.3 for a summary of the

---

TABLE **7.3** Advantages and Disadvantages of Using ANOVA to Analyze the Results of a Correlational Study

Advantages	Disadvantages

• Allows you to perform two important analyses easily.
  1. You can do more than look at the simple relationships between two variables. Instead, you can look at the relationship among three or more variables at once.
  2. You could determine whether the relationship between variables is nonlinear.
	• You have less power than testing the statistical correlation coefficient because ANOVA doesn't use actual scores. Instead, it uses much less detailed information. For example, if you use a two-level ANOVA, you are recording only whether the score is in the top half or the bottom half of the distribution. Furthermore, you can examine both complex relationships among variables and nonlinear relationships without ANOVA.

• You can minimize the problem of losing detail by dividing scores into more groups. That is, you are not limited to just comparing the top half versus the bottom half. Instead, you could compare the top fifth versus the second fifth, versus the third fifth, versus the fourth fifth, versus the bottom fifth. Because you would be entering more detailed information into your analysis, you would have reasonable power.	• You still do not have as much detail and power as if you had used participants' actual scores.
• Provides a convenient way to analyze data.	• It may not be so convenient if you have an unequal number of participants in each group. In that case, you would have what is called an unbalanced ANOVA. Some computer programs can't accurately compute statistics for an unbalanced ANOVA.
• Is a familiar way to analyze data.	• Because it is a conventional way to analyze experimental data, people may falsely conclude that significant results mean that one variable causes changes in the other.

advantages and disadvantages of using ANOVA to analyze the results of correlational research; for more information about multiple regression, see Appendix E).

■ CONCLUDING REMARKS

In this chapter, you have learned how to conduct several types of descriptive research. You have seen that although descriptive research cannot answer "why" (cause–effect) questions, it can answer "what," "when," and "where" questions. Furthermore, you have seen that such questions can be grounded in theory and can involve more than simple relationships between two variables. Although you have learned a great deal about descriptive research, you have not learned about the most common method of doing descriptive research: asking questions. Therefore, the next chapter is devoted to showing you how to conduct surveys.

SUMMARY

1. Descriptive research allows you to describe behavior accurately. The key to descriptive research is to measure and record your variables accurately using a representative sample.

2. Although descriptive research cannot tell you whether one variable causes changes in another, it may suggest cause–effect (causal) hypotheses that you could test in an experiment.

3. Description is an important goal of science. Description also paves the way for prediction.

4. Ex post facto research uses data that you collected before you came up with your hypothesis.

5. Archival research uses data collected and sometimes coded by someone else.

6. With both ex post facto and archival research, data may not have been measured, collected, or coded in a way appropriate for testing your hypothesis.

7. Observational methods are used in both correlational and experimental research.

8. In both naturalistic observation and participant observation, the researcher must be careful that the observer does not affect the observed and that coding is objective.

9. Using preexisting, validated tests in your correlational research may increase the construct validity of your study. As with all research, the external validity of testing research depends on the representativeness of the sample.

10. Using a scatterplot to graph your correlational data will tell you the direction of the relationship (positive or negative) and give you an idea of the strength of the relationship.

11. Correlational coefficients give you one number that represents the direction of the relationship (positive or negative). These numbers range from –1.00 to +1.00.

12. A positive correlation between two variables indicates that if a participant scores high on one of the variables, the participant will probably also score high on the other.

13. A negative correlation between two variables indicates that if a participant scores high on one of the variables, the participant will probably score low on the other variable.

14. A zero correlation between two variables indicates there is no relationship between how a participant scores on one variable and how that participant will score on another variable. The further a correlation coefficient is away from zero, the stronger the relationship. Thus, a –.4 correlation is stronger than a +.3.

15. By squaring the correlation coefficient, you get the coefficient of determination, which tells you the strength of the relationship between two variables. The coefficient of determination can range from 0 (no relationship) to 1 (perfect relationship). Note that the coefficient of determination of both a −1 and a +1 correlation coefficient is +1.

16. If your results are based on a random sample, you may want to use inferential statistics to analyze your data.

17. Remember, statistical significance means only that your results can be generalized to the population from which you randomly sampled. Statistical significance does *not* mean that you have found a cause–effect relationship.

18. Beware of doing too many tests of significance. Remember, if you do 100 tests and use a .05 level of significance, 5 of those tests might be significant by chance alone.

19. You may obtain null (nonsignificant) results even though your variables are related. Common culprits are insufficient number of observations, nonlinear relationships, restriction of range, and insensitive measures.

KEY TERMS

95% confidence interval *(p. 248)*
archival data *(p. 234)*
coefficient of determination *(p. 260)*
content analysis *(p. 236)*
correlation coefficient *(p. 256)*
ex post facto research *(p. 232)*
frequency distribution *(p. 245)*

illusory correlations *(p. 231)*
instrumentation bias *(p. 238)*
laboratory observation *(p. 241)*
median split *(p. 252)*
median *(p. 246)*
mode *(p. 246)*
moderator variable *(p. 270)*
naturalistic observation *(p. 241)*
negative correlation *(p. 255)*
nonreactive measure *(p. 239)*

participant observation *(p. 241)*
positive correlation *(p. 253)*
restriction of range *(p. 267)*
scatterplot *(p. 253)*
standard deviation *(p. 245)*
standard error of the mean *(p. 249)*
zero correlation *(p. 255)*

EXERCISES

1. Steinberg and Dornbusch (1991) found that there is a positive correlation between cutting class and hours per week that adolescents work. In addition, they find a negative correlation between grade-point average and number of hours worked.
   a. In your own words, describe what the relationship is between class-cutting and hours per week that adolescents work.
   b. In your own words, describe what the relationship is between grade-point average and hours per week that adolescents work.
   c. What conclusions can you draw about the *effects* of work? Why?

   d. If you had been analyzing their data, what analysis would you use? Why?

2. Steinberg and Dornbusch (1991) also reported that the correlation between hours of employment and interest in school was statistically significant. Specifically, they reported that $r(3,989) = -.06, p < .001$. [Note that the $r(3,989)$ means that they had 3,989 participants in their study.] Interpret this finding.

3. Brown (1991) found that a measure of aerobic fitness correlated +.28 with a self-report measure of how much people exercised. He also found that the measure of aerobic fitness correlated −.41 with resting heart rate. Is resting heart rate or self-report

of exercise more closely related to the aerobic fitness measure?

4. In the same study, gender was coded as 1 = male, 2 = female. The correlation between gender and aerobic fitness was −.58, which was statistically significant at the $p < .01$ level.
   a. In this study, were men or women more fit?
   b. What would the correlation have been if gender had been coded as 1 = female and 2 = male?
   c. From the information here, can you conclude that one gender tends to be more aerobically fit than the other? Why or why not?

5. Suppose you wanted to see whether men differed from women in terms of the self-descriptions they put in personal ads. How would you get your sample of ads? How would you code your ads? That is, what would your content analysis scheme look like?

6. Suppose that a physician looked at 26 instances of crib death in a certain town and found that some of these deaths were due to parents suffocating their children. As a result, the physician concluded that most crib deaths in this country are due not to problems in brain development, but to parental abuse and neglect. What problems do you have with the physician's conclusions?

7. Researchers began by looking at how a sample of 5-year-olds were treated by their parents. Thirty-six years later, when the participants were 41-year-olds, the study examined the degree to which these individuals were socially accomplished. The investigators then looked at the relationship between childrearing practices when the child was 5 and how socially accomplished the person was at 41 (Franz, McClelland, & Weinberger, 1991). They concluded that having a warm and affectionate father or mother was significantly associated with "adult social accomplishment."
   a. What advantages does this prospective study have over a study that asks 41-year-olds to reflect back on their childhood?
   b. How would you measure adult social accomplishment?
   c. How would you measure parental warmth? Why?
   d. Assume, for the moment, that the study clearly established a relationship between parenting practices and adult social accomplishment. Could we then conclude that parenting practices account for (cause) adult social accomplishment? Why or why not?
   e. Imagine that the researchers had failed to find a significant relationship between the variables of adult social accomplishment and parental warmth. What might have caused their results to fail to reach significance?

WEB RESOURCES

1. Go to the Chapter 7 section of the book's student website and
   1. Look over the concept map of the key terms.
   2. Test yourself on the key terms.
   3. Take the Chapter 7 Practice Quiz.
2. Get a better sense of what descriptive research is like by using the "Participate in a Descriptive Study" link.
3. Become more comfortable with correlation coefficients by computing correlation coefficients using a statistical calculator, accessible from the "Statistical Calculator" link.
4. Get a better sense of the coefficient of determination by clicking on the "Coefficient of Determination" link.

Survey Research

A fool can ask more questions in an hour than a wise man can answer in seven years.
—English Proverb

A prudent question is one half of wisdom.
—Francis Bacon

CHAPTER OVERVIEW

If you want to know *why* people do what they do or think what they think, you should use an *experimental* design. If, on the other hand, you want to know *what* people are thinking, feeling, or doing, you should use a *nonexperimental* design, such as a **survey**.

To conduct a successful survey, you must meet three objectives. First, you must know what your research hypotheses are so that you know what you want to measure. Second, your questionnaire, test, or interview must accurately measure the thoughts, feelings, or behaviors that you want to measure. Third, you must be able to generalize your results to a certain, specific group. This group, called a **population**, could be anything from all U.S. citizens to all students in your research methods class.

Survey research that fails to meet these three objectives will be flawed. Thus, there are three ways survey research can go wrong.

First, survey research may be flawed because the researchers did not know what they wanted to find out. If you don't know what you're looking for, you probably won't find it. Instead, you will probably be overwhelmed by irrelevant data.

Second, survey research may be flawed because the questionnaire, test, or interview measure has poor construct validity. This occurs when

1. The questions demand knowledge that your respondents don't have.
2. The questions hint at the answers the researcher wants to hear, leading respondents to lie.
3. The respondents misinterpret the questions.
4. The researcher misinterprets or miscodes respondents' answers.

Third, survey research may have little external validity because the people who were questioned do not represent the target population. For example, a telephone survey of U.S. citizens that obtained its sample from phone books might underrepresent college students and overrepresent adults over 65 (Blumberg & Luke, 2008).

As you can see, there is more to survey research than asking whatever questions you want to whomever you want. Instead, survey research, like all research, requires careful planning. You must determine whether the survey design is appropriate for your research problem. Then, you must decide what questions you are going to ask, why you are going to ask those questions, to whom you are going to ask those questions, how you are going to ask those questions, and how you are going to analyze the answers to those questions.

Unfortunately, few people engage in the careful planning necessary to conduct sound survey research. Consequently, even though the survey is by far the most commonly used research method, it is also the most commonly abused. By reading this chapter, you can become one of the few people who know how to conduct sound and ethical survey research.

QUESTIONS TO ASK BEFORE DOING SURVEY RESEARCH

The most obvious—but least asked—question in survey research is, "Should I use a survey?" To answer this question correctly, you must answer these five questions:

1. What is my hypothesis?
2. Will I know what to do with the data after I have collected them?
3. Am I interested in either describing or predicting behavior—or do I want to make cause–effect statements?
4. Can I trust respondents' answers?
5. Do my results apply only to those people who responded to the survey, or do the results apply to a larger group?

What Is Your Hypothesis?

The first question to ask is, "What is my hypothesis?" Because good research begins with a good hypothesis, you might think that everyone would ask this question. Unfortunately, many inexperienced researchers try to write their survey questions without clear research questions. What they haven't learned is that you can't ask pertinent questions if you don't know what you want to know. Therefore, before you write your first survey question, make sure you have a clear hypothesis on which to base your questions.

Do Your Questions Relate to Your Hypothesis?

Having a hypothesis doesn't do you much good unless you are disciplined enough to focus your questions on that hypothesis. If you don't focus your questions on your hypothesis, you may end up with an overwhelming amount of data—and still not find out what you wanted to know. For example, the now-defunct United States Football League (USFL) spent millions of dollars on surveys to find out whether it should be a spring or fall league. Despite the fact that it took more than 20 books to summarize the survey results, the surveys did not answer the research question ("Injury Quiets," 1984). So don't be seduced by how easy it is to ask a question. Instead, ask questions that address the purpose of your research.

Asking useful questions involves two steps. First, determine what analyses you plan to do *before* you administer the questionnaire. You can do this by constructing a table like Table 8.1, the table we used to help develop the survey displayed in Box 8.1. If you don't plan on doing any analyses

TABLE **8.1** Table to Determine the Value of Including Questions From Box 8.1 in the Final Survey

Question Number	Purpose(s) of Question	Predictions Regarding Question	Analyses to Test Prediction
1	1. Qualify. 2. See whether sample reflects the population.	Percent of instructors in sample will be the same as in the population.	Compare percentages of sample at each rank with school's report of the total faculty at each rank.
2–3	Find out text messaging habits without asking a leading question.	1. Average number of text messages will be fewer than 25 per week. 2. Professors who text message will be more sympathetic to students than professors who don't text message. 3. Female faculty members will text message more than male faculty members will. 4. Younger faculty members will text message more than older faculty members will.	1. Compute mean and confidence intervals. 2. Correlate Question 3 with the sum of Questions 6–11. (Graph data to see whether there is a curvilinear relationship.) 3. Correlate Question 3 with Question 16. 4. Correlate Question 3 with Question 15.
4–5	Engage respondent and help set up the next set of questions.		
6–11	Scale to measure attitudes toward students.	1. Faculty members will have positive attitudes toward students. 2. See predictions made under Questions 2–3.	1. Compute average and confidence intervals for Question 11 and for sum of scale (sum of items 6–11). See whether the mean is significantly above the scale's midpoint. 2. See analyses described under Questions 2–3.
12–16	See if sample reflects the population.	Sample will reflect the population.	Compare sample's demographic characteristics against the demographic characteristics of the school's faculty.

BOX **8.1**   **Sample Telephone Survey**

Hello, my name is _____. I am conducting a survey for my Research Design class at Bromo Tech. Your name was drawn as a part of a random sample of university faculty. I would greatly appreciate it if you would answer a few questions about your job and your use of text messaging. The survey should take only 5 minutes. You can skip any questions you wish and you can terminate the interview at any time. If you agree to participate, your answers will be kept confidential. Will you help me?

1. **What is your position at Bromo Tech?** (read as an open-ended question)

    _____ Instructor
    _____ Assistant Professor
    _____ Associate Professor
    _____ Full Professor
    _____ Other (If other, terminate interview)

2. **Do you text message?**

    _____ Yes
    _____ No (put "0" in slot for Question 3, and skip to 4)

3. **How many text messages do you send in a typical week?** (read as open-ended question)

    _____ (Write number, then put a check next to the appropriate box. If exact number is not given, read categories and check appropriate box.)
    _____ <10
    _____ 10–50
    _____ 51–100
    _____ 101–150
    _____ 151–200
    _____ >201

    Please indicate how much you agree or disagree with the following statements. State whether you strongly agree (SA), agree (A), are undecided (U), disagree (D), or strongly disagree (SD).

4. **Text messaging has made the job of the average professor less stressful.**

    SA   A   U   D   SD

5. **Text messaging has made the average student's life less stressful.**

    SA   A   U   D   SD

6. **College is stressful for students.**

    SA   A   U   D   SD

7. **Colleges need to spend more time on students' emotional development.**

    SA   A   U   D   SD

8. **Colleges need to spend more time on students' physical development.**

    SA   A   U   D   SD

(Continued)

BOX 8.1 (Continued)

9. College students should be allowed to postpone tests when they are sick.

   SA   A   U   D   SD

10. College students work hard on their studies.

   SA   A   U   D   SD

11. I like college students.

   SA   A   U   D   SD

Demographics
Finally, I have just a few more questions to ensure that we get opinions from a variety of people.

12. **How long have you been teaching at Bromo Tech?**[*]

   _____ (Write years, then check the appropriate box. If exact years are not given, read categories
   and check appropriate box.)
   _____ 0–4 years
   _____ 5–9 years
   _____ 10–14 years
   _____ 15–19 years
   _____ 20 or more years

13. **What department do you teach in?**

   _____ Anthropology
   _____ Art
   _____ Biology
   _____ Business
   _____ Chemistry
   _____ English
   _____ History
   _____ Math
   _____ Physical education
   _____ Physics
   _____ Political science
   _____ Psychology
   _____ Sociology
   _____ Other _____

14. **What is the highest academic degree you have earned?**

   _____ BA/ BS
   _____ MA/ MD
   _____ PhD/ EdD/PsyD
   _____ Other _____

[*]Questions 12–15 can be read as open-ended questions.

15. **How old are you?**

_____ (Write age, then check the appropriate box. If exact years not given, read categories and check appropriate box.)

_____ <25

_____ 26–34

_____ 35–44

_____ 45–54

_____ 55–64

_____ >65

_____ Refused

Thank you for your help.

Note: Complete the following after the interview is finished. Do not read item 16 (below) to the participant.

16. **Gender (don't ask)**

_____ Male

_____ Female

involving responses to a specific question and the question serves no other purpose, get rid of that question.

Second, with the remaining questions, imagine participants responding in a variety of ways. For example, you may graph the results you predict and results that are completely opposite of what you would predict. If you find that no pattern of answers to a question would provide useful information, eliminate that question. Thus, when doing a survey to test hypotheses, eliminate a question if you determine that no matter how participants answer the question, it wouldn't disprove any of your hypotheses. Similarly, when doing a survey to help an organization, eliminate a question if you determine that no matter how participants answer the question, it wouldn't change how that organization runs its business.

Do You Have a Cause–Effect Hypothesis?

If your questions focus on a research hypothesis, your survey will be able to address that hypothesis—as long as you do _not_ have a cause–effect hypothesis. To do survey research, you must have a **descriptive hypothesis**: a hypothesis about a group's characteristics or about the correlations between variables. Usually, you will test one of the following four types of descriptive hypotheses.

First, you may do a survey to find out how many people have a certain characteristic or support a certain position. For example, a social worker may do a survey to find out what percentage of adolescents in the community have contemplated suicide. Similarly, a politician may do a survey to find out what percentage of the voters in her district support a certain position. Such surveys can reveal interesting information. For example, one survey of formerly obese individuals found that every individual surveyed "would rather have some disability than be obese again" (Kolata, 2007, p. 69).

Second, you may do a survey to develop a detailed profile of certain groups. You might use surveys to develop a list of differences between those who support gun control and those who don't or between college students who are happy and those who aren't. For example, Diener and Seligman (2002) found that very happy college students were more outgoing than other students.

Third, you may do a survey to examine the relationships between two or more variables. For example:

- Davis, Shaver, and Vernon (2004) used a survey to test the hypothesis that attachment styles (being securely attached, being anxiously attached, or being insecurely attached) are related to sex drive and to reasons for having sex.
- Haselton, Buss, Oubaid, and Angleitner (2005) used surveys to test the hypothesis that how upset people will be when their partner lies to them will be related both to the type of lie and to gender. Specifically, the researchers hypothesized that women would be more upset by men lying about their income, whereas men will be more upset about women lying about their past sexual history.
- Oishi, Diener, and Lucas (2007) used surveys to test the hypothesis that, beyond a certain level, happiness is not associated with financial and educational success.
- Lippa (2006) used surveys to test the hypothesis that, for most heterosexual men, increased sex drive is associated with increased sexual attraction to women, but that, for most heterosexual women, increased sex drive is associated with increased sexual attraction to both men and women.
- Swann and Rentfrow (2001) used surveys to test the hypothesis that blirtatiousness—the degree to which a person tends to quickly respond to others by saying whatever thoughts pop into the person's head—is positively correlated with self-esteem and impulsivity but negatively correlated with shyness.

Fourth, you might want to describe people's intentions so that you can predict their behavior. For example, news organizations conduct surveys to predict how people will vote in an election, and market researchers conduct surveys to find out what products people will buy.

As we have discussed, the survey is a useful tool for finding out what people plan to do. However, the survey is *not* a useful tool for finding out *why* people do what they do. Like all nonexperimental designs, the survey design does not allow you to establish causality. Therefore, if you have a cause–effect hypothesis, do not use a survey.

To illustrate why you cannot make causal inferences from a survey design, let's imagine that you find that professors are more sympathetic toward students than college administrators are. In that case, you cannot say that being an administrator causes people to have less sympathy for students. It could be that professors who didn't like students became administrators; or, it could be that some other factor (like being bossy) causes one to be an administrator and that factor is also associated with having less sympathy toward students (see Figure 8.1).

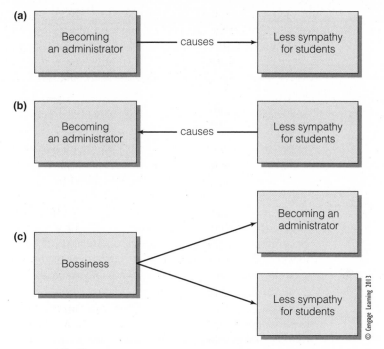

FIGURE **8.1** Correlation does not establish causality.

As you can see, finding that professors who became administrators are less sympathetic to students than other professors is not proof that becoming an administrator causes a loss of sympathy for students. There are at least two other possibilities.

Even among students who realize that nonexperimental methods cannot establish causality, some try to use survey methods to establish causality. They argue that all they have to do to establish causality is ask people why they behaved in a certain manner. However, those students are wrong: People do not necessarily know the causes of their behavior. For example, Nisbett and Wilson (1977) demonstrated that when participants are shown a row of identical television sets, participants will prefer the TV farthest to the right. Yet, when participants are asked why they preferred the TV on the right, nobody says, "I like it because it's on the right."

Can Self-Report Provide Accurate Answers?

Nisbett and Wilson's (1977) research illustrates a general problem with questioning people: People's answers may not reflect the truth. People's self-reports may be inaccurate for four reasons:

1. Participants never knew, and never will know, the answer to your question.
2. Participants no longer remember the information needed to correctly answer your question.
3. Participants do not yet know the correct answer to your question.
4. Participants know the correct answer to your question, but they don't want to give you the correct answer.

Are You Asking People to Tell More Than They Know?

When you ask people why they did something or how they feel about something, they often don't know. As research has shown, people are, in a sense, "strangers to themselves" (Haidt, 2006; Wilson, 2002). For example, people may not know why they like (or dislike) broccoli, why they call "heads" more often than "tails," why they usually wake up right before their alarm goes off, or why they find some comedians funnier than others. In short, although asking people questions about why they behave the way they do is interesting, you can't accept their answers at face value.

Are You Asking More Than Participants Can Accurately Remember?

As Nisbett and Wilson's (1977) study showed, when people do not know the real cause, they make up a reason—and they believe that reason. Similarly, even though people have forgotten certain facts, they may still think they remember. For example, obese people tend to underreport what they have eaten, and students tend to overreport how much they study. Both groups are surprised when they actually record their own behavior (R. L. Williams & Long, 1983).

Because memory is error prone, you should be careful when interpreting responses that place heavy demands on participants' memories. If you aren't skeptical about the meaning of those responses, be assured that your critics will be. Indeed, one of the most commonly heard criticisms of research is that the results are questionable because they are based on **retrospective self-reports**: participants' statements about their past behavior.

Are You Asking Participants to Look Into a Crystal Ball?

As bad as people are at remembering the past, they can be even worse about predicting the future. Thus, asking people "What would you do in _____ situation?" may make for interesting conversation, but the answers will often have little to do with what people would actually do if they were in that situation (S. J. Sherman, 1980). For instance, although people who are asked what they would do if they were in a conformity experiment typically report that they would not conform, 60–80% of actual participants do conform. For example, nearly two thirds of Milgram's participants went along with an experimenter's orders to give high-intensity shocks to another person (Milgram, 1974)—and recent replications of the study suggest the same proportion would conform today (Burger, 2007).

As you might expect, people are particularly bad at predicting how they will react to a situation they haven't experienced. For example, when a committee surveyed high school girls in Illinois to help it decide whether establishing high school volleyball teams in that state would be a waste of time, the surveys indicated that girls had virtually no interest in the sport. But when the committee went ahead and set up leagues anyway, volleyball became the most popular girls' high school sport in that state (Brennan, 2005).

If Participants Know, Will They Tell?

To this point, we have discussed cases in which participants aren't giving you the right answer because they don't know the right answer. However, even when participants know the right answer, they may not share that answer with you because they want to impress you, they want to please you, or they don't want to think about the question.

Social Desirability Bias.   If you ask participants questions, you need to be concerned about **social desirability bias**: participants understating, exaggerating, or lying to give an answer that will make them look good. For instance, if you went by U.S. adults' survey responses, 40% of U.S. adults regularly attended religious services in the early 1990s. However, research looking at actual attendance suggests that actual attendance was about half that (Hadaway, Marler, & Chaves, 1993). Participants are most likely to commit the social desirability bias when the survey is not anonymous and the question is extremely personal (e.g., "Have you cheated on your spouse?").

Obeying Demand Characteristics.   Sometimes, participants will give you the answer they think you want to hear. Their behavior may be similar to yours when, after having a lousy meal, the server asks you, "Was everything okay?" In such a case, rather than telling the server everything was lousy and ruining his day, you say what you think he wants to hear—"Yes, everything was okay." In technical terms, you are obeying the **demand characteristics** of the situation.

Following Response Sets.   Rather than think about what answer you want, participants may hardly think about their answers—or your questions—at all. Instead, participants may follow a **response set**: a habit of responding in a certain, set way, regardless of what the question says. Participants who use the "agree" or "strongly agree" option in response to every rating scale question are following a "*yea-saying*" response set. Participants who respond "disagree" or "strongly disagree" to every statement are following a "*naysaying*" response set. Participants who always choose the "neutral" or "neither agree nor disagree" option are following the *central tendency* response set.

To Whom Will Your Results Apply?

Even if you have a good set of questions that are all focused on your hypotheses, and you can get accurate answers to those questions, your work is probably not done. Usually, you want to generalize your results beyond the people who responded to your survey. For example, you might survey a couple of classes at your university, not because you want to know what the people in those particular classes believe, but because you hope those classes are a representative sample of your college as a whole. But are they? Unfortunately, they probably aren't. For example, if you selected a first-year English course because "everybody has to take it," your sample may exclude seniors. As you will see, obtaining an unbiased sample is difficult.

Even if you start out with an unbiased sample, by the end of the study, your sample may become biased because not everyone in your sample will fill out your questionnaire. In fact, if you do a mail or e-mail survey, don't be surprised if only 5% of your sample returns the survey. Unfortunately, the 5% who responded are probably not typical of your population. Those 5% probably feel more strongly about the issue than the 95% who did not bother to respond. Because so many people refuse to take part in surveys and because the few who do respond are different from those who do not respond, one of the most serious threats to the survey design's external validity is **nonresponse bias**: members of the original sample refusing to participate in the study, resulting in a biased sample.

Conclusions About the Advantages and Disadvantages of Survey Research

In summary, surveys can be valuable. A survey can be a fast and inexpensive way to collect a lot of information about a sample's attitudes, beliefs, and self-reported behaviors.

Although surveys can be valuable, recognize that if participants' self-reports are inaccurate, the survey will have poor construct validity. If the sample is biased, the survey will have poor external validity. Finally, no matter what, the survey will have poor internal validity because it cannot reveal why something happened: If you want to know what causes a certain effect, don't use a survey design.

THE ADVANTAGES AND DISADVANTAGES OF DIFFERENT SURVEY INSTRUMENTS

If you decide that a survey is the best approach for your research question, then you need to decide what type of survey instrument you are going to use. You can choose between two main types of survey instruments: (1) **questionnaire** surveys, in which participants read the questions and then write their responses, and (2) **interview** surveys, in which participants hear the questions and then speak their responses.

Written Instruments

If you are considering a questionnaire survey, you have three options: self-administered questionnaires, investigator-administered questionnaires, and psychological tests. In this section, we will discuss the advantages and disadvantages of these three written instruments.

Self-Administered Questionnaires

A **self-administered questionnaires**, as the name suggests, is filled out by participants in the absence of an investigator. Behavioral scientists, as well as manufacturers, special-interest groups, and magazine publishers, all use self-administered questionnaires. You probably have seen some of these questionnaires in your mail, on the Internet (see Table 8.2), at restaurant tables, and in magazines.

Self-administered questionnaires have two main advantages. First, self-administered questionnaires are easily distributed to a large number of people. Second, self-administered questionnaires often allow anonymity.

TABLE **8.2** Advantages and Disadvantages of Web Surveys—and Strategies for Dealing With the Disadvantages

Characteristic	Advantages	Disadvantages	Comments/Solutions
Participants can be anonymous.	• Less social desirability bias (Gosling, Vazire, Srivasta, & John, 2004; Lin, 2004).	• Participants may be underage, may take the survey several times, or may even take it as a group (Nosek et al., 2002). • Because a passer-by could see responses while the participant is taking the survey, responses may not be anonymous (Nosek et al., 2002).	• You can delete surveys that come from the same IP address—or ones that come from the same IP address and have similar characteristics (Gosling et al., 2004). • You can ask participants whether they have taken the survey before (Gosling et al., 2004). • You can ask participants whether they are alone at the computer (Gosling et al., 2004). • Participants don't have to be anonymous: You can require them to register for the study (Nosek et al., 2002). • Internet surveys seem to get the same pattern of results as paper-and-pencil surveys. You can compare your results to an off-line sample (Gosling et al., 2004).
Researcher exerts less control over participants.	• Fewer ethical problems that are due to researcher influencing participant to continue the study (Nosek, Banaji, & Greenwald, 2002).	• Participants might be distracted while filling out the survey and not pay attention to the questions. • Participants might not complete the survey. • Participants might not get the reassurance or debriefing they need (Nosek et al., 2002).	• Research finds that the reliability of a measure when administered over the Internet is equivalent to its reliability when administered via paper and pencil (Gosling, Vazire, Srivasta, & John, 2004; Miller et al., 2002; Riva, Teruzzi, & Anolli, 2003). Thus, Internet participants seem to be taking the survey questions as seriously as other participants. • Keeping the survey short can decrease the dropout rate.

(Continued)

TABLE **8.2** Continued

Characteristic	Advantages	Disadvantages	Comments/Solutions
Can survey anyone with a computer.	• Geography is not a boundary (Gosling, Vazire, Srivasta, & John, 2004). • Large samples possible, meaning that statistical inferences can be made about even groups that make up only a small percentage of the population (Gosling, Vazire, Srivasta, & John, 2004). • Can target groups that have special interests or characteristics by targeting members of online discussion groups (Nosek, Banaji, & Greenwald, 2002).	• Older people are less likely to be sampled (Gosling, Vazire, Srivasta, & John, 2004). • Web samples are usually not representative of the population (Gosling, Vazire, Srivasta, & John, 2004).	• Web samples are often more representative samples than the samples used in most laboratory research (Gosling, Vazire, Srivasta, & John, 2004).

© Cengage Learning 2013

Allowing respondents to be anonymous may be important if you want honest answers to highly personal questions.

Using a self-administered questionnaire can be a cheap and easy way to get honest answers from thousands of people. However, using a self-administered questionnaire has at least two major drawbacks.

First, surveys that rely on self-administered questionnaires usually have a low return rate. Because the few individuals who return the questionnaire may not be typical of the people you tried to survey, you may have a biased sample. In other words, nonresponse bias is a serious problem with self-administered questionnaires.

Second, because the researcher and the respondent are not interacting, problems with the questionnaire can't be corrected. Thus, if the survey contains an ambiguous question, the researcher can't help the respondent understand the question. For example, suppose we ask people to rate the degree to which they agree with the statement, "College students work hard." One respondent might think this question refers to a job a student might hold in addition to school. Another respondent might interpret this to mean, "Students work hard at their studies." Because respondent and researcher are not interacting, the researcher will have no idea that these two respondents are, in a sense, answering two different questions.

Investigator-Administered Questionnaires

To avoid the self-administered questionnaire's weaknesses, some researchers use the investigator-administered questionnaire. The investigator-administered questionnaire is filled out in the presence of a researcher.

Investigator-administered questionnaires share many of the advantages of the self-administered questionnaire. With both types of measures, many respondents can be surveyed at the same time. With both types of measures, surveys can be conducted in a variety of locations, including the lab, the street, in class, over the phone, and at respondents' homes.

A major advantage of having an investigator present is that the investigator can clarify questions for the respondent. In addition, the investigator's presence encourages participants to respond. As a result, surveys that use investigator-administered questionnaires have a higher response rate than surveys using self-administered questionnaires.

Unfortunately, the investigator's presence may do more than just increase response rates. The investigator-administered questionnaire may reduce perceived anonymity. Because such respondents feel their answers are less anonymous, respondents to investigator-administered surveys may be less open and honest than respondents to self-administered surveys (see Figure 8.2).

Psychological Tests: Borrowing From the Best

An extremely refined form of the investigator-administered questionnaire is the psychological test. Whereas questionnaires are often developed in a matter of days, psychological tests are painstakingly developed over months, years, and, in some cases, decades. Nevertheless, the distinction between questionnaires and tests is sometimes blurred.

One reason there is not always a clear-cut difference between questionnaires and tests is that questionnaires often incorporate questions from psychological tests. For example, in a study of people's concern about body weight, Pliner, Chaiken, and Flett (1990) incorporated two psychological tests into their questionnaire: Garner and Garfinkel's (1979) Eating Attitudes Test (EAT) and Janis and Field's (1959) Feeling of Social Inadequacy Scale.

FIGURE **8.2** Participant bias in an investigator-administered survey.

Even if you do not include a test as part of your questionnaire, try to incorporate the best aspects of psychological tests into your questionnaire. To make your questionnaire as valid as a test, try to follow these seven steps:

1. Pretest your questionnaire (as Schwarz and Oyserman [2001] point out, although it would be best to have volunteers fill out the questionnaire and then interview them about what they were thinking as they answered each question, you can, at the very least, answer your own questionnaire).

2. Standardize the way you administer the questionnaire.

3. Balance out the effects of response-set biases, such as "yea-saying" (always agreeing) and "nay-saying" (always disagreeing) by asking the same question in a variety of ways. For example, you might ask, "How much do you like the president?" as well as, "How much do you dislike the president?"

4. When possible, use "objective" questions (such as multiple-choice questions) that do not require the person scoring the test to interpret the participant's responses.

5. Prevent scorer bias on those questions that do require the scorer to interpret responses by (a) developing a detailed scoring key for such responses and (b) not letting scorers know the identity of the respondent. If the hypothesis is that male respondents will be more aggressive, for example, do not let coders know whether the survey they are scoring is a man's or a woman's.

6. Make a case for your measure's reliability. About a month after you surveyed your respondents, administer the survey to them again, and see whether they score similarly both times. Finding a strong positive correlation between the two times of measurement would suggest that scores reflect some stable characteristic rather than random error.

7. Make a case for your measure's validity by correlating it with measures that do not depend on self-report. For example, Steinberg and Dornbusch (1991) justified using self-reported grade-point average (GPA) rather than actual grade-point average by establishing that previous research had shown that school-reported and self-reported GPA were highly correlated.

Written Instruments: A Summary

To review, an investigator-administered survey is generally better than a self-administered survey because administering the survey gives you higher response rates and more control over how the questionnaire is administered. If you follow our seven additional steps to make your questionnaire more like a test, your investigator-administered questionnaire may have almost as much construct validity as a psychological test has.

Interviews

At one level, there is very little difference between the questionnaire and the interview. In both cases, the investigator is interested in participants' responses to questions. The only difference is that, in an interview, rather than having respondents provide written answers to written questions, the interviewer records respondents' spoken answers to spoken questions. As subtle as this difference is, it still has important consequences.

One important consequence is that interviews are more time consuming than questionnaires. Whereas you can administer a written questionnaire to many people at once, you should not interview more than one person at a time. If you interviewed more than one participant at a time, what one participant said might depend on what other participants had already said. For example, participants might go along with the group rather than disclosing their true opinions.

Because interviews are more time consuming than questionnaires, they are also more expensive. However, some researchers think interviews are worth the extra expense.

Advantages of Interviews

The added expense of the interview buys you additional interaction with the participant. This additional interaction lets you clarify questions that the respondents don't understand and lets you follow up on responses you do not understand or did not expect—a tremendous asset in exploratory studies in which you have not yet identified all the important variables. The additional personal interaction may also increase your response rate.

Two Methodological Disadvantages of Interviews

Unfortunately, the personal nature of the interview creates two major problems. First, there is the problem of **interviewer bias:** The interviewer may influence respondents' responses by verbally or nonverbally signaling approval of "correct" answers.

Second, participants may try to impress the interviewer. As a result, rather than telling the truth, participants may give socially desirable responses that would make the interviewer like them or think well of them. Thus, answers may be tainted by the social desirability bias (de Leeuw, 1992).

Advantages of Telephone Interviews

Psychologists have found that the telephone interview is less affected by interviewer bias and social desirability bias than the personal interview. Furthermore, in some cases, the telephone interview may have fewer problems with sampling bias than other survey methods.

Because the telephone interviewer can't see the respondents, the interviewer cannot bias respondents' responses via subtle visual cues such as frowns, smiles, and eye contact. Furthermore, by monitoring and tape recording the interviews, you can discourage interviewers from saying anything that might bias respondents' answers. For example, you could prevent interviewers from changing the wording or order of questions or from giving more enthusiastic and positive verbal feedback (e.g., "Great!") for answers that support your hypothesis and less enthusiastic feedback (e.g., "Okay") for answers that do not support your hypothesis.

Because the telephone interviewer can't see the participants, participants feel more anonymous, which, in turn, appears to reduce desirability bias

(Groves & Kahn, 1979). Thus, thanks to the lack of nonverbal cues, the telephone survey may be less vulnerable to both interviewer bias and respondent biases than the personal interview.

The telephone survey also reduces sampling bias by making it easy to get a representative sample. If you have a list of your population's phone numbers, you can randomly select numbers from that list. If you don't have a list of the population's phone numbers, you may still be able to get a random sample of your population by **random digit dialing**: taking the area code and the 3-digit prefixes that you are interested in and then adding random digits to the end to create 10-digit phone numbers. Note, however, that with random digit dialing, you call many fax numbers and disconnected or unused numbers. Furthermore, you will not contact people who either have no phone or whose only phone is a cell phone (although it would be technologically possible to call people's cell phones, the ethical problems—from participants having to pay for the cost of the call, to participants being minors, to participants feeling their privacy has been invaded, to participants divulging private information in a public place, to a participant potentially getting in a car accident while answering your call—are immense; Lavrakas, Shuttles, Steeh, & Fienberg, 2007).

Thus far, we have discussed two basic advantages of telephone interviews. First, partly because there are no visual, nonverbal cues, the telephone survey is superior to the personal interview for reducing both respondent biases and interviewer biases. Second, because it is easy to get a large random sample, the telephone survey may give you the best sample of any survey method. However, the main reason for the popularity of the telephone interview is practicality: The telephone survey is more convenient, less time consuming, and cheaper than the personal interview.

Disadvantages of Telephone Interviews

Although there are many advantages to using the telephone interview, you should be aware of its four most serious limitations. First, as with any survey method, there is the possibility of sampling bias. Even if you followed proper random sampling techniques, telephone interviews are limited to those households with landline phones. Although this limitation may not seem serious, realize that many households have cell phones instead of landline phones, and some people do not have any kind of phone (Blumberg & Luke, 2008). Furthermore, if you are drawing your random sample from listed phone numbers, realize that many people (more than 25% of U.S. households) have unlisted numbers (Dillman, 2000).

Second, as with any survey method, nonresponse bias can be a problem. Some people will, after screening their calls through an answering machine or through "caller ID," choose not to respond to your survey. Indeed, one study reported that 25% of men between the ages of 25 and 34 screen all their calls (Honomichl, 1990). Even when a person does answer the phone, he or she may refuse to answer your questions. In fact, some people get angry when they receive a phone call regarding a telephone survey. We have been yelled at on more than one occasion by people who believe that a call to request a telephone interview is a violation of their privacy.

Third, telephone surveys limit you to asking simple and short questions. Rather than focusing on answering your questions, participants' attention may be focused on the television show they are watching, the ice cream that is melting, or the baby who is crying.

Fourth, by using the telephone survey, you limit yourself to learning only what participants tell you. You can't see anything for yourself. Thus, if you want to know what race the respondent is, you must ask. You can't see the respondent or the respondent's environment—and the respondent knows you can't. Therefore, a 70-year-old bachelor living in a shack could tell you he's a 35-year-old millionaire with a wife and five kids. He knows you have no easy way of verifying his story.

How to Conduct a Telephone Survey

After weighing the pros and cons of the different surveys (see Table 8.3), you may decide to conduct a telephone survey. How should you go about it?

Your first step is to determine what population (what particular group) you wish to sample from and figure out a way to get all of their phone numbers. Often, your population is conveniently represented in a telephone book, membership directory, or campus directory. Once you obtain the telephone numbers, you are ready to draw a random sample from your population. (Later in this chapter, you will learn how to draw a random sample.)

When you draw your sample, pull more names than you actually plan to survey. You won't be able to reach everyone, so you'll need some alternate names. Usually, we draw 25% more names than we actually plan on interviewing.

Next, do what any good survey researcher would do. That is, as you'll see in the next section ("Planning a Survey"), (1) decide whether to ask your questions as essay questions, multiple-choice questions, or some other format; (2) edit your questions; and (3) put your questions in a logical order.

After editing your questions and putting them in the right order, further refine your survey. Start by having a friend read the survey to you. Often, you will find that some of your questions don't "sound" right. Edit them so they sound better. Then, conduct some practice telephone interviews. For example, interview a friend on the phone. The practice interviews may show you that you need to refine your questions further to make them easier to understand. Much of this editing will involve shortening the questions.

Once you've made sure your questions are clear and concise, concentrate on keeping your voice neutral and slow. Try not to let your tone of voice signal that you want participants to give certain answers. Tape yourself reading the questions, and play it back. Is your voice hinting at the answer you want participants to give? If not, you're ready to begin calling participants— provided you have (a) taken proper steps to preserve the anonymity of respondents and the confidentiality of their responses, (b) weighed the benefits of the survey against the costs to participants, and (c) received approval from your professor and either your school's institutional review board (IRB) or your department's ethics committee.

If you get a busy signal or a phone isn't answered, try again later. Usually, you should phone a person six to eight times at different times of the day before replacing that person with an alternate name or number.

TABLE **8.3** Comparing Different Ways of Administering Questionnaires

	Personal Interview	Phone Interview	Investigator Administered (to a Group)	Self-Administered (Includes Mail Surveys, E-mail Surveys, Web Surveys)
Quality of Answers				
Interviewer bias	Facial expressions and tone of voice could affect responses. However, having a supervisor monitor interview sessions or videotaping them or watching them through a one-way mirror) could prevent and detect interviewer bias.	Tone of voice could affect responses. However, having a supervisor monitor interview sessions or taping sessions could prevent and detect interviewer bias.	Minimal interaction with investigator, so little chance of interviewer bias.	No interviewer, no interviewer bias.
Social desirability bias	Participant may want to impress interviewer.	Participants often feel anonymous.	Participants often feel anonymous.[a]	Participants are anonymous.
Problems due to participants misunderstanding questions	Interviewer can clarify questions.	Interviewer can clarify questions.	In a group setting, participants are unlikely to ask questions even when they don't understand.	There may not even be an opportunity for participants to ask the meaning of a question.
Potential for fraud (people filling out multiple questionnaires)	Not a problem.	Not a problem.	Not a problem.	Potential for filling out multiple surveys.

	Personal Interview	Phone Interview	Investigator Administered (to a Group)	Self-Administered (Includes Mail Surveys, E-mail Surveys, Web Surveys)
Sampling Issues				
Geographical diversity	Difficult.	Easy if you can make long distance calls	Difficult.	Easy to distribute to a wide area.
Cheaply contacting a representative sample.	Expensive to get a large sample.	Long-distance calls are affordable.[b]	Group administered surveys are cheaper than individually administered surveys.	Very cheap because there is no need to have someone administer the survey.[c]
Getting a high response rate (Avoiding nonresponse bias)	Good: People respond to the personal approach.	If you precede your call with a letter, you may get a high response rate. If at first you get a low response rate, try calling again and again.		Potential problem.
Ethical Issues			Group debriefing may be less effective than individual debriefing.	May be unable to debrief participants.

[a]When administering a measure to a group, you should take steps to make participants feel that their responses cannot be seen by other participants. These steps may include providing a cover sheet, separating participants by at least one desk, allowing participants to put their completed questionnaire in an unmarked envelope, and having participants put their questionnaires into a box (rather than handing the questionnaire to another participant or to the researcher).

[b]If you use random-digit dialing, you get a better sample than you would if you used most other methods; however, some groups will still be underrepresented (e.g., people who do not have phones and, because of the ethical problems involved in calling people's cell phones, people whose only phone is a cell phone).

[c]The quality of your initial sample will depend on the method you use. If you use a mail survey, you will probably have a pretty good sample, although you will underrepresent the homeless and people who have recently moved. If you use a web survey, you will underrepresent the poor and people over 50.

When you do reach a person, identify yourself and ask for the person on the list—or, if you don't have a list, randomly select a member of the household who meets your criteria (e.g., "I would like to speak to the adult living in the house who most recently had a birthday."). Note that if you survey whoever happens to answer the phone, you will bias your sample (because women are more likely to answer the phone than men and because more outgoing people are more likely to answer the phone than less outgoing people).

Once you are talking with the appropriate person, briefly introduce the study. Tell the person

- the general purpose of the study (but do not bias the respondents' answers by stating specific objectives or specific hypotheses)
- the topics that will be covered
- the sponsor of the survey
- the average amount of time it takes to complete the interview (you learned this when you practiced giving your survey to a friend)
- the steps that you are taking to safeguard the respondent's confidentiality
- that the respondent is free to skip any question
- that the respondent can quit the interview at any time

After providing this introduction, ask the person whether she or he is willing to participate. If the person agrees, ask each question slowly and clearly. Be prepared to repeat and clarify questions.

Once the survey is completed, thank your respondent. Then, offer to answer any questions. Usually, you should give participants the option of being mailed a summary of the survey results.

PLANNING A SURVEY

Before conducting a telephone survey—or any other type of survey—you need to do some careful planning. In this section, we will take you through the necessary steps in developing and executing your survey.

Deciding on a Research Question

As with all psychological research, the first step in designing a survey is to have a clear research question. You need a hypothesis to guide you if you are to develop a cohesive and useful set of survey questions. Writing a survey without a hypothesis to unify it is like writing a story without a plot: In the end, all you have is a set of disjointed facts that tell you nothing.

Part of developing a clear research question is specifying your target population: all the members of the group that you want to generalize to. Knowing your population will help you word your questions, pretest your questions, and obtain a representative sample of your population.

Not only do you want a clear research question but you also want an important one. Therefore, before you write your first survey question, justify why your research question is important. You should be able to answer at least one of these questions:

1. What information will the survey provide?
2. What practical implications could the survey results have?

Choosing the Format of Your Questions

You've decided that you can use a survey to answer your research question. In addition, you've decided what kind of survey instrument (questionnaire or interview) will give you the best answer to your question. Now, you are ready to decide what types of questions to use.

Fixed-Alternative Questions

You might decide to use **fixed-alternative questions**: questions in which respondents have to choose between two or more answers. Your survey might include several types of fixed-alternative questions: true–false, multiple-choice, and rating scale.

Nominal-Dichotomous Items. Sometimes, fixed-alternative questions ask respondents to tell the researcher whether they belong to a certain category. For example, participants may be asked to categorize themselves according to gender, race, or religion. Because these questions do not tell us about how much of a quality a participant has but instead only whether the person has a given quality, these questions yield nominal data. (For more on nominal data, see Chapter 6.)

Dichotomous questions—questions that allow only two responses (usually "yes" or "no")—also give you nominal (qualitative) data because they ask whether a person has a given quality. Often, respondents are asked whether they are members of a category (e.g., "Are you employed?" or "Are you married?").

Sometimes, several dichotomous questions are asked at once. Take the question: "Are you African American, Hispanic, Asian, or Caucasian (non-Hispanic)?" Note that this question could be rephrased as several dichotomous questions: "Are you African American?" ("yes" or "no"), "Are you Hispanic?" ("yes" or "no"), and so on. The information is still dichotomous—participants either claim to belong to a category or they don't. Consequently, the information you get from these questions is still categorical, qualitative information. If you code African American as a "1," Hispanic as "2," and so on, there is no logical order to your numbers. Higher numbers would not stand for having more of a quality. In other words, different numbers stand for different types (different qualities) rather than for different amounts (quantities).

The fact that **nominal-dichotomous items** present participants with only two—usually very different—options has at least two advantages. First, respondents often find it easier to choose between two points (such as, "Are you (1) for animal research or (2) against animal research?"), than between 13 points (e.g., "Rate how favorably you feel toward animal research on a

13-point scale."). Second, when there are only two very different options, respondents and investigators are likely to have similar interpretations of the options. Therefore, a well-constructed dichotomous item can provide reliable and valid data.

Although there are advantages of offering only two choices, there are also disadvantages. One disadvantage of nominal-dichotomous items is that some respondents will think that their viewpoint is not represented by the two alternatives given. To illustrate this point, consider the following question:

"Do you think abortion should continue to be legal in the United States?" ("yes" or "no").

How would people who are ambivalent toward abortion respond? How would people who are fervently opposed to legalized abortion feel about not being allowed to express the depth of their feelings?

If you have artificially limited your respondents to two alternatives, your respondents may not be the only ones irritated by the fact that your alternatives prevent them from accurately expressing their opinions. You, too, should be annoyed—because by depriving yourself of information about subtle differences among respondents, you deprive yourself of **power**: the ability to find relationships among variables.

Likert-Type and Interval Items. One way to give yourself power is to use **Likert-type items**. Likert-type items typically ask participants to respond to a statement by choosing *strongly disagree* (scored a 1), *disagree* (scored a 2), *undecided* (3), *agree* (4), or *strongly agree* (5).

Traditionally, most psychologists have assumed that a participant who strongly agrees (a 5) and a participant who merely agrees (a 4) differ by as much, in terms of how they feel, as a participant who is undecided (a 3) differs from someone who disagrees (a 2). That is, participants who differ by the same distance on the scale (e.g., 1 point), supposedly differ by the same amount psychologically. In other words, Likert-type scales are assumed to yield interval data (To learn more about interval data, see Chapter 6). Questions 4–11 in Box 8.1 are examples of Likert-type, interval scale items.

Likert-type items are extremely useful in questionnaire construction. Whereas dichotomous items allow respondents only to agree or disagree, Likert-type items give respondents the freedom to strongly agree, agree, be neutral, disagree, or strongly disagree. Thus, Likert-type items yield more information than nominal-dichotomous items. Furthermore, because Likert-type items yield interval data, responses to Likert-type items can be analyzed by more powerful statistical tests than nominal-dichotomous items.

The major disadvantage of Likert-type items is that some respondents may resist the fixed-alternative nature of the question. One approach to this problem is to have a "don't know" option. That way, respondents won't feel forced into an answer that doesn't reflect their true position. In an interview, you can often get around the problem by reading a Likert-type question (e.g., "On a 1, *strongly disagree*, to 5, *strongly agree*, scale, do you think college is stressful for students?") as if it were an open-ended question (e.g., "Do you think college is stressful for students?") and then recording the participant's answer under the appropriate alternative. As you can see, many of the questions in Box 8.1 could be read like open-ended items.

Using Likert-Type Items to Create Summated Scores. If you have several Likert-type items that are designed to measure the same variable (such as liking for students), you can sum (add up) each respondent's answers to all those questions to get a total score for each respondent on that variable. For example, consider Questions 6–11 in Box 8.1. For each of those questions, a "5" indicates a high degree of liking for students, whereas a "1" indicates a low level of liking for students. Therefore, you might add (sum) the answers (scores) for each of those questions to produce a **summated score** for student liking. Suppose you obtained the following pattern of responses from one professor:

Question 6 = 1 (strongly disagree)

Question 7 = 2 (disagree)

Question 8 = 1 (strongly disagree)

Question 9 = 3 (undecided)

Question 10 = 1 (strongly disagree)

Question 11 = 2 (disagree)

Then, the summated score (total score for liking students) would be 10 (because $1 + 2 + 1 + 3 + 1 + 2 = 10$).

There are two statistical advantages to using summated scores. First, just as a 50-question multiple-choice test is more reliable (less influenced by random error) than a 1-question multiple-choice test, a score based on several questions is more reliable than a score based on a single question. Second, analyses are often simpler for summated scores. If we summed the responses for Questions 6–11 in Box 8.1, we could compare professors who text message to those who don't on "student liking" by doing one t test.[1] Without a summated score, you would have to perform six separate t tests, and then correct the t test for the effects of having done multiple analyses.[2]

Conclusions About Fixed-Alternative Items. You can use fixed-alternative questions for more than asking respondents whether they belong to a certain category, support a certain position, or do a certain behavior. You can use fixed-alternative questions to ask respondents how strongly respondents

---

[1] To compute a t, you would subtract your two group means and then divide by the standard error of the differences. To calculate the standard error of the differences by hand, you can use the formula: standard error of the mean $= \sqrt{s_1^2/N_1 + s_2^2/N_2}$. In that formula, $s =$ standard deviation, $N =$ number of participants, 1 means the symbol refers to group 1, and 2 means the symbol refers to group 2. Alternatively, you may follow the more detailed set of instructions in Appendix E.

[2] Alternatively, you could use a more complex statistical procedure, such as multivariate analysis of variance (MANOVA).

believe in a certain position. For example, a question might ask, "How much do you agree or disagree with the following statement?" (The fixed alternatives could be strongly agree, agree, disagree, strongly disagree.) Similarly, fixed-alternative questions can ask how much of a certain behavior the person did. For instance, "How many days a week do you study?" (The fixed alternatives could be a. 0, b. 1, c. 2, d. 3, e. 4, f. 5, g. 6, h. 7.) If asked the right way, these "how many" and "how much" questions can yield interval data (as discussed in Chapter 6, interval data allow you to compare participants in terms of how *much* of a quality they have).

Unfortunately, many of these "how much" and "how many" questions are *not* asked the right way; thus, they do not yield interval data. For example, when asking respondents about their grade-point averages, some researchers make 1 = 0.00–0.99, 2 = 1.0–1.69, 3 = 1.7–2.29, 4 = 2.3–2.7, 5 = 2.8–4.0. Note that the response options do not cover equal intervals: The interval covered by option "4" is .4, whereas the range of grade-point averages (GPAs) covered by option "5" is 1.2. Because there aren't equal intervals between response options, averaging participants' responses is meaningless.

A better choice of options would be 1 = 0.00–0.99, 2 = 1.00–1.99, 3 = 2.00–2.99, and 4 = 3.00–4.00. Probably the best thing to do would be to abandon the fixed-response format and just ask participants the open-ended question, "What is your grade-point average?"

Open-Ended Questions

Before discussing other situations in which you might want to use **open-ended questions**, let's distinguish open-ended questions from fixed-alternative questions. Whereas fixed-response items may resemble multiple-choice questions, open-ended questions may resemble fill-in-the-blank, short-answer, or essay questions. Whereas fixed-response questions force participants to choose among several researcher-determined response options, open-ended questions allow participants to respond in their own words. There are two major advantages of letting participants respond in their own words.

First, you avoid putting words in participants' mouths. To illustrate how fixed alternatives may influence participants, consider the question "What is the most important thing for children to prepare them for life?" When the question was asked as a fixed-alternative question, almost two-thirds of respondents chose the alternative "To think for themselves." However, when the question was asked as an open-ended question, fewer than 1 in 20 respondents gave any response resembling "To think for themselves" (Schwarz, 1999).

Second, open-ended questions may let you discover the beliefs behind the respondents' answers to the fixed-alternative questions. In some cases, open-ended questions may reveal that there are no beliefs behind a participant's answers to the fixed-alternative questions. That is, you might find that although the respondent is dutifully checking and circling responses, the respondent really doesn't know anything about the topic.

In other cases, asking open-ended questions allows you to discover that respondents making the same ratings have different opinions. For example, consider two professors who respond to Question 11 in Box 8.1, "I like college students" with "undecided." Open-ended questions may allow you to discover that one professor circles "undecided" because he is new to the college and doesn't know, whereas the other professor circles "undecided" because she has mixed feelings about students. Without asking open-ended questions, you would not have known that these two respondents have different reasons for giving the same response.

Although there are two major advantages to letting respondents answer in their own words, there are also two major disadvantages. First, open-ended questions are harder for participants to answer. Because of the difficulty of generating their own responses, participants will often skip open-ended questions. Second, answers to open-ended questions are hard to score. Answers may be so varied that you won't see an obvious way to code them. If you aren't careful, the coding strategy you finally adopt will be arbitrary.

To help you come up with a logical and systematic method of coding open-ended questions, try to come up with a content analysis scheme (see Chapter 7) *before* you start collecting data. Once you have done a content analysis, you may convert the information from your open-ended questions into numbers. If you rated answers to a question on a 1 (*not at all aggressive*) to 5 (*extremely aggressive*) scale, you would analyze these quantitative ratings as interval data. If you coded responses in terms of whether ideas about loyalty were mentioned (not mentioned = 0, mentioned = 1), you would analyze these qualitative, categorical data as nominal data (to learn more about nominal data, see Chapter 6).

Choosing the Format of Your Survey

If you use an interview, in addition to deciding on the format of your questions, you also need to decide on the format of your interview. You have a choice between three interview formats: structured, semistructured, and unstructured.

Structured Interview

In psychological research, the most popular interview format is the **structured interview**: an interview in which all respondents are asked a standard list of questions in a standard order. The structured interview is popular because the structure reduces the risk of interviewer bias and increases reliability. To build on the strengths of the structured interview, consider asking only fixed-alternative questions: By using a standard list of questions, you reduce the risk of interviewer bias—and, by using fixed-alternative questions, you obtain easily interpretable responses.

Semistructured Interview

In psychological research, a less popular interview format is the **semistructured interview**. Like the structured interview, the semistructured interview is constructed around a core of standard questions. Unlike the structured interview, however, the interviewer may expand on any question in order to explore a given response in greater depth.

Like the structured interview, the semistructured interview tells you how respondents answered the standard questions. In addition, the semistructured interview allows the investigator to ask additional questions to follow up on any interesting or unexpected answers to the standard questions.

Unfortunately, the advantage of being able to follow up on questions is usually outweighed by two major disadvantages. First, data from the follow-up questions are difficult to interpret because different participants are asked different questions. One can't compare how Participant 1 and Participant 2 answered follow-up Question 6c if Participant 1 was the only person asked that follow-up question.

Second, even the answers from the standard questions are difficult to interpret because the standard questions were not asked in the same standard way to all participants. Participant 1 might have answered Question 2 right after Question 1, whereas Participant 2 answered Question 2 after answering Question 1 and 10 minutes of follow-up questions. Those follow-up questions might shape the answers to Question 2 (Schwarz, 1999). Thus, in giving the interviewer more freedom to follow up answers, you may be giving the interviewer more freedom to bias the results. In other words, by deciding which answers to probe and how to probe them, the interviewer may affect what participants say in response to the standard questions.

Given the disadvantages of the semistructured interview, when should it be used? Perhaps the best time to use it is when you are conducting a pilot (preliminary) study so that you can better formulate your research question. For instance, you may know a few questions you want to ask, but you also know that, for the most part, you "don't really know enough to know what to ask." The standard questions may give you some interpretable data, from which you may be able to get some tentative answers to the specific questions you do have. From the answers to the follow-up questions, you may get some ideas for other questions you could ask in your next survey.

In short, if you do not yet know enough about your respondents or a certain topic area to create a good structured interview, you may want to first conduct a semistructured interview. What you learn from the results of that interview may enable you to generate a good set of questions that you can then use in either a structured interview or a questionnaire.

Unstructured Interview

The **unstructured interview** is popular in the media, in the analyst's office, and with inexperienced researchers. In the unstructured interview, interviewers have objectives that they believe can be best met without an imposed structure. Therefore, there isn't a set of standard questions. The interviewer is free to ask what she wants, how she wants to, and the respondent is free to answer how he pleases. Without standardization, the information is extremely vulnerable to interviewer bias and is usually too disorganized for meaningful analysis.

Because of these problems, the unstructured interview is best used as an exploratory device. As a research tool for reaping meaningful and accurate information, the unstructured survey is limited. As a tool for a beginning researcher, the unstructured interview is virtually worthless.

Editing Questions: Nine Mistakes to Avoid

Now that you have probably decided that you will use either a structured interview or a questionnaire, it's time to focus on your questions. Although asking questions is a part of everyday life, asking good survey questions is not. Therefore, in this section, you will learn how to avoid nine mistakes people often make when writing questions.

1. Avoid Leading Questions: Ask, Don't Answer

Remember, your aim is to get accurate information, not to get agreement with your beliefs. Therefore, don't ask **leading questions**: questions that clearly lead participants to the answer you want. For example, don't ask the question, "You disapprove of the biased, horrible way that television news covers the abortion issue, don't you?" Instead ask, "Do you approve or disapprove of the way television news shows cover the abortion issue?"

2. Avoid Questions That Are Loaded With Social Desirability

Don't ask questions that have a socially correct answer, such as, "Do you donate money to worthwhile causes?" Generally, the answers to such questions cannot be trusted because participants will respond with the socially desirable answer. Such questions may also contaminate participants' responses to subsequent questions because such questions may arouse respondents' suspicions. For instance, the respondent may think, "They said there were no right or wrong answers. They said they just wanted my opinion. But obviously, there are right and wrong answers to this survey." Or, the respondent may think, "They knew I would give that answer. Anyone would give that answer. This survey is starting to feel like one of those 'surveys' used by people who try to sell you something. What are they trying to sell?"

3. Avoid Double-Barreled Questions: "No ands or buts about it"

You wouldn't think of asking a respondent more than one question at the same time. But that's exactly what happens when you ask a **double-barreled question**: more than one question packed into a single question (e.g., "How much do you agree with the following statement: 'Colleges need to spend more time on students' emotional *and* physical development'?"). The responses to this question are uninterpretable because you don't know whether participants were responding to the first statement, "Colleges need to spend more time on students' emotional development," the second statement, "Colleges need to spend more time on students' physical development," or both statements.

As you can see, the conjunction *and* made the question double-barreled. Almost all double-barreled questions are joined by *and* or some other conjunction. So when looking over your questions, look suspiciously at all *ands, ors, nors,* and *buts.*

4. Avoid Long Questions: Short Is Good

Short questions are less likely to be double-barreled. Furthermore, short questions are easier to understand. A useful guideline is to keep most of your questions under 10 words and all your questions under 20 words.

5. Avoid Negations: No and Not Are Bad

The appearance of a negation, such as *no* or *not*, in a questionnaire item increases the possibility of misinterpretation. This is probably because it takes more time to process and interpret a negation than a positively stated item. To illustrate, compare the next two questions: "Do you not like it when students don't study?" versus "Do you like it when students study?"

6. Avoid Irrelevant Questions

Make sure your questions seem relevant to your participants and that your questions are relevant to your research question. For example, "Do you eat fondue?" is irrelevant to the research question, "Are professors who use text messaging more sympathetic to students?"

Although there are many reasons not to ask irrelevant questions, the most important reason is that such questions annoy respondents. If you ask an irrelevant question, many respondents will conclude that you are either incompetent or disrespectful. Because they have lost respect for you, they will be less likely to give accurate answers to the rest of your questions. In fact, they may even refuse to continue with the survey.

7. Avoid Poorly Worded Response Options

From your experiences with multiple-choice tests, you are keenly aware that the response options are part of the question. The options you choose will affect the answers that participants give (Schwarz, 1999). Therefore, you should carefully consider how to word each option and how many options you will include.

As a general rule, the more options you provide, the greater your ability to detect subtle differences between participants' answers. According to this rule, if you use a 1-to-7 scale, you may find differences that you would have failed to find had you used a 1-to-3 scale.

However, like most rules, this one has exceptions. If you give participants too many options, participants may be overwhelmed. Likewise, if the options are too similar, participants may be confused. The easiest way to determine how many options are appropriate is to pretest your questions.

8. Avoid Big Words

Your task is not to impress respondents with your large vocabulary. Instead, your task is to make sure respondents understand you; therefore, use simple words and avoid jargon.

9. Avoid Words and Terms That May Be Misinterpreted

To make sure that participants know exactly what you are talking about, take three steps. First, avoid abbreviations and slang terms. Abbreviations that are meaningful to you may be meaningless to some respondents. Similarly, slang terms often have different meanings to different groups. Thus, if you want to know people's attitudes toward marijuana, use the word "marijuana" rather than a slang term like "dope." Dope may be interpreted as meaning marijuana, heroin, or all drugs.

Second, be specific. If you want to know whether your respondents like college students, don't ask, "How do you feel about students?" Instead, ask, "Do you like college students?"

Third, pretest the questions on members of your target population. Often, the only way to find out that a question or term will be misinterpreted is by asking members of your target group what they think the question means. For example, through extensive pretesting, you might find that a seemingly straightforward question such as, "Should Pittsburgh increase coke production?" may be interpreted in at least five different ways:

1. Should Pittsburgh increase cocaine production?
2. Should Pittsburgh increase coal production?
3. Should Pittsburgh increase steel production?
4. Should Pittsburgh increase soft drink production?
5. Should Pittsburgh increase Coca-Cola production?

Similarly, if you were asking about sex, your participants may or may not consider masturbation and oral sex as instances of sex. If pretesting shows that participants will not interpret the word the way you intend, you may need to use another word or you may need to define the term (e.g., "For the purpose of this survey, consider sex to include masturbation, oral sex, and sexual intercourse.").

In conclusion, even if you carefully evaluate and edit each question, there are some problems that you can discover only by having people try to answer your questions. Therefore, *pretesting questions is one of the most important steps in developing questions.*

Sequencing Questions

Once you have edited and pretested your questions, you need to decide in what order to ask them. Ordering questions is important because the sequence of questions can influence results (Krosnick & Schuman, 1988; Schwarz & Oyserman, 2001). To appropriately sequence questions, follow these five rules:

1. Put the least personal questions first.
2. Qualify early.
3. Be aware of response sets.
4. Keep similar questions together.
5. Put demographic questions last.

1. Put the Least Personal Questions First

Participants are often tense or anxious at the beginning of a survey. They don't know what to expect. They don't know whether they should continue the survey. Indeed, they may be looking for an excuse to quit the survey. Thus, if the first question is extremely personal, participants may decide to withdraw from the survey. Even if participants don't withdraw, starting them out with a personal question may put them on the defensive for the entire survey. If, on the other hand, your first questions are simple, interesting, innocuous, and nonthreatening, participants may relax and feel comfortable enough to respond frankly to personal questions.[3]

Putting the most sensitive questions at the end of your survey will not only give you more honest responses but also more responses. To illustrate, suppose that you have a 20-item survey in which all but one of the questions are relatively innocuous. If you put the sensitive item first, respondents may quit the survey immediately. Because this item was the first question you asked, you have gathered no information whatsoever. If, on the other hand, you put the sensitive item last, respondents may answer the question because they have a deeper involvement with both you and the survey than they did at the beginning. Furthermore, even if they quit, you still have their responses to 19 of the 20 questions.

2. Qualify Early

If people must meet certain qualifications to be asked certain questions, find out if your participant has those qualifications before asking her those questions. In other words, don't ask people questions that don't apply to them. There is no need to waste their time—and yours—by collecting useless information. Participants don't like having to repeatedly answer questions by saying, "Doesn't apply."

The survey in Box 8.1 begins with a simple qualifying question: "What is your position at Bromo Tech?" This question establishes the presence of two qualifications for the survey: (1) that the person is a professor, and (2) the person teaches at Bromo Tech. If a respondent doesn't meet these qualifications, the survey is terminated. By terminating the survey early in the interview, we save our time, the respondent's time, and our client's money.

3. Be Aware of Response Sets

If all your questions have the same response options, some participants may lock onto one of those options. For example, if each question has the alternatives, "Strongly Agree, Agree, Neutral, Disagree, Strongly Disagree," some respondents may circle the option "Neutral" for every question. By always checking the neutral option, they can get the questionnaire over with as soon as possible.

To avoid the neutral response set, you may want to eliminate the neutral option. However, you will still be vulnerable to response sets because the neutral response set isn't the only response bias. As we mentioned earlier,

---

[3] Not everyone agrees with this rule. For example, Dillman (1978) suggests that surveys should start with questions that hook the respondents' interest. If you are having trouble getting people to participate in your survey, you might consider Dillman's advice. However, when we have carefully explained the purpose of the survey before administering it (in accordance with the principle of informed consent), participants who start the survey usually finish it.

there are a variety of response sets, including the "yea-saying" (always agreeing) and the "nay-saying" (always disagreeing) biases.

One of the most common ways of dealing with response sets is to alternate the way you phrase the questions. You might ask respondents to strongly agree, agree, disagree, or strongly disagree to the statement, "Most students work hard on their studies." Then, later in the questionnaire, you could ask them to strongly agree, agree, disagree, or strongly disagree with the statement, "Most students are lazy when it comes to their studies."

4. Keep Similar Questions Together

There are three reasons why you get more accurate responses when you keep related questions together.

First, your participants will perceive the survey to be organized and professional. Therefore, they will take the survey seriously.

Second, participants will be less likely to misunderstand your questions. You minimize the problem of participants thinking that you are asking about one thing when you are really asking about another topic.

Third, because you ask all the related questions together, participants are already thinking about the topic before you ask the question. Because they are already thinking about the topic, they can respond quickly and accurately. If respondents aren't thinking about the topic before you ask the question, it may take some respondents a while to think of the answer to the question. At best, this makes for some long pauses. At worst, respondents will avoid long pauses by saying they don't know or by making up an answer.

5. Put Demographic Questions Last

In addition to writing items that directly address your research question, you should ask some questions that will reveal your sample's demographics: statistics relating to the age, sex, education level, and other characteristics of the group or its members. In our survey of college professors (see Box 8.1), we asked four demographic questions (Questions 12–15).

By comparing our sample's responses to these demographic questions with our population's demographics, we may be able to detect problems with our sample. For example, we can look in the college catalog or go to the personnel office to find out what percentage of the population we are interested in (all teachers at Bromo Tech) are men. Then, we can compare our sample demographics to these population demographics. If we found that 75% of the faculty are men, but that only 25% of our sample were men, we would know that our sample wasn't representative of the faculty.

Note that we, like most researchers, put the demographic questions last (Questions 12–15). We put them last for two reasons. First, respondents are initially suspicious of questions that do not clearly relate to the purpose of the survey. Second, people seem increasingly reluctant to provide demographic data. To reduce suspiciousness and increase openness, we try to put our respondents at ease before we ask demographic questions.

Putting the Final Touches on Your Survey Instrument

You've written your questions, carefully sequenced them, and pretested them. Now, you should carefully proofread and pretest your questionnaire to make sure that it is accurate, easy to read, and easy to score.

Obviously, participants are more likely to take your research seriously if your questionnaire looks professional. Therefore, your final copy of the questionnaire should be free of smudges and spelling errors. The spaces between questions should be uniform.

Even though the questionnaire is neatly typed, certain key words may have been scrambled and omitted. At best, these scrambled or missing words could cause embarrassment. At worst, they would cause you to lose data. Therefore, not only should you proofread the questionnaire to ensure that the form of the questionnaire looks professional but you should also pretest the questionnaire to ensure that the content is professional.

Once you have thoroughly checked and rechecked both the form and the content of the questionnaire, you should fill out the questionnaire and then code your own responses. Then, you should consider three strategies for making coding easier.

1. Put the answer blocks in the left margin. This will allow you to score each page quickly because you can go straight down the page without shifting your gaze from left to right and without having to filter out extraneous information (see Box 8.1).
2. Have respondents put their answers on a separate answer sheet. With an answer sheet, you don't have to look through and around questions to find the answers. The answer sheet is an especially good idea when your questionnaire is longer than one page because the answer sheet saves you the trouble of turning pages.
3. Have participants put their responses on a coding sheet that can be scored by computer. Computer scoring is less time consuming, less tedious, and more accurate than hand scoring.

Choosing a Sampling Strategy

You have decided what questions you will ask and how you will ask them. You know why you are asking the questions: Your questions will answer a question you have about your population. Your next step is to decide who, of all the people in your population, will be in your sample (as the Ghostbusters put it, "Who you gonna call?"). If your population is extremely small (all art history teachers at your school), you may decide to survey every member of your population. Usually, however, your population is so large that you can't easily survey everyone. Therefore, instead of surveying the entire population, you will survey a sample of people from that population. Whether you acquire your sample by a probability sampling method such as pure random sampling or proportionate stratified random sampling, or whether you use a nonprobability sampling method such as convenience sampling or quota sampling, your goal is to get a sample that is representative of your population.

Random Sampling

In **random sampling,** each member of the population has an equal chance of being selected. Furthermore, the selection of respondents is independent: The selection of a given person has no influence on the selection or exclusion of other members of the population from the sample. For example, your having selected Sam doesn't have any effect on whether you will select Mary.

Obtaining a Random Sample. To select a random sample for a survey, you would first identify every member of your population. Next, you would go to a random numbers table and use that table to assign each member of the population a random number. Then, you would rank each member from lowest to highest based on the size of his or her random number. Thus, if a person were assigned the random number 00000, that person would be the first person on the list, whereas a person assigned the number 99999 would be the last person on the list. You would select your sample by selecting names from the beginning of this list until you got the sample size you needed. If you needed 100 respondents, you would select the first 100 names on the list.

As you can imagine, random sampling can be time consuming. First, you have to identify every member of the population—and have a way of contacting them. Identifying every member of the population can be a chore, depending on your population. Obtaining their contact information can be a real nightmare—especially if you are trying to get their e-mail addresses. If you are interested in a student sample, then a trip to the registrar's office might yield a list of all currently enrolled students and their phone numbers. In fact, most schools can generate a computerized random sample of students for you. If your population is your local community, the local telephone book may help you assess much of that population. However, realize that the phone book will leave out people who can't afford or choose not to have phones, people who use only their cell phones, people who have recently moved, and people with unlisted numbers. If you have the money, you can avoid many of these problems by purchasing phone lists from marketing research firms.

After you've identified the population and obtained the best list of that population you can get, you have to assign random numbers to your potential respondents. Just the first step—assigning random numbers to all members of a population—can be cumbersome and time consuming. Imagine assigning 1 million random numbers to names! But after that's done, you still have to order the names based on these random numbers to determine whom you will sample. Fortunately, computers can eliminate many of the headaches of random sampling—especially if you can find a computer file or database that already has all the names of everybody in your population.

Despite the hassles involved with random sampling, researchers willingly use it because random sampling allows them to generalize the results of one study to a larger population. To be more specific, you can use inferential statistics to infer the characteristics of a population from a random sample of that population.

Determining an Appropriate Sample Size. As you know, your random sample may differ from the population by chance. For example, although 51% of the people in your population are women, perhaps only 49% of the people in your random sample will be women. You also know that you can reduce random sampling error by increasing your sample size. In other words, a random sample of 10,000 will tend to reflect the population more accurately than a random sample of 10. However, surveying 10,000 people may cost more time and energy than the added accuracy it buys. To determine how many people you will need to randomly sample, consult Table 8.4.

Proportionate Stratified Random Sampling

What if you can't afford to survey as many people as Table 8.4 says you need? Then, if you use pure random sampling, random sampling error may cause your sample to be less representative than you would like. With pure

---

TABLE **8.4** Required Sample Size as a Function of Population Size and Desired Accuracy (Within 5%, 3%, or 1%) at the 95% Confidence Level

	Sampling Error		
	5%	3%	1%
Size of the Population	**Minimum Sample Size Required**		
50	44	48	50
100	79	92	99
200	132	169	196
500	217	343	476
1,000	278	521	907
2,000	322	705	1,661
5,000	357	894	3,311
10,000	370	982	4,950
20,000	377	1,033	6,578
50,000	381	1,066	8,195
100,000	383	1,077	8,926
1,000,000	384	1,088	9,706
100,000,000	384	1,089	9,800

Example of how this table works: If you are sampling from a population that consists of 50 people and you want to be 95% confident that your results will be within 5% of the true percentage in the population, you need to randomly sample at least 44 people.

Note: Table provided by David Van Amburg of MarketSource, Inc.

random sampling, the only defense you have against random sampling error is a large sample size.

With **proportionate stratified random sampling**, on the other hand, you don't leave the representativeness of your sample entirely to chance. Instead, you make sure that the sample is similar to the population in certain respects. For example, if you know that the population is 75% male and 25% female, you make sure your sample is 75% male and 25% female.[4] You would accomplish this goal by dividing your population (stratum) into two subpopulations, or substrata. One substratum would consist of the population's men; the other substratum would consist of the population's women. Next, you would decide on how many respondents you would sample from each substratum (e.g., you might sample 75 from the male stratum and 25 from the female stratum). Finally, you would draw random samples from each substratum. In this last step, the only difference between proportionate stratified random sampling and basic random sampling is that you are collecting two random samples from two substrata (e.g., male professors and female professors), rather than one sample from the main population (e.g., professors).

By using proportionate stratified random sampling, you have all the advantages of random sampling, but you don't need to sample nearly as many people. Thus, thanks to proportionate stratified sampling, the Gallup Poll can predict the outcome of U.S. presidential elections based on samples of only 300 people.[5] Furthermore, a proportionate stratified random sample ensures that your sample matches the population on certain key variables.

---

[4] If you are going to do stratified random sampling, typically you will do proportionate random sampling. That is, if the first stratum comprises ¾ of the population and the second was ¼ of the population, ¾ of your total sample would be from the first population and ¼ would be from the second population. In other words, the size of your sample from the first stratum would be 3 times as big as your sample from the second stratum. However, there are cases in which you would not do proportionate random sampling. For example, suppose that you wanted a sample of at least 100 persons from a certain subgroup (stratum) so that you could make relatively precise statements about that subgroup, but that subgroup (stratum) made up a tiny percentage of the population (e.g., 1%). If you used proportionate sampling, to get 100 people from a subgroup that made up only 1% of the population, you would need a total sample of 10,000 ($10,000 \times .01 = 100$) people (9,900 of which would be from the majority group). In such a case, you would probably use disproportionate random sampling. For example, you might sample 100 from your 1% group and 100 from your 99% group. To make estimates of the total population's behavior, you have to correct for oversampling from the 1% group. For example, if the average rating on a -50 (extremely dissatisfied) to $+50$ (extremely satisfied) scale was -20 for your 1% group and $+20$ for your 99% group, you should not estimate the population satisfaction at 0. Instead, you should give each response from a member of your 99% stratum 99 times more weight than you would give a response from the 1% group. Consequently, you would multiply the average of the 99% group by 99, multiply the average of the 1% group by 1, add those two results, and then divide by 100 to get your estimate of the population's satisfaction: $+19.6$ ([(99 × 20) + (1 × −20)]/100).

[5] As of this writing, the Gallup poll has successfully predicted the winner in the last eight presidential elections and has usually been accurate within a percentage point in predicting the actual percentage the losing candidate will get.

Convenience Sampling

In **convenience sampling** (also called accidental sampling, haphazard sampling, and nonprobability sampling), you simply sample people who are easy to survey. Convenience surveys are very common. Newspapers ask people to e-mail their responses to a survey question, and radio stations ask people to call in their reactions to a question. Even television stations sometimes ask viewers to express their views by text messaging or by filling out a survey on the station's website.

To see how you would get a convenience sample, suppose that you were given 1 week to get 1,000 responses to a questionnaire. What would you do? Provided you had approval from your school's institutional review board (IRB), you might (a) go to areas where you would expect to find lots of people, such as a shopping mall; (b) ask your professors if you could do a survey in their classes; (c) put an ad in the newspaper, offering people money if they would respond to a questionnaire; or (d) put your survey on the Internet.

Although you can use convenience sampling techniques to get a relatively large sample quickly, you do not know whether the sample represents your population. Your best bet is that it does *not* (see Figure 8.3). In fact, if your respondents are actively volunteering to be in your survey, you can bet that your sample is biased. People who call in to radio shows, answer surveys posted on Internet sites such as Facebook, or respond to ads asking for

"We need a bit more science in our market surveys."

FIGURE **8.3** Convenience samples are biased samples.

Reprinted by permission of CartoonStock./cartoonstock.com

people to be in a survey do not represent a significant portion of the population: people without the time or desire to respond to such surveys.

Quota Sampling

Quota sampling is designed to make your convenience sample more representative of the population. Like proportionate stratified random sampling, quota sampling is designed to guarantee that your sample matched the population on certain characteristics. For instance, you might make sure that 25% of your sample was female, or that 20% of your sample was Hispanic.

Unlike proportionate stratified random sampling, however, quota sampling doesn't involve random sampling. So, even though you met your quotas, your sample may not reflect the population at all. For example, you may meet your 20% quota of Hispanics by hanging around a hotel where there is a convention of high school Spanish teachers; obviously, the Hispanics in your survey would not be representative of the Hispanics in your community.

Conclusions About Sampling Techniques

If we were to rank sampling techniques in terms of their ability to produce representative samples, the rankings would be

1. proportionate stratified random sampling,
2. random sampling,
3. quota sampling, and
4. convenience sampling.

To get samples that represent your population, we recommend that you use either simple random sampling or proportionate stratified random sampling because random samples have three big advantages over other sampling techniques: (1) they are not biased, (2) their margin of error can be calculated, and (3) if they are of adequate size, they are fairly accurate. To appreciate how accurate random samples are relative to other methods, note that a sample of 400 people using either of these random sampling techniques will get you a more representative sample of a large population than a sample of a million people using nonrandom sampling. Unfortunately, Alfred Kinsey, a biologist turned sex researcher, did not appreciate this fact. Because he stubbornly used nonrandom sampling, Kinsey collected much more data than he needed to and was able to draw far fewer legitimate conclusions than he should have been able to.

Note, however, that random sampling will not be accurate unless you have an accurate list of your population—and you may not be able to get such a list. For example, you may want to randomly sample people who will vote in the next election, but the list of such voters does not exist. You can get a list of registered voters, you can get a list of voters who have voted in the last few elections, but you cannot get a list of only those people who will vote in the next election.

Although most polling organizations do a decent job of maintaining a good list of the population, some use poor lists. One of the most infamous cases of working from a poor list was the poll that led to the headline you may have seen in a history book: "Dewey beats Truman." The problem was that the list the polling company worked from was compiled from telephone books and automobile registrations. Back in 1936, the wealthy were much more likely to have phones, to have cars, and to be Republicans. Thus, the poll of Republicans yielded a strong preference for the Republican candidate (Dewey) rather than the actual winner (Truman).

However, even if you have a perfect list and draw a perfect sample from that list, you may not end up with a perfect sample because of nonresponse bias (see Figure 8.4). In other words, your sample will not represent members of the population who choose not to respond. Nonresponse bias is so powerful that it can even cause exit polls of voters to be inaccurate. For example, in the 2004 election, Bush voters were less likely to participate in exit polls than Kerry voters, thus making it appear that Kerry had defeated Bush.

There are two things you can do about the bias caused by nonresponse. First, you can get your response rate so high that nonresponse is not a big problem. For instance, by mailing out information to participants in advance, by keeping the survey short, and by calling people back repeatedly, some telephone interviewers have obtained response rates of 97% or better.

Second, keep detailed records on the people who refused. If possible, unobtrusively record their sex, race, and estimated age. By knowing who is dropping out of your sample, you may know to whom your results don't apply.

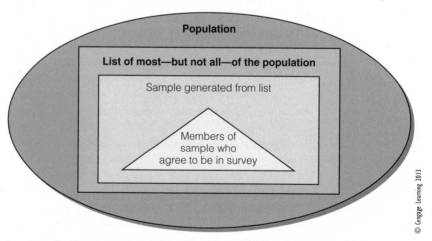

© Cengage Learning 2013

FIGURE **8.4** The challenge of capturing the population.

We would like to measure the population: all the members of a particular group. However, we usually do not start with the actual population. Instead, we start with a list of the population members—a list that is usually incomplete. Then, we usually take a sample from that list, a sample that is not a perfect sample of that list. Finally, we get people who agree to and actually do fill out the survey—a group that usually is a biased subgroup of our sample. Thus, our respondents are usually not a perfect reflection of the population.

ADMINISTERING THE SURVEY

You have your survey questions. You've carefully sequenced your questions, and you've determined your sampling technique. You have weighed the benefits and the risks of doing your survey, and you have taken steps to maintain your participants' confidentiality. You have had your study approved by your professor and either your department's ethics committee or your school's institutional review board (IRB). Now it's time for you to actually administer your survey.

As with any research, you must follow APA's ethical guidelines (APA, 2002) and conduct yourself professionally (see Appendix D: "Practical Tips for Conducting an Ethical and Valid Study"). For example, participants should always be greeted. If participants can't be greeted in person (e.g., you have a mail questionnaire), the questionnaire should be accompanied by a written greeting: a cover letter. In this written greeting, you should introduce yourself, explain the nature of your study, and request the participant's help—just as you would if you were greeting the participant in person. In your greeting, inform participants about (a) anything that would affect their decision to participate, such as how long the survey will take, whether the questions deal with any sensitive issues, and whether participants will be compensated for their participation; (b) their right to skip questions they don't want to answer; (c) their right to quit the study at any point; and (d) steps you will take to keep their responses confidential.

As with any other study, your instructions should be as clear as possible. Thus, if you are administering the questionnaire, you should probably repeat or restate the questionnaire's written instructions.

After participants complete the survey, they should be thanked and debriefed about the survey's purpose. At the end of a mail questionnaire, you should thank your participants, give them any additional instructions, and give them the opportunity to be debriefed. For example, you might write, "Please mail your questionnaire in the enclosed envelope. To find out more about the survey, put a check mark in the upper left-hand corner of the questionnaire, and we will send you a summary of the results once the data have been analyzed. If you wish to talk to me about the study, please call me at (your phone number or your department's phone number). Thank you for your participation."

Finally, as in all studies, you should be careful to ensure your participants' confidentiality. Before the survey begins, you and any other people working on the survey should sign a statement that they will not discuss participants' responses. As the data are collected, you must ensure that their responses are kept private. If you are interviewing participants, for instance, you must interview them in a private place where their responses will not be overheard. If you are having them fill out a questionnaire in a group setting, you should use a cover page and spread out participants so that they do not see one another's responses. If possible, you should not have participants put their names on the survey.

After participants respond to your survey, you must store and dispose of data in a way that keeps their data private. For example, you must store the information in a secure place (e.g., a locked file cabinet). If participants' names or other identifying information are on the cover sheet, you should probably either destroy the cover sheet as soon as possible or store the cover sheet in one place and the rest of the survey data in another place.

ANALYZING SURVEY DATA

Once you have collected your survey data, you need to analyze them. In this section, we will show you how to summarize and make inferences from your data.

Summarizing Data

The first step in analyzing survey data is to determine what data are relevant to your hypotheses. Once you know what data you want, you need to summarize those data. How you summarize your data will depend on what kind of data (e.g., nominal, ordinal, interval, or ratio) you have.

The type of data you have will depend on the type of questions you ask. When you ask rating scale questions or when you ask people to quantify their behavior ("How many text messages do you send in a typical a week?"), you probably can assume that your data are interval scale data.

If, on the other hand, you ask questions that people have to answer either "yes" or "no" ("Do you text message?" "Do you like students?"), or questions that call on people to put themselves into qualitatively different categories ("Are you male or female?"), you have nominal data.

Summarizing Interval Data

If all you need to know is the typical response to an interval scale question (e.g., the average of respondents' answers to the question, "How many text messages do you send in a typical week?"), all you need to calculate is the mean and standard deviation for that question.[6]

Summarizing Relationships Between Pairs of Variables. Rather than being interested only in the average response to a question, you will probably also want to know about the relationship between the answer to one question and the answers to other questions—in other words, the relationship between two or more variables. To begin to explore such a relationship, you will usually want to construct tables of means. For example, because we expected that there would be a relationship between text messaging and sympathy for

---

[6] If you need help computing these statistics, you can (a) use a web calculator to do so, (b) use the formula for the mean (total divided by number of scores) and the formula for the standard deviation (For each score, subtract it from the mean, square the result, add up all those squared terms, divide that total by one less than the number of scores, and then take the square root.), or (c) follow the more detailed instructions in Appendix E.

TABLE **8.5** Table of Means and Interactions

Table of Means on Question 11: "I Like College Students" Broken Down by Text Messaging Status

Text messaging status	
Yes	No
4.0	3.0

Average score on a 1 (*strongly disagree*) to 5 (*strongly agree*) scale.

Table of Means on Question 11: "I Like College Students" Broken Down by Gender

Gender	
Men	Women
3.25	3.75

Average score on a 1 (*strongly disagree*) to 5 (*strongly agree*) scale.

Table of Means on Question 11: "I Like College Students" Broken Down by Gender and Text Messaging Status

Gender	Text messaging status	
	Yes	No
Men	3.5	3.0
Women	4.5	3.0

Average score on a 1 (*strongly disagree*) to 5 (*strongly agree*) scale.

students, we compared the average sympathy for students of professors who text messaged to the average sympathy for students of professors who did not text message (see the top of Table 8.5). To supplement your tables of means, you may want to compute a correlation coefficient[7] to get an idea of the strength of the relationship between your two variables.[8]

---

[7] Technically, the name of the correlation you would compute would be called the point biserial correlation. There is a special formula you can use specifically for calculating the point biserial *r*. However, if you use the formula for the Pearson *r* or if you have a computer calculate the Pearson *r*, you will obtain the correct value for the correlation coefficient (Cohen & Cohen, 1983).

[8] To compute the Pearson *r* by hand, you can use the formula:

$$N \times \Sigma(X \times Y) - (\Sigma X) \times (\Sigma Y) / \sqrt{\left([N \times (\Sigma X^2) - (\Sigma X)^2] \times [N \times (\Sigma Y^2) - (\Sigma Y)^2] \right)}$$

Describing Complex Relationships Among Three or More Variables. Once you have looked at relationships between pairs of variables (e.g., text messaging and sympathy, gender and sympathy, gender and text messaging), you may want to see how three or more variables are related. The easiest way to compare three or more variables is to construct a table of means, as we have done at the bottom of Table 8.5. As you can see, this 2 × 2 table of means allows us to look at how both text messaging and gender are related to sympathy.

Summarizing Ordinal and Nominal Data

If your data are not interval scale data, don't summarize your data by computing means. For example, if you code 1 = man, 2 = woman, do not say, "the mean gender in my study was 1.41."

Similarly, if you are having participants *rank* several choices, do not say that the mean rank for Option B was 2.2. To understand why not, imagine that five people were ranking three options (Semon, 1990). Option A was ranked as second best by all five people (the rankings were "2-2-2-2-2"). Option B, on the other hand, was ranked best by two people and ranked third best by three people (the rankings were "1-1-3-3-3"). The mean rank for Option A is 2.0; the mean rank for Option B is 2.2 (Semon, 1990). Thus, according to the mean, A is assumed to be better liked (because it is closest to the average rank of 1.0, which would mean first choice).

In this case, however, the mean is misleading (Semon, 1990). The mean gives the edge to A because the mean assumes that the difference between being a second choice and being a third choice is the same as the difference between being a first choice and being a second choice. As you know, this is not the case. There is usually a considerable drop-off between one's first (favorite, best) choice and the one's second choice (runner-up, second best), but not such a great difference between one's second and third choices (Semon, 1990). For example, you may find an enormous drop-off between your liking of your best friend and your second best friend or between your favorite football team and your second favorite football team.

To go back to our example of Options A and B, recall that A's average rank was better than B's. However, because two people ranked B best and nobody ranked A first, we could argue that Option B was better liked (Semon, 1990). The moral of this example is that if you do not have interval data, do not use means to summarize your data. Instead, use frequencies or percentages.

Summarizing Relationships Between Pairs of Variables. To look at relationships among nominal variables, use a table to compare the different groups' responses. You could use a table of percentages to display the percentage of people belonging to one category (e.g., those belonging to the category "women professors") who also belong to a second category (e.g., "text messagers"). Alternatively, you could use a *frequency* table to display the *number* of people belonging to one category who also belong to a second

TABLE **8.6** Tables of Nominal Data

Text Messaging by Gender

	Gender	
Text Messaging	**Men**	**Women**
Yes	(A)	(B)
	20	15
No	(C)	(D)
	55	10

Text Messaging by Gender and Academic Department

Text messaging by gender			Text messaging by gender		
	Gender			Gender	
Text Messaging	**Men**	**Women**	**Text Messaging**	**Men**	**Women**
Yes	10	10	Yes	20	20
No	80	0	No	40	20
Physical science professors			Social science professors		

category. As you can see from Table 8.6, a frequency table can help you visualize similarities and differences between groups.

If you want to compute a measure to quantify how closely two nominal variables are related, you can calculate a correlational coefficient called the *phi coefficient*.[9] Like most correlation coefficients, phi ranges from −1 (perfect negative correlation) to +1 (perfect positive correlation).

Describing Complex Relationships Among Three or More Variables. If you want to look at how three or more variables are related, do not use the phi coefficient. Instead, construct tables of frequencies, as we have done in Table 8.6. These two 2 × 2 tables of frequencies do for our ordinal data what the 2 × 2 table of means did for our interval data—allow us to look at three variables at once.

---

[9] Suppose you have two variables (e.g., employment status and obesity) and that you code both variables using just "0's" and "1's," with "0" meaning does not belong to the category (e.g., not employed, not obese) and "1" meaning belongs to the category (e.g., employed, obese). In that case, you can calculate the phi coefficient using the Pearson *r* formula:

$$N \times \Sigma(X \times Y) - (\Sigma X) \times (\Sigma Y)/\sqrt{([N \times (\Sigma X^2) - (\Sigma X)^2] \times [N \times (\Sigma Y^2) - (\Sigma Y)^2])}$$

For step-by-step instructions on how to compute phi, see Appendix E.

Using Inferential Statistics

In addition to using descriptive statistics to describe the characteristics of your sample, you may wish to use inferential statistics. Inferential statistics may allow you to generalize the results of your sample to the population it represents. There are two main reasons why you might want to use inferential statistics.

First, you might want to use inferential statistics to estimate certain **parameters** (characteristics of the population) such as the population mean for how many text messages professors send. For example, if you wanted to use the average number of text messages professors in your sample said they sent each week to estimate the average number of text messages all Bromo Tech professors sent each week, you would be using **parameter estimation**.

Second, you might want to determine whether the relationship you found between two or more variables would hold in the population. For instance, you might want to determine whether text messaging and student sympathy are related in the population. Because you are deciding whether to reject the **null hypothesis** (that the variables are *not* related in the population), this use of inferential statistics is called **hypothesis testing**.

In hypothesis testing, the researcher determines how likely it is that the obtained results would occur if the null hypothesis is true. If the results are extremely unlikely given that the null hypothesis is true, the null hypothesis is rejected. Typically, "extremely unlikely" means that the *probability* of finding such a result given that the null hypothesis is true is less than 5 in 100 ($p < .05$). If the null hypothesis is rejected, the relationship between variables is declared **statistically significant**: probably not due to chance alone; reliable.

Parameter Estimation With Interval Data

As we just mentioned, one reason for using inferential statistics would be to estimate population parameters. From our survey of text messaging and student sympathy, we might want to estimate one parameter: the amount of sympathy the average professor at our school has for students.

Our best guess of the amount of sympathy the average professor at our school has for students is the average amount of sympathy the average professor in our sample has for students. This guess may be inaccurate because the average for our sample may differ from the average in the population. Therefore, it is often useful to establish a range in which the population mean is likely to fall. For example, you may want to establish a **95% confidence interval**: a range in which you can be 95% sure that the population mean falls.

You can establish 95% confidence intervals for any population mean from the sample mean, if you know the **standard error of the mean**.[10] You establish the lower limit of your confidence interval by subtracting

---

[10] The standard error of the mean equals the standard deviation (*SD*) divided by the square root of the number of participants. Thus, if the *SD* is 8 and the sample size is 100, the standard error of the mean would be $8/\sqrt{100} = 8/10 = .8$ For more on the standard error, see either Chapter 7 or Appendix E.

approximately 2 standard errors from the sample mean. Then, you establish the upper limit of your confidence interval by adding approximately 2 standard errors[11] to the sample mean. Thus, if the average sympathy rating for all the professors in our sample was 3.0 and the standard error was .5, we could be 95% confident that the true population mean was somewhere between 2.0 and 4.0.[12]

Hypothesis Testing With Interval Data

You can also use statistics to see if there are significant differences between groups. That is, we might want to know if the differences between groups that we observe in our sample also apply to the population at large.

Testing Relationships Between Two Variables. By using a *t* test,[13] we could test whether the differences in sympathy we observed between professors who text message and professors who don't text message were too large to be due to sampling error alone and thus probably represented a true difference.

The *t* test between means is not the only way to determine whether there is a relationship between text messaging and student sympathy. We could also see whether a relationship exists by determining whether the correlation coefficient between those two variables was significant.[14]

If you were comparing more than one pair of variables, you could do several *t* tests or test the significance of several correlations (for more about these analyses, see **Chapter 7 or the Chapter 7 section of www. studyRDE.com). In either case, you should correct for doing more than a

---

[11] To determine precisely what you should multiply the standard error by, look at the .05 significance column of Table F-1 (in Appendix F) in the row corresponding to one less than your sample size. If your sample size is less than 61, you will have to multiply the standard error by a number larger than 2. If your sample size is greater than 61, multiplying by 2 will give you a larger confidence interval than you need: You would be more than 95% confident that the true population mean is within that interval. Usually, the number of standard errors will vary from 1.96 to 2.776. To be more precise, the exact number will depend on your degrees of freedom (*df*)—and your *df* will be 1 less than your number of participants. (e.g., if you have a mean based on 11 participants' scores, your *df* will be 10.) Once you have calculated your *df*, go to the *t* table (Table F-1) in Appendix F. In that table, you will look under the .05 column (it starts with 12.706) and find the entry corresponding to your *df*. Thus, if you have a *df* of 10, you would multiply your standard error by 2.228; if you had a *df* of 120, you would multiply your standard error by 1.98.

[12] If we had 61 participants (see the previous footnote).

[13] To compute a *t*, you would subtract your 2 group means and then divide by the standard error of the differences. To calculate the standard error of the differences by hand, you can (a) use the formula: standard error of the mean = $\sqrt{s_1^2/N_1 + s_2^2/N_2}$, where s = standard deviation and N = number of participants, and where 1 refers to group 1 and 2 refers to group 2; or (b) follow the more detailed set of instructions in Appendix E. To learn more about *t*, see either Chapter 10 or Appendix E.

[14] The formula for this test is

$$t = \frac{r \times \sqrt{(N-2)}}{\sqrt{(1 - [r \times r])}}$$

To find out more about the test, see Appendix E.

single statistical test. One way to correct for doing more than one test is to use a more stringent significance level than the conventional .05 level, such as a .01 significance level. Note that the more tests you do, the more stringent your significance level should be. For example, if you looked at 5 comparisons, you might use the .01 level; if you looked at 50 comparisons, you might use the .001 level.

To understand why you should correct for doing multiple tests, imagine that you are betting on coin flips. You win if you get a heads. If you flip a coin once, it's fair to say that there's a 50% chance of getting a heads. However, if you flip a coin three times and declare victory if any of those flips come up heads, it's not fair to claim that you only had a 50% chance of winning. Similarly, a .05 significance level implies that you only have a 5% chance of getting those results by chance alone. This false alarm rate (Type 1 error rate) of .05 applies only if you are doing only one test: If you are doing 100 tests and none of your variables are related, it would not be unusual for you to get 5 false alarms (because $.05 \times 100 = 5$).

Testing Relationships Among More Than Two Variables. Suppose you wanted to look at more than two variables at once. For example, suppose you wanted to explore the relationship between text messaging, gender, and sympathy summarized in Table 8.5. You might especially be interested to see whether gender was a moderator variable—whether it qualified, modified, or changed the relationship between text messaging and sympathy. For example, you might ask, "Is text messaging a better predictor of student sympathy for women or for men?" To answer questions involving moderator variables, you might analyze your data using analysis of variance (ANOVA).[15]

If you are dealing with multiple predictors, ANOVA is probably the simplest analysis you can do. To learn about more sophisticated analyses that you might use or that you may encounter as you read research articles, see Box 8.2.

Using Inferential Statistics With Nominal Data

Just as inferential statistics can be applied to interval data, inferential statistics can be applied to nominal data. Indeed, if you do research as part of your job, you may be more likely to do parameter estimation and hypothesis testing with nominal data than with interval data.

Parameter Estimation With Nominal Data. You might start by doing some basic parameter estimation, such as estimating the percentage of people who have some characteristic. If you used random sampling and chose your sample size according to the first column of Table 8.4, you can be 95% confident that your sample percentages are within 5% of the population's percentages. In that case, if you found that 35% of your participants were women, you could be 95% confident that 30–40% of your population were women.

---

[15] To learn more about ANOVA, see Chapter 11 or Appendix E.

BOX **8.2** Advanced Analyses

If your data include multiple measures or multiple predictors, you should consider using analyses designed to deal with multiple variables. For example, suppose that, rather than trying to determine whether a single predictor (text messaging) predicted the answers to a single variable (sympathy), you were trying to determine whether two predictors (text messaging and gender) predicted the answers to two dependent measures: (1) responses to Question 4 and (2) responses to Question 10.

Multiple ANOVAs and Multivariate Analysis of Variance

To determine whether text messaging and gender predict the answers to Question 4, you could do an ANOVA with testing and gender as your predictors and Question 4 as your dependent variable. If you also wanted to know whether texting and gender predict the answer to Question 10, you would do another ANOVA that again used testing and gender as predictors but used Question 10 as the dependent measure. If you perform multiple ANOVAs, you should correct your significance level for the number of ANOVAs you computed, just as you would if you computed multiple *t* tests (see this chapter's section "Testing Relationships Between Two Variables").[1]

Factor Analysis

In an earlier example, we avoided the problem of doing analyses on multiple measures (e.g., separate analyses for Questions 6, 7, 8, 9, and 10) by deciding that the sum of answers to Questions 6 through 10 would be our student sympathy measure. Although combining the answers to those five questions into one measure simplified our analyses, a critic might question whether those five questions actually measured the same underlying variable. In published research, most investigators would do a **factor analysis** to make the case that those five items were indeed measuring the same underlying factor (to learn more about factor analysis, see Appendix E).

Multiple Regression

Even if a survey researcher's sole goal is to predict responses to a single question (e.g., Question 11), analyzing the results may be complicated. For example, suppose your sole goal was to predict respondents' answers to Question 11: "I like students." Although you are trying to predict responses to only one question, you have many potential predictors (e.g., professor rank, professor experience, text messaging, gender, answers to Questions 4–10, and answers to Questions 12–15). If you want to know (a) how best to use these predictors to come up with an accurate prediction of how people will respond to Question 11, (b) how accurate that prediction will be, or (c) how important each of these predictors are, you should probably use multiple regression to analyze your data (to learn more about multiple regression, see Appendix E).

Structural Equation Modeling

Structural equation modeling (SEM) often has two elements: (1) a measurement model that, like factor analysis, specifies how an observed measure (e.g., answers to some test questions) correlates with

---

[1]Some believe the way to make sure that your actual significance level (your chance of making a Type 1 error: the error of declaring that a relationship between variables is statistically significant when, in reality, there is no relationship between the variables) is equal to your stated significance level when doing multiple ANOVAs is to do a multivariate analysis of variance (MANOVA) on your data first. If the overall MANOVA is significant, many believe you can then do your specific ANOVAs. However, as Huberty and Morris (1989) point out, such a strategy may not be an effective way to reduce Type 1 errors. In our example, there would be benefits to using MANOVA to look at the effect of text messaging on answers to Question 4 and 10. A significant main effect for text messaging would suggest that the text messaging predicts the answers to at least one of those two survey questions. A significant interaction between text messaging and question would suggest that the correlation between text messaging and Question 4 is different from the correlation between text messaging and Question 10.

(Continued)

BOX **8.2** **(Continued)**

a hypothetical, unobserved factor (e.g., shyness), and (2) a cause–effect model that specifies which variables are causes and which are effects (Kline, 1998). However, not all SEMs involve both aspects.

Factor analysis, for example, is an SEM technique that does not look for cause–effect relationships between variables. Instead, it focuses exclusively on establishing a measurement model: a model that specifies how scores on a measure—called indicators, observed variables, or manifest variables—are related to some invisible (latent), underlying factor. Therefore, in factor analysis, researchers collect scores on several indicators of each hypothetical variable and then create a measurement model that specifies how these observed measures (e.g., test items or test scores) are associated with an unobservable, hypothetical latent factor (e.g., creativity or love) that is not directly observed.

Thus, if we thought our survey was measuring two different things (e.g., creativity and assertiveness), we would hope that our factor analysis would support the idea that our questions were measuring two factors (in technical lingo, we would hope that "those two factors accounted for over 60% of the variability in scores"). We would also hope that our creativity questions all correlated with each other, so that we could infer that the creativity questions all correlated with (loaded on) the same factor (construct), and that our assertiveness questions correlated with each other and that they loaded on a different construct. Thus, Figure 8.5 would support the idea that we had two factors (the two circles), one measured by Questions 1–3, another measured by Questions 4–6.

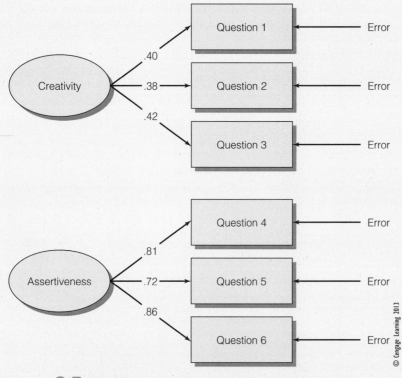

© Cengage Learning 2013

FIGURE **8.5** Possible results from a factor analysis.

Path analysis, on the other hand, does not have a measurement model. It does not try to make connections between observed scores and some invisible construct. Instead, path analysis focuses exclusively on trying to establish cause–effect paths between the observed, measured variables. One use of path analysis is to test a hypothesis that one variable mediates the effect of another.

For example, suppose we measure A, an attitude (e.g., liking research methods), B, a behavior (e.g., studying research methods), and C, a consequence (e.g., grade in research methods). One possibility is that A (liking research methods) has a direct effect on C (grades)—an A → C model. Suppose, however, that the investigator hypothesizes a model in which liking research methods (A) affects one's grade in research methods by increasing how much one studies (B). That is, the hypothesis is that A's (liking's) effect on C (grades) is indirect and mediated by B (studying). In other words, the investigator is proposing that the causal path is best described as A → B → C.

To test this or any causal path, the researcher using path analysis uses multiple regression to estimate the strength of the paths between the variables. (Path analysis relies so much on multiple regression that one expert calls path analysis "multiple regression with pictures" [B. M. Byrne, 2004].) If the direct path between A → C is strong and the indirect path (A → B → C)[2] is weak, the researcher would conclude that A's effect on C is direct (e.g., liking directly improves grades). If, on the other hand, the indirect path (A → B → C) is strong and the direct A → C path is weak, the researcher might conclude that A's effect on C is mediated through B (e.g., liking leads to studying, which leads to better grades).

In short, most structural equation models are more complex than either path analysis or factor analysis. Most SEMs are more complicated than path analysis because, rather than confining themselves to observed variables, they test relationships between unobserved (latent) factors. In a sense, because most SEMs use multiple indicators of each hypothetical factor, most SEMs incorporate a factor analysis. However, most SEMs are more complicated than factor analysis because most SEMs test not only a relationship between a hypothetical factor and measures of that factor but also try to determine how one hypothetical factor *causes* a change in another.

[2]If you aren't given the strength of the indirect path (A → B → C), you can calculate it by multiplying the A → B path coefficient by the B → C coefficient. So, if the A → B path was .4 and the B → C path was .2, the A → B → C path would be .08. (because .4 × .2 = .08)

Hypothesis Testing With Nominal Data. After estimating the percentages of the population that had certain characteristics, you might go on to look for differences between groups. In that case, you would use significance tests to determine whether differences between sample percentages reflect differences in population percentages. For example, in your sample, you may find that more men than women text message. Is that a relationship that holds in your population—or is that pattern due to random sampling error? To rule out the possibility that the pattern is an artifact of random sampling error, use a statistical test. But instead of using a *t* test, as you would with interval data, you would use a test that doesn't require you to have interval data: the **chi-square (χ^2) test**.[16]

---

[16]To compute a chi square, you first calculate the expected number of observations that should be in each cell by taking the row total for that cell, multiplying it by the column total, and then dividing by the total number of observations. Then, for each cell, you take the actual total for the cell, subtract the expected score for that cell, square the difference, and then divide the difference by the expected score. Now that you have a result for each cell, add up all those results to get your chi square. To see whether your chi square is significant, go to Table F-2 in Appendix F. For more detailed instructions and an example of how to do a chi-square test, see Appendix E.

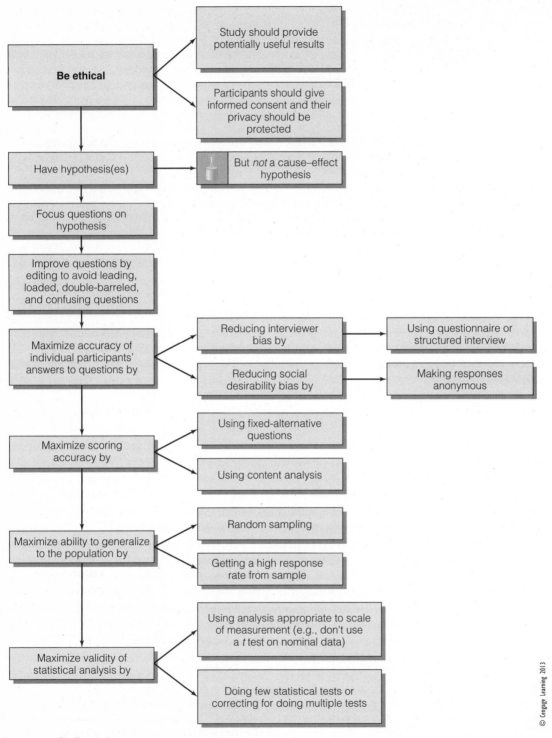

FIGURE **8.6** Guidelines for survey research.

© Cengage Learning 2013

If you are performing more than one chi-square test, you should correct for the number of analyses performed by raising your significance level to compensate for doing multiple analyses. For example, if you are comparing five chi-squares, you should use a .01 significance level rather than a .05 significance level.

CONCLUDING REMARKS

In this chapter, you learned the essence of good survey research. Early in the chapter, you were introduced to the applications and limitations of survey research. You saw the advantages and disadvantages of different survey formats, as well as the strengths and weaknesses of different kinds of questions. After learning how to write a survey, you learned how to administer, score, and analyze survey data. If you apply what you have learned in this chapter (see Figure 8.6), you will be a skilled survey researcher.

SUMMARY

1. Surveys can help you describe what people are thinking, feeling, or doing.
2. Surveys allow you to gather information from a large sample with less effort and expense than most other data-gathering techniques.
3. In a survey, it is important to ask only relevant questions.
4. Don't accept respondents' answers as truth. People don't always tell the truth or even know what the "truth" is.
5. Surveys yield only correlational data. You cannot draw cause–effect conclusions from correlational data.
6. There are two main drawbacks to self-administered questionnaires: (1) They have a low return rate, and (2) respondents may misinterpret questions.
7. Investigator-administered questionnaires have a higher response rate than self-administered questionnaires.
8. Interviews are especially useful for exploratory studies. However, interviews are expensive, and the interviewer may bias participants' responses.
9. Telephone surveys have higher response rates, are easier to administer, and offer greater anonymity than personal interviews.
10. Your first step in survey research is to have a hypothesis.
11. There are three basic question formats: nominal-dichotomous, Likert-type, and open-ended.
12. Structured surveys are more useful than unstructured surveys.
13. In survey research, you want to ask the right people the right questions.
14. To ask the "right people," you need a representative sample. To get a representative sample, you must first know what your population (the group that you want to generalize to) is. Once you know your population, you can try to get a representative sample by using either random or proportionate stratified random sampling. Unfortunately, getting your random sample may be hampered by nonresponse bias.
15. To ask good questions, (1) make sure they relate to your hypotheses; (2) edit them so they are short, clear, and unbiased; and (3) pretest them.
16. Careful attention should be placed on sequencing questions. Keep similar questions together and put personal questions last.
17. Be aware of response biases, such as a tendency of participants to agree with

statements or the tendency to answer questions in a way that puts the participant in a positive light.
18. Spending a little time deciding how to code your questionnaire before you administer it can save a great deal of time later on.
19. Both random and proportionate stratified random sampling allow you to make statistical inferences from your data.
20. Participants in survey research should be treated with the same respect as human participants in any other kind of study.

KEY TERMS

95% confidence interval (p. 320)
chi-square (χ^2) test (p. 325)
convenience sampling (p. 312)
demand characteristics (p. 285)
demographics (p. 280)
descriptive hypothesis (p. 281)
dichotomous questions (p. 297)
double-barreled question (p. 303)
factor analysis (p. 323)
fixed-alternative questions (p. 297)
hypothesis testing (p. 320)
interview (p. 286)
interviewer bias (p. 291)

leading questions (p. 303)
likert-type items (p. 298)
nominal-dichotomous items (p. 297)
nonresponse bias (p. 286)
null hypothesis (p. 320)
open-ended questions (p. 300)
parameter estimation (p. 320)
parameters (p. 320)
population (p. 276)
power (p. 298)
proportionate stratified random sampling (p. 311)
questionnaire (p. 286)
quota sampling (p. 313)
random digit dialing (p. 292)

random sampling (p. 309)
response set (p. 285)
retrospective self-reports (p. 284)
self-administered questionnaires (p. 286)
semistructured interview (p. 301)
social desirability bias (p. 285)
standard error of the mean (p. 320)
statistically significant (p. 320)
structured interview (p. 301)
summated score (p. 299)
survey (p. 276)
unstructured interview (p. 302)

EXERCISES

1. You probably have participated in many surveys. For one of those surveys, answer the following questions:
   a. What was the topic of the survey?
   b. What do you think the hypothesis was?
   c. Did they use an oral interview or a written questionnaire? Do you think they made the right choice? Why or why not?
2. State a hypothesis that can be tested by administering a survey. Why is a survey a good way to test your hypothesis? (If you are having trouble generating a hypothesis, Omarzu [2004] suggests thinking of doing a survey that would provide useful information to your school or to the psychology department.)
3. Is an interview or a questionnaire the best way to test your hypothesis? Why?

4. For the three basic question formats, list their advantages and disadvantages.
5. Write three nominal-dichotomous questions that might help you test your hypothesis.
6. Write three Likert-type questions that might help you test your hypothesis.
7. A Gallup/CNN poll asked, "How likely do you think it is that Democrats in the Senate would attempt to block Bush's nominee for inappropriate political reasons." Which two of this chapter's nine tips for writing questions did this question violate? Rewrite the question to improve its validity.
8. A former president of the Association for Psychological Science wrote, "Sampling ain't simple" (Gernsbacher, 2007, p. 13). Explain why that is a true statement. What questions

would you ask of a sample to determine how much to trust that sample?

9. Why can you make statistical inferences from data obtained from a random sample?

10. Why might having participants sign informed consent statements (a statement that they had been informed of the nature of the study, the risks and benefits of the study, the participants' right to refuse to be in the study, the participants' right to quit the survey at any point, and the participants' right to confidentiality) make a survey research study less ethical? (Hints: Under what circumstances does the APA ethical code not require informed consent for surveys [see Appendix D]? Under what circumstances would requiring informed consent reduce the value of the survey without providing any benefits to participants?)

 ## WEB RESOURCES

1. Go to the Chapter 8 section of the book's student website and
   a. Look over the concept map of the key terms.
   b. Test yourself on the key terms.
   c. Take the Chapter 8 Practice Quiz.
2. If you are ready to draft a method section, click on the "Method Section Tips" link.
3. If you want to have a better understanding of correlation coefficients, click on the "Correlator" link.

4. Use the sample data and the statistical calculators available from the "Evaluate a Questionnaire" link to evaluate the reliability and construct validity of a questionnaire.
5. Use the sample data and the statistical calculators available from the "Analyzing Results" link to practice analyzing and interpreting data from a survey. If you wish, you can also use that link to find out how to use multiple regression to analyze survey responses.

Internal Validity

Any person armed with an understanding of causation has the power to change, alter, repair, and control.
—Neal Roese

True wisdom consists in tracing effects to their causes.
—Oliver Goldsmith

CHAPTER OVERVIEW

This chapter is about **internal validity**: establishing that a factor causes an effect. Internal validity is important for two reasons. First, to know what makes people tick, we need to know what *causes* them to behave the way they do. Thus, good answers to "why" questions start with "be**cause**," and good explanations are **caus**al explanations. Second, to help people, we need to know what treatments are **effect**ive. So, if you need to determine whether a treatment, action, intervention, training program, lecture, or therapy works, you need to conduct a study that has internal validity. For example, you would need a study with internal validity to determine whether

- piano lessons increase IQ scores
- listeners will be more persuaded by hearing a weak argument when listeners are sitting down than by hearing the same argument when they are standing up
- students will do better on an exam if it is printed on blue rather than white paper
- a restaurant server's manner (e.g., squatting down next to a customer as opposed to standing up, smiling an open-mouthed smile as opposed to a closed-mouth smile) increases the amount of money the server gets in tips
- music will cause shoppers to go through the store faster
- sugar will make young children more active
- students will have higher test scores when taught in classrooms that have windows
- keeping a log of what one should be grateful for will make people score higher on a happiness test
- full-spectrum lighting will increase people's scores on a mood scale

People want studies to have internal validity, but few studies deliver it. To establish that your study has internal validity, you and your study must meet three challenges. Each challenge builds on, and is more difficult than, the one before it.

First, because changes in the cause must be accompanied by changes in its effect, you must establish that changes in the alleged cause are *related* to changes in the outcome variable. For example, if you are going to show that sugar causes children to be more active, you first need to show that when children have more sugar, they are more active. Similarly, if you are going to show that writing about things one is thankful for increases happiness, you need to establish that people who write about things for which they are thankful are happier than people who do not write about things for which they are thankful or that people are happier when they write about things they are thankful for than when they don't. You should be able to determine whether differences in one factor are accompanied by differences in the outcome variable by measuring both variables and using the appropriate statistics. For example, you might be able to establish that the average happiness score of the group asked to write about what they are grateful for is significantly (reliably) different from the average happiness score of the group not assigned that task.

Second, because the cause must come before its effect, you must establish that changes in the treatment came *before* changes in the outcome variable. If you administer a treatment, you will usually be able to establish that changes in the treatment came before—and are followed by—changes in the outcome variable. For example, if you have the treatment group participants write about things they are thankful for and then measure mood, you will usually be able to make the case that participants wrote about what they were grateful for before—not after—their happiness increased. Note that if you did not manipulate a treatment—for example, if you just counted how many grateful entries people had in their diaries—it could be that happiness caused gratitude rather than gratitude causing happiness. In other words, if you don't manipulate the treatment, what you think is a cause may actually be an effect.

Third, because many nontreatment factors may have caused the changes in your outcome variable, you must establish that the treatment (writing about what they are grateful for) is the only factor responsible for the effect (higher scores on the happiness measure). Put another way, your final challenge is to show that the difference in the outcome measure (higher happiness scores) is not due to **extraneous factors (variables)**: anything other than the treatment. Because meeting this last challenge is so tricky, ruling out extraneous factors is the focus of this chapter.

In an ideal world, you would rule out extraneous factors by eliminating them from your study, thereby making sure that they couldn't be responsible for your results. You might dream of two perfect ways to get rid of extraneous factors:

1. *The ideal version of the two-group design.* Find two identical groups; treat them identically, except that you give only one of the groups the treatment; then compare the treatment group to the no-treatment group.

2. *The ideal version of the pretest–posttest (before-after) design.* Find some participants; give them the outcome measure; make sure that nothing in their lives changes, except that they get the treatment; then give them the outcome measure again.

In practice, however, it is impossible to conduct a study using the ideal version of either the two-group design or the pretest-posttest design because it is impossible to keep everything the same except for the treatment. That is, the actual, implemented versions of these designs can't eliminate all extraneous variables. Because neither design can eliminate all extraneous variables, neither can *prove* that a treatment caused an effect (despite what many infomercials imply).

In this chapter, you will learn why these two approaches fail to establish internal validity. Along the way, you will learn about Campbell and Stanley's (1963) eight threats to internal validity:

1. *Selection:* Treatment and no-treatment group scores differ because the groups were different on the outcome measure even before you administered the treatment. As the old saying goes, you are "comparing apples with oranges," but you are claiming that your treatment—rather than the fact that the apples and oranges were different to start with—made the apples different from the oranges.

2. *Selection by maturation interaction:* Treatment and no-treatment groups, although scoring similarly on the outcome measure at the start of study, were different in ways that would inevitably make them grow apart by the end of the study. In a sense, you are matching a group of tree seeds with a group of pea seeds on size, fertilizing the tree seeds, waiting a few years, and then claiming that the trees are bigger than the peas because of the fertilizer.

3. *Regression:* Participants who have extreme scores on the pretest are likely to have less extreme scores on the posttest. If you are not careful, you may confuse this tendency of things to return to normal for a treatment effect. Why do extreme scores on a pretest become less extreme on a retest? The key is to understand that all scores are affected by random error—so that when you are looking at extreme scores, you are looking at scores that are being pushed toward the extreme by random error, and so they are likely to regress back toward the mean. Note that regression applies to your "scores" in real life: When you hit bottom, there is no place (for random error

to push scores) but up. Similarly, when you "hit top" (e.g., you get 100% on an exam), there is no way for random error to push your score higher.

4. *Mortality (Attrition, Treatment-Related Participant Loss, Differential Dropout)*: The average score in the treatment condition differs from the average score in the no-treatment conditions because more participants who would have scored low dropped out of the treatment condition than dropped out of the no-treatment condition. By comparing the average score for an elite subgroup of treatment survivors to everyone in the no-treatment condition, you are, in a sense, comparing the cream of one crop against the average of another crop—but claiming that the crops are different because of your treatment.

5. *History*: Factors *outside* the participant, other than the treatment, cause the person to change. Because "life happens"—the world does not stop while a treatment is being administered—history is often a serious threat to a study's internal validity.

6. *Maturation*: Factors *inside* the participant (e.g., brain development) cause the participant to change. Because—even without treatment—individuals are constantly growing, changing, and maturing, maturation is often a serious threat to a study's internal validity.

7. *(Re)Testing*: Taking the pretest changed the participants—and thus changed their scores on the posttest (e.g., practice on the test made participants better on the retest).

8. *Instrumentation*: Posttest scores were different from pretest scores because the posttest *measuring instrument* was not the same, not scored the same, or not administered the same as the pretest measure. With instrumentation, the participant may stay the same, but the participant's score may change.

By the end of this chapter, you will know enough about those eight threats to

1. detect their presence in research that erroneously claims to prove that a certain factor has an effect and

2. take steps to prevent these threats from corrupting the internal validity of your research.

PROBLEMS WITH TWO-GROUP DESIGNS

To begin our exploration of Campbell and Stanley's eight threats to validity, let's examine the first approach for ruling out extraneous variables: obtaining two identical groups. Specifically, suppose you obtain two groups of participants and treat them identically, except that only one of the groups receives the treatment (e.g., writing about events for which they are grateful). Then, you give both groups a happiness scale and note that they have different levels of happiness.

Why We Never Have Identical Groups

What do you conclude? If the groups were identical before you introduced the treatment, you would correctly conclude that the treatment caused the groups to differ. However, you cannot assume that the groups were identical before you introduced the treatment. Therefore, the difference in scores could be due to **selection** (also called **selection bias**): having groups that were different from one another before the study began. *Selection bias is usually the most serious threat to the internal validity of a two-group design.*

Self-Assignment to Group Produces Selection Bias

How can you avoid selection bias? A first step toward avoiding selection is to prevent **self-selection**: participants choosing what condition they want to be in. You want to avoid self-selection because it leaves you with groups that you know differ in at least one way—one group chose the treatment, whereas the other chose to avoid the treatment—and that probably also differ in ways that you do not know about.

Sometimes the effects of self-selection are obvious. For example, suppose you compare two groups—one group offers to stay after work to attend a seminar on "Helping Your Company"; the other does not. If you later find that the group who attended the seminar is more committed to the company than the group who did not, you can't conclude that the effect is due to the seminar: After all, the groups probably differed in commitment before the study began.

Sometimes the effects of self-selection are not as obvious. For instance, what if you let participants choose whether they get to be in the "gratitude" condition or in the "no-gratitude" condition? If you find that the gratitude group is happier than the no-gratitude group, you still can't conclude that the effect was due to the gratitude manipulation. People who prefer to write about what they are thankful for may already be happier than people who prefer not to write about what they are thankful for. You may not know exactly how participants who choose one condition differ from those who choose another condition. But you do know that they differ at the beginning of the study—and that those differences may cause the groups to differ at the end of the study.

Researcher Assignment to Group Produces Selection Bias

We've seen that letting participants assign themselves to a group creates unequal, nonequivalent groups. However, if you assign participants to groups, you might yourself unintentionally bias your study. For example, you might "stack the deck" in favor of your hypothesis by unconsciously putting all the smiling participants in the gratitude condition (treatment group) and all the frowning participants in the no-gratitude condition (no-treatment group).

Arbitrary Assignment to Group Produces Selection Bias: Choosing Groups Based on Their Differences Results in Having Groups That Are Different

To avoid the bias of stacking the deck, you might assign participants to groups on the basis of some arbitrary rule. For example, why not assign students on the right-hand side of the room to the no-treatment group and assign students on the left side of the room to the treatment group? The answer is simple: because the groups are not equal. At the very least, the groups differ in that one group prefers the right side, whereas the other group prefers the left side. The groups probably also differ in many other ways. For instance, if the left side of the room is near the windows and the right side is near the door, we can list at least four additional potential differences between "left-siders" and "right-siders":

1. People sitting on the left side of the room may be more energetic (they chose to walk the width of the room to find a seat).
2. People sitting on the left side of the room may be early-arrivers (students who came in late would tend to sit on the right side so they would not disrupt class by crossing the width of the room).
3. People sitting on the left side may be more interested in the outdoors (they chose to be close to the windows).
4. People sitting on the left side may have chosen those seats to get a better view of the professor's performance (if the professor shows the typical right-hander's tendency of turning to the right, which would be the students' left).

You can probably come up with many other differences between left-siders and right-siders in a particular class. But the point is that the groups definitely differ in at least one respect (choice of side of room), and they almost certainly differ in numerous other respects (see Figure 9.1).

What's true for the arbitrary rule of assigning participants to groups on the basis of where they sit is true for any other arbitrary rule. Thus, any researchers who assign participants on the basis of an arbitrary rule (e.g., the first-arriving participants assigned to the treatment group, people whose last names begin with a letter between A and L in the treatment group, etc.) make their research vulnerable to selection bias.

One infamous example of how arbitrary assignment can produce misleading research was Brady's (1958) "executive monkey" study. In that study, Brady tested monkeys in pairs. Each pair consisted of an "executive monkey" and a "worker monkey." The executive monkey controlled a switch that, if pressed at the right time, would prevent both monkeys from getting a shock. Brady found that the executive monkeys were more likely to get ulcers than the worker monkeys.

a. The rule of choosing "every other person" to get the treatment is not random. The problem with this rule is most obvious when applied to situations in which people are encouraged to line up "boy/girl."

b. The arbitrary rule of assigning the front of the class to one treatment and the back of the class to no treatment does not work. Ask any teacher! The two groups are definitely different.

FIGURE **9.1** Arbitrary assignment to groups produces selection bias.

c. Assigning by left side versus right side ruins an attention study's internal validity. Students on the window-side of the room may be sitting there because they want to look out the window or at the clock. The students on the other side of the room may be sitting there to avoid distractions.

FIGURE **9.1** (Continued)

Although the study seemed to suggest that human executives deserve their high salaries because their responsibilities give them stress and ulcers, later research showed the opposite: Individuals who do not have control (like the worker monkeys) are more likely than individuals who have control (like executive monkeys) to be stressed and get ulcers (Seligman, 1975). The problem with Brady's research was selection bias—he assigned the monkeys who learned how to use the switch the fastest to be the executive monkeys. This arbitrary assignment was a big mistake, probably because the monkeys who learned to use the switch the fastest were those who were most upset by the shocks.

In conclusion, arbitrarily assigning participants to groups does not work because you are assigning participants to groups based on their differences. Your groups can't be equal when you are deliberately ensuring that they are different on some variable—even if that variable (e.g., how fast one learns to use a switch, preference for side of the room, etc.) seems unimportant.

Problems With Matching on Multiple Variables

If you can assign participants in a way that guarantees they are different, why can't you assign participants in a way that guarantees they are identical? In other words, why not use **matching**: trying to choose groups in such a way that the groups are identical on key variables?

The Impossibility of Perfectly Matching Individual Participants: Identical Participants Do Not Exist. In the abstract, matching seems like an easy, foolproof way of making sure that your two groups are equal. In practice, however, matching is neither easy nor foolproof. Imagine trying to find two people who match on every characteristic and then assigning one to the no-treatment condition and the other to the treatment condition. It would be impossible. Even identical twins would not be exactly alike—they have different first names and different experiences.

The Impossibility of Matching Groups on Every Variable: There Are Too Many Variables. Obviously, you can't create the situation in which each member of the treatment group has a clone in the no-treatment group. Nor can you get two groups that, on the average, match on every variable. Try as you might, there will always be some variable on which you had not matched—and that variable might be important. Even if you created two groups that had the same average age, same average intelligence, same average income, same average height, and same average weight, there would still be thousands of variables on which the groups might differ. The groups might differ in how they felt on the day of the study, how they were getting along with their parents, how many books they had read, their overall health, and so forth.

The Impossibility of Matching Groups on Every Relevant Variable. You know you can't match your no-treatment and treatment groups on every single characteristic, but do you *need* to make the groups identical in every respect? No—you need them to be identical only with respect to the variable you want to measure. For example, suppose you were studying happiness. Then, all you would need to do is match your groups on every characteristic that will influence their scores on your happiness measure.

Unfortunately, there is a big problem with this "solution." Matching only on those factors that influence the key variable of happiness is impossible, because there are thousands of factors that influence happiness. You cannot know every single characteristic that influences happiness—after all, if you knew everything about happiness, you would not be doing a study to find out about happiness!

Matching on Pretest Scores Rules Out Selection But Leads to Two Other Problems

Instead of matching participants on every characteristic that affects the variable you want to measure, why not match participants on the variable you want to measure? In your case, why not match participants on happiness scores? Before you assign participants to groups, test people on the happiness scale (what researchers call a "pretest"). Next, match your groups so that the treatment group and no-treatment group have the same average pretest score. Finally, at the end of the study, test the participants again, giving participants what researchers call a "posttest." If you find a difference between your groups on the posttest, then you should be positive that the treatment worked, right? Wrong!

Even if the treatment had no effect whatsoever, two groups that scored the same on the pretest could still differ on the posttest. As you will see in the next two sections, groups that are similar on a pretest measure may not be equivalent for two reasons: (1) selection by maturation interactions and (2) regression.

Selection by Maturation Interactions: Participants Who Are Similar Now May Grow Apart. The first reason matching on pretest scores doesn't work is that there might be a **selection by maturation interaction:** The groups started out the same on the pretest but afterward naturally developed at different rates or in different directions. Selection by maturation interactions occur when participants who are similar in one way differ in other ways—and those differences cause the participants to grow apart.

To visualize the strong impact that selection by maturation interaction can have, imagine you studied some 4th-grade boys and girls. You put all the boys in one group. Then, you had them lift weights. You saw that the average weight they could lift was 40 lbs (18.14 kg). You then picked a group of 4th-grade girls who could also lift 40 lbs. Thus, your groups are equivalent on the pretest. Then, you introduced the treatment: strength pills. You gave the boys strength pills for 8 years. When both groups were in the 12th grade, you measured their strength. You found that the boys were much stronger than the girls. Although this difference might be due to the strength pills, the difference might be due to the boys naturally developing greater strength than the girls. In other words, the difference may be due to failing to match on a variable (gender) that influences muscular maturation.

In addition to growing apart because of different rates of physical maturation, groups may also grow apart because of different rates of social, emotional, or intellectual maturation. To illustrate this point, let's examine a situation in which the two groups are probably changing in different ways on virtually every aspect of development.

Suppose a researcher matched a group of 19-year-old employees with a group of 66-year-old employees on job performance. The researcher then enrolled the 19-year-olds in a training program. When the researcher compared the groups 2 years later, the researcher found that the 19-year-olds were performing better than the 66-year-olds. Why?

Although the difference could have been due to training, it may have had nothing to do with the training. Instead, the difference may have been due to (1) the 19-year-olds' productivity increasing as they learned their new jobs and (2) the 66-year-olds' productivity naturally declining as this group anticipates retirement. Therefore, the apparent treatment effect may really be a selection by maturation interaction.

You may be saying to yourself that you would never make the mistake of matching 19-year-olds and 66-year-olds on pretest scores. If so, we're glad. You intuitively know that you can't make groups equivalent by merely matching on pretest scores. We would caution you, however, to realize that age is not the only—or even the most important—variable that affects maturation.[1] Many factors, such as intelligence, motivation, and health, affect maturation. Thus, if you are going to match on pretest scores, you must also match on all of the variables that might affect maturation. Otherwise, you run the risk of a selection by maturation interaction.

To repeat, *one reason matching on pretest scores is incomplete is that participants who are the same on a characteristic at one time will not necessarily be the same on that characteristic at a later time.* Many factors affect how a participant will change from pretest to posttest. If the groups are not matched on all those factors, two groups that started out the same on the pretest may naturally grow apart. Consequently, what the average person thinks is a treatment effect may really be a selection by maturation interaction.

Regression—and How Groups With the Same Score on a Pretest May Score Differently on a Retest.

If you were somehow able to match on pretest scores and all other relevant variables, you would be able to rule out selection by maturation. However, your matched groups still might not be equal on the pretest variable.

How could your groups not be equal if you measured them and made sure that they were equal? The problem is that because measurement is not perfect, *measuring groups as equal does not mean they are equal.*

Even though we tend to assume that measurement is perfect, it is not. For example, if a police officer stops you for speeding, the officer might say, "You were going 75." Or the officer might say, "I clocked you at 75." The officer's two statements are very different. You may have been going 40 mph and the radar mis-timed you (radars have clocked trees as going over 100 mph), or you may have been going 95 mph. In any event, you probably were not going at exactly the speed that the officer recorded due to variations in the accuracy of the radar. Even in this age of advanced technology, something as simple as measuring someone's height is not immune to measurement error. In fact, one of the authors fluctuates between 5 ft. 5 in. (165 cm) and 5 ft. 8 in. (172.7 cm), depending on which physician's office measures her. If measurements of variables as easy to measure as height are contaminated with random error, measurements of psychological variables—variables that are harder to measure than height—will also be affected by random measurement error.

Because of random measurement error, a measure of an individual's height, weight, mood, free-throw shooting ability, or almost anything else might be inaccurate. Thus, two individuals having the same score on a measure might actually differ on the variable being measured. For example, if you tested

---

[1] Note that, contrary to ageist stereotypes, we might find that the older workers improved more than the younger workers. That is, older workers are much more productive and involved than many people assume. Indeed, ageism is probably why our poor researcher was forced to do such a flawed study. Perhaps the researcher was able to get management to invest in training for the younger workers but not for the older workers. In essence, the researcher used the older workers as a comparison group only because management gave her no choice.

free-throw shooting ability by having people shoot two free throws, both a good and a poor free-throw shooter could score 50% on your measure.

Although random error might cause two individuals who differ to have the same scores, could random error cause two groups that differ to have the same average score? At first, you might think the answer would be "no." You might reason that because random error is, by definition, unsystematic and unbiased, it should affect each group to about the same extent. Because random error tends to balance out, it would seem unlikely, for example, that random measurement error would inflate the free-throw shooting percentage of individuals in the treatment group but deflate the free-throw percentage of the individuals in the no-treatment group. Yet, even though your first reaction is reasonable, it is a mistake to make it an absolute rule because random error may in fact have one effect on the treatment group and another on the no-treatment group.

Given that random error tends to balance out, how could random error have one effect on the treatment group and another effect on the no-treatment group? To answer this question, imagine a group of extremely high scorers and a group of extremely low scorers. For the purpose of this example, let's imagine having hundreds of people each shoot five free throws. From those hundreds, we will select two groups of free-throw shooters: (1) a group in which all members hit all five free throws, and (2) a group in which all members missed all five free throws.

Why is the extremely high-scoring group doing so well? It's unlikely that these scores reflect each individual's true ability. Indeed, none of the people who hit all five foul shots are really 100% free-throw shooters. It's more likely that most of these are good free-throw shooters, but they are also benefiting from some good fortune. A few may be average or even poor foul shooters whose scores are being pushed up by random error (even Shaq has hit five free throws in a row). One thing we know for sure—nobody in this group had random error push down their free-throw percentage. In short, the average score of this group has been pushed up by random error.

Now, let's look at the group of extremely low scorers. Why are they scoring so low? Perhaps all of them are 0% foul shooters. It is more likely, however, that many are poor to average foul shooters experiencing a run of bad luck. One thing we know for sure—nobody in this group had random error inflate his or her free-throw percentage. In short, the average score of this group has probably been pushed down by random error.

What will happen if we retest both groups? The first group will tend to do a bit worse than before: Their average will not be 100%. On the pretest, random error pushed their scores in only one direction—up. That probably won't happen on the retest. Instead, random error will probably push some of their scores up and some of their scores down. As a result, their scores will revert to more normal levels on the retest. Similarly, the second group will tend to score at more normal levels on the retest: Their average will probably not be 0%. On the pretest, random error pushed their scores in only one direction—down. That probably won't happen two times in a row. As we have seen, the 0% group will do better on the retest, but the 100%

group will do worse. Put another way, both groups' average scores become less extreme on the retest.

Why does each group's average score become less extreme on the retest? In other words, why do their scores *revert back to more normal levels*? The short answer is that on the retest, each group's average score is less influenced by random error. The long answer is that (1) the groups were initially selected because of their extreme pretest scores; (2) their extreme pretest scores were due, in part, to random error pushing their scores in one direction; and (3) random error, which by its very nature is inconsistent, probably won't push all the groups' scores in that same direction on the retest.

Thus far, we have considered the case in which two groups who score much differently on the pretest (0% versus 100% on a foul-shot test) might appear to grow more similar on a retest. But how could two groups (1) seem to be similar on a pretest and then (2) seem to grow apart on the retest? For example, how could two groups that hit 60% of their free throws on the pretest end up scoring very differently on the retest? The key to seeing how this illusion would work is to realize that extreme scores are only extreme relative to their group's average.

To illustrate, suppose we have a large group of 90% career free-throw shooters and a large group of 30% career free-throw shooters. We then have people from each group shoot 10 free throws. We find that several from each group shoot 60% (6 out of 10) on our pretest. For the career 30% free-throw shooters, 60% is extremely good. For the career 90% free-throw shooters, 60% is extremely bad.

We now have two groups that each shot 60% on our pretest. The first group was taken from extreme low scorers from the 90% group, whereas the second group was taken from extreme high scorers from the 30% group. The two groups match on pretest *scores*, but this matching is just a mirage due to random measurement error. On the posttest, this mirage will disappear because participants' scores will be affected by chance to a more usual (and lesser) degree. The first group will score closer to its average score of 90% and the second group will score closer to its average score of 30%. In technical terminology, both groups will exhibit what scientists call **regression toward the mean (also called regression effect, regression artifact, statistical regression, reverting to average, and regression)**: the tendency for scores that are extremely unusual to revert back to more normal levels on the retest.

As you might imagine, regression toward the mean could mimic a treatment effect. If, in our free-throw shooting example, you administered a treatment between the pretest and the posttest, people might mistakenly believe that the treatment was responsible for the groups scoring differently on the posttest. For example, if you yelled at the first group after their poor (for them) pretest performance, people might think that your yelling is what caused them to do better on the posttest.

Regression toward the mean also explains why many parents believe that punishment is more effective than it is. After children have behaved unusually badly, their behavior will tend to revert to more normal (better) levels (regression to the mean). But those parents who have punished their children usually do not say, "Well, the behavior would have improved anyway because of regression toward the mean." Instead, they tend to say, "Punishing her made her behave better."

Regression toward the mean also tricks some parents, teachers, and bosses into believing that praise actually harms performance. They note that when they reward outstanding performance, the next performance tends to be not be as good. Consequently, they decide that the praise worsened performance by making the praised person overconfident. They have failed to realize that, as Rosenzweig (2007) puts it, "Nothing recedes like success" (p. 105). In other words, they have been tricked by regression toward the mean.

Although regression toward the mean is tricky enough by itself to fool most of the people most of the time, sometimes it has help. A deceiving swindler might intentionally use regression toward the mean to make it look like a worthless treatment had a positive effect. The key would be to intentionally take advantage of random measurement error to make it look like two dissimilar groups were really similar on the pretest (e.g., the "new diet" group would be made up of people who had been underweight until recently, whereas the comparison group would be made up of people who had been overweight all their lives).

Unfortunately, a researcher might also unintentionally use random measurement error to match two groups. To see how random error might make two unequal groups *look* equal, suppose a researcher who works for a continuing care retirement community (CCRC) wants to test a memory-improvement program. The researcher decides she wants to provide an intervention for those residents who score between the 40th and 45th percentiles for older adults on the Wechsler Memory Scale because she believes this group will benefit most from her treatment. The researcher needs to find two groups whose scores fall within this range, give one group the memory training, and see whether the training group scores better on the posttest than the no-training group.

In her CCRC, there are three levels of care: independent living, assisted living, and nursing care. The researcher decides to focus on the assisted living residents because she believes that those residents will be most likely to contain individuals who are healthy and who score somewhat below average (50th percentile is average) on the memory scale.

She administers the pretest but finds only eight assisted-living residents who score between the 40th and 45th percentiles. She knows that she needs more participants. She decides to use these eight residents as her treatment group and looks elsewhere for her no-treatment group. She rules out the nursing-care residents because she wants the groups to be equivalent in terms of health and activity level. Instead, she tests independent-living residents and finds eight who score within the 40–45 percentile range.

At the end of the study, the researcher gives both groups the memory test again (the posttest). When she looks at the results, she is horrified: The no-treatment group (the eight independent-living residents) scores much higher on the posttest than the treatment group (the eight assisted-living residents). On closer examination, she finds that the scores of the independent-living residents increased from pretest to posttest, whereas the scores of the assisted-living residents decreased from pretest to posttest.

What happened? Did the true level of memory functioning improve for the independent-living residents even though they received no memory training? No. Did the training program actually decrease the memory functioning of those residents in assisted living? No.

What happened was that the investigator selected scores that were likely to be heavily contaminated with random measurement error. To understand how this occurred, think about what would cause healthy older adults who are capable of independent living—a group that would average well above the 50th percentile—to score in the 40th–45th percentile on a memory test. These scores, which would be uncharacteristically low for them, might be due to some unusual event, such as the flu or to jet lag following a vacation abroad. If they scored in the 45th percentile on the pretest because of illness or jet lag, would it be likely that they would score this low on the posttest? No, chances are that they would not be suffering from jet lag on the posttest. Consequently, chances are that their posttest scores will be a more accurate reflection of their true memory ability, and thus be higher.

Not only did the researcher select independent-living participants whose scores were likely to be loaded with random measurement error but the researcher also selected assisted-living participants whose scores were likely to be loaded with random measurement error. People requiring assisted living are much more likely to suffer from health problems that will directly (cardio-vascular disease, mild dementia) or indirectly (medication side effects) decrease their memory ability to below the 40th percentile.

Consider how a person in assisted living could score in the 45th percentile. What would cause them to score above their true score? Probably some form of luck would be involved. Just as you may occasionally get lucky guessing on a multiple-choice test, perhaps a few people in assisted living might get lucky on a memory test. That is, if you tested 200 people in assisted living, 8 might score near average on memory function just by chance. But would these same 8 be lucky a second time? It is a good bet that they would not. Instead, their second score should be a more accurate reflection of their true score. Consequently, when retested, they would get lower scores than they did the first time.

Conclusions About Matching on Pretest Scores. In short, matching on pretest scores does not make your groups equal for two reasons. First, matching on pretest scores is *incomplete* because participants who are the same on a characteristic at the time of the pretest may not be the same on that characteristic at the time of the posttest (e.g., if you were to rank the happiness of all the students in your high school, that ranking might change if you were to re-rank them 4 years later). To get groups who will be the same on that characteristic at pretest and posttest, you need to match the groups not only on that characteristic but also on every other variable that might affect how participants will change on that characteristic. If you do not match on all those other characteristics, you may have two groups that started out the same, but naturally grew apart—no thanks to the treatment. In that case, what you believe is a treatment effect may really be a selection by maturation effect.

Second, matching on pretest scores is *imperfect* because scores are not perfect indicators of what they are supposed to be measuring. Instead, pretest scores are contaminated by random error. Because of random measurement error, it's possible to get two groups that match on pretest scores but that differ on the pretest variable. That is, random error may create the illusion that two dissimilar groups are similar. Specifically, this illusion of similarity is caused by choosing those participants whose scores had, on the pretest, been blown in a certain direction by random error. However, because random error is inconsistent, the winds of chance will probably not blow those scores in the same direction on the posttest. Consequently, on the posttest, the illusion will disappear. If two groups that only appeared to be similar on the pretest reveal their true differences during the posttest, a naïve observer may believe that the groups "became different" because of the treatment.

Mortality (Attrition): Differential Drop Out. Even if our groups were identical to start with, they might not stay that way because of **mortality** (**attrition, treatment-related participant loss, differential dropout**): one condition being affected more than another by participants dropping out of the study. Like selection, mortality can make the participants in one group systematically different from participants in the other. But whereas selection makes groups differ by affecting who *enters* each group, mortality makes groups differ by affecting who *exits* each group.

To understand the threat posed by mortality, suppose we have designed a program for high-risk youths. To test the program's effectiveness, we put 40 at-risk youths into our intense training program and compare them to a no-treatment group consisting of 40 other at-risk youths. We find that youths who complete our training program are much more likely to get a good job than the youths in the no-training group. However, 75% of the treatment group youths drop out of our rigorous program. Thus, we are comparing the 10 elite survivors of the treatment group against everyone in the no-treatment group. Consequently, our training program's apparent "success" may simply be due to comparing the best of one group against everyone in the other group.

Conclusions About Two-Group Designs: Selection Is Not Easily Overcome

In the previous example, mortality seriously threatened the validity of our study. Even if the groups had been the same to start with, they were not the same at the end of the study.

But groups will not be the same at the start of the study, no matter what we try (see Box 9.1). Even if we use matching, our groups may still be different before the treatment is introduced (see Figure 9.2). In short, differences between the treatment and no-treatment groups at the end of the study may have nothing to do with the treatment and instead be due to the groups being different before the treatment was introduced.

BOX **9.1**   **Six Flawed Strategies for Getting Two Identical Groups—And Why They Don't Work**

1. Self-assignment causes selection bias.
2. Researcher assignment can cause selection bias.
3. Arbitrary assignment to a group causes selection bias by making the groups differ in at least one respect.
4. Matching on every variable is impossible.
5. Matching participants on all relevant variables is impossible.
6. Matching on pretest score is flawed because pretest scores are not perfect predictors of posttest scores. Even if the treatment doesn't work, two groups matched on pretest scores may differ on the posttest because
   a. they differ on the pretest variable: Their pretest scores only match due to random error (regression); or
   b. they differ on variables that cause them to grow apart on the posttest (selection by maturation).

© Cengage Learning 2013

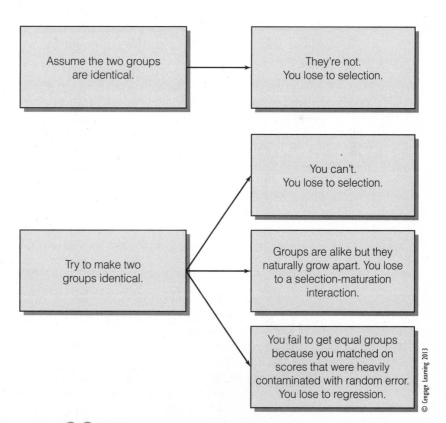

FIGURE **9.2** Making two groups identical: a game you can't win.

PROBLEMS WITH THE PRETEST–POSTTEST (BEFORE–AFTER) DESIGN

Because it is impossible to get two different groups of participants who are identical, you might decide not to compare one group of participants who receives the treatment with a different group who does not receive the treatment. Instead, you might decide to compare each participant **before** that participant gets the treatment with that same participant **after** that participant gets the treatment. In technical terminology, this before–after design is called a **pretest–posttest design**: a design in which you measure each participant before giving the treatment (the pretest), then administer the treatment, and then measure each participant again (the posttest).

By making sure that the participants in the treatment group are the same participants who were in the no-treatment group, you **eliminate the threat of selection bias** (and you also eliminate the threat of selection bias by maturation interactions).

At first glance, the pretest–posttest design seems to be a perfect way to establish internal validity. However, the pretest–posttest design can have internal validity only if the treatment is the sole reason that posttest scores differ from pretest scores. Unfortunately, the treatment is not the only reason that participants' scores may change from pretest to posttest.

Three Reasons Participants May Change Between Pretest and Posttest

Even without the treatment, participants may change over time. Specifically, participants may change from pretest to posttest because of three factors having nothing to do with the treatment: (1) maturation, (2) history, and (3) testing.

1. Maturation: Participants Change on Their Own

A participant may change between the time of the pretest and the time of the posttest as a result of **maturation**: the natural biological or developmental changes that occur inside the participant (see Figure 9.3). People are constantly

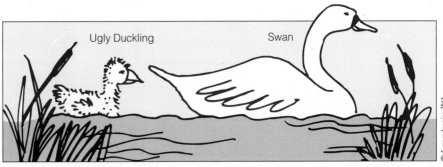

FIGURE **9.3** A happy case of maturation.

changing. From one moment to the next, they may become more bored, more hungry, or more tired. From one month to the next, they will grow older, and they may mature.

To see how maturation might masquerade as a treatment effect, suppose you institute a safe-driving program for young adults. You start your study with a group of 20-year-olds, show them videos about the dangers of risky driving, and measure them again when they are 25. You find that when they are 25 they take fewer risks than when they were 20. Your problem is that you do not know whether the safe-driving program or natural development is responsible for the change. Similarly, if you give a baby 10 years of memory training, you will find that her memory improves. However, this difference is probably due to maturation rather than to the training.

Even without treatment, many physical and psychological conditions improve over time. However, if a treatment is administered, "treatment, not time, may get the credit" (Painter, 2008, p. 8D). So, when listening to stories about how someone allegedly recovered due to some miracle treatment, remember the old saying: "If you have a cold and go to the doctor, it will take you 7 days to get well; if you don't go to the doctor, it will take you a whole week."

2. History: Environment Changes Participants

In addition to changing because of events that occur inside the participant, the participant may change because of events—other than the treatment—that occur in the outside world. Thus, even if the treatment has no effect, a participant may change between pretest and posttest because of **history**: any change in the participant's *environment* that has nothing to do with the treatment but has a systematic effect on a condition's average score (see Figure 9.4). History can involve events as important and far-reaching as a world war or as relatively unimportant and limited as a campus rumor.

To understand the kinds of events that can be labeled "history" and how history can bias a study, suppose two social psychologists have a treatment (an ad) they think will change how Americans feel about space exploration. However, between pretest and posttest, a spacecraft explodes. The change in attitudes may be due to the explosion (history) rather than to their ad. Or, suppose an investigator was examining the effect of diet on maze-running speed. However, between pretest and posttest, the heat went off in the rat room, and the rats nearly froze to death. As you can see from these examples, events that happen in a participant's life (history) between the pretest and the posttest can cause changes that could be mistaken for treatment effects.

3. (Re)Testing: Measuring Participants Changes Participants

One event that always occurs between the start of the pretest and the start of the posttest is the pretest itself. If the pretest changes participants (e.g., it motivates them to learn what is on the test or it makes them better at taking the test by giving them practice on the test), you have a **testing effect**. For example, if your instructor gave you the same test twice, you would score better the second time around. Your improvement would be due to finding

FIGURE **9.4** A mythical case of history.

out and remembering the answers to questions you missed. Because of the testing effect, people who have taken many intelligence tests (for example, children of clinical psychologists) may score very high on IQ tests regardless of their true intelligence. (Because of the testing effect, you should take the sample quizzes on this text's website—as Roediger and Karpicke [2006] point out, "Testing is a powerful means of improving learning, not just assessing it" [p. 249].)

The (re)testing effect is not limited to knowledge tests: It can occur with any measure. To illustrate, let's look at a pretest that has nothing to do with knowledge. Suppose we were to ask people their opinions about Greenland entering the World Bank. Would we get a different answer the second time we asked this question? Yes, because the very action of asking for their opinion may cause them to think about the issue more and to develop or change their opinion. In short, whether you are measuring a person's attitudes, cholesterol, exercise habits, or almost anything else, your measurements may stimulate the person to change.

Three Measurement Changes That May Cause Scores—But Not Participants—to Change Between Pretest and Posttest

Obviously, participants' scores may change because participants have changed. What is less obvious is that participants' scores may change because how participants are measured has changed. As you will soon see, even when the participants themselves have not changed, their scores may change due to (1) instrumentation, (2) regression, and (3) mortality.

1. Instrumentation: Changes in How Participants Are Measured

One reason a participant's score may change from pretest to posttest is **instrumentation**: changes in the measuring instrument—or in how it is administered or scored—that cause scores to change. But why would the measuring instrument used for the posttest be different from the one used during the pretest?

Sometimes, changes in the measuring instrument are unintentional. Suppose you are measuring aggression using the most changeable measuring instrument possible: the human rater. As the study progresses, raters may broaden their definition of aggression. Consequently, raters may give participants higher posttest scores on aggression, even though participants' behavior has not changed.

Unfortunately, there are many ways that raters could change between pretesting and posttesting. Raters could become more conscientious, less conscientious, more lenient, less lenient, and so forth. Any of these changes could cause an instrumentation effect.

Sometimes, changes in the instrument occur because the researcher is trying to make the posttest better than the pretest. For example, the researcher may retype the original questionnaire to make the scales look nicer, to fix typographical errors, or to eliminate bad questions. Unfortunately, these changes, no matter how minor they may seem and no matter how logical they may be, can cause instrumentation effects. Thus, the time to refine your measure is before—not while—you conduct your study.

2. Regression Revisited: Changes in How Random Error Affects Measurements

Even if the measuring instrument is the same for both the pretest and posttest, the degree to which random measurement error affects scores may differ from pretest to posttest. Consequently, regression toward the mean can sabotage pretest–posttest designs.

To show how regression can destroy a pretest–posttest design's internal validity just as easily as it can destroy a two-group design's, think back to the researcher who was investigating the effects of a memory-training program in older adults. Suppose that she had decided not to compare the eight highest-scoring assisted-living residents with a group of independent-living residents. Instead, after having the eight assisted-living residents who scored highest on the pretest complete the training program, suppose she had re-administered the memory test. What would she have observed?

As before, she would have observed that the assisted-living residents' memory scores dropped from pretest to posttest. This drop would not be due to the training program robbing patients of memories. Rather, the posttest scores would more accurately reflect the patients' poor memories. The posttest scores would be lower than the pretest scores only because the pretest scores were inflated by random measurement error.

The pretest scores were destined to be inflated by measurement error because the investigators selected only those residents whose scores were extremely high (for their group). Extreme scores tend to have extreme amounts of measurement error.

To understand why extreme scores tend to have extreme amounts of measurement error, realize that a participant's score is a function of two things: the participant's true characteristics and measurement error. Thus, an extreme score may be extreme because measurement error is making the score extreme. To take a concrete example, let's consider three possibilities for a student getting a perfect score on an exam:

1. The student is a perfect student.
2. The student is a very good student and had some good luck.
3. The student is an average or below-average student but got incredibly lucky.

As you can see, if you study a group of people who got perfect scores on the last exam, you are probably studying a group of people whose scores were inflated by measurement error. If participants were measured again, random error would probably be less generous. (After all, random error could not be more generous. There's only one place to go from a perfect score—down.) Therefore, if you were to give those participants a treatment (memory training) and then look at their scores on the next exam, you would probably be disappointed. The group that averaged 100% on the first test might average "only" 96% on the second test.

In the case we just described, regression's presence is relatively obvious because people recognize that test scores are influenced by random error. Note, however, that almost any measure is influenced by random error—and any measure that is influenced by random error is potentially vulnerable to regression toward the mean. For example, suppose you are trying to make inferences about a participant's typical behavior from a sample of that participant's behavior. If the behavior you observe is not typical of the participant's behavior, you have measurement error. Even if you measured the *observed* behavior perfectly, you have measurement error because you have not measured the participant's *typical* behavior perfectly.

To see how a sample of behavior may not be typical of normal behavior, let's look at a coin's behavior. Suppose you find a coin that comes up heads 6 times in a row. Although you have accurately recorded that the coin came up heads 6 times in a row, you might be making a mistake if you concluded that the coin was biased toward heads. In fact, if you were to flip the coin 10 more times, you probably would not get 10 more heads. Instead, you would probably get something close to 5 heads and 5 tails.

Coins are not the only things to exhibit erratic behavior. Every behavior is inconsistent and therefore prone to atypical streaks. For example, suppose you watch someone shoot baskets. You accurately observe that she made 5 out of 5 shots. Based on these observations, you may conclude that she is a great shooter. However, you may be wrong. Perhaps if you had observed her shooting on a different day, you would have seen her make only one of five shots.

To illustrate how this subtle form of measurement error can lead to regression toward the mean, suppose a person who had been happy most of her life feels depressed. This depression is so unlike her that she seeks therapy. Before starting the therapy, the psychologist gives her a personality test. The test verifies that she is depressed. After a couple of sessions, she is feeling better. In fact, according to the personality test, she is no longer depressed. Who could blame the psychologist for feeling proud?

But has the psychologist changed the client's personality? No, the client is just behaving in a way consistent with her normal personality. The previous measurements were contaminated by events that had nothing to do with her personality. Perhaps her depressed manner reflected a string of bad fortune: getting food poisoning, her cat running away, and being audited by the IRS. As this string of bad luck ended and her luck returned to normal, her mood returned to normal.

Regression toward the mean is such a clever impersonator of a treatment effect that regression fools most of the people most of the time. Many people swear that something really helped them when they had "hit bottom." The baseball player who recovers from a terrible slump believes that hypnosis was the cure; the owner whose business was at an all-time low believes that a new manager turned the business around; and a man who was at an all-time emotional low feels that his new girlfriend turned him around. What these people fail to take into account is that things are simply reverting back to normal (regressing toward the mean).

3. Mortality (Attrition): Changes in How Many Participants Are Measured

The last reason that you could find differences between the average pretest score and the average posttest score would be that you were measuring fewer participants at posttest than you were at pretest. If the lower scoring participants are dropping out, the increase in the average score from pretest to posttest may not be due to your treatment.

To illustrate how much of an impact mortality can have on a pretest–posttest study, imagine that you are studying the effect of diet on memory in older adults. You pretest your participants, give them your new diet, and test them again. You find that the average posttest score is higher than the average pretest score. However, if the pretest average is based on 100 participants and the posttest average is based on 70 participants, your results may well be due to mortality. Specifically, the reason posttest scores are higher than pretest scores may be that the people who scored poorly on the pretest are no longer around for the posttest.

Although death is the most dramatic way to lose participants, it is not the most common way. Usually, attrition results from participants quitting the study, failing to follow directions, or moving away.

Note that not all attrition is equal. For example, if you are losing participants due to their moving away, it is possible that you are losing just as many low scorers as high scorers and that this attrition has little systematic effect on posttest scores. If, on the other hand, you are losing participants who can't or won't stay on your demanding treatment program, you are probably losing the low-scoring participants—and that treatment-related participant loss may be your average posttest score's gain.

CONCLUSIONS ABOUT TRYING TO KEEP EVERYTHING EXCEPT THE TREATMENT CONSTANT

We tried to create a situation in which we manipulated the treatment while keeping everything else constant. However, nothing we tried worked.

When we tried to compare a treatment group with a no-treatment group, we had to worry that our groups were not identical before the study started (**selection**). Even when we matched our groups, we realized that the groups might not be identical because

1. we could not match groups on every characteristic (and those unmatched factors might cause the groups to grow apart), and
2. we could not match groups based on participants' actual characteristics, so we had to match them based on imperfect measures of those characteristics.

Because we could not get equivalent groups at the start of the study, we did not dwell on the additional problems of keeping groups equivalent. That is, we did not stress the **mortality** problem that would result if, for example, more participants dropped out of the treatment group than out of the no-treatment group.

Because of the problems with comparing a treatment group against a no-treatment group (see Table 9.1), we tried to measure the same group before and after giving them the treatment. Although this pretest–posttest tactic got rid of two threats to the two-group study's internal validity (selection bias and selection by maturation), it did not get rid of two others: mortality and regression. Furthermore, it introduced four others (see Figure 9.5).

TABLE **9.1** Questions to Ask When Examining a Two-Group (Treatment Versus No-Treatment) Study

Selection	Were groups equal before the study began?
Selection by maturation interaction	Even without the treatment, would the groups have grown apart?
Regression effects	Even if the groups appeared equivalent before the study began, was this apparent equivalence merely a temporary illusion created by random measurement error?
Mortality	Did more participants drop out of one group than dropped out of the other group?

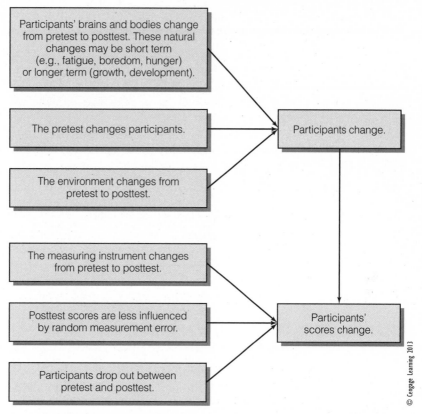

FIGURE **9.5** The impossible dream: Making sure the only thing that could make participants' scores change in a pretest–posttest design is the treatment.

In short, as Table 9.2 shows, participants may change from pretest to posttest for three reasons that have nothing to do with the treatment:

1. natural development (maturation)
2. other things in their lives changing (history)
3. learning from the pretest (testing)

In addition, participants may *appear* to change from pretest to posttest for three reasons:

1. the posttest measure was a different instrument (or was scored or administered differently) than the pretest measure (instrumentation)
2. their pretest scores were unduly influenced by chance (setting up regression toward the mean)
3. participants dropped out of the study, so that the posttest group is a select subgroup of the pretest group (mortality)

You would like to say that the treatment was the only factor that could cause the scores to change from pretest to posttest, but that's not easy to do.

TABLE **9.2** Questions to Ask When Examining a Pretest–Posttest (Before–After) Study

Maturation	Could the before–after (pretest–posttest) differences have been due to natural changes resulting from participants becoming older?
History	Could other events in the participants' lives or in the world have caused the pretest–posttest differences?
Testing	Could participants have scored differently on the posttest because of the practice and experience they got on the pretest?
Instrumentation	Were participants measured with the same instrument, in the same way, both times?
Regression	Were participants selected for their extreme pretest scores? Participants who get extreme scores will often get less extreme scores the second time around.
Mortality	Did everyone who took the pretest stick around for the posttest—or is the posttest group a more select group than the pretest group?

© Cengage Learning 2013

RULING OUT EXTRANEOUS VARIABLES

Why couldn't we eliminate extraneous variables? Was it because we used improper research techniques? No—as you will see in later chapters, matching participants and testing participants before and after treatment are useful research techniques.

We couldn't eliminate extraneous variables because it can't be done. Keeping everything the same is impossible. Imagine, in our ever-changing world, trying to make sure that only one thing in a participant's life changed!

Accounting for Extraneous Variables

Fortunately, you do not have to eliminate extraneous variables to rule out their effects. As you will learn in Chapter 10, you can combine random assignment and statistics to rule out the effects of extraneous variables. (Random assignment involves using a random process, such as a coin flip, to determine which treatment a participant receives. In the simplest case, random assignment results in half the participants receiving the treatment and half receiving no treatment.)

Even without using random assignment, you can still try to rule out the effects of extraneous variables. In a sense, tracking down a treatment's effect without using random assignment is much like a detective tracking down a murderer. Just as the detective is confronted with more than one suspect for a murder, you are confronted with more than one suspect for an effect. Just as the detective can't make the suspects disappear, you can't eliminate extraneous factors. However, like the detective, you can use logic to rule out some suspects.

Before you can begin to account for the actions of every suspicious extraneous variable, you have to know "who" each of these variables is. At first glance, identifying all of the thousands of variables that might account for the relationship between the treatment and the effect seems as impossible as eliminating all those variables.

Identifying Extraneous Variables

Fortunately, identifying the extraneous variables is not as difficult as it first appears because every one of these thousands of factors falls into eight categories: Campbell and Stanley's (1963) eight threats to validity. Thus, you really have only eight suspects: selection, history, maturation, testing, regression, mortality, instrumentation, and selection by maturation. If you can show that none of these eight threats were responsible for the effect (Chapter 14 shows you how), you can conclude that the treatment was responsible.

THE RELATIONSHIP BETWEEN INTERNAL AND EXTERNAL VALIDITY

If you rule out all eight threats, you have established *internal validity*—you have demonstrated that a *factor causes an effect* in your particular study. But you have not demonstrated that you can generalize your results outside your particular study. Internal validity alone does not guarantee that an investigator repeating the study using different participants (patients hospitalized with depression instead of college students) or using a different setting (a library instead of a lab) would obtain the same results. If you want to *generalize* your results to different places, people, and time periods, you need *external validity*.

If internal validity does not guarantee external validity, why bother with establishing internal validity? One answer is that you may not care about external validity. Instead of wanting to generalize your results, you may only want to show that a certain treatment causes an effect in a certain group in a certain setting. To understand why you might be so focused on internal validity, let's look at two types of researchers who have that focus.

First, investigators trying to isolate a process that would help us understand how something (the brain, vision, memory, or reading) works may not care about external validity. Indeed, to isolate and understand a particular process, they might deliberately use an artificial environment (e.g., a brain-imaging chamber, a Skinner box). Note that precisely because the process does not operate in isolation in real life, the investigator would not expect the study's results to replicate in a real-life setting—just as a physicist would not expect a study done in a vacuum to turn out the same way in a real-life setting (Stanovich, 2007).

Second, some therapists may want to show that with their patients, in their hospital, giving the patients an exercise program reduces patients' alcohol consumption. The therapists may not care whether the treatment would work with other kinds of patients at other hospitals (external validity). They only care that they have a method that works for them. However, few people are so single-minded that they are totally unconcerned with external validity.

Given that most researchers are concerned about external validity, you might think that most researchers would take many steps to maximize their study's external validity. However, for at least three reasons, researchers often take relatively few steps targeted specifically at boosting their study's external validity.

First, results from internally valid experiments tend to generalize (Anderson & Bushman, 1997; Anderson, Lindsay, & Bushman, 1999). That is, if an experiment showing that a factor has an effect is replicated (repeated) with a different group of participants or in a different setting, the replication will usually also find that the factor has an effect. As Anderson et al. wrote, "The psychological laboratory has generally produced psychological truths rather than trivialities" (p. 3).

Second, if other researchers using other types of participants and other settings all replicate the findings of the original study, these replications make a strong case for the finding's external validity. Indeed, replications by other researchers usually produce stronger evidence that a finding has external validity than anything the original researcher can do.

Third, the things that the original researcher would do to improve a study's external validity may reduce its internal validity (see Table 9.3). Or, to look at it another way, the steps a researcher might take to improve internal validity may end up reducing the study's external validity. For instance, to reduce the problem of selection bias, you might use twins as your participants. Although using twins as participants could increase internal validity by reducing differences between your treatment and no-treatment groups, it might hurt the generalizability of your study: Your results might apply only to twins.

Similarly, you might reduce the threat of history by testing your participants in a situation such as a lab where they are isolated from nontreatment factors. This highly controlled situation may increase internal validity because the treatment was one of the only things to change during the study. However, you would have to wonder whether the treatment would have the same effect outside this artificial situation. For example, would the results generalize to real life, where the factors from which you isolated your participants would come into play?

TABLE **9.3** Classic Conflicts Between the Goals of Internal and External Validity

Tactic Used to Help Establish Internal Validity	Tactic's Impact on External Validity
Use participants who are very similar to each other to reduce the effects of selection. For example, study only twins or study only rats.	Studying such a narrowly defined group raises questions about the degree to which the results can be generalized to different participant populations.
Study participants in a highly controlled laboratory setting to reduce the effects of extraneous factors such as history.	Studying participants in an isolated, controlled environment, such as a lab, raises questions about the extent to which the results might generalize to more complex, real-life settings.

CONCLUDING REMARKS

As you have seen, internal validity and external validity are sometimes in conflict. The same procedures that increase internal validity may decrease external validity. Fortunately, however, internal validity and external validity are not necessarily incompatible. Indeed, as you will see in future chapters, you can do studies that have both.

If you want to establish both internal and external validity, many would argue that you should first establish internal validity. After all, before you can establish that a factor causes an effect in most situations, you must show that the factor causes an effect in at least one situation.

But how can you establish internal validity? In this chapter, we tried two basic approaches (the no-treatment/treatment group design and the pretest–posttest design), and both failed. In the next chapter, you will learn the easiest and most automatic way to establish internal validity: the simple experiment.

SUMMARY

1. If you observe an effect in a study that has internal validity, you know what caused that effect.

2. Campbell and Stanley (1963) described eight major threats to internal validity: selection, selection by maturation interaction, regression, maturation, history, testing, mortality, and instrumentation.

3. When you compare a treatment group to a no-treatment group, beware of two non-treatment reasons your groups could differ: (1) the groups being different even before you start the study (selection) and (2) the groups becoming different because of treatment-related participant loss (mortality).

4. To reduce selection bias, participants should never get to choose what amount of treatment they get. In addition, participants' characteristics, attitudes, or behaviors should have nothing to do with whether they are put in the treatment rather than in the no-treatment group.

5. It is impossible to match two groups of participants so that they are identical in every respect: Participants simply differ in too many ways.

6. Even matching participants on pretest scores is not perfect because pretest scores are not perfect predictors of posttest scores.

7. If groups that score the same on the pretest (e.g., happiness) differ in ways (e.g., optimism about getting older) that will cause the groups to grow apart by the time they take the posttest, there is a selection by maturation interaction. That interaction may be mistaken for a treatment effect.

8. If groups *score* the same on the pretest, that does not mean they are the same on the pretest variable. Instead, random measurement error may be temporarily making two groups that are different look similar. As this illusion of similarity disappears on the posttest, the groups will score differently—and some people may mistakenly believe that the change in scores was due to the treatment.

9. In the pretest–posttest design, you measure individuals before they receive the treatment and again after they receive the treatment.

10. Unlike the two-group design, the pretest–posttest design is not vulnerable to selection or selection-maturation interactions. However, like the two-group design, it is vulnerable to mortality and regression. In addition, it is vulnerable to testing, history, maturation, and instrumentation.

11. Regression can occur in the pretest–posttest design because the participant may have

gotten the treatment when he or she had "hit bottom." Consequently, there was no place to go but up.

12. Maturation refers to inner, biological changes that occur in people merely as a result of time. In some cases, becoming more mature—not the treatment—accounts for people changing from pretest to posttest.

13. History refers to outside events—other than the treatment—that may influence participants' scores. Events that occur in the participants' world between pretest and posttest can cause participants to change from pretest to posttest.

14. Testing effect refers to the fact that taking a pretest may affect performance on a posttest.

15. Instrumentation occurs when the measuring instrument used in the posttest is different from the one used in the pretest.

16. External validity is the degree to which the results from a study can be generalized to other types of participants and settings.

17. Internal and external validity are different but not necessarily incompatible. For example, if men dropped out of your treatment group but not your no-treatment group, that might hurt your internal validity. If all the men in both groups quit your study, that would hurt your external validity, but not your internal validity. If nobody dropped out of either group, that would be good for both your internal and external validity.

KEY TERMS

extraneous
    factors *(p. 332)*
history *(p. 349)*
instrumentation (bias) *(p. 351)*
internal validity *(p. 331)*
matching *(p. 338)*

maturation *(p. 348)*
mortality *(p. 346)*
pretest–posttest design
    (p. 348)
regression (toward the mean)
    (p. 343)

selection (or selection bias)
    (p. 335)
selection by maturation
    interaction *(p. 340)*
testing effect *(p. 349)*

EXERCISES

1. What questions would you ask a researcher who said that the no-treatment and treatment groups were identical before the start of the study?

2. In all of the following cases, the researcher wants to make cause–effect statements. What threats to internal validity is the researcher apparently overlooking?

    a. Employees are interviewed on job satisfaction. Bosses undergo a 3-week training program. When employees are reinterviewed, dissatisfaction seems to be even higher. Therefore, the researcher concludes that the training program caused further employee dissatisfaction.

    b. After completing a voluntary workshop on improving the company's image, workers are surveyed. Those who attended the workshop are now more committed than those in the no-treatment group who did not make the workshop. Therefore, the researcher concludes that the workshop made workers more committed.

    c. After a 6-month training program, employee productivity improves. Therefore, the researcher concludes that the training program caused increased productivity.

    d. Morale is at an all-time low. As a result, the company hires a "humor consultant." A month later, workers are surveyed and morale has improved. Therefore, the researcher concludes that the consultant improved morale.

    e. Two groups of workers are matched on commitment to the company. One group is asked to attend a 2-week workshop on improving the company's image; the other is the no-treatment group. Workers who complete the workshop are more committed than those in the no-treatment group. Therefore, the researcher concludes that the workshop made workers more committed.

3. A hypnotist claims that hypnosis can cause increases in strength. To "prove" this claim, the hypnotist has participants see how many times they can squeeze a hand-grip in 2 minutes. Then, he hypnotizes them and has them practice for 2 weeks. At the end of 2 weeks, they can squeeze the hand-grips together many more times than they could at the beginning. Other than hypnosis, what could have caused this effect?

4. How could a quack psychologist or "healthcare expert" take advantage of regression toward the mean to make it appear that certain phony treatments actually worked? Why should a baseball team's general manager consider regression toward the mean when considering a trade for a player who made the All-Star team last season?

5. How could a participant's score on an ability test change even though the person's actual ability had not?

6. Suppose a memory researcher administers a memory test to a group of residents at a nursing home. He finds grade-school students who score the same as the older patients on the memory pretest. He then administers an experimental memory drug to the older patients. A year later, he gives both groups a posttest.
   a. If the researcher finds that the older patients now have a worse memory than the grade-school students, what can the researcher conclude? Why?
   b. If the researcher finds that the older patients now have a better memory than the grade-school students, what can the researcher conclude? Why?

7. Suppose there is a correlation between the use of night-lights in an infant's room an increased incidence of nearsightedness later. What might account for this relationship?

8. What is the difference between
   a. testing and instrumentation?
   b. history and maturation?

9. Suppose a researcher reports that a certain argument strategy has an effect, but only on those participants who hold extreme attitudes. Why might the researcher be mistaken about the effects of the persuasive strategy? (Hint: Whereas magic caused Cinderella to return to normal when the clock struck 12, this causes scores return to normal on retesting.)

10. What is the difference between internal and external validity?

WEB RESOURCES

1. Go to the Chapter 9 section of the book's student website and
   a. Look over the concept map of the key terms.
   b. Test yourself on the key terms.
   c. Take the Chapter 9 Practice Quiz.

CHAPTER 10

The Simple Experiment

What you have is an experience, not an experiment.
—**R. A. Fisher**

Happy is the person who gets to know the reasons for things.
—**Virgil**

CHAPTER OVERVIEW

Why do people behave the way they do? How can we help people change? To answer these questions, we must be able to isolate the underlying causes of behavior, and to do that, we must design a study that has **internal validity**: the ability to determine whether a factor causes an effect.

This chapter introduces you to one of the easiest ways to establish that a factor causes an effect: the simple experiment. You will start by learning the basic logic behind the simple experiment. Then, you will learn how to weigh statistical, ethical, and validity issues in order to design a useful simple experiment. Finally, you will learn how to interpret the results of such an experiment.

LOGIC AND TERMINOLOGY

Suppose that a researcher has a list of people who have volunteered to be in her simple experiment. Suppose she puts a + mark by half of the names and a √ mark by the other half in such a way that the people with + marks do not differ in any systematic way from the people with √ marks. For example, the average heights, happiness, and even the hula skills of the two halves are similar. Then, the experimenter schedules individual testing sessions for each person on the list. The first person on the list comes in, receives a treatment, is tested, is debriefed, and leaves. The second person on the list comes in... and so on through all of the participants. During these individual sessions, the experimenter gives one treatment to participants who have a "+" next to their names and a different treatment to participants who have a √ next to their names. For example, the experimenter might

- Assign the + individuals and the √ individuals to different *types* of activities (e.g., the + individuals play a violent video game whereas the √ individuals play a nonviolent video game).
- Assign the + individuals and the √ individuals to different *amounts* of an activity (e.g., the + individuals meditate for 30 minutes, whereas the √ individuals meditate for 10 minutes).
- Appear one way (e.g., formally dressed) to the + individuals and another way (e.g., casually dressed) to the √ individuals.

- Have confederates (people who pretend to be participants but who are actually the researcher's assistants) behave one way (e.g., agreeing with the participant) when interacting with + individuals and another way (e.g., disagreeing with the participant) when interacting with √ individuals.
- Have a certain object (e.g., a mirror or a gun) in the testing room when + individuals are tested but not when √ individuals are tested.
- Make the testing room's environment more intense on a certain dimension (e.g., how hot it is, how loud it is, how bright it is, how strongly citrus-scented it is, or how negatively charged the air in it is) when + individuals are tested and less intense on that dimension when √ individuals are tested.
- Give the + individuals and √ individuals different instructions ("memorize these words by repeating them over and over" versus "make a sentence out of these words," or "keep a log of what you have to be grateful for" versus "keep a log of hassles you encounter in your daily life").
- Give + individuals and √ individuals different printed stimuli (e.g., graphic cigarette warning labels showing pictures of diseased lungs versus written cigarette warning labels stating that smoking may cause lung cancer; concrete and easy to visualize words [e.g., "bell"] versus abstract and hard to visualize [e.g., "liberty"], exams printed on blue paper versus exams printed on white paper).
- Give + individuals and √ individuals different contexts for interpreting stimuli (the researcher may vary the gender, age, attractiveness, or background of the person whose job application is being judged).
- Give + individuals and √ individuals different scenarios. Sometimes, these scenarios describe the same situations, but do so in different words (e.g., "Valerie and I are best friends" versus "We are best friends" or "You can have $5.00 now or $6.20 in a month" versus "You can have $5.00 now and $0 in a month or $0 now and $6.20 in a month"). Sometimes, the situation described in the scenarios differs in one key respect (e.g., gender, race, or job experience of characters; the possible or likely causes of an event [e.g., the person was—or was not—drunk, the disease could—or could not—be transmitted through sexual contact]).
- Give + individuals and √ individuals different feedback (e.g., "the test suggests you are outgoing" versus "the test suggests you are shy," "the test suggests you will spend much of your future alone" versus "the test suggests you will spend much of your future with friends and loved ones," or "you did well on the task" versus "your performance on the task was average").
- Give the + individuals and √ individuals different chemicals (sugar-sweetened lemonade versus artificially sweetened lemonade, caffeinated versus decaffeinated colas).

As you can imagine, the real trick to doing an experiment is figuring out how to put the + and √ marks next to the participants' names so that you start with two sets of individuals who do not differ in any systematic way.

Once you have two groups that do not differ in any systematic way, the rest will be easy. Just administer a treatment to one group (the treatment group) and don't administer it to the other (the no-treatment group). If, at the end of the experiment, the two groups differ significantly, you could conclude that the treatment—the only systematic difference between the groups—caused that significant difference.

But how do you set up a situation in which the only *systematic* difference between the no-treatment and the treatment groups is the treatment? The answer is **independent random assignment**.

With independent random assignment, a process similar to determining what treatment the participant will receive based on a coin flip, each participant—regardless of that participant's characteristics and regardless of who else has been assigned to receive the treatment—*has an equal chance of being assigned to either the treatment or no-treatment group*. Thus, random assignment is not haphazard or arbitrary assignment: It is a special type of assignment that takes care to ensure that every participant has an equal chance of being assigned to any of the experiment's groups.

If we can successfully provide each participant an equal chance of being assigned to either group, there should be *no systematic differences* between the groups at the start of the experiment—and, at the end of the experiment, there should be only one systematic difference between them: the treatment. Therefore, if the individuals getting one treatment behave significantly differently from the individuals getting the other treatment, we can assume that the treatment is responsible for this difference. Thus, if we can randomly assign participants to groups, we can make cause–effect statements. In other words, random assignment is what gives experiments the ability to make cause–effect statements.

Because random assignment is so important to the experiment, experiments are sometimes called "randomized trials." Because random assignment is so important to making cause–effect statements, the first question researchers ask about a study that makes a cause–effect claim is, "Were participants randomly assigned?"

Because random assignment is so important, let's review its two key aspects. First, we *randomly*—rather than arbitrarily or haphazardly—divide our participants into two similar halves that do not differ in any systematic way. Second, we *assign* one of those halves to get one treatment and the other half to get a different treatment. For example, half may be allowed to choose the deadlines for their term papers, whereas the other half are not; or half the participants are given a violent video game to play, whereas the other half are given a nonviolent video game.

To understand why random assignment is so important, suppose you had all your participants roll a die. You ask half of them (the "low group") to aim for a low number (e.g., a "1") and the other half (the "high group") to aim for a high number (e.g., a "6"). Suppose that, on the average, the low group rolled a "3" and the high group rolled a "4." Although the high group's score is higher than the low group's, you do not know whether this is a consistent, reliable result. You have merely observed one sample of many possible samples. If you repeated the experiment, you would probably get a different set of results. So, your question would not be "Did one group roll higher numbers than the other group?" (After all, you can clearly see that one group did roll higher numbers than the other.) Instead, your question would be "How likely is it (e.g., 30 out of 100 times, 5 out of 100 times, 1 in a 1000 times) to get this result if instructions actually have no effect on what people roll?"

To answer this question, you would need to (a) randomly assign participants to groups so that any nontreatment differences between groups are due to random error and (b) have a good model of random error so that you can tell how likely it is that random error alone would produce the difference you observe. Fortunately, statisticians have developed a good model of how random error behaves, so all you need to worry about is randomly assigning participants.

But how do you randomly assign participants?[1] You know you shouldn't impulsively, casually, arbitrarily, erratically, or haphazardly decide who is in which group because that would not give each participant an equal chance of being in each group. (So, the "eenie meenie minie moe" and "one potato–two potato" methods are out.) You shouldn't try to assign participants based on a pattern that looks random to you because people are very bad at recognizing, much less deliberately creating, random patterns (Mlodiow, 2008).

You might think that you could use a computer to randomly assign participants. However, computers have difficulty producing random sequences (Klarreich, 2004), and some computer programs are better than others (Cooper, 2011). So, using a computer does not guarantee proper random assignment. You might think that you could flip a coin for each participant: If the coin comes up heads, the participant gets the treatment; if the coin comes up tails, the participant does not get the treatment. Coin-flipping is better than many methods, but coin-flipping is not perfectly random because "a tossed coin is slightly more likely to land on the face it started out on than on the opposite face" (Klarreich, 2004, p. 363).

So how do you ensure that each participant has exactly a 50% chance of being in one group and a 50% chance of being in the other group? The solution is to use a random numbers table to assign participants to condition (Wilkinson & the Task Force on Statistical Inference, 1999). To learn how to use a random numbers table (and to see that random assignment is a careful, systematic process), follow the instructions in Box 10.1.

Experimental Hypothesis: The Treatment Has an Effect

If you do not randomly assign your participants to two groups, you do not have a simple experiment. However, before you randomly assign participants, you should have an **experimental hypothesis**: a prediction that varying the treatment will *cause* an effect. To generate an experimental hypothesis, you must predict that the treatment and no-treatment groups will differ because

---

[1] Instead of using pure independent random assignment, researchers typically use independent random assignment with the restriction that an equal number of participants must be in each group.

BOX **10.1**   **Randomly Assigning Participants to Two Groups**

There are many ways to randomly assign participants to groups. Your professor may prefer another method. However, following these steps guarantees random assignment and an equal number of participants in each group.

Step 1: On the top of a sheet of paper, make two columns. Title the first "Control Group." Title the second "Experimental Group." Under the group names, draw a line for each participant you will need. Thus, if you were planning to use eight participants (four in each group), you would draw four lines under each group name.

Control Group	Experimental Group
_____	_____
_____	_____
_____	_____
_____	_____

Step 2: Turn to a random numbers table, like the one at the end of this box (or the one in Appendix F). Roll a die to determine which column in the table you will use. Make a note in that column so that others could check your methods (Wilkinson & the Task Force on Statistical Inference, 1999).

Step 3: Assign the first number in the column to the first space under Control Group, the second number to the second space, and so on. When you have filled all the spaces for the control group, place the next number under the first space under Experimental Group and continue until you have filled all the spaces. Thus, if you used the random numbers table at the end of this box and you rolled a "5," you would start at the top of the fifth column of that table (the column starting with the number 81647), and your sheet of paper would look like this:

Control Group	Experimental Group
81647	06121
30995	27756
76393	98872
07856	18876

Step 4: At the end of each control group score, write down a "C." At the end of each experimental group score, write down an "E." In this example, our sheet would now look like this:

Control Group	Experimental Group
81647C	06121E
30995C	27756E
76393C	98872E
07856C	18876E

Step 5: Rank these numbers from lowest to highest. Then, on a second piece of paper, put the lowest number on the top line, the second lowest number on the next line, and so on. In this example, your page would look like this:

06121E	30995C
07856C	76393C

(Continued)

BOX **10.1** **(Continued)**

18876E	81647C
27756E	98872E

Step 6: Label the top line "Participant 1," the second line "Participant 2," and so forth. The first participant who shows up will be in the condition specified on the top line, the second participant who shows up will be in the condition specified by the second line, and so forth. In this example, the first participant will be in the experimental group, the second in the control group, the third and fourth in the experimental group, the fifth, sixth, and seventh in the control group, and the eighth in the experimental group. Thus, our sheet of paper would look like this:

Participant Number 1 = 06121E
Participant Number 2 = 07856C
Participant Number 3 = 18876E
Participant Number 4 = 27756E
Participant Number 5 = 30995C
Participant Number 6 = 76393C
Participant Number 7 = 81647C
Participant Number 8 = 98872E

Step 7: To avoid confusion, recopy your list, but make two changes. First, delete the random numbers. Second, write out "Experimental" and "Control." In this example, your recopied list would look like the following:

Participant Number 1 = Experimental
Participant Number 2 = Control
Participant Number 3 = Experimental
Participant Number 4 = Experimental
Participant Number 5 = Control
Participant Number 6 = Control
Participant Number 7 = Control
Participant Number 8 = Experimental

Random Numbers Table

			Column			
Row	1	2	3	4	5	6
1	10480	15011	01536	02011	81647	69179
2	22368	46573	25595	85393	30995	89198
3	24130	48360	22527	97265	76393	64809
4	42167	93093	06243	61680	07856	16376
5	37570	39975	81837	76656	06121	91782
6	77921	06907	11008	42751	27756	53498
7	99562	72905	56420	69994	98872	31016
8	96301	91977	05463	07972	18876	20922

of the treatment's effect. For example, you might hypothesize that participants getting 3 hours of full-spectrum light will be happier than those getting no full-spectrum light because full-spectrum light causes increases in happiness.

Although you can make a wide variety of experimental hypotheses (e.g., you could hypothesize that participants forced to trade their lottery tickets would be unhappier than those who were not forced to trade their lottery tickets or that participants asked to describe their relationship with their friend with "My friend and I _____" sentences would feel less happy with the relationship than people asked to describe their relationship with "We _____" sentences), realize that not all hypotheses are cause–effect hypotheses. Sometimes, hypotheses involve describing what happens rather than finding out what makes things happen. If you generate a hypothesis that is *not* a cause–effect statement, it is *not* an experimental hypothesis. Thus, if you hypothesize that men are more romantic than women, you do not have an experimental hypothesis. Similarly, if you predict that athletes will be more assertive than nonathletes, you do not have an experimental hypothesis. In short, to have an experimental hypothesis, you must predict that varying the *treatment* (some factor that *you manipulate*) will *cause* an *effect*.

Null Hypothesis: The Treatment Does Not Have an Effect

Once you have an experimental (cause–effect) hypothesis, pit it against the **null hypothesis**: the hypothesis that varying the treatment has *no* effect. The null hypothesis essentially states that any difference you observe between the treatment and no-treatment group scores could be due to chance. Therefore, if our experimental hypothesis was that getting 3 hours of full-spectrum lighting will cause people to be happier, the null hypothesis would be that varying full-spectrum lighting will have *no* demonstrated effect on happiness.

If your results show that the difference between groups is probably not due to chance, you can reject the null hypothesis. By rejecting the null hypothesis, you tentatively accept the experimental hypothesis: You conclude that the treatment has an effect.

But what happens if you fail to demonstrate conclusively that the treatment has an effect? Can you say that there is no effect for full-spectrum lighting? No, you can only say that you *failed* to prove beyond a reasonable doubt that full-spectrum lighting causes a change in happiness. In other words, you're back to where you were before you began the study: You do not know whether full-spectrum lighting causes a change in happiness.[2]

---

[2] Those of you who are intimately familiar with confidence intervals may realize that null results do not necessarily send the researcher back to square one. Admittedly, we do not know whether the effect is greater than zero, but we could use confidence intervals to estimate a range in which the effect size probably lies. That is, before the study, we may have no idea of the potential size of the effect. We might think the effect would be anywhere between −100 units and +100 units. However, based on the data collected in the study, we could estimate, with 95% confidence, that the effect is in a certain range. For example, we might find, at the 95% level of confidence, that the effect is somewhere in the range between −1 units and +3 units.

To reiterate a key point, *the failure to find a treatment effect doesn't mean that the treatment has no effect.* If you had looked more carefully, you might have found the effect.

To help yourself remember that you can't prove the null hypothesis, think of the null hypothesis as saying, "The difference between conditions *may* be due to chance." Even if you could prove that "The difference may be due to chance," what would you have you proved? Certainly, you would not have proved that the difference *is* due to chance.

Conclusions About Experimental and Null Hypotheses

In summary, you have learned four important points about experimental and null hypotheses:

1. The experimental hypothesis is that varying the treatment has an effect.
2. The null hypothesis is that varying the treatment has no effect.
3. If you reject the null hypothesis, you can tentatively accept the hypothesis that the treatment has an effect.
4. If you fail to reject the null hypothesis, you can't draw any conclusions.

To remember these four key points, think about these hypotheses in the context of a criminal trial. In a trial, the *experimental* hypothesis is that the defendant *did cause* the crime; the *null* hypothesis is that the defendant *did not* commit the crime. The prosecutor tries to disprove the null hypothesis so that the jury will accept the experimental hypothesis. In other words, the prosecutor tries to *disprove*, beyond a reasonable doubt, the hypothesis that the defendant is not guilty of the crime. If the jury decides that the null hypothesis is highly unlikely, they reject it and find the defendant guilty. If, on the other hand, they still have reasonable doubt, they fail to reject the null hypothesis and vote "not guilty." Note that their "not guilty" verdict is not an "innocent" verdict. Instead, it is a verdict reflecting that they are not sure, beyond a reasonable doubt, that the null hypothesis is false. In other words, it is a "not proved" verdict.

Manipulating the Independent Variable

Once you have your hypotheses, your next step is to manipulate the treatment by giving (assigning) some participant one treatment and giving other participants another. In any experiment, "participants are presented with the same general scenario (e.g., rating photographs of potential dating partners), but at least *one aspect of this general scenario is manipulated*" (Ickes, 2003, p. 22). By treating participants the same except that you vary the treatment, you can isolate the treatment's effect, as was done in the following classic experiments:

- In the first study showing that leading questions could bias eyewitness testimony, Loftus (1975) had students watch a film of a car accident and then gave students a questionnaire. The manipulation was whether the first question on the questionnaire was "How fast was Car A going when it ran the stop sign?"—a misleading question because Car A did *not* run the stop sign—*or* "How fast was Car A going when it turned right?"—a question that was not misleading because the car *did* turn right.

- In the first study showing that people's entire impressions of another person could be greatly influenced by a single trait, Asch (1946) had participants think about a person who was described as either (a) "intelligent, skillful, industrious, *warm*, determined, practical, cautious" *or* (b) "intelligent, skillful, industrious, *cold*, determined, practical, cautious."
- In the first study showing that sex role stereotypes affect how people perceive infants, Condry and Condry (1976) had all participants use a form to rate the same baby. The only difference between how participants were treated was whether the infant rating form listed the infant's name (a) as "David" and sex as "male" *or* (b) as "Dana" and sex as "female."
- In the first study showing that the pronouns people use when they describe their closest relationships affect how people see those relationships, Fitzsimons and Kay (2004) had all participants rate their relationship with their closest same-sex friend after writing five sentences about that friend. The only difference between groups was that one group was told to begin each sentence with "We," and was given the example, "We have known each other for 2 years," whereas the other group was told to begin each sentence with "(Insert friend's name) and I," and given the example, "John and I have known each other for 2 years."

To understand how you would manipulate a treatment, let's go back to trying to test the hypothesis about the effect of full-spectrum lighting on mood. To do this, you must vary the amount of light people get—and the amount should be independent of (should not depend on or be affected by) the individual's personal characteristics. To be specific, the amount of full-spectrum light participants receive should be determined by independent random assignment. Because the amount *varies* between the treatment group and the no-treatment group, because it varies *independently* of each participant's characteristics, and because it is determined by *independent* random assignment, the amount of full-spectrum lighting (the experimental intervention) is the **independent variable**. (Note that participants are people—not variables. Thus, participants are neither independent variables nor dependent variables. Participants may, however, be in the experimental group or in the control group.)

In simple experiments, there are two **levels of the independent variable**: different values (amounts or types) of the treatment variable. Those two levels can be *amounts* of treatment (e.g., 1 hour of lighting versus 2 hours of lighting) or they can be *types* of treatment (e.g., lighting versus psychotherapy). In our lighting experiment, participants are randomly assigned to one of the following two levels of the independent variable: (1) 3 hours of full-spectrum lighting and (2) no full-spectrum lighting.

Experimental and Control Groups: Similar, but Treated Differently

The participants who are randomly assigned to get the higher level of the treatment (3 hours of full-spectrum light) are usually called the **experimental group**. The participants who are randomly assigned to get a lower level of the treatment (in this case, no treatment) are usually called the **control group**. Thus, in our example, the experimental group is the treatment group and the control group is the no-treatment group.

The control group is a comparison group. We *compare* the experimental (treatment) group participants to the control (no-treatment) group participants to see whether the treatment had an effect. If the treatment group scores the same as the comparison group, we would suspect that the treatment group would have scored that way even without the treatment. If, on the other hand, the treatment group scores differently than the control group, we would suspect that the treatment had an effect. For example, Ariely (2007) gave experimental group participants a chance to cheat. After taking a 50-item test, all participants transferred their answers from their tests to an answer sheet. For participants in the experimental group, the answer sheets already had the correct answers marked. Experimental group participants then shredded their tests and handed in their answer sheets. In this condition, students averaged about 36 questions correct. Did they cheat—and, if they did, how could Ariely possibly know? The only way to find out whether the experimental group cheated was to compare their scores to control group participants who were not allowed to cheat. Those control participants answered only about 33 questions correctly. By using statistics to compare the experimental group to the control group, Ariely found,that the experimental group answered significantly (reliably) more questions than the control group. Thus, he concluded that the experimental group cheated. Note that his conclusion—like that of any experimenter who uses a control group—only makes sense if the groups were equivalent at the start of the experiment. Thus, experimenters need to make sure that there are no systematic differences between the groups before the experimenter gives the different groups different levels of the independent variable.

As the terms "experimental *group*" and "control *group*" imply, you should have several participants (preferably more than 30) in each of your conditions. The more participants you have, the more likely it is that your two groups will be similar at the start of the experiment. Conversely, the fewer participants you have, the less likely it is that your groups will be similar before you administer the treatment. For example, if you are doing an experiment to evaluate the effect of a strength pill and have only two participants (a 6 ft. 4 in., 280-lb [1.9 m, 127 kg] professional body builder and a 5 ft. 1 in., 88-lb [1.5 m, 40 kg] person recovering from a long illness), random assignment will not have the opportunity to make your "groups" equivalent. Consequently, your control group would not be a fair comparison group.

The Value of Independence: Why Control and Experimental Groups Shouldn't Be Called "Groups"

Although we have noted that the experimental and control groups should have several participants in each "group," these "groups" should not already *be* groups and should not be allowed to *become* groups. Thus, to conduct an experiment, you do *not* find two groups of participants and then randomly assign one group to be the experimental group and the other to be the control group.

Why You Should Not Choose Two Preexisting Groups

To see why not, suppose you were doing a study involving 10,000 janitors at a Los Angeles company and 10,000 managers at a New York company. You have 20,000 people in your experiment: one of the largest experiments in history. Then, you flip a coin and—on the basis of that single coin flip—assign the LA janitors to no treatment and the New York managers to treatment. Even though you have 10,000 participants in each group, your treatment and no-treatment groups differ in at least two systematic ways (where they live and what they do) before the study begins. Your random assignment is no more successful in making your groups similar than it was when you had only two participants. Consequently, to get random assignment to equalize your groups, you need to assign each participant **independently**: individually, without regard to how previous participants were assigned.

Why You Should Not Let Your "Groups" Become Groups

Your concern with independence does not stop at assignment. After you have independently assigned participants to condition, you want each of your participants to remain independent. To maintain independence, *do not test the control participants in one group session and the experimental participants in a separate group session.* Having one testing session for the control group and a second session for the experimental group hurts independence in two ways.

First, when participants are tested in groups, they may become group members who influence each other's responses rather than independent individuals. For example, instead of giving their own individual, independent responses, participants might respond as a conforming mob.

As an example of the perils of letting participants interact, imagine that you are doing an ESP experiment. In the control group, only 30 of the 60 participants correctly guessed that the coin would turn up heads. In the experimental group, on the other hand, all 60 participants correctly guessed that the coin would turn up heads. Had each experimental group participant made his or her decision independently, such results would rarely[3] happen by chance. Thus, we would conclude that the treatment had an effect. However, if all the experimental group members talked to one another and made a group decision, they were not acting as 60 individual participants but as one group. In that case, the results would not be so impressive: Because all 60 experimental participants acted as one, the chances of all of them correctly guessing the coin flip were the same as the chances of one person correctly guessing a coin flip: 1 in 2 (50%).

Although this example shows what can happen when participants are tested in groups and allowed to interact freely, interaction can disturb independence even when group discussion is prohibited. Participants may influence one another through inadvertent outcries (laughs, exclamations like, "Oh no!") or through subtle nonverbal cues. In our lighting–happiness

---

[3] To be more precise, it should happen with a probability of $(1/2)^{60}$, which is less than .000000000000000009% of the time.

experiment, if we tested all the participants in a single group session, one participant who is crying uncontrollably might cause the entire experimental group to be unhappy, thereby leading us to falsely conclude that the lighting caused unhappiness. If, on the other hand, we tested each participant individually, the unhappy participant's behavior would not affect anyone else's responses.

The second reason for not testing all the experimental participants in one session and all the control participants in another is that such group testing turns the inevitable, random differences between testing sessions into systematic effects. For instance, suppose that when the experimental group was tested, there was a distraction in the hall, but there was no such distraction while the control group was tested. Like the treatment, this distraction was presented to all the experimental group participants, but to none of the control group participants. Thus, if the distraction did have an effect, its effect might be mistaken for a treatment effect. If, on the other hand, participants were tested individually, it is unlikely that only the experimental participants would be exposed to distractions. Instead, distractions would have a chance to even out so that participants in both groups would be almost equally affected by distractions.

But what if you are sure you won't have distractions? Even then, the sessions will differ in ways unrelated to the treatment. If you manage to test the participants at the same time, you'll have to use different experimenters and different testing rooms. If you manage to use the same experimenter and testing room, you'll have to test the groups at different times. Consequently, if you find a significant difference between your groups, you will have trouble interpreting those results. Specifically, you have to ask, "Is the significant difference due to the groups getting different levels of the treatment or to the groups being tested under different conditions (e.g., having different experimenters or being tested at different times of day)?"

To avoid these problems in interpreting your results, make sure that the only consistent difference between your participants is that some receive one level of the independent variable whereas others receive a different level of the independent variable. To make sure that the treatment manipulation is the only consistent difference between your conditions, randomly assign each of your participants individually and then test your participants individually (or in small groups) so that random differences between testing sessions have a chance to even out. If you must run participants in large groups, do not run groups made up exclusively of *either* experimental or control participants. Instead, run groups made up of *both* control and experimental participants.

Note that running participants individually will not always guarantee independence. For example, suppose that you are trying to see whether a citrus scent (e.g., the smell of lemons) makes people more alert. It is a hassle to set up your research room to add the lemon smell to the room (for experimental group participants) and then to air it out to take away the smell (for the control participants). Therefore, you decide to run all the control participants in the morning, add lemon scent to the room, and then test all the experimental participants in the afternoon. Even if you test each participant individually, there is a systematic difference between your conditions that has nothing to do with scent: Control participants were all tested in the morning, whereas experimental participants were all tested in the afternoon. In short, to avoid ruining your study, maintain independence even if maintaining independence is inconvenient.

The Value of Assignment (Manipulating the Treatment)

We have focused on the importance of independence to independent random assignment. Independence helps us start the experiment with two "groups" of participants that do not differ in any systematic way. But assignment is also a very important aspect of independent random assignment.

Random Assignment Makes the Treatment the Only Systematic Difference Between Groups

Random assignment to treatment group helps ensure that the only systematic difference between the groups is the treatment. With random assignment, our groups will be equivalent on the nontreatment variables we know about as well as on the (many) nontreatment variables we don't know about.

In our experiment, random assignment makes it so that one random sample of participants (the experimental group) is assigned to receive a high level of the independent variable whereas the other random sample of participants (the control group) is assigned to receive a low level of the independent variable. If, at the end of the study, the groups differed by more than would be expected by chance, we could say that the difference was due to the only nonchance difference between them: the treatment.

Without Random Assignment You Do Not Have a Simple Experiment

As you have seen, the key to the simple experiment is random assignment. Therefore, *if you cannot randomly assign participants to your different groups, you cannot do a simple experiment.* Because you cannot randomly assign participants to have certain personal characteristics, simple experiments cannot be used to study the effects of pre-assigned participant characteristics such as gender, race, personality, and intelligence.[4] For example, it makes no sense to assign a man to be a woman, a 7 ft. 2 in. (218 cm) person to be short, or a shy person to be outgoing.

Admittedly, limiting yourself to variables that can be randomly assigned limits the variables you can study (but, as you can see by referring back to Tip 4 on page 87, it does not limit you as much as you may think). However, to make cause–effect statements, you need variables that can be randomly assigned. For example, suppose you want to know whether staring at the back of people's heads will make them look at you. To test this hypothesis, you obviously have to stare at the back of their heads *before*

---

[4] You can, however, use experiments to investigate how participants react to people who vary in terms of these characteristics. For example, you can have an experiment in which participants read the same story except that one group is told that the story was written by a man, whereas the other group is told that the story was written by a woman. Similarly, you can randomly determine, for each participant, whether the participant interacts with a male or female experimenter.

they look at you: Causes must come *before* effects. The *assignment* aspect of random assignment allows you to guarantee that your treatment manipulation (e.g., staring at the back of their heads) comes before they look at you—guaranteeing that staring is not an effect, but rather a possible cause. The *random* aspect of random assignment allows you to randomize other variables so that the treatment (your staring) is the only systematic difference between the "stare" and "no-stare" groups. Because staring is the only consistent difference between the groups, if one group does look at you significantly more than the other group, you can say that the difference is due to your staring rather than to some other cause.

To better understand why we need to be able to randomly assign participants, let's imagine that you try to look at the effects of lighting on mood without using random assignment. Suppose you get a group of people who use light therapy and compare them to a group of people who do not use light therapy. What would be wrong with that?

The problem is that you are selecting two groups of people who you know are different in at least one way, and then you are assuming that they don't differ in any other respect. The assumption that the groups are identical in every other respect is probably wrong. The light therapy group probably feels more depressed, lives in colder climates, is more receptive to new ideas, and is richer than the other group (because light boxes are expensive).

Because the groups differ in many ways other than in terms of the "treatment," it would be foolish to say that the treatment—rather than one of the many other differences between the groups—is what caused the groups to score differently on the happiness measure. For example, if the group of light users is more depressed than our sample of nonusers, we could not conclude that the lighting caused their depression. After all, the lighting is more likely to be a partial cure for—rather than a cause of—their depression.

But what if the group of lighting users is less depressed? Even then, we could not conclude that the lighting is causing an effect. Lighting users may be less depressed because they are richer, have more spare time, or differ in some other way from those who don't use lights. In short, if you do not randomly *assign* participants to groups, you cannot conclude anything about the effects of a treatment.

If, on the other hand, you start with one group of participants and then randomly assign half to full-spectrum lighting and half to normal lighting, interpreting differences between the groups would be much simpler. Because the groups probably were similar before the treatment was introduced, large group differences in happiness are probably due to the only systematic difference between them—the lighting.

Collecting the Dependent Variable

Before you can determine whether the lighting caused the experimental group to be happier than the control group, you must measure each participant's happiness. You know that each person's happiness will be somewhat *dependent* on the individual's personality, and you predict that his or her score on the happiness *variable* will also be *dependent* on the lighting. Therefore, scores on the happiness measure are your **dependent variable.**

Because the dependent variable is what the participant does that you *measure*, the dependent variable is also called the **dependent measure**.

The Statistical Significance Decision: Deciding Whether to Declare That a Difference Is Not a Coincidence

After measuring the dependent variable, you will want to compare the experimental group's happiness scores to the control group's. One way to make this comparison is to subtract the average of the happiness scores for the control (comparison) group from the average of the experimental group's happiness scores.

Unfortunately, knowing how much the groups differ doesn't tell you how much of an effect the treatment had. After all, even if the treatment had no effect, nontreatment factors would probably still make the groups differ. In other words, even if the treatment had no effect, the groups may differ due to random error.

How can you determine that the difference between groups is due to something more than random error? To determine the probability that the difference is just due to chance alone, you need to use **inferential statistics**: the science of chance, the science of making inferences about a population (the larger group that you are interested in) from a random sample of that population.

Statistically Significant Results: Declaring That the Treatment Has an Effect

If, after using statistics, you find that the difference between your groups is greater than could be expected if only chance were at work, your results are **statistically significant**: statistically *reliable*. The term *statistical significance* means that you are sure, beyond a reasonable doubt, that the difference you observed is due to more than mere coincidence.

What is a reasonable doubt? Usually, before researchers commit themselves to saying that the treatment has an effect, they want a 5% probability ($p = .05$) or less ($p < .05$) that they would get such a pattern of results when there really was no effect of the treatment. Consequently, you may hear researchers say that their results were "significant at the point-oh-five level" and, in journal articles, you will often see statements like, "the results were statistically significant ($p = .04$)."

To review, if you do a **simple experiment**—a study in which you independently and randomly assign participants to one of two conditions, you will probably find that the treatment group mean is different from the control group mean. Such a difference is not, by itself, evidence of the treatment's effect. Indeed, because random assignment does not create identical groups, you would expect the two group means to differ to some extent. Therefore, the question is not "Is there a difference between the group means?" but rather "Is the difference between the group means a *reliable* one—one bigger than would be expected if only random factors were at work?" To answer that question, you need to use inferential statistics.

By using inferential statistics, you might find that if only chance factors were at work (i.e., if the independent variable had no effect), you would get a difference as large as that less than 5% of the time ($p < .05$). If differences as big or bigger than what you found occur less than 5% of the time by chance alone when the null hypothesis is true, you would probably conclude that the null hypothesis is not true. To state your conclusion more formally, you might say that "the results are statistically significant at the .05 level." By "statistically significant," you mean that because it's unlikely that the difference between your groups is due to chance alone, you conclude that some of the difference was due to the treatment.

With statistically significant results, you would be relatively confident that if you repeated the study, you would get the same pattern of results—the independent variable would again cause a similar type of change in the scores on the dependent variable. In short, statistical significance suggests that the results are reliable and replicable.

Statistically Significant Effects May Be Small

Statistical significance, however, does not mean that the results are significant in the sense of being large. Just because a difference is statistically significant—reliably different from zero—doesn't mean the difference is large. Even a tiny difference can be statistically reliable. For example, if you flipped a coin 5,000 times and it came up heads 51% of the time, this result would be significantly different from 50% heads.

Statistically Significant Results May Be Insignificant (Trivial)

Nor does statistical significance mean that the results are significant in the sense of being important. If you have a meaningless hypothesis, you may have results that are statistically significant but scientifically meaningless.

Statistically Significant Results May Refute Your Experimental Hypothesis

Finally, statistically significant results do not necessarily support your hypothesis. For example, suppose your hypothesis is that the treatment improves behavior. A statistically significant effect for the treatment would mean that the treatment had an effect. But did the treatment improve behavior or make it worse? To find out, you have to look at the group averages to see whether the treatment group or no-treatment group is better off.

Summary of the Limitations of Statistically Significant Results

In short, statistically significant results tell you nothing about the direction, size, or importance of the treatment effect (see Box 10.2). Because of the

---

BOX 10.2 Limits of Statistical Significance

Statistically significant differences are

1. probably not due to chance alone
2. not necessarily large
3. not necessarily in the direction you predicted
4. not necessarily important

limitations of statistical significance, the American Psychological Association appointed a task force to determine whether significance testing should be eliminated. The task force did "not support any action that could be interpreted as banning the use of null significance testing or p values in psychological research and publication" (American Psychological Association, 1996b). However, the task force did recommend that, in addition to reporting whether the results were statistically significant, authors should provide information about the direction and size of effects.

Null Results: Why We Can't Draw Conclusions From Nonsignificant Results

You now know how to interpret statistically significant results. But what if your results are *not* statistically significant? That is, what if you can't reject the null hypothesis that the difference between your groups could be due to chance? Then, you have *failed* to reject the null hypothesis; therefore, your results would be described as "not significant."

As the phrase "not significant" suggests, you can't draw any conclusions from such findings. With **nonsignificant results** (also called **null results**), you do not know whether the treatment has an effect that you failed to find or whether the treatment really has no effect (see Figure 10.1).

Nonsignificant results are analogous to a "not guilty" verdict: Is the defendant innocent, or did the prosecutor just present a poor case? Often, defendants get off, not because of overwhelming proof of their innocence, but because of lack of conclusive proof of their guilt.

You have seen that nonsignificant results neither confirm nor refute that the treatment had an effect. Unfortunately, you will find some people acting like null results are proof that the treatment has an effect—whereas others will act like null results are proof that the treatment has no effect (see Table 10.1).

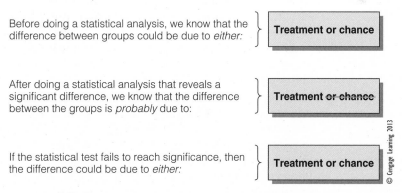

FIGURE 10.1 The meaning of statistical significance.

If the results are statistically significant, we can conclude that the difference between the groups is not due entirely to chance and therefore some of the difference must be due to the treatment. However, if the results are not statistically significant, the results could be due to chance or treatment. Put another way, we don't know any more than we did before we subjected the results to statistical analysis.

TABLE **10.1** Common Errors in Discussing Null Results

Statement	Flaw
"The results were not significant. Therefore, the independent variable had no effect."	"Not that I found" is not the same as proving "there isn't any."
"The treatment had an effect, even though the results are not significant."	"Not significant" means that you failed to find an effect. Therefore, the statement could be translated as, "I didn't find an effect for the treatment, but I believe that I really did."

© Cengage Learning 2013

Nonsignificant Results Are Not Significant

All too often, people act as if nonsignificant results are really significant. They may say, "The difference between my groups shows that the treatment had an effect, even though the difference is not significant." Reread the previous quote because you're sure to see it again: It's one of the most common contradictory statements that researchers make. People making this statement are really saying, "The difference is due to the treatment, even though I've found no evidence that the difference isn't just due to chance." What scientists hear these people saying is this: "Even though the significance test says that results are inconclusive, I want to prove my hypothesis so badly that I'm going to ignore the significance test (and science)."

Null Results Do Not Prove the Null Hypothesis: "I Didn't Find It" Doesn't Mean It Doesn't Exist

As we have just discussed, some people act like null results secretly prove the experimental hypothesis. On the other hand, some people make the opposite mistake: They incorrectly assume that null results prove the null hypothesis. That is, they falsely conclude that null results prove that the treatment had no effect. Some individuals make this mistake because they think the term "null results" implies that the results prove the null hypothesis. Those people would be better off thinking of null results as "no results" than to think that null results support the null hypothesis.

Thinking that nonsignificant results support the null hypothesis is a mistake because it overlooks the difficulty of conclusively proving that a treatment has an effect. People should realize that not finding something is not the same as proving that the thing does not exist. After all, people often fail to find things that clearly exist, such as books that are in the library, items that are in the grocery store, and keys that are on the table in front of them.

Even in highly systematic investigations, failing to find something doesn't mean the thing does not exist. For example, in 70% of all murder investigations, investigators do not find a single identifiable print at the murder scene—not even the victim's. Thus, the failure to find the suspect's fingerprints at the scene is hardly proof that the suspect is innocent. For essentially the same reasons, the failure to find an effect is not proof that there is no effect. In short, just as a "not guilty" verdict doesn't mean the defendant did not cause the crime, a "not significant" verdict does not mean that the treatment did not have an effect.

Summary of the "Ideal" Simple Experiment

Thus far, we have said that the simple experiment gives you an easy way to determine whether a factor causes an effect. If you can randomly assign participants to either a treatment or no-treatment group, all you have to do is find out whether your results are statistically significant. If your results are statistically significant, your treatment probably had an effect. No method allows you to account for the effects of nontreatment variables with as little effort as random assignment.

ERRORS IN DETERMINING WHETHER RESULTS ARE STATISTICALLY SIGNIFICANT

There is one drawback to random assignment: Differences between groups may be due to chance rather than to the treatment. Admittedly, statistical tests—by allowing you to predict the extent to which chance may cause the groups to differ—minimize this drawback. Statistical tests, however, do not allow you to perfectly predict chance all of the time. Therefore, you may err by either underestimating or overestimating the extent to which chance is causing your groups to differ (see Table 10.2).

Type 1 Errors: "Crying Wolf"

If you underestimate the role of chance in your study, you may make a **Type 1 error**: mistaking a chance difference for a real difference. In the simple experiment, you would make a Type 1 error if you mistook a chance difference between your experimental and control groups for a treatment effect. More specifically, you would make a Type 1 error if you declared that a difference between your groups was statistically significant, when the treatment really didn't have an effect. Type 1 errors, then, are statistical false alarms. In nonresearch settings, examples of Type 1 errors include

- a jury convicting an innocent person because they mistake a series of coincidences as evidence of guilt

TABLE **10.2** Possible Outcomes of Statistical Significance Decision

Statistical Significance Decision	Real State of Affairs	
	Treatment Has an Effect	**Treatment Does Not Have an Effect**
Significant: Reject the null hypothesis	Correct decision	Type 1 error
Not significant: Do not reject the null hypothesis	Type 2 error	Correct decision

- a person responding to a false alarm, such as thinking that the phone is ringing when it's not or thinking that an alarm is going off when it's not
- a physician making a "false positive" medical diagnosis, such as telling a woman she is pregnant when she isn't
- a fire alarm going off when there is no fire

Reducing Your Risk of a Type 1 Error by Using a .05 Significance Level

What can you do about Type 1 errors? There is *only one thing you can do: You can decide what risk of a Type 1 error you are willing to take.* Usually, experimenters decide that they are going to take less than a 5% risk of making a Type 1 error. For example, they say their results must be significant at the .05 level ($p < .05$) before they declare that their results are significant. In other words, if the null hypothesis is true, they will only "cry wolf" 5% of the time. Although they are comfortable with the odds of their making a Type 1 error being less than 5 in 100, why take even that risk? Why not take less than a 1% risk? Why take any risk at all?

Accepting the Risk of a Type 1 Error

To understand why researchers risk making Type 1 errors, imagine you are betting with someone who is flipping a coin. For all 10 flips, she calls "heads." She wins most of the 10 flips.

Let's suppose that you will refuse to pay up if you have statistical proof that she is cheating. However, you do not want to make the Type 1 error of attributing her results to cheating (using a biased coin) when the results are due only to luck. How many of the 10 flips does she have to win before you "prove" that she is cheating?

To help you answer this question, we looked up the chances of getting 8, 9, or 10 heads in 10 flips of a fair coin.[5] Those probabilities are as follows:

Event	Probability Expressed in Percentages	Probability Expressed in Decimal Form
Chances of 8 or more heads	5.47%	.0547
Chances of 9 or more heads	1.08%	.0108
Chances of 10 heads	0.1%	.001

© Cengage Learning 2013

From these probabilities, you can see that you can't have complete, absolute proof that she is cheating. Thus, if you insist on taking 0% risk of falsely accusing her (you want to be absolutely 100% sure), you would not call her a cheat— even if she got 10 heads in a row. As you can see from the probabilities we listed, it is very unlikely (.1% chance), but still possible, that she could get 10 heads in a row, purely by chance alone. Consequently, if you are going to accuse her of cheating, you are going to have to take some risk of making a false accusation.

---

[5] You do not need to know how to calculate these percentages.

If you were willing to take more than a 0% risk but were unwilling to take even a 1% risk of falsely accusing her (you wanted to be more than 99% sure), you would call her a cheat if all 10 flips turned up heads—but not if 9 of the flips were heads. If you were willing to take a 2% risk of falsely accusing her (you wanted to be 98% sure), you would call her a cheat if either 9 or 10 of the flips turned up heads. Finally, if you were willing to take a 6% risk of falsely accusing her (you would settle for being 94% sure), you could refuse to pay up if she got 8 or more heads.

This betting example gives you a clue about what happens when you set your risk of making a Type 1 error. When you determine your risk of making a Type 1 error, you are indirectly determining how much the groups must differ before you will declare that difference statistically significant. If you are willing to take a relatively large risk of mistaking a difference that is due only to chance for a treatment effect, you may declare a relatively small difference statistically significant. If, on the other hand, you are willing to take only a tiny risk of mistakenly declaring a chance difference statistically significant, you must require that the difference between groups be relatively large before you are willing to call it statistically significant. In other words, all other things being equal, the larger the difference must be before you declare it significant, the less likely it is that you will make a Type 1 error. To take an extreme example of this principle, if you would not declare even the biggest possible difference between your groups statistically significant, you would never make a Type 1 error (just as if you never yelled "wolf," you would never falsely cry "wolf").

Type 2 Errors: "Failing to Announce the Wolf"

The problem with not taking any risk of making a Type 1 error is that, if the treatment did have an effect, you would be unable to detect it. You would be like the person who deactivated a smoke alarm to stop its false alarms but then slept through a fire because the alarm didn't go off. That is, by trying to be very sure that a difference is due to treatment and not to chance, you may set yourself up to make a **Type 2 error**: overlooking a genuine treatment effect because you think the differences between conditions might be due to chance. Examples of Type 2 errors in nonresearch situations include

- a jury letting a criminal go free because they wanted to be sure beyond any doubt and they realized that it was possible that the evidence against the defendant was due to numerous, unlikely coincidences
- a person failing to hear the phone ring
- a radar detector failing to detect a speed trap
- a physician making a "false negative" medical diagnosis, such as failing to detect that a woman was pregnant
- a fire alarm not going off when there is a fire

In short, whereas Type 1 errors are statistical false alarms that mistake a coincidence for a real difference, Type 2 errors are statistical alarms failing to alert us that the difference we see is more than a coincidence. Because the errors are opposites, choosing to reduce your risk of making a Type 1 error by being overly cautious about declaring a result statistically significant will increase your risk of making a Type 2 error. In the extreme case, if you were never willing to risk making a Type 1 error, you would make a Type 2 error every time the treatment did have an effect. Thus, because you want to detect real treatment effects, you will take a risk of making a Type 1 error. In other words, you will take some risk of Type 1 errors so that your study has **power**: the ability to find real differences and declare those differences statistically significant; or, put another way, the ability to avoid making Type 2 errors.[6]

The Need to Prevent Type 2 Errors: Why You Want the Power to Find Significant Differences

Although increasing your risk of making a Type 1 error will increase your power, you can increase power without increasing your risk of making a Type 1 error. Unfortunately, many people don't do what it takes to have power.

If you don't do what it takes to have power, your study may be doomed: Even if your treatment has an effect, you will fail to find that effect statistically significant. In a way, looking for a significant difference between your groups with an underpowered experiment is like looking for differences between cells with an underpowered microscope.

As you might imagine, conducting a low-powered experiment often leads to frustration over not finding anything. Beginning researchers frequently frustrate themselves by conducting such low-powered experiments. (We know we did.) Why do beginning researchers often fail to design sufficiently powerful experiments?

STATISTICS AND THE DESIGN OF THE SIMPLE EXPERIMENT

One reason inexperienced researchers fail to design powerful experiments is that they simply do not think about power—a "sin" that many professional researchers also commit (Cohen, 1990). But even when novice researchers do think about power, they often think that it is a statistical concept and therefore has nothing to do with design of experiments. Admittedly, power *is* a statistical concept. However, *statistical concepts should influence the design of research*. Just as a bridge builder should consider engineering principles

---

[6] In a sense, power (defined as 1.00 − the probability of making a Type 2 error) and Type 2 errors are opposites. *Power* refers to the chances (given that the treatment really does have a certain effect) of *finding* a significant treatment effect, whereas the probability of a Type 2 error refers to the chances (given that the treatment really does have a certain effect) of *failing to find* a significant treatment effect. If you plug numbers into the formula "1.00 − power = chances of making a Type 2 error," you can see that power and Type 2 errors are inversely related. For example, if power is 1, you have a 0% chance of making a Type 2 error (because 1.00 − 1.00 = 0%). Conversely, if the treatment has an effect and power is 0, you have a 100% chance of making a Type 2 error (because 1.00 − 0 = 100%). Often, power is around .40, meaning that, if the treatment has an effect, the researcher has a 40% (.40) chance of finding that effect and a 60% chance of not finding that effect (because 1.00 − .40 = 60%).

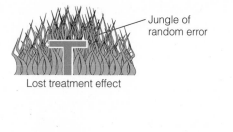

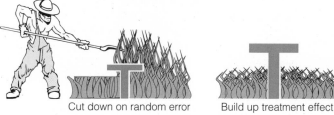

FIGURE **10.2** Cutting down on random error and building up the treatment effect: Two ways to avoid losing your treatment effect in a "jungle" of random error.

when designing a bridge, a researcher should consider statistical principles when designing a study. If you consider statistical power when designing your study, your study should have enough power to find the differences that you are looking for—if those differences really exist.

Power and the Design of the Simple Experiment

To have enough power, you must reduce the risk of chance differences hiding the treatment effect. As you can see from Figure 10.2, two ways to stop random error from overwhelming your treatment effect are (1) reduce the effects of random error and (2) increase the size of the treatment effect.

Reduce the Effect of Random Error

One of the most obvious ways to reduce the effects of random error is to reduce the potential sources of random error. The major sources of random error are random differences between testing situations, random measurement error, random differences between participants, and sloppy coding of data.

Standardize Procedures and Use Reliable Measures. Because a major source of random error is random variation in the testing situation, you can reduce random error by standardizing your experiment. Standardization consists of keeping the testing environment and the experimental procedures as constant as possible. Thus, to improve power, you might want the noise level, illumination level, temperature, and other conditions of testing to be the same for each participant. Furthermore, you would want to treat all your experimental group participants identically and treat all your control group participants identically. In addition to reducing random error by standardizing procedures, you should also reduce random error by using a reliable dependent measure (for more about how reliable measures boost power, see Chapter 6).

The desire for both reliable measures and strict standardization makes some psychologists love both instruments and the laboratory. Under the lab's carefully regulated conditions, experimenters can create powerful and sensitive experiments.

Other experimenters, however, reject the laboratory setting in favor of real-world settings. By using real-world settings, they can more easily make a case for their study's external validity. The price they pay for leaving the laboratory is that they are no longer able to keep many nontreatment variables (temperature, distractions, noise level, etc.) constant. These variables, free to vary wildly, create a jungle of random error that may hide the treatment's effect.

Because of the large variability in real-world settings and the difficulties of using sensitive measures in the field, even die-hard field experimenters may first look for a treatment's effect in the lab. Only after they have found that the treatment has an effect in the lab will they try to detect the treatment's effect in the field.

Study Participants Who Are All Alike. Like differences between testing sessions, differences between participants can hide treatment effects. Even if the treatment effect causes a large difference between your groups, you may overlook that effect, mistakenly believing that the difference between your groups is due to your participants being years apart in age and worlds apart in terms of their experiences.

To decrease the chances that between-subject differences will mask the treatment's effect, choose participants who are similar to one another. For instance, select participants who are the same gender, same age, and have the same IQ—or, study rats instead of humans. With rats, you can select subjects that have grown up in the same environment, have similar genes, and even have the same birthday. By studying homogeneous subjects under standardized situations, rat researchers can detect very subtle treatment effects.

Code Data Carefully. Obviously, sloppy coding of the data can sabotage the most sensitively designed study. So, why do we mention this obvious fact?

We mention it because careful coding is a cheap way to increase power. If you increase power by using nonhuman animals as participants, you may lose the ability to generalize to humans. If you increase power by using a lab experiment rather than a field experiment, you may lose some of your ability to generalize to real-world settings. But careful coding costs you nothing—except for a little time spent rechecking the coding of your data.

Let Random Error Balance Out. Thus far, we have talked about reducing the effects of random error by reducing the amount of random error. But you can reduce the *effects* of random error on your data without reducing the *amount* of random error in your data.

The key is to give random error more chances to balance out. To remind yourself that chance does balance out in the long run, imagine flipping a fair coin. If you flipped it six times, you might get five tails and one head—five times as many tails as heads. However, if you flipped it 1,000 times, you would end up with almost as many heads as tails.

Similarly, if you use five participants in each group, your groups probably won't be equivalent before the experiment begins. Thus, even if you found large differences between the groups at the end of the study, you might have to say that the differences could be due to chance alone. However, if you use 60 participants in each group, your groups should be equivalent before the study begins. Consequently, a treatment effect that would be undetected if you used 5 participants per group might be statistically significant if you used 60 participants per group. In short, to take advantage of the fact that random error balances out, boost your study's power by studying more participants.

Create Larger Effects: Bigger Effects Are Easier to See

Until now, we have talked about increasing power by making our experiment more sensitive to small differences. Specifically, we have talked about two ways of preventing the "noise" caused by random error from making us unable to "hear" the treatment effect: (1) reducing the amount of random error and (2) giving random error a chance to balance out. However, we have left out one obvious way to increase our experiment's ability to detect the effect: making the effect louder (bigger) and thus easier to hear.

As you might imagine, bigger effects are easier to find. But how do we create bigger effects? Your best bet for increasing the size of the effect is to give the control group participants a very low level of the independent variable while giving the experimental group a very high level of the independent variable. Hence, to have adequate power in the lighting experiment, rather than giving the control group 1 hour of full-spectrum light and the experimental group 2 hours, you might give the control group no full-spectrum light and the experimental group 4 hours of full-spectrum light.

To see how researchers can maximize the chances of finding an effect by giving the experimental and control groups widely different levels of treatment, let's consider an experiment by T. D. Wilson and Schooler (1991). Wilson and Schooler wanted to determine whether thinking about the advantages and disadvantages of a choice could hurt one's ability to make the right choice. In one experiment, they had participants rate their preference for the taste of several fruit-flavored jams. Half the participants rated their preferences after completing a "filler" questionnaire asking them to list reasons why they chose their major. The other half rated their preferences after completing a questionnaire asking them to "analyze why you feel the way you do about each jam in order to prepare yourself for your evaluations." As Wilson and Schooler predicted, the participants who thought about why they liked or didn't like the jam made less accurate ratings (ratings that differed more from experts' ratings) than those who did not consider how they felt about the jam.

Although the finding that one can think too much about a choice is intriguing, we want to emphasize another aspect of Wilson and Schooler's study: the difference between the amount of time experimental participants reflected on jams versus the amount of time that control participants reflected on jams. Note that the researchers did not ask the control group to do any reflection whatsoever about the jams. To reiterate, Wilson and Schooler did not have the control group do a moderate amount of reflection and the experimental group do slightly more reflection. If they had, Wilson and Schooler might have failed to find a statistically significant effect.

Conclusions About How Statistical Considerations Impact Design Decisions

By now, you can probably appreciate why Fisher said, "To consult a statistician after an experiment is finished is often merely to ask him to conduct a post mortem examination. He can perhaps say what the experiment died of." The reason you should think about statistics before you do an experiment is that statistical considerations influence virtually every aspect of the design process (see Table 10.3). For example, statistical considerations even dictate what kind of hypothesis you can test. Because you cannot accept the null hypothesis, the only hypotheses that you can hope to support are hypotheses that the groups will differ. Therefore, you cannot do a simple experiment to prove that two treatments have the same effect or that a certain treatment will be just as ineffective as no treatment.

Not only do statistical considerations dictate what types of hypotheses you can have, they also mandate how you should assign your participants. Specifically, if you do not assign your participants to groups using independent random assignment, you do not have a valid experiment.

Statistical considerations also dictate how you should treat your participants. You will not have a valid experiment if you let participants influence one another's responses or if you do anything else that would violate the statistical requirement that individual participants' responses must be independent.

TABLE **10.3** Implications of Statistics for the Simple Experiment

Statistical Concern/Requirement	Implications for Designing the Simple Experiment
Observations must be independent.	You must use independent random assignment and, ideally, you will test participants individually.
Groups must differ for only two reasons—random differences and the independent variable.	You must randomly assign participants to groups.
It is impossible to accept the null hypothesis.	You cannot use the experiment to prove that a treatment has no effect or to prove that two treatments have identical effects.
You need enough power to find a significant effect.	You should 1. Standardize procedures. 2. Use sensitive, reliable dependent variables. 3. Code data carefully. 4. Use homogeneous participants. 5. Use many participants. 6. Use extreme levels of the independent variable.

Even when statistics are not dictating what you must do, they are suggesting what you should do. To avoid making Type 2 errors (that is, to have adequate power), you should do the following:

1. Standardize your procedures.
2. Use sensitive and reliable dependent measures.
3. Carefully code your data.
4. Use homogeneous participants.
5. Use many participants.
6. Use extreme levels of the independent variable.

NONSTATISTICAL CONSIDERATIONS AND THE DESIGN OF THE SIMPLE EXPERIMENT

Statistical issues are not the only issues that you should consider when designing a simple experiment. If you considered only statistical power, you could harm your participants, as well as your experiment's external and construct validity. Therefore, in addition to statistical issues such as power, you must also consider external validity, construct validity, and ethical issues.

External Validity Versus Power

Many of the things you can do to improve your study's power may hurt your study's external validity (generalizability). For example, using a laboratory setting, participants who are all alike, and extreme levels of the independent variable all improve power, but may also reduce external validity.

By using a lab experiment to stop unwanted variables from varying, you may have more power to find an effect. However, by preventing unwanted variables from varying, you may hurt your ability to generalize your results to real life—where those unwanted variables *do* vary.

By using a homogeneous set of participants (18-year-old, White males with IQs between 120 and 125), you reduce between-subject differences, thereby enhancing your ability to find treatment effects. However, because you used such a restricted sample, you would not be as able to generalize your results to the average American as a researcher whose participants were a random sample of Americans would.

Finally, by using extreme levels of the independent variable, you may be able to find a significant effect for your independent variable. If you use extreme levels, though, you may be like the person who used a sledgehammer to determine the effects of hammers—you don't know the effect of realistic, naturally occurring levels of the treatment variable.

Construct Validity Versus Power

Your efforts to improve power may hurt not only your experiment's external validity but also its construct validity. For example, suppose you had two choices for your measure. The first is a 100-point rating scale that is sensitive and reliable. However, the measure is vulnerable to subject bias: If participants guess your hypothesis, they can easily circle the rating they think you want them to. The second is a measure that is not very reliable or sensitive, but it is a measure that participants can't easily fake. If power were your only concern, you would pick the first measure, despite its vulnerability to subject bias. With it, you are more likely to find a statistically significant effect. However, because construct validity should be an important concern, many researchers would suggest that you pick the second measure.

If you sought only statistical power, you might compromise the construct validity of your independent variable manipulation. For instance, to maximize your chances of getting a significant effect for full-spectrum lighting, you would give the experimental group full-spectrum lighting and make the control group an **empty control group**: a group that gets no kind of treatment. Compared to the empty control group, the treatment group

1. receives a gift (the lights) from the experimenter
2. gets more interaction with, and attention from, the experimenter (as the experimenter checks participants to make sure they are using the lights)
3. adopts more of a routine than the controls (using the lights every morning from 6:00 a.m. to 8:00 a.m.)
4. has higher expectations of getting better (because they have more of a sense of being helped) than the controls

As a result of all these differences, you would have a good chance of finding a significant difference between the two groups. Unfortunately, if you find a significant effect, it's hard to say that the effect is due to the full-spectrum lighting and not due to any of these other side effects of your manipulation.[7]

To minimize these side effects of the treatment manipulation, and to increase your construct validity, you might give your control group a **placebo treatment**: a substance or treatment that has no effect. Thus, rather than using a no-light condition, you might expose the control group to light from an ordinary 75-watt incandescent light bulb. You would further reduce the chances of bias if you made both the experimenters and participants **blind (masked)**: unaware of which kind of treatment the participant was getting. If you make the researcher who interacts with the participants blind, that researcher will not bias the results in favor of the experimental hypothesis. Similarly, by making participants blind, you make it less likely that participants will bias the results in favor of the hypothesis.

In short, the use of placebos, the use of **single blinds** (in which either the participant or the experimenter is unaware of which treatment the participant received), and the use of **double blinds** (in which both the participant and the experimenter are blind) all may reduce the chances that you will obtain a significant effect. However, if you use these procedures and still find a significant effect, you can be relatively confident that the treatment variable—rather than some side effect of the treatment manipulation—is causing the effect.

---

[7]The problem of using an empty control group is even more apparent in research on the effect of surgery. For example, if a researcher finds that rats receiving brain surgery run a maze slower than a group of rats not receiving an operation, the researcher should not conclude that the surgery's effect was due to removing a part of the brain that plays a role in maze-running.

You have seen that what is good for power may harm construct validity, and vice versa. But what trade-offs should you make? To make that decision, you might find it helpful to see what trade-offs professional experimenters make between power and construct validity. Do experienced experimenters use empty control groups to get significant effects? Or, do they avoid empty control groups to improve their construct validity? Do they avoid blind procedures to improve power? Or, do they use blind procedures to improve construct validity?

Often, experimenters decide to sacrifice power for construct validity. For example, in their jam experiment, Wilson and Schooler did not have an empty control group. In other words, their control group did not simply sit around doing nothing while the experimental group filled out the questionnaire analyzing how they felt about a jam. Instead, the control group also completed a questionnaire. The questionnaire was a "filler questionnaire" about their reasons for choosing a major. If Wilson and Schooler had used an empty control group, critics could have argued that it was the act of filling out a questionnaire—not the act of reflection—that caused the treatment group to make less accurate ratings than the controls. For example, critics could have argued that the controls' memory for the jams was fresher because they were not distracted by the task of filling out a questionnaire.

To prevent critics from arguing that the experimenters influenced participants' ratings, Wilson and Schooler made the experimenters blind. To implement the blind technique, Wilson and Schooler employed two experimenters. The first experimenter had participants (a) taste the jams and (b) fill out either the control group (filler) questionnaire or the experimental group (reasons) questionnaire. After introducing the participants to Experimenter 2, Experimenter 1 left the room. Then, Experimenter 2—who was unaware of (blind to) whether the participants had filled out the reasons or the filler questionnaire—had participants rate the quality of the jams.

Ethics Versus Power

As you have seen, increasing a study's power may conflict with both external and construct validity. In addition, increasing power may conflict with ethical considerations. For example, suppose you want to use extreme levels of the independent variable (food deprivation) to ensure large differences in the motivation of your animals. In that case, you need to weigh the benefits of having a powerful manipulation against ethical concerns, such as the comfort and health of your subjects (for more about ethical concerns, see Chapter 2 and Appendix D).

Ethical concerns determine not only how you treat the experimental group but also how you treat the control group. Just as it might be unethical to administer a potentially harmful stimulus to your experimental participants, it also might be unethical to withhold a potentially helpful treatment from your control participants. For instance, it might be ethically questionable to withhold a possible cure for depression from your controls. Therefore, rather than maximizing power by completely depriving the control group of a treatment, ethical concerns may dictate that you give the control group a moderate dose of the treatment. (For a summary of the conflicts between power and other goals, see Table 10.4.)

TABLE **10.4** Conflicts Between Power and Other Research Goals

Action to Help Power	How Action Might Harm Other Goals
Use a homogeneous group of participants to reduce random error due to participants.	May hurt your ability to generalize to other groups.
Test participants under controlled laboratory conditions to reduce the effects of extraneous variables.	1. May hurt your ability to generalize to real-life situations where extraneous variables are present. 2. Artificiality *may* hurt construct validity. If the setting is so artificial that participants are constantly aware that what they are doing is not real and just an experiment, they may *act* to please the experimenter rather than expressing their true reactions to the treatment.
Use artificially high or low levels of the independent variables to get big differences between groups.	1. You may be unable to generalize to realistic levels of the independent variable. 2. May be unethical.
Use an empty control group to maximize the chance of getting a significant difference between the groups.	Construct validity is threatened because the significant difference may be due to the participants' expectations rather than to the independent variable.
Test many participants to balance out the effects of random error.	Expensive and time-consuming.

© Cengage Learning 2013

ANALYZING DATA FROM THE SIMPLE EXPERIMENT: BASIC LOGIC

After carefully weighing both statistical and nonstatistical considerations, you should be able to design a simple experiment that would test your experimental hypothesis in an ethical and internally valid manner. If, after consulting with your professor, you conduct that experiment, you will have data to analyze.

To understand how you are going to analyze your data, remember why you did the simple experiment. You did it to find out whether the treatment would have an effect on a unique **population**: all the participants who took part in your experiment. More specifically, you wanted to know the answer to the hypothetical question: "If I had put all my participants in the experimental condition, would they have scored differently than if I had put all of them in the control condition?" To answer this question, you need to know the averages of two populations:

Average of Population #1—what the average score on the dependent measure would have been if all your participants had been in the control group.

Average of Population #2—what the average score on the dependent measure would have been if all your participants had been in the experimental group.

Unfortunately, you cannot measure both of these populations. If you put all your participants in the control condition, you won't know how they would have scored in the experimental condition. If, on the other hand, you put all your participants in the experimental condition, you won't know how they would have scored in the control condition.

Estimating What You Want to Know: You Have Sample Means But You Want to Know the Population Means

You can't directly get the population averages you want, so you do the next best thing—you estimate them. You can estimate them because, thanks to independent random assignment, you split all your participants (your population of participants) into two random samples. That is, you started the experiment with two random samples from your original population of participants. These two "chips off the same block" were the control group and the experimental group (see Figure 10.3).

The average score of the random sample of your participants who received the treatment (the experimental group) is an estimate of what the average score would have been if all your participants received the treatment. The average score of the random sample of participants who received no treatment (the control group) is an estimate of what the average score would have been if all of your participants had been in the control condition.

Calculating Sample Means: Getting Your Estimates

Even though only half your participants were in the experimental group, you will assume that the experimental group is a fair sample of your entire population of participants. Thus, the experimental group's average score

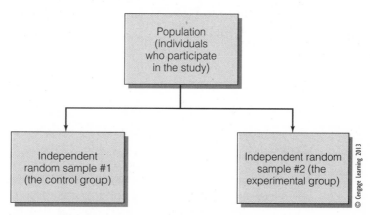

FIGURE **10.3** The control group and the experimental group are two samples drawn from the same population.

Problem: If the average score for the experimental group is different from the average score for the control group, is this difference due to (a) the two groups receiving different treatments or (b) To random error related to sampling? (Two random samples from the same population may differ.)

should be a good estimate of what the average score would have been if all your participants had been in the experimental group. Similarly, you will assume that the control group's average score is a good estimate of what the average score would have been if all your participants had been in the control group. Therefore, the first step in analyzing your data will be to calculate the average score for each group. Usually, the average you will calculate is the **mean**: the result of adding up all the scores and then dividing by the number of scores (e.g., the mean of 3 and 5 is 4 because $3 + 5 = 8$ and $8/2 = 4$).

Comparing Sample Means: How to Compare Two Imperfect Estimates

Once you have your two sample means, you can compare them. Before talking about how to compare them, let's understand why we are comparing the means. We are comparing the sample means because we know that, before we administered the treatment, both groups represented a random sample of the population consisting of every participant who took part in the study. Thus, at the end of the experiment, if the treatment had no effect, the control and experimental groups would both still be random samples from that population.

As you know, two random samples from the same population will probably be similar to each other. For instance, two random samples of the entire population of New York City should be similar to each other, two random samples from the entire population of students at your school should be similar to each other, and two random samples from the entire group of participants who took part in your study should be similar to each other. Consequently, if the treatment has no effect, at the end of the experiment, the experimental and control groups should be similar to each other.

Why We Must Do More Than Subtract the Means From Each Other

Because two random samples from the same population should be similar to each other, you might think all we need to do is subtract the control group mean from the experimental group mean to find the effect. But such is not the case: Even if the treatment has no effect, the means for the control group and experimental group will rarely be identical. To illustrate, suppose that Dr. N. Ept made a serious mistake while trying to do a double-blind study. Specifically, Dr. N. Ept succeeded in not letting his assistants know whether the participants were getting the real treatment or a placebo, but failed in that all the participants got the placebo. In other words, both groups started and ended up being random samples of the same population—participants who did not get the treatment. Even in such a case, the two groups will probably have different means.

How Random Error Affects Data From the Simple Experiment

Dr. N. Ept's study illustrates an important point: Even if groups are random samples of the same population, they may still differ because of random error. You are probably aware of random error from reading about public opinion polls that admit to a certain degree of sampling error.

To help you see how random error could affect the results of a simple experiment, let's simulate conducting a small-scale experiment. Be warned that this simulation won't show us what would typically happen in an experiment. Instead, this simulation is rigged to demonstrate the worst random error can do. Nevertheless, the simulation does demonstrate a fundamental truth: Random error alone can create groups that differ substantially from each other.

To conduct this simulation, assume that you have the following four participants, who would tend to score as follows:

Abby	10
John	20
Mary	70
Paul	40

Now use Box 10.1 to randomly assign each participant to either the experimental or control group. Then, get an average for each group. Repeat this process several times. If you do this, you will simulate what happens when you do an experiment and the treatment has no effect.

As doing this simulation will reveal, which participants end up in which group varies greatly depending on where on the random numbers table you happen to start—and there are many different places you could start. Not all of these possible ways of splitting participants into control and experimental groups are going to produce identical groups. Indeed, you may even find that random assignment sometimes results in having all men in the experimental group and all women in the control group.

In summary, the control and experimental groups start off as random samples of your participants. At the start of the study, these groups are not identical. Instead, they will probably merely be similar. Occasionally, however, they may start off being fairly different. If they start off as different, then they may score differently on the dependent measure task at the end of the experiment—even when the treatment has no effect. Thus, even if the treatment had no effect, random error might make the experimental group score differently (either higher or lower) than the control group.

Because random error can affect the results of a study, you need to understand random error to understand the results of a study. More specifically, to interpret the results of a simple experiment, you need to understand two important statistical principles:

1. Random error affects individual scores.
2. Random error may also cause group means to differ.

Fortunately, as you will soon see, you already intuitively understand both of these principles.

Random Error Makes Scores Within a Group Differ

To see that you intuitively grasp the first principle (that random error affects individual scores), consider the following scores:

Control	Experimental
70	80
70	80
70	80

© Cengage Learning 2013

Is there something strange about these data? Most students we show these data to realize that these data are faked. Students are suspicious of these data because scores within each group do *not* vary: There are no within-groups differences in this experiment. These data make it look like the only thing that affects scores is the treatment. With real data, however, scores would be affected by factors other than the treatment. Consequently, the scores within each group would vary. That is, there would be what statisticians call *within-groups variability*.

When asked to be more specific about why they think the data are faked, students point out that there are at least two reasons why scores within each group should differ. First, participants within each group differ from each other, so their scores would reflect those differences. That is, because participants in the control group aren't all clones of each other, their scores won't all be the same. Likewise, because participants in the experimental group aren't all identical, their scores shouldn't all be identical.

Second, even if a group's participants were all identical, random measurement errors alone would prevent participants from getting identical scores. For instance, even if the control group participants were clones, participants' scores would probably vary due to the measure's less-than-perfect reliability. Similarly, even if all the experimental group participants were identical, their scores would not be identical because many random factors—from random variations in how the experimenter treated each participant to random errors in coding of the data—would cause scores within the experimental group to differ.

In summary, most students have an intuitive understanding that there will be differences within each group (within-groups variability), and these differences are due to factors completely unrelated to the treatment. To be more specific, these differences are due to random error caused by such factors as individual differences, random measurement error, and imperfect standardization.

Random Error Can Make Group Means Differ

To see whether you intuitively grasp the second principle (random error may cause group means to differ from each other), consider the following data:

Control	Experimental
70	70
80	80
70	100

© Cengage Learning 2013

Do you think the experimental group is scoring significantly higher than the control group? Most students wisely say "no." They realize that if the participant who scored "100" had been randomly assigned to the control group rather than the experimental group, the results may have been completely different. Thus, even though the group means differ, the difference may not be due to the treatment. Instead, the difference between these two group means could be entirely due to random error.

As you have just seen, even if the treatment has no effect, random error may cause the experimental group mean to differ from the control group mean. Therefore, we cannot say that there is a treatment effect just because there is a difference between the experimental group's average score and the control group's. Instead, if we are going to find evidence for a treatment effect, we need a difference between our groups that is "too big" to be due to random error alone.

When Is a Difference Between Means Too Big to Be Due to Random Error?

What will help us determine whether the difference between group means is too big to be due to random error alone? In other words, what will help us determine that the treatment had a statistically significant (reliable) effect?

To answer the question of how we determine whether the treatment had a statistically significant effect, we'll look at three sets of experiments. Let's begin with the two experiments tabled below. Which of the following two experiments do you think is more likely to reveal a significant treatment effect?

Experiment A		Experiment B	
Control	Experimental	Control	Experimental
70	70	70	80
71	73	71	81
72	72	72	82

© Cengage Learning 2013

Bigger Differences Are Less Likely to Be Due to Chance Alone

If you picked Experiment B, you're right! All other things being equal, bigger differences are more likely to be "too big to be due to chance alone" than smaller differences. Therefore, bigger differences are more likely to reflect a treatment effect. Smaller differences, on the other hand, provide less evidence of a treatment effect.

To appreciate the fact that small differences provide less evidence of a treatment effect, let's consider an extreme case. Specifically, let's think about the case where the difference between groups is as small as possible: zero. In that case, the control and experimental groups would have identical means. If the treatment group's mean is the same as the no-treatment group's mean, there's no evidence of a treatment effect.

"Too Big to Be Due to Chance" Partly Depends on How Big "Chance" Is

You have seen that the difference between means is one factor that affects whether a result is statistically significant. All other things being equal, bigger differences are more likely to be significant.

The size of the difference isn't the only factor that determines whether a result is too big to be due to chance. To illustrate this fact, compare the two experiments below. Then, ask yourself, is Experiment A or Experiment B more likely to reveal a significant treatment effect? That is, in which experiment is the difference more likely to be too big to be due to chance?

Experiment A		Experiment B	
Control	Experimental	Control	Experimental
68	78	70	70
70	80	80	80
72	82	60	90

© Cengage Learning 2013

Differences Within Groups Tell You How Big Chance Is

In both experiments, the difference between the experimental and control group mean is 10. Therefore, you can't tell which difference is more likely to be too big to be due to chance just by seeing which experiment has a bigger difference between group means. Instead, to make the right choice, you have to figure out the answer to this question: "In which experiment is chance alone a less likely explanation for the 10-point difference?"

To help you answer this question, we'll give you a hint. The key to answering this question correctly is to look at the extent to which scores vary within each group. The more variability within a group, the more random error is influencing scores. All other things being equal, the more random error makes individual scores within a group differ from one another (i.e., the bigger the within-groups variability), the more random error will tend to make group means differ from each other.

Now that you've had a hint, which experiment did you pick as being more likely to be significant? If you picked Experiment A, you're correct!

If you were asked why you picked A instead of B, you might say something like the following: "In Experiment B, the experimental group may be scoring higher than the control group merely because the participant who scored a 90 randomly ended up in the experimental group rather than in the control group. Consequently, in Experiment B, the difference between the groups could easily be due to random error."

Such an explanation is accurate, but too modest. Let's list the four steps of your reasoning:

1. You realized that there was more variability within groups in Experiment B than in Experiment A. That is, in Experiment B relative to Experiment A, (1) control group scores were further from the control group mean, and (2) experimental group scores were further from the experimental group mean.

2. You recognized that within-groups variability could not be due to the treatment. You realized that the differences among participants' scores within the control group could not be due to the treatment because none of those participants received the treatment. You also realized that the differences among scores within the experimental group could not be due to the treatment because every participant in the experimental group received the same treatment. Therefore, when scores within a group vary, these differences must be due to nontreatment factors such as individual differences.

3. You realized that random assignment turned the variability due to nontreatment factors (such as individual differences) into random error. Thus, you realized that the greater within-groups variability in Experiment B meant there was more random error in Experiment B than in Experiment A.

4. You realized that the same random error that caused differences within groups could cause differences between groups. Thus, the more random error is blowing the scores within each group away from each other, the more random error could be blowing two group means from the same population away from each other.

As you have seen, all other things being equal, the *larger* the *differences between* your *group means*, the *more likely* the results are to be *statistically significant*. As you have also seen, the *smaller* the *differences* among scores *within* each of your *groups* (i.e., the less your individual scores are influenced by random error), the *more likely* your results are to be *statistically significant*. Thus, you have learned two of the three factors that determine whether a difference is significant. To find out what the third factor is, compare Experiments A and B below. Which is more likely to produce a significant result?

Experiment A		Experiment B	
Control	Experimental	Control	Experimental
68	70	68	70
70	72	70	72
72	74	72	74
		68	70
		70	72
		72	74
		68	70
		70	72
		72	74

© Cengage Learning 2013

In both experiments, the group means are equally far apart, so you can't look at group differences to figure out which experiment is more likely to be significant. In both experiments, the random variability within each group is the same; therefore, looking at within-groups variability will not help you figure out which experiment is more likely to be significant. Which one do you choose?

With Larger Samples, Random Error Tends to Balance Out

If you chose Experiment B, you're correct! Experiment B is the right choice because it had more participants. In Experiment B, it's less likely that random error alone would cause the groups to differ by much because *with large enough samples, random error tends to balance out to zero.* If you flip a coin 4 times, you will often get either 75% heads or 75% tails. That is, random error alone will probably cause a deviation of 25% or more from the true value of 50% heads. If, on the other hand, you flip a coin 4,000 times, you will almost never get more than 51% heads or fewer than 49% heads. Because 4,000 flips gives random error an opportunity to balance out, random error will almost never cause a deviation of even 1% from the true value.

Just as having more coin flips allows more opportunities for the effects of random error to balance out, having more participants allows more opportunities for random error to balance out. Thus, Experiment B, by having more participants, does a better job than Experiment A at allowing the effects of random error to balance out. Consequently, it's less likely that random error alone would cause Experiment B's groups to differ by a large amount. Therefore, a difference between the control group mean and the treatment group mean that would be big enough to be statistically significant (reliable) in Experiment B might *not* be significant in Experiment A.

ANALYZING THE RESULTS OF THE SIMPLE EXPERIMENT: THE *t* TEST

To determine whether a difference between two group means is significant, researchers often use either ANOVA[8] (analysis of variance, a technique we will discuss in the next chapter) or the ***t* test** (to see how to do a *t* test, you can use the formula in Table 10.5 or consult Appendix E).[9] The basic idea behind the *t* test is to see whether the difference between two groups is larger than would be expected by random error alone. Thus, the *t* test takes

---

[8] The logic of ANOVA is similar to that of the *t* test. Indeed, for a simple experiment, the *p* value for the ANOVA test will be exactly the same as the *p* value from the *t* test. Thus, if the *t* test is statistically significant (*p* is less than .05), the ANOVA test will also be statistically significant (*p* will be less than .05). In addition, for the simple experiment, you can get the value of the ANOVA test statistic (called "F") by squaring your *t* value. Thus, if *t* is 2, *F* will be 4. To learn more about ANOVA, see the next chapter or see Appendix E.

[9] Although *t* test and ANOVA analyses are commonly used, they are criticized. The problem is that both *t* tests and ANOVA tell us only whether a result is statistically significant—and, as we discussed earlier, nonsignificant results don't tell you anything and significant results don't tell you anything about the size of your effect. Therefore, many argue that, rather than using significance tests, researchers should use confidence intervals. For more on the statistical significance controversy, see Box 1 in Appendix E. For more about confidence intervals, see Appendix E.

TABLE **10.5** Basic Idea of the *t* Test

General Idea	Formula
Top of *t* ratio: Obtain observed difference (between two group means)	$$t = \frac{\text{Group 1 Mean} - \text{Group 2 Mean}}{\sqrt{\dfrac{S_1^2}{N_1} + \dfrac{S_2^2}{N_2}}}$$
Bottom of *t* ratio: Estimate difference expected by chance (using the standard error of the difference between means)	where S_1 = standard deviation of Group 1, S_2 = standard deviation of Group 2, N_1 = number of participants in Group 1, and N_2 = number of participants in Group 2. The standard deviation can be calculated by the formula $$S = \sqrt{(\Sigma X - M)^2/(N-1)}$$ where X stands for the individual scores, M is the sample mean, and N is the number of scores.

Notes:

1. A large *t* value is likely to be statistically significant. That is, a large *t* (above 2.6) is likely to result in a *p* value smaller than .05.
2. *t* will tend to be large when
   a. The difference between experimental group mean and the control group mean is large.
   b. The standard error of the difference is small. The standard error of the difference will tend to be small when
      i. The standard deviations of the groups are small (scores in the control group tended to stay close to the control group mean, scores in the experimental group tended to stay close to the experimental group mean).
      ii. The groups are large.

the difference between the group means and divides that difference by an index of the extent to which random error alone might cause the groups to differ. To be more precise, *t* equals the difference between means divided by the standard error of the difference between means (see Table 10.5).

Making Sense of the Results of a *t* Test

Once you have obtained your *t* value, you should calculate the degrees of freedom for that *t*. To calculate degrees of freedom, subtract 2 from the number of participants. Thus, if you had 32 participants, you should have 30 degrees of freedom.

If you calculate *t* by hand, you need to compare your calculated *t* to a value in a *t* table (you could use Table 1 in Appendix F) to determine whether your *t* ratio is significant. To use the *t* table in Appendix F, you need to know how many degrees of freedom *(df)* you have. For example, if you had data from 32 participants, you would look at the *t* table in Appendix F under the row labeled "30 *df*." When comparing the *t* ratio you calculated to the value in the table, act like your *t* ratio is positive even if your *t* value is actually negative (e.g., treat −3 as if it were +3). In other words, take the absolute value of your *t* ratio.

If the absolute value of your t ratio is *not* bigger than the number in the table, your results are *not* statistically significant at the .05 level. If, on the other hand, the absolute value of your t ratio is bigger than the number in the table, your results are statistically significant at the $p < .05$ level.

If you had a computer calculate t for you, make sure that the degrees of freedom *(df)* for t are two fewer than the number of participants. For example, if you thought you entered scores for 32 participants but your computer says $df = 18$, you know there is a problem because the computer is acting as though you entered only 20 scores.

If you had a computer calculate t for you, it might provide you with only the t, the degrees of freedom, and the p value, as in the following case: $df = 8$; $t = 4$; and $p = .0039$

$$df = 8, t = 4, \text{ and } p = .0039$$

From the *df* of 8, you know that the t test was calculated based on scores from 10 participants ($10 - 2 = 8$). From the p value of less than .05, you know the results are statistically significant at the .05 level. That is, you know that if the null hypothesis were true, the chances of your obtaining differences between groups that were as big as or bigger than what you observed would be less than 5 in 100.

Many computer programs will provide you with more information than the df, t, and p values. Some will provide you with what might seem like an overwhelming amount of information, such as the following:

1. $df = 8$, $t = 4$, and Sig. (2-tailed) $= .0039$
2. Mean difference $= 4.00$
3. 95% CI of this difference: 1.69 to 6.31
4. Group 1 mean $= 11.00$; Group 1 $SD = 1.58$; $SEM = 0.71$
5. Group 2 mean $= 7.00$; Group 2 $SD = 1.58$; $SEM = 0.71$

The first line tells you that the t test was calculated based on scores from 10 participants ($10 - 2 = 8$, the *df*) and that the results were statistically significant.

The second line tells you that the Group 1 mean was 4 units bigger than the Group 2 mean. The third line tells you that you can be 95% confident that the true difference between the means is between 1.69 units and 6.31 units. (To learn more about how the confidence interval [CI] was calculated, see Box 10.3.)

The fourth line describes Group 1's data, and the fifth line describes Group 2's data. Both of those lines start by providing the group's average score (the mean) followed by a measure of how spread out the group's scores are: the standard deviation (*SD*). Be concerned if the *SD* of either group is extremely high—a high *SD* may mean that you have entered a wrong value (e.g., when entering responses from a 1-to-5 scale, you once typed a "55" instead of a "5"). Both lines end with their group's standard error of the mean (*SEM*): an indicator of how far off the group's sample mean is likely to be from the actual population mean. If either group's *SEM* is large, your experiment probably has little power, and you probably failed to find a significant effect.

BOX **10.3**   **Beyond Statistical Significance: Obtaining Information About Effect Size**

Your study's *t* value gives you almost everything you need to know to determine whether your results are statistically significant. However, you may also want to know whether your results are practically significant: big enough to care about.

Using *t* to Estimate the Treatment's Average Effect: Confidence Intervals

One way to estimate effect size is to take advantage of information you used when you computed your *t*. Let's start by looking at the top of the *t* ratio: the difference between the mean of the no-treatment group and the mean of the treatment group. The top of the *t* ratio is an estimate of the treatment effect. Thus, if the treatment group scores 2 points higher than the no-treatment group, our best estimate is that the treatment improved scores by 2 points.

Unfortunately, our best estimate is almost certainly wrong: We have almost no confidence that the treatment effect is exactly 2.000. We would be more confident of being right if we said that the treatment effect was somewhere between 1 and 3 points. We would be even more confident of being right if we said that the real effect was somewhere between 0 and 4 points.

Rather than merely saying that we are confident that the real effect is within a certain range, it would be better to be able to say how confident we are that the real effect is within a certain range. For example, we would like to be able to say that we are "95% confident that the effect of the treatment is between 1 and 3 points."

Fortunately, we can specify that we are 95% confident that the real effect is within a certain range. That is, we can compute a 95% confidence interval. All the information we need to compute a 95% confidence interval is the information we had when we determined whether our results were significant. Specifically, we need the difference between the two means (the top of our *t* ratio), the standard error of the difference (the bottom of our *t* ratio), and the critical value of *t* at the .05 level (what we used to see whether the *t* value we calculated was significant). (You can find the critical value of *t* by looking in the *t* table—Table 1 of Appendix F— at the intersection of the ".05" column and the row corresponding to your experiment's degrees of freedom. For example, if you had data from 42 participants, the value would be 2.021.)

The first step to building on our confidence interval is to look at the difference between the treatment group mean and the no-treatment group mean. To find that difference, we can just look at the top of the *t* ratio. In this example, that difference is 2. We realize that our observed difference is our best single guess of the treatment's effect. Furthermore, we realize that the treatment's real effect is just as likely to be higher than that guess as it is to be lower than that guess. Consequently, the observed difference will be the middle point of our confidence interval.

To get our confidence interval's upper value, we will add a number to that middle point. Specifically, we will add the number we get by multiplying the standard error of the difference (the bottom of our *t* value) by the critical value of *t*. For example, if the standard error of the difference was 1, and the critical value of *t* was 2.021, we would add 2.021 (the standard error of the difference [1] × the critical value of *t* [2.021] = 2.021). Thus, in this case, we would add 2.021 to 2 (the observed difference between means) to obtain the top of our confidence interval: 4.021.

To get the bottom of our confidence interval, we reverse the process. We will again start with 2 (the difference between our means). This time, however, we will subtract, rather than add, 2.021 (the product of multiplying the standard error by the critical *t* value) from 2. Therefore, the lower value of our interval would be −0.021 [2 − (1 × 2.021) = 2 − 2.021 = −0.021].

As the result of our calculations, we could say that we were 95% confident that the true effect was in the interval ranging from −0.021 to 4.021. By examining this interval, we can form two conclusions. First, we cannot confidently say that the treatment effect has any effect because 0 (zero effect, no effect) was within our interval. Second, we see that because our confidence interval is large, our study lacks power and precision. Therefore, we may want to repeat the study in a way that shrinks the confidence

(Continued)

BOX **10.3** (Continued)

interval (e.g., using more participants, using more reliable measures, or using more standardized procedures) so that we can more precisely estimate the treatment's effect.

To see the value of increasing the power of the study, imagine that we repeated the study using 62 participants instead of 42. If such a study again found a 2-point difference between our groups, we could probably be 95% confident that the true effect was between .35 and 3.6.[1] Because this interval would be narrower than the original interval (which went from −0.021 to +4.021), we would have a better idea of what the true effect is. Because this interval would not include zero, we could confidently say that the treatment did have some effect.

Note another lesson from this example: Even though the first study's results were not statistically significant (because we could not say that the treatment effect was significantly different from zero) and the second study's results were significant (because we could say that the treatment effect was significantly different from zero), the two studies do not contradict each other. The difference in the results is that the second study, by virtue of its greater power, allows you to make a better case that the treatment effect is greater than zero.

Using t to Compute Other Measures of Effect Size: Cohen's d and r^2

In the previous section, you learned how to provide a range that, 95 times out of 100, will contain the treatment's average effect. However, even if you knew precisely what the average effect of the treatment was, you would not know all you should know about the treatment's effect size. For example, suppose you know that the average effect was 2. Is 2 a small effect? If your participants' scores range from 0 to 100, a difference between your control group and experimental group of 2 units might be a relatively small effect. If, on the other hand, scores within your groups vary by less than 1 unit, a treatment effect of 2 units would be a relatively large effect. Therefore, to know the relative size of an effect, you need an effect size measure that takes into account the variability of the scores.

One popular effect size measure is **Cohen's d**, which tells you how big your effect size is using a standard unit: the standard deviation. Thus, whereas your measured difference between means would be 1000X larger if you measured in milliseconds rather than in seconds, Cohen's d would be the same. For example, if one mean was 2 standard deviations above the other, Cohen's d would be 2 regardless of whether you measured in milliseconds, seconds, or hours.

A Cohen's d of 2 would be very large. Usually, social scientists view a d of 0.2 as indicating a small effect, a d of 0.5 as indicating a medium effect, and a d of 0.8 as indicating a large effect.

If you had the same number of participants in each group, you can calculate Cohen's d from your t value by using the following formula: Cohen's $d = 2t/\sqrt{df}$ Thus, if t is 3 and df is 9, Cohen's d will be $(2 \times 3)/\sqrt{9} = 6/3 = 2$.

Instead of talking about effect size in terms of the average difference between your two groups, we could look at the correlation between your treatment variable and your outcome variable. Often, researchers measure the relationship between your treatment variable and your dependent variable by squaring the correlation (r) between the treatment and the dependent variable. The result will be a measure, called the **coefficient of determination**, that can range from 0 (no relationship) to 1.00 (perfect relationship). Usually, social scientists view a coefficient of determination of .01 as small, of .09 as moderate, and of .25 as large (for more about the coefficient of determination, see Chapter 7).

If you have computed d, you can compute the coefficient of determination (r^2) by using the following formula: $r^2 = d^2/(d^2 + 4)$. To see the relationships among the mean difference, Cohen's d, and the coefficient of determination, see Table 10.6.

---

[1]When we calculated this confidence interval, we assumed that the standard deviations (an index of the extent to which participants' scores differ from the mean, in which a 0 would mean that nobody's score differed from the mean) within each of your groups would be the same as they were in the original study. If your procedures were more standardized when you repeated the study, the standard deviations might be smaller and so your intervals might be even smaller than what we projected.

TABLE 10.6 Relationships Among Different Effect Size Measures

	Information from the *t* Test			Effect Size Measures *t*	
t	Degrees of Freedom	Mean Difference (example with low variability in scores)	Mean Difference (example with moderate variability in scores)	*d*	r^2 (also called h^2)
2	9	2	4.7	1.33	.31
2	16	1.4	3.7	1.0	.20
2	25	1.2	3.0	0.8	.14
2	36	1.0	2.5	0.67	.10
2	49	0.8	2.2	0.57	.08
2	64	0.7	1.9	0.50	.06
2	81	0.7	1.7	0.44	.05
2	100	0.6	1.5	0.40	.04

© Cengage Learning 2013

Suppose that your experiment is powerful enough to find that your effect is statistically significant at the $p < .05$ level. In that case, because there's less than a 5% chance that the difference between your groups is solely due to chance, you can be reasonably sure that some of the difference is due to your treatment.

To learn about the size of your treatment's effect, you might want to use Box 10.3 to compute an index of effect size such as Cohen's *d*. For example, suppose your computer analysis presented the following results:

1. $df = 30$, $t = 3.10$, and $p = .004$
2. Mean difference = 3.46
3. 95% CI of this difference: 1.57 to 5.35; *SED* = 1.12
4. Group 1 mean = 8.12; Group 1 *SD* = 3.0; *SEM* = 0.75
5. Group 2 mean = 4.66; Group 2 *SD* = 3.32; *SEM* = 0.83

Using that data and Box 10.3, you would be able to determine that Cohen's *d* was 1.13.

Then, you could write up your results as follows:[10] "As predicted, the experimental group recalled significantly more words ($M = 8.12$, $SD = 3.0$) than the control group ($M = 4.66$, $SD = 3.32$), $t(30) = 3.10$, $p = .004$, $d = 1.13$."

You could include even more information: APA strongly encourages researchers to supplement significance tests with means, standard deviations, confidence intervals, effect size estimates, and confidence intervals for effect

---

[10] *M* stands for mean, *SD* stands for standard deviation (a measure of the variability of the scores; the bigger the SD, the more spread out the scores are and the less the scores cluster around the mean), and *d* stands for Cohen's *d* (a measure of effect size). *SD* will usually be calculated as part of computing *t* (for more about *SD*, see Appendix E). To learn how to compute *d*, see Box 10.3.

sizes. However, at the very least, you should say something like this: "As predicted, the experimental group recalled significantly more words ($M = 8.12$) than the control group ($M = 4.66$), $t(30) = 3.10$, $p = .004$."

In conclusion, you must do more than report that your results are statistically significant. Indeed, largely because some researchers have focused only on whether their results are statistically significant, a few researchers have suggested that statistical significance testing be banned (for more on the statistical significance controversy, see Box 1 in Appendix E). Although only a few people agree that statistical significance testing should be banned, almost everyone agrees that researchers need to do more than report p values.

Assumptions of the t Test

The validity of any p values you obtain from any significance test will depend on how well you meet the assumptions of that statistical test. Faulty assumptions will lead to faulty conclusions. For the t test, two of these assumptions are especially important: (1) having at least interval scale data and (2) having independent observations.

Two Critical Assumptions

When determining whether one group's mean score is significantly larger than the other's, the t test assumes that groups with higher means have more of the quality you are measuring than groups with lower means. Because only interval and ratio scale data allow you to compute such "meaningful means," you must be able to assume that you have either interval scale or ratio scale data (for a review of interval and ratio scale data, see Chapter 6).

Because you cannot compute meaningful means on either qualitative data or ranked data, you cannot do a t test on those data. You cannot compute meaningful means on qualitative (nominal, categorical) data because scores relate to categories rather than amounts. With qualitative (nominal) data, 1 might equal "nodded head," 2 might equal "gazed intently," and 3 might equal "blinked eyes." With such nominal data, computing a mean (e.g., the mean response was 1.8) would be meaningless.

With ranked and other ordinal data, the numbers have an order, but they still don't refer to specific amounts and so means can be meaningless and misleading. For example, although averaging the ranks of second- and third-place finishers in a race would result in the same mean rank (2.5) as averaging the ranks of the first- and fourth-place finishers, the average times of the first- and fourth-place finishers could be much faster or much slower than the average of the times of the second- and third-place finishers.

Although having either nominal or ordinal data prevents you from comparing group means with a t test, you can still compare two groups using tests, such as the Mann-Whitney U test (for ordinal data) and the chi-square test (for either nominal or ordinal data), that do not involve comparing means. (For more on these tests, see Appendix E.)

The second assumption you must meet to perform a legitimate t test is that your observations must be independent. Specifically, (a) participants must be assigned independently (e.g., individually, so that the assignment of Mary to the experimental group has no effect on whether John is assigned to the experimental group); (b) participants must respond independently (e.g., no participant's response influences any other participant's response); and (c) participants must be tested independently so that, other than the treatment, there is no systematic difference between how experimental and control group participants are treated.

If you followed our advice and independently and randomly assigned each participant to either the experimental or the control conditions, and then ran participants individually (or in small groups or in larger groups that mixed experimental and control participants), your observations are independent. If, however, your observations are not independent, you cannot legitimately do a conventional independent groups *t* test. Indeed, violating independence often means that the data from your study are unanalyzable and thus worthless.

To reiterate, to do a meaningful independent *t* test in a simple experiment, your data must meet two key assumptions: You must have at least interval-scale data, and you must have used independently assigned participants to groups. In addition to these two pivotal assumptions, the *t* test makes two less vital assumptions (see Table 10.7).

Two Less Critical Assumptions

First, the *t* test assumes that the individual scores in the population from which your sample means were drawn are **normally distributed**: Half the scores are below the average score; half are above; the average score is the most common score; about 2/3 of the scores are within one standard deviation of the mean; about 19/20 of the scores are within two standard deviations of the mean; and if you were to plot how often each score occurred, your plot would resemble a bell-shaped curve.

The reason for this assumption is that if the individual scores in the population are normally distributed, the distribution of sample means based

---

TABLE **10.7** Effects of Violating the *t* Test's Assumptions

Assumption	Consequences of Violating Assumption
Observations are independent (participants are independently assigned and participants do not influence one another's responses).	Serious violation; probably nothing can be done to salvage your study.
Data are interval or ratio scale (e.g., numbers must not represent qualitative categories, nor may they represent ranks [first, second, third, etc.]).	Do not use a *t* test. However, you may be able to use another statistical test (e.g., Mann-Whitney U, Chi-square).
The population from which your sample means was drawn is normally distributed.	If the study used more than 30 participants per group, this is not a serious problem. If, however, fewer participants were used, you may decide to use a different statistical test.
Scores in both conditions have the same variance.	Usually not a serious problem.

on those scores will also tend to be normally distributed.[11] The assumption that individual scores are normally distributed is usually nothing to worry about because most distributions are normally distributed.

But what if the individual scores aren't normally distributed? Even then, your sample means probably will be normally distributed—provided you have more than 30 participants per group. That is, as the **central limit theorem** states, with large enough samples (and 30 per group is usually large enough), the distribution of sample means will be normally distributed, regardless of how individual scores are distributed.

To understand why the central limit theorem works, realize that if you take numerous large samples from the same population, your sample means will differ from one another for only one reason: random error. Because random error is normally distributed, the distributions of sample means will be normally distributed—regardless of the shape of the underlying population.

The *t* test's second less critical assumption is that the variability of scores within your experimental group will be about the same as the variability of scores within your control group. To be more precise, the assumption is that scores in both conditions will have the same variance.[12] Usually, the penalty for violating the assumption of equal variances is not severe. Specifically, if you have unequal variances, it won't seriously affect the results of your *t* test, as long as one variance isn't more than 2½ times larger than the other.

■ QUESTIONS RAISED BY RESULTS

Obviously, if you violate key assumptions of the *t* test, people should question your results. But even if you don't violate any of the *t* test's assumptions, your results will raise questions—and this is true whether or not your results are statistically significant.

---

[11] Why do we have to assume that the distribution of sample means is normally distributed? We need to know precisely how the sample means are distributed to establish how likely it is that the two sample means could differ by as much as they did by chance alone. In other words, if we are wrong about how the sample means are distributed, our *p* value—our estimate of the probability of the sample means differing by as much as they did if their population means were the same—would be wrong.

[12] To get the variance for a group, square that group's standard deviation (*SD*). If you used a computer to get your *t*, the computer program probably displayed each group's *SD*. If you calculated the *t* by hand, you probably calculated each group's *SD* as part of those calculations. Some computer programs will do a statistical test such as Levene's Test for Equality of Variance to tell you how reasonable it is to assume that the groups have the same variance. If the *p* value for the Levene's Test for Equality of Variance is statistically significant, it means that the variances are probably different: It does *not* mean that the treatment has an effect. If the variances are significantly different, instead of a conventional *t* test, you may want to do Welch's test instead. Some programs will also calculate two *t* values for you: one assuming equal variances, one not making that assumption.

Questions Raised by Nonsignificant Results

Nonsignificant results raise questions because the null hypothesis cannot be proven. Therefore, null results inspire questions about the experiment's power such as the following:

1. Did you have enough participants?
2. Were the participants homogeneous enough?
3. Was the experiment sufficiently standardized?
4. Were the data coded carefully?
5. Was the dependent variable sensitive and reliable enough?
6. Would you have found an effect if you had chosen two different levels of the independent variable?

Questions Raised by Significant Results

If your results are statistically significant, it means you found an effect for your treatment. So, there's no need to question your study's power. However, a significant effect raises other questions. Sometimes, questions are raised because statistical significance doesn't tell us how big the effect is (see Box 10.3).

Sometimes, questions are raised because the experimenter sacrificed construct or external validity to obtain adequate power. For example, if you used an empty control group, you have questionable construct validity. Consequently, one question would be: "Does your significant treatment effect represent an effect for the construct you tried to manipulate or would a placebo treatment have had the same effect?" Or, if you used an extremely homogeneous group of participants, the external validity of your study might be questioned. For instance, skeptics might ask: "Do your results apply to other kinds of participants?" Thus, skeptics might want you to increase the external validity of your study by repeating it with a more representative sample. Specifically, they might want you to first use random sampling to obtain a representative group of participants and then randomly assign those participants to either the control or experimental group.

At other times, questions are raised because of a serious limitation of the simple experiment: It can study only two levels of a single independent variable. Because of this, there are two important questions you can ask of any simple experiment:

1. To what extent do the results apply to levels of the independent variable that were not tested?
2. To what extent could the presence of other variables modify (strengthen, weaken, or reverse) the treatment's effect?

■ CONCLUDING REMARKS

As you have seen, the results of a simple experiment always raise questions. Although results from any research study raise questions, some questions raised by the results of the simple experiment occur because the simple experiment is limited to studying only two levels of a single variable. If the logic of the simple experiment could be used to create designs that would study several levels of several independent variables, such designs could answer several questions at once. Fortunately, as you will see in Chapters 11 and 12, the logic of the simple experiment can be extended to produce experimental designs that will allow you to answer several research questions with a single experiment.

SUMMARY

1. Psychologists want to know the causes of behavior so that they can understand people and help people change. Only experimental methods allow us to isolate the causes of an effect.

2. Studies that don't manipulate a treatment are not experiments.

3. Many variables, such as participant's age, participant's gender, and participant's personality, can't be manipulated. Therefore, many variables can't be studied using an experiment. Put another way, variables that can't be randomly assigned (but instead are pre-assigned or self-assigned) can't be studied using an experiment.

4. The simple experiment is the easiest way to establish that a treatment causes an effect. That is, simple experiments have internal validity.

5. The experimental hypothesis states that the treatment will cause an effect.

6. The null hypothesis, on the other hand, states that the treatment will not cause an effect.

7. With the null hypothesis, you only have two options: You can reject it, or you can fail to reject it. You can never accept the null hypothesis.

8. Typically, in the simple experiment, you administer a low level of the independent (treatment) variable to some of your participants (the comparison or control group) and a higher level of the independent variable to the rest of your participants (the experimental group). Near the end of the experimental session, you observe how each participant scores on the dependent variable: a measure of the participant's behavior.

9. To establish causality with a simple experiment, participants' responses must be independent. Because of the need for independence, your experimental and control groups are not really groups. Instead, these "groups" are sets of individuals who may have never met and may have only one thing is common: Individuals belonging to the same "group" received the same treatment.

10. Random assignment is different from arbitrary assignment or haphazard assignment. In the simple experiment, it involves using a systematic process to ensure that every participant has a 50% chance of being assigned to the experimental group and a 50% chance of being assigned to the control group.

11. Independent random assignment is the cornerstone of the simple experiment: Without it, you don't have internal validity; without that, you do not have an experiment.

12. Independent random assignment has three aspects: (a) assignment—individuals are assigned to receive a certain treatment, (b) randomness—the assignment system gives each individual an equal chance of being assigned to any of the groups, and (c) independence—assignment is individual, so that one individual's assignment to a group has no effect on any other person's assignment.

13. Independent random assignment is necessary because it is the only way to make sure that the only differences between your groups can be due to only two things: chance and the treatment.

14. Independent random assignment makes it likely that your control group is a fair comparison group. Therefore, if you use random assignment, the control and experimental groups will probably be equivalent before you introduce the treatment.

15. Because participants' scores must be independent, control groups and experimental groups are not really groups. For example, control group participants typically do not see, meet, or interact with other control group participants. Likewise, experimental group participants typically do not know of each other's existence.

16. Your goal in using independent random assignment is to create two samples that accurately represent your entire population of participants. You use the mean of the control group as an estimate of what would have happened if all your participants had been in the control group. You use the experimental group mean as an estimate of what the mean would have been if all your participants had been in the experimental group. You use inferential statistics to determine whether the difference you observed between these samples reflects a real difference between the two populations.

17. The *t* test tries to answer the question, "Does the treatment have an effect?" In other words, would participants have scored differently had they all been in the experimental group than if they had all been in the control group?

18. If the results of the *t* test are statistically significant, the difference between your groups is greater than would be expected by chance (random error) alone. Therefore, you reject the null hypothesis and conclude that your treatment has an effect. Note, however, that statistical significance does not tell you that your results are big, important, or of any practical significance.

19. There are two kinds of errors you might make when attempting to decide whether a result is statistically significant. Type 1 errors occur when you mistake a chance difference for a treatment effect. Before the study starts, you choose your "false alarm" risk (risk of making a Type 1 error). Most researchers decide to take a 5% risk. Type 2 errors occur when you fail to realize that the difference between your groups is due to more than just chance. In a sense, Type 2 errors involve overlooking a genuine treatment effect.

20. By reducing your risk of making a Type 1 error, you increase your risk of making a Type 2 error. That is, by reducing your chances of falsely "crying wolf" when there is no treatment effect, you increase your chances of failing to yell "wolf" when there really is a treatment effect.

21. Because Type 2 errors can easily occur, nonsignificant results are inconclusive results.

22. To prevent Type 2 errors, (a) reduce random error, (b) use many participants to balance out the effects of random error, and (c) try to increase the size of your treatment effect.

23. You can decide how much of a risk of making a Type 1 error you will take, but there's no way you can design your experiment to reduce that risk. In contrast, it is hard to determine your risk of making a Type 2 error, but there are many ways you can design your study to reduce your risk of making such errors.

24. If your experiment minimizes the risk of making Type 2 errors, your experiment has power. (Power is rarely having to say that you made a Type 2 error.) In the simple experiment, *power* refers to the ability to obtain statistically significant results when your independent variable really does have an effect.

25. Sometimes, efforts to improve power may hurt the study's external validity. For example, to get power, researchers may use a highly controlled lab setting rather than a real-life setting. Likewise, power-hungry researchers may study participants who are very similar to each other rather than a wide range of participants.

26. Occasionally, efforts to improve power may hurt the study's construct validity.

27. Using placebo treatments, single blinds, and double blinds can improve your study's construct validity.

28. Ethical concerns may temper your search for power—or even cause you to decide not to conduct your experiment.

29. Because of random error, you cannot determine whether your treatment had

an effect simply by subtracting your experimental group mean from your control group mean. Instead, you must determine whether the difference between your group means could be due to random error.

30. The *t* test is a common way to analyze data from a simple experiment. However, you should not use the *t* test if your data do not meet the assumptions of that test.

31. The *t* test involves dividing the difference between means by an estimate of the degree to which the groups would differ when the treatment had no effect. More specifically, the formula for the *t* test is: (Mean 1 – Mean 2)/ standard error of the difference.

32. The degrees of freedom for a two-group between-subjects *t* test are 2 less than the total number of participants.

KEY TERMS

blind (masked) *(p. 390)*
central limit theorem *(p. 408)*
coefficient of determination *(p. 404)*
Cohen's *d (p. 404)*
control group *(p. 371)*
dependent variable (dependent measure) *(p. 376)*
double blinds *(p. 390)*
empty control group *(p. 390)*
experimental group *(p. 371)*
experimental hypothesis *(p. 366)*

independent random assignment *(p. 365)*
independent variable *(p. 371)*
independently, independence *(p. 373)*
inferential statistics *(p. 377)*
internal validity *(p. 363)*
levels of the independent variable *(p. 371)*
mean *(p. 394)*
normally distributed, normally distribution *(p. 407)*

null hypothesis *(p. 369)*
null results (nonsignificant results) *(p. 379)*
p < .05 level *(p. 377)*
placebo treatment *(p. 390)*
populations *(p. 392)*
power *(p. 384)*
simple experiment *(p. 377)*
single blinds *(p. 390)*
statistically significant *(p. 377)*
t test *(p. 400)*
Type 1 error *(p. 381)*
Type 2 error *(p. 383)*

EXERCISES

1. A professor has a class of 40 students. Half of the students chose to take a test after every chapter (chapter test condition) outside of class. The other half of the students chose to take in-class "unit tests." Unit tests covered four chapters. The professor finds no statistically significant differences between the groups on their scores on a comprehensive final exam. The professor then concludes that type of testing does not affect performance.
   a. Is this an experiment?
   b. Is the professor's conclusion reasonable? Why or why not?

2. Participants are randomly assigned to meditation or no-meditation condition. The meditation group meditates three times a week; the no-meditation group is an empty control group. The meditation group reports being significantly more energetic than the nomeditation group.
   a. Why might the results of this experiment be less clear-cut than they appear?
   b. How would you improve this experiment?

3. Theresa fails to find a significant difference between her control group and her experimental group—*t*(10) = 2.11, which is not significant.
   a. Given that her results are not significant, what—if anything—would you advise her to conclude?
   b. What would you advise her to do? (Hint: You know that her *t* test, based on 10 degrees of freedom, was not significant. What does the fact that she has 10 degrees of freedom tell you about her study's sample size, and what does that sample size suggest about her study's power?)

4. A training program significantly improves worker performance. What should you know before advising a company to invest in such a training program?

5. Jerry's control group is the football team; his experimental group is the baseball team. He assigned the groups to condition using random assignment. Is there a problem with Jerry's experiment? If so, what is it? Why is it a problem?

6. Students were randomly assigned to two different strategies of studying for an exam. One group used visual imagery, the other group was told to study the normal way. The visual imagery group scores 88% on the test as compared to 76% for the control group. This difference was not significant.

   a. What, if anything, can the experimenter conclude?

   b. If the difference had been significant, what would you have concluded?

   c. "To be sure that they are studying the way they should, why don't you have the imagery people form one study group and have the control group form another study group?" Is this good advice? Why or why not?

   d. "Just get a sample of students who typically use imagery and compare them to a sample of students who don't use imagery. That will do the same thing as random assignment." Is this good advice? Why or why not?

   e. "Setting up the room for the different conditions takes time. Let's save some time by running all the experimental participants in the morning and all the control participants in the afternoon. We will still have independence because we will randomly assign participants to a condition and because we will test participants individually." Why is this bad advice?

7. Bob and Judy are doing the same study, except that Bob has decided to put his risk of a Type 1 error at .05 whereas Judy has put her risk of a Type 1 error at .01.

   a. If Judy has 22 participants in her study, what t value would she need to get significant results? (Hint: Review main point 32, then consult Table 1 in Appendix F.)

   b. If Bob has 22 participants in his study, what t value would he need to get significant results? (Hint: Review main point 32, then consult Table 1 in Appendix F.)

   c. Who is more likely to make a Type 1 error? Why?

   d. Who is more likely to make a Type 2 error? Why?

8. Gerald randomly assigned participants to receive their test on either yellow or blue paper. Gerald's dependent measure is the order in which people turned in their exam (first, second, third, etc.). Can Gerald use a t test on his data? Why or why not? What would you advise Gerald to do in future studies?

9. Are the results of Experiment A or Experiment B more likely to be significant? Why?

Experiment A	
Control Group	**Experimental Group**
3	4
4	5
5	6

Experiment B	
Control Group	**Experimental Group**
0	0
4	5
8	10

10. Are the results of Experiment A or Experiment B more likely to be significant? Why?

Experiment A		Experiment B	
Control Group	**Experimental Group**	**Control Group**	**Experimental Group**
3	4	3	4
4	5	4	5
5	6	5	6
		3	4
		4	5
		5	6
		3	4
		4	5
		5	6

© Cengage Learning 2013

WEB RESOURCES

1. Go to the Chapter 10 section of the book's student website and
   a. Look over the concept map of the key terms.
   b. Test yourself on the key terms.
   c. Take the Chapter 10 Practice Quiz.
   d. Do the interactive end-of-chapter exercises.
2. Do a *t* test using a statistical calculator by going to the "Statistical Calculator" link.
3. Find out how to conduct a field experiment by reading "Web Appendix: Field Experiments."
4. If you want to write your method section, use the "Tips on Writing a Method Section" link.
5. If you want to write up the results of a simple experiment, click on the "Tips for Writing Results" link.

CHAPTER 11

Expanding the Simple Experiment:
The Multiple-Group Experiment

Perhaps too much of everything is as bad as too little.
—**Edna Ferber**

Scientific principles and laws do not lie on the surface of nature. They are hidden, and must be wrested from nature by an active and elaborate technique of inquiry.
—**John Dewey**

CHAPTER OVERVIEW

We devoted Chapter 10 to the simple experiment: the design that involves *randomly assigning* participants to two groups. The simple experiment is internally valid and easy to conduct. However, it is limited in that you can study only two values of an independent variable.

In this chapter, you will see why you might want to do an experiment that allows you to study more than two values of an independent variable. Then, you will see how to randomly assign participants to three or more groups so that you can do such an experiment. Finally, you will learn how to analyze data from such a multiple-group experiment.

THE ADVANTAGES OF USING MORE THAN TWO VALUES OF AN INDEPENDENT VARIABLE

The simple experiment is ideal if an investigator wants to compare a single treatment group to a single no-treatment control group. However, as you will see, investigators often want to do more than compare two groups.

Comparing More Than Two Kinds of Treatments

We do not live in a world where there are only two flavors of ice cream, only two types of music, and only two opinions on how to solve any particular problem. Because people often choose between more than two options, investigators often compare more than two different kinds of treatments.

For instance, to decide how police should respond to a domestic dispute, investigators compared three different strategies: (1) arrest a member of the couple, (2) send one member away for a cooling off period, and (3) give advice and mediate the dispute (Sherman & Berk, 1984). Clearly, investigators could not compare three different treatments in one simple, two-group experiment. Therefore, instead of randomly assigning participants to two different groups, they randomly assigned participants to three different groups. (To learn how to randomly assign participants to more than two groups, see Box 11.1.)

BOX **11.1**   Randomly Assigning Participants to More Than Two Groups

Step 1 Across the top of a piece of paper write down your conditions. Under each condition draw a line for each participant you will need.

Group 1	Group 2	Group 3
——	——	——
——	——	——
——	——	——
——	——	——

Step 2 Turn to a random numbers table (there's one in Table 6, Appendix F). Roll a die to determine which column in the table you will use.

Step 3 Assign the first number in the column to the first space under Group 1, the second number to the second space, and so forth. When you have filled the spaces for Group 1, put the next number under the first space under Group 2. Similarly, when you fill all the spaces under Group 2, place the next number in the first space under Group 3.

Group 1	Group 2	Group 3
12	20	63
39	2	64
53	37	95
29	1	18

Step 4 Replace the lowest random number with "Participant 1," the second lowest random number with "Participant 2," and so on. Thus, in this example, your first two participants would be in Group 2, and your third participant would be in Group 1.

© Cengage Learning 2013

In another case of attacking an applied problem, Cialdini (2005) saw something we all see—a well-intentioned, written request to do something good—and wondered what most of us have wondered: Would wording the request differently make it more effective? Specifically, he questioned the effectiveness of hotel room signs that urge guests to conserve water by reusing towels. Typically, those signs state that reusing towels will preserve the environment and help the hotel donate money to an environmental cause. Cialdini believed he could design more effective signs by using scientifically established principles of persuasion—and he could think of at least two psychological principles that he could use.

First, he could use an approach that applied the principle that people tend to do what they believe others do. Thus, he created a sign stating that 75% of guests reuse their towels.

Second, he could use an approach that applied the principle that people tend to repay a favor. Thus, he created a sign stating that the hotel had already donated money to protect the environment on behalf of the hotel guests and wanted to recover that expense.

To test his two solutions against conventional practice, Cialdini needed at least three groups: (1) a group that got the conventional hotel sign, (2) a "most other people are doing it" group, and (3) a "repay a favor" group. As Cialdini suspected, both the "repay a favor" and the "others are doing it" groups reused their towels much more than the group that saw the sign hotels typically used.

Clearly, Cialdini could not compare three groups in a single, two-group experiment. Thus, he used a multiple-group experiment. Similarly, Nairne, Thompson, and Pandeirada (2007) hypothesized that people are best able to remember information when they rate its relevance to their survival. To see whether the survival rating task was the best rating task for helping participants recall information, Nairne et al. used a multiple-group experiment to compare their rating task to other rating tasks that help memory (e.g., rating how pleasurable the word is, rating how personally relevant the word is). In short, if, like Cialdini or Nairne and his colleagues, you want to compare more than two treatments, you should use a multiple-group experiment.

Comparing Two Kinds of Treatments With No Treatment

Even when you are interested in comparing only two types of treatments, you may be better off using a multiple-group experiment. To understand why, let's consider the following research finding: For certain kinds of back problems, people going to a chiropractor end up better off than those going for back surgery. Although an interesting finding, it leaves many questions unanswered. For example, is either treatment better than nothing? We don't know because the researchers didn't compare either treatment to a no-treatment control condition. It could be that both treatments are worse than nothing and chiropractic treatment is merely the lesser of two evils. On the other hand, both treatments could be substantially better than no treatment and chiropractic could be the greater of two goods.

Similarly, if we compared two untested psychological treatments in a simple experiment, we would know only which is better than the other: We would not know whether the better one was the less harmful of two "bad" treatments or the more effective of two "good" treatments. Plus, we would not know whether the lesser of the two treatments was (1) moderately harmful, (2) neither harmful nor helpful, or (3) mildly helpful. However, by using a three-group experiment that has a no-treatment control group, we would be able to judge not only how effective the two treatments were relative to each other but also their overall, general effectiveness. Consider the following examples of how adding a no-treatment control group helps us know what effect the treatments had.

• In a classic experiment, Loftus (1975) found that leading questions distorted participants' memories of a filmed car accident. All participants watched a film of a car accident, completed a test booklet that contained questions about the film, and a week later, answered some more questions about the film. But participants were not treated identically because not all participants got the same test booklet. Instead, each participant was randomly assigned to receive one of the following three test booklets:

1. The "presume" booklet contained 40 questions asked of all participants, plus 5 additional questions that asked whether certain

objects—objects that were *not* in the film—were seen in the film. These 5 additional questions were leading questions: questions suggesting that the object was shown in the film (e.g., "Did you see *the* school bus in the film?").

2. The "mention but don't presume" booklet contained 40 questions asked of all participants, plus 5 additional questions that asked whether certain objects—objects that were *not* in the film—were seen in the film. This booklet was the same as the "presume" booklet except that the 5 additional questions did *not* suggest that the item was shown in the film (e.g., "Did you see *a* school bus in the film?").

3. The control booklet contained only the 40 questions asked of all participants.

   Note that without a control group, Loftus would not have known whether the difference between the nonleading question and leading question group was due to (a) the nonleading question condition sharpening memory or (b) the leading question condition distorting memory.

- Crusco and Wetzel (1984) looked at the effects of having servers touch restaurant customers on the tips that servers received. Had they only compared hand-touching with shoulder-touching, they would not have known whether touching had an effect. Thanks to the no-touch control group, they learned that both kinds of touching increase tipping.

- Anderson, Carnagey, and Eubanks (2003) looked at the effects of violent lyrics on aggressive thoughts. Had they used only nonviolent and violent songs, they would not have known whether nonviolent songs reduced aggressive thoughts or whether violent songs increased aggressive thoughts. Thanks to the no-song control condition, they learned that violent lyrics increased aggressive thoughts.

- Strayer and Drews (2008) looked at the effects of cell phones on driving. Had they only compared the driving performance of drivers who use hand-held cell phones to drivers who use hands-free cell phones, they would not have found an effect for cell phones. However, thanks to a no cell phone control group, they learned that cell phone use impairs driving.

Comparing More Than Two Amounts of an Independent Variable to Increase External Validity

In the simple experiment, you are limited to two amounts of your independent variable. However, we do not live in a world where variables come in only two amounts. If we did, other people would be either friendly or unfriendly, attractive or unattractive, like us or unlike us, and we would be either rewarded or punished, included or excluded, and in complete control or have no control. Instead, we live in a world where situations vary not so much in terms of whether a quality (e.g., noise) is present but rather the degree to which that quality is present.

Not only that, but we live in a world where more is not always better. Sometimes, too little of some factor can be bad, too much can be bad, but (to paraphrase the littlest of the three bears) a medium amount is just right. In such cases, a simple, two-valued experiment can lead us astray.

Why You Need More Than Two Amounts of an Independent Variable to Know the Functional Relationship Between Your Variables

To see how simple experiments can be misleading, suppose that a low amount of exercise leads to a poor mood, a moderate amount of exercise leads to a good mood, and a high amount of exercise leads to a poor mood. Such an upside-down "U"-shaped relationship is plotted in Figure 11.1a. As you can see, if we did a multiple-group experiment, we would uncover the true relationship between exercise and mood. However, if we did a simple experiment, our findings might be misleading. For example, if we did the simple experiment depicted in

- Figure 11.1b, we might conclude that exercise *increases* mood
- Figure 11.1c, we might conclude that exercise *decreases* mood
- Figure 11.1d, we might conclude that exercise does not affect mood

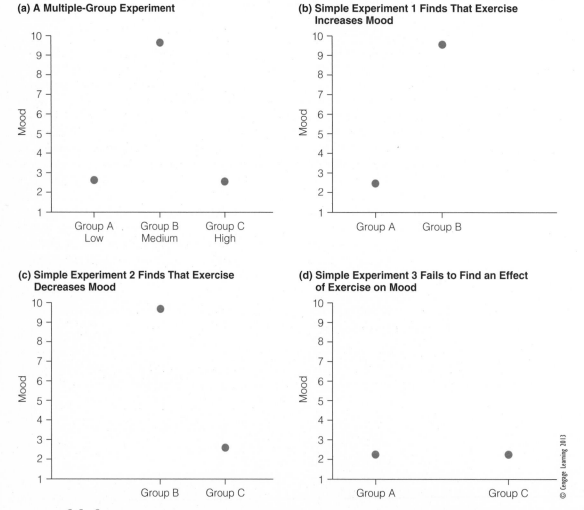

FIGURE **11.1** How a multiple-group experiment can give you a more accurate picture of a relationship than a simple experiment.

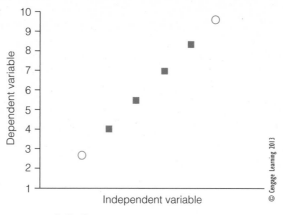

FIGURE **11.2** Linear relationship between two points.

As you have just seen, if a researcher is to make accurate statements about the effects of different levels of an independent variable, the researcher must know the shape of the independent and dependent variables' **functional relationship**: how scores on the outcome variable change as a function of changing the amount of the treatment.

If you are going to map the shape of a functional relationship accurately, you need more than the two data points that a simple experiment provides. To appreciate that knowing only two points does not allow you to know the *shape* of the functional relationship between two variables, consider Figure 11.2. From the two known data points (the empty circles), can you tell the shape of the variable's functional relationship?

No, you can't. Perhaps the relationship is a **linear relationship**: one that is represented by a straight *line*. A straight line does fit your two points. However, maybe your relationship is *not* linear: As you can see from Figure 11.3, many other curved lines also fit your two points.

Because lines of many different shapes can be drawn between the two points representing a simple experiment's two group means, the simple experiment does not help you discover the functional relationship between the variables. Thus, if your simple experiment indicated that 100 minutes of exercise produced a better mood than 0 minutes of exercise, you would still be clueless about the functional relationship between exercise and mood. Therefore, if we asked you about the effects of 70 minutes of exercise on mood, you could do little more than guess. If you assumed that the exercise–mood relationship is linear, you would guess that exercising 70 minutes a day would be (a) better than no exercise and (b) worse than exercising 100 minutes a day. But if your assumption of a linear relationship is wrong (and it well could be), your guess would be wrong.

To get a line on the functional relationship between variables, you need to know more than two points. Therefore, suppose you expanded the simple experiment into a multilevel experiment by adding a group that gets 50 minutes of exercise a day. Then, you would have a much clearer idea of

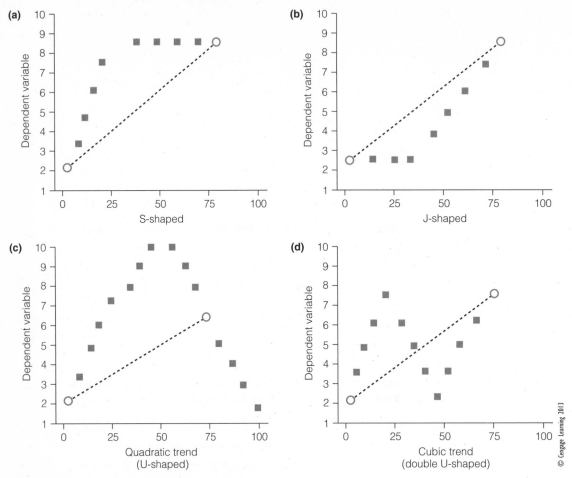

FIGURE **11.3** Some possible nonlinear relationships.

Note: The circles represent the known data points. Each dotted line represents the straight line that the experimenter would draw between the two data points. The boxes between the circles are what might happen at a given level of the independent variable, depending on whether the relationship between the variables is characterized by (a) an S-shaped (negatively accelerated) trend, (b) a J-shaped (positively accelerated) trend, (c) a U-shaped (quadratic) trend, or (d) a double U-shaped (cubic) trend. Note that in all cases, the simple experimenter would not see the curve.

the functional relationship between exercise and happiness. As you can see in Figure 11.4 on page 423, using three levels can help you identify the functional relationship among variables. If the relationship is linear, you should be able to draw a straight line through your three points. If the relationship is U-shaped, you'll be able to draw a "U" through your three points.

Because you can get a good picture of the functional relationship when you use three levels of the independent variable, you can make accurate predictions about unexplored levels of the independent variable. For example, if

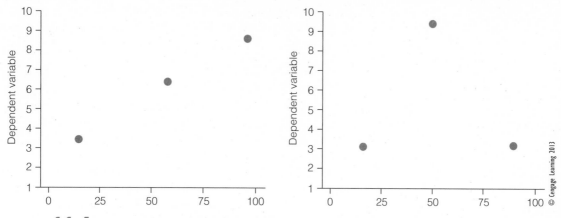

FIGURE **11.4** Having three levels of the independent variable (three data points) helps you identify the shape of the functional relationship.

With these three points, we can be relatively confident that the relationship is linear (fits a straight line). Most nonlinear relationships (see Figure 11.3) would not produce data that would fit these three data points.

If we had these three data points, we could be relatively confident that the relationship is curvilinear (fits a curved line). Specifically, we would suspect that we had a quadratic relationship: a relationship shaped like a "U" or an upside-down "U."

the functional relationship between exercise and happiness was linear, you might obtain the following pattern of results:

Group 1	0 minutes of exercise per day	0.0 self-rating of happiness
Group 2	50 minutes of exercise per day	5.0 self-rating of happiness
Group 3	100 minutes of exercise per day	10.0 self-rating of happiness

In that case, you could confidently predict that 70 minutes of exercise would be less beneficial for increasing happiness than 100 minutes of exercise.

If, on the other hand, the relationship was S-shaped (as in Figure 11.3), you might get the following pattern of results:

Group 1	0 minutes of exercise per day	0.0 self-rating of happiness
Group 2	50 minutes of exercise per day	10.0 self-rating of happiness
Group 3	100 minutes of exercise per day	10.0 self-rating of happiness

In that case, you would predict that a person who exercised 70 minutes would do as well as someone exercising 100 minutes a day.

Why You May Not Need More Than Four Amounts of an Independent Variable to Know the Functional Relationship Between Your Variables

The more groups you use, the more accurately you can pin down the shape of the functional relationship. Yet, despite this fact, you do not need to use numerous levels of the independent variable. Why? Because nature prefers simple patterns. That is, most functional relationships are linear (straight lines), and few are more complex than U-shaped functions. Consequently, you will rarely need more than four levels of the independent variable to pin down a functional relationship. In fact, you will usually need no more than three carefully chosen levels.

Review: Why We Want to Know the Functional Relationship Between the Independent and Dependent Variable

In summary, knowing the functional relationship between two variables is almost as important as knowing that a relationship exists. If you want to give practical advice, you should be able to say more than: "If you exercise 100 minutes a day, you will be happier than someone who exercises 0 minutes a day." Who exercises exactly 100 minutes a day? You want to be able to generalize your results so that you can tell people the effects of exercising 50 minutes, 56 minutes, 75 minutes, and so forth. Yet, you do not want to test the effects of every possible amount of exercise a person might do. Instead, you want to test only a handful of exercise levels. If you choose these levels carefully, you will be able to map the functional relationship between the variables. Mapping the functional relationship, in turn, will allow you to make educated predictions about the effects of treatment levels that you have not directly tested.

When applying psychology, you need to know the functional relationship so you can know how much of a therapy or other treatment to administer. How much is too little? At what point is additional treatment not worth it? How much is too much? If you know the answers to these questions, not only do you avoid wasting your time and your client's time on unnecessary treatments, but you also free up time and resources to help a client who needs it (Tashiro & Mortensen, 2006).

Examples of Multiple Group Experiments That Uncovered Functional Relationships

When mapping functional relationships, psychologists often manipulate independent variables that have names starting with "number of," such as number of others, number of milligrams of a drug, or number of seconds of exposure to a stimulus. You may be inspired by studies like the following classics.

- In Asch's (1955) line-judging experiments, he led participants to believe that they were part of a group that was participating in a visual perception study. The participant's job was to pick the line on the right that matched the line on the left. In reality, the experiment was a social influence experiment, and the other members of the group were really confederates (assistants of the experimenter). Asch wanted to know whether the size of group would affect people's conformity to the group. He found that as group size went from two to five, participants were more likely to conform. However, he found that increasing the group size beyond seven actually decreased the chances that participants would go along with the group.
- In Latané, Williams, and Harkins's (1979) social loafing experiments, investigators wanted to know how loafing would change as a function

of group size. They found that adding two members to a group increased loafing, but that adding two members increased loafing more in smaller groups than in larger groups.

- In Milgram, Bickman, and Berkowitz's (1969) conformity experiment, confederates looked up at the sixth floor window of an office building. Because the researchers were interested in the effects of group size on conformity, the researchers had 1, 2, 3, 5, 10, or 15 confederates look up at the office building. Then, they counted the number of people passing by who also looked up. They found that the bigger their initial group, the stronger the group's influence.
- In Darley and Latané's (1968) study of helping behavior, participants thought they were talking via intercom to either one, two, or five other participants (actually, the participant was alone—the voices came from a tape-recording) when one of them had a seizure. They found that the *more* people participants thought were in the group, the *less* likely participants were to help.
- In Middlemist, Knowles, and Matter's (1976) urinal study, researchers found that the closer a confederate was standing to a participant, the longer it took for the participant to begin urinating.
- In Ambady and Rosenthal's (1993) "thin slices" experiments, participants watched—with the sound off—three video clips of a professor. The clips varied in length: One group saw 2-second clips, a second group saw 5-second clips, and a third group saw 10-second clips. The researchers found that participants in all three groups gave the professor the same ratings as students who sat in the professor's class all term gave that professor.
- In Basson, McInnes, Smith, Hodgson, and Koppiker's (2002) study of the effect of Viagra on women, the 10-mg, 50-mg, and 100-mg doses of Viagra were not more effective in increasing sexual response than a placebo.

How To Get More Than Two Amounts of the Treatment So You Can Uncover Functional Relationships

Although it is easy to map functional relationships when the name of your independent variable starts with "number of," realize that you can map functional relationships between most variables because—with a little work—most variables can be quantified.

If you can manipulate a variable between two extreme levels (e.g., low and high), you can probably also manipulate it in between those extremes (e.g., medium). To illustrate, consider a variable that is not obviously quantitative: similarity. Byrne (1961) manipulated similarity from 0% to 100% by (a) making participants believe they were seeing another student's responses to an attitude survey and then (b) varying the proportion of responses that matched the participant's attitudes from 0% to 100%.

If your independent variable involves exposing participants to a stimulus, you can usually quantify your manipulation. Specifically, you can usually (a) produce several variations of the stimulus, (b) have volunteers rate each of those variations on a 0 to 10 scale, and then (c) use, as your stimuli, the variations that have the scale values that you want. For example, suppose you wanted to manipulate physical attractiveness by showing participants photos of people who varied in attractiveness. You could take a photo of an attractive person, then (a) get some less attractive photos of the person by using a computerized photo editing program to mess with the original picture, (b) have some volunteers rate the attractiveness of each photo on a 0-to-10 scale, and (c) use, as your three stimuli, the photos that were consistently rated 4, 6, and 8, respectively. Realize that this scaling strategy is not just for pictures: If your manipulation was "severity of a crime," "legitimacy of an excuse," or almost anything else, you could still use this scaling strategy.

Even without scaling your independent variable, you may still be able to order the levels of your manipulation from least to most and then see whether more of the manipulation creates more of an effect. For example, in one study (Risen & Gilovich, 2007) all participants were to imagine the following scenario:

> You bought a lottery ticket from a student who was organizing a lottery. However, on the day of the drawing, you left your money at home and so had no money to buy lunch. To buy lunch, you sold your ticket back to the student who sold it to you.

Then, one third of the participants were told to imagine that the lottery ticket was eventually purchased by their best friend, one third were told to imagine that the ticket was purchased by a stranger, and one third were told to imagine that the ticket was purchased by their "ex-friend and least favorite person at school" (p. 16). As Risen and Gilovich predicted, the *less* the participants liked the person who would eventually own their original ticket, the *more* likely participants thought that ticket would be the winning ticket. Conversely, Young, Nussbaum, and Monin (2007) showed that if a disease could be spread by sexual contact, people were reluctant to have themselves tested for it, and this reluctance was unaffected by whether sexual contagion was an unlikely (e.g., only 5% of the cases were due to sexual contact) or likely (e.g., 90% of the cases were due to sexual contact) cause of the disease.

Using Multiple Groups to Improve Construct Validity

You have seen that multilevel experiments—because their results can generalize to a wider range of treatment levels—can have more external validity than simple experiments. In this section, you will learn that multilevel experiments can also have more construct validity than simple experiments.

Confounding Variables Often Hurt the Simple Experiment's Construct Validity

In Chapter 10, you saw that, thanks to random assignment, simple experiments are able to rule out the effects of variables that have nothing to do with the treatment manipulation. For example, because participant variables such as gender, race, and personality are randomly assigned, the treatment group should not have significantly more women than men. Thus, it is unlikely that what a researcher doing a simple experiment thinks is a treatment effect is really a gender effect.

Because simple experiments have internal validity, a statistically significant difference between the control group and the experimental group at the end of the experiment probably means that the treatment manipulation had an effect. That is, the groups behaved differently because the researcher treated them differently.

Impurities in a Manipulation May Introduce Confounding Variables That Hurt The Simple Experiment's Construct Validity.

But in what way(s) did the researcher treat the groups differently? In other words, we have the following construct validity question: What variable(s) did the treatment manipulate?

If a manipulation manipulates more than one variable, its construct validity is questionable. For example, if an "exercise manipulation" also manipulates social support, the construct validity of that exercise manipulation is questionable. In such a case, the simple experiment might lead us to conclude that exercise had an effect when social support (a side effect of the treatment manipulation) was responsible for the effect.

Ideally, your treatment would be a pure manipulation that creates one—and only one—systematic difference between the experimental group and the control group. Thus, ideally, a simple experiment would always have good construct validity. In reality, however, it is rare to have a perfect manipulation. Instead, the treatment manipulation usually produces several differences between how the experimental and control groups are treated.

For example, suppose that a simple experiment apparently found that the "attractive" defendant was more likely to get a light sentence than the "unattractive" defendant. We would know that the "attractiveness" manipulation had an effect. However, it could be that in addition to manipulating attractiveness, the researchers also manipulated perceived wealth. Thus, wealth, rather than attractiveness, might account for the manipulation's effect. Specifically, people may be less likely to give wealthy defendants long sentences.

Because of impurities in manipulations, researchers using simple experiments often end up knowing that the treatment manipulation had an effect, but not knowing whether the treatment had an effect because it manipulated (a) the variable they wanted to manipulate or (b) some other variable that they did not want to manipulate. In short, simple experiments may lack construct validity because the independent variable manipulation is contaminated by variables that are unintentionally manipulated along with the treatment. In technical terminology, the manipulation's construct validity is weakened by **confounding variables**: variables, other than the independent variable, that may be responsible for the differences between your conditions.

The following example[1] illustrates the general problem of confounded manipulations. Imagine being in a classroom that has five light switches, and you want to know what the middle light switch does. Assume that in the "control" condition, all the light switches are off. In the "experimental" condition, you want to flick the middle switch. However, because it is dark, you accidentally flick on the middle three switches. As the lights come on,

---

[1] We are indebted to an anonymous reviewer for this example and other advice about confounding variables.

the janitor bursts into the room, and your "experiment" is finished. What can you conclude?

You can conclude that your manipulation of the light switches had an effect. That is, your study has internal validity. But, because you manipulated more than just the middle light switch, you can't say that you know what the middle light switch did. Put another way, if you were to call your manipulation a "manipulation of the middle switch," your manipulation would lack construct validity.

Because of confounding variables, it is often hard to know what it is about the treatment that caused the effect. In real life, variables are often confounded. For example, your friend may know she got a hangover from drinking too much wine, but not know whether it was the alcohol in the wine, the preservatives in the wine, or something else about the wine that produced the awful sensations. A few years ago, a couple of our students joked that they could easily test the hypothesis that alcohol was responsible. All they needed us to do was donate enough money to buy mass quantities of a pure manipulation of alcohol—180 proof, totally devoid of impurities. These students understood how confounding variables can contaminate real-life manipulations—and how confounding variables can make it hard to know what it was about the manipulation that caused the effect.

Multiple Groups May Help You Have Fewer Differences Between Your Groups and Thereby Isolate the Source of a Treatment's Effect. Having a multiple-group experiment can allow you to know the source of a treatment's effect. For example, if you wanted to look at the effects of cell phones on driving behavior, you could have a no cell phone group, a cell phone group, and a cell phone with headset group. By comparing the regular cell phone group to the headset group, you might be able to see whether reaching for the phone was a source of the cell phone users' driving problems (Strayer & Drews, 2008). To see how having more than two groups has helped researchers track down the source of a treatment's effect, consider the following examples.

- Gesn and Ickes (1999) found that participants who saw a video of another person did a passable job at knowing what that person was thinking. But why? Was it the person's words—or was it their nonverbal signals? To find out, Gesn and Ickes compared one group that heard only the words (audio only) to another group that got only the nonverbal signals. (The nonverbal group saw video of the person accompanied by a filtered sound track that allowed participants to hear the pitch, loudness, and rhythm of the person's speech, but not the actual words.) Gesn and Ickes found that the words, rather than nonverbal signals, were what helped participants figure out what the person was thinking. Specifically, whereas the audio-only group did nearly as well as the normal video group, the video with filtered audio group did very poorly.
- Langer, Blank, and Chanowitz (1978) had their assistant get into lines to use the copier and then ask one of three questions:
  1. Can I cut in front of you?
  2. Can I cut in front of you because I'm in a rush?
  3. Can I cut in front of you because I want to make a copy?

The researchers found that 60% of participants in the no-excuse condition let the assistants cut in front, 94% of the participants in the good-excuse condition let the assistants cut in, and 93% of the participants in the poor-excuse condition let the assistants cut in front of them. By having both a no-excuse control group and a good-excuse control group, the researchers were able to establish that it was (a) important to have an excuse but (b) the quality of the excuse was unimportant.

- In the false memory study we discussed earlier, Loftus (1975) included a control group who, like the experimental group, was asked questions about objects that weren't in the film, but who, unlike the experimental group, were not asked questions that implied that those objects were in the film (e.g., the control group might be asked "Did you see a red stop sign?" whereas the experimental group would be asked, "Did you see the red stop sign?"). The fact that this control group did not have false memories allowed Loftus to discover that the false memories in the leading question condition were caused by suggesting that the object was present—and not by the mere mention of the false object.

- Lee, Frederick, and Ariely (2006) found that people told that they were about to drink some beer that had vinegar added to it rated the beer more negatively than participants not told about the vinegar. One possibility for this finding is that participants merely obeyed demand characteristics: Participants might expect that the experimenter wanted them to give low ratings to vinegar-tainted beer. Fortunately, Lee et al. were able to rule out this possibility because they had a control group that was told about the vinegar *after* tasting the beer—and that "after" group rated the beer as positively as the group that didn't know about the vinegar. Consequently, the researchers were able to conclude that knowing about the vinegar *beforehand* changed how the beer tasted to participants.

- Baumeister, DeWall, Ciarocco, and Twenge (2005) found that participants believing they would spend the future alone exhibited less self-control than participants believing they would spend the future with friends. However, this finding could mean either that social rejection leads to less self-control or that expecting unpleasant outcomes leads to less self-control. Therefore, Baumeister et al. added a control group of participants who were led to expect an unpleasant, injury-riddled future. That "misfortune" group did not show a low level of self-control, suggesting that it was rejection, not negative events, that caused the lowered self-control.

To understand how confounding variables can contaminate a simple experiment, let's go back to the simple experiment on the effects of exercise that we proposed earlier in this chapter. You will recall that the experimental group got 100 minutes of exercise class per day, whereas the control group got nothing. Clearly, the experimental group participants were treated differently from the control group participants. The groups didn't differ merely in terms of the independent variable (exercise). They also differed in terms of several other (confounding) variables: The exercise group received more attention and had more structured social activities than the control group.

Hypothesis-Guessing Often Hurts the Construct Validity of Simple Experiments That Use Empty Control Groups.

Furthermore, participants in the experimental group knew they were getting a treatment, whereas participants in the control group knew they were not receiving any special treatment. If experimental group participants suspected that the exercise program should have an effect, the exercise program may appear to have an effect—even if exercise does not really improve mood. In other words, the construct validity of the study might be ruined because the experimental group participants engaged in **hypothesis-guessing**: they *guessed* and played along with the *hypothesis*.

Because of the impurities (confounding variables) of this exercise manipulation, you cannot say that the difference between groups is due to exercise by itself. Although all manipulations have impurities, this study's most obvious—and avoidable—impurities stem from having an **empty control group**: a group that gets no treatment, not even a placebo (a placebo is a treatment that doesn't have an effect, other than possibly by changing a participants' expectations). Thus, if you chose to use a placebo control group instead of the empty control group, you could reduce the impact of confounding variables.

Increasing Construct Validity Through Multiple Control Groups

Choosing the placebo control group over the empty control group does, however, often come at a cost. Often, it would be better to have both control groups.

To see how hard it can be to choose between an empty control group and a placebo group, consider the studies comparing the effect of antidepressant drugs to the effect of a placebo. If those simple experiments had compared groups getting antidepressants to groups getting no treatment, those studies, because they would have given the drugs credit for any placebo effects, would have grossly overestimated the effectiveness of antidepressant drugs (Kirsch, Moore, Scoboria, & Nicholls, 2002). However, because those studies did not use empty control groups, they don't tell us the difference between getting the drug and receiving no treatment. Given that patients will be often choosing between drug treatment and no treatment (Moerman, 2002), the lack of an empty control group is a problem. It would have been nice to have compared the antidepressant group to both an empty control group as well as to a placebo group.

The Value of a Placebo Group.

To take another example of the difficulty of choosing between a placebo group and an empty control group, let's go back to the problem of examining the effects of exercise on mood. If you use an empty control group that has nothing done to its participants, interpreting your results may be difficult. More specifically, if the exercise group does better than this "left alone" group, the results could be due to hypothesis-guessing (e.g., participants in the exercise condition figuring out that exercise should boost their mood) or to any number of confounding variables (such as socializing with other students in the class, being put into a structured routine, etc.).

If, on the other hand, you use a placebo-treatment group (for example, meditation classes), you would control for *some* confounding variables. For example, both your treatment and placebo groups would be assigned to a structured routine. Now, however, your problem is that you only know how the treatment compares to the placebo: You do not know how it compares to no treatment. Consequently, you won't know what the treatment's effect is.

The Value of an Empty Control Group: "Placebos" May Not Be Placebos. You won't know what the effect of your treatment is because you do not know what the effect of your placebo treatment is. Ideally, you would like to believe that your placebo treatment has no effect. In that case, if the treatment group does worse than the placebo group, the treatment is harmful; if the treatment group does better, the treatment is helpful.

If, however, what you hope is a purely placebo treatment turns out to be a treatment that really does have an effect, you are going to have trouble evaluating the effect of your treatment. For example, suppose you find that the exercise group is more depressed than the meditation group. Could you conclude that exercise increases depression? No, because it might be that although exercise reduces depression, meditation reduces it more. Conversely, if you found that the exercise group is less depressed than the meditation group, you could not automatically conclude that exercise decreases depression. It may be that meditation increases depression greatly, and exercise increases depression only moderately: Exercise may merely be the lesser of two evils.

To find out whether exercise increases or decreases depression, you need to compare the exercise group to a no-treatment group. Thus, if you were interested in the effects of exercise on depression, you have two options: (1) Use a simple experiment and make the hard choice between an empty control group and a placebo group, or (2) use a multiple-group experiment so that you can include both an empty control group and a placebo control group.

Using Multiple Imperfect Control Groups to Compensate for Not Having the Perfect Control Group. Even if you are sure you do not want to use an empty control group, you may still need more than one control group because you will probably not have the perfect control group. Instead, you may have several groups, each of which controls for some confounding variables but not for others. If you were to do a simple experiment, you may have to decide which of several control groups to use. Choosing one control group—when you realize you need more than one—is frustrating. It would be better to be able to use as many as you need.

But how often do you need more than one control group? More often than you might think. In fact, even professional psychologists sometimes underestimate the need for control groups. Indeed, many professional researchers get their research articles rejected because a reviewer concluded that they failed to include enough good control groups (Fiske & Fogg, 1990).

You often need more than one control group so that your study will have adequate construct validity. Even with a poor control group, your study has internal validity: You know that the treatment group scored differently than the control group. But what is it about the treatment that is causing the effect? Without good control group(s), you may think that one aspect of your treatment (the exercise) is causing the effect, when the difference is really due to some other aspect of your treatment (the socializing that occurs during exercise).

To illustrate how even a good control group may still differ from the experimental group in several ways having nothing to do with the independent variable, consider the meditation control group. The meditation control group has several advantages over the empty control group. For example, if the exercise group was less depressed than a meditation control group, we could be confident that this difference was not due to hypothesis-guessing, engaging in structured activities, or being distracted from worrisome thoughts for awhile. Both groups received a "treatment," both engaged in structured activities, and both were distracted for the same length of time.

The groups, however, may differ in that the exercise group did a more social type of activity, listened to louder and more upbeat music, and interacted with a more energetic and enthusiastic instructor. Therefore, the exercise group may be in a better mood for at least three reasons having nothing to do with exercise: (1) the social interaction with their exercise partners, (2) the upbeat music, and (3) the upbeat instructor.

To rule out all these possibilities, you might use several control groups. For instance, to control for the "social activity" and the "energetic model" explanations, you might add a group that went to a no-credit acting class taught by an enthusiastic professor. To control for the music explanation, you might add a control group that listened to music or perhaps even watched aerobic dance videos. By using all of these control groups, you may be able to rule out the effects of confounding variables.

ANALYZING DATA FROM MULTIPLE-GROUP EXPERIMENTS

You have just learned that multiple control groups, by controlling for multiple impurities in your manipulation, may give you more construct validity than one control group. Earlier, you learned that multiple treatment groups, by letting you use low, medium, and high amounts of the treatment, may give you more external validity than an experiment that lets you compare only two different treatment amounts. Before that, you learned that, in a world in which we are usually not limited to two treatment options, the multiple-group experiment allows you to compare more of those options than a two-group experiment does. In short, you have learned at least three good reasons to conduct a multiple-group experiment instead of a simple experiment:

1. to improve construct validity,
2. to improve external validity, and
3. to compare several treatments at once.

However, before you conduct a multiple-group experiment, you should understand how it will be analyzed because the way that it will be analyzed has implications for (a) what treatment groups you should use, (b) how many participants you should have, and even (c) what your hypothesis should be.

Even if you never conduct a multiple-group experiment, you will read articles that report results of such experiments. To understand those articles, you must understand the logic and vocabulary used in analyzing them. Therefore, we will devote the next sections to helping you understand how to understand how multiple-group experiments are analyzed.

Analyzing Results From the Multiple-Group Experiment: An Intuitive Overview

As a first step to understanding how to analyze the results of multiple-group experiments, let's look at data from three experiments that compared the effects of no-treatment, meditation, and aerobic exercise on happiness. All of these experiments had 12 participants rate their feelings of happiness on a 0-to-100 (*not at all happy* to *very happy*) scale. Here are the results of Experiment A:

	No-treatment	Meditation	Exercise
	50	51	53
	51	53	53
	52	52	54
	51	52	52
Group Means	51	52	53

© Cengage Learning 2013

Compare these results to the results of Experiment B:

	No-treatment	Meditation	Exercise
	40	60	78
	42	60	82
	38	58	80
	40	62	80
Group Means	40	60	80

© Cengage Learning 2013

Suppose that either Experiment A or Experiment B found that varying the treatment had a reliable effect. Which experiment—A or B—do you think found that the treatment made a difference? If you say B, why do you give B as your answer? You answer B because there is a *bigger difference between the groups* in Experiment B than in Experiment A. That is, the group means for Experiment B are further apart than the group means for Experiment A. Group B's means being further apart—what statisticians call greater **variability between group means**—lead you to think that Experiment B is more likely to be the study that obtained significant results.

Why do you think the experiment with greater variability between means is more likely to be statistically significant? You probably have two reasons.

First, you intuitively realize that to find a treatment effect, you need between-group variability. That is, to argue that the treatment makes a difference, you must show that groups getting different treatments have different group means. After all, if the no-treatment, meditation, and exercise groups all had the same means, you couldn't argue that the treatment made a difference.

Second, you intuitively realize that a small difference between group means might easily be due to chance (rather than to the treatment), but that a larger difference is less likely to be due entirely to chance.[2] In other words, you realize that the more variability there is between group means, the more likely it is that at least some of that variability is due to treatment.

Now, compare Experiment B with Experiment C. Here are the results of Experiment C:

	No-treatment	Meditation	Exercise
	10	10	100
	80	90	80
	60	60	60
	10	80	80
Group Means	40	60	80

© Cengage Learning 2013

Which experiment do you think provides stronger evidence of a treatment effect—Experiment B or Experiment C? Both experiments have the same amount of variability between group means. Therefore, unlike in our first example, you cannot choose using the rule of choosing the experiment with the means that differ the most. Yet, once again, you will pick Experiment B. Why?

You will pick Experiment B because you are concerned about one aspect of Experiment C: scores of individuals belonging to the same group being very different. In technical terminology, you are concerned about the extreme amount of variability within each group.

You realize the only reason scores within a group vary is random error. (If participants in the same treatment group get different scores, those different scores can't be due to the treatment. Instead, the differences in scores must be due to nontreatment variables, such as individual differences. In a randomized experiment, nontreatment variables produce random error.) Thus, you see that Experiment C is more affected by random error than Experiment B.

The large amount of random error in Experiment C (as revealed by the *within-groups* variability) bothers you because you realize that this random error—rather than the treatment—might be the reason the groups differ from

---

[2] Similarly, if your favorite team lost by one point, you might blame luck. However, if your team lost by 30 points, you would be less likely to say that bad luck alone was responsible for the defeat.

one another. That is, the same random error that makes individual scores within a group differ from each other might also make the group means differ from each other.[3] For example, if the participant who scored 100 had been assigned to the no-treatment group instead of to the meditation group, the results of the study might be quite different.

In Experiment B, on the other hand, you don't have to worry much about random error because the small amount of within-group variability indicates that there is virtually no random error in that experiment's data. Therefore, in Experiment B, you feel fairly confident that random error is *not* causing the group means to differ from one another. Instead, you believe that the means differ from one another because of the treatment.

Intuitively then, you understand the three most important principles behind analyzing the results of a multiple-group experiment. Specifically, you realize the following:

1. *Within-groups variability*—differences in scores within a group—is not due to the treatment, but instead is due to *random error*. For example, differences between no-treatment group participants can't be due to treatment because none of those participants are getting the treatment. Similarly, differences within a treatment group can't be due to the treatment because everyone in the group is getting the same treatment. Thus, differences among group members must be due to random factors such as individual differences and random measurement error.

2. *Between-groups variability*—differences between group means—is not entirely due to the treatment. Admittedly, if the treatment has an effect, the means of groups getting different levels of treatment should differ from one another. However, even if the treatment has no effect, random error will probably make the group means differ from one another. Thus, between-group variability is affected by both *random error and treatment effects*.

3. If you compare between-group variability (which is affected by *both* random error and the treatment) to within-group variability (which is affected *only* by random error), you may be able to determine whether the treatment had an effect.

Analyzing Results From the Multiple-Group Experiment: A Closer Look

You now have a general idea of how to analyze data from a multiple-group study. To better understand the logic and vocabulary used in these

---

[3] To get a sense of how random sampling error might cause the group means to differ, randomly sample two scores from the no-treatment group (scores are in the table on page 434). Compute the mean of this group. If you do this several times, you will get different means. These different means can't be due to a treatment effect because none of the participants in any of your samples are receiving the treatment. The reason you are getting different means even though you are sampling the same group is random sampling error. Fortunately, statistics can help us determine how likely it is that the differences between group means are entirely due to random error.

analyses—a must if you are to understand an author's or a computer's report of such an analysis—read the next few sections.

Within-Groups Variability: A Pure Measure of Error

As you already know, within-groups variability does not reflect the effects of treatment. Instead, it reflects the effects of random error. For example, because all the participants in the meditation group are getting the same treatment (meditation), any differences among those participants' scores can't be due to their getting different treatments. Instead, the differences among scores of meditation group participants are due to such random factors as individual differences, unreliability of the measure, and lack of standardization. Similarly, differences among the scores of participants in the no-treatment group are due not to treatment, but to irrelevant random factors. The same is true for differences within the exercise group. Thus, calculating within-groups variability will tell us the extent to which chance causes individual scores to differ from each other.

To measure this within-groups variability, we first look at the variability of the scores within each group. To be more specific, we calculate an index of variability called the variance. If we have three groups, we could calculate the variance within each group. Each of these three within-group variances would be an independent estimate of how much random error is affecting the study. However, we do not need three different estimates of random error—we just need one good one. To end up with one estimate of variability due to random error, we average all three within-group variances to come up with the best estimate of random variability—the **within-groups variance**.

The within-groups variance gives us a good idea of the average extent to which random error causes individual scores within a group to differ from each other. However, what we want to know is the extent to which random error is likely to cause group means to differ from each other.

Fortunately, we can use the within-groups variance to estimate the extent to which random error is likely to cause group means to differ from each other. Partly because within-groups variance gives us an index of the degree to which random *error* alone may cause your group means to differ, within-groups variance is often referred to as **error variance**.

Between-Groups Variability: Differences Between Group Means

Why You Need to Calculate the Between-Groups Variance. Once you have an index of the degree to which your groups could vary from each other due to chance alone (the within-groups variance), the next step is to get an index of the degree to which your groups actually vary from one another. It is at this step where it becomes clear that you cannot use a *t* test to analyze data from a multiple-group experiment. When using a *t* test, you determine the degree to which the groups differ from one another in a straightforward manner: You subtract the average score of Group 1 from the average score of Group 2.

Subtraction works well when you want to know the difference between two groups, but it does not work well when you want to know the difference between more than two groups because you can subtract only two means at a time. Thus, if you want to use subtraction and have three groups, which two groups do you compare? Group 1 with Group 2? Or, Group 2 with Group 3? Or, Group 1 with Group 3?

You might answer this question by saying "all of the above." If so, you are saying that, with three groups, you would do three t tests: one comparing Group 1 against Group 2, a second comparing Group 1 against Group 3, and a third comparing Group 2 against Group 3. However, that's not allowed!

An analogy will help you understand why you cannot use multiple t tests. Suppose a stranger comes up to you with a proposition: "Let's bet on coin flips. If I get a 'head,' you give me a dollar. If I don't, I give you a dollar." You agree. He then proceeds to flip three coins at once and then makes you pay up if even one of the coins comes up heads. Why is this unfair? This is unfair because he misled you: You thought he was going to flip only one coin at a time, so you thought he had only a 50% chance of winning. But because he's flipping three coins at a time, his chances of getting at least one head are much better than 50%.[4]

When you do multiple t tests, you are doing basically the same thing as the coin hustler. You start by telling people the odds that a single t test will be significant due to chance alone. For example, if you use conventional significance levels, you would tell people that if the treatment has no effect, the odds of getting a statistically significant result for a particular t test are less than 5 in 100. In other words, you are claiming that your chance of making a Type 1 error (your false alarm rate) is no more than 5%.

Then, just as the hustler gave himself more than a 50% chance of winning by flipping more than one coin, you give yourself a more than 5% chance of getting a statistically significant result by doing more than one t test. The 5% chance you quoted would hold only if you had done a single t test. If you are using t tests to compare three groups, you will do three t tests, which means the chances of at least one turning out significant by chance alone are much more than 5%.[5]

So far, we've talked about the problems of using a t test when you have a three-group experiment. What happens if your experiment has more than three groups? Then, the t test becomes even more deceptive (just as the coin hustler would be cheating even more if he flipped more than three coins at a time). The more groups you use in your experiment, the greater the difference between the significance level you report and the actual odds of at least one t test being significant by chance (Hays, 1981).

To give you an idea of how great the difference between your stated significance level and the actual odds can be, suppose you had six levels of the independent variable. To compare all six groups with one another, you would need to do 15 t tests. If you did that and used a .05 significance level, the probability of getting at least one significant effect by chance alone would be more than 50%: Your risk of making a Type 1 error would be 10 times greater than you were claiming it was!

---

[4] To be more precise, his chances of getting at least one head are 87.5%.

[5] To be more precise, your chances are 14.26%.

As you have seen, the *t* test is not useful for analyzing data from the multiple-group experiment because it measures the degree to which groups differ (vary) by using subtraction—and you can only subtract two group averages at a time. To calculate the degree to which more than two group means *vary*, you need to calculate a *variance* between those means.

The between-groups variance indicates the extent to which the group means vary (differ). Thus, if all your groups have the same mean, between-groups variance would be zero because there would be no (zero) differences between your group means. If, on the other hand, there are large differences between the group means, between-group variance will be large.

Between-Groups Variance = Random Error Plus (Possibly) Treatment Effects.
So, the size of the between-groups variance depends on the extent to which the group means differ. But what affects the extent to which the group means differ? As you saw earlier, there are two factors.

One factor is random error. Even if the treatment has no effect, random error alone will usually cause the group means to differ. If the experiment uses an unreliable measure, few participants, and poorly standardized procedures, random error alone may cause large differences between the group means. Even if the experiment uses a reliable measure, many participants, and highly standardized procedures, random error alone would still tend to cause small differences between the group means.

The second factor that *may* affect the extent to which the groups differ from each other is the treatment effect. If the treatment has an effect, the differences between the group means should be greater than when the treatment doesn't have an effect. Because of the treatment effect's influence on the size of the between-groups variance, the between-groups variance is often called **treatment variance**.

Note, however, that "treatment variance" is a misleading name for between-groups variance. Such a name implies that between-group variance is due only to the treatment, but that implication is misleading in two ways. First, if the treatment has no effect, between-groups variance is not affected by the treatment. Second, regardless of whether the treatment has an effect, between-group variance will usually be affected by random error. Indeed, because of random error, the treatment groups will often have different means even when the treatment has no effect.

To recap, when there is a treatment effect, the between-group variance is the sum of two quantities: an estimate of random error plus an estimate of treatment effects. Therefore, if the treatment has an effect, between-groups variance (which is affected by the treatment plus random error) should be larger than the within-groups variance (which is affected only by random error).

Comparing Between-Group Variance to Within-Groups Variance: Are the Differences Between Groups Due to More Than Random Error?

Once you have the between-groups variance (an estimate of random error plus any treatment effects) and the within-groups variance (an estimate of random error), the next step is to compare the two variances. If the between-groups variance is larger than the within-groups variance, some of the between-groups variance *may* be due to a treatment effect—as illustrated by the following table:

	If the treatment has no effect, the variance is an estimate of the study's	If the treatment has an effect, the variance is an estimate of the study's
Between-group variance (differences between the means of the different treatment groups)	Random error	Random error + treatment effect
Within-group variance (differences between participants who are in the same group)	Random error	Random error

© Cengage Learning 2013

Put another way, the task of teasing out a treatment's effect is similar to—but more solvable than—the problem faced by those who wanted to measure the weight of the soul. If they obtained the person's weight, that weight did not tell them the weight of the person's soul because that weight represented the combined weight of the body and (possibly) the soul. To solve the problem of not being able to being able to weigh the soul directly, they weighed a person right before death (weight of body and soul) and compared that weight to the weight right after death (body alone). If the live body weighed more than the dead body, the difference would be due to whatever left the body at death—or to measurement error.

Similarly, if you look at differences between treatment group means, those differences do not tell you the effect of the treatment because those differences represent the combined effects of random error and (possibly) the treatment effect. To solve the problem of not being able to see pure treatment effect, you compare the difference between groups (an estimate of the combined effects of random error and treatment effects) to differences within groups (an estimate of random error alone). If the between-groups variance is greater than the within-groups variance, the difference may be due to the treatment effect—or to an error in estimating the effects of random error.

The statistical *analysis* that allows you to compare the between-groups *variance* to the within-groups *variance* to determine whether the treatment had an effect is called **analysis of variance (ANOVA)**.

When doing an ANOVA, you compare two variances by *dividing* the between-groups variance by the within-groups variance. That is, you set up the following ratio:

$$\frac{\text{between-groups variance}}{\text{within-groups variance}}$$

Instead of using the term *variance,* you are more likely to see the term *mean square (MS)*. Thus, you are more likely to read about authors setting up the following ratio:

$$\frac{\text{mean square between groups}}{\text{mean square within groups}}$$

Note that authors tend to leave off the word *groups*. As a result, you are likely to see the ratio described as

$$\frac{\text{mean square between}}{\text{mean square within}}$$

In addition, authors often abbreviate mean square as *MS*. Thus, you may see

$$\frac{MS \text{ between}}{MS \text{ within}}$$

To shorten the expression even further, authors may abbreviate mean square between as MSB and mean square within as MSW. Therefore, you are likely to see the ratio of the variances described as

$$\frac{MSB}{MSW}$$

To complicate things further, authors may not use the terms *between* or *within*. Rather than use a name that refers to how these variances were calculated (looking at differences *between* group means for *MS between* and looking at differences *within* groups for *MS within*), authors may instead use a name that refers to what these variances estimate. Thus, because between-groups variance is, in part, an estimate of treatment effects, authors may refer to mean square between as mean square *treatment* (*MST*). Similarly, because within-groups variance is an estimate of the degree to which random *error* is affecting estimates of the treatment group means, authors may refer to mean square *within* as mean square *error* (*MSE*).

Regardless of what names or abbreviations authors give the two variances, the ratio of the between-groups variance to the within-groups variance is called the *F* ratio. Consequently, the following three ratios are all *F* ratios:

$$\frac{MSB}{MSW} = \frac{MS \text{ treatment}}{MS \text{ error}} = \frac{MST}{MSE}$$

In ANOVA summary tables, terms are shortened even more. Thus, when scanning computer printouts or when reading articles, you may see tables resembling the one below:

Source	Mean Square	F
Treatment	10	2
Error	5	

© Cengage Learning 2013

Why an *F* Must Be Larger Than 1 to Show That the Treatment Had an Effect. Conceptually, the *F* ratio can be portrayed as follows:

$$F = \frac{\text{random error} + \text{possible treatment effect}}{\text{random error}}$$

By examining this conceptual formula, you can see that the *F* ratio will rarely be much less than 1. To illustrate, imagine that the null hypothesis is true: There is no (zero) treatment effect. In that case, the formula is (random error + 0)/random error, which reduces to random error/random error. As you know, if you divide a number by itself (e.g., 5/5, 8/8), you get 1.[6]

You now know that if the null hypothesis were true, the F ratio would be approximately 1.00.[7] That is,

$$F = \frac{\text{random error} + 0}{\text{random error}} = \frac{\text{random error}}{\text{random error}} \approx 1.00$$

To be more precise, if the null hypothesis were true,

$$F = \frac{\text{estimate of random error} + 0}{\text{estimate of random error}} = \frac{\text{estimate of random error}}{\text{estimate of random error}} \approx 1$$

This formula tell us two things about what happens when the treatment has no effect:

1. Because a number divided by itself equals 1, F will tend to be around 1.
2. Because we are dealing with estimates that are not perfectly accurate, F will not always be exactly 1. For example, when the null is true, F may be less than 1 or it may be bigger than 1 (e.g., around 4).

But what would happen to the F ratio if the treatment had an effect? To answer this question, let's look at what a treatment effect would do to the top and the bottom half of the F ratio.

If the treatment has an effect, the top of the F ratio—the between-groups variance—should get bigger. Intuitively, you realize that the differences between groups getting different treatments will be greater when the treatment has an effect than when it does not. After all, when there is a treatment effect, not only is random error making the group means differ (as it was even when the treatment did not have an effect), but the treatment is also making the group means differ.

We just explained that a treatment effect increases the *top* of the F ratio, but what does a treatment effect do to the *bottom* of the F ratio? Nothing. Regardless of whether there is a treatment effect, the bottom of the F ratio, the within-groups variance, always represents only random error: With or without a treatment effect, a group's scores (e.g., the scores of the participants in the control group) differ from one another solely because of random error.

Let's now use our knowledge of how treatment effects influence the top and bottom parts of the F ratio to understand how treatment effects influence the entire F ratio. When there is a treatment effect, the differences between group means are due not only to random error (the only thing that affects within-groups variance) but also to the treatment's effect. Consequently,

---

[6] The only exception is that 0/0 = 0.

[7] If you get an F below 1.00, it indicates that you have found no evidence of a treatment effect. Indeed, in the literature, you will often find statements such as, "There were no other significant results, all $Fs < 1$." If you get an F substantially below 1.00, you may want to check to be sure you did not make a computational error. If your F is negative, you have made a computational error: F can't be less than 0.

when there is a treatment effect, the between-groups variance (an index of random error plus treatment effect) should be larger than the within-groups variance (an index of random error alone). Put more mathematically, when there is a treatment effect, you would expect the ratio of between-groups variance to within-groups variance to be greater than 1. Specifically,

$$F = \frac{\text{between groups variance (treatment} + \text{random error)}}{\text{within groups variance (random error)}} > 1,$$

when the treatment has an effect.

Why You Need to Use an *F* Table to Determine Whether an *F* Above 1 Is Significant. *Not all Fs above 1.00 are statistically significant*, however. Admittedly, if there is no treatment effect, we are essentially dividing one estimate of random error by another. In such a case, if our estimates were perfect, we would get 1. However, our two estimates of random error are not perfect. Thus, by chance alone, an *F* could be bigger than 1. Therefore, to determine whether an *F* ratio is enough above 1.00 to indicate that there is a significant (reliable) difference between your groups, you need to consult an *F* table, like the one in Appendix F.

Calculating Degrees of Freedom. To use the *F* table, you need to know two degrees of freedom: one for the top of the *F* ratio (between-groups variance, MS treatment) and one for the bottom of the *F* ratio (within-groups variance, MS error).

Calculating the degrees of freedom for the top of the *F* ratio (between-groups variance) is simple: It's just one less than the number of values of the independent variable. So, if you have three values of the independent variable (no-treatment, meditation, and exercise), you have 2 (3 − 1) degrees of freedom. If you had four values of the independent variable (e.g., no-treatment, meditation, archery, aerobic exercise), you would have 3 (4 − 1) degrees of freedom. Thus, for the experiments we have discussed in this chapter, *the degrees of freedom for the between-groups variance equals the number of groups 1.*

Computing the degrees of freedom for the bottom of the *F* ratio (within-groups variance) is also easy. The formula is N (number of participants) − G (groups). Thus, if there are 20 participants and 2 groups, the degrees of freedom = 18 (20 − 2 = 18).[8]

Let's now apply this formula to some multiple-group experiments. If we have 33 participants and 3 groups, the *df* for the error term = 30 (because 33 − 3 = 30). If we had 30 participants and 5 groups, the *df* error would = 25 (because 30 − 5 = 25). To repeat, the simplest way of computing the error *df* for the experiments we discussed in this chapter is to use the formula N − G, where N = total number of participants and G = total number of groups (see Table 11.1).

---

[8] As you may recall, you could have used this N − G formula to get the degrees of freedom for the *t* test described in Chapter 10. However, because the *t* test always compares 2 groups, people often memorize the formula N − 2 for the *t* test instead of the more general formula N − G.

TABLE **11.1** Calculating Degrees of Freedom

Source of Variance (*SV*)	Calculation of *df*
Treatment (between groups)	Number of groups − 1 (*G* − 1)
Within subjects (error variance)	Number of participants minus number of groups (*N* − *G*)
Total	*N* − 1

© Cengage Learning 2013

Once you know the degrees of freedom, find the entry in the $p < .05$ *F* table (Table 3 of Appendix F) that corresponds to those degrees of freedom. If your *F* ratio is larger than the value listed, the results are statistically significant at the $p < .05$ level.

Making Sense of an ANOVA Summary Table or Computer Printout.   Usually, you will not have to look up *F* values in an *F* table. Instead, you will have a computer calculate *F* and look it up in a table for you. However, if you had a computer calculate *F* for you, you should make sure that the degrees of freedom on the printout are correct. If they are not, do not trust that analysis: It is wrong. If the printout shows more degrees of freedom for the treatment than it should, you probably entered a wrong number for a participant's condition (e.g., in a 3-group experiment, you entered "4" for a participant's group number). If the printout shows fewer degrees of freedom than it should, you probably didn't enter all your data.

The degrees of freedom are not the only thing that you need to understand about the printout. To see how to make sense of the printout, let's look at a sample of an analysis of variance (ANOVA) summary table:

Source of Variance	Sum of Squares (*SS*)	*df*	MS	*F*	*p*
Treatment (Between)	88	2	44	44	<.001
Error (Within)	12	12	1		
Total	100	14			

© Cengage Learning 2013

The first column, the source of variance column, may sometimes have only the heading "Source." The two main sources of variance will be your (1) treatment (which may be labeled as "Treatment," "Between groups," "BG," "Groups," "Between," "Model," or the actual name of your independent variable) and (2) random error (which may be labeled as "Error," "Within groups," "WG," or "Within").

The second column, the sum of squares column, may be labeled "Sum of Squares," "SS," or "Type III Sum of Squares." Note that if you add the sum of squares treatment to the sum of squares error, you will get the sum of squares total.

The third column, the degrees of freedom column, is often labeled with the abbreviation *df*. As we mentioned earlier, you should check the *df* column to make sure that the analysis is based on the right number of treatment groups and the right number of participants.

From the *df* column in our ANOVA table, we know two things. First, because the formula for the *df* treatment is G − 1 and because the treatment *df* is 2, we know that a 3 group ANOVA has been calculated (because 3 [groups] − 1 = 2 [*df*]). Second, because the formula for total *df* is N − 1 (number of participants − 1) and because the total *df* is 14, we know that the ANOVA is based on data from 15 participants (because 15 [participants] − 1 = 14 [*df*]).

The fourth column, the Mean Square column, is often labeled with the abbreviation *MS*. The *MS* Treatment will be the *SS* Treatment divided by the *df* Treatment. Note that if the *MS* Treatment is *not* bigger than *MS* Error, the results will *not* be statistically significant.

The fifth column contains the *F* ratio. The *F* ratio is the *MS* Treatment divided by *MS* Error. In the table above *F* is 44 because 44 (*MST*) divided by 1 (*MSE*) = 44.

The sixth column, the *p* value column, tells you how likely it would be to get differences between the groups this large or larger if the null hypothesis (the null hypothesis is that the treatment has no effect) were true. In this case, *p* is less than .05, suggesting that it is unlikely that you would obtain these results if the null hypothesis were true. Traditionally, such results would be called "statistically significant." An author might start to summarize the results of such an ANOVA by writing, "Consistent with the hypothesis, the treatment had an effect, $F(2, 12) = 44$, $p < .001$."

What Statistical Significance Means in ANOVA—And What It Doesn't Mean

If your results are statistically significant, what does that mean? *Statistical significance means that you can reject the null hypothesis.* In the multiple-group experiment, the null hypothesis is that the differences among all your group means are due to chance. That is, all your groups are essentially the same. Rejecting this hypothesis means that, because of treatment effects, all your groups are *not* the same. Note that this is like hearing that four people each bought a dollar lottery ticket and that, after the drawing, those tickets are no longer all worth the same amount. In that case, you would wonder what the average difference in value was (one dollar? 30 million dollars?) and how many of the four had winning tickets. Similarly, on learning from the ANOVA that not all the groups were the same, you might ask two questions:

1. "By how much do the groups differ?"
2. Which groups differ from each other—just two of them (which two?), most of them? (which ones?), or all of them?"

To begin to answer the "How large is the effect?" question, you could just look to see how big the differences were between the means. For example, looking at such differences suggests that the effect of antidepressants on relieving depression is only to increase scores by 2 points on a 50-point scale (Kirsch, Moore, Scoboria, & Nicholls, 2002).

One problem with looking only at the mean differences to determine effect size is that you might not know what a 2-unit difference represents: Is a 2-unit difference a large difference or a small difference? The problem is that the size of a 2-unit difference depends on many things, including what your units are and what you are measuring. For example, if your 2-unit difference is a 2-second difference in a reaction time task, that difference is enormous. If, on the other hand, the 2-unit difference is a 2-millisecond difference in the time it takes to run a marathon, that difference is infinitesimal. Similarly, whereas a 2-point difference on a 10-point classroom quiz might be meaningful, a 2-point difference on the 800-point Math SAT would be meaningless. To make comparing differences more useful, we need some standard way of comparing differences.

One way to have a standard against which we can compare differences—even when those differences involve different units and different tasks—is to divide the average group difference by the group's standard deviation. By dividing the mean difference by the standard deviation, we can talk about effects in terms of the same standard (deviation) units. With that system, an effect size of 1 tells us that the treatment made a difference of 1 standard deviation.

Another strategy for looking at the strength of a treatment effect is to describe the relationship on a 0 (*no relationship*) to 1 (*perfect relationship*) scale. For example, you might compute **eta squared** (η^2): an estimate of effect size that ranges from 0 to 1 and is comparable to *r* squared (r^2).[9]

Computing eta squared from an ANOVA summary table is simple: Just divide the Sum of Squares Treatment by the Sum of Squares total. For example, in our ANOVA table, *SS* treatment was 88 and *SS* total was 100; therefore, eta squared was .88 (because 88/100 = .88). Thus, an author might start to describe such results by writing, "Consistent with the hypothesis, the treatment had an effect, $F(2, 12) = 44$, $p < .001$, $\eta^2 = 0.88$." Note that you would normally not get such a large eta squared. Indeed, social scientists tend to view any eta squared (or *r* squared) of .25 or above to be large (.09 to .25 is considered moderate; less than .09 is considered small).

In addition to finding out how big the treatment's effect is, you should also answer our second question: "Which groups differ from each other?" Remember, that, a significant *F* means that not all the groups are the same, *but it does not tell you which groups differ*. For example, if you had a three-group experiment, a significant ANOVA would tell you that the following statement is false: Group 1 = Group 2 = Group 3. In other words, at least one of the following three statements is false:

Group 1 = Group 2; Group 1 = Group 3; and Group 2 = Group 3.

---

[9] To learn about *r* squared, review our section titled "Coefficient of Determination" in Chapter 7, see Box 10.2 in Chapter 10, or look at Appendix E.

But which one(s) are false? It could be that

Group 1 ≠ Group 2; Group 1 = Group 3; Group 2 = Group 3.

Or that

Group 1 = Group 2; Group 1 ≠ Group 3; Group 2 = Group 3

Or that

Group 1 = Group 2; Group 1 = Group 3; Group 2 ≠ Group 3

Or that

Group 1 ≠ Group 2; Group 1 ≠ Group 3; Group 2 = Group 3.

Or that

Group 1 ≠ Group 2; Group 1 = Group 3; Group 2 ≠ Group 3

Or that

Group 1 = Group 2; Group 1 ≠ Group 3; Group 2 ≠ Group 3

Or that

Group 1 ≠ Group 2; Group 1 ≠ Group 3; Group 2 ≠ Group 3.

Therefore, once you have performed an F test to determine that at least some of your groups differ, you need to do additional tests to determine which of your groups differ from one another.

Beyond ANOVA: When and Why We Do Follow-up Tests

You might think that all you would have to do to determine which groups differ is compare group means. Some group means, however, may differ from others solely due to chance. To determine which group differences are due to treatment effects, you need to do additional tests. These additional, more specific tests are called post hoc t tests.

Post Hoc t Tests Among Group Means: Which Groups Differ?   At this point, you may be saying that you wanted to do t tests all along. Before you complain to us, please hear our two-pronged defense.

First, you can go in and do **post hoc tests** only *after* you get a significant F. That is, you can't legitimately use follow-up tests to ask "which of the groups differ" until you first establish that at least some of the groups do indeed differ. To do post hoc tests without finding a significant F is considered statistical malpractice. Such behavior would be like a physician doing a specific test to find out which strain of hepatitis you had after doing a general test that was negative for hepatitis. At best, the test will not turn up anything, and your only problem will be the expense and pain of an unnecessary test. At worst, the test results will be misleading because the test is being used under the wrong circumstances. Consequently, you may end up being treated for a hepatitis you do not have. Analogously, a good researcher does not ask which groups differ from one another unless the more general, overall analysis of variance test has first established that at least some of the groups do indeed differ.[10]

Second, post hoc tests are not the same as conventional t tests. Unlike conventional t tests, post hoc t tests are designed to correct for the fact that you are doing more than two comparisons. As we mentioned earlier, doing more than one t test at the $p = .05$ level and claiming that you have only a 5% risk of making a *Type 1* error is like flipping more than one coin at a time and claiming that the odds of getting a "heads" are only 50%. In both cases, the odds of getting the result you hope for are much greater than the odds you are stating. Thus, we cannot simply do an ordinary t test. Instead, we must correct for the number of comparisons we are making. Post hoc t tests take into consideration how many tests are being done and make the necessary corrections.

At this point, we will not require you to know how to do post hoc tests. (If you want to know how to conduct a post hoc test, see Appendix F.) You should, however, be aware that if you choose to do a multiple-group experiment, you should be prepared to do post hoc analyses.

You should also be aware that if you read a journal article describing the results of a multiple-group experiment, you may read the results of post hoc tests. For example, you may read about a Bonferroni t test, Tukey test, Scheffe test, Dunnett test, Newman-Keuls test, Duncan, or an LSD test. When reading about the results of such tests, do not panic: The author is merely reporting the results of a post hoc test to determine which means differ from one another.

When and Why to Do a Post Hoc Trend Analysis. Rather than wanting to know which particular groups differ from one another, you may want to know the shape of the functional relationship between the independent and dependent variables so that you could either (a) better generalize to levels of the treatment that were not tested or (b) test a theory that predicts a certain functional relationship. If you want to know the shape of the functional relationship, instead of following up a significant main effect with post hoc tests between group means, follow up the significant effect with a **post hoc trend analysis**.

But why should you do a trend analysis to determine the shape of the functional relationship between your independent and dependent variables?

---

[10] Although everyone agrees that you need to do an ANOVA before doing a post hoc t test, not everyone agrees that you need to do an ANOVA before doing other tests. Indeed, Robert Rosenthal (1992) argued that researchers should almost never do the general, overall F test. Instead, he argued that if you have specific predictions about which groups differ, you should do normal t tests to compare those group means. Those t tests are often called "planned comparisons" because the researcher planned to make those comparisons before collecting data. Planned comparisons involving t tests are sometimes also called "*a priori t* tests" ("a priori" means in advance) to emphasize that the t tests were done before peeking at the data. Sometimes, planned comparisons will be called "planned contrasts." One planned contrast that you will see when the researcher is trying to determine whether the two experimental groups differ from the control group or whether the two control groups differ from the experimental group is called the "two vs. one" contrast.

Can't you see this relationship by simply graphing the group means? Yes and no. Yes, graphing your sample's means allows you to see the pattern in the data produced by your experiment. No, graphing does not tell you that the pattern you observe represents the true relationship between the variables because your pattern could be due to random error (e.g., if even one mean is thrown off by random error, that one misplaced mean could make a linear relationship look nonlinear). So, just as you needed statistics to tell you if the difference between two groups was significant (even though you could easily see whether one mean was higher than the other), you need statistics to know if the pattern you observe in your data (a straight line, a curved line, a combination of a curve and a straight line, etc.) would occur if you repeated the experiment. The statistical test you need to determine whether the pattern in your data reflects a *reliable* functional relationship is a post hoc trend analysis.

Computing a post hoc trend analysis is easy. You can either follow the simple directions in Appendix F or use a computer program that does the analysis for you. Because post hoc trend analyses are easy to do and because you don't do them until after you have collected your data, you might be tempted to forget about post hoc trend analysis until it comes time to analyze your data. Don't make that mistake.

If you do not think about post hoc trend analysis before you do your experiment, you will probably be unable to do a valid post hoc trend analysis after you do your experiment. Therefore, if you think that you might want to know about the functional relationship between the variables in your experiment, you should keep three facts in mind *before* conducting that experiment (see Box 11.2).

First, to do a post hoc trend analysis, you should have selected levels of your independent variable that increase proportionally. For example, if you were using three levels of a drug, you would not use 5 mg, 6 mg, and 200 mg. Instead, you might use 10 mg, 20 mg, and 30 mg or 10 mg, 100 mg, and 1000 mg.

Second, you must have at least an interval scale measure of your dependent variable. Your map of the functional relationship can't be accurate unless your measure of the dependent variable is to scale. If you tried to find the relationship between the loudness of the music playing on participants' personal stereos and distance walked, you would have to measure distance

---

BOX **11.2** Requirements for Conducting a Valid Post Hoc Trend Analysis

1. Your independent variable must have a statistically significant effect.
2. Your independent variable must be quantitative, and the levels used in the experiment should vary from one another by some constant amount or proportion.
3. Your dependent variable must yield interval or ratio-scale data so that your map of the functional relationship will be to scale.
4. The number of trends you can look for is one less than the number of levels of your independent variable.

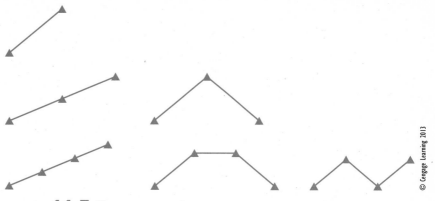

FIGURE **11.5** Note: more points = more possible trends.

Note: With 2 points, you can only see which straight line fits your data. With 3 points, you can see how well a straight line and lines with 1 bend in them fit your data. With 4 points, you can see how well a straight line, lines with 1 bend in them, and even lines with 2 bends in them fit your data.

by number of meters walked rather than by blocks walked (unless all your blocks are the same length). In short, you can't do a trend analysis if you have ordinal or nominal data.

Third, the more levels of the independent variable you have, the more trends you can look for. Specifically, the number of trends you can examine is one less than the number of levels you have. If you had only two levels, you can test only for straight lines (linear component). If you have three groups, you can test for both straight lines (linear component) and for U-shaped curves (quadratic component). With four levels, you can test for straight lines, U-shaped curves, and double U-shaped lines (cubic component). Thus, if you are expecting a double U-shaped curve, you must use at least four levels of the independent variable.

If you think about graphing your data, you can see why the "number of trends = number of levels – 1" rule is true (see Figure 11.5). If you had only two levels, you would have only two points on your graph (representing your two means). With only two levels, you could only see how well different straight lines fit those two points. Your two points, no matter what they were, would not form the outline of a curved line. If you had three levels, you would have three points, and you could see whether those three points lined up in a straight line or whether the formed a line with a bend in it. If you had four points, you could see whether those four points formed a straight line, a line with one bend in it, or a line with two bends in it.

CONCLUDING REMARKS

By using a multiple-group experiment rather than a simple experiment, you can ask more refined questions. For example, you can go beyond asking, "Is there an effect?" to asking "What is the nature of the functional relationship?"

By using a multiple-group experiment rather than a simple experiment, you can get more valid answers to your questions. For example, by using appropriate control groups, you can learn not only that the treatment manipulation worked but also why it worked.

Although adding more levels of the treatment is a powerful way to expand the simple experiment, an even more powerful way to expand the simple experiment is to add independent variables. As you will see in the next chapter, adding independent variables not only increases construct and external validity but also opens up a whole new arena of research questions.

SUMMARY

1. The multiple-group experiment is more sensitive to nonlinear relationships than the simple experiment. Consequently, it is more likely to obtain significant treatment effects and to accurately map the functional relationship between your independent and dependent variables.

2. Knowing the functional relationship allows more accurate predictions about the effects of unexplored levels of the independent variable.

3. To use the multiple-group experiment to discover the functional relationship, you should select your levels of the independent variable carefully, and your dependent measure must provide at least interval scale data.

4. Multiple-group experiments may have more construct validity than a simple experiment because they can have multiple control groups and multiple treatment groups.

5. To analyze a multiple-group experiment, you first have to conduct an analysis of variance (ANOVA). An ANOVA will produce an F ratio.

6. An F ratio is a ratio of between-groups variance to within-groups variance.

7. Random error will make different treatment groups differ from each other. If the treatment has an effect, the treatment will also cause the groups to differ from each other. In other words, between-groups variance is due to random error and may also

be due to treatment effects. Because it may be affected by treatment effects, between-groups variance is often called treatment variance.

8. Scores within a treatment group differ from each other for only one reason: random error. That is, the treatment cannot be responsible for variability within each treatment group. Therefore, within-groups variance is an estimate of the degree to which random error affects the data. Consequently, another term for within-groups variance is error variance.

9. The F test is designed to see whether the difference between the group means is greater than would be expected by chance. It involves dividing the between-groups variance (an estimate of random error plus possible treatment effects) by the within-groups variance (an estimate of random error). If the F is 1 or less, there is no evidence that the treatment has had an effect. If the F is larger than 1, you need to look in an F table (under the right degrees of freedom) to see whether the F is significant.

10. The first degrees of freedom (between groups/ treatment) equals the number of groups minus one, abbreviated $G - 1$. The second degrees of freedom (within groups/error) equals the number of participants minus the number of groups, abbreviated $N - G$. Thus, if you had 5 groups and 40 participants, you would look at the F table under 4 ($5 - 1$) and 35 ($40 - 5$) degrees of freedom.

11. You are most likely to get a significant *F* if between-group variability is large (your group means are very different from each other) and within-groups variability is small (within each group, most scores are close to each other).

12. If you get a significant *F*, you know that the groups are not all the same. If you have more than two groups, you have to find out which groups differ. To find out which groups are different, do not just look at the means to see which differences are biggest. Instead, do post hoc tests to find out which groups are reliably different.

13. The following table summarizes the mathematics of an ANOVA table.

Source of Variance (SV)	Sum of Squares (SS)	Degrees of Freedom (df)	Mean Square (MS)	F
Treatment (T)	SST	Levels of T − 1	SST/df T	MST/MSE
Error (E)	SSE	Participants − Groups	SSE/df E	
Total	SST + SSE	Participants − 1		

© Cengage Learning 2013

KEY TERMS

analysis of variance (ANOVA) *(p. 439)*

confounding variables *(p. 427)*

empty control group *(p. 430)*

error variance *(p. 436)*

eta squared (η^2) *(p. 445)*

F ratio *(p. 440)*

functional relationship *(p. 421)*

hypothesis-guessing *(p. 430)*

linear relationship *(p. 421)*

post hoc tests *(p. 446)*

post hoc trend analysis *(p. 447)*

treatment variance *(p. 438)*

variability between group means *(p. 433)*

within-groups variance *(p. 436)*

EXERCISES

1. A researcher randomly assigns each member of a statistics class to one of two groups. In one group, each student is assigned a tutor who is available to meet with the student 20 minutes before each class. The other group is a control group not assigned a tutor. Suppose the researcher finds that the tutored group scores significantly better on exams.

   a. Can the researcher conclude that the experimental group students learned statistical information from tutoring sessions that enabled them to perform better on the exam? Why or why not?

   b. What changes would you recommend in the study?

2. Suppose people living in homes for older adults were randomly assigned to two groups: a no-treatment group and a transcendental meditation (TM) group. Transcendental meditation involves more than sitting with eyes closed. The technique involves both "a meaningless sound selected for its value in facilitating the transcending, or settling-down, process and a specific procedure for using it mentally without effort again to facilitate transcending" (Alexander, Langer, Newman, Chandler, & Davies, 1989, p. 953). The TM group was given instruction in how to perform the technique; then "they met with their instructors half an hour each week to verify that they were meditating

correctly and regularly. They were to practice their program 20 minutes twice daily (morning and afternoon) sitting comfortably in their own room with eyes closed and using a timepiece to ensure correct length of practice" (Alexander et al., 1989, p. 953).

Suppose that the TM group performed significantly better than other groups on a mental health measure.[11]

a. Could the researcher conclude that it was the transcendental meditation that caused the effect?

b. What besides the specific aspects of TM could cause the difference between the two groups?

c. What control groups would you add?

d. Suppose you added these control groups and then got a significant F for the treatment variable? What could you conclude? Why?

3. Assume you want to test the effectiveness of a new kind of therapy. This therapy involves screaming and hugging people in group sessions followed by individual meetings with a therapist. What control group(s) would you use? Why?

4. Assume a researcher is looking at the relationship between caffeine consumption and sense of humor.

a. How many levels of caffeine should the researcher use? Why?

b. What levels would you choose? Why?

c. If a graph of the data suggests a curvilinear relationship, can the researcher assume that the functional relationship between the independent and dependent variables is curvilinear? Why or why not?

d. Suppose the researcher used the following four levels of caffeine: 0 mg, 20 mg, 25 mg, and 26 mg. Can the researcher easily do a trend analysis? Why or why not?

e. Suppose the researcher ranked participants based on their sense of humor. That is, the person who laughed least got a score of 1, the person who laughed second-least scored a 2, and so on. Can the researcher use these data to do a trend analysis? Why or why not?

f. If a researcher used four levels of caffeine, how many trends can the researcher look for? What are the treatment's degrees of freedom?

g. If the researcher used three levels of caffeine and 30 participants, what are the degrees of freedom for the treatment? What are the degrees of freedom for the error term?

h. Suppose the F is 3.34. Referring to the degrees of freedom you obtained in your answer to "g" (above) and to Table 3 (Appendix F), are the results statistically significant? Can the researcher look for linear and quadratic trends?

5. A computer analysis reports that $F(6, 23) = 2.54$. The analysis is telling you that the F ratio was 2.54, and the degrees of freedom for the top part of the F ratio = 6 and the degrees of freedom for the bottom part = 23.

a. How many groups did the researcher study?

b. How many participants were in the experiment?

c. Is this result statistically significant at the .05 level? (Refer to Table 3 of Appendix F.)

6. A friend gives you the following Fs and significance levels. On what basis would you want these Fs—or significance levels—rechecked?

a. $F(2, 63) = .04$, not significant

b. $F(3, 85) = -1.70$, not significant

c. $F(1, 120) = 52.8$, not significant

d. $F(5, 70) = 1.00$, significant

---

[11] A modification of this study was actually done. The study included appropriate control groups.

7. Complete the following table. (Hint: See main point #13.)

Source of Variance (SV)	Sum of Squares (SS)	Degrees of Freedom (df)	Mean Square (MS)	F
Treatment (T) 3 levels of treatment	180	—	—	—
Error (E), also known as within-groups variance	80	8	—	

© Cengage Learning 2013

8. Complete the following table. (Hint: See main point #13.)

Source of Variance (SV)	Sum of Squares (SS)	Degrees of Freedom (df)	Mean Square (MS)	F
Treatment (T) (between groups variance)	50	5	—	
Error (E), (within-groups variance)	100	—	—	—
Total	—	30	—	

© Cengage Learning 2013

9. A study compares the effect of having a snack, taking a 10-minute walk, or getting no treatment on energy levels. Sixty participants are randomly assigned to a condition and then asked to rate their energy level on a 0 (not at all energetic) to 10 (very energetic) scale. The mean for the "do nothing" group is 6.0, for having a snack 7.0, and for walking 7.8. The F ratio is 6.27.
   a. Graph the means.
   b. Are the results statistically significant?
   c. If so, what conclusions can you draw? Why?
   d. What additional analyses should you do? Why?
   e. How would you extend this study?
   f. If you only knew the group means, could you do an ANOVA? Why or why not?

WEB RESOURCES

1. Go to the Chapter 11 section of the book's student website and
   a. Look over the concept map of the key terms.
   b. Test yourself on the key terms.
   c. Take the Chapter 11 Practice Quiz.
2. Do an analysis of variance using a statistical calculator by going to the "Statistical Calculator" link.
3. If you want to write your method section, use the "Tips on Writing a Method Section" link.
4. If you want to write up the results of a one-factor, between-participants experiment, click on the "Tips for Writing Results" link.

CHAPTER 12

Expanding the Experiment:

Factorial Designs

I'm an earth sign, she was a water sign—together we made mud.
—Woody Allen

The pure and simple truth is rarely pure and never simple.
—Oscar Wilde

CHAPTER OVERVIEW

If your professor asked you to design an experiment that tested whether exercise affected happiness, you might propose an experiment in which participants were randomly assigned to either sit down or to exercise. Suppose your friend proposed the same experiment, except that your friend's simple experiment was to be conducted outside whereas yours was to be conducted inside.

What would happen if you joined forces with your friend? If you combined your ideas properly, your new experiment would independently vary two **factors** (potential causes, independent variables): (1) whether participants were indoors or outdoors and (2) whether participants were sitting down or exercising.

As you can see from Figure 12.1, one way to visualize doing this study is that you would randomly assign each participant twice: first to be either indoors or outdoors and then to either sit or to exercise. Consequently, you have four types of participants: (1) participants who are indoors *and* sit down, (2) participants who are indoors *and* exercise, (3) participants who are outdoors *and* sit down; and (4) participants who are outdoors *and* exercise.

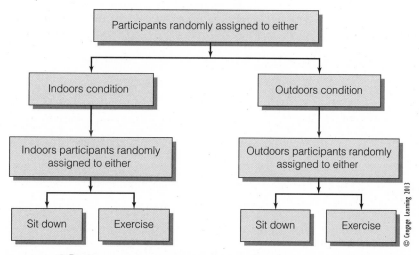

FIGURE **12.1** A factorial experiment.

In technical terminology, you would do a **factorial experiment** an experiment that studies the effects of two or more independent variables (*factors*) on at least one dependent measure. To be more specific, you have the simplest form of a factorial experiment: the 2 × 2. This particular experiment is a 2 (exertion level: sitting or exercising) × 2 (location: indoors or outdoors) experiment. The "×"—spoken as "by"—indicates that each level of the first factor is crossed (combined) with each level of the second factor. Consequently, each participant will be randomly assigned to one level of the first factor (e.g., sitting or exercise) *and* to one level of the second factor (e.g., indoors or outdoors).

Because a factorial experiment must manipulate at least two factors (independent variables), the following experiment that looks at the effect of exercise on mood is not a factorial experiment. Although it involves two variables (exercise and mood), it manipulates only one factor (exercise). Thus, rather than being a factorial experiment, it is a two-level, one-factor experiment.

Level of Exercise Factor	
Sitting (First level of the exercise factor)	**Exercising (Second level of the exercise factor)**
Group 1: Participants assigned to sit	Group 2: Participants assigned to run

© Cengage Learning 2013

Because a factorial experiment must have two manipulated factors (independent variables) and must cross its factors, the following experiment is not a factorial experiment.

Level of Exercise/Location Factor	
Sitting inside (First level of exercise/location factor)	**Running outside (Second level of exercise/location factor)**
Group 1: Participants assigned to sit inside	Group 2: Participants assigned to run outside

© Cengage Learning 2013

At first glance, you might think that this is a factorial experiment because you think it manipulates two variables: exertion (sitting vs. running) and location (indoors versus outdoors). However, those two variables are not manipulated as two separate, independent factors. Instead, those two variables have been fused into one, two-level factor (sitting inside versus running outside). Because the two variables are not manipulated separately, you can't figure out their separate effects. Thus, if Group 2 was in a better mood than Group 1, all you would know is that the combination of running and being outside puts people in a better mood than the combination of sitting and being inside. You would not know whether running helps mood more than sitting, and you would not know whether being outside helps mood more than being inside.

To know the individual effects of each of those variables, you would need to manipulate each variable independently. That is, you would need to set up a situation in which being assigned to get one level of one factor (e.g., being indoors) should have no effect on (should be *independent* of) whether a participant is assigned to a certain level of another factor (e.g., sitting). To make the conditions independent (e.g., to make being assigned to be indoors independent of being assigned to sit), you need to cross all your levels of the first factor with all your levels of the second factor, and then randomly assign each participant to one of those four treatment combinations, as in the following 2 × 2 factorial experiment.

Level (Type) of Setting	Level of Exercise	
	1. Sitting	**2. Exercising**
1. Indoors	Indoors *and* sitting participants	Indoors *and* exercising participants
2. Outdoors	Outdoors *and* sitting participants	Outdoors *and* exercising participants

© Cengage Learning 2013

Because factorial experiments must have at least *two* factors, because each level of one factor must be crossed (×) with each level of the other factor(s), and because each factor must have at least *two* levels, a between-subject factorial experiment must have at least *four (2 × 2)* groups.

Not all four-group experiments are factorial experiments, however. For example, the following four-group experiment is not a factorial experiment because it manipulates only one factor: exercise. Participants sit, walk, jog, *or* sprint. They are not assigned to different *combinations* of treatments.

Level of Exercise Factor			
1. Sitting	**2. Walking**	**3. Jogging**	**4. Sprinting**
Participants assigned to *sit*	Participants assigned to *walk*	Participants assigned to *jog*	Participants assigned to *sprint*

© Cengage Learning 2013

You now know the difference between a 2 × 2 factorial experiment and a single-factor experiment. In the next section, you will learn about different types of factorial experiments. Then, after learning why you should do factorial experiments, you will learn how to design them and how to make sense of their results.

TYPES OF FACTORIAL EXPERIMENTS

Although all factorial experiments must include *at least* two levels of two factors, factorial experiments can have more than two levels of each factor and can have more than two factors. To let people know how many levels each factor has, researchers use terminology similar to what builders use. When a builder refers to a "2 by 4," the builder means a board for which the first dimension (thickness) is 2 inches and the second dimension (width) is 4 inches. Similarly, when a researcher refers to a "2 × 4," the researcher means that the first experimental factor has 2 levels and the second experimental factor has 4 levels. Thus, if we had two levels of location and four levels of exertion, we might have a 2 (location: indoors or outdoors) × 4 (degree of exertion: lying down, sitting, walking, or running) factorial experiment. As you can see from the following table, this 2 × 4 design involves eight (2 × 4) groups of participants.

Level (Type) of Setting Factor	Level of Exertion Factor			
	1. Lying down	2. Sitting	3. Walking	4. Running
1. Indoors	Group 1: Indoors/lying down	Group 2: Indoors/sitting	Group 3: Indoors/walking	Group 4: Indoors/running
2. Outdoors	Group 5: Outdoors/lying down	Group 6: Outdoors/sitting	Group 7: Outdoors/walking	Group 8: Outdoors/running

© Cengage Learning 2013

Suppose that, instead of adding more levels of exertion, we wanted to add another factor (e.g., music) to our original 2 × 2 factorial experiment. Then, as you can see from the table below, we would have a 2 × 2 × 2 design composed of eight (2 × 2 × 2) groups.

Setting Factor	Level of Exertion		Level of Exertion	
	Sitting	Running	Sitting	Running
Indoors	Group 1: Indoors/sitting/no music participants	Group 2: Indoors/running/no music participants	Group 5: Indoors/sitting/ music participants	Group 7: Indoors/running/ music participants
Outdoors	Group 3: Outdoors/ sitting/no music participants	Group 4: Outdoors/running/ no music participants	Group 6: Outdoors/sitting/ music participants	Group 8: Outdoors/running/ music participants
	No Music Conditions		**Music Conditions**	

© Cengage Learning 2013

Factorial experiments are not limited to eight groups. Indeed, the factorial experiment can be as complicated as the researcher can handle. Thus, a researcher could design a 2 × 2 × 3 × 6 factorial experiment—or something even more complicated. We, however, will spend the rest of this chapter on something much simpler: the 2 × 2 between-subjects factorial.

THE 2 × 2 FACTORIAL EXPERIMENT: USING GRAPHS TO SEE THE THREE PATTERNS IT CAN DETECT

You now know that you could create a 2 × 2 factorial experiment by sticking two simple experiments together. "But," you might ask, "what can I find out by using a factorial experiment that I couldn't find out by just doing a simple experiment?" To know what you could find out, you need to know how to represent the results—and the predicted results—of a factorial experiment.

To see how to represent your predictions and findings, let's return to our 2 (Exertion: sitting or running) × 2 (Setting: indoors or outdoors) experiment. Because we essentially combined two simple experiments into one, it may not surprise you that, to graph data from this study, we would essentially combine two graphs into one. That is, if we can graph your original simple experiment by using one line (see Figure 12.2a):

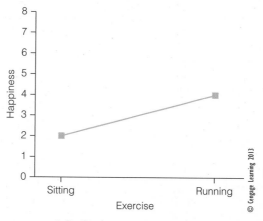

FIGURE **12.2a** A graph of your original simple experiment.

and graph your friend's original simple experiment by using another line (see Figure 12.2b):

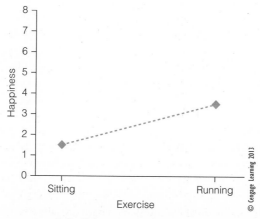

FIGURE **12.2b** A graph of your friend's original simple experiment.

then, we can graph the combined experiment by using both of those lines (see Figure 12.2c):

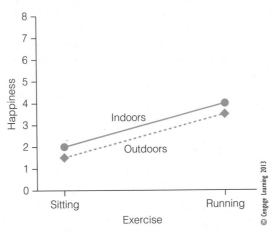

FIGURE **12.2c** A graph of your factorial experiment.

Now that you know how to graph a 2 × 2 experiment, what can you do with that knowledge? Before you do a 2 × 2 experiment, you could use that knowledge to make a graph of your predicted results. After you do a 2 × 2 experiment, you could use that knowledge to graph your actual results. In both cases, your graph will display information about three effects that you could not possibly uncover with a simple experiment. But what are those three effects—and how do you find them in a graph?

1. *Running's average (overall) effect.* With your original simple experiment, you could see running's effect only on participants who were indoors. With your friend's simple experiment, you could see running's effect only on participants who were outdoors. With your factorial experiment, on the other hand, you could see whether running, on the average, has an effect in a study that looks at both participants who are indoors and participants who are outdoors. Thus, the first thing you can find out by doing a factorial experiment instead of your original simple experiment is whether participants, *generally*—when scores are *averaged* across both indoors and outdoors participants—are in a better mood when running than when sitting.

You can see the average effect of running versus sitting by comparing the left (sitting) side of the graph to the right (running) side of the graph. You are probably comfortable comparing the left and right side of a one-line graph. Indeed, you probably realize that, in a simple experiment, comparing the left (sitting) side of the graph to the right (running side) merely involves recognizing one of three patterns. First, if the line slopes up as it goes from sitting to running (as in Figure 12.3a), then running makes participants' moods go up more than sitting does. Second, if the line slopes down as it goes from sitting to running (as in Figure 12.3b), then running, as compared to sitting, makes

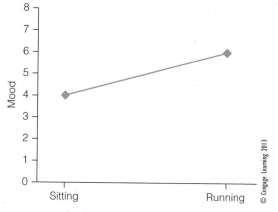

FIGURE **12.3a** Running, relative to sitting, seems to make mood go up.

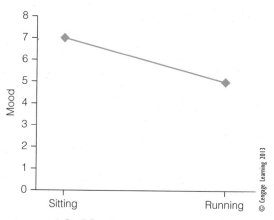

FIGURE **12.3b** Running, relative to sitting, seems to make mood go down.

participants' moods go down. Third, if the line goes neither up nor down (as in Figure 12.3c), the sitting/running variable does not appear to have a noticeable effect.

The problem you might have when interpreting the graph of a factorial experiment is that it has two lines (one for indoors and one for outdoors) when you really just want one line: a line that represents both indoors and outdoors participants. Suppose you had such a line: a line that combined (averaged) the two lines into one. Then, you would know how to make sense of your graph. If that line went up as it moved to the running side of the graph, running (on the average) made mood go up relative to sitting; if that line went down as it moved to the running side of the graph, running made mood go down relative to sitting. But how do you get such a line?

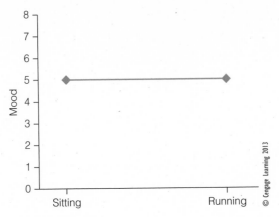

FIGURE **12.3c** Running, relative to sitting, seems not to make mood go up or down.

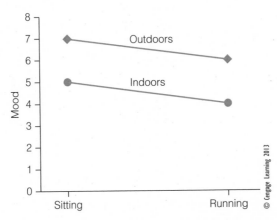

FIGURE **12.4** Running, relative to sitting, seems to make mood go down.

Making such a "combined/averaged" line is often so easy that you can do it in your head. For example, in the figure above (Figure 12.4), the lines are so similar, that you can easily mentally combine the lines into one—and know that the combined line is going down, indicating that mood, on the average, goes down in the running condition.

If you are having trouble averaging the lines in your head, you can average the lines on paper in three easy steps. First, make a dot halfway between the left end of the top line and the left end of the bottom line (see Figure 12.5a). Second, make a dot halfway between the right end of the top line and the right end of the bottom line (see Figure 12.5b). Third, draw a line between those two dots (see Figure 12.5c).

FIGURE **12.5a** Step 1 in combining the lines from a 2 × 2 experiment.

FIGURE **12.5b** Step 2 in combining the lines from a 2 × 2 experiment.

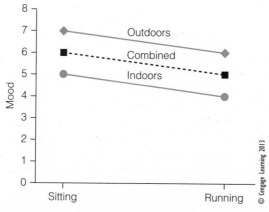

FIGURE **12.5c** Step 3 in combining the lines from a 2 × 2 experiment.

2. *Indoors–outdoors factor's average (overall) effect.* In addition to seeing the average effect of running, the factorial experiment lets you see whether the indoors/outdoors factor had an effect. In your simple experiment, you did not vary setting: All your participants were indoors. Consequently, you could not learn whether setting had an effect. In the factorial experiment, you vary setting, so you can see its effect.

From your graph, you can see whether outdoors participants are, on the average, happier than indoor participants by seeing whether one line is higher than the other. If the outdoors line is higher (see Figure 12.6a), outdoors participants are happier; if the indoors line is higher, the indoors participants are happier. Often, just a glance at the lines will tell you which is higher. For example, in Figure 12.6b, which line is higher? Which group is happier?

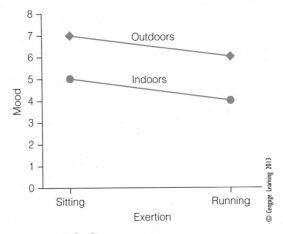

FIGURE **12.6a** Compare the height of these two lines.

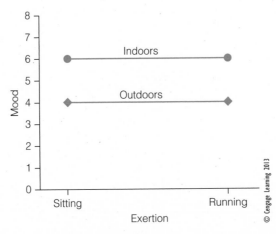

FIGURE **12.6b** Compare the height of these two lines.

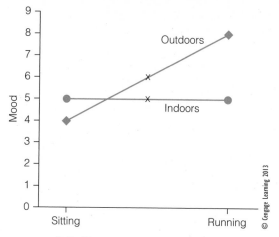

FIGURE **12.7** Compare the height of these two lines.

If you are having trouble telling which line is higher, put an "x" in the middle of each line (see Figure 12.7). If the outdoor line's midpoint is above the indoor line's midpoint, outdoor participants are happier. If the midpoint of the indoor line is above the outdoor line's midpoint, the indoors participants are happier. If neither midpoint is above the other, there is no evidence that the indoors–outdoors variable has an effect.

3. Running's different effect on different groups. The third thing that you can learn from a factorial experiment is whether varying exercise (sitting vs. running) has a different effect for participants who are indoors than for participants who are outdoors. For example, suppose that, when participants are inside, they would prefer to sit, but when they are outside, they would prefer to run. If you did your simple experiment inside, you would conclude that running decreases mood. If you did your simple experiment outside, you would conclude that running increases mood. However, if you did a factorial experiment, you would discover that running's effect *depended* on whether participants were inside or outside.

You can see whether the effect of running is different for indoors participants than for outdoors participants by seeing whether the lines have *different slopes*. Specifically, if the lines have the same slope—the lines are parallel (the two lines almost form an "equals" sign)—the effect of varying exercise is the same (i.e., equal or parallel) for both indoors and outdoors participants (as it is in Figures 12.6a and 12.6b). If, on the other hand, the lines are not parallel—they cross, they go away from each other (perhaps even forming an "<" sign), they come toward each other (perhaps even forming a ">" sign), or one line has a steeper slope that the other—then, the effect of varying the exercise appears to be different for indoor participants than it is for outdoor participants. Figure 12.8 shows some examples of graphs in which the two lines are not parallel.

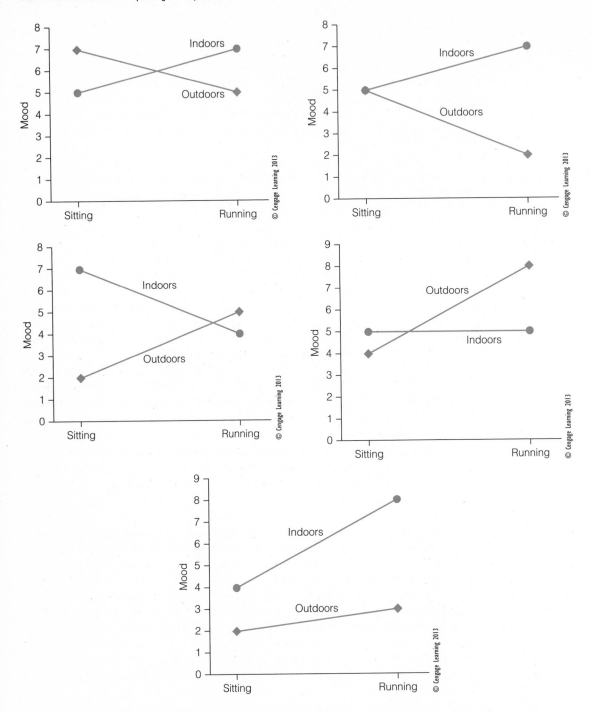

FIGURE **12.8** Five examples of graphs suggesting that running's effect is different indoors than outdoors.

The 2 × 2 Factorial Experiment: Using Tables to See the Three Patterns It Can Detect

To better understand how a 2 × 2 factorial experiment works, let's turn to an actual 2 × 2 experiment: Pronin and Wegner's (2007) experiment on manic thinking. In that experiment, the researchers were interested in seeing whether getting participants' thoughts to race would boost participants' moods—and whether this boost in mood would occur even when people were thinking negative thoughts.

To manipulate what participants thought, Pronin and Wegner had participants read aloud 60 statements that were either uplifting or depressing. Participants randomly assigned to the uplifting statements groups read a neutral statement—"Today is no better or worse than another day"—and then read statements that became increasingly positive. For example, the second statement participants in the uplifting statements group read was "I do feel pretty good today, though," whereas the last statement they read was "Wow! I feel great!" Participants randomly assigned to the depressing statements read the same first statement as the uplifting statements group ("Today is no better or worse than any other day") but then read statements that became increasingly negative. For example, the second statement they read was "However, I feel a little low today," and the last statement they read was "I want to go to sleep and never wake up."

To control how fast participants were thinking, Pronin and Wegner made participants read the statements either very quickly or very slowly. Specifically, although both fast and slow condition participants read statements aloud from a PowerPoint® presentation, the PowerPoint® presentation went nearly four times as fast in the fast condition as in the slow condition.

Because it was a 2 × 2, the factors were crossed. Consequently, half of the participants in the uplifting statements condition were randomly assigned to read the statements quickly (about twice as fast as students would normally read those statements) whereas the other half were to read the statements slowly (about half as fast as students would normally read those statements). Similarly, half the participants in the depressing statements condition were randomly assigned to read the statements quickly, whereas the other half were randomly assigned to read the statements slowly.

After the participants read the statements, they filled out several scales, one of which was a mood scale. Then, participants were debriefed and dismissed.

If you were to repeat the Pronin and Wegner's 2 (statement type: negative or positive) × 2 (statement speed: slow or fast), you might randomly assign participants twice: first to either negative statements or positive statements and then to either a slow or fast presentation. The result should be that one-fourth of your participants were in each of the four groups described in the following table:

Group 1: Gets negative statements and slow presentation	**Group 2:** Gets negative statements and fast presentation
Group 3: Gets positive statements and slow presentation	**Group 4:** Gets positive statements and fast presentation

© Cengage Learning 2013

Each Column and Each Row of the 2 × 2 Factorial Is Like a Simple Experiment

You could view each *row* of the 2 × 2 factorial as a simple experiment. With that view, you would see Pronin and Wagner's 2 × 2 factorial experiment as two simple experiments, both of which looked at whether participants are in better moods when statements are presented quickly than when statements are presented slowly. That is, as you can see from the following table, both "experiments" compare *slow presentation* to *fast* presentation.

Simple Experiment 1	Group 1	Group 2
(Effect of slow vs. fast presentation for negative statements)	*Negative statements* and *slow presentation* group	*Negative statements* and *fast presentation* group

Simple Experiment 2	Group 3	Group 4
(Effect of slow vs. fast presentation for positive statements)	positive statements and slow presentation group	positive statements and fast presentation group

© Cengage Learning 2013

You could also view each *column* of the 2 × 2 factorial as a simple experiment. With that perspective, you would Pronin and Wagner's 2 × 2 factorial experiment as two different simple experiments, both of which looked at whether participants are in a better mood after reading positive statements than after reading negative statements (see the following table).

Simple Experiment 3	Simple Experiment 4
Group 1	**Group 2**
Negative statements Slow presentation group	*Negative statements* Fast presentation group
Group 3	**Group 4**
Positive statements Slow presentation group	*Positive statements* Fast presentation group

© Cengage Learning 2013

If you looked at both the rows and the columns, you would see that the factorial experiment contains four simple experiments (see the following table).

	Column Containing Simple Experiment 3	Column Containing Simple Experiment 4
	Group 1	Group 2
Row Containing Simple Experiment 1	Negative statements Slow presentation group	Negative statements *Fast presentation* group
	Group 3	Group 4
Row Containing Simple Experiment 2	Positive statements Slow presentation group	Positive statements *Fast presentation* group

© Cengage Learning 2013

The 2 × 2 Yields Four Simple Main Effects

As the previous table shows, each row and each column of the 2 × 2 is essentially a simple experiment. As that table also shows, the 2 × 2 contains 2 rows and 2 columns, and therefore essentially contains four (2 + 2) simple experiments. If we did those four different experiments separately, we would be able to test for four separate effects. Because if we essentially do those four experiments all at once in a 2 × 2 design, we would be able to test for those four separate effects. All we would have to do is look at each "experiment" separately. That is, rather than look at the table as a whole, we need to look at each row and each column separately. To see how you would split your 2 × 2 factorial into four slices (each representing a simple experiment), imagine you have the following 2 × 2 table of means:

	Slow presentation groups	*Fast* presentation groups
Negative statement groups	<u>4</u> (the mean of the negative statement, <u>slow</u> presentation group)	6 (the mean of the negative statement, *fast* presentation group)
Positive statement groups	<u>12</u> (the mean of the positive statement, <u>slow</u> presentation group)	14 (the mean of the positive statement, *fast* presentation group)

© Cengage Learning 2013

For now, don't look at this table as a whole. Instead, start by looking only at the top row:

	Slow presentation groups	*Fast* presentation groups
Negative statement groups	<u>4</u> (the mean of the negative statement, <u>slow</u> presentation group)	6 (the mean of the negative statement, *fast* presentation group)

© Cengage Learning 2013

The top row looks like a simple, two-group experiment in which all participants read negative statements, but some read them at a <u>slow</u> rate and some read them at a *fast* rate. To whom do these results apply? The negative statement groups. What is the difference between how these groups are treated? Presentation rate—One gets a powerpoint presentation in which the slides are presented at a <u>slow</u> rate; the other gets a presentation in which the slides are presented at a *fast* rate. What is the difference between the group's means? The difference is 2 (6 − 4). Thus, for the negative statement groups, our estimate of the difference made by varying presentation rate is 2.

Next, look only at the second row:

	Slow presentation groups	*Fast* presentation groups
Positive statement groups	<u>12</u> (the mean of the positive statement, <u>slow</u> presentation group)	14 (the mean of the positive statement, *fast* presentation group)

© Cengage Learning 2013

This row looks like a two-group experiment in which all participants read positive statements, but some read them at a <u>slow</u> rate and some read them at a *fast* rate. What is the difference in how these groups are treated? Again, it's presentation rate—You are comparing <u>slow</u> presentation, positive statements group with the *fast* presentation, positive statements group. What do the two groups have in common? They are both positive statement groups. What is the difference between the group's means? It is 2 (14 − 12). So, our estimate of the difference made by presenting the slides at different rates to the positive statement groups is 2.

Now, let's focus on the first column:

	<u>Slow</u> presentation groups
Negative statement groups	<u>4</u> (the mean of the negative statement, <u>slow</u> presentation group)
Positive statement groups	<u>12</u> (the mean of the positive statement, <u>slow</u> presentation group)

© Cengage Learning 2013

The first column looks like another two-group experiment. In this one, all participants saw a slow slide show, but some read negative statements whereas others read positive statements. What do the groups have in common? They are both *slow* presentation groups. What is the difference between how the groups are treated? Statement type: One group read **negative** statements whereas the other group read **positive** statements. What is the difference between the group means? It is <u>8</u> (<u>12</u> − <u>4</u>). So, for the *slow* presentation groups, the effect of presenting different types of statements is estimated to be 8.

Finally, for our last slice of the table, let's focus on the second column:

	Fast presentation groups
Negative statement groups	6 (the mean of the **negative** statement, *fast* presentation group)
Positive statement groups	14 (the mean of the **positive** statement, *fast* presentation group)

© Cengage Learning 2013

The second column looks like yet another two-group experiment. In this one, all participants saw a *fast* slide show, but some read negative statements whereas others read positive statements. What do the groups have in common? They are both *fast* presentation groups. What is the difference between how the groups are treated? Statement type: One group read **negative** statement, whereas the other group read **positive** statements. What is the difference between the groups means? It is 8 (14 − 6). So, for *fast* presentation groups, the difference made by giving different groups different statements is estimated to be 8.

Let's review the three general principles you have just learned. First, in any row of a factorial design (e.g., in the negative statements row), one factor is kept constant (e.g., negative statements) while another factor varies (e.g., speed of presentation). In other words, all the groups in the same row have one thing in common (e.g., all were presented with negative statements), but those groups differ from each other in one way (e.g., the words were presented slowly to some participants and quickly to others).

Second, in any column of a factorial design (e.g., in the slow presentation column), one factor is kept constant (e.g., slow presentation) while another varies (e.g., whether the statements are positive or negative). In other words, although all the groups in the same column received the same level of one treatment (e.g., slow presentation), different groups received different levels of another treatment (e.g., some received positive statements, whereas others received negative statements).

Third, to tell whether giving participants different levels of a factor makes a difference, you need to compare participants who have received different levels of that factor. This may sound obvious, but it has some less than obvious implications for looking at the means, rows, and columns of a 2 × 2 table.

For looking at means, the implication is that *looking at any single mean in a table tells you nothing about a factor's effect*. A mean only has meaning when compared to other means.

For looking at rows and columns, the implication is that *if a level of a variable (e.g., "negative statements") is the label for a row or column, comparing the means within that row or column will tell you nothing about that variable's (e.g., statement type [negative vs. positive]) effect*. For example, looking at the negative statements row (i.e., the participants who all read negative statements) will not tell you anything about the effect of reading different types of statements because you will only be looking at participants who have read negative statements. If you want to know the effect of reading different types of statements, you need to compare participants who have read negative statements to participants who have read positive statements. You can compare participants who have read negative statements to participants who have read positive statements by looking at the slow presentation column or by looking at the fast presentation column. By the same logic, looking at the slow presentation column (i.e., all the participants who saw the slow slide show) does not tell you anything about the effect of presentation speed because all those participants were presented with the words at the same slow speed. To learn about the effect of varying presentation speed, you must look at rows that contain groups that vary in terms of the speed at which the information was presented: the negative statements row and the positive statements row.

If you look at the effect of varying presentation speed for the groups receiving negative statements, you are looking for presentation speed's **simple main effect** (also called a **simple effect**)—the effect of varying one factor for the groups/conditions receiving one particular level of the other factor—for the negative statement groups. If you look at the effect of presentation speed for the positive statement groups, you are looking for presentation speed's *simple main effect* for the positive statement groups.

Because any factor in a factorial experiment will have at least two simple main effects, you cannot refer to "**the** simple main effect of" a factor. For example, if you referred to "the simple main effect of presentation speed," nobody but you would know whether you were talking about the simple main effect of presentation speed for the positive statement groups or whether you were talking about the simple main effect of presentation speed for the negative statement groups.

In short, a 2 × 2 experiment, like four simple experiments, gives you four simple main effects: 2 from its 2 rows and 2 from its 2 columns. To know which simple main effect a row refers to, just fill in the blanks in the following sentence: "Subtracting the means in the (insert name of row [e.g., *"negative statements"*]) _____ row from each other produces an estimate of the simple main effect of the (insert name of your column variable [e.g., *"presentation speed"*]) _____ for the (insert name of row [e.g., *"negative statements"*]) _____ groups. Similarly, to know which simple main effect a column refers to, just fill in the blanks in the following sentence: "Subtracting the means in the (insert name of column [e.g., *"slow presentation"*]) _____ column from each other produces an estimate of the simple main effect of the (insert name of your row variable [e.g., *"statement type"*]) _____ for the (insert name of column [e.g., *slow presentation*]) _____ groups.

The 2 × 2 Yields Two Pairs of Simple Main Effects—One Pair for Each Factor

We have shown you that the 2 × 2 can yield four simple main effects—and that these four simple main effects are essentially the same four effects you would get if you did four separate simple experiments. However, doing 2 × 2 does not result in your getting four completely unrelated simple main effects. Instead, you get two *pairs* of simple main effects: one pair belonging to your first factor and one pair belonging to your second factor. For example, in the Pronin and Wagner study, one pair of simple main effects belongs to the statement type factor (the simple main effect of statement type [positive vs. negative] in the slow presentation condition and the simple main effect of statement type in the fast presentation condition). The other pair belongs to the presentation speed factor (the simple main effect of speed in the negative statements condition and the simple main effect of speed in the positive statements condition).

To capitalize on the two pairs of simple main effects that the 2 × 2 produces, researchers' analyses focus on three things:

1. combining (*averaging*) the first factor's pair of simple main effects to estimate the first factor's overall, average effect
2. combining (*averaging*) the second factor's pair of simple main effects to estimate the second factor's overall, average effect
3. contrasting (*subtracting*) a factor's pair of simple main effects to determine whether the factor has one effect on one group of participants but a different effect on a different group of participants

Averaging a Treatment's Simple Main Effects Lets You Estimate the Overall Main Effect: The Average Effect of Varying a Factor

To combine a factor's simple main effects, you average them. The *average* of a factor's two simple main effects allows you to estimate the factor's **overall main effect**: the average effect of varying that factor.

In the 2 (speed of thought: slow or fast) × 2 (type of thought: positive or negative) experiment, the researcher would average the two simple main effects of speed to get an estimate of the overall main effect for speed. To illustrate, suppose the simple main effect of presentation speed was +2 in the negative statements condition (the fast presentation, negative statements participants scored 2 points higher on the mood scale than the slow presentation, negative statements participants). Furthermore, suppose that the simple main effect of presentation speed was +4 in the positive statement conditions (the fast presentation, positive statements participants scored 4 points higher on the mood scale than the slow presentation, positive statements participants). In that case, the estimate for the overall main effect of presentation speed would be 3 (because the average of 2 and 4 is 3).

Similarly, to estimate the overall main effect for statement type (negative vs. positive), the researcher would average the two statement type simple main effects. If the overall statement type effect was statistically significant, it would mean that, on the average, participants who read negative statements were in a different mood than the participants who read positive statements.

One reason researchers emphasize overall main effects is convenience. It is easier to talk about a factor's one overall main effect than about its two simple main effects.

A more important reason for averaging the factor's two simple main effects into an overall main effect is that it allows us to make *general* statements about that factor's effects. To appreciate this advantage, consider the advantage of averaging the two simple main effects of speed. Because we combined two simple main effects, we are not limited to saying that speeding up thoughts improves mood if you are already thinking positive thoughts. Instead, we can say that, on the average, across conditions that varied from participants thinking negative thoughts to participants thinking positive thoughts, participants who thought faster were in better moods.

Subtracting a Treatment's Simple Main Effects Lets You Estimate the Interaction

But what if the simple main effect for speed of thought is different in the negative thought condition than in the positive thought condition? In that case,

1. You should *not* make a general statement about the effects of thought speed without mentioning that the effect of speeding up thought changes depending on whether the person is thinking negative thoughts or positive thoughts.
2. You should be happy that you can compare thought speed's simple main effects with each other because that comparison lets you know that the effect of speeding up thoughts depends on whether the person is thinking negative thoughts or positive thoughts.

By comparing speed's two simple main effects (the speed simple main effect for the negative statements condition and the speed simple main effect for the positive statements condition), you would be able to tell whether the effect of speeding up thoughts *depended on* whether participants were thinking positive or negative thoughts. If, for example, you found that that speeding up thoughts had a negative effect in the negative statements condition, but had a positive effect in the positive statements condition, you could say that the effect of speeding up thoughts depends on the type of statements participants read.

If the simple main effects of speed differ *depending* on the type of statement (i.e., the simple main effect of speed is different for the positive statements group than for the negative statements group), there is a speed × statement type of **interaction:** *the effect of a factor depends on the level of another factor*. If, on the other hand, speed's simple main effects do not differ from each other (speed has the same effect in the negative statements condition as it has in the positive statements condition), you do not have an interaction.

To review, one way—the simplest way—to see whether you have an interaction is to look at a factor's simple main effects. If the same factor's simple main effects are the same (e.g., both are 0 or both are 2), you do not have an interaction. Similarly, if one factor's simple main effects are both 2 and the other factor's simple main effects are both 7, you do not have an interaction. If, on the other hand, the *same* factor's simple main effects are significantly different from each other, you have an interaction.

Another way to see whether you have an interaction, is to see whether the effect of combining two factors is the same as the sum of their individual effects. For example, suppose your study obtained the following overall main effects. First, on the average, participants who have seen positive statements score 1 point higher on the mood scale than participants who have seen negative statements. Second, on the average, participants who saw a fast presentation score 2 points higher on the mood scale than participants who saw a slow presentation. Given these findings, you might guess that the positive, fast statement group would score 3 points higher than the negative, slow statements group. That is, you would add 1 (the positive statements effect) plus 2 (the fast statement effect) to get 3.

If you did not have an interaction, your guess of 3 would be a good one. The positive, fast statement group would score about 3 points higher than the negative, slow statements group.

If you had an interaction, however, your guess would be a bad one. The positive, fast statement group would not score about 3 points higher than the negative, slow statements group. Instead, it would either score substantially higher or lower than 3. Thus, whenever people say that the whole is different from the sum of its parts, they are referring to an interaction. (To review the difference between simple main effects, overall main effects, and interactions, see Table 12.1).

Why You Want to Look for Interactions:
The Importance of Moderating Variables

Interactions are important and common (see Table 12.2). Treatments will tend to have one effect on one group but another effect on another group. For example, eating grapefruit is good for most people, but not for people who are taking certain kinds of medications. For those people, eating grapefruit

TABLE **12.1** Simple Main Effects, Overall Main Effects, and Interactions

Simple Main Effects

Definition	The effects of one independent variable at a specific level of a second independent variable. The simple main effect could have been obtained by doing a simple, two-group experiment.
How to Estimate	Depends on which simple main effect you want to estimate. To estimate one the 2 × 2's four simple main effects, you could subtract the first row's two means (for instance, subtract the average for the slow presentation, negative thoughts group from the average for the fast presentation, negative thoughts group).
Question Addressed	What is the effect of the thought speed in the negative statements condition?

Overall Main Effect

Definition	The average effect of a treatment.
How to Estimate	Average a treatment's simple main effects. If the average of the two simple main effects is significantly different from zero, there is an overall main effect.
Question Addressed	What is the average effect of speeding up thoughts in this study?

Interaction

Definition	The effect of a treatment is different, depending on the level of a second independent variable. That is, the effect of a variable is uneven across conditions.
How to Estimate	Look at the differences between the same treatment's simple main effects. If the treatment's simple main effects are the same, there is no interaction. If, however, the treatment's two simple main effects differ significantly, there is an interaction.
Question Addressed	Does speeding up thoughts have a different effect on those who read negative statements than it has on those who read positive statements?

© Cengage Learning 2013

may kill them. For them, the positive main effect for eating grapefruit is unimportant relative to the dangerous grapefruit × drug interaction.[1]

Interactions do not have to be dangerous. The only requirement for an interaction is that the effect of combining treatments is different from the sum of their individual effects. For example, there is an interesting interaction involving the stimulants caffeine and nicotine. Nicotine, by itself, increases arousal. Caffeine, by itself, increases arousal. So, you might guess that drinking coffee and smoking would greatly increase arousal. However, for people who have a lot of nicotine in their system, caffeine actually *reduces* physiological arousal: The person who has smoked several cigarettes can wind down by drinking a caffeinated cola.[2]

---

[1] A popular and effective allergy medicine was taken off the market because of this deadly interaction.

[2] We are indebted to an anonymous reviewer for this example.

TABLE **12.2** Ways of Thinking About Interactions

Viewpoint	How Viewpoint Relates to Interactions
Chemical Reactions	Lighting a match, in itself, is not dangerous. Having gasoline around is not, in itself, dangerous. However, the *combination* of lighting a match in the presence of gasoline is explosive. Because the explosive effects of combining gas and lighting a match are different from simply adding their separate, individual effects, gasoline and matches interact.
Personal Relationships	John likes most people. Mary is liked by most people. *But* John dislikes Mary. Based only on their individual tendencies, we would expect John to like Mary. Apparently, however, like gasoline and matches, the combination of their personalities produces a negative outcome.
Sports	A team is not the sum of its parts. The addition of a player may do more for the team than the player's abilities would suggest—or the addition may help the team much less than would be expected because the addition upsets team "chemistry." In other words, the player's skills and personality may interact with those of the other players on the team. Knowing the interaction between the team and the player—how the two will mesh together—may be almost as important as knowing the player's abilities. Good pitchers get batters out. Poor hitters are easier to get out than good hitters are. However, sometimes a poor hitter may have a good pitcher's "number" because the pitcher's strengths match the hitter's strengths. Similarly, some "poor" pitchers are very effective against some of the league's best batters. Managers who can take advantage of these interactions can win more games than would be expected by knowing only the talents of the individual team members.
Prescription Drugs	Drug A may be a good, useful drug. Drug B may also be a good, useful drug. However, taking Drug A and B together may result in harm or death. Increasingly, doctors and pharmacists have to be aware of not only the effects of drugs in isolation but also of their combined effects. Ignorance of these interactions can result in deaths and in malpractice suits.
Making General Statements	Interactions indicate that you cannot talk about the effects of one variable without mentioning that the effect of that variable depends on a second variable. Therefore, if you have an interaction and are discussing a factor's effect, you need to say "but," "except when," "depending on," "only under certain conditions." Indeed, you will often see results sections say that the main effect was "qualified by a _____ interaction" or "the effect of the _____ variable was different depending on the level of (the other) variable."

Visually	If you graph an interaction, the lines will not be parallel. That is, the lines either already cross or if they were extended, they would eventually cross.
Mathematically	If you have an interaction, the effect of combining the variables is not the same as adding their two effects. Rather, the effect is better captured as the result of multiplying the two effects. That is, when you add 2 to a number, you know the number will increase by 2, regardless of what the number is. However, when you multiply a number by 2, the effect will depend on the other number. When doubling a number, the effect is quite different when the number to be doubled is 4 than when it is 1,000 or than when it is −40. To take another example of the effect of multiplication, consider the multiplicative effects of interest rates on your financial condition. If interest rates go up, that will have a big, positive effect on your financial situation if you have lots of money in the bank; a small, positive effect if you have little money in the bank; and a negative effect on your finances if you owe money to the bank (you will have to pay more interest on your debt).

Interactions do not have to involve reversing the treatment's original effect. To have an interaction, all that is required is that the effect of combining the treatments has an effect that is different from the sum of their individual effects. Because there are many ways that a combination of treatments can have an effect that is different from the sum of their individual effects, there are many kinds of interactions. To give you an idea of the variety of interactions that are possible, we will quickly show you concrete examples of seven different kinds of drug interactions.

To imagine the first three, suppose that we had two drugs, each of which individually boosted mood by an average of 1 point on a 10-point scale. In that case, if using both treatments together improved mood by 2 points, we would not have an interaction. We would not have an interaction because combining the factors had the same effect as we would expect from adding their individual effects together. In such a case, you could say that 1 (the effect of the first factor) + 1 (the effect of the second factor) = 2 (the effect you get when you use both factors).

If, however, combining the treatments produced any increase of mood that was significantly more than or significantly less than 2 points, you would have an interaction. In other words, you could say you had an interaction if 1(the main effect of the first factor) + 1 (the main effect of the second factor) ≠ 2 (the effect you get when you use both factors together)—if you interpret "≠" as meaning "is reliably different from." Thus, if combining the treatments had any of these three effects, you might have an interaction:

1. Together, they boost mood by 3 points (1 + 1 = 3: a "they make each other better" effect).

2. Together, they boost mood by only 1 point (1 + 1 = 1: a "one is enough" effect).
3. Together, they don't boost mood (1 + 1 = 0: a "better apart" effect).

To imagine the next two interactions, assume that neither drug has an effect. If each drug, by itself, has no effect and if taking both of them together also has no effect, you would not have an interaction. In that case, adding up the individual main effects of each factor (0 + 0) would result in their combined effect (0).

If, however, the combined effect was substantially bigger or smaller than 0, you would have an interaction. For example, you might have an interaction if

4. Together, they boost mood (0 + 0 = 3: a "two nothings make a something" effect).
5. Together, they reduce mood (0 + 0 = −2: an "it takes two to harm" effect).

To imagine the next two interactions, assume that one of the treatments had no effect (0) and the other had a strong effect (+2). In that case, the only way you would not have an interaction would be if combining them together boosted mood by about 2 (0 + 2) points. Thus, if combining the treatments had the following effects, you might have an interaction:

6. Combined, the two treatments had no effect (0 + 2 = 0; as though the neutral treatment deactivated the effective treatment).
7. Combined, the two treatments boosted mood by 4 points (0 + 2 = 4; as though the neutral treatment acted as a catalyst for the effective treatment).

As you have seen, there are many kinds of interactions. Although there are many kinds of interactions, you do not need to know the different kinds to know whether you have an interaction. Instead, all you need to know is that *if your factors' combined effect is reliably different from the sum of their individual effects, you have an interaction.*

Until now, we have focused our discussion of real-life interactions on drug interactions for two reasons. First, drug interactions are easy to visualize (although thinking of Grandpa being whisked to the hospital because he took Viagra with his blood pressure medicine may be something you don't want to visualize). Second, given that drug interactions are responsible for thousands of deaths a year in the U.S. alone, you have heard the term "drug interactions."

However, we do not wish to imply that drug interactions are the only interactions people know about or think about. Indeed, in social situations, most people suspect that the effect of their actions will depend on (will *interact* with) other factors. For example,

• Most people know that telling someone "congratulations" will have a good effect if she has just been promoted but a bad effect if she has just been fired.

- Most people suspect that, under some conditions, it pays to accuse others of something, but under some conditions, accusing others may backfire.

Research supports the popular notion that some treatments will have one effect on one group of participants, but a different effect on another group. For example, Rucker and Petty (2003) found that, of the two groups of participants who read about an employee who had a *bad* work ethic, the group that learned that the employee had accused his coworkers of having a bad work ethic liked the employee *more* than did the group that did not learn of the employee making such accusations. On the other hand, of the two groups of participants who read about an employee who had a *good* work ethic, the participants who learned that the employee had accused his coworkers of having a bad work ethic liked the employee *less* than did the participants who were not told that the employee had made any accusations. Thus, there was an employee reputation × accusation interaction.

You have seen that interactions—the effects of a combination of treatments being different from the sum of those factors' individual effects—may involve biological factors (e.g., drugs) and may involve social factors. Realize that interactions can involve any variables—even physical variables such as noise and lighting. For instance, consider the effects of two manipulated variables: (1) noise level and (2) perception of control. If you make a group of participants believe they have no control over the noise level in the room, increasing the noise level seriously harms performance. But for participants led to believe that they could control the noise level, increasing the noise level does *not* harm performance. Thus, noise level interacts with perceived control (Glass & Singer, 1972).

Because of this interaction between noise level and perceived control, you cannot simply say that noise hurts performance. You have to say that the effect of noise level on performance *depends* on (is moderated by) perceived control. In other words, rather than stating a simple rule about the effects of noise, you have to state a more complex rule. This complex rule puts qualifications on the statement that noise hurts performance. Specifically, the statement that noise hurts performance will be qualified by some phrase such as "depending on," "but only if," or "however, that holds only under certain conditions." In short, as Gernsbacher (2007) puts it, if the rule suggested by a main effect is like the spelling rule "*i* before *e*," the rule describing an interaction is more like "*i* before *e except after c*." Note that both in the case of spelling and real life, the rule described by the interaction is not as simple as the overall main effect, but it is more accurate. Thus, as Stanovich (2007) points out, interactions encourage us to go beyond simplistic "either/or" thinking (e.g., is your performance due to your personality or your environment) to "and" thinking (e.g., how is your performance affected by your personality, the environment, and the interaction between your personality and the environment).

Because the concept of interaction is so important, let's consider one more example. As a general rule, we can say that getting within 12 inches (30 cm) of another person in the general U.S. culture will make that person uncomfortable. Thus, the main effect of getting physically closer to someone is to produce a negative mood. However, what if the person who comes that close is extremely attractive? Then, getting closer may elicit positive feelings. Because the effect of interpersonal distance is moderated by attractiveness, we can say that there is an interaction between distance and attractiveness.

In short, you now know two facts about interactions. First, if there is an interaction involving your treatment, it suggests that the treatment has one effect under one set of conditions but another effect under another set of conditions. Second, interactions play an important role in real life because in real life, the right answer often depends on interactions between factors.

Interesting Questions in Modern Psychology Are Often Questions About Interactions

As psychology has progressed, psychologists have focused increasingly more attention on interactions. One reason psychologists focus on interactions is that psychologists have already discovered the main effects of many variables. We know how most individual variables act in isolation. Now, it is time to go to the next step—addressing the question, "What is the effect of combining these variables?" Put another way, once we learn what the general effect of a variable is, we want to find out what specific conditions may modify (moderate) this general, overall effect. Consequently, in Chapter 3 (pages 95–99), we encouraged you to generate research ideas that involved moderating variables. In other words, we encouraged you to do what many psychologists do—focus on interactions rather than main effects.

Another reason psychologists focus on interactions is that interactions are common. Consequently, psychologists now frame general problems and issues in terms of interactions. Rather than asking, "What is the (main) effect of personality and what is the (main) effect of the situation?" psychologists are now asking, "How do personality and the situation interact?" Asking this question has led to research indicating that some people are more influenced by situational influences than others (Snyder, 1984).

Similarly, rather than looking exclusively at the main effects of heredity and the main effects of environment, many scientists are looking at the interaction between heredity and environment. In other words, rather than asking, "What is the effect of a certain environment?" they are asking, "Are the effects of a certain environment different for some people than for others?"

Looking for these interactions sometimes produces remarkable findings. For example, psychologists have found that certain children may thrive in an environment that would harm children who had inherited a different genetic predisposition (Plomin, 1993). Eventually, such research may lead to new ways of educating parents. For instance, rather than telling parents the one right way to discipline children, parent education may involve teaching parents to identify their child's genetic predispositions and then alter their parenting strategies to fit that predisposition. In short, much of the recent research in psychology has involved asking questions that relate to interactions, such as "Under what conditions do rewards hurt motivation?"

External Validity Questions Are Questions About Interactions

We do not mean to imply that the interest in interactions is an entirely new phenomenon. Anyone interested in external validity is interested in interactions. If you are concerned that a treatment won't work on a certain type of person (women, minorities, retired adults), you are concerned about a treatment × type of person interaction. If you are concerned that a treatment that worked in one setting (a hospital) won't have the same effect in a different setting (a school), you are concerned about a treatment × setting interaction. If you are concerned that a treatment won't have the same effect in another culture, you are concerned about a treatment × culture interaction. If you are concerned that the superiority of one treatment over another will diminish over time, you are concerned about a treatment × time interaction. In summary, determining the external validity of your findings is often a matter of determining whether your treatment interacts with time, setting, culture, or type of participant.

Questions in Applied Psychology Are Often Questions About Interactions

Understandably, applied psychologists have always been interested in interactions. One of the founders of applied psychology, Walter Dill Scott, was fascinated by the fact that some people will like an advertisement that others will hate. Therefore, he investigated personality × type of ad interactions.

Most applied psychologists have shared Scott's interest in determining which treatments work on which type of people. For example, therapists know that a therapeutic approach (behavior therapy, drug therapy) that works well for some patients (e.g., individuals with phobias) may not work as well for others (e.g., individuals who are depressed). In other words, good therapists know about treatment × type of patient interactions.

In conclusion, the applied psychologist is keenly interested in interactions. When clients pay for advice, they do not want the expert to know only about main effects. That is, they do not want the expert to stop at saying, "My recommended course of action works in the average case, and so it may work for you." Instead, clients may quiz the expert about interactions involving the expert's proposed treatment. For example, they may ask, "Are there circumstances in which this treatment might make things worse—and does my case fit those circumstances?" To answer this question—that is, to know when a treatment will be helpful and when it will be harmful—the expert must know about the interactions involving that treatment.

Examples of Questions You Can Answer Using the 2 × 2 Factorial Experiment

Now that you have a general understanding of main effects and interactions, let's apply this knowledge to a specific experiment. If you were to replicate Pronin and Wegner's (2007) 2 (statement type: positive statements vs. negative statements) × 2 (speed: slow vs. fast) experiment we described earlier, you would look for three different kinds of effects (see Table 12.3).

First, you could look at the main effect of statement type: statement type's *average* effect. You could estimate the overall main effect for statement type by

TABLE **12.3** Questions Addressed by a 2 × 2 Experiment

Effect	Question Addressed
Overall main effect for speed	"On the average, does varying speed have an effect?"
Overall main effect for statement type	"On the average, does varying statement type have an effect?"
Interaction between speed and statement type	"Does the effect of speed differ *depending on* what type of statements (positive vs. negative) participants read?"
	Put another way,
	"Does the effect of statement type (positive vs. negative) differ *depending on* whether participants are in the slow vs. fast condition?"

© Cengage Learning 2013

averaging the two statement type simple main effects (the simple main effect for statement type for the slow presentation groups and the simple main effect for statement type for the fast presentation groups). If, on the average, positive statement participants were in a reliably better (or worse) mood than participants who read negative statements, you would have a statement type main effect.

Second, you could look at the main effect of speed: speed's average effect. You could estimate the overall main effect for speed by *averaging* the two speed simple main effects (the simple main effect for speed for the negative statement groups and the simple main effect for speed for the positive statement groups). If, on the average, participants who were in the fast-thought groups were in a reliably better (or worse) mood than participants in the slow-thought conditions, you would have a speed main effect.

Third, you could look at the interaction between speed and statement type: the extent to which speed's effect *differs* depending on what type of statement participants read. You could estimate the interaction effect by *subtracting* speed's two simple main effects from each other to get the *difference* between them. If there is no difference between speed's two simple main effects, there is no interaction: Speed's simple main effects are both the same, and the effect of speed does not depend on type of statement type. Without an interaction, if speed boosts mood by 2 points in the positive-statement conditions, it also boosts mood by 2 points in the negative statement-conditions. If, on the other hand, there is an interaction, the two simple main effects of speed will be different. Thus, if the following occurred, you would have an interaction:

- Speed amplified (multiplied) the effect of the positive/negative statements manipulation. Consequently, fast speed increased mood for the positive statement groups but decreased mood for the negative statement groups (e.g., the simple main effect for speed was +5 in the positive statements group, but −5 in the negative statements group).
- Speed muted (weakened) the positive/negative statements manipulation. Consequently, fast speed decreased mood in the positive-statement groups but increased mood in the negative-statement groups (e.g., the simple main effect for speed was −4 in the positive statements group, but +4 in the negative statements group).

- Speed had no effect in the negative conditions, but had a strong effect in the positive conditions.
- Speed had no effect in the positive conditions, but had a strong effect in the negative conditions.

To review, a significant main effect for statement type would mean that, on the average, varying statement type had an effect on mood. A significant main effect for speed would mean that, on the average, varying speed had an effect on mood. Finally, a significant interaction would mean that the combination of statement type and speed produces an effect that is different (more, less, or opposite) from what you would expect from knowing only statement type's and speed's separate effects.

To illustrate that an interaction indicates that the combination of factors has an effect that is different from the sum of the factors' overall main effects, imagine the following situation. Suppose the average effect of positive statements was to boost mood by 2 points and the average effect of fast presentation was to boost mood by 3 points. In addition, suppose we asked you to guess how much better a mood the participants who had the advantages of both receiving positive statements as well as a fast presentation speed (the positive statements/fast presentation group) were in relative to the participants who had neither of these advantages (the negative statements/slow presentation participants). In that case, you might, after adding up the effects of positive statements (+2) and fast statements (+3), say "5." In other words, you would guess that, in this case, $2 + 3 = 5$. If there is no interaction, your guess would be right.

But if there is an interaction, your guess would be wrong. The positive statements/fast presentation participants would *not* have a mood that averaged 5 points higher than the mean for the negative statement/slow presentation participants.

With an interaction, the positive/statements group mean might be much less than 5 points higher than the negative/slow presentation group—or it might be much more. In short, if you had a statement type × speed interaction, you couldn't predict the mood of the positive statements/fast presentation group merely by adding the statement type effects to the speed effects.

As you can imagine, significant interactions force scientists to answer such questions as, "Does working in groups cause people to loaf?" by saying, "Yes, but it depends on ..." or "It's a little more complicated than that." Psychologists do not give these kinds of responses to make the world seem more complicated than it is.

On the contrary, psychologists would love to give simple answers. Like all scientists, psychologists prefer simple, elegant explanations that involve few principles to more complex explanations. Therefore, psychologists would love to report main effects that are not qualified by interactions. Psychologists would like to say that speeding up people's thoughts always increases mood. However, if interactions occur, scientists have the obligation to report them— and in the real world, interactions abound. Only the person who says "Give me a match; I want to see if my gas tank is empty" is unaware of the pervasiveness of interactions. Most of us realize that when variables combine, the effects are different from what you would expect from knowing only their individual, independent effects.

Because we live in a world where we are exposed to more than one variable at a time and because the variables we are exposed to often interact, you may be compelled to do an experiment that captures some of this complexity. But how would you describe the results from such a factorial experiment?

POTENTIAL RESULTS OF A 2 × 2 FACTORIAL EXPERIMENT

You would describe the results of a 2 × 2 factorial experiment in terms of three different independent outcomes: (1) whether you had an overall main effect for your first independent variable, (2) whether you had an overall main effect for your second independent variable, and (3) whether you had an interaction. For example, you might report no main effect for your first variable, an effect for your second variable, and an interaction.

To understand that these three outcomes are independent, consider three other outcomes that are independent: three independent coin flips. Just as you can get a "heads" on your third coin flip regardless of the outcome of the first two flips, you can get an interaction, regardless of whether you have any main effects. Just as there are eight possible patterns of heads and tails you can get from three coin flips (HHH, HHT, HTH, HTT, TTT, TTH, THT, THH), there are eight different patterns of effects and noneffects you can get from a 2 × 2 (see Table 12.4).

General Approach for Deciphering a 2 x 2's Table of Means

If you did a study, how would you know which of these patterns of results you obtained? At some point, you would need to do a statistical analysis, such as an analysis of variance (ANOVA). Without such a statistical analysis, the patterns you observed in your data might be due to random error rather than to statistically reliable effects. Either before or after doing such an analysis, however, you would probably like to see what patterns exist in your data. Therefore, you might calculate the mean response for each group and then make a table of those means.

TABLE **12.4** Eight Potential Outcomes of a 2 × 2 Factorial Experiment

1. A main effect for variable 1	No main effect for variable 2	No interaction
2. No main effect for variable 1	A main effect for variable 2	No interaction
3. **A main effect for variable 1**	**A main effect for variable 2**	**No interaction**
4. A main effect for variable 1	A main effect for variable 2	An interaction
5. **No main effect for variable 1**	**No main effect for variable 2**	**An interaction**
6. A main effect for variable 1	No main effect for variable 2	An interaction
7. No main effect for variable 1	A main effect for variable 2	An interaction
8. No main effect for variable 1	No main effect for variable 2	No interaction

Note that having (or not having) a main effect has no effect on whether you will have an interaction. Similarly, having two main effects has no effect on whether you will have an interaction.

In the next sections, we will go through many examples of how to decode tables of means. You will see that you can decode any 2 × 2 table using the same five-step strategy:

1. Find the four simple main effects.
2. Group those four simple main effects into two pairs of main effects: one pair for each factor.
3. Average the first factor's pair of main effects to get an estimate of the first factor's overall main effect.
4. Average the second factor's pair of main effects to get an estimate of the second factor's overall main effect.
5. Subtract the first factor's two simple main effects from each other to get an estimate of the interaction's effect.

If you need help with any of these steps, please see Figure 12.9.

One Main Effect and No Interaction

Let's start by supposing you replicate the Pronin and Wegner (2007) experiment we discussed earlier. Using a 2 (positive statements vs. negative statements) × 2 (slow speed vs. fast speed) factorial experiment, suppose you found results like the ones displayed in Table 12.5. To understand your results, you might start looking at the experiment as though it were four separate simple experiments. Thus, if you look only at the first row, it is just like you are looking at the effects of speed in a simple experiment in which all participants read negative statements.

As you can see from the first row of Table 12.5, the slow speed/negative statements group was in the same mood (6) as the fast speed/negative statements group. Thus, varying speed had no noticeable effect in the negative statements condition.

To find out what happened in the positive statements groups, look at the second row. Note that looking at the second row is just like looking at a simple experiment that varied speed (while making all the participants read positive statements). As you can see by the fact that both the slow presentation and the fast presentation groups scored the same on the mood scale (8), varying speed had no noticeable effect in the positive statement condition.

Averaging the effect of speed over both the negative statements and the positive statements conditions, you find that speed's average (overall) effect was zero. Put another way, the slow speed groups' scores, on the average, were the same as the high speed groups'. Thus, there was no overall main effect for the speed manipulation.

Looking at the columns tells you about the effect of varying whether statements were negative or positive. For example, looking at the first column is like looking at a simple experiment that varied statement type (while having all participants read the statements slowly). As you can see, the positive statements group scores an average of 2 points *higher* (8 − 6 = 2) than the negative statements group. Thus, there may be a simple main effect for statement type in the slow speed condition.

Looking at the second column shows you the effect of statement type for the fast-speed participants. In a way, looking at the second column is like

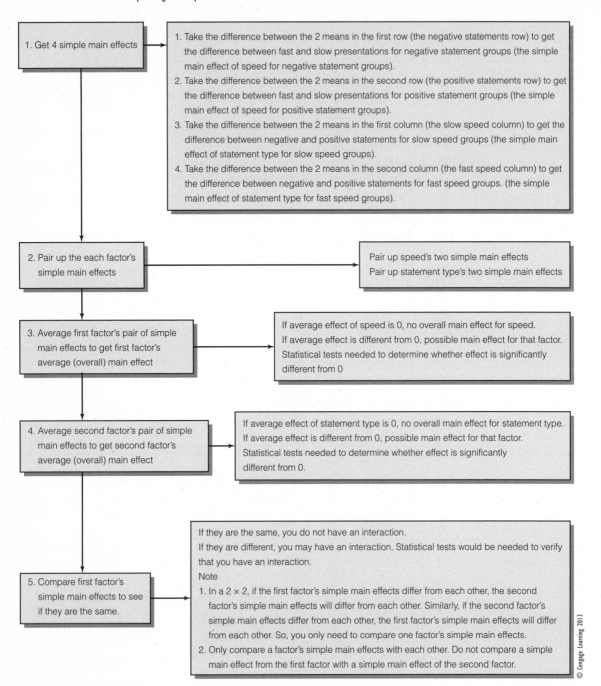

FIGURE **12.9** Five steps to decoding a 2 × 2 table of means.

TABLE **12.5** Main Effect for Statement Type, No Interaction

	Slow Speed	Fast Speed	Speed Simple Main Effects
Negative statements	6	6	0 ($\underline{6} - 6 = 0$)
Positive statements	8	8	0 ($\underline{8} - \underline{8} = 0$)

Statement type simple main effects   2 ($8 - 6 = 2$)   2 ($\underline{8} - \underline{6} = 2$)

Averaging a treatment's simple main effects gives us the treatment's overall main effect:

Simple main effect of *Statement type* in the slow presentation condition	2
Simple main effect of *Statement type* in the fast presentation condition	2
Average effect (overall main effect) of *Statement type*	4/2 = 2
Simple main effect of SPEED in the negative statements condition	0
Simple main effect of SPEED in the positive statements condition	0
Average effect (overall main effect) of SPEED	0/2 = 0

Comparing a treatment's simple main effects tells us whether there is an interaction:

Because there are no differences between statement type's two simple main effects (both are 2), there is no interaction. In other words, because the effect of statement type is not affected by the speed with which the statements are presented, there is no interaction.

looking at a simple experiment that manipulated statement type (while having all participants read the statements quickly). As you can see, the positive statement group scores an average of 2 points *higher* on the mood scale than the negative statement group ($\underline{8} - \underline{6} = 2$). Thus, there may be a simple main effect for statement type in the fast-speed condition.

Because statement type increases mood for both the slow-speed and the fast-speed participants, there seems to be an overall main effect for statement type. Our best estimate of this average effect of statement type is that positive statements increase mood 2 points more than negative statements do.[3]

Because statement type's effect does *not* differ depending on speed condition, there is *no* interaction between statement type and speed. Specifically, there is no interaction because positive statements increase mood by the same number of points (2) in the slow-statements condition as they do in the fast-statements condition.

Although making tables of means is a useful way to summarize data, perhaps the easiest way to interpret the results of a factorial experiment is to graph the means. To see how graphing can help you interpret your data,

---

[3] Because of random error, you don't know what the effect actually is. Indeed, without using statistical tests, you can't claim that you have a significant main effect or an interaction. However, because our purpose in this section is to teach you how to interpret tables and graphs and because the tables and graphs you will see in journal articles will almost always be accompanied by a statistical analysis, we will pretend—in this section—that any differences between means are statistically significant and due entirely to treatment effects.

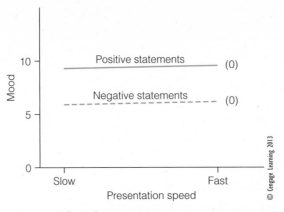

FIGURE **12.10** Main effect for statement type, no interaction.

Note: Numbers in parentheses represent the speed simple main effects. Thus, the simple main effect of speed was 0 in both the positive statements condition and the negative statements condition.

graph the data in Table 12.5. Before you plot your data, start by beginning to make a graph of a simple experiment that manipulates speed. Once you have a vertical *y*-axis labeled "Mood," and a horizontal *x*-axis that has labels for both slow presentation and fast presentation, you are ready to plot your data. Start by plotting two points representing the two means from the top row. Next, draw a line between those points and label that line "Negative statements." Then, plot the bottom row's two means. Draw a line between those two points and label that line "Positive statements." Your graph should look something like Figure 12.10. If it doesn't, please consult Box 12.1.

Figure 12.10 confirms what you saw in Table 12.5. Negative statements decreased mood relative to positive statements, as shown by the negative statements participants' line being below the positive statements participants' line. Speed did *not* affect mood, as shown by the fact that both lines stay perfectly level as they go from slow presentation (left) side to fast presentation (right) side of the graph.

Finally, there is no interaction between speed and statement type on mood, as shown by the fact that the lines are parallel.[4] The lines have the same slope because speed is having the same effect on the positive-statements

---

[4] If you have a bar graph instead of a line graph, you can't simply look to see if the lines are parallel because there are no lines. Instead, the key is to see whether the relationship between the dark bar and the light bar on the left side of the graph is the same as the relationship between the dark bar and the light bar on the right side of the graph. For example, if, on the left side of the graph, the dark bar is taller than the light bar, but on the right side of the graph, the dark bar is shorter than the light bar, you may have an interaction. Alternatively, you may convert the bar graph into a line graph by (a) drawing one line from the top, right corner of the first dark bar to the top, left corner of the other dark bar; and (b) drawing a second line from the top, right corner of the first light bar to the top, left corner of the other light bar.

BOX 12.1   Turning a 2 × 2 Table Into a Graph

If you have never graphed a 2 × 2 before, you may need some help. How can you graph three variables (the two factors and the dependent variable) on a two-dimensional piece of paper? The short answer is that you need to use two lines instead of one.

To see how to make such a graph, get a sheet of notebook paper and a ruler. Starting near the left edge of the sheet, draw a 4-inch (10.16 cm) line straight down the page. This vertical line is called the y-axis. The y-axis corresponds to scores on the dependent measure. In this case, your dependent measure is mood. So, label the y-axis "Mood."

Now that you have a yardstick (the y-axis) for mood, your next step is to put marks on that yardstick. Having these marks will make it easier for you to plot the means accurately. Start marking the y-axis by putting a little hash mark on the very bottom of the y-axis. Label this mark "0." Half an inch above this mark, put another mark. Label the mark "5." Keep making marks until you get to 20.

Your next step is to draw a horizontal line that goes from the bottom of the y-axis to the right side of the page. (If you are using lined paper, you may be able to trace over one of the paper's lines.) This horizontal line is called the x-axis. On the x-axis, you should put one of your independent variables. It usually doesn't matter which independent variable you put on the x-axis. However, some people believe you should put the moderator variable on the x-axis. If you don't have a moderator variable, those same people believe you should put the factor you consider most important on the x-axis. For the sake of this example, put "Presentation speed" about an inch below the middle of the x-axis. Then, put a mark on the left-hand side of the x-axis and label this mark "Slow." Next, put a mark on the right side of the x-axis and label it "Fast."

You are now ready to plot the means in the first row of Table 12.5. Once you have plotted those 2 means, draw a straight line between them. Label that line "Negative statements." Next, plot the 2 means in the right column of Table 12.5. Then, draw a line between those two points. Label this second line (which should be above your first line) "Positive statements." Your graph should look something like Figure 12.10.

group as it is on the negative-statements group. In this case, speed is having no (0) effect on either group.

Note that if you graph your data, you need to see only whether the lines are parallel to know whether you have an interaction. *If your lines are parallel, you do not have an interaction.* If, on the other hand, your lines have different slopes, you *may* have an interaction.[5]

Instead of having no interaction and a main effect for statement type, you could have no interaction and a main effect for speed. This pattern of results is shown in Table 12.6. From the top row, you can see that in the negative-statements groups, fast presentation increased mood by 5 points (10 − 5 = 5). Looking at the bottom row, you see that in the positive statements groups, fast presentation also increased mood scores by 5 points (10 − 5 = 5). By averaging the effect of speed over both the negative statements and the positive statements conditions, you could estimate that speed's average effect, the overall main effect of speed, was 5.

---

[5] Remember that because of random error, we don't know what the effect actually is. To know whether we had an interaction, we would need to do a statistical significance test.

TABLE **12.6** Main Effect for Speed, No Interaction

	Slow Speed	Fast Speed	**Speed Simple Main Effects**
Negative statements	5	<u>10</u>	5 (<u>10</u> − 5 = 5)
Positive statements	5	<u>10</u>	5 (<u>10</u> − 5 = 5)

Statement type simple main effects   0 (5 − 5 = 0)   0 (<u>10</u> − <u>10</u> = 0)

Averaging a treatment's simple main effects gives us the treatment's overall main effect:

Simple main effect of *Statement type* in the slow presentation condition	0
Simple main effect of *Statement type* in the fast presentation condition	<u>0</u>
Average effect (overall main effect) of *Statement type*	0/2 = 2
Simple main effect of SPEED in the negative statements condition	5
Simple main effect of SPEED in the positive statements condition	<u>5</u>
Average effect (overall main effect) of SPEED	10/2 = 5

Comparing a treatment's simple main effects tells us whether there is an interaction:

Because there are no differences between statement type's two simple main effects (both are 0), there is no interaction. In other words, because the effect of statement type is not affected by the speed with which the statements are presented, there is no interaction.

Whereas looking at the rows tells you about the effects of speed, looking at the columns tells you about the effect of statement type. Looking at the first column tells you about the effect of statement type in the slow presentation conditions. In the slow presentation conditions, the negative statement participants were in the same mood as the positive statements participants (both averaged 5 on the mood scale). Thus, there was no simple main effect of statement type in the slow presentation conditions.

Looking at the second column (the fast presentation column) tells you about the effect of statement type in the fast conditions. You can see that, in the fast presentation condition, the negative-statement participants were in the same mood as positive-statements participants (both averaged 10 on the mood scale). Thus, there was no simple main effect for statement type in the fast presentation condition.

To determine the overall main effect of statement type, compute the average of the two statement-type simple main effects. Because there was no (zero) observed effect for varying statement type in both the slow-presentation condition (the first column) and the fast-presentation condition (the second column), there is no (zero) overall main effect for varying statement type.

To determine whether there is a statement type × speed interaction, you could subtract the statement type simple main effects from each other (0 − 0 = 0). Or, you could subtract the speed simple main effects from each other (5 − 5 = 0). Either way, the result is zero, suggesting that you don't have a speed × statement type interaction. You do not have an interaction because the effect of speed is not affected by the statement-type variable: Increasing presentation speed increases mood by 5 points, regardless of whether statements are positive or negative.

Two Main Effects and No Interaction

Table 12.7 reflects another pattern of effects you might obtain. From the first row, you can see that, in the negative-statements groups, fast statements increased mood scores by 4 points ($8 - 4$). Looking at the second row, you see that, in the positive-statements groups, speed also increased mood scores by 4 points ($\underline{10} - 6$). Averaging the effect of speed over all the statement type conditions, you find that the average effect of speed (the overall main of speed) was to increase mood scores by 4 points.

Looking at the columns tells you about the effect of varying statement type. The first column tells you about what happens in the slow-presentation conditions. As you can see, in the slow-presentation conditions, the participants who read positive statements averaged 2 points higher ($6 - 4$) on the mood scale than those who read negative statements. Looking at the second column, you see that, in the fast-presentation conditions, participants who read positive statements score, on the average, 2 ($\underline{10} - \underline{8}$) points higher on the mood scale than participants who read negative statements. Because positive statements increase mood in both the slow-presentation and the fast-presentation groups, it appears that there is a statement type main effect.

TABLE **12.7** Main Effect for Speed and Statement Type, No Interaction

	Slow Speed	Fast Speed	Speed Simple Main Effects
Negative statements	4	$\underline{8}$	4 ($\underline{8} - 4 = 4$)
Positive statements	6	$\underline{10}$	4 ($\underline{10} - 6 = 4$)

Statement type simple main effects   2 ($6 - 4 = 2$)   2 ($\underline{10} - \underline{8} = 2$)

Averaging a treatment's simple main effects gives us the treatment's overall main effect:

Simple main effect of *Statement type* in the slow presentation condition	2
Simple main effect of *Statement type* in the fast presentation condition	$\underline{2}$
Average effect (overall main effect) of *Statement type*	$4/2 = 2$
Simple main effect of SPEED in the negative statements condition	4
Simple main effect of SPEED in the positive statements condition	$\underline{4}$
Average effect (overall main effect) of SPEED	$8/2 = 4$

Comparing a treatment's simple main effects tells us whether there is an interaction:

Because there are no differences between statement type's two simple main effects (both are 2), there is no interaction. In other words, because the effect of statement type is not affected by the speed with which the statements are presented, there is no interaction.

Comparing the two columns tells you that there is *no* interaction because the effect of statement type is unaffected by speed. As Table 12.7 demonstrates, the effect of statement type is independent of (does not depend on) speed. In this case, positive statements increase mood by 2 points, regardless of whether participants are in the slow or fast thought condition.

To look at this lack of statement type × speed interaction from a different perspective, look at the rows. Comparing the rows shows you that the effect of speed is unaffected by the type (positive or negative) of statement. Specifically, fast statements increase mood by 4 points for both the negative statement groups and the positive statements groups.

We have shown you two ways to use a table of means (like Table 12.7) to determine whether you have an interaction: (1) by comparing (subtracting) the simple main effects of the two rows, and (2) by comparing (subtracting) the simple main effects of the two columns. There is a third way. If either of a factor's simple main effects is the same as that factor's overall main effect, you do *not* have an interaction. Thus, in the current example, we know there is no interaction because the simple main effect of fast statements in the positive statements conditions (4) is the same as the overall main effect of fast statements (4).

Although a table of means gives you valuable information, you may understand your data better if you graph the means. To appreciate this point, look at a graph of Table 12.7's means: Figure 12.11. As you can see from the positive-statements line being above the negative-statements line, positive statements increased mood relative to negative statements. As you can see from both lines sloping upward as they go from the slow statements (left) side to fast statements (right) side of Figure 12.11, fast statements, relative to slow statements, increased mood. Finally, as you can see from the

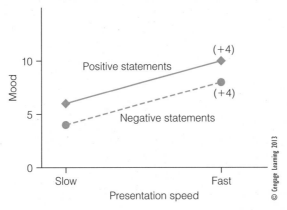

FIGURE **12.11** Main effect for statement type and speed, no interaction.

Note: Numbers in parentheses represent the simple main effects of speed. Thus, the simple main effect of speed was +4 in both the positive statements condition and in the negative statements condition.

parallel lines, there is no interaction between speed and statement type. The lines are parallel because speed affects the negative statements groups the same (parallel) way that it affects the positive statements groups.

Two Main Effects and an Interaction

Now imagine that you got a very different set of results from your statement type–speed study. For example, suppose you found the results in Table 12.8.

As the table shows, you have main effects for both speed and statement type. The average effect of fast statements is to *decrease* mood scores by 3 points, and the average effect of positive statements is to *increase* mood scores by 5.

Although, on the average, fast statements have an effect, the specific effect of fast statements varies depending on whether participants read negative or positive statements. In the positive-statements condition, fast statements, relative to slow statements, *increased mood* by 2 points (12 vs. 10). In the negative statements condition, on the other hand, fast statements *decreased* mood by 8 points (12 vs. 20). Because the effect of speed differs depending on statement type, there is an interaction.

To see this interaction, look at Figure 12.12a. As you can see, the lines are not parallel because the slope of the negative-statements line is different from the slope of the positive-statements line. This difference in slope indicates that the effect of speed is different for the negative-statements groups

---

TABLE **12.8** Main Effect for Speed and Statement Type, and a (Crossover) Interaction

	Slow Speed	Fast Speed	Speed Simple Main Effects
Negative statements	10	12	2 (12 − 10 = 2)
Positive statements	20	12	−8 (12 − 20 = −8)

Statement type simple main effects   10 (20 − 10 = 10)   0 (12 − 12 = 0)

Averaging a treatment's simple main effects gives us the treatment's overall main effect:

Simple main effect of *Statement type* in the slow presentation condition   10

Simple main effect of *Statement type* in the fast presentation condition   0

   Average effect (overall main effect) of *Statement type*   10/2 = 5

Simple main effect of SPEED in the negative statements condition   2

Simple main effect of SPEED in the positive statements condition   −8

   Average effect (overall main effect) of SPEED   −6/2 = −3

Comparing a treatment's simple main effects tells us whether there is an interaction:

Because there are differences between statement type's two simple main effects (one is 10, one is 0), there is an interaction. In other words, because the effect of statement type is affected by the speed with which the statements are presented, there is an interaction.

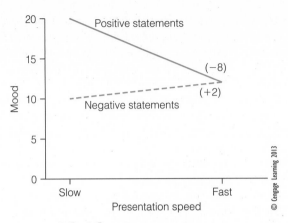

FIGURE **12.12a** Main effects for statement type and speed, and a crossover (disordinal) interaction.

Note: Numbers in parentheses represent the simple main effects of speed. Thus, the simple main effect of speed was −8 in the positive statements condition and was +2 in the negative statements condition.

than for the positive-statements groups. In this case, the negative-statements line slopes upward (indicating that negative-statements participants are in a *better* mood in the fast-statements condition than in the slow-statements condition), whereas the positive-statements line slopes downward (indicating that positive-statements participants are in a *worse* mood in the fast condition than in the slow condition). When the lines slope in opposite directions—indicating that the effect a treatment has with one group of participants is opposite from that treatment's effect on the other group of participants—the interaction is often called a **crossover interaction** (because the lines often *cross*).

Crossover interactions are due to the combination of treatments having effects that are different from the sum of the treatments' individual effects. In crossover interactions, the factor is having one *kind* of effect for some groups and a different kind of effect for other groups.

Crossover interactions can't be due to your measure being poor at distinguishing between different amounts of a variable. That is, even a crude measurement system that is not good at determining how much more of a variable one group has than another will not create the illusion that a factor is having a different *kind* of effect in one condition than another.

To illustrate that even a crude measure's flaws will not make you think that two things are changing in different directions when they are not, suppose you and your friend go on a diet. If you both gained weight on the diet, would any valid scale make it look like you lost weight whereas your friend gained weight? No. If both of you lost weight on the diet, would any valid scale make it look like you lost weight whereas your friend gained weight? No. If neither of you lost weight, would any valid scale, no matter how crude, make it look like you lost weight whereas your friend gained weight? No. Thus, even if you had an ordinal measure of weight, that measure would not create the illusion that a diet was having one kind of effect on you (e.g., making it look like you lost weight) but having a different kind of effect on your friend (e.g., making it look like your friend gained weight). Similarly, an ordinal measure couldn't create the illusion that a factor was having one *kind* of effect on one group and the opposite effect on another group. Because crossover interactions can't be due to having an ordinal level of measurement, crossover interactions are also called **disordinal interactions**: interactions that cannot be due to ordinal data but instead represent a "true" interaction.

Although having an ordinal level of measurement can't create crossover interactions, having an ordinal level of measurement can be responsible for another type of interaction. To see how, consider Figure 12.12b, in which both lines slope downward but the negative statements line slopes downward more sharply than the positive statements line. As you can see from Figure 12.12b, the lines are not parallel—and, therefore, there is an interaction.

What does this interaction mean? It could mean that negative statements participants were *more* affected by the fast-thought manipulation than the positive-statements participants were. That is, such an interaction could be due to the treatment having more of an effect in one condition than in another—but there is another possibility.

In this case, for you to say that the treatment had *more* of an effect in one condition than another, you must be able to say that the difference between a 15 and an 8 *on the variable you are measuring* is more than the difference between a 20 and an 18. You can legitimately say that—if your measure is an interval- or ratio-scale measure. However, if your measure is

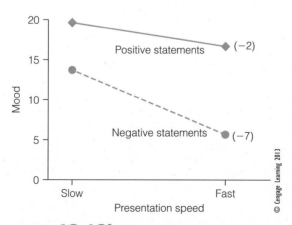

FIGURE **12.12b** Main effects for statement type and speed and an ordinal interaction.

Note: Numbers in parentheses represent the simple main effects of speed. Thus, the simple main effect of speed was only −2 in the positive statements condition but was −7 in the negative statements condition.

ordinal, you can't legitimately say that—and it is almost always possible that your measure might be ordinal. Consequently, any time a treatment *seems* to have more of an effect on some groups than on others, realize that you may be seeing an *illusion* caused by thinking you had interval measurement when you really had *ordinal measurement.*

To imagine how ordinal measurement could create such an illusion, suppose the mood score was based on participants selecting the adjective that best describes them. If checking "omnipotent" was scored as "20," checking "superior" was scored as "18," checking "powerful" was scored as "12," and checking "influential" was scored as "5," this measure may be ordinal. With such an ordinal measure, although going from 12 to 5 is clearly more of a decrease in *scores* than going from 20 to 18, going from 12 to 5 (from powerful to influential) may *not* be more of a difference in *actual mood* than going from 20 to 18 (from omnipotent to merely superior).

You could argue that we should not have used a measure that was so clearly ordinal. True, but realize that even measures that seem interval are not unquestionably interval. To illustrate, suppose you are very competitive with a friend about how much each of you is improving. Now, suppose she goes from a 62 on the first exam to a 70 on the second; you go from an 86 to a 92. She has clearly improved more in terms of points. But if you are using test scores as a measure of who has improved more in terms of knowledge, who has improved more? We couldn't say. Similarly, if, over the semester, she goes from Level 1 to Level 3 on the stair machine while you go from Level 3 to Level 4, who has increased their fitness more? We couldn't say.

In summary, even if you believe you have an interval measure, you may actually have an ordinal measure. An ordinal measure may make a treatment that has the same effect on all groups look like it has more of an effect for some groups than for others. Because interactions that *appear* to be due to a treatment having *more* of an effect in one condition than in another could actually be an illusion caused by having *ordinal* data, such interactions are called *ordinal interactions.*

An Interaction and No Main Effects

You have seen that you can have main effects with interactions, but can you have interactions without main effects? To answer this question, consider the data in Table 12.9 and Figure 12.13a.

From the graph (Figure 12.13a), you can see that the lines are not parallel. Instead, the lines actually cross. In this case, the crossover interaction is due to speed having one kind of effect (increasing mood) in the negative statements condition, but having an opposite effect (decreasing mood) in the positive statements condition. (In this case, "X" marks the crossover interaction. However, graphs of crossover interactions don't always look like Xs. As you can see from Figure 12.13b, a graph of a crossover interaction sometimes looks like a sideways "V" rather than an "X.")

Although you have an interaction between statement type and speed, you do not have a main effect for either statement type or speed. As you can tell by looking at Table 12.9, the slow-presentation groups have the same average

TABLE **12.9** No Main Effects for Speed or Statement With a (Crossover) Interaction

	Slow Speed	Fast Speed	Speed Simple Main Effects
Negative statements	10	<u>15</u>	+5 (<u>15</u> − 10 = 5)
Positive statements	15	<u>10</u>	−5 (<u>10</u> − 15 = −5)

Statement type simple main effects   +5 (15 − 10 = 5)   −5 (10 − <u>15</u> = −5)

Averaging a treatment's simple main effects gives us the treatment's overall main effect:

Simple main effect of *Statement type* in the slow presentation condition	+5
Simple main effect of *Statement type* in the fast presentation condition	−<u>5</u>
Average effect (overall main effect) of *Statement type*	0/2 = 0
Simple main effect of Sᴘᴇᴇᴅ in the negative statements condition	+5
Simple main effect of Sᴘᴇᴇᴅ in the positive statements condition	−<u>5</u>
Average effect (overall main effect) of Sᴘᴇᴇᴅ	0/2 = 0

Comparing a treatment's simple main effects tells us whether there is an interaction:

Because there are differences between statement type's two simple main effects (one is +5, one is −5), there is an interaction. In other words, because the effect of statement type *depends on* the speed with which the statements are presented, there is an interaction.

© Cengage Learning 2013

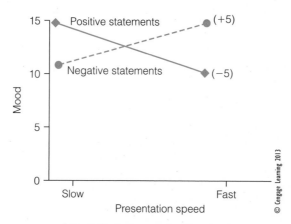

FIGURE **12.13a** No main effects and a crossover interaction: the classic "X"-shaped pattern.

Note: Numbers in parentheses represent the simple main effects of speed. Thus, the simple main effect of speed was −5 in the positive statements condition but +5 in the negative statements condition.

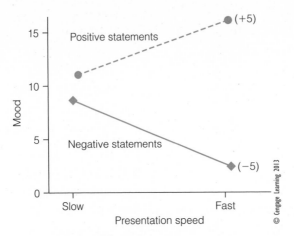

FIGURE **12.13b** No main effects and a crossover interaction: the classic "V"-shaped pattern.

Note: Numbers in parentheses represent the simple main effects of speed. Thus, the simple main effect of speed was +5 in the positive statements condition but −5 in the negative statements condition.

mood as the fast-presentation groups. Therefore, there isn't a speed main effect. Similarly, because the negative statements groups have the same average mood as the positive statements groups, there isn't a statement type main effect.

Thus, you would have to say that neither statement type nor speed has a main effect. Yet, you would not want to say that neither statement type nor speed has any effect. Instead, you would either say that (a) statement type has an effect, but its effect *depends* on the speed at which the statements are presented, or (b) speed has an effect, but its effect *depends* on whether the statements are positive or negative.

Regardless of whether you emphasize the effect of statement type (as in the first statement) or the effect of speed (as in the second statement), you cannot talk about the effect of one variable without talking about the other. In short, if you have an interaction, the effect of one variable depends on the other—even when you don't have any main effects.

An Interaction and One Main Effect

You have seen that you can have no main effects and an interaction. You have also seen that you can have two main effects and an interaction. Can you also have one main effect and an interaction? Yes—such a pattern of results is listed in Table 12.10 and graphed in Figure 12.14.

As Table 12.10 reveals, the average effect of varying statement type is zero. (The −2 effect of statement type in the slow condition is cancelled out by the +2 effect of statement type in the fast condition.) The average effect of varying speed, on the other hand, is to increase mood scores by 2. Note, however, that speed's effect is uneven. In the negative-statements condition,

TABLE **12.10** Main Effect for Speed With an Interaction

	Slow Speed	Fast Speed	Speed Simple Main Effects
Negative statements	10	<u>10</u>	0 (<u>10</u> − 10 = 0)
Positive statements	8	<u>12</u>	4 (<u>12</u> − 8 = 4)

Statement type simple main effects   −2 (8 − 10 = −2)   +2 (<u>12</u> − <u>10</u> = +2)

Averaging a treatment's simple main effects gives us the treatment's overall main effect:

Simple main effect of **Statement type** in the slow presentation condition	−2
Simple main effect of **Statement type** in the fast presentation condition	+2
Average effect (overall main effect) of **Statement type**	0/2 = 0
Simple main effect of Speed in the negative statements condition	0
Simple main effect of Speed in the positive statements condition	<u>4</u>
Average effect (overall main effect) of Speed	4/2 = 2

Comparing a treatment's simple main effects tells us whether there is an interaction:

Because there are differences between statement type's two simple main effects (one is −2, the other is +2), there is an interaction. In other words, because the effect of statement type is affected by the speed with which the statements are presented, there is an interaction.

© Cengage Learning 2013

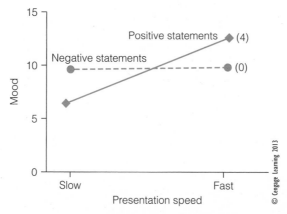

FIGURE **12.14** Main effect for speed with an interaction.

Note: Numbers in parentheses represent the simple main effects of speed. Thus, the simple main effect of speed was 4 in the positive statements condition but 0 in the negative statements condition. Because the simple main effect of speed differs depending on statement type, there is an interaction.

fast statements have no observable effect (10 − <u>10</u> = 0). But in the positive-statements condition, speed has an effect (<u>12</u> − 8 = 4). Because the effect of speed differs depending on statement type, there is a speed × statement type interaction.

Figure 12.14 tells the same story. By looking at that figure, you realize there may be an interaction because the lines are not parallel. They are not parallel because the effect of speed is dramatic in the positive statements conditions but undetectable in the negative statements conditions.

Whereas you can glance at Figure 12.14 and instantly see the interaction, seeing the main effects requires more mental visualization. If there is a main effect for statement type, one of the statement type lines should, on the average, be higher than the other. When one line is always above the other, it is easy to tell whether there seems to be a main effect. In this case, however, the lines cross—making it hard to tell whether one line is, on the average, above the other. If you get a ruler and mark the midpoint of each line, you will see that the midpoint of both lines is at the same spot. Or, you may realize that the negative statements line is below the positive statements line just as often and to the same extent as it is above the positive statements line. In either case, you would conclude that there is no main effect for statement type.

To determine whether there is a main effect for speed, you could mentally combine the two lines. If you do that, you would "see" that this combined line slopes upward, indicating a positive main effect for speed. (If you can't visualize such a line, you can create one in three steps. First, take a ruler and put a point halfway between the left ends of the two lines [i.e., a point halfway between the two slow statements points]. Second, put a point halfway between the right ends of the two lines [i.e., a point halfway between the two fast statements points]. Third, draw a line between the two points you just drew.) Alternatively, you could reason that because the positive-statements line slopes upward and the negative-statements line stays level, the average of the two lines would be to slope upward.

If you prefer not to think about lines at all, convert the graph into a table of means. To practice, take Figure 12.14 and see if you can convert it into a table resembling Table 12.10. Once you have your table of means, you will be able to see that the average for the fast-statements groups is higher than the average for the slow-statements groups.

No Main Effects and No Interaction

The last pattern of results you could obtain is to get no statistically significant results. That is, you could fail to find a statement-type effect, fail to find a speed effect, and fail to obtain an interaction between statement type and speed. An example of such a dull set of findings (possibly caused by a lack of power) is listed in Table 12.11.

TABLE **12.11** No Main Effects and No Interaction

	Slow Speed	Fast Speed
Negative statements	12	<u>12</u>
Positive statements	12	<u>12</u>

ANALYZING RESULTS FROM A FACTORIAL EXPERIMENT

You can now graph and describe all the possible patterns of results from a 2 × 2 experiment. But how would you analyze your results to determine whether a main effect or an interaction is significant?

You would probably use analysis of variance (ANOVA) to analyze your data. Using ANOVA to analyze a factorial experiment is similar to using ANOVA to analyze data from a single-factor experiment. In both cases, you are trying to figure out whether the difference you observed is due to more than just random error. The main difference is that instead of testing for one main effect, you will be testing for two main effects and an interaction. Thus, your ANOVA summary table might look like this:

Source Of Variance	Sum of Squares (SS)	df	Mean Square (MS)	F
Speed Main Effect (A)	900	1	900	9.00
Statement Main Effect (B)	200	1	200	2.00
Interaction (A × B)	100	1	100	1.00
Error (Within Groups)	3600	36	100	
Total	4800	39		

© Cengage Learning 2013

What Degrees of Freedom Tell You

Despite the fact that this ANOVA table has two more sources of variance than the ANOVA for the multiple-group experiment described in Chapter 11, most of the rules that apply to the ANOVA table for that design also apply to the table for a factorial design (see Box 12.2). In terms of degrees of freedom, you can still use the two rules we discussed in Chapter 11:

1. The number of treatment levels is one more than the treatment's degrees of freedom. Because the ANOVA summary table above states that the degrees of freedom for speed is 1, we know that the study used two levels of speed. Likewise, because the degrees of freedom for statement type is 1, we know the study used two statement types. Thus, the ANOVA summary table tells us that the study used a 2 × 2 design.
2. The total number of participants is one more than the total degrees of freedom. Therefore, because the ANOVA table states that the total degrees of freedom was 39, we know that there were 40 (39 + 1) participants in the experiment.

The only new rule is for the interaction's degrees of freedom. To calculate the interaction term's degrees of freedom, multiply the degrees of freedom for the main effects making up that interaction. For a 2 × 2 experiment, that would be 1 (df for first main effect) × 1 (df for second main effect) = 1. For a 2 × 3 experiment, that would be 1 (the df for the first main effect) × 2 (the df for the second main effect) = 2.

BOX 12.2 The Mathematics of an ANOVA Summary Table for Between-Subjects Factorial Designs

1. Degrees of freedom (df) for a main effect equal 1 less than the number of levels of that factor. If there are three levels of a factor (low, medium, high), that factor has 2 df.
2. Degrees of freedom for an interaction equal the product of the df of the factors making up that effect. If you have an interaction between a factor that has 1 df and a factor that has 2 df, that interaction has 2 df (because $1 \times 2 = 2$).
3. To get the total degrees of freedom, subtract 1 from the number of participants. Therefore, if you have 60 participants, the total degrees of freedom should be 59 ($60 - 1$).
4. To get the df for the error term, determine how many groups you had. Then, subtract the number of groups from the number of participants. In a 2×2, you have four (2×2) groups. Therefore, if you had 60 participants, your df error is 56 ($60 - 4$). If you had a 3×2, you would have six (3×2) groups. Therefore, the df error would be 54 ($60 - 6$). Another way to get the df error is to (a) add up the df for all the main effects and interactions and then (b) subtract that sum from the total degrees of freedom. Thus, if you had 1 df for the first main effect, 1 df for the second main effect, 1 df for the interaction, the sum of the df for your main effects and interactions would be **3** ($1 + 1 + 1$). You would then subtract that sum (3) from the df total. Thus, if the df total was 59, your error term would be 56 ($59 - 3$).
5. To get the mean square for any effect, get the sum of squares for that effect and then divide by that effect's df. If an effect's sum of squares was 300, and its df was 3, its mean square would be 100 (because $300/3 = 100$). If the effect's sum of squares was 300, and its df was 1, its mean square would be 300 (because $300/1 = 300$).
6. To get the F for any effect, get its mean square and divide it by the mean square error. If an effect's mean square was 100, and the mean square error was 50, the F for that effect would be 2 (because $100/50 = 2$).

What F and p Values Tell You

To determine whether an effect was significant, you look at the p value for the effect. If the p value is less than .05, the effect is statistically significant. If you do not have the p values, compare the F for that effect to the value given in the F table (see Table 3 in Appendix F) under the appropriate number of degrees of freedom. If your obtained F is larger than the value in the table, the effect is statistically significant.

What Main Effects Tell You: On the Average, the Factor Had an Effect

Usually, you will want to start your inspection of the ANOVA results by seeing whether any of your overall main effects are significant. If you have a significant effect for a factor, the overall effect of that factor is either to increase or to decrease scores on the dependent measure. If you have a significant main effect, your next step would be to find out whether this main effect is qualified by an interaction.

If the interaction was not significant, your conclusions are simple and straightforward. Having no interactions means there are no "ifs" or "buts" about your main effects. That is, you have not found anything that would lead you to say that the main effect occurs only under certain conditions.

For instance, if you have a main effect for statement type and no interactions, statement type had the same kind of effect throughout your experiment—no matter the speed at which participants read those statements. When you don't have interactions, you can just talk about the overall main effects. Thus, your Results section might resemble the following:

> A 2 (statement type: positive statements, negative statements) × 2 (speed: slow, fast) between-subjects ANOVA was conducted to assess the effects of statement type and speed on mood. Contrary to our hypothesis, this analysis did not find that the positive-statements group was in a better mood ($M = 11.8$) than the negative-statements group ($M = 12.2$), $F(1, 48) = 2.14$, *ns*. However, the analysis did reveal the expected main effect for speed, with participants in the fast-presentation groups scoring higher on mood ($M = 16$) than participants in the slow-presentation groups ($M = 8$), $F(1, 48) = 4.21$, $p = .04$, $r_{\text{effect size}} = .12$. The speed main effect was not qualified by a speed × statement type interaction, $F(1, 48) = 1.42$, *ns*.

If, on the other hand, you did have an interaction, you would replace the last sentence with something like the following:

> These findings are qualified, however, by a significant speed × statement type interaction, $F(1, 48) = 4.60$, $p = .04$, $\eta^2 = .08$. In the positive-statements conditions, the participants in the slow-presentation condition scored almost as high on the mood scale ($M = 16.1$, $SD = 3.33$) as participants in the fast-presentation condition ($M = 16.3$, $SD = 3.46$). However, in the negative-statements conditions, participants in the slow-presentation condition were in a worse mood ($M = 6.11$, $SD = 3.11$) than participants in the fast-presentation condition ($M = 10.1$, $SD = 3.22$).

What Interactions Usually Tell You: Combining Factors Leads to Effects That Differ From the Sum of the Individual Main Effects

As you just saw, when you have a significant interaction, describing the results is more complicated than when you don't have a significant interaction. Rather than talk about the means for each row or column as a whole (e.g., the mean for fast presentations groups), you need to talk about each individual cell mean (e.g., the mean for the slow-presentation/negative-statement group). You can't just talk about one factor's effect without also stating that the factor's effect depends on (is moderated by, is qualified by) a second variable.

At a more concrete level, having an interaction means that a treatment factor has a different effect on one group of participants than on another. In our statement type–speed example, having an interaction would mean that the simple main effect of statement type in the slow-statements condition is different from the simple main effect of statement type in the fast statements condition. In that case, because statement type's simple main effects would differ, rather than talking only about statement type's general, average, overall main effect, you would talk about the specific, individual, simple main effects that make up that overall main effect.

Before you can talk about those simple main effects, however, you must understand them. The easiest way to understand the pattern of the simple main effects—and thus understand the interaction—is to graph them.[6] In addition to looking at the slope of each line, examine the relationship between your lines to see why they aren't parallel.

If the lines are sloping in different directions, you have a disordinal interaction and you know that the interaction is not merely an artifact of having ordinal data. Therefore, you know that the treatment has one kind of effect in one condition and a different kind of effect in another.

If, on the other hand, both lines are sloping in the same direction but one is steeper than the other, you have an ordinal interaction. With an ordinal interaction, your interaction may be an artifact of having ordinal data. In other words, if you had used a measure that produced interval data, your interaction might disappear. Because your interaction may be due to having an ordinal measure, you can't be confident that the interaction is due to the treatment having a stronger effect on one group than on another.

PUTTING THE 2 × 2 FACTORIAL EXPERIMENT TO WORK

You now understand the logic behind the 2 × 2 design and can interpret the analysis of a 2 × 2 experiment. In the next sections, you will see how you can use the 2 × 2 to produce research that is more interesting, has greater construct validity, and has greater external validity than research produced by a simple experiment.

Looking at the Combined Effects of Variables That Are Combined in Real Life

Suppose you are aware of research showing that driving while talking on cell phones impairs driving performance and that you are also aware that driving while drunk impairs driving performance, but you are unaware of any research looking at the combined effects of both these factors. Then, if you think a study examining both factors would have practical implications (some people use cell phones while driving drunk) or theoretical implications (to see whether inattention is the mechanism for both), you might propose a study that looked at both factors at once (you would use a driving simulator rather than having people actually drive). Similarly, you could look at how driving performance was affected by the interaction of cell phone use with any of the following variables: sleep deprivation, caffeine, number of passengers in the car, or driving conditions.

Ruling Out Demand Characteristics

Suppose you design a simple experiment in which half of your participants think about their own death and the other half think about going to the dentist. You expect that participants made to think about death are more likely

---

[6] Interactions suggest that, rather than looking at the overall main effects, you should look at the individual simple main effects. One way to understand an interaction is to do statistical analyses on the individual simple main effects. The computations for these tests are simple. However, there are some relatively subtle issues involved in deciding which test to use.

to have happy thoughts than people made to think about going to the dentist. A friend criticizes your proposal, suggesting that your findings would just be the result of participants playing along with your hypothesis. To test that possibility, you could add two more groups to your study: a group that imagines how they would feel if they were in the death-salience condition and a group that imagines how they would feel if they were in the dental-pain condition (you are now proposing a replication of DeWall & Baumeister, 2007). If the pattern of results for the groups that really experienced the treatment is different from the pattern of results for the groups that role-played receiving the treatment, you would show that your hypothesis was not as intuitive as your friend believed. Note that all simple experiments involve comparing two levels of treatment (e.g., treatment 1 vs. treatment 2), and that you could convert most of those experiments into 2 (treatment 1 vs. treatment 2) × 2 (imagined vs. direct experience) experiments just by adding two groups that imagine—rather that actually—experience the treatments.

Adding a Replication Factor to Increase Generalizability

The generalizability of results from a single simple experiment can always be questioned. Critics ask questions such as, "Would the results have been different if a different experimenter had performed the study?" and "Would the results have been different if a different manipulation had been used?" Often, the researcher's answer to these critics is to do a **systematic replication**: a study that varies from the original only in some minor aspect, such as using different experimenters or different stimulus materials.

For example, Morris (1986) found that students learned more from a lecture presented in a rock-video format than from a conventional lecture. However, Morris used only one lecture and one rock video. Obviously, we would have more confidence in his results if he had used more than one conventional lecture and one rock-video lecture.

Morris would have benefited from doing a 2 × 2 experiment. Because the 2 × 2 factorial design is like doing two simple experiments at once, Morris could have (1) obtained his original findings and (2) replicated them with a different set of stimulus materials. Specifically, in addition to manipulating the factor of presentation type (conventional lecture vs. rock-video lecture), he could also have manipulated the replication factor of **stimulus sets**: the particular stimulus materials shown to one or more groups of participants. For example, he could have done a 2 (presentation type [conventional lecture vs. rock-video format]) × 2 (stimulus sets [material about Shakespeare vs. material about economics]) study. Because psychologists often want to show that the manipulation's effect can occur with more than just one particular stimulus set, experimenters routinely include stimulus sets as a replication factor in their experiments.[7]

---

[7] However, psychologists have not all agreed that the traditional, fixed-effects analysis of variance should be used to analyze such studies (see Clark, 1973; Cohen, 1976; Coleman, 1979; Kenny & Smith, 1980; Richter & Seay, 1987; Wickens & Keppel, 1983; Wike & Church, 1976).

Stimulus sets are not the only replication factor that researchers use. Some researchers employ more than one experimenter to run the study and then use experimenter as a factor in the design.

Some of these researchers use experimenter as a factor to show the generality of their results. Specifically, they want to show that certain experimenter characteristics (gender, attractiveness, status) do not alter the treatment's effect.

Other researchers use experimenters as a factor to establish that the experimenters are not biasing the results. For instance, Ranieri and Zeiss (1984) were worried that experimenters might unintentionally influence participants' responses to their experiment's dependent measure: a self-report scale of mood. Therefore, they used three experimenters and randomly assigned participants to experimenter. If different experimenters had obtained different patterns of results, Ranieri and Zeiss would have suspected that the results might be due to experimenter effects rather than to the manipulation itself.

Thus far, we have discussed instances in which the investigator's goal in using the factorial design was to increase the generalizability of the experimental results. Thus, in a study that uses stimulus set as a replication factor, researchers hope that the treatment × stimulus set interaction will not be significant. Similarly, most researchers who use experimenter as a factor hope that there will not be a treatment × experimenter interaction.

Using an Interaction to Find an Exception to the Rule: Looking at a Potential Moderating Factor

Often, however, researchers are interested in finding an interaction. For example, you may read about a study's results and say to yourself, "But I bet that would not happen under _____ conditions." In that case, you should do a study in which you essentially repeat the original experiment except that you add what you believe will be a moderating factor that will interact with the treatment.

To see how a moderating factor experiment would work, let's look at a study by Jackson and Williams (1985). Although aware of the phenomenon of social loafing—individuals don't work as hard on tasks when they work in groups as when they work alone—Jackson and Williams felt that social loafing would not occur on extremely difficult tasks. Therefore, they did a study, which, like most social-loafing studies, manipulated whether participants worked alone or in groups. In addition, they added what they thought would be a moderating factor—whether the task was easy or difficult (e.g., whether participants completed a simple maze or a challenging maze).

As expected, and as other studies had shown, social loafing occurred. But, social loafing occurred only when the task was easy. When the task was difficult, the reverse of social loafing occurred: Participants worked harder in groups than alone. This interaction between task difficulty and number of workers confirmed Jackson and Williams's hypothesis that task difficulty moderated social loafing (see Figure 12.15).

To see how you could take advantage of Jackson and Williams's research strategy, let's review what they did. With part of their study, they replicated

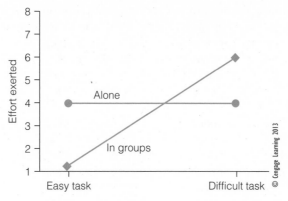

FIGURE **12.15** Interaction between task difficulty and number of coworkers on effort.

Note: Effort was scored on a 1-to-7 scale, with higher numbers indicating more effort.

an existing finding (the social-loafing main effect). With the other part, they tested whether another variable would moderate (interact with) the social-loafing main effect. If you like this strategy of proposing a study that tests both a safe prediction (e.g., a replication) and a risky prediction (e.g., an untested interaction), consider a moderating factor study. Note that this strategy works well if you have an idea about how to neutralize a bad effect (e.g., a training program that would reduce frustration's effect of increasing aggression) or intensify a good effect (e.g., instructions that may improve the positive effects of a placebo). For more tips on designing a moderating-factor study, see pages 95–99.

Using Interactions to Create New Rules

Although we have discussed looking for an interaction to find an exception to an existing rule, some interactions do more than complicate existing rules. Some interactions reveal new rules. Consider Tversky's (1973) 2 × 2 factorial experiment. She randomly assigned students to one of four conditions:

1. Student expected a multiple-choice test and received a multiple-choice test.
2. Student expected a multiple-choice test and received an essay test.
3. Student expected an essay test and received a multiple-choice test.
4. Student expected an essay test and received an essay test.

She found an interaction between type of test expected and test received. Her interaction showed that participants did better when they got the *same* kind of test they expected (see Figure 12.16).

Similarly, a researcher might find an interaction between mood (happy, sad) at the time of learning and mood (happy, sad) at the time of recall. The

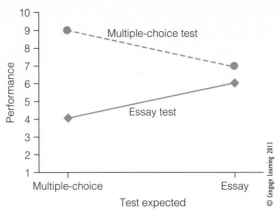

FIGURE **12.16** The effect of expectations and type of test on performance.

interaction might reveal that recall was best when participants were in the *same* mood at the time of learning as they were at the time of recall. As you can see, the 2 × 2 experiment may be useful for you if you are interested in assessing the effects of *similarity*.

Conclusions About Putting the 2 × 2 Factorial Experiment to Work

As you have seen, expanding a simple experiment into a 2 × 2 experiment allows you to test more—and more interesting—hypotheses. You can look at the main effect of the factor you would have studied with the simple experiment, plus the main effect of an additional factor, plus the interaction between those two factors. In many cases, the hypothesis involving the interaction may be the most interesting.

HYBRID DESIGNS: FACTORIAL DESIGNS THAT ALLOW YOU TO STUDY NONEXPERIMENTAL VARIABLES

Rather than converting a simple experiment into a 2 × 2 experiment by adding a second experimental factor, you could convert a simple experiment into a 2 × 2 hybrid design by adding a nonexperimental factor. The nonexperimental factor could be any variable that you cannot randomly assign, such as age, gender, or personality type.

Hybrid Designs' Key Limitation: They Do Not Allow Cause–Effect Statements Regarding the Nonexperimental Factor

In such a hybrid 2 × 2 design, you could make cause–effect statements about the effects of the experimental factor, but *you could not make any cause–effect statements regarding the nonexperimental factor*. Thus, although the

TABLE **12.12** The Hybrid Design: A Cross Between an Experiment and a Nonexperiment

	Men	Women	Gender Simple Main "Effects"
Negative statements	10	<u>12</u>	2 (<u>12</u> − 10 = 2)
Positive statements	8	<u>14</u>	6 (<u>14</u> − 8 = 6)
Statement type simple main effects	−2 (8 − 10 = −2)	+2 (<u>14</u> − <u>12</u> = +2)	

Averaging a factor's simple main effects gives us the factor's overall main effect:

Simple main effect of *Statement type* for men	−2
Simple main effect of *Statement type* for women	+<u>2</u>
Average effect (overall main effect) of *Statement type*	0/2 = 0
Simple main "effect" of **Gender** in the negative statements condition	2
Simple main "effect" of **Gender** in the positive statements condition	<u>6</u>
Average "effect" (overall main effect) of **Gender**	8/2 = 4

Comparing a treatment's simple main effects tells us whether there is an interaction:

Because there are differences between statement type's two simple main effects (i.e., −2 is different from +2), there is an interaction. In other words, because the effect of statement type is different for men than for women, there is a statement type × gender interaction

Note that the hybrid 2 × 2 design answers two questions that the simple experiment does not:

1. Do male and female participants differ on the dependent variable? (Answered by the gender main effect)
2. Does the effect of statement type differ depending on which group (men or women) we are examining? (Answered by the gender × treatment interaction)

study described in Table 12.12 includes gender of participant as a variable, the study does not allow us to say anything about the *effects* of a participant's gender.

You can't make cause–effect (causal) statements regarding the effects of the participant's gender because your two groups may differ not only in terms of gender but also in hundreds of other ways. For example, they may differ in terms of college major, age, self-esteem, religiosity, parental support, or loneliness. Any one of the hundreds of potential differences between the groups might be responsible for the difference in behavior between the two groups. Therefore, you cannot legitimately say that gender differences—rather than any of these other differences—caused your two groups to behave differently.

To help emphasize that *you can make causal statements only about those independent variables that you randomly assign*, randomly assigned variables are often called "true" independent variables or "strong" independent variables. In contrast, predictor variables that are not randomly assigned are called "weak" independent variables to highlight the fact that you can't determine whether they have an effect.

Reasons to Use Hybrid Designs

If you cannot make causal statements about the nonexperimental factor, why would you want to add a nonexperimental variable to your simple experiment? The most obvious and exciting reason is that you are interested in that nonexperimental variable.

To see how adding a nonexperimental variable (age of participant, introvert–extrovert, etc.) can spice up a simple experiment, consider the following simple experiment: Participants are either angered or not angered in a problem-solving task by a confederate who poses as another participant. Later, participants get an opportunity to punish or reward the confederate. Obviously, we would expect participants to punish the confederate more when they had been angered. This simple experiment, in itself, would not be very interesting.

Holmes and Will (1985) added a nonexperimental factor to this study— whether participants were Type A or Type B personalities. (People with Type A personalities are thought to be tense, hostile, and aggressive, whereas people with Type B personalities are thought to be more relaxed and less aggressive.) The results of this study were intriguing: If participants had *not* been angered, Type A participants were more likely to punish the confederate than Type B participants. However, if participants had been angered, Type A and Type B participants behaved similarly (see Figure 12.17).

Likewise, Hill (1991) could have done a relatively uninteresting simple experiment. He could have determined whether research participants are more likely to want to talk to a stranger if that stranger is supposed to be "warm" than if the stranger supposedly lacks warmth. The finding that people prefer to affiliate with nice people would not have been startling.

Fortunately, Hill conducted a more interesting study by adding another variable: need for affiliation. He found that participants who were

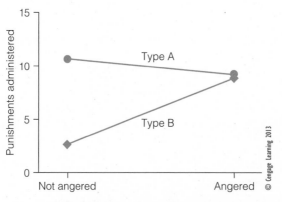

FIGURE **12.17** The effect of being angered on the aggressiveness of type a and type b personality types.

Source: Adapted from Holmes, D. S., & Will, M. J. (1985). Expression of interpersonal aggression by angered and nonangered persons with Type A and Type B behavior patterns by D. S. Holmes and M. J. Will, 1985, *Journal of Personality and Social Psychology, 48,* 723–727.

high in need for affiliation were very likely to want to interact with an allegedly warm stranger, but very unlikely to want to interact with a stranger who allegedly lacked warmth. For low need-for-affiliation participants, on the other hand, the alleged warmth of the stranger made little difference.

As you have seen, adding a nonexperimental factor can make a study more interesting. As you will see in the next sections, you can add a nonexperimental variable to a simple experiment for most of the same reasons you would add an experimental variable: to increase the generalizability of the findings, to look for a similarity effect, and to look for a moderating factor. In addition, you may add a nonexperimental factor to increase your chances of finding a significant effect for your experimental factor.

Increasing Generalizability

You could increase the generalizability of a simple experiment that used only men as participants by (a) using both men and women as participants and then (b) making gender of the participant a factor in your design. This design would allow you to determine whether the effect held for both men and women. For example, researchers (Crusco & Wetzel, 1984) wondered whether restaurant servers' "Midas touch"—touching customers results in bigger tips—holds for both men and women customers. (It does.)

Some effects do not generalize across genders. For example, whereas men were *more* likely to say "yes" to a stranger's request to have sex than to say "yes" to a stranger's request to go on a date, women were *much less* likely to say "yes" to a stranger's request to have sex than to say "yes" to a stranger's request to go on a date (Clark & Hatfield, 2003).

In addition to seeing whether an effect generalizes across genders, you could see whether an effect generalizes across age, experience, or personality. For example, researchers have found that sleep-deprived younger drivers benefit more from a short nap than older drivers (Sagaspe et al., 2007); that both police officers and experienced judges are more likely to think that a videotaped confession is voluntary when the camera recording the confession is focused more on the suspect than on the detective (Lassiter, Diamond, Schmidt, & Elek, 2007); and that, on math problems, people who normally do well in math are more likely to choke under pressure than people who normally do not do so well (Beilock & Carr, 2005).

Studying Effects of Similarity: The Matched Factors Design

If you were interested in similarity, you might include some participant characteristic (gender, status, etc.) as a factor in your design, while manipulating the comparable (matching) experimenter or confederate factor. For example, if you were studying helping behavior, you could use style of dress of the participant (well-dressed/casual) and style of dress of the confederate as factors in your design. You might find this interaction: Well-dressed participants were more likely to help confederates who were well-dressed, but casually dressed participants were more likely to help confederates who were casually dressed. This interaction would suggest that similarity of dress influences helping behavior (see Figure 12.18).

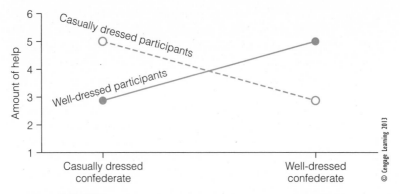

FIGURE **12.18** A hybrid design in which an interaction represents a similarity effect.

Finding an Exception to the Rule: The Moderating Factors Design

Looking for the effects of similarity is not the only reason you would want to examine interactions involving participant characteristics. As we mentioned earlier, you might look at interactions involving participants to see whether a treatment that works with one type of person is as effective with another type of person. The treatment could be any intervention—from a therapy technique to a teaching style.

For instance, if you thought that intelligence would be a moderating variable for the effectiveness of computerized instruction, you might use intelligence as a factor in your design. To do this, you would first give your participants an IQ test and then divide them into two groups (above-average intelligence and below-average intelligence). Next, you would randomly assign the high-intelligence group to condition so that half of them were in computerized instruction and half were in lecture instruction. You would do the same for the low-intelligence group.

This hybrid study might reveal some interesting findings. Suppose you found that computerized instruction substantially increases learning for low-IQ children but slightly decreases learning for high-IQ children. If you had done only a simple experiment, you might have found a significant positive effect for computerized instruction. On that basis, you might have made the terrible mistake of recommending that computerized instruction be used to teach all children.

Boosting Power: The Blocked Design

Suppose you were solely interested in seeing whether instructional technique had an effect and you had no interest in either IQ or the interaction between IQ and instructional technique. Even then, you might still include IQ as a factor in your experiment. Specifically, before the study begins, you might divide your participants into two *blocks* (groups): the low-IQ block and the high-IQ block. Then, you would randomly assign each member of the high-IQ group to instruction condition, thereby ensuring that half of the high-IQ participants are assigned to the computerized instruction condition and half are assigned to the lecture condition. Next, you would randomly assign each member of the low-IQ block to instruction condition.

In other words, you would do exactly the same study that we just recommended you do if you were looking at IQ as a moderating factor. However, this study would be called a **blocked design**: a factorial design in which, to boost power, participants are first divided into groups (blocks) on a participant variable (e.g., low-IQ block and high-IQ block) that is highly correlated with the dependent measure, and then participants from each block are randomly assigned to experimental condition.

The difference between doing this blocked design and doing the moderating factors study we just described is not *what* you are doing, but *why* you are doing it. If you are using a blocked design, you do not care about your blocking variable, and you do not care about the interaction between your blocking variable and your treatment. You are using the blocking variable solely to boost your chances of finding a statistically significant effect for your treatment.

To understand how the blocking variable will increase your chances of finding the treatment's effect, you first have to understand that just like decreasing the amount of dust on a microscope's lens increases your chances of seeing differences between cells, decreasing error variance increases your chances of seeing differences between treatment conditions. Then, you have to understand that blocked designs reduce error variance.

To understand how blocked designs reduce error variance, realize what error variance is—variability that is not accounted for in your study. If you use a simple experiment, individual differences in IQ are not accounted for; consequently, any variations in scores due to individual differences in IQ contribute to error variance. If, on the other hand, you use a blocked design that blocks on IQ, you account for some of the variance due to individual differences in IQ, thereby reducing your error variance. In a sense, you use your blocking variable to soak up variance that would otherwise be error variance. By shrinking the error variance, you make your treatment's effect easier to spot.

CONCLUDING REMARKS

We hope that you understand how factorial designs can help you refine your existing research ideas and generate new research ideas. We know that understanding factorial designs, one of the most common research methods in psychology, will increase your ability to read, understand, and evaluate other people's research.

SUMMARY

1. Factorial experiments allow you to look at the effects of more than one independent variable at a time.

2. The simplest factorial experiment is the one that looks at the effects of only two levels of two independent variables: the 2 × 2 ("two by two") experiment.

3. The 2 × 2 produces four simple main effects—two for each factor.

4. When looking at a table of means for a 2 × 2, realize that the individual means don't tell you anything about effects. To see if a factor makes a difference, you need to take the difference between means of groups that differ on that factor. The first step to seeing if a factor makes a difference is to estimate its simple main effects. You can estimate your experiment's four simple main effects by (1) taking the differences between the means in the first row, (2) taking the differences between the means in the second row, (3) taking the differences between the means in the first column, and (4) taking the differences between the means in the second column.

5. Averaging a factor's pair of simple main effects gives you an estimate of that factor's overall main effect.

6. In addition to allowing you to see the average effects of two factors in one experiment, the 2 × 2 experiment allows you to see whether the factors' combined effects are different from the sum of their individual effects.

7. Whenever the effect of combining two independent variables is different from the sum of their individual effects, you have an interaction. That is, if a factor's simple main effects differ from each other, you have an interaction. An interaction means that one independent variable's effect depends on the level of a second (moderating) variable. For example, the independent variable may have one effect when the second factor is absent and a different effect when the second factor is present.

8. Subtracting a factor's two simple main effects gives you an estimate of the interaction. If the factor's two simple main effects are the same, you do not have an interaction.

9. Interactions often indicate that a general rule does not always apply. For instance, a treatment × distraction interaction indicates that the treatment does not have the same effect on people who are being distracted as on people who are not being distracted.

10. Interactions can most easily be observed by graphing your data. If your two lines aren't parallel (i.e., they have different slopes), you may have an interaction. You need statistical tests to determine whether the difference in slopes you observed is a statistically reliable difference.

11. A significant interaction usually qualifies main effects. Thus, if you find a significant interaction, you can't talk about your main effects without referring to the interaction. Put another way, because the interaction means that a factor's overall main effect is different from its simple main effects and that its simple main effects differ from each other, an interaction often forces you to talk about the individual simple main effects. If, on the other hand, you don't have an interaction, you can just talk about your overall main effects: You do not need to talk about your simple main effects.

12. You can have an interaction without having a main effect—and having two main effects doesn't mean you will have an interaction.

13. When looking at a graph of a 2 × 2, realize that the two different levels of one factor are represented by two different lines and that the two levels of the other factor are represented by the left and right sides of the graph (i.e., the "x" axis). If one line is above the other, there may be an effect for the variable represented by the lines. If, on the average,

the lines slope up or down, there may be an effect for the factor represented by the "*x*" axis. If the lines have different slopes, you may have an interaction.

14. Sometimes, an interaction represents similarity. For instance, in a 2 (place of learning: basement or top floor) × 2 (place of testing: basement or top floor) factorial experiment, an interaction may reveal that it is best to be tested in the *same* place you learned the information.

15. The following summarizes the mathematics of an ANOVA summary table for a factorial design:

Source of Variance (*SV*)	Sum of Squares (*SS*)	Degrees of Freedom (*df*)	Mean Square (*MS*)	F
A	*SS A*	Levels of *A*–1	*SSA/df A*	*MSA/MSE*
B	*SS B*	Levels of *B*–1	*SSB/df B*	*MSB/MSE*
A × *B* Interaction	*SS (A* × *B)*	*df A* × *df B*	*SS/df A* × *B*	*MS(A* × *B)/MSE*
Error	*SSE*	Participants – Groups	*SSE/df E*	
Total	*SS A* + *SS B* + *SS(AXB)* + *SSE*	Participants –1		

© Cengage Learning 2013

16. With the hybrid factorial design, you can look at an experimental factor and a factor that you do not manipulate (personality, gender, age) in the same study. However, because you did not manipulate the nonexperimental factor, *you cannot make any cause–effect statements about your nonexperimental factor.*

17. Once you have an idea for a simple experiment, you can easily expand that idea into an idea for a factorial experiment. For example, you could add a replication factor (such as stimulus set) to try to establish the generalizability of your treatment's effect. In that case, you would not be expecting a significant interaction. Alternatively, if you wanted to show that the treatment didn't have the same effect under all circumstances, you could add a potential moderating variable. In that case, you would be expecting a significant interaction between the treatment and the factor that you believe will moderate its effect.

18. If you have a nonmanipulated factor (e.g., participant's age), you can look at differences between groups on this factor. However, even though these differences are called main effects of the factor, do not make the mistake of thinking that these differences represent effects of the factor. Remember, *if you do not randomly assign a factor, you cannot make cause–effect statements about that factor.*

KEY TERMS

blocked design *(p. 513)*
crossover (disordinal) interaction *(p. 494)*
factorial experiments *(p. 456)*

factors *(p. 455)*
interaction *(p. 474)*
ordinal interaction *(p. 496)*
overall main effect (p. 472)

simple main effect *(p. 471)*
stimulus sets *(p. 505)*
systematic replication *(p. 505)*

EXERCISES

1. Match the following terms to their definitions.
   a. interaction _____ The average effect of a factor
   b. simple main effect _____ A factor's effect on one group of participants
   c. overall main effect _____ A factor's effect on one group of participants being different from its effect on another group of participants

2. What is the difference between
   a. a simple main effect and an overall main effect?
   b. an overall main effect and an interaction?

3. Can you have an interaction without a main effect? Why or why not?

4. Suppose an experimenter looked at the status of speaker and rate of speech on attitude change and summarized the experiment's results in the following table. Describe the pattern of those results in terms of main effects and interactions. Assume that all differences are statistically significant.

	Status of Speaker	
Rate of Speech	Low Status	High Status
Slow	10	15
Fast	20	30
	Attitude Change	

5. Describe the pattern of results in the following table in terms of main effects and interactions. Assume that all differences are statistically significant.

	Status of Speaker	
Rate of Speech	Low Status	High Status
Slow	10	15
Fast	20	25
	Attitude Change	

6. Forty participants receive a placebo. The other forty receive a drug that blocks the effect of endorphins (pain-relieving substances, similar to morphine, that are produced by the brain). Half the placebo group and half the drug group get acupuncture. Then, all participants are asked to rate the pain of various shocks on a 1-to-10 (*not at all painful* to *very painful*) scale. The results are as follows: placebo, no acupuncture group, 7.2; placebo, acupuncture group, 3.3; drug, no acupuncture group, 7.2; drug and acupuncture group, 3.3.
   a. Graph the results. (If you need help graphing the results, see Box 12.1 or use the tools on this chapter's website.)
   b. Describe the results in terms of main effects and interactions (making a table of the data may help).
   c. What conclusions would you draw?

7. The following table is an incomplete ANOVA summary table of a study looking at the effects of similarity and attractiveness on liking. Complete the table. (Hint: If you are having trouble, consult Box 12.2 or the sample ANOVA summary table in Summary point 15.) Then, answer these three questions.
   a. How many participants were used in the study?
   b. How many levels of similarity were used?
   c. How many levels of attractiveness were used?

SV	SS	df	MS	F
Similarity *(S)*	10	1	—	—
Attractiveness *(A)*	—	2	20	—
S × *A* interaction	400	—	200	—
Error	540	54	—	
Total	990	59		

8. A professor does a simple experiment. In that experiment, the professor finds that students who are given lecture notes do better than students who are not given lecture notes. Imagine that you are asked to replicate the

professor's simple experiment as a 2 × 2 factorial.

    a. What variable would you add to change the simple experiment into a 2 × 2?

    b. Graph your predictions.

    c. Describe your predictions in terms of main effects and interactions.

9. A lab experiment on motivation yielded the following results:

Group	Productivity
No financial bonus, no encouragement	25%
No financial bonus, encouragement	90%
Financial bonus, no encouragement	90%
Financial bonus, encouragement	90%

    a. Make a 2 × 2 table of these data.

    b. Graph these data (for help with graphing, see Box 12.1, p. 489).

    c. Describe the results in terms of main effects and interactions. Assume that all differences are statistically significant.

    d. Interpret the results.

10. A memory researcher looks at the effects of processing time and rehearsal strategy on memory.

Group	Percent Correct
Short exposure, simple strategy	20%
Short exposure, complex strategy	15%
Long exposure, simple strategy	25%
Long exposure, complex strategy	80%

    a. Graph these data. (For help, see Box 12-1 or this book's website.)

    b. Describe the results in terms of main effects and interactions. Assume that all differences are statistically significant.

    c. Interpret the results.

11. Suppose a researcher wanted to know whether lecturing was more effective than group discussion for teaching basic facts. The researcher did a study and obtained the following results:

Source of Variance	SS	DF	MS	F
Teaching (T)	10	1	10	5
Introversion/ Extroversion (I)	20	1	20	10
$T \times I$	50	1	50	25
Error	100	50	2	

    a. What does the interaction seem to indicate?

    b. Even if there had been no interaction between teaching and extroversion, would there be any value in including the introversion–extroversion variable? Explain.

    c. What, if anything, can you conclude about the *effects* of introversion on learning?

WEB RESOURCES

1. Go to the Chapter 12 section of the book's student website and
    a. Look over the concept map of the key terms.
    b. Test yourself on the key terms.
    c. Take the Chapter 12 Practice Quiz.

2. Use the Chapter 12 website to practice
    a. interpreting ANOVA tables and
    b. interpreting graphs of results of factorial experiments.

3. Do an ANOVA using a statistical calculator by going to the "Statistical Calculator" link.

CHAPTER 13

Matched Pairs, Within-Subjects, and Mixed Designs

The art of being wise is the art of knowing what to overlook.
—William James

CHAPTER OVERVIEW

In Chapters 10, 11, and 12, you learned that you could perform an internally valid experiment by independently and randomly assigning participants to groups. Although you understand that randomly assigning participants to groups produces internally valid experiments, you may still have two basic reservations about between-subjects designs.

First, you may believe that these designs are wasteful in terms of the number of participants they require. For example, in the simple experiment, each participant is either in the control group *or* in the experimental group. If each participant was in both the control group *and* the experimental group, one participant could do the job of two.

Second, you may be concerned that, in between-subjects designs, it is hard to tell that a difference between groups at the end of the experiment is due to the treatment because the groups were probably different before the experiment started. To illustrate, suppose you hypothesize that playing a multiplayer role-playing video game makes one more cooperative than playing a video game in which an individual player plays against the computer. In addition, suppose that your hypothesis is correct—but the effect is small. If you use a simple, two-group, between-subjects experiment and the group assigned to play the role-playing video game scores slightly higher on cooperativeness than the comparison group, you couldn't claim that the treatment made the groups slightly different. Instead, you would have to admit that this small difference might be due to the groups being different at the start of the experiment. For example, it may be that most of the highly cooperative participants were randomly assigned to the multiplayer online role-playing game group, whereas most of the highly competitive participants were assigned to the comparison group. But if you used each participant as his or her own control, the difference that the treatment created could not be dismissed as being due to random differences between two groups. Consequently, the treatment's small effect might be detected.

Because pure randomized between-group experiments often require too many participants and have too much random error, they often have too little power. In this chapter, you will learn about designs that get you more power with fewer participants.

You will start by learning about a special type of between-subjects design: the **matched-pairs design**. Like the simple experiment, the matched pairs design involves comparing individuals who have received one treatment with individuals who have received a different treatment. Like the simple experiment, random assignment determines which participant gets which treatment. Thus, like the simple experiment, the matched pairs design has internal validity because participants are randomly assigned to condition.

Unlike a simple experiment, however, participants are paired up based on their scores on a key characteristic *before* they are randomly assigned. For example, in our study of the effects of video games on cooperation, we might give our 10 participants a test of cooperativeness and then rank our participants from 1 to 10 based on their scores on that test. We would then make ranks 1 and 2 a pair, 3 and 4 a pair, 5 and 6 a pair, 7 and 8 a pair, and 9 and 10 a pair. Then, we would randomly determine, separately for each pair, which member of the pair would play the individual game version of the game and which member would play the multiple player version. Thus, for each pair, we could compare the member who played the individual version with the member who played the multiple player version.

If the members of our pairs are well-matched, the individuals who were assigned to the multiple player group will be similar to the individuals who were assigned to the comparison group. Thus, matching will make it less likely that differences between groups that existed before we administered the treatment would hide differences between the groups that were caused by the treatment.

Rather than compare one group with another group, we could compare participants with themselves in a **within-subjects design** (also called a "**repeated measures design**"). In a within-subjects design, you could have each participant play both the individual and multiple-player version of the game and measure each participant's cooperativeness after they played each version. That way, you could compare how cooperative a participant was after playing the individual game with how cooperative that same participant was after playing the multiple-player version. Because you are getting two scores for each participant and because each participant is being compared to himself or herself rather than being compared to a different person, your within-subjects experiment could have impressive power.

One problem with your within-subjects experiment is that participants may react differently on the first trial than on the last trial. As a result, if you always administer the treatments in the same sequence, what you think is a treatment effect ("they respond best to treatment A") may be an order effect (they really respond best to the first treatment they get). To deal with order effects, within-subjects researchers typically try one of two strategies to mix up the sequence in which participants get the treatment.

One strategy researchers use to ensure that not every participant gets the treatments in the same sequence is to use a random process, like a coin flip, to determine the sequence of the treatments a participant gets. For example, if the coin flipped for a participant came up "heads," that participant might receive the treatments in the following sequence: individual-player game first, multiple-player game second; if the coin came up "tails," the participant might receive this sequence: multiple-player game first, individual-player game second. Such a design would be called a randomized within-subjects design.

One problem with the randomized within-subjects design is that one sequence may be over-represented. For example, suppose you have 10 participants and you assign them to sequence based on a coin flip. If the coin comes up heads 7 out of 10 times, the "individual-player game first, multiple-player game second" sequence will occur more than twice as often as the "multiple-player game first, individual-player game second" sequence.

To make sure that no one sequence of treatments is over-represented, you might randomly assign participants in a way that guarantees that half will get the "individual-player game first, multiple-player game second" sequence and half will get "multiple-player game first, individual-player game second" sequence. That is, you might use a **counterbalanced design**: a design in which each participant gets every level of the treatment, but different groups of participants get the treatment levels in different sequences.

Note that because different *groups* of participants receive different treatment sequences, the counterbalanced design is not a pure within-subjects design. Instead, the counterbalanced design is a type of **mixed design**: a design in which at least one factor is a within-subjects factor, and at least one factor is a between-subjects factor.

In mixed designs, all participants get all levels of the within-subjects factor(s), but different participants get different levels of the between-subjects factor(s). For example, you might use a mixed design in which all participants played both the individual-player video game and the multiple-player video game, but some participants played the games in a hot room, whereas others played the game in a normal temperature room.

After learning about all these designs, you will learn how to weigh the trade-offs involved in choosing among them. Thus, by the end of this chapter, you will be better able to choose the best experimental design for your research problem.

THE MATCHED-PAIRS DESIGN

If you do not have enough participants to do a powerful simple experiment, you might use a design, such as a matched-pairs design, that requires fewer participants. As you will see, the **matched-pairs design** combines the best aspects of matching and random assignment: It uses *matching* to increase *power*, and it uses *random assignment* to establish *internal validity*.

Procedure

In the matched-pairs design, you first measure your participants on a variable that correlates with the dependent measure. For example, if you were doing a memory experiment, you might first give all your participants a memory test.

After getting scores from all your participants on your measure of the matching variable, you would consider the two highest scorers to be a matched pair, the next two highest scorers to be another matched pair, and you would continue until you had matched up all the participants. This would give you pairs of participants with similar scores on the matching variable. Finally, you would randomly assign one member of each pair to the control group and the other member to the experimental group. For example, you might assign random numbers to all the participants and then put the member of the pair with the higher random number in the experimental condition and put the member with the lower random number in the control condition.

Considerations in Using Matched-Pairs Designs

You now have a general idea of how to conduct a matched-pairs experiment. You also know how it compares to a simple experiment: Unlike a simple experiment, it uses matching; like a simple experiment, it uses random assignment (see Table 13.1). But should you use a matched-pairs experiment instead of a simple experiment? When considering a matched-pairs design, you ask four questions:

1. Will matching be practical?
2. Will matching give you more power?
3. Will matching harm external validity?
4. Will matching harm construct validity?

TABLE **13.1** Comparing the Matched Design with the Simple Experiment

Matched Design	Simple Experiment
First, *match* participants on key characteristics.	*No matching.*
Then, *randomly assign* each member of the pair to condition.	*Randomly assign* participants to condition.

© Cengage Learning 2013

Finding an Effective Matching Variable

As we suggested earlier, you can make effective use of the matched-pairs design only if you can create pairs in which the members of each pair are very similar to each other in terms of the dependent measure. The most direct way to get such pairs is to start your study by giving all the participants the dependent measure as a pretest and then matching participants based on their pretest scores. Thus, in a memory experiment, participants could be matched based on scores on an earlier memory test; in a maze-running experiment, subjects could be matched based on scores on an earlier maze-running trial.

If you cannot match on pretest scores, you may have to search the research literature (see Web Appendix B) to find a matching variable. If you are extremely lucky, you will find that other researchers have done studies using your same dependent variable and have effectively used a certain measure to match up their participants. More likely, however, you will find out what variables correlate with your dependent measure. Unfortunately, after doing your library research, you may find (a) that there are no variables that have a strong, documented relationship with performance on the dependent measure or (b) that there are good matching variables, but for ethical or practical reasons you cannot use them.

Power—If You Do a Good Job of Matching

You want to find an appropriate matching variable so that your study will have adequate **power**: the ability to find differences between conditions. After all, the reason you would use a matched-pairs design rather than another between-subjects design is to avoid the power problems that plague researchers who use pure between-subjects designs.

Researchers who rely exclusively on random assignment to make groups similar lose power because individual differences between participants hide treatment effects. Specifically, because participants differ from each other, between-subjects researchers can't assume that the treatment group and the no-treatment group are extremely similar before the start of the experiment. Consequently, if the groups differ at the end of the experiment, these researchers may not know whether this difference is due to the treatment or to the groups being different before the experiment began. Indeed, if a simple experiment has fewer than 30 participants, even a large difference between the treatment and no-treatment groups could be entirely due to random error.

If matching makes your groups extremely similar to each other before the experiment begins, then there isn't much random error due to individual differences to hide your treatment effects. Therefore, the small difference that would not be statistically significant with a simple experiment may be significant with a matched-pairs design.

As we just said, the matched-pairs design does a good job of finding treatment effects because it reduces the factor that hides treatment effects: random error. But how, specifically, does reducing random error result in finding a statistically significant treatment effect? The two-part answer to that question is that (1) reducing random error results in larger *t* values and (2) larger *t* values are more likely to be statistically significant.

Why would the *t* value be larger in a matched-pairs design? Recall that the *t* value equals the difference between the means of the two conditions *divided by* an estimate of *random error* (the standard error of the difference). So, *with less random error*, the difference between groups is divided by less, and so the *t value becomes larger* (and thus more likely to be statistically significant). For example, suppose you do a matched-pairs study and a simple experiment and that in both cases, your treatment group is 6 milliseconds faster than your no-treatment group. If the standard error of the difference for a simple experiment is 6 milliseconds, then a difference of 6 milliseconds between conditions would yield a *t* value of 1.0 (because 6/6 = 1.0). That *t* value is too low to be statistically significant. However, if a matched-pairs design reduced random error so much that the standard error of the difference was only 1, then that same difference of 6 milliseconds would yield a *t* value of 6.0 (because 6/1 = 6.0). That *t* value would be statistically significant. In other words, if matching limits the effects of individual differences, you may be able to find even relatively small treatment effects.

But what if matching fails to reduce random error? For example, suppose a researcher matched participants on an irrelevant variable such as shoe size. In that case, the random differences between the two groups would be the same as if the researcher had done a simple experiment. Because matching hasn't reduced the study's random error, the *t* value will be roughly the same as it would have been in the simple experiment. In other words, because matching hasn't reduced the random differences between the treatment and no-treatment group, the matched-pairs experiment would not be more powerful than the simple experiment. In fact, even though the matched pairs experiment would produce roughly the same *t* value as the simple experiment, the matched-pairs experiment would be *less powerful* than the simple experiment.

To understand why poor matching leads to a matched-pairs design that is less powerful than a simple experiment, you need to know understand two facts: (1) matched-pairs designs have half the degrees of freedom of a same-sized simple experiment, and (2) all other things being equal, fewer degrees of freedom means less power. So, let's look at some examples that will help you see the implications of these two facts.

By using a matched-pairs design instead of a simple experiment, you lose half your degrees of freedom because, whereas degrees of freedom for a simple experiment equals number of *participants*−2, the degrees of freedom for a matched-pairs study equals number of *pairs*−1. Thus, if you used 20 participants in a simple experiment, you would have 18 degrees of freedom (two fewer than the number of participants). But if you used 20 participants (10 pairs) in a matched-pairs design, you would have only 9 degrees of freedom (one fewer than the number of pairs).

Critical Values of *t*

	Level of Significance for Two-Tailed *t* Test
df	.05
1	12.706
9	2.262
18	2.101
60	2.000
120	1.980

© Cengage Learning 2013

Losing degrees of freedom can cause you to lose power. As you can see by looking at this mini *t* table, the fewer degrees of freedom you have, the larger your *t* value must be to reach significance. For example, with 18 degrees of freedom (what you'd have if you tested 20 participants in a simple experiment), you would need only a *t* value of 2.101 for your results to be statistically significant at the .05 level. On the other hand, with 9 degrees of freedom (what you'd have if you tested 20 participants [10 pairs of participants] in a matched-pairs experiment), your *t* value would have to be at least 2.262 to be statistically significant at the .05 level. So, a difference between your treatment conditions that would have been big enough to be statistically significant if you had used a simple experiment might not be statistically significant with a matched-pairs design in which you matched on a variable that did not correlate with your measure. Thus, if you obtain the same *t* value with the matched-pairs design as you would have obtained with a simple experiment, the matched-pairs design costs you power.

If your matching is any good, however, you should not get the same *t* value with a matched-pairs design as with a simple experiment. Instead, you will get a larger *t* value with a matched-pairs design because you have reduced a factor that shrinks *t* values—random error due to differences between participants. Usually, the increase in the size of the *t* value will more than compensate for the degrees of freedom you will lose. Thus, as long as you can match participants on a relevant variable, you will get more power by switching from a simple experiment to a matched-pairs design.

External Validity

Power is not the only consideration in deciding to use a matched-pairs design. You may use—or avoid—matching for reasons of external validity.

Matched-Pairs Designs May Have Better External Validity Than an Equally Powerful Simple Experiment. A matched-pairs design may have more external validity than an equally powerful simple experiment. Why? Because unlike the simple experiment, the matched-pairs design can have power without limiting who can be in the experiment.

To obtain adequate power, a researcher using a simple experiment may have to severely restrict the kind of individual who can be in the study. That is, to reduce the degree to which differences between participants create random differences between treatment and no-treatment groups, the experimenter may be forced to use participants who are all very similar. For example, to create a simple experiment that would be as powerful as a matched-pairs design, an experimenter might need to limit participants to male, albino rats between 180 and 185 days of age. Another researcher might attempt to reduce random error due to individual differences by limiting participants to middle-class college women with IQs between 115 and 120.

With a matched-pairs design, on the other hand, you can minimize random differences between the treatment and no-treatment groups without studying participants who are all alike. Because you reduce random error by matching up the participants you do have rather than by limiting the kinds of participants you can have, you can have a powerful matched-pairs design that still allows you to generalize your results to a broad population.

Matched-Pairs Designs May Have Less External Validity Than a Simple Experiment. Matched-pairs designs, however, do not always have better external validity than simple experiments. For example, if participants drop out of the study between the time they are tested on the matching variable and the time they are to perform the experiment, matching will reduce the generalizability of your results. For instance, suppose you start off with 16 matched pairs, but end up with only 10 pairs. In that case, your experiment's external validity is compromised because your results may not apply to individuals resembling the participants who dropped out of your experiment.

Even if participants do not drop out, matching may still harm external validity because your results generalize only to situations in which individuals perform the matching task before getting the treatment. To illustrate, imagine that an experimenter uses a matched-pairs design to examine the effect of caffeine on anxiety. In that experiment, participants take an anxiety test, then either consume caffeine (the experimental group) or do not consume caffeine (the control group), and then take the anxiety test again. Suppose that the participants receiving caffeine become more anxious than those not receiving caffeine.

Can the investigator generalize her results to people who have not taken an anxiety test before consuming caffeine? No, it may be that caffeine increases anxiety only when it is consumed after taking an anxiety test. For example, taking the anxiety test may make participants so concerned about their level of anxiety that they interpret any increase in arousal as an increase in anxiety. Because of the anxiety test, the arousal produced by caffeine—which might ordinarily be interpreted as invigorating—is interpreted as anxiety.

Construct Validity: Using a Matched-Pairs Design May Help Participants Figure out the Hypothesis

In the caffeine study we just discussed, taking the anxiety test before and after the treatment might make participants aware that the experimenter is looking at the effects of a drug on anxiety. The participants' awareness of the hypothesis may harm the study's construct validity. For example, if participants believe that the hypothesis is that the drug will increase anxiety, they may act more anxious to help the researcher prove the hypothesis.

However, the fact that participants guess the hypothesis does not, by itself, ruin the experiment's construct validity. For instance, if you used a treatment condition and a placebo condition, it does not matter whether participants think that taking a pill is supposed to increase anxiety. Because both groups have the same expectations ("The pill I took will increase my anxiety"), having that expectation would not cause the treatment group to differ from the placebo group. Therefore, a significant difference between groups would have to be due to the treatment (the drug in the treatment group's pill).

If, on the other hand, your independent variable manipulation has poor construct validity, matching will make your manipulation's weaknesses more damaging. To see how *matching can magnify a manipulation's weaknesses*, imagine that the caffeine study used an empty control group (nothing was given to the participants who did not receive the treatment). The experimental group participants fill out an anxiety measure, take a caffeine pill, and then fill out another anxiety measure. The experimental group participants might think that the pill is supposed to increase their anxiety level, thereby causing them to be more anxious—or at least, to report being more anxious. The control group participants, not having been given a pill, would not expect to become more anxious. Consequently, a significant difference between the groups might be due to the two groups acting on different beliefs about what the researchers expected—rather than to any ingredient in the pill.

Analysis of Data: Use a Within-Subjects, Rather Than a Between-Subjects, *t* Test

We have talked about how matching, by making your study powerful, can help you find a significant difference. We have also warned you about external validity and construct validity problems that should make you cautious when interpreting such a significant difference. But how do you know whether you have a significant difference?

You should *not* use a regular between-subjects *t* test (also known as an independent groups *t* test). That test compares the overall, average score of the treatment group with the overall, average score of the no-treatment group.

You need a test that will allow you to compare the score of one member of a matched pair directly with the score of the other member—and to make that comparison for each of your pairs. If you have ratio or interval scale data,[1] you can make those comparisons using the within-subjects *t* test (also known as the **dependent groups *t* test**).[2] If you plan to do a dependent groups *t* test by hand, see Appendix E. If you plan to have a computer do a dependent groups *t* test for you, see Box 13.1.

---

[1] If you have only ordinal data, you should use the sign test. If you don't know what type of data you have, consult Chapter 5.

[2] You can also analyze such data using a within-subjects ANOVA (see Box 13.2).

BOX **13.1**   **Using the Computer to Conduct a Dependent Groups *t* Test**

When looking for a computer program to do an analysis on a matched-pairs design or on a two-condition within-subjects design, be aware of three facts. First, you should not do an independent groups *t* test (also known as a between-subjects *t* test). Instead, you should you a dependent groups *t* test. Second, the dependent *t* test goes by at least five other names: (1) *t* test for correlated samples, (2) *t* test for dependent samples, (3) *t* test for paired samples, (4) repeated-measures *t* test, and (5) within-subjects *t*. Third, you are not limited to using a dependent groups *t* test. For example, you could do a within-subjects analysis of variance (see Box 13.2).

If you use a dependent groups *t* test, the computer should provide you with at least three sets of information. First, it should tell you the number of observations you had in each condition. Thus, if you had four scores for condition 1, it should tell you that "*n*" for condition 1 was 4. Second, it should give you the mean (*M*) and the standard deviation (*SD*) for each condition. Third, it should give you the *t* value, the degrees of freedom (*df*) for the test, and the two-tailed probability (*p*) of obtaining a difference at least that great or greater between your two means if the null hypothesis were true. For example, a printout might look like the following:

Condition 1		**Condition 2**
n	4	4
M	6.25	2.5
SD	0.95	1.29

t	*df*	two-tailed *p*
15	3	.006

You might report such results this way:[a] "As predicted, significantly more words were recalled in the treatment condition (*M* = 6.25, *SD* = 0.95) than in the control condition (*M* = 2.5, *SD* = 1.29), *t*(3) = 15.0, p < .001."

[a]*M* stands for mean, *SD* stands for standard deviation (a measure of the variability of the scores), and *p* stands for the probability of obtaining a difference between conditions at least that large if the treatment had no effect. *SD* will usually be calculated as part of computing *t* (for more about *SD*, see Appendix E).

Conclusions About the Matched-Pairs Design

In summary, the matched-pairs design's weaknesses stem from the matching (see Table 13.2). If you can't find an effective matching variable, matching may hurt power. If matching alerts participants to the purpose of your experiment, matching may hurt your construct validity. If participants drop out of the experiment between the time they are measured on the matching variable and the time they are to be given the treatment, matching costs you the ability to generalize your results to the participants who dropped out. Finally, even if participants do not get suspicious and do not drop out, matching still costs you time and energy.

Although matching has its costs, matching usually offers one big advantage—power without restricting your subject population. Because the matched-pairs design combines the power of matching with the internal validity–promoting properties of random assignment, the matched-pairs design is hard to beat when you can study only a few participants.

BOX **13.2** **Using the Computer to Conduct a Within-Subjects Analysis of Variance**

If you had conducted a matched-pairs study or a two-condition within-subjects study, you could analyze your data using a dependent groups *t* test (see Box 13.1) or by using a within-subjects ANOVA. If you use a within-subjects ANOVA instead of a dependent *t* test, you will get similar results. For example, if we had used a within-subjects ANOVA on the data we analyzed in Box 13.1, a computer printout of that analysis would look like this:

Descriptive Statistics

	Condition 1	Condition 2
n	4	4
M	6.25	2.5
SD	0.95	1.29

Within-Subjects ANOVA Table

Source	SS	df	MS	F	p
Treatment	27.68	1	27.68	225	.0006
Error	0.37	3	0.123		

   If you compare this ANOVA printout with the within-subjects *t* printout in Box 13.1, you will note three similarities. First, the table listing the descriptive statistics in the within-subjects ANOVA printout is identical to the table listing the descriptive statistics in the within-subjects *t* test printout. Regardless of whether you use a within-subjects *t* test or a within-subjects ANOVA, the computer reports the same number of observations per condition, the same average for each condition, and the same variability of scores within each condition.

   Second, the *p* value for the treatment (.0006) in the within-subjects ANOVA table is the same as the *p* in the within-subjects *t* test. Both tests are equally likely to find a significant result.

   Third, the *df* error for the ANOVA (3) is the same as the *df* for the *t*. In both cases, *df error equals the number of pairs of scores minus 1*.

   Even the differences between the printouts reveal similarities. For example, the within-subjects ANOVA's *F* value (225) is the dependent groups *t* value (15) squared.

   Given the similarities between the two types of analyses, you probably will not be surprised to learn that they would be written up similarly. Thus, you might report the above-described results as follows. "As predicted, significantly more words were recalled in the treatment condition (*M* = 6.25, *SD* = 0.95) than in the control condition (*M* = 2.5, *SD* = 1.29), *F* (1, 3) = 225.0, *p* < .001."

   If you had more than two levels of your independent variable, you could not use a within-subjects *t* test to analyze your data. You could, however, analyze such data with a multiple-level within-subjects ANOVA.

   If you were to analyze such data with a multiple-level within-subjects ANOVA, your printout might resemble the printout of a two-level within-subjects ANOVA. Indeed, the most noticeable difference would be that your degrees of freedom would be different. For example, if you had 3 levels of the treatment, your treatment *df* would be 2.

(Continued)

BOX **13.2** **(Continued)**

As we have suggested, if you switch from looking at the printout of a two-level within-subjects design to looking at the printout of a three-level within-subjects design, you probably will not see a big difference. However, if you switch from looking at the printout from one computer program to another, you may notice a big difference. For example, in one program, a three-level, within-subjects ANOVA printout might look like the following printout.

Within-Subjects ANOVA Table

Source	SS	df	MS	F	p
Treatment	12.133	2	6.067	26	.0001
Error	1.867	8	0.233		

However, the same analysis in another program might look like the table below—minus the footnotes. We added the footnotes to help you decipher the table.

Tests of Within-Subjects Effects

	Measure				
Source	Type III Sum of Squares (SS)[a]	df	Mean Square[b]	F[c]	SIG.[d]
Treatment	12.133	2[e]	6.067	26	.000
Error (Treatment)	1.867	8	.233		

[a]Treat this column like the previous table's sum of squares (SS) column.
[b]Mean Square is calculated by dividing the Sum of Squares by the df.
[c]$F = MS$ for the effect divided by MS error. The bigger F is, the more likely the results are to be statistically significant.
[d]This column represents how likely it is that one would obtain a result this large or larger if the null hypothesis were true. Traditionally, when the value in this column is less than .05, the results are considered "statistically significant."
[e]If there are 2 degrees of freedom (df), then there must be three levels of the "Treatment" variable.

In yet another program, the printout might look like the following table—minus the footnotes. (We added the footnotes to help you decipher the table.)

General Linear Models Procedure Repeated Measures Analysis of Variance Univariate Tests of Hypotheses for Within-Subjects Effects Source: Treatment

df	Type III Sum of Squares (SS)	Mean Square	F Value	P R > F[a]	Geisser Greenhouse Epsilon Prob Level[b] (G–T)
2	12.33	6.067	26	0.0001	0.0001

[a]The value in this column corresponds to the p value or significance level that most programs give you.
[b]The probability value in this column or in the next column should be used if certain assumptions of the within-subjects ANOVA have been violated.

© Cengage Learning 2013

TABLE **13.2** Advantages and Disadvantages of Matching

Advantages	Disadvantages
More power because matching reduces the effects of differences between participants.	Matching makes more work for the researcher.
Power is not bought at the cost of restricting the subject population. Thus, results may, in some cases, be generalized to a wide variety of participants.	Matching may alert participants to the experimental hypothesis.
	Results cannot be generalized to participants who drop out after the matching task.
	The results may not apply to individuals who have not been exposed to the matching task prior to getting the treatment.

© Cengage Learning 2013

WITHIN-SUBJECTS (REPEATED MEASURES) DESIGNS

One set of designs that can beat the matched-pairs design, at least in terms of power, are the **within-subjects designs** (also called **repeated-measures designs**). In all within-subjects designs, each participant receives all the levels or types of the treatment that the experimenter administers, and the participant is measured after receiving each level or type of treatment. In the simplest case, each participant would receive only two levels of treatment: no treatment and the treatment. For example, a participant might complete the dependent-measure task (e.g., take an aggression test), get a treatment (e.g., play a violent video game), and repeat the dependent-measure task again (e.g., retake the aggression test). The experimenter would estimate the effect of the treatment by comparing how each participant scored right after receiving the treatment (e.g., after playing a violent video game) with how that same participant scored before receiving the treatment (e.g., before playing the violent video game).

Considerations in Using Within-Subjects Designs

You now have a general idea of how a within-subjects (repeated-measures) experiment differs from a between-subjects design (for a review, see Table 13.3). But what do you have to gain—or lose—by using a within-subjects design instead of a between-subjects design? As you'll soon see, by using a within-subjects design instead of a between-subjects design, you will gain power, and you may gain external validity; however, you may lose internal and construct validity.

Increased Power

Despite potential problems with the within-subjects design's internal and construct validity, the within-subjects design is extremely popular because it increases power in two ways.

TABLE **13.3** Comparing Three Designs

	Between-Subjects	**Matched-Pairs Design**	**Within-Subjects**
Role of random assignment	Randomly assign participants to treatment condition.	Randomly assign members of each pair to condition.	Randomly assign to sequence of treatment conditions.
Approach to dealing with the problem that differences between participants may cause differences between the treatment and notreatment conditions.	Allow random assignment and statistics to account for any differences between conditions that could be due to individual differences.	Use matching to reduce the extent to which differences between conditions could be due to individual differences. Then, use random assignment and statistics to deal with the effects of individual differences that were not eliminated by matching.	Avoid the problem of individual differences causing differences between conditions by comparing each participant's performance in one condition with his or her performance in the other condition(s).
How analyzed	Between-subjects t test, between-subjects ANOVA	Within-subjects t test, within-subjects ANOVA	Within-subjects t test, within-subjects ANOVA, MANOVA

© Cengage Learning 2013

The first way is similar to how the matched-pairs design increases power—by reducing random error due to individual differences. Specifically, both the matched-pairs experimenter and the within-subjects experimenter are concerned that, when you compare two groups that have received different treatments, you can't easily tell whether differences between the groups are due to treatment making the groups different or to the groups being different to start with. As you may recall, the matched-pairs experimenter tries to solve this problem by comparing pairs of participants in which both members of the pair are similar to start with. Within-subjects experimenters go one step farther: They compare scores from participants who were identical to start with. That is, they compare each participant's score after receiving one treatment with that same participant's score after receiving a different treatment.

The second way the within-subjects design increases power is by increasing the number of observations you get from each participant. The more observations you have, the more random error will tend to balance out; the more random error balances out, the more power you will have. With between-subjects designs, the only way you can get more observations is to get more participants because you can only get one observation per participant. But in a within-subjects experiment, you get at least two scores out of each participant. In the simplest case, your participants serve double duty by being in both the control and experimental conditions. In more complex within-subjects experiments, your participants might do triple duty, quadruple duty, or even more than quadruple duty. For example, in a study of how men's muscularity affected women's ratings of men, Frederick and Haselton

(2007) had participants do octuple duty. Specifically, to test their hypothesis that muscularity—up to a point—would increase attractiveness ratings, Frederick and Haselton had women rate the attractiveness of eight drawings that varied in muscularity. If Frederick and Haselton had used a purely between-subjects design, each participant would have rated only one drawing. However, because Frederick and Haselton used a within-subjects design, each participant could rate all eight figures.

Order Effects May Harm Internal Validity

As you intuitively realize, the main advantage of within-subjects designs is their impressive power. By comparing each participant with himself or herself, even subtle treatment effects may be statistically significant.

However, as you may also intuitively realize, the problem with comparing participants with themselves is that people are not perfectly consistent: The same person may act, think, or feel one way at one time and a different way at a different time. Because participants may change over time, the **order** (first or last) in which an event occurs within a sequence of events may have a strong impact on how participants react to an event. For example, the lecture that might have been fascinating had it been the first lecture you heard that day might be boring if it is your fourth class of the day. Because order affects responses, if a participant reacts differently to the first treatment than to the last, we have a dilemma: Do we have a treatment effect or an order effect?

To get a better idea of how **order (trial) effects** can complicate within-subjects experiments, let's examine a within-subjects experiment. Imagine being a participant in a within-subjects experiment where you take a drug (e.g., caffeine), play a video game, take a second drug (e.g., aspirin), and play the video game again.

If you perform differently on the video game the second time around, can the experimenters say that the second drug has a different effect than the first drug? No. The experimenters can't safely make conclusions about the difference between the two drugs because they are comparing your performance on trial 1, when you had been exposed to only one treatment (drug 1), to your performance on trial 2, by which time you had been exposed to three unofficial "treatments": (1) drug 1, (2) playing the game, and (3) drug 2 (see Table 13.4).

TABLE **13.4** In a Within-Subjects Design, the Treatment May Not Be the Only Factor Being Manipulated

	Events That Occur Before Being Tested	
	Drug 1 Condition	**Drug 2 Condition**
Between-subjects experiment	Get drug 1	Get drug 2
Within-subjects design	Get drug 1	Get drug 1
		Play video game
		Get drug 2

Four Sources of Order Effects

In the next few sections, you will see how being exposed these other "treatments" may affect your performance and thus hurt a within-subjects experiment's internal validity. We will start by showing you how the variable of *order* (first trial vs. second trial) may affect your performance. Specifically, we will look at four nontreatment reasons why you may perform differently on the task after the second treatment:

1. You may do better after the second treatment because you are seeing the dependent-measure task for the second time—and familiarity with the task makes you better at it. For example, the practice you got playing the game after taking the first drug may help you when you play the game again.
2. You may do worse after the second treatment because you are seeing the dependent-measure task for the second time—and you are now bored by it.
3. You may score differently after the second treatment because you are experiencing some delayed effects of the first treatment.
4. You may score differently after the second treatment because, by the time you see two of the treatment manipulations and see the dependent measure task twice, you have figured out the experimental hypothesis.

In summary, you need to be aware that the order in which participants get a treatment may affect the results. Thus, Treatment A may *appear* to have one kind of effect when it comes first, but may *appear* to have a different kind of effect when it comes second.

Practice Effects

If you perform better after the second treatment than you did after the first treatment, your improvement may merely reflect **practice effects**: You may have learned from the first trial. The first trial, in effect, trained you how to play the video game—although that wasn't the researcher's plan. Not surprisingly, practice effects are common: Participants often perform better as they warm up to the experimental environment and get familiar with the experimental task. Unfortunately, rather than seeing that you improved because of practice, the researcher may mistakenly believe that you improved due to the treatment.

Fatigue Effects

Instead of experience helping your performance (practice effects), experience may hurt your performance (**fatigue effects**). You may do worse on later trials merely because you are becoming tired or less enthusiastic as the experiment goes on. Unfortunately, a researcher might interpret your fatigue as a treatment effect.

Like practice effects, fatigue effects have nothing to do with any of the treatments participants receive. Like practice effects, fatigue effects are often due to becoming familiar with the dependent-measure task. However, instead of performance getting better as you learn the game, performance is getting worse as you get bored with the game. Thus, you could consider fatigue effects to be negative practice effects.

Treatment Carryover Effects

If you perform differently on the second trial, it could be due to the first treatment—if that treatment has some delayed effects. In technical terms, the effects of an earlier *treatment* that *carry over* to affect responses on later trials are called **carry-over (treatment carryover) effects**.

To see how treatment carryover effects might affect a within-subjects design, suppose that you were a participant in a study that examined the effects of drugs on video game performance. On Trial 1, the researcher gave you a drug and then measured your video game performance. On Trial 2, the researcher gave you a placebo and then measured your video game performance. If your performance was worst in the placebo (no-drug) condition, the researcher might think that your better performance on the first trial was due to the drug improving your performance. The researcher, however, could be wrong. Your poor performance in the placebo condition may be due to carryover effects from the previous treatment: You may just be starting to feel certain effects of the drug that you consumed during an earlier trial. Depending on the drug and the time between the trials, you may be feeling either "high" or hung-over.

Sensitization Effects

In addition to practice effects, fatigue effects, and carryover effects, you might perform differently after the second treatment due to **sensitization**. Sensitization occurs when, after getting several different treatments and performing the dependent variable task several times, participants realize (become *sensitive* to) what the independent and dependent variables are, and thus, during the latter parts of the experiment, they guess the experimental hypothesis and play along with it. For example, by the third trial of the video game experiment, a participant is likely to realize that the experiment had something to do with the effects of drugs on video game performance.

Note that sensitization has two effects. First, it threatens *construct validity* because participants figure out what the hypothesis is and thus may be acting to support the hypothesis rather than reacting to the treatment. Second, it threatens internal validity because it makes participants behave differently during the last trial (when they know the hypothesis) than they did during the first trial (when they did not know the hypothesis).

Review of the Four Sources of Order Effects

You have seen that because of practice, fatigue, carryover, and sensitization, the sequence in which participants receive the treatments could affect the results. For example, suppose participants all received the treatments in this sequence: Treatment A first, Treatment B second, and Treatment C last. Even if none of the treatments had an effect, the effect of order (first vs. second vs. last) might make it look like the treatments had different effects.

If practice effects caused participants to do better on the last trial, participants would do best on the trial in which they received Treatment C. In that case, even if none of the treatments had an effect, the investigator might mistakenly believe that Treatment C improves performance.

If, on the other hand, fatigue effects caused participants to perform the worst on the last treatment condition, participants would do worst on the trial in which they received Treatment C. In that case, even if none of the treatments had an effect, the investigator might mistakenly believe that Treatment C decreases performance.

Treatment carryover effects might also affect performance on the last trial. For example, if the effect of Treatment B is helpful but delayed, it might help performance on the last trial. If, on the other hand, the effect of Treatment B is harmful but delayed, it might harm performance on the last trial. Thus, even if Treatment C has no effect, the investigator might mistakenly believe that Treatment C is harmful (if Treatment B's delayed effect is harmful) or that Treatment C is helpful (if Treatment B's delayed effect is helpful).

Sensitization might also create the illusion that Treatment C has an effect. The participants were most naïve about the experimental hypothesis when receiving the first treatment (Treatment A) and least naïve when receiving the last treatment (Treatment C). Thus, the ability of the participant to play along with the hypothesis increased as the study went on. Changes in the ability to play along with the hypothesis may create order effects that could masquerade as treatment effects.

Dealing With Order Effects

You have seen that (a) the sources of order effects are practice, fatigue, carryover, and sensitization; and that (b) order effects threaten the internal validity of a within-subjects design. How can you use this knowledge to prevent order effects from threatening your experiment's internal validity?

Minimizing Each of the Individual Sources of Order Effects

Perhaps the best place to start to reduce the effect of order is to attack the four root causes of order effects: practice, fatigue, carryover, and sensitization.

Minimizing Practice Effects. To minimize the effects of practice, you can give participants extensive practice before the experiment begins. For example, if you are studying maze running and you have the rats run the maze 100 times before you start administering treatments, they've probably learned as much from practice as they can. Therefore, it's unlikely that the rats will benefit greatly from the limited practice they get during the experiment.

Minimizing Fatigue Effects. You can reduce fatigue effects by making the experiment interesting, brief, and undemanding. If the experiment ends before participants get bored or tired, you will not have to worry about fatigue effects.

Minimizing Treatment Carryover Effects. You can reduce carryover effects by lengthening the time between treatments. That way, the effect of earlier treatments will wear off before the participant receives the next treatment. For instance, if you were looking at the effects of drugs on how well rats run a maze, you might reduce treatment carryover effects by spacing your treatments a week apart (for example, antidepressant pill, wait a week, anti-anxiety pill, wait a week, placebo).

Minimizing Sensitization Effects. You can reduce sensitization by preventing participants from noticing that you are varying anything (Greenwald, 1976). For example, suppose you were studying the effects of different levels of full-spectrum light on typing performance. In that case, there would be three ways that you could prevent sensitization.

First, you could use very similar levels of the treatment in all your conditions. By using slightly different amounts of full-spectrum light, participants may not realize that you are actually varying amount of light.

Second, you could change the level of the treatment so gradually that participants do not notice. For example, while you gave participants a short break in between trials, you could change the lighting level watt by watt until it reached the desired level (Greenwald, 1976).

Third, you might be able to reduce sensitization effects by using good placebo treatments. Thus, if you were looking at the effects of full-spectrum light, you would not use darkness as the control condition. Instead, you would use light from a normal bulb as the control condition.

One General Strategy for Reducing Order Effects: Use Fewer Treatments

To this point, we have given you some strategies to reduce practice effects, to reduce fatigue effects, to reduce carryover effects, and to reduce sensitization (see Table 13.5 for a review). However, by reducing the number of experimental conditions, you can reduce all four causes of order effects at once because there will be fewer opportunities for them to affect your study.

TABLE **13.5** Order Effects and How to Minimize Their Impact

Effect	Example	Ways to Reduce Impact
Practice Effects	Getting better on the task due to becoming more familiar with the task or with the research situation.	Give extensive practice and warm-up before introducing the treatment.
Fatigue Effects	Getting tired as the study wears on.	Keep study brief, interesting.
Carryover Effects	Effects of one treatment lingering and affecting responses on later trials.	Use few levels of treatment. Allow sufficient time between treatments for treatment effects to wear off.
Sensitization	As a result of getting many different levels of the independent variable, the participant—during the latter part of the study—becomes aware of what the treatment is and what the hypothesis is.	Use subtly different levels of the treatment. Gradually change treatment levels. Use few treatment levels.

To see how fewer conditions leads to fewer order-effect problems, compare a within-subjects experiment that has 11 conditions with one that has only 2 conditions. In the 11-condition experiment, participants have 10 opportunities to practice on the dependent-measure task before they get the last treatment; in the 2-condition experiment, participants only have one opportunity for practice. The 11-condition participants have 11 conditions to fatigue them; 2-condition participants only have 2. In the 11-condition experiment, there are 10 treatments that could carry over to the last trial; in the 2-condition experiment there is only 1. Finally, in the 11-condition experiment, participants have 11 chances to figure out the hypothesis; in the 2-condition experiment, they only have 2 chances.

Mixing Up Sequences to Try to Balance Out Order Effects: Randomizing and Counterbalancing

Although you can take steps to reduce the impact of order, you can never be sure that you have eliminated its impact. Therefore, if you gave all your participants Treatment A first and Treatment B second, you could not be sure that the difference between the average of the Treatment A scores and the average of the Treatment B scores was due to a treatment effect. Instead, the difference could simply be due to an *order* (*trials*: first vs. second) effect.

To avoid confusing an order (trials) effect for a treatment effect, you should not give all your participants the same sequence of treatments. For example, in a two-condition study, you should not give all of your participants the treatments in this sequence: Treatment A first, Treatment B second. Instead, some participants should get this treatment sequence: Treatment B first and then Treatment A.

To understand how you could mix things up, imagine that you were doing an informal, 30-day study to find out whether working out was more effective in boosting your mood than taking a nap. You know that you shouldn't always take a nap, fill out a mood scale, work out, and then fill out a mood scale. So, you would take steps to prevent yourself from getting the same nap-workout sequence each day.

You might give yourself different sequences by flipping a coin every afternoon: "heads" you take a nap, then work out; "tails, you work out, and then take a nap. As you'll soon see, this approach is similar to what you would do in a randomized within-subjects design.

If you decided to do something more elaborate, you might say that, in the next 30 days, you will work out and then take a nap on 15 days, and you will take a nap and then work out on 15 days—and you will let chance determine which 15 days you work out first and which 15 days you nap first. As you'll see, this more elaborate approach is similar to what you would do in a counterbalanced design.

RANDOMIZED WITHIN-SUBJECTS DESIGNS

You can ensure that not all participants get the same sequence of treatments by randomly determining, for each participant, which treatment they get first, which treatment they get second, and so on. In the simplest case, if you flip a coin and it comes up "heads," the participant gets Treatment A, then

Treatment B; if it comes up "tails," the participant gets Treatment B, then Treatment A. If you use this randomization strategy to sequence each participant's series of treatments, you have a randomized within-subjects design.

Procedure

The **randomized within-subjects design** is very similar to the matched-pairs design. Indeed, most of the differences between the two-condition, randomized, within-subjects experiment and matched-pairs experiment stem from a single difference: In the within-subjects experiment, you get a pair of scores from a single participant, whereas in the matched-pairs design, you get a pair of scores from a matched pair of participants. That is, in the matched-pairs case, each participant only gets one treatment, but in the within-subjects experiment, each participant gets two treatments.

Other than each participant receiving more than one treatment, the two designs are remarkably similar. The matched-pairs researcher randomly determines, for each pair, who will get what treatment. In some pairs, the first member will get Treatment A, whereas the second member will get Treatment B; in other pairs, the first member will get Treatment B, whereas the second member will get Treatment A.

The within-subjects researcher randomly determines, for each individual, the sequence of the treatments. For some individuals, the first treatment will be Treatment A (and the second treatment will be Treatment B); for other individuals, the first treatment will be Treatment B (and the second treatment will be Treatment A). In short, whereas the matched-pairs experimenter randomly assigns members of pairs to different treatments, the within-subjects experimenter randomly assigns individual participants to different sequences of treatments.

To see the similarities and differences between the matched-pairs and within-subjects designs, imagine that you are interested in whether observers' judgments about other people are influenced by irrelevant information. Specifically, you want to see whether pseudo-relevant information (information that seems relevant but really isn't relevant) affects whether observers see others as passive or assertive. Therefore, you produce pseudo-relevant descriptions ("Bill has a 3.2 GPA and is thinking about majoring in psychology") and "clearly irrelevant" descriptions ("Bob found 20 cents in a pay phone in the student union when he went to make a phone call").

In a matched-pairs design, you would match participants—probably based on how assertively they tend to rate people. Then, one member of the pair would read a "pseudo-relevant" description while the other read a "clearly irrelevant" description. After reading the information, each participant would rate the assertiveness of the student he or she read about on a 9-point scale ranging from "very passive" to "very assertive."

In a randomized within-subjects design, on the other hand, each participant would read both "pseudo-relevant" and "clearly irrelevant" descriptions. After reading the information, they would rate the assertiveness of each of these students on a 9-point scale ranging from "very passive" to "very assertive." Thus, each participant would provide data for both the "pseudo-relevant" condition and the "clearly irrelevant" condition. The sequence of the descriptions would be randomized, with some sequences having the pseudo-relevant description first and others having the clearly irrelevant description first.

Hilton and Fein (1989) conducted such a randomized within-subjects experiment and found that participants judged the students described by pseudo-relevant information as more assertive than students described by clearly irrelevant information. Consequently, Hilton and Fein concluded that even irrelevant information affects our judgments about people.

Analysis of Data

To analyze data from the two-condition within-subjects design, you can use the same dependent groups *t* test that you used to analyze matched-pairs designs.[3] The only difference is that instead of comparing each member of the pair with the other member of that pair, you compare each participant with him- or herself. Because the dependent groups *t* test can be used to analyze data from a within-subjects design, it is often called the *within-subjects t test*.

You do not have to use a within-subjects *t* test. For example, instead of using a within-subjects *t* test (see Box 13.1), you could use a within-subjects analysis of variance (see Box 13.2).

Conclusions About Randomized Within-Subjects Designs

As you might expect from two designs that can be analyzed with the same technique, the randomized within-subjects design and the matched-pairs design are very similar. In terms of procedures, the only real difference is that the matched-pairs experimenter randomly assigns members of pairs to treatments, whereas the randomized within-subjects experimenter randomly assigns individual participants to sequences of treatments. Because both designs have impressive power, both should be seriously considered if participants are scarce.

The randomized within-subjects design, however, has some unique strengths and weaknesses stemming from the fact that it collects more than one observation per participant (see Table 13.6). Because it uses individual participants (rather than matched pairs) as their own controls, the randomized within-subjects design is more powerful than the matched-pairs design—and more useful when you want to generalize your results to real-life situations in which individuals get more than one "treatment." Thus, if you were studying the effects of political ads, you might use a within-subjects design because, in real life, a person is likely to be exposed to more than one political ad about a candidate (Greenwald, 1976).

Although there are benefits to collecting more than one observation per participant, having to contend with order effects (practice, fatigue, carryover, and sensitization) is a major drawback. As we have suggested, you can try to minimize order effects, and you can *hope* that randomization will balance out the sequence of your treatments so that each condition comes first about the same number of times as it comes last.

---

[3] If you have more than two conditions, you cannot use a *t* test. Instead, you must use either within-subjects analysis of variance (ANOVA) or multivariate analysis of variance (MANOVA).

TABLE **13.6** Comparing the Matched-Pairs Design With the Within-Subjects Design

Matched-Pairs Design	Within-Subjects Design
Powerful	More powerful
Order effects are *not* a problem.	Order effects are a serious problem.
Matching may lead participants to guess the hypothesis.	Being exposed to the dependent measure task more than once and to more than one level of the treatment may lead participants to guess the hypothesis.
Uses random assignment to balance out differences between participants	Uses randomization to balance out order effects
Useful for assessing variables that vary between subjects in real life	Useful for assessing variables that vary within subjects in real life

© Cengage Learning 2013

USING COUNTERBALANCING IN WITHIN-SUBJECTS DESIGNS

Instead of merely hoping that chance will make it so that the same number of participants receive each sequence, why not make sure? For example, if you have two treatments (A and B), you could randomly assign participants in such a way that half got the A-B sequence and half got the B-A sequence. When you use administer a series of sequences in a way that balances out order effects, you are using counterbalancing. In the next sections, we will address two key questions about using counterbalanced sequences to balance out order effects. First, what sequences should you use to balance out order effects? Second, should you give each participant all those sequences—or should you give different participants different sequences?

Complete counterbalancing versus Latin Square Counterbalancing. What sequences of treatments should you use? One possibility is to use every possible treatment sequence. For example, if you had two treatments, your sequences would be A-B and B-A. Including every possible sequence is called **complete counterbalancing**.

One problem with complete counterbalancing is that, as you add treatment levels, you greatly expand the number of possible sequences you have. For example, if you have 2 treatment levels, you have 2 sequences; if you have 4 treatment levels, you have 24 sequences; and if you have 5 treatment levels, you have 120 sequences. Consequently, if you have many treatment levels, you will probably find complete counterbalancing impractical. Therefore, you may decide to use some form of partial counterbalancing. For example, if you had four treatments (A, B, C, and D), you could use two sequences: a sequence (e.g., A-B-C-D) and its reverse (D-C-B-A).

A better form of partial counterbalancing would be to use Latin Square counterbalancing (see Box 13.3). The genius of Latin Square counterbalancing is that it allows you to select a small set of sequences that share two characteristics with complete counterbalancing.

First, Latin Square counterbalancing gives you a set of sequences that, like complete counterbalancing, guarantees that every condition occurs in every position equally often (e.g., Treatment A occurs first just as often as it occurs second, third, and fourth—and the same is true of all your treatments). The simplest way to guarantee that every condition occurs in every position equally often is to have each condition appear *once* in each position. Thus, if you had four treatments, treatment A would appear first in one sequence, second in one sequence, third in one sequence, and fourth in one sequence—and the same would be true of treatments B, C, and D.

BOX 13.3 Latin Square Designs: The ABCs of Counterbalancing Complex Designs

You have seen an example of the simplest form of counterbalancing in which one group of participants gets Treatment A followed by Treatment B (A-B) and a second group gets Treatment B followed by Treatment A (B-A). This simple form of counterbalancing is called A-B, B-A counterbalancing. Note that even this simple form of counterbalancing accomplishes two goals.

First, it guarantees that every condition occurs in every position equally often. Thus, in A-B, B-A counterbalancing, A occurs first half the time and last half the time. The same is true for B: For half the participants, B is the first treatment they receive; for the other half, B is the last treatment they receive.

Second, each condition precedes every other condition just as many times as it follows that condition. That is, in A-B, B-A counterbalancing, A comes before B once and comes after B once. This symmetry is called *balance*.

Although achieving these two objectives of counterbalancing is easy with only two conditions, with more conditions, counterbalancing becomes more complex. For example, with four conditions (A, B, C, D) you would have four groups. To determine what order the groups will go through the conditions, you would consult the following 4 × 4 Latin Square:

	Position			
	1	**2**	**3**	**4**
Group 1	A	B	D	C
Group 2	B	C	A	D
Group 3	C	D	B	A
Group 4	D	A	C	B

In this 4 × 4 complete Latin Square, Treatment A occurs in all four positions (first, second, third, and fourth), as do Treatments, B, C, and D. In addition, the square has balance. As you can see from looking at the square, every letter precedes every other letter twice and follows every other letter twice. For example, if you just look at Treatments A and D, you see that A comes before D twice (in Groups 1 and 2) and follows D twice (in Groups 3 and 4).

Balance is relatively easy to achieve for 2, 4, 6, 8, or even 16 conditions. But, what if you have 3 conditions? Immediately you recognize that with a 3 × 3 Latin Square, A cannot precede B the same number of times as it follows B. Condition A can either precede B twice and follow it once or precede it once and follow it twice. Thus, with an uneven number of conditions, you cannot create a balanced Latin Square.

One approach to achieving balance when you have an uneven number of treatment levels is to add or subtract a level so you have an even number of levels. However, adding a level may greatly increase the number of sequences and groups you need. Subtracting a level, on the other hand, may cause you to lose vital information. Therefore, you may not wish to alter your study to obtain an even number of levels. Fortunately, you can achieve balance with an uneven number of treatment levels by using two Latin Squares.* For instance, consider the 3 × 3 squares below.

If you randomly assign subjects to six groups, as outlined above, you ensure balance. See for yourself that if you take any two conditions, one condition will precede the other three times and will be preceded by the other condition three times.

	Square 1 Position				**Square 2 Position**		
	1	**2**	**3**		**1**	**2**	**3**
Group 1	A	B	C	Group 4	C	B	A
Group 2	B	C	A	Group 5	A	C	B
Group 3	C	A	B	Group 6	B	A	C

*Another option is to use incomplete Latin Square designs. However, the discussion of incomplete Latin Square designs is beyond the scope of this book.

© Cengage Learning 2013

Second, Latin Squares counterbalancing makes sure that each condition comes before every other condition just as many times as it comes after that condition (e.g., if Treatment A comes before Treatment B in two of the sequences, Treatment A comes after Treatment B in two of the sequences). That is, A would occur before B in half the sequences and after B in half the sequences—and the same would be true for any pair of treatments.

As we have said, Latin Square counterbalancing gives you a set of sequences that, like complete counterbalancing, control for order effects by meeting two criteria: (1) each condition occurs in each position equally often and (2) each condition comes before every other condition half the time (see Box 13.3). The advantage of Latin Square counterbalancing over complete counterbalancing is that Latin Square counterbalancing requires fewer sequences. For example, suppose you had 4 levels of a treatment. With complete counterbalancing, you would use 24 sequences; with Latin Square counterbalancing, you would need only 4 sequences. Thus, if you are using more than 3 levels of treatment, you will probably use Latin Square counterbalancing rather than complete counterbalancing. (If you are using 2 or 3 levels of treatment, complete counterbalancing and Latin Square counterbalancing are the same.)

The Counterbalanced Design: Randomly Assigning Different Groups Different Sequences. Once you have your Latin Square counterbalanced sequences, do you give each participant all those sequences—or do you one sequence to some participants and another sequence to other participants? For example, if you were comparing two treatments (A and B), you might be able to give every participant the series A-B, B-A. That way, if participants got consistently better as the experiment went on, that order effect would not help Treatment A more than Treatment B.

Although it is possible to give each participant the same series of sequences, a much more common tactic is to give different participants different sequences by using what is called a **counterbalanced design**. In the counterbalanced design, as in a randomized within-subjects design, each participant gets the same treatments. In counterbalanced design, as in a randomized within-subjects design, not all participants will receive those treatments in the same sequence: One participant may get one sequence of treatments, whereas another participant may get the treatments in a different sequence. Thus, the counterbalanced design and the randomized within-subjects design have much in common. However, the randomized within-subjects design and the counterbalanced mixed design differ in terms of how they decide how many participants get each sequence.

The randomized within-subjects design lets chance determine both what sequence a participant will get and how many other participants will get that sequence. To illustrate, suppose you do a randomized within-subjects experiment and have two conditions (A and B). What sequence will a participant get? In the pure, randomized within-subjects design, the sequence depends entirely on chance. Because the sequence of treatments is randomly determined for each participant, a participant is equally likely to get either of the two possible sequences of the two treatments (A-B or B-A). How many participants will get a particular sequence? Again, that is entirely due to chance. Thus, if you do a randomized within-subjects experiment with eight participants, it is possible that most of those participants would see treatment A on the first trial.

The counterbalanced design, on the other hand, is not so trusting of chance. In a counterbalanced design, you decide what sequences you are going to use (usually by consulting a Latin Square, see Box 13.3), and you make sure that an equal number of participants receive each of the sequences you selected. To make sure that the same number of participants get each of your sequences, randomly assign participants just the way you would for a between-subjects experiment. The only difference is that instead of assigning participants to receive different treatments, you will assign participants to receive different sequences of treatments.

Now that you have a general understanding of how counterbalancing makes sure that *routine order effects* are balanced out,[4] let's see how you

---

[4] In football, for example, teams change sides every quarter and this usually balances out the effects of wind. However, if the wind shifts in the fourth quarter, counterbalancing fails to balance out the effects of wind. Similarly, if basketball teams change sides at the end of every half (as in international rules), but a rim gets bent (or fixed) during halftime, counterbalancing has failed to balance out the effects of different baskets.

could actually do a counterbalanced, within-subjects experiment. We will start by showing how you could study a two-level factor, then we will move to more complex designs.

Procedure

If you were to use a counterbalanced mixed design to study a two-level factor, you would randomly assign half of your participants to receive Treatment A first and Treatment B second, whereas the other half would receive Treatment B first and Treatment A second. By randomly assigning your participants to these counterbalanced sequences, most order effects will be neutralized. For example, if participants tend to do better on the second trial, this will not help Treatment A more than Treatment B because both occur in the second position equally often.

Advantages and Disadvantages of Counterbalancing

By using a counterbalanced mixed design, you have not merely balanced out routine order effects. You have also changed your study from one in which you had manipulated only one variable (the treatment, a within-subjects factor) to one in which you have also manipulated the between-subjects factor of counterbalancing sequence.

You could ignore the fact that you had varied counterbalancing sequence. In that case, you would analyze your experiment the same way you would analyze a pure within-subjects design. However, most experts would argue that you should examine the effect of the counterbalancing factor—even though doing so causes two problems: a minor problem and a major problem.

Disadvantages of Adding and Studying the Counterbalancing Factor

A minor problem with trying to find effects related to your counterbalancing factor (e.g., if you want to examine the difference between the group getting the A-B sequence and the group getting the B-A sequence) is that your statistical analysis is now more complex. Rather than using the dependent (within-groups) *t* test, you now have to use a mixed analysis of variance. This would be a major problem if you had to compute statistics by hand. However, because computers can do these analyses for you, this disadvantage really is minor.

The major problem with studying the between-subjects factor of counterbalancing sequence is that you need more participants than you would if you treated your study as a pure within-subjects design. To have enough power to see whether the group getting the A–B sequence has higher average scores than the group getting the B–A sequence, you will need at least 30 participants in each group.[5]

Advantages of Adding a Counterbalancing Factor

The disadvantage of needing more participants is often offset by being able to discover more effects. With the two-condition within-subjects experiment, you

---

[5] In most cases, 30 participants per group is too few. Usually, researchers should have at least 60 participants per group (Cohen, 1990).

BOX 13.4   A 2 × 2 Counterbalanced Design

The members of the first group get a list of words, are asked to form images of these words, and are asked to recall these words. Then, they get a second list of words, are asked to form a sentence with these words, and are asked to recall the words.

The members of the second group get a list of words, are asked to form a sentence with these words, and are asked to recall these words. Then, they get a second list of words, are asked to form images of those words, and are asked to recall those words.

Group 1	
First Task	Second Task
Form Images	Form Sentences

Group 2	
First Task	Second Task
Form Sentences	Form Images

Questions this study can address include the following:

1. Do people recall more when asked to form sentences than when asked to form images?
2. Do Group 1 participants recall more words than Group 2 participants? In other words, is one sequence of using the two different memory strategies better than the other?
3. Do people do better on the first list of words they see than on the second? That is, does practice help or hurt?

can find only one effect (the treatment main effect). By adding the two-level factor of counterbalancing sequence, you converted the two-condition experiment into a 2 (the within-subjects factor of treatment) × 2 (the between-subjects factor of counterbalancing sequence) experiment, thus giving you more information. Specifically, you can look for two main effects and an interaction (see Box 13.4). By looking at these three effects, you can find out three things.

First, as was the case with the pure within-subjects design, by looking at the treatment main effect, you can find out whether the treatment had an effect. In the experiment described in Box 13.4, you can look at the treatment main effect to find out whether forming images of words is a more effective memory strategy than making sentences out of the words.

Second, by looking at the counterbalancing-sequence main effect, you find out whether the group of participants getting one sequence of treatments (A–B) did better than the participants getting the other (B–A) sequence. In the experiment described in Box 13.4, the question is, "Did Group 1 (who formed images first and then formed sentences) recall more words than Group 2 (who formed sentences first and then formed images)?"

Third, by looking at the *treatment × counterbalancing interaction*, you find out whether participants score differently on their first trial than on their second. Looking at the treatment × counterbalancing interaction allows you to detect what some people call a "*trials effect*" and what others call an "*order effect*."

But how can looking at an interaction tell you that participants score differently on the first trial than on the second? After all, significant interactions usually indicate exceptions to general rules rather than indicating a general rule, such as "participants do better on the first trial."

The first step to seeing why a significant treatment × counterbalancing interaction tells you that participants score differently on the first trial than on the second is to imagine such an interaction. Suppose that participants who get Treatment A *first,* score highest after receiving Treatment A, *but* participants who get Treatment B *first,* score highest after receiving Treatment B. At one level, this is an interaction: The rule that participants score highest when receiving Treatment A only holds when participants receive Treatment A first. However, the cause of this interaction is an order (trials) effect: Participants score highest on the first trial.

To get a clearer idea of what a counterbalanced study can tell us, let's look at data from the memory experiment we mentioned earlier. In that experiment, each participant learned one list of words by making a sentence out of the list and learned one list of words by forming mental images. Thus, like a within-subjects design, each participant's performance under one treatment condition (sentences) was compared with that same participant's performance under another treatment condition (images).

Like a two-group between-subjects design, participants were randomly assigned to one of two groups. As would be expected from a counterbalanced design, the groups differed in terms of the counterbalanced sequence in which they received the treatments. Half the participants (the group getting the sentence–image sequence) formed sentences for the first list, then formed images to recall the second list. The other half (the group getting the image–sentence sequence) formed images to recall the first list, then formed sentences to recall the second list.

Now that you have a basic understanding of the study's design, let's examine the study's results. To do so, look at both the table of means for that study (Table 13.7) and the analysis of variance summary table (Table 13.8).

By looking at Table 13.8, we see that the main effect for the between-subjects factor of counterbalanced sequence is not significant. As Table 13.7 shows, members of both groups recalled, on the average, 14 words in the course of the experiment. Participants getting the treatment sequence A–B did not, on the average, recall more words than participants getting the sequence B–A.

Next, we see that the within-subjects factor of the memory strategy factor was also not significant. Because participants recalled the same number of words in the imagery condition (7) as they did in the sentence condition (7), we have no evidence that one strategy is superior to the other. Thus, there is no treatment effect.

Finally, we have a significant interaction of memory strategy and group sequence. By looking at Table 13.7, we see that this interaction is caused by

TABLE **13.7** Table of Means for a Counterbalanced Memory Experiment

	Memory Strategy		
Group's Sequence	**Images**	**Sentences**	**Images–Sentences Difference**
Group 1 (images first, sentences second)	<u>8</u>	<u>6</u>	+2
Group 2 (sentences first, images second)	6	8	−2
	14/2 = 7	14/2 = 7	Strategy Main Effect = 0

Counterbalancing Main Effect = 0

On the average, participants in both groups remembered a total of 14 words (8 in one condition, 6 in another)

Strategy Effect = 0

     Average recalled in image condition was 7 ([<u>8</u> + 6]/2).

     Average recalled in sentence condition was 7([<u>6</u> + 8]/2).

Order Effect = +2

     Participants remember the first list best.

     They averaged 8 words on the first list, 6 on the second.

The order (first vs. second) effect is revealed by an *interaction* involving counterbalancing *group* and rehearsal *strategy*.

That is, Group 1 did better in the image condition (<u>8</u> to <u>6</u>), but Group 2 did better in the sentence condition (8 to 6).

© Cengage Learning 2013

TABLE **13.8** ANOVA Summary Table for a Counterbalanced Design

	Analysis of Variance Table				
Source	**SS**	**df**	**MS**	**F**	**p**
Group Sequence (counterbalancing: Group 1 vs. Group 2)	0	1	0	0	*n.s.*[*]
Between-Subjects Error Term	44	22	2		
Memory Strategy (Treatment A vs. Treatment B)	0	1	0	0	*n.s.*
Interaction Between Memory Strategy and Group Sequence (effect of order—first vs. second list)	10	1	10	10	$p < .01$
Within-Subjects Error Term	23	23	1.0		

[a]*n.s.* is an abbreviation for not statistically significant.

Note: "*p*" values in an ANOVA summary table indicate the probability that the researchers could get differences between their conditions that were this big even if the variables were not related. That is, the *p* values tell you the probability that the difference between the groups could occur due to chance alone. Thus, the smaller the *p* value, the less likely the results are due only to chance—and the more likely that the variables really are related.

© Cengage Learning 2013

the fact that Group 1 (which gets images first) recalled more words in the imagery condition whereas Group 2 (which gets sentences first) recalled more words in the sentences condition. In other words, participants did better on the first list than on the second.

What does this order (trials) effect mean? If the researchers were not careful in their selection of lists, the order effect could merely reflect the first list being made up of words that were easier to recall than the second list. The researchers, however, presumably did not make that mistake.[6] Therefore, if the experiment was properly conducted, the order effect must reflect either the effects of practice, fatigue, treatment carryover, or sensitization. In this case, it probably reflects the fact that the practice participants get on the first list hurts their memory for the second list. Psychologists do not consider this negative practice effect a nuisance. On the contrary, this negative practice effect is one of the most important and most widely investigated facts of memory—proactive interference.

Now that you understand the three effects (two main effects and the treatment × counterbalancing interaction) that you can find with a 2 × 2 counterbalanced design, let's look at an experiment where the researcher is interested in all three effects. Suppose that Mary Jones, a politician, produces two commercials: an emotional commercial and a rational commercial. She hires a psychologist to find out which commercial is most effective so she'll know which one to give more airtime. The researcher uses a counterbalanced design to address the question (see Table 13.9).

TABLE **13.9** Effects Revealed by a 2 × 2 Counterbalanced Design

Group 1

First Ad	Second Ad
Emotional Ad	Rational Ad

Group 2

First Ad	Second Ad
Rational Ad	Emotional Ad

Questions Addressed by the Design:

1. Is the rational ad more effective than the emotional ad? (main effect of the within-subjects factor of type of ad)
2. Is it better to show the emotional ad and then the rational ad or the rational ad and then the emotional ad? (main effect of the between-subjects factor of counterbalancing sequence)
3. Are attitudes more favorable toward the candidate after seeing the second ad than after seeing the first? (ad by counterbalancing interaction)

© Cengage Learning 2013

---

[6] There are at least three ways to avoid this mistake: (a) extensively pretest the lists to make sure that both are equally memorable, (b) consult the literature to find lists that are equally memorable, and (c) counterbalance lists so that, across participants, each list occurred equally often under each instructional condition. The third approach is probably the best.

By looking at the treatment main effect, the researcher is able to answer the original question, "Which ad is more effective?" By looking at the counterbalancing sequence main effect, the researcher is able to find out whether one sequence of showing the ads is better than another, thus enabling him to answer the question, "Should we show the emotional ad first and then the rational ad or should we show the ads in the opposite sequence?" Finally, by looking at the ad × counterbalancing interaction, the researcher is able to determine if there is an order (trials) effect, leading him to be able to answer the question, "Do participants feel more favorable toward the candidate after they've seen the second ad?" Obviously, he would hope that voters would rate the candidate higher after seeing the second ad than they did after seeing the first ad.

Let's suppose that all three effects were statistically significant and the means were as follows:

	Type of Ad	
	Emotional Ad	Rational Ad
Group 1: (emotional–rational sequence)	<u>4</u>	6
Group 2: (rational–emotional sequence)	8	<u>7</u>

Note: Scores are rating of the candidate on a 1 (strongly disapprove of) to 9 (strongly approve of) scale.

© Cengage Learning 2013

As you can see from comparing the emotional ad column with the rational ad column, the treatment main effect is due to the rational ad, on the average, being more effective than the emotional ad. As you can see from comparing the Group 1 row with the Group 2 row, Group 2 likes the candidate more than Group 1. Thus, the between-groups counterbalancing sequence main effect suggests that it would be better to present the ads in the Rational–Emotional sequence (Group 2's sequence) than in the Emotional–Rational sequence (Group 1's sequence).

To help you find the order effect in this table, we have underlined the mean for the ad that each group saw first. Thus, we underlined 4 because Group 1 saw the emotional ad first, and we underlined 7 because Group 2 saw the rational ad first. By recognizing that 4 + 7 is less than 8 + 6, you could determine that scores were lower on the first trial than on the second. To make it easier to see the order effect, you should rearrange the table so that the columns represent "Order of Ads" rather than "Type of Ad." Your new table would look like this:

	Order of Ads	
	First Ad	Second Ad
Group 1: (emotional–rational sequence)	<u>4</u>	6
Group 2: (rational–emotional sequence)	<u>7</u>	8

© Cengage Learning 2013

As you can see from this table, the order effect reveals that people like the candidate more after the second ad. The ads *do* build on each other.

It's possible, however, that the consultant would not have obtained an order effect. For example, suppose the consultant obtained the following pattern of results:

	Type of Ad	
	Emotional Ad	Rational Ad
Group 1: (emotional–rational sequence)	<u>5</u>	6
Group 2: (rational–emotional sequence)	5	<u>6</u>

© Cengage Learning 2013

In this case, both Group 1 participants and Group 2 participants rate the candidate one point higher after seeing the rational ad than after seeing the emotional ad. Thus, there is no treatment by counterbalancing interaction. Because there is no treatment × counterbalancing interaction, there is no order effect. An easier way to see that there was no order effect would be to create the following table.

	Order of Ads	
	First Ad	Second Ad
Group 1: (emotional–rational sequence)	<u>5</u>	6
Group 2: (rational–emotional sequence)	<u>6</u>	5

© Cengage Learning 2013

With these data, the consultant would probably decide to just use the rational ad.

Instead of obtaining no order effect, the consultant could have obtained an order effect such that people always rated the candidate worse after the second ad. For example, suppose the consultant obtained the following results:

	Order of Ads	
	First Ad	Second Ad
Group 1: (emotional–rational sequence)	<u>5</u>	4
Group 2: (rational–emotional sequence)	<u>6</u>	4

© Cengage Learning 2013

If the consultant obtained these results, he would take a long, hard look at the ads. It may be that both ads are making people dislike the candidate, or it may be that the combination of these two ads does not work. Seeing both ads may reduce liking for the candidate by making her seem inconsistent. For example, one ad may suggest that she supports increased military spending while the other may suggest that she opposes increased military spending.

Conclusions About Counterbalanced Within-Subjects Designs

As you can see from this last example, the counterbalanced design does more than balance out routine order effects. It also tells you about the impact of both trials (order: first vs. second) and sequence (e.g., rational then emotional ad vs. emotional ad then rational ad). Therefore, you should use counterbalanced designs when

1. You want to make sure that routine order effects are balanced out.
2. You are interested in sequence effects.
3. You are interested in order (trials) effects.

You will usually want to balance out order effects because you don't want order effects to destroy your study's internal validity. That is, you want a significant treatment main effect to be due to the treatment, rather than to order effects.

You will often be interested in **sequence effects** because real life is often a sequence of treatments (Greenwald, 1976). That is, most of us are not assigned to receive either praise or criticism; to see either ads for a candidate or against a candidate; to experience only success or failure, pleasure or pain; and so on. Instead, we usually receive both praise and criticism, see ads for and against a candidate, and experience both success and failure. Counterbalanced designs allow us to understand the effects of receiving different sequences of these "treatments." In counterbalanced designs, the main effect for the between-subjects factor of counterbalancing sequence can help you answer questions like the following:

- Would it be better to eat and then exercise—or to exercise and then eat?
- Would it be better to meditate and then study—or to study and then meditate?
- If you are going to compliment and criticize a friend, would you be better off to criticize, then praise—or to praise, then criticize?

Order (trials) effects, on the other hand, will probably interest you if you can control whether a particular event will be first or last in a series of events. Thus, you might be interested in using a counterbalanced design to find out whether it's best to be the first or the last person interviewed for a job. Or, if you want to do well in one particular course (research methods, of course), should you study the material for that course first or last? To find out about these order effects, you'd use a counterbalanced design and look at the treatment × counterbalancing interaction.

CHOOSING THE RIGHT DESIGN

If you want to compare two levels of an independent variable, you have several designs you can use: matched pairs, within-subjects designs, counterbalanced designs, and the simple between-subjects design. To help you choose among these designs, we will briefly summarize the ideal situation for using each design.

Choosing a Design When You Have One Independent Variable
The matched-groups design is ideal when

1. You can readily obtain participants' scores on the matching variable without arousing their suspicions about the purpose of the experiment.

2. The matching variable correlates highly with the dependent measure.
3. Participants are scarce.

The randomized within-subjects design is ideal when

1. Sensitization, practice, fatigue, or carryover effects are not problems;
2. You want a powerful design;
3. Participants are scarce, and
4. You want to generalize your results to real-life situations, and in real life, individuals tend to be exposed to both levels of the treatment.

The 2 × 2 counterbalanced design is ideal when

1. You want to balance out the effects of order.
2. You are interested in order effects, sequence effects, or both.
3. You have enough participants to meet the requirement of a counterbalanced design.
4. You are not concerned that being exposed to both treatment levels will alert participants to the purpose of the experiment.

The pure between-subjects design is ideal when

1. You think fatigue, practice, sensitization, or carryover effects could affect the results.
2. You have access to a relatively large number of participants.
3. You want to generalize your results to real-life situations, and in real life, individuals tend to receive either one treatment or the other, but not both.

Choosing a Design When You Have More Than One Independent Variable

Thus far, we have discussed how to choose a design when you are studying the effects of a single variable (see Table 13.10). Often, however, you may want to investigate the effects of two or more variables.

In that case, you would appear to have three choices: a between-subjects factorial design, a within-subjects factorial design, and a counterbalanced design. However, counterbalancing becomes less attractive—especially for the beginning researcher—as the design becomes more complicated. Thus, beginning researchers who plan on manipulating two independent variables usually are choosing between a two-factor within-subjects design and a two-factor between-subjects design.

Using a Within-Subjects Factorial Design

You should use a pure within-subjects design when

1. You can handle the statistics (you will have to use within-subjects analysis of variance or multivariate analysis of variance).
2. Sensitization, practice, fatigue, and carryover effects are not problems.
3. You are concerned about power.
4. In real-life situations, people are exposed to all your different combinations of treatments.

TABLE **13.10** Ideal Situations for Different Designs

Simple Experiment	Matched Groups	Within-Subjects	Counterbalanced Design
Participants are plentiful.	Participants are very scarce.	Participants are very scarce.	Participants are somewhat scarce.
Order effects could be a problem.	Order effects could be a problem.	Order effects are not a problem.	Want to assess order effects or order effects can be balanced out.
Power isn't vital.	Power is vital.	Power is vital.	Power is vital.
In real life, people usually only get one or the other treatment, rarely get both.	In real life, people usually only get one or the other treatment, rarely get both.	In real life, people usually get both treatments, rarely get only one or the other.	In real life, people usually get both treatments, rarely get only one or the other.
Multiple exposure to dependent measure will tip participants off about hypothesis.	Exposure to matching variable will *not* tip participants off about hypothesis.	Multiple exposure to dependent measure will *not* tip participants off about hypothesis.	Multiple exposure to dependent measure will *not* tip participants off about hypothesis.
Exposure to different levels of the independent variable will tip participants off about hypothesis.	Exposure to different levels of the independent variable will tip participants off about hypothesis. Matching variable is easy to collect and correlates highly with the dependent measure.	Exposure to different levels of the independent variable will *not* tip participants off about hypothesis.	Exposure to different levels of the independent variable will *not* tip participants off about hypothesis.

© Cengage Learning 2013

Using a Between-Subjects Factorial Design

On the other hand, you should use a between-subjects design when

1. You are worried about the statistics of a complex within-subjects design.
2. You are worried that order effects would destroy the internal validity of a within-subjects design.
3. You are not worried about power.
4. In real-life situations, people are exposed to either one combination of treatments or another.

Using a Mixed Design

Sometimes, however, you will find it difficult to choose between a completely within-subjects design and a completely between-subjects design. For example, consider the following two cases.

Case 1: You are studying the effects of brain lesions and practice on how well rats run mazes. On the one hand, you do not want to use a completely within-subjects design because you consider brain damage to occur "between subjects"

13.11 Ideal Situations for Making a Factor Between or Within

Should a Factor Be a Between-Subjects Factor or a Within-Subjects Factor?

Make Factor Between Subjects	Make Factor Within Subjects
Order effects pose problems.	Order effects are not a problem.
Lack of power is *not* a concern.	Lack of power is a serious concern.
You want to generalize the results to situations in which participants receive either one treatment or another.	You want to generalize the results to situations in which participants receive all levels of the treatment.

© Cengage Learning 2013

in real life (because some individuals suffer brain damage and others do not). On the other hand, you do not want to use a completely between-subjects design because you think that practice occurs "within subjects" in real life (because all individuals get practice and, over time, the amount of practice an individual gets increases).

Case 2: You are studying the effects of subliminal messages and electroconvulsive therapy on depression. You expect that if subliminal messages have any effect, it will be so small that only a within-subjects design could detect it. However, you feel that electroconvulsive shock should not be studied in a within-subjects design because of huge carryover effects (see Table 13.11).

Fortunately, in these cases, you are not forced to choose between a totally within-subjects factorial and a totally between-subjects factorial. As you know from our discussion of counterbalanced designs, you can do a study in which one factor is varied between subjects and the other is varied within subjects. Such designs, called **mixed designs**, are analyzed using a mixed analysis of variance. (To learn how to interpret the results of a mixed analysis of variance, see Box 13.5.)

In both Case 1 and Case 2, the mixed design allows us to have both internal validity and power. In Case 1, we could make lesions a between-subjects variable by randomly assigning half the participants to get lesions and half not. That way we do not have to worry about carryover effects from the brain lesions. We could make *practice* a within-subjects variable by having each participant run the maze three times. Consequently, we have the power to detect subtle differences due to practice (see Table 13.12 and Figure 13.1).

In Case 2, we could make ECS therapy a between-subjects variable by randomly assigning half the participants to get electroconvulsive (ECS) therapy and half not. That way, we do not have to worry about carryover effects from the ECS. Then, we would expose all participants to a variety of subliminal messages, some designed to boost mood and some to be neutral. By comparing the average overall depression scores from the ECS therapy group to that of the no-ECS group, we could assess the effect of ECS. By comparing participants' scores following the "positive" subliminal messages to their scores following "neutral" subliminal messages, we could detect even rather subtle effects of subliminal messages.

BOX 13.5 Not Getting Mixed Up About Mixed Designs

If you use a mixed design, you will probably have a computer analyze your data for you. Often, both entering the data and interpreting the printout are straightforward. For example, suppose you had two groups (one received Treatment X, the other Treatment Y), had each participant go through three trials, and collected the following data:

Participant	Group	Trial 1	Trial 2	Trial 3
Steve	X	1	3	7
Mary	X	2	4	6
Todd	X	3	6	7
Melissa	X	4	5	7
Tom	Y	4	5	7
Amy	Y	5	4	7
Rob	Y	4	5	6
Kara	Y	4	4	7

You might input the data as follows:

Group	Trial 1	Trial 2	Trial 3
1	1	3	7
1	2	4	6
1	3	6	7
1	4	5	7
2	4	5	7
2	5	4	7
2	4	5	6
2	4	4	7

Your printout might be relatively straightforward and resemble the following:

	T 1 Mean	T 2 Mean	T 2 Mean
Group 1	2.5	4.5	6.75
Group 2	4.25	4.5	6.75
Total	3.375	4.5	6.75

Between Ss

Source	SS	df	MS	F	p
A	2.04	1	2.04	1.69	.24
Error term	7.25	6	1.21		

Within Ss					
B	47.25	2	23.63	47.26	<.001
A × B	4.08	2	2.04	4.08	.044
Error term	6.0	12	.5		

However, in some programs, entering your data and interpreting the printout can be more complicated. To make sure that the computer has done the analysis you expected, check your printout carefully.

If your printout contains only one error term, the computer is analyzing your data as if you have a completely between-subjects design. If you take the *MS* for any treatment or interaction and divide it by your one and only *MSE,* you will get the *F* for that effect.

If, on the other hand, every main effect and every interaction has its own error term, the computer is analyzing your data as if you have a completely within-subjects design. In that case, if you have three effects (two main effects and an interaction effect), you will have three error terms.

Even if the computer seems to be analyzing your study as a mixed design, check the computer printout to be sure that it has correctly identified which factors are within and which are between. Start by looking at the degrees of freedom for all your main effects. If your between-subjects factor(s) have more levels than your within-subjects factor(s), then the degrees of freedom for your between-subjects main effect should be larger than the degrees of freedom for your within-subjects main effect. In any event, make sure that the *df* for each of your variable's main effects is one fewer than the number of levels of that variable. For example, if you have 4 levels of the between variable and 2 levels of the within variable, be sure that the degrees of freedom for the between variable is 3 and that the degrees of freedom for the within variable is 1.

Next, focus on your between-subjects factor(s). All between-subjects main effects—and all interactions that involve only between-subjects factors—should be tested against a single error term. To check on this, divide the *MS* for each between-factors main effect and each exclusively between-factors interaction by the *MS* for the between-subjects error term. In every case, you should get the same *F* that is reported in the printout.

To double-check that the computer correctly identified all the between-subjects variables, add up the degrees of freedom for all the between-subjects main effects, the *df* for the interactions that involved only between-subjects factors, and the *df* for the between-subjects error term. The total of these degrees of freedom should be one fewer than the number of participants.

Next, check the within factors. Each within-subjects main effect and each interaction that involves only within-subjects factors should be tested against a different error term.

Finally, look at interactions in which at least one variable is a between factor and at least one variable is a within factor. To find the appropriate error term for these interactions, attend only to the within-subjects factors: Ignore the between-subjects factors. If A is a within factor and B is a between factor and you see an A × B interaction, this interaction should be tested against the same error term that A is tested against. If A is a within factor and B and C are between factors, the error term for the A × B × C interaction should still be the same error term that was used for testing A. If it is not, there is a mix-up about which of your factors are within and which are between.

In a mixed design, you are able to test not only the main effects of two treatments but also the interaction of those treatments. In Case 1, the interesting effects will probably involve the interaction rather than the two main effects. That is, we would not be terribly surprised to find a main effect for

TABLE **13.12** Analysis of Variance Summary Table for a Mixed Design

Source of Variance	df	SS	MS	F	p
Brain Lesion	1	51.0	*51.0*	*10.0*	.0068
Between-Subjects Error	14	72.4	<u>5.1</u>		
Trials	2	26.6	13.3	11.1	.0003
Lesions × Trials	2	13.7	6.8	5.7	.0083
Within-Subjects Error	28	33.6	<u>1.2</u>		

Note: The mean square error for the within-subjects term is much smaller than the between-subjects error term (1.2 to 5.1), giving the design tremendous power for detecting within-subjects effects. This table corresponds to the graph in Figure 13.1.

© Cengage Learning 2013

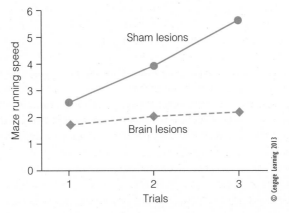

FIGURE **13.1** An interaction in a mixed design.

lesion, telling us that the brain-lesioned rats performed worse.[7] Nor would we be surprised to find a main effect for practice, telling us that participants improve with practice. However, we would be interested in knowing about the practice × lesion interaction. A significant practice × lesion interaction would tell us that one group of rats was benefiting from practice more than another. In this case, as you can see from Figure 13.1, the nonlesion group benefits most from practice. In Case 2, although we would be interested in both the ECS and subliminal message main effects, we might be most interested in the interaction between ECS and subliminal messages: Such an interaction would tell us whether the ECS group was more influenced by the subliminal messages than the no-ECS group.

---

[7] The lesion main effect would be especially unsurprising if our control group didn't get any surgery. However, such empty control groups are rare. Typically, the control group would be a "sham lesion" control group that got brain surgery and was treated the same as the treatment group except that, instead of being injected with a chemical that would destroy (lesion) part of the brain, they would be injected with a harmless saline solution.

In many mixed designs, both a main effect and the interaction will be of interest. For example, Hebl and Mannix (2003) found a between-subjects main effect indicating that participants who saw a picture of a male job applicant sitting next to an overweight woman rated the job applicant more harshly than participants who saw a picture of the same man sitting next to an average-weight woman. This between-subjects main effect was of interest. The interaction between this main effect and the within-subjects variable of rating dimension (willingness to hire applicant, applicant's professional qualities, applicant's interpersonal skills) was also of interest because Hebl and Mannix wanted to see whether being seen with an overweight woman influenced hiring judgments more than it affected judgments about the applicant's interpersonal skills.

Note the problems Hebl and Mannix would have had in interpreting their results if they had used either a completely within-subjects or a completely between-subjects design. If they had used a completely within-subjects design, each participant would rate the applicant both (1) after seeing the applicant in the presence of an overweight woman and (2) after seeing the applicant in the presence of an average-weight woman. Participants would have found the study strange and would probably have figured out the hypothesis, thereby making the weight-of-woman main effect hard to interpret.

If Hebl and Mannix had used a completely between-subjects design, one group of participants would make hiring judgments, another group would make interpersonal skills judgments, and yet another group would make judgments about the applicant's professional qualities. Because each participant would be providing one set of ratings rather than the three sets that Hebl and Mannix's participants did, each participant in a between-subjects design would be providing only one third as much data as the participants in Hebl and Mannix's actual study. Because participants would be providing less data, the study would have been less powerful than Hebl and Mannix's actual study. Thus, if Hebl and Mannix had used a completely between-subjects design and failed to find an effect for the interaction, a scientist reading their work would wonder whether they would have succeeded in finding an interaction had they used a more powerful design.

As you can see from Hebl and Mannix's study and from our two hypothetical cases (Case 1 and Case 2), the mixed design has two major strengths. First, it allows you to examine the effects of two independent variables and their interaction. Second, instead of trading off the needs of one independent variable for the needs of another, you are able to give both independent variables the design they need. Because of its versatility, the mixed design is one of the most popular experimental designs.

CONCLUDING REMARKS

This chapter has expanded your ability to read about and conduct research. When reading reports of either within-subjects or mixed designs, you now know to ask

1. whether the multiple measures and manipulations may have led participants to figure out the hypothesis,

2. what steps (e.g., counterbalancing) were taken to reduce order effects (practice, fatigue, carryover, and sensitization)—and whether those steps were sufficient to ensure the study's internal validity, and

3. whether a between-subjects design might have been more internally valid.

When planning, conducting, or analyzing research, you now can

1. Do experiments to determine the effect of a treatment and have a reasonable chance of finding the treatment effect even if the effect is small and you can study only a few participants.

2. Replicate between-groups experiments that failed to find an effect with a more powerful design that is more likely to find an effect.

3. Use counterbalancing to control for order effects.

4. Take steps to minimize practice, fatigue, carryover, and sensitization, thereby minimizing order effects.

5. Do research assessing the effects of order (trials) and the effect of interactions involving trials (e.g., does the effect of one treatment get stronger when it is repeatedly presented, whereas the effect of another treatment weakens with repeated exposures?).

6. Do research to determine the effect of different treatment sequences (e.g., is it more effective to have cognitive therapy followed by antidepressants or to have antidepressants followed by cognitive therapy?).

7. Determine whether you should use a pure between-subjects experiment, a matched-pairs experiment, a within-subjects design, or a mixed design.

8. Interpret computer printouts of analysis of variance (ANOVA) analyses of within-subjects as well as mixed designs.

SUMMARY

1. The matched-pairs design uses matching to reduce the effects of random differences between participants and uses random assignment and statistics to account for the remaining effects of random error. Because of random assignment, the matched-pairs design has internal validity. Because of matching, the matched-pairs design has power.

2. Because the matched-pairs design gives you power without limiting the kind of participant you can use, you may be able to generalize your results to a broader population than if you had used a simple experiment.

3. The matched-pairs design's weaknesses stem from matching: Matching may be time-consuming, ineffective, change how participants react to the treatment, and make participants drop out of the study.

4. Within-subjects designs are also known as repeated-measures designs.

5. The two-condition within-subjects design gives you two scores per participant.

6. The within-subjects design increases power by eliminating random error due to individual differences and by increasing the number of observations that you obtain from each participant. As a result, the within-subjects design is the most powerful experimental design. However, within-subjects designs have more internal validity and construct validity problems than between-subjects designs.

7. Both the matched-pairs design and the two-condition pure within-subjects design can be analyzed by the dependent groups t test. Complex within-subjects designs require more complex analyses. Specifically, they should be

analyzed by within-subjects analysis of variance (ANOVA) or by multivariate analysis of variance (MANOVA).

8. Because of practice, fatigue, carryover, and sensitization effects, the participant may respond one way if receiving a treatment first and a different way if receiving the treatment last. Because of these order effects, a within-subjects design may mislead you about a treatment's real effect.

9. To reduce the effects of order, you should randomly determine the sequence in which each participant will get the treatments or use a counterbalanced design.

10. In the counterbalanced design, participants are randomly assigned to systematically varying sequences of conditions to ensure that routine order effects are balanced out. The keys are to (a) find a set of sequences that do balance each other out and (b) randomly assign participants in such a way that an equal number of participants get each sequence.

11. Order effects (often called *trials effects*) are different from sequence effects. *Order effects* refer to whether participants respond differently on one trial (e.g., the first) than on some other trial (e.g., the last). Order is a within-subjects factor in a counterbalanced design because every participant gets some treatment first and another treatment second.

12. Order effects can be detected by looking at the treatment × counterbalancing sequence interaction.

13. *Sequence effects* refer to whether participants respond differently to getting a series of treatments in one sequence than getting the treatments in a different sequence. For example, the group of participants who get the treatments arranged in the sequence Treatment A, then Treatment B may have higher overall average scores than the group of participants who get the treatments arranged in the sequence Treatment B, then Treatment A. Sequence is a between-subjects factor.

14. A counterbalanced design allows you to see whether (a) the treatment had an effect (by looking at the treatment main effect), (b) a group getting one sequence of treatments did better than group(s) getting a different sequence of treatments (by looking at the counterbalancing main effect), and (c) trials/order had an effect (by looking at the counterbalancing × treatment interaction).

15. If you include the between-subjects factor of counterbalancing in your analyses, counterbalanced designs require more participants than pure within-subjects designs.

16. If you want to compare two levels of an independent variable, you can use a matched-pairs design, a within-subjects design, a counterbalanced design, or a simple between-subjects design.

17. Mixed designs have both a within- and a between-subjects factor. Counterbalanced designs are one form of a mixed design.

18. Mixed designs should be analyzed with a mixed analysis of variance or a multivariate analysis of variance.

KEY TERMS

carry-over (treatment carryover) effects *(p. 535)*
counterbalanced within-subjects design *(p. 544)*
dependent groups *t* test *(p. 527)*
fatigue effects *(p. 534)*

matched-pairs design *(p. 520)*
mixed designs *(p. 555)*
order *(p. 533)*
order (trial) effects *(p. 533)*
power *(p. 523)*
practice effects *(p. 534)*

randomized within-subjects design *(p. 539)*
sensitization *(p. 535)*
sequence effects *(p. 552)*
within-subjects designs *(repeated-measures designs) (p. 531)*

EXERCISES

1. What feature of the matched-pairs design makes it
   a. an internally valid design?
   b. a powerful design?

2. A researcher uses a simple between-subjects experiment involving 10 participants to examine the effects of memory strategy (repetition vs. imagery) on memory.
   a. Do you think the researcher will find a significant effect? Why or why not?
   b. What design would you recommend?
   c. If the researcher had used a matched-pairs study involving 10 participants, would the study have more power? Why? How many degrees of freedom would the researcher have? What type of matching task would you suggest? Why?

3. An investigator wants to find out whether hearing jokes will allow a person to persevere longer on a frustrating task. The researcher matches participants based on their reaction to a frustrating task. Of the 30 original participants, 5 quit the study after going through the "frustration pretest." Beyond the ethical problems, what problems are there in using a matched-pairs design in this situation?

4. What problems would there be in using a within-subjects design to study the "humor-perseverance" study (discussed in question 3)? Would a counterbalanced design solve these problems? Why or why not?

5. Why are within-subjects designs more powerful than matched-pairs designs?

6. Two researchers hypothesize that spatial problems will be solved more quickly when the problems are presented to participants' left visual fields than when stimuli are presented to participants' right visual fields. (They reason that messages seen in the left visual field go directly to the right brain, which is often assumed to be better at processing spatial information.) Conversely, they believe verbal tasks will be performed more quickly when stimuli are presented to participants' right visual fields than when the tasks are presented to participants' left visual

fields. What design would you recommend? Why?

7. A student hypothesizes that alcohol level will affect sense of humor. Specifically, the student has two hypotheses. First, the more people drink, the more they will laugh at slapstick humor. Second, the more people drink, the less they will laugh at other forms of humor. What design would you recommend the student use? Why?

8. You want to determine whether caffeine, a snack, or a brief walk has a more beneficial effect on mood. What design would you use? Why?

9. Using a driving simulator and a within-subjects design, you want to compare the differences between driving unimpaired, driving while talking on a cell phone, and driving while legally intoxicated.
   a. Which order effects do you have to worry about? Why?
   b. To what degree would counterbalancing solve the problems caused by order effects?
   c. How would you try to prevent order effects from harming the validity of your study?

10. A researcher wants to know whether music lessons increase scores on IQ subtests and whether music lessons have more of an effect on some subtests (e.g., more of an effect on math than on vocabulary) than others.
    a. Would you make music lessons a between- or within-subjects factor? Why?
    b. Would you make subtests a between- or within-subjects factor? Why?
    c. If the researcher did an analysis of variance (ANOVA) on the data, the researcher would obtain three effects. Name those three effects.
    d. What effect would the researcher look for to determine whether music lessons increase scores on IQ subtests?
    e. What effect would the researcher look for to determine whether music lessons have more of an effect on math subtests than on vocabulary subtests?

WEB RESOURCES

Go to the Chapter 13 section of the book's student website and

1. Look over the concept map of the key terms.
2. Test your self on the key terms.
3. Take the Chapter 13 Practice Quiz.
4. Download the Chapter 13 tutorial to practice
   a. distinguishing between order and sequence effects

b. interpreting printouts from within-subjects designs
   c. choosing among designs
5. Do an analysis on data from a within-subjects design using a statistical calculator by going to the "Statistical Calculator" link.

Single-*n* Designs and Quasi-Experiments

Real life is messy.
—Anonymous

The average human has about one breast and one testicle.
—Statistics 101

CHAPTER OVERVIEW

To solve real-world problems, applied psychologists must identify the problems'
causes. One powerful tool applied psychologists use to identify causes is the
randomized experiment. However, when applied psychologists cannot randomly
assign participants to treatment, they turn to two other types of studies: single-*n*
designs and quasi-experiments. In this chapter, you will learn about these two types
of studies and about how they compare to the randomized experiment. After
reading this chapter, you will be able to design a study to determine the effect
of a real-life treatment.

INFERRING CAUSALITY IN RANDOMIZED EXPERIMENTS

Whether you use a randomized experiment or any other design, you must
satisfy three criteria if you are to infer that one variable (e.g., smiling at
others) causes a change in another variable (others helping you). Specifically,
you must establish

1. covariation (that changes in the treatment are associated with changes
    in behavior);
2. temporal precedence (that changes in the treatment occur before changes
    in behavior); and
3. that the change in behavior is not due to something other than the
    treatment.

Establishing Covariation: Finding a Relationship Between Changes in the Suspected Cause and Changes in the Outcome Measure

Before you can show that the treatment causes a change in behavior, you
must first establish **covariation**: that changes in the treatment are accompanied by changes in the behavior. Therefore, to show that smiling causes
people to help you, you must show that people are more helpful to you
when you smile than when you do not.

In the randomized experiment, you would establish covariation by seeing
whether the amount of help you received when you smiled was greater than
when you did not smile. If the average amount of helping was the same in
both groups, you would not have covariation. Because you would not have

covariation (variations in smiling would not correspond with variations in helping), you would not conclude that the treatment had an effect. If, on the other hand, you received more help in the smiling condition than in the no-smile condition, you would have covariation.

Establishing Temporal Precedence: Showing That Changes in the Suspected Cause Come Before Changes in the Outcome Measure

Establishing covariation, by itself, does not establish causality. You must also establish **temporal precedence**: that the treatment comes before the change in behavior. In other words, you must show that you smile at others before they help you. Otherwise, it may be that you react with a smile after people help you. Without temporal precedence, you can't determine which variable is the cause and which is the effect. Thus, one reason correlational designs fail to establish that changes in the "first" variable caused changes in the "second" variable is that such designs often don't allow you to know which variable changed first. For example, if we find that successful companies have employees with high morale, we don't know that high morale causes success: After all, it could be that success causes high morale (Rosenzweig, 2007).

In a randomized experiment, you automatically establish that the treatment comes before the change in behavior (temporal precedence) by manipulating the treatment. You always present the independent variable (smiling) before you present the dependent measure task (giving participants an opportunity to help).

Battling Spuriousness: Showing That Changes in the Outcome Measure Are Not Due to Something Other Than the Suspected Cause

In addition to establishing temporal precedence (that the cause came before the effect), you must show that the covariation you observed could be due only to the treatment. Ideally, you would do this by showing that the treatment is the only thing that varies. Therefore, to show that your smiling causes others to help you, you must show that everything—except for your smiling—is the same during the times that you smile and the times that you do not smile.

The Value of Battling Spuriousness

It's difficult to prove that the only difference between the times when you get help and times when you don't is your smile. But without such proof, you can't say that your smiling causes people to be more helpful. Why not? Because you might be smiling more when the weather is nice or when you are with your friends. These same conditions (being with friends, nice weather) may be the reason you are getting help—your smile may have nothing to do with it. If you cannot be sure that everything else was the same, the relationship between smiling and helpfulness may be a **spurious relationship**: a statistical relationship between two variables that is not due to one of the variables influencing the other but instead is due to both variables being influenced by some third variable. Because correlational designs do not rule out spuriousness, you can't use those designs to make cause–effect statements.

Battling Spuriousness Without Keeping All Nontreatment Variables Constant

In the randomized experiment, you do not keep everything—except for the treatment variable—constant. There are some nontreatment variables, such as individual differences, that you can't control. There may be other nontreatment variables that you choose not to control. For example, you may decide to do your experiment in a real-world setting where you can't keep temperature, noise, and other factors constant.

How do you deal with these nontreatment variables that aren't being controlled? You use random assignment so that these uncontrolled variables are now random variables. As you will see in the next two sections, there are two advantages of converting nontreatment variables into random variables: (1) Random variables should not influence one group significantly more than another, and (2) statistics can be used to estimate the effects of random variables.

Random Variables Affect All Groups (Almost) Equally. One advantage of random assignment is that the nontreatment variables should not substantially affect one group more than the other. Random assignment should spread those variables more or less equally into each of your groups, just as an electric mixer should distribute ingredients fairly equally to both sides of the bowl. With random assignment, your conditions will be equivalent except for the effects of the independent variable and the chance impact of random variables. Therefore, as a result of random assignment, only random variables stand in the way of keeping irrelevant variables constant.

Statistics Can Help You Estimate the Effects of Random Variables. If you could remove those random variables, you would be able to keep everything constant, thereby isolating the treatment as the cause of the change in behavior. Unfortunately, in the randomized experiment, you cannot keep nontreatment variables constant and you cannot remove them. However, you can use statistics to estimate their effects: If the difference between groups is greater than the estimated effects of random variables, the results are declared "statistically significant."

If you find a statistically significant effect for your treatment variable, you can argue that your treatment variable causes a change in scores on the dependent measure. However, you may be wrong. Even with statistics, you can't perfectly estimate the effects of random variables 100% of the time. If you underestimate the effects of random variables in your study, then you may falsely label a chance difference as a treatment effect. In technical terminology, you may make a Type 1 error.

Fortunately, before you do the study, you establish what your chances are of making a Type 1 error. Usually, most investigators make the chances of committing a Type 1 error fairly remote. Specifically, most investigators set the probability of mistaking chance variation as a genuine treatment effect at less than 5 in 100 ($p < .05$).

SINGLE-*n* DESIGNS

Can we make reasonable inferences about the causes of an effect without random assignment? Yes. Psychological pioneers such as Wundt, Helmholtz, Ebbinghaus, Fechner, and Skinner often did so by conducting studies that involved intensively studying a single participant.

Although these researchers were intensively studying individual participants, their research did not involve clinical case histories. They did not look back at events that happened in an individual's life, try to determine that a particular event came before the individual started acting in a certain way, and then trust that no other event could be responsible for the participant acting that way. Instead, these pioneers isolated the cause of the key behavior by controlling the events that occurred in the participant's life.

To get a general sense of how their approach differs from the case study approach (a detailed description of an individual), consider Skinner's experimental study of superstitious behavior. Skinner did not find a person who already engaged in superstitious behavior and then try to identify which of the person's thousands of genes, millions of life experiences, or billions of interactions between events and genes was responsible for the superstitious behavior. Nor did he look for people who claimed to have become less superstitious after an event (e.g., after going through primal scream therapy) and then assume that no other event within the person (becoming more mature) or outside the person (experiencing success) could be responsible for the change.

Instead, in a highly controlled environment (a Skinner box), Skinner induced and then eliminated superstitious behavior. Skinner arranged for a hungry pigeon to receive food "at regular intervals with no reference whatsoever to the bird's behavior" (Skinner, 1948, p. 168). Skinner was able to induce behavior that could be called superstitious, extinguish such behavior, and recondition it. He also demonstrated that behavior that could be described as superstitious was most likely when there was a 15-second interval between reinforcements. By systematically introducing and withdrawing the treatment, by observing changes in the behavior following changes in the treatment, and by not allowing other changes in the pigeon's environment, Skinner was able to make a convincing case that the treatment was the cause of the pigeon's superstitious behavior.

Note that the approach of a single-*n* researcher such as Skinner is like that of a physicist performing an experiment whereas the approach of a case study researcher is like that of a naturalist observing a rare specimen. Specifically, Skinner and other researchers who use the single-*n* approach are able to do three things researchers who use case studies cannot (see Table 14.1).

First, Skinner was able to establish covariation because he was able to see how behavior changed as he repeatedly introduced and removed the treatment. Thus, he was able to establish a reliable, coincidence-free connection between the treatment and the superstitious behavior. The case study researcher, on the other hand, does not establish covariation. For example, suppose such a researcher studied the case of the brother of a Russian leader

TABLE **14.1** Designs and Causality

	Establish Covariation	Manipulate Treatment to Establish Treatment Comes Before Effect	Rule out other Explanations for Relationship
Case studies			
Correlational study	√		
Quasi-experimental study	√	√	?
Randomized experiment	√	√	√
Single-*n* studies	√	√	?

© Cengage Learning 2013

who apparently recovered from a mental illness after being thrown into an icy river (Henderson, 1985). Even if the story is true, the researcher has not established a reliable connection between being thrown into a cold river and mental health improvement.

Second, Skinner was able to establish that the connection between rewards and superstitious behavior was a cause–effect connection because he was able to rule out nontreatment factors. By showing that, before he introduced the rewards, the pigeon rarely produced the behavior, Skinner established that he had kept relevant nontreatment factors relatively constant. The case study researcher, on the other hand, cannot rule out the effects of other factors.

Third, Skinner and others can easily replicate (repeat) the study to verify the findings. If skeptical others obtain similar results, we can be more confident of the reliability of the original study's findings (to learn more about why researchers are skeptical of case studies, see Table 14.2).

Like researchers using experimental designs, researchers using single-*n* designs establish that the cause comes before the effect (temporal precedence) by introducing the treatment variable before presenting the dependent-measure task. Thus, like a researcher using a randomized experiment, a researcher using a single-*n* design would smile at the participant *before* giving the participant an opportunity to help. Like researchers using experimental designs, researchers using single-*n* designs establish covariation by comparing the different treatment conditions (comparing the amount of help received in the smiling vs. no-smiling conditions). However, unlike researchers using experimental designs, researchers using single-*n* designs do not rely on randomization and statistical tests to rule out the effects of nontreatment factors (rule out spuriousness).

Instead, the single-*n* researcher strives to keep nontreatment factors constant. That is, rather than letting nontreatment factors vary and then statistically accounting for the effects of those variables, single-*n* researchers try to stop nontreatment factors from varying, thereby isolating the treatment's effect (see Table 14.3).

Battling Spuriousness by Keeping Nontreatment Factors Constant: The A–B Design

To understand how single-*n* researchers keep nontreatment factors constant, let's examine the simplest **single-*n* design**, the A–B design. In the **A–B design**,

TABLE **14.2** Why Case Studies Are Not (Scientifically) Convincing

Question	Problem	Example/Explanation

Did it happen?	Informal observation can be flawed due to
	1. Misperception
		• Thousands of people have reported seeing aliens.
		• People have "observed" that
		  • rotten food turned into insects;
		  • the sun revolved around the earth;
		  • planets had round orbits;
		  • the earth was flat.

	2. Misremembering	• People often think they were worse off than they were. • Nurses "remember" more psychiatric patients being admitted during full moons.
	3. Misreporting	• Exaggerating or lying about an improvement, especially if the person likes the researcher or therapist. Technically, this is called "obeying demand characteristics."
Does it happen consistently?	We may be unable to replicate a case study, thus making it difficult to know how often the event occurs.	• Correlations in psychology are rarely 1.0, meaning that psychological rules have exceptions. Therefore, we do not know whether a case is an example of what typically happens or an example of an exception to the typical case. Thus, psychologists are not convinced by "I know someone who…" evidence. • If we just looked at people who made the biggest returns on their investments—lottery winners, gamblers, stock market speculators—we would conclude that the way to make money was to take large risks (Rosenzweig, 2007). However, our case study research would be misleading because we would not be looking at all those who lost money on such risky investments. Thus, despite the many cases of people who have won the lottery, no responsible financial consultant would tell a retired person to invest his or her savings in lottery tickets.
Why did it happen?	Because there is no control group, we cannot rule out other explanations for the findings, such as 1. Coincidence	• Psychological problems may appear, disappear, and improve over time. If a treatment is administered during one of these times, the treatment may get credit—or blame—for the change (Painter, 2008). For example, after being

Question	Problem	Example/Explanation
		subjected to medical treatments now known to be harmful (e.g., having an operation to put asbestos over one's heart, eating lizard dung), some people get better.
		• One week, you receive an e-mail telling you a stock is going up. It does. The next week, you receive an e-mail telling you another stock is going down. It does. The next week, you receive an e-mail telling you another stock is going up. It does. Although you might be impressed, the "expert" stock picker has used a simple trick. The trick starts by sending out thousands of e-mails, half predicting that a stock will go up, half predicting that a stock will go down. Then, the trickster waits until the stock goes up or down. He doesn't send any more e-mail to the people to whom he sent the wrong prediction. To the people to whom he sent the correct prediction, he will send half of them a prediction that another stock will go up and the other half a "down" prediction. He repeats this several times. Then, he asks the people to whom he has sent a series of accurate predictions to invest their money with him (Stanovich, 2007). Note that although this dishonest scheme is taking advantage of coincidence, even the success of legitimate investment gurus can be explained by coincidence (Mlodinow, 2008; Whyte, 2005).
	2. Researchers unintentionally and nonverbally telling participants what to do.	• People thought that Clever Hans ("the mathematical horse") tapped out correct answers to math questions because he knew math. In actuality, he knew body language: He stopped tapping when people stopped looking at his feet.
		• People thought that severely impaired autistic children could communicate complex thoughts using facilitated communication—a technique in which a helper "steadied" their hands as they typed. The effect turned out to be a Ouija board type effect, similar enough to Clever Hans that Wegner, Fuller, and Sparrow (2003), described it as a case of "Clever Hands."
	3. Participant bias	• If people believe that a treatment will work, they will tend to feel better after receiving that treatment. Physicians know the power of the placebo effect.
		• Participants may change other behaviors or attitudes when taking the treatment—and those other changes may cause the desired results (Painter, 2008). For example, a person may start taking vitamins and start exercising at the same time, but credit the weight loss to the vitamins rather than to the exercise (Painter, 2008).

TABLE **14.3** How Different Designs Infer Causality

Requirement	Randomized Experiments	A–B Single-*n* Design
Temporal Precedence (treatment came before changes in scores)	Introduce treatment before there is a change in the dependent variable.	Introduce treatment before there is a change in the dependent variable.
Covariation (different treatment conditions score differently on measure)	Observe difference between treatment and control conditions.	Observe difference between conditions A (baseline) and B (posttreatment behavior).
Accounting for Irrelevant Variables (determining that the change in behavior is not due to nontreatment factors)	1. Use independent random assignment to make sure all irrelevant factors vary randomly rather than systematically. 2. Then, use statistics to account for effects of these random factors. If the difference between groups is greater than would be expected as a result of these random factors, the difference is assumed to be the effect of the one nonrandom, systematically varied factor: the treatment.	1. Eliminate between-subjects variables by using only one participant. 2. Control relevant environmental factors. Demonstrate that those factors have been controlled by establishing a stable baseline. Then, introduce treatment. If change occurs, that change is assumed to be due to the treatment.

© Cengage Learning 2013

as in all single-*n* designs, the researcher studies a single participant and tries to make sure that the participant's behavior on the dependent-measure task occurs at a consistent rate. If we were studying the effects of rewards on how often a chicken pecked at a bar, we would first make sure the chicken was pecking at a constant rate. The process of ensuring that the behavior occurs at a steady, consistent rate is called establishing a **stable baseline**. This first step, the baseline behavior, is designated as A. Next, the researcher introduces the treatment and then compares posttreatment behavior (B) with baseline behavior (A).

As with all single-*n* designs, the A–B design strives to keep everything but the treatment constant. Specifically, the A–B design tries to make sure that differences between the conditions are not due to either of the two basic types of nontreatment variability: (a) between-subjects variability unrelated to the treatment and (b) within-subjects variability unrelated to the treatment.

As with all single-*n* designs, the A–B design makes sure that between subjects variability can't cause the difference between treatment conditions. The difference between scores in the A condition and the B condition can't possibly be due to differences between participants because the participant in A was the same individual who was in the B condition.

Within-subjects variability, however, is a problem. An individual's moods and behaviors may naturally vary from moment to moment. Thus, in a sense, the same participant could be a different participant during the A phase of the study than during the B phase. So, how does the single-*n* researcher know that the treatment, rather than natural within-subjects variability, is responsible for the change in the participant's behavior?

The single-*n* researcher is confident that the difference between notreatment and treatment conditions is not due to random within-subject variability because she has established a stable baseline. The **baseline** shows that the participant's behavior is not varying.

But how does a single-*n* researcher obtain a stable baseline? To obtain a stable baseline, the single-*n* researcher must hold constant all those variables that might affect the participant's responses.

If the researcher does not know what the relevant variables are, the participant's environment must be kept as constant as possible. Consequently, the researcher might perform the study under highly controlled conditions in a soundproof laboratory.

If the researcher knows what the relevant variables are, only the relevant variables need to be kept constant. Thus, if a researcher knew that parental praise was the only relevant variable in increasing studying behavior, the researcher would need to control only that one variable. However, the researcher usually does not know which variables can be safely ignored. Psychology has not advanced to the state where we can catalog what variables affect and don't affect every possible response.

The researcher looks at the baseline to check whether she has succeeded at controlling key variables. If the baseline is not stable, the researcher continues to control variables until the behavior becomes stable.

But what if a researcher cannot achieve a stable baseline? Then, the researcher planning to use an A–B design has a problem: Changes in behavior that occur after the treatment is introduced may be due to something other than the treatment. Consequently, the researcher might not know whether the change in behavior is due to (a) normal fluctuations in the participant's behavior or (b) treatment effects.

There is still hope for the researcher who can't achieve a stable baseline. As you can see from Figure 14.1, if the participants' behavior changes dramatically after the treatment is introduced, A–B researchers can make a convincing case that the results are not due to normal baseline fluctuations.

Although it is difficult to achieve a stable baseline, we should point out that single-*n* researchers often do achieve it. They are especially successful when they put a simple organism (e.g., a pigeon) in a simple environment (e.g., a Skinner box) and have it perform a simple behavior (e.g., peck a disk).

To this point, you have seen how the single-*n* researcher using an A–B design can hold individual difference variables and relevant environmental variables constant. But how does the researcher know that the difference between conditions is not due to **maturation**: natural biological changes in the organism, such as those due to development or fatigue?

The single-*n* researcher may limit maturation by choosing an organism that she knows won't mature substantially during the course of the study. She might use a pigeon or a rat because the extent of their maturation as it relates to certain tasks (bar pressing and pecking) is well documented.

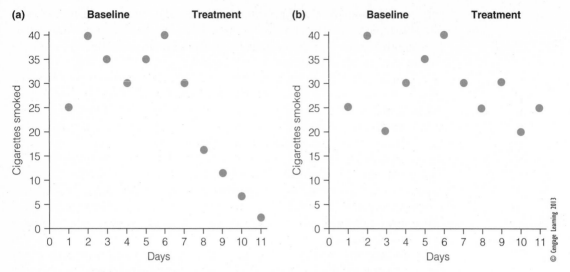

FIGURE **14.1** One behavior modification program appears to reduce a client's cigarette smoking, one doesn't.

Note:
In Figure 14-1a, the decrease in cigarettes smoked after the treatment was introduced on Day 7 seems to be due to the treatment. In Figure 14-1b, on the other hand, it is unclear whether the decrease in cigarettes smoked after Day 7 is due to anything more than normal fluctuations in the participant's behavior.

Or, as you will soon see, the researcher may use a design that will allow her to account for maturation. But before looking at a design that accounts for maturation, let's look at an example of the A–B design.

In an early study of the effects of psychoactive drugs, Blough (1957) wanted to study the impact of LSD on a pigeon's visual perception. His first step was to place the pigeon in a highly controlled environment—a Skinner box—equipped with a light that lit up a spot.

By varying how bright the light was, Blough could make the spot easier—or more difficult—to see. To determine whether the pigeon could see the spot, the pigeon was conditioned to peck at disk "1" when the spot was visible and to peck at disk "2" when the spot was not visible.

Before Blough administered his independent variable (LSD), he had to make sure that no other variables were influencing the pigeon's behavior. To do this, he had to keep all the relevant variables in the pigeon's environment constant. Therefore, he placed the pigeon in the Skinner box and carefully observed the pigeon's behavior. If he had succeeded in eliminating all non-treatment variables, the pigeon's behavior would be relatively stable—the relationship between pecking and illumination would be constant. If he had failed, he would have observed erratic fluctuations in the pigeon's pecking.

Once the pigeon's behavior was stable, Blough was ready to introduce the independent variable, LSD. After administering the LSD, Blough compared the pigeon's behavior after the treatment (B), to its behavior before the treatment (A). Blough found that after taking the LSD, the pigeon experienced decreased visual ability. Specifically, the pigeons pecked at disk 2 (cannot see spot) under a level of illumination that—prior to treatment—*always* led to a peck at disk 1. Because Blough had ensured that nontreatment variables were not influencing the pigeon's behavior, he concluded that the LSD was the sole cause of the decrease in visual ability.

Blough's study was exceptional because he knew that the pigeon's behavior on this task normally wouldn't change much over time. In studies with other kinds of participants or tasks, the researcher would not know whether participants would change, develop, or learn over time. Therefore, most researchers are not so confident that they have controlled all the important variables. In fact, as you will soon see, researchers know that two potentially important nontreatment variables have changed from measurement at baseline (A) to measurement after administering the treatment (B).

First, because the posttest occurs after the pretest, participants have had more practice on the posttest task. In technical terminology, their improved performance may be due to **testing:** the effects of doing the dependent measure task on subsequent performance on that task. For example, the practice a participant gets doing the task during the A phase may help the participant do better during the B phase.

Second, because the posttest occurs after the pretest, changes from pretest to posttest may be due to maturation. For instance, the participant's behavior may have changed over time due to fatigue, boredom, or development.

Variations on the A–B Design

Because psychologists want to know that their results are due to the treatment rather than to testing or maturation, single-*n* researchers rarely use the A–B design. Instead, they use variations on the A–B design such as the reversal design, psychophysical designs, and the multiple-baseline design.

The Reversal Design: Giving and Taking Away

In the **reversal design**, also known as the **A–B–A design** and the **A–B–A reversal design,** the researcher measures behavior (A), then administers the treatment and measures behavior (B), and then withdraws the treatment and measures behavior again (A).

To see why the A–B–A design is superior to the A–B design, consider one in a series of classic single-*n* studies demonstrating that behavior modification was an effective therapy for patients in mental hospitals (Ayllon & Azrin, 1968). In a mental hospital, Ayllon and Azrin worked with individuals who had been diagnosed as psychotic to see if a token economy was an effective way of increasing socially appropriate behavior. In a typical study, Ayllon and Azrin first identified an appropriate behavior (e.g., feeding oneself). Next, the researchers observed how often a certain patient performed that behavior. This phase of collecting baseline data for a patient could be labeled A. They then attempted to reinforce that behavior with a "token." Like money, the token could be exchanged for desirable outcomes such as candy, movies, social interaction, or privacy. During the treatment phase (labeled B), Ayllon and Azrin gave the patient tokens for each instance of the socially appropriate behavior and measured the behavior. They found that the patient performed more socially appropriate behaviors after the tokens were introduced. Therefore, a token economy increases socially appropriate behavior, right?

If Ayllon and Azrin's (1968) study had ended here, you could not be confident about that conclusion. Remember, with an A–B design, you don't know whether a change in behavior is due to maturation, testing, or the treatment.

Fortunately, Ayllon and Azrin (1968) expanded the A–B design to an A–B–A design by stopping the treatment while continuing to observe their patient's behavior. After removing the treatment, the incidence of socially appropriate behavior decreased. Consequently, they were able to determine that the treatment (tokens) increased socially appropriate behavior.

If, after withdrawing the treatment, socially appropriate behavior had continued to increase, they would not have concluded that the increase in socially appropriate behavior was due to the treatment. Instead, they would have concluded that the increase could be due to maturation or testing.

We should point out that the results were not quite as neat as we described. Admittedly, they found that socially appropriate behavior increased when they introduced the tokens and decreased when they stopped giving out tokens. Removing the tokens, however, did not cause the behavior to fall back all the way to baseline levels. Instead, the behavior fell back to near-baseline levels.

If tokens caused the effect, shouldn't their withdrawal cause the behavior to fall to baseline rather than near baseline? Admittedly, if the dependent measure (the rate of socially appropriate behavior) returned to baseline level, it would help make the case that the treatment had an effect. However, most behaviors won't return to baseline after you withdraw the treatment because of

1. maturation effects
2. testing effects
3. **carryover effects**: the treatment's effects persisting even after the treatment has been removed

Because of these three effects, you might be willing to say that the treatment had an effect, even if the behavior did not return to baseline. For example, you might be willing to say that the treatment had an effect if the participant's behavior was substantially different during treatment phase (B) than during either the pretreatment (A) and posttreatment (A) conditions (see Figure 14.2).

Unfortunately, even if posttreatment behavior returns to baseline, if the effects of practice or maturation are cyclical, your claim that the treatment caused an effect could be wrong. For instance, suppose performance was affected by menstrual cycles. Performance might be good during the pretreatment phase (before menstruation), poor during the treatment phase (during menstruation), and good during the posttreatment phase (after menstruation). Although such an unsteady effect of maturation or testing would be unlikely, it is possible.[1]

---

[1] A similar problem could result if the individual you studied regularly went through periods of depression followed by periods of normal mood followed by depression (cyclical depression).

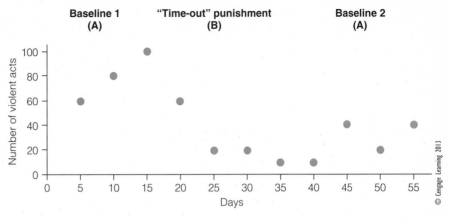

FIGURE **14.2** Results from A–B–A design: number of violent acts performed by jim during the no-punishment and "time-out" punishment phases.

Note:
Even though posttreatment violence did not revert back to pretreatment levels, a strong case can still be made that the time-out punishment reduced Jim's violent behavior.

To rule out the possibility that apparent treatment effects are due to some simple cyclical pattern involving either maturation or practice, you might extend the A–B–A design. For example, you might make it an A–B–A–B design. Ayllon and Azrin (1968) expanded their design to an A–B–A–B design and found that reintroduction of the token rewards led to an increase in the socially appropriate behavior.

Expanding beyond even the A–B–A–B design allows you to rule out the possibility of an even more complicated maturational or practice cycle. The more you expand the design, the less likely that maturation or practice would increase performance every time the treatment is introduced, but never increase performance when the treatment is removed. Thus, it would be very hard to describe a cycle of maturation and practice effects that could mimic treatment effects in an A–B–A–B–A–B–A–B–A–B–A–B–A–B design.

Psychophysical Designs

Psychophysical designs extend the A–B–A–B–A design. In psychophysical designs, participants are asked to judge stimuli. For instance, they may be asked to rate whether one light is brighter than another, one weight is heavier than another, or one picture is more attractive than another. The idea is to see how variations in the stimulus relate to variations in judgments. Because the dependent variable is *psycho*logical judgment and the independent variable is often some variation of a stimulus's *physi*cal characteristic (loudness, intensity, etc.), the name *psychophysics* is appropriate.

Because a participant can make psychophysical judgments quickly, a participant in a psychophysical experiment will be asked to make many judgments. Indeed, in one psychophysical experiment, each participant made 67,000 judgments!

With so many judgments, you might worry about maturation effects. Participants might get tired as the research session goes on—and on.

In addition, you might be concerned about treatment carryover effects. Specifically, you might worry that earlier stimuli may affect ratings of later stimuli. Suppose you were rating how heavy you thought a 50-pound weight was. If the last 10 weights you had judged were all about 100 pounds, you might tend to rate 50 pounds as light. However, if the last 10 weights had all been around 10 pounds, you might tend to rate 50 pounds as heavy. Similarly, if you were judging how wealthy a person making $50,000 was, your rating would be affected by whether the previous people you had judged had been multimillionaires or poverty stricken (Wedell & Parducci, 1988).

Because of treatment carryover and maturation, the order of the treatments may affect the results. To deal with potential order effects, researchers often follow the advice of Gustav Fechner—psychophysics' inventor—by presenting each stimulus more than once and counterbalancing the order in which they present the stimuli. For example, if the researcher was interested in ratings of two stimuli (A and B), the researcher would present Stimulus A before Stimulus B half the time; the other half of the time, Stimulus B comes before Stimulus A. If Stimulus A receives different ratings when it is presented first than when it is presented last, the researchers know there are order effects. However, thanks to counterbalancing, these order effects should not make the average of Stimulus A's ratings different from the average of Stimulus B's.

In summary, maturation, testing, and carryover may cause order effects. To deal with these order effects, psychophysical designs often use three techniques:

1. multiple ratings,
2. averaging, and
3. counterbalancing.

The Multiple-Baseline Design

Another single-*n* design that rules out the effects of maturation, testing, and carryover is the multiple-baseline design. In a typical **multiple-baseline design**, you collect baselines for several key behaviors. For example, you might collect baselines for a child making her bed, putting away her toys, washing her hands, and vacuuming her room. Then, you would reinforce one of those key behaviors. If the behavior being reinforced (putting away her toys) increases, you might suspect that reinforcement is causing the behavior to increase.

Unfortunately, the increase in the desired behavior might be due to the child becoming more mature or due to some other nontreatment effect. To see whether the child's improvement in behavior is due to maturation or some other nontreatment factor, you would look at her performance on the other tasks. If those tasks are still being performed at baseline level, then nontreatment factors such as maturation and testing are not improving performance on those tasks and are probably also not increasing the particular behavior you decided to reinforce. Therefore, you would be relatively confident that the improvement in putting away toys was due to reinforcement.

To be even more confident that the reinforcement is causing the change in behavior, you would reinforce a second behavior (washing hands) and compare it against the other nonreinforced behaviors. You would continue the process until you had reinforced all the behaviors, hoping to find that when you reinforced hand washing, hand washing increased—but that no other behavior increased. Similarly, when you reinforced tooth brushing, you would hope tooth brushing—and only tooth brushing—increased. If increases in behavior coincided perfectly with reinforcement, you would be confident that reinforcement was responsible for the increases in behavior (see Figure 14.3).

Evaluation of Single-*n* Designs

You have now examined some of the more popular single-*n* designs. Before leaving these designs, let's see how they stand up on three important criteria: internal, construct, and external validity.

Internal Validity

One strategy the single-*n* researcher uses to achieve internal validity is to keep many relevant variables constant. The single-*n* researcher holds individual difference variables constant by studying a single participant and may hold environmental variables constant by placing that participant in a highly controlled environment. For example, the single-*n* researcher may study a single rat pressing a bar inside a soundproof Skinner box.

Like the within-subjects researcher (see Chapter 13), the single-*n* researcher must worry that the changes in the participant's behavior could be due to the participant naturally changing over time (maturation) or due to the participant getting practice on the dependent-measure task (testing). Not surprisingly, within-subjects and single-*n* researchers may adapt similar strategies to deal with the threats of maturation and testing.

Both within-subjects and single-*n* researchers may try to rule out maturation by keeping their study so short that there is not enough time for maturation to occur. Both may try to reduce the effects of testing by giving participants extensive practice on the task before introducing the treatment, thereby reducing the chances that participants will benefit from any additional practice they get during the research study.

You don't have to take the single-*n* researchers' word that participants got enough practice. By showing that the response rate is stable before the treatment is introduced (the stable baseline), single-*n* researchers show that neither the practice nor anything else is causing the participant to improve during the latter part of the pretreatment phase.

Like the within-subjects experimenter, the single-*n* experimenter must be concerned about treatment carryover effects. Because of carryover, investigators using an A–B–A design frequently find that participants do not return to the original baseline. These carryover problems multiply when you use more levels of the independent variable and/or when you use more than one independent variable. Because carryover effects are a serious concern, most single-*n* researchers minimize carryover's complications by doing studies that

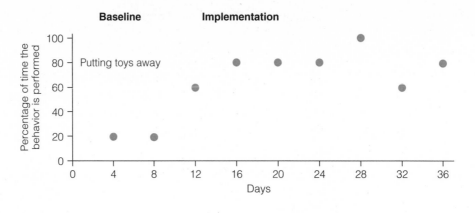

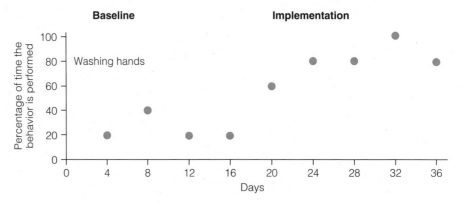

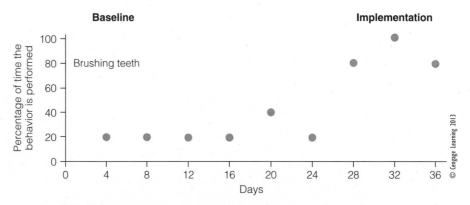

FIGURE **14.3** Hypothetical data from a multiple-baseline design.

have only two levels of a single independent variable. That is, rather than use an A–B–C–D– E–F–G–G–F–E–D–C–B–A design, most single-*n* researchers only use A–B–A–B designs. However, as you will soon see, internal validity concerns are not the only reason for simpler designs. Construct validity concerns also lead to choosing simpler designs.

TABLE **14.4** Similarities Between Within-Subjects Experiments and Single-*n* Designs

Problem	Single-*n* Experiment	Within-Subjects Design
Practice effects may harm internal validity.	Provide extensive practice before introducing treatment.	Provide extensive practice before introducing treatment.
Fatigue or maturation may harm internal validity.	Keep study brief.	Keep study brief.
Assorted order effects may harm internal validity.	Counterbalance sequence of treatments.	Counterbalance sequence and randomly assign participants to different sequences.
Carryover effects may harm internal validity.	Use few levels and few variables. Wait a long time between treatments.	Use few levels and few variables. Wait a long time between treatments.
Participants may learn what the study is about (sensitization), thus harming construct validity.	1. Use placebo treatments. 2. Use few levels of treatment. 3. Gradually increase or decrease intensity of treatment.	1. Use placebo treatments. 2. Use few levels of treatment. 3. Gradually increase or decrease intensity of treatment.

© Cengage Learning 2013

Construct Validity

Although there are similarities between the single-*n* researcher and the within-subjects researcher in how they deal with internal validity concerns, those researchers have even more in common when they attack threats to construct validity (see Table 14.4). For both researchers, *sensitization* (participants figuring out the hypothesis because they have been exposed to several levels of the treatment) poses a serious problem, and both researchers use the same solutions. Specifically, both try to reduce the effects of sensitization by

1. using placebo treatments,
2. using very few levels of treatment, and
3. making the difference between the treatment conditions so subtle that participants don't realize that anything has changed (such as gradually varying the loudness of a stimulus).

External Validity

You may be satisfied with both the single-*n* design's internal validity and its construct validity. However, you probably question its external validity because you are concerned about (a) generalizing from a sample of one to most people and (b) generalizing from research conducted in highly controlled circumstances to real life.

To reduce your concerns about generalizing from a single participant, the single-*n* researcher would make four points. First, although it is risky to generalize from the sample of one that the single-*n* researcher uses, it is also risky to generalize from the nonrepresentative samples that most other experimenters use. If the participants who volunteer for a multiple-participant study are not a representative sample of any recognizable group, it is difficult to argue that such a study has more generalizability than a single-*n* study (Dermer & Hoch, 1999).

Second, even if a multiple-participant (multiple-*n*) experiment used a representative sample, the results may apply to groups but not to individuals. Just as a study may find that the average American family has 2.2 children even though no individual American family has 2.2 children, multiple-*n* experiments may find general truths that do not apply to any individual. For example, suppose that your treatment helps half the people but hurts the other half. Your multiple-*n* experiment might find no effect for the treatment (Dermer & Hoch, 1999).

Third, single-*n* researchers establish the external validity of their findings by replicating their studies. By demonstrating that the treatment has the same effect on each individual studied, they provide some evidence that the effect generalizes across individuals (Dermer & Hoch, 1999). In contrast, note that replicating a multiple-*n* experiment numerous times might fail to establish that the results apply to most people. For example, one could use multiple large random samples to replicate a multiple-*n* experiment numerous times, consistently obtain an average treatment effect, and fail to realize that the effect occurred only for certain types of participants.

Fourth, when single-*n* researchers investigate universal, fundamental processes, the results obtained from one individual can be generalized to the entire species. For example, the results of classical conditioning experiments performed on a single individual can be generalized to other members of that species.

To reduce your concerns about generalizing from research conducted in highly controlled circumstances, the single-*n* researcher would make three points. First, lab studies tend to have excellent external validity (Anderson, Lindsay, & Bushman, 1999).

Second, the setting to which you want to generalize may be just as highly controlled as the lab setting. For instance, you may want to generalize the results to clients in a biofeedback lab.

Third, not all single-*n* studies are conducted in lab settings. Many times, the setting is the real world. Single-*n* studies have been done in homes, schools, and businesses.

In short, the single-*n* design does not get high marks for external validity. However, under some circumstances, the results from a study using a single-*n* design may have a high degree of generalizability.

Conclusions About Single-*n* Designs

We have evaluated the single-*n* designs in terms of internal, construct, and external validity. Overall, single-*n* designs, although not possessing the internal validity of a true, randomized experiment, have some internal validity. Generally, single-*n* designs can have adequate construct validity. Thus, often the decision about whether to use a single-*n* design comes down to whether the researcher is worried about external validity. Consequently, single-*n* designs are most useful under two circumstances: (a) when the researcher does not need to show that the results generalize to other individuals and (b) when the researcher can argue that the results from one participant generalize to other individuals.

In some applied situations, the investigator is interested in the causes of one particular individual's behavior—not in generalizing the results to others. Suppose you were trying to change your own behavior or the behavior of a family pet. Or, suppose a therapist is treating a client and wants to see if the treatment is having a measurable effect on that particular patient. In all these cases, the single-n design would be the best way to evaluate the effect of the treatment.

In some situations, generalizing the results from one participant to a larger group may be reasonable. For example, suppose that you want to make statements about fundamental, universal processes that we understand fairly well. Then, according to single-n researchers, you should use a single-n design. After all, it would be wasteful to study many participants if the treatment has the same effect on everyone. Because all people tend to respond similarly to reinforcements, the single-n design is commonly used in behavior modification research. Likewise, because everyone seems to respond similarly to psychophysical manipulations, the single-n design is also a popular alternative to the randomized experiment in psychophysical research.

QUASI-EXPERIMENTS

Another popular alternative to the randomized experiment is the **quasi-experiment**. Like true experiments, quasi-experiments involve administering a treatment. Unlike true experiments, though, participants are not randomly assigned to treatment (Cook & Campbell, 1979).[2]

Ideally, researchers could determine a treatment's effect by randomly assigning participants to different treatments. Researchers, however, do not run the world.

People who do run the world usually won't relinquish their power to researchers. Those in control want to decide who gets which treatment, rather than letting researchers use random assignment to determine who gets which treatment. Judges usually like to decide what sentence to give, rather than leaving it up to random assignment. Parents want to determine whether their children should watch violent television, rather than leaving it up to random assignment. Bosses usually want to choose who gets training. Cable companies probably want to decide who to serve based on geography and income rather than on random assignment.

Even when money and power aren't issues, some people object that random assignment is not fair. On one hand, this argument seems absurd.

---

[2] According to this definition, single-n designs are quasi-experiments. However, people usually think of single-n designs as being different from quasi-experiments because, relative to other quasi-experiments, single-n experiments study fewer participants under more controlled conditions.

What could be fairer than allowing everyone who wants a treatment an equal chance at it? On the other hand, a good case can be made that the treatment should be given to the people who are the most needy or the most qualified.

For a variety of reasons, researchers are often unable to randomly assign participants to condition. Even when researchers can randomly assign, the internal validity of those studies may be weak because the random assignment does not stick (Ehrenberg, Brewer, Gamoran, & Williams, 2001). The random assignment may not stick because

1. participants assigned to receive one condition may get themselves re-assigned so they can receive what they consider the better treatment (e.g., the drug rather than the placebo, the enrichment program rather than the ordinary program) or
2. participants drop out of one group much more than another (Ehrenberg et al., 2001).

Even if a researcher finds a situation in which the researcher can (a) randomly assign participants and (b) get the assignment to stick, that situation is probably not typical. Consequently, there may be questions about that field experiment's external validity (Ehrenberg et al., 2001).

As you have seen, researchers wishing to use random assignment to evaluate the effects of real-world treatments face three problems: (1) the powers that be may prohibit the study; (2) participants may reassign themselves to condition, thereby ruining the study's internal validity; and (3) studying participants who fully cooperate with random assignment may involve studying a nonrepresentative group of participants, thereby harming the study's external validity. Consequently, when evaluating the effects of many real-world treatments—from therapy, to training programs, to introducing new technology to social programs—using quasi-experimental designs is often the researcher's best option.

Because quasi-experimental designs are so useful for assessing the effects of real-life treatments, we will devote the rest of this chapter to these designs. We will begin by discussing the general logic behind quasi-experimental designs. Then, we will take a more detailed look at some popular quasi-experimental designs.

Battling Spuriousness by Accounting for—Rather Than Controlling—Nontreatment Factors

Like experimenters, quasi-experimenters try to establish temporal precedence by showing that the change in participants occurred *after* the researchers administered the treatment. Also like experimenters, quasi-experimenters assess covariation by comparing treatment vs. nontreatment conditions. However, unlike experimenters, quasi-experimenters do not rule out spuriousness by randomizing the effects of nontreatment factors and then statistically controlling for those random effects. Furthermore, unlike single-*n* researchers, quasi-experimenters do not rule out spuriousness by keeping nontreatment factors constant.

Identifying Nontreatment Factors: The Value of Campbell and Stanley's Spurious Eight

The challenge in quasi-experiments is to rule out the effects of nontreatment variables without either the aid of random assignment or the ability to control nontreatment variables. The first step in meeting this challenge is to identify all the variables other than your treatment that might account for the change in participants' scores. After you have identified those nontreatment factors, you will try to demonstrate that those nontreatment factors did not account for the change in participants' scores so that you can argue that your treatment caused the effect.

To identify every possible nontreatment factor that could threaten your study's internal validity might seem like an unmanageable task. However, Campbell and Stanley (1963) made the task manageable by discovering that all these potential threats to internal validity fall into eight general categories. Thus, rather than dealing with an almost infinite number of specific threats to internal validity, researchers can focus on the following eight general threats to internal validity:

1. **Testing:** apparent treatment effects that are really due to participants having learned from the pretest. For example, practice on the pretest may improve performance on the posttest.
2. **Maturation:** apparent treatment effects that are really due to natural biological changes—from changes due to growing and developing to changes due to becoming more tired or more hungry.
3. **History:** apparent treatment effects that are really due to events in the outside world that are unrelated to the treatment.
4. **Instrumentation:** apparent treatment effects that are really due to changes in the measuring instrument. For example, the researcher may use a revised version of the measure on the retest.
5. **Regression (regression toward the mean, statistical regression):** apparent treatment effects that are really due to the tendency for participants who receive extreme scores on the pretest to receive less extreme scores on the posttest.
6. **Mortality (attrition):** apparent treatment effects that are really due to participants dropping out of the study. For instance, suppose that participants who would score poorly drop out of the treatment condition, but not out of the no-treatment condition. In that case, the treatment group would score higher than the no-treatment group, even if the treatment had no effect.
7. **Selection:** apparent treatment effects that are really due to the different treatment groups being different from each other before the study started.
8. **Selection-maturation interaction:** apparent treatment effects that are really due to groups that scored similarly on the pretest naturally growing apart and therefore scoring differently from each other on the posttest.

As you will soon see, the eight threats to validity fall into three general categories. First, there are those environmental and physiological events—other than the treatment—that cause individuals to change. Second, there are errors in measurement that cause changes in individuals' *scores*. Third, there are problems related to the fact that treatment and no-treatment groups—because different individuals are in the two groups—may differ from each other even when the treatment has no effect.

Three Reasons Individuals Change Even Without Treatment. The first three threats to validity—testing, maturation, and history—include all the nontreatment factors that can cause individual participants to change. The first two—testing and maturation—are threats we talked about in terms of the single-*n* design. As you may recall, we were concerned that testing (the participant learning from performing the dependent measure task several times) might cause the participant's behavior to change between the A and B phases of an A–B design. We were also concerned that maturation (any changes in the participant's internal, physiological environment, such as changes due to growing old or becoming hungry) might cause the participant's behavior to change. Maturation is a concern because many conditions improve over time (Painter, 2008). However, because we could isolate the participant from the larger world and could keep the laboratory environment constant, we were—unlike the quasi-experimenter—unconcerned about history (any nontreatment changes in the external environment).

How Measurement Errors Can Look Like a Treatment Effect. The next two threats, instrumentation and statistical regression, can cause participants in the treatment conditions to have different scores than they did in the no-treatment condition even though the participants themselves have not changed. With instrumentation, participants are tested with one measuring instrument in one condition and a different measuring instrument in another condition (this would include a revised version of an original measure). No wonder their scores are different!

Statistical regression is harder to spot. To understand statistical regression (also called regression and regression toward the mean), remember that most scores contain some random error. Usually, however, random error's net effect on the overall average score of a group is zero because the scores that random error pushes upward are balanced out by the scores that random error pulls downward.

But what if we select only those participants whose scores have been pushed way up by random error? When we retest them, their scores will go down. Their scores going down might fool us into thinking they had really changed. In fact, all that has happened is that random error isn't going to push up all these scores again (just as lightning is unlikely to strike the same place twice). Instead, this second time, random error will push some scores up, some scores down, and have almost no effect on the remaining scores.

You might wonder how we could select scores that have been pushed up by random error. One way is to select extreme scores. For example, if we select only those people who got 100% on an exam, we know that random error did not decrease their scores. However, random error (lucky guesses, the scorer failing to see a question that was missed) could have increased those scores. Thus, if we give these people a test again, their scores are likely to go down. This tendency for extreme scorers to score less extremely when they are retested is called regression toward the mean.

Regression toward the mean is a powerful effect. Whether watching a baseball player on a hitting streak (or in a hitting slump), watching the economy, or observing a patient, you will find that extreme events tend to revert back to more normal levels.

Until now, we have talked about factors that could change an individual's scores. We explained that a participant in the treatment condition may change for reasons having nothing to do with the treatment (maturation, testing, history). We have also talked about how an individual's score can change, even though the individual doesn't really change (instrumentation, regression). In some cases, these changes in individual participants' scores could cause a treatment group to score differently from a no-treatment group.

Three Differences Between Treatment and No-Treatment Groups That Have Nothing to Do With the Treatment. Even when the individual scores are accurate and unaffected by treatment-irrelevant influences, the treatment group may differ from the no-treatment group simply because the participants in the treatment group have different characteristics than those in the no-treatment group. The three treatment-irrelevant factors that could cause participants in the treatment condition to systematically differ from participants in the no-treatment condition are mortality, selection, and selection-maturation.

Mortality (attrition) would be a problem if your poor performers have dropped out of the treatment condition. In that case, your treatment condition scores would be higher than your no-treatment condition scores—even if your treatment had no effect.

Selection would be a problem if you were comparing groups that were different before the study began. As the saying goes, "that's not fair—you're comparing apples and oranges." If your treatment group started out being different from your no-treatment group, the differences between your group's scores at the end of the study may not be due to the treatment. Therefore, you shouldn't conclude that the difference in scores between your two groups is due to the treatment. Even if you selected two groups who scored similarly on your measure before you introduced the treatment, you can't conclude that they would have scored similarly at the end of the study because of *selection-maturation interactions*: Groups that scored similarly in the pretest may naturally mature at different rates.

Using Logic to Combat the Spurious Eight

Once you have identified the threats to internal validity, you must determine which threats are automatically ruled out by the design and which threats you can eliminate through logic (see Table 14.5). Quasi-experimental designs differ in their ability to automatically rule out the eight threats to internal validity. Some designs rule out most of these threats; some rule out only a few. Yet, even with a quasi-experimental design that automatically rules out only a few of these threats, you may occasionally be able to infer causality.

TABLE **14.5** Steps Quasi-Experimenters May Take to Minimize Threats to Internal Validity

Threats	Precautions
History	Isolate participants from external events during the course of the study.
Maturation	Conduct the study in a short period to minimize the opportunities for maturation. Use participants who are maturing at slow rates.
Testing	Only test participants once.
	Give participants extensive practice on task prior to collecting data so that they won't benefit substantially from practice they obtain during the study.
	Know what testing effects are (from past data) and subtract out those effects. Use different versions of the test to decrease the testing effect.
Instrumentation	Administer same measure, the same way, every time.
Mortality	Use rewards, innocuous treatments, and brief treatments to keep participants from dropping out of the study.
	Use placebo treatments or subtly different levels of the treatment so that participants won't be more likely to drop out of the treatment condition.
	Make sure participants understand instructions so that participants aren't thrown out for failing to follow directions.
Regression	Don't choose participants on basis of extreme scores.
	Use reliable measures.
Selection	Match on all relevant variables.
	Don't use designs that involve comparing one group of participants with another.
Selection Interactions	Match on all relevant variables, not just on pretest scores. In addition, use tips from earlier in this table to reduce the effects of variables—such as history and maturation—that might interact with selection. In other words, reducing the role of maturation will also tend to reduce selection by maturation interactions.

© Cengage Learning 2013

To illustrate the potential usefulness of quasi-experimental designs, we will start by looking at a design that most people would not even consider to be in the same class as a quasi-experimental design: the **pretest–posttest design**. As the name suggests, you test one group of participants, administer a treatment, and then retest them.

This design does not rule out many threats automatically; hence, its low status as a design. However, because you are comparing individuals against themselves, it does automatically rule out selection and selection-maturation interactions.

Although the pretest–posttest design does not automatically rule out mortality, instrumentation, regression, maturation, history, and testing, you may still be able to rule out these threats. If nobody dropped out of your study, mortality (attrition) is not a problem. If you were careful enough to use the same measure and administer it in the same way, instrumentation is not a problem. If there were only a few minutes between the pretest and posttest, history is unlikely.

If there were only a few minutes between pretest and posttest, maturation is also unlikely. About the only maturation that could occur in a short period would be boredom or fatigue. Thus, if performance was better on the posttest than on the pretest, then you could rule out boredom and fatigue—and thus maturation.

You might even be able to rule out regression. The key to ruling out regression is to realize that regression occurs when extreme pretest scores that were inflated (or deflated) by random error revert back to more average scores on the retest. Therefore, to rule out regression, you need to make the case that random error had not inflated (or deflated) pretest scores by establishing either that

1. your measure was so reliable (so free of random error) that random error would have little impact on pretest scores or
2. participants in the study did not have pretest scores that were extreme.

Thus far, in this particular study, you have been able to rule out every threat except testing—and you might even be able to rule out testing. For instance, if participants did not know they had been observed (e.g., you unobtrusively recorded how long they gazed into each other's eyes), testing should not be a problem. Or, if you used a standardized test, you might know how much people tend to improve when they take the test the second time. If your participants improved substantially more than people typically improve upon retesting, you could rule out the testing effect as the explanation for your results.

As you have seen, the pretest–posttest design, by itself, has poor internal validity because it automatically eliminates only a few threats to internal validity. But, as you have seen, you may be able to use your wits to rule out the remaining threats and thereby infer causality (see Table 14.6 for a

TABLE **14.6** How to Deal With the Threats to Internal Validity if You Must Use a Pretest–Posttest Design

Threat	How to Deal with it
Selection	Automatically eliminated because participants are tested against themselves.
Selection by Maturation	Automatically eliminated because participants are tested against themselves.
Mortality	Not a problem if participants don't drop out. Conduct study over short period of time and use an undemanding treatment.
Instrumentation	Standardize the way you administer the measure.
Regression	Do not select participants based on extreme scores. Use a reliable measure.
Maturation	Minimize the time between pretest and posttest.
History	Minimize the time between pretest and posttest.
Testing	Use an unobtrusive measure. Have data from previous studies about how much participants' scores tend to change from test to retest.

review). Furthermore, as you will soon see, by extending the pretest–posttest design, you can create a quasi-experimental design that eliminates most threats to internal validity: the time-series design.

Time-Series Designs

Like the pretest–posttest design, the **time-series design** tests and retests the same participants. However, rather than use a single pretest and a single posttest, the time-series design uses several pretests and posttests. Thus, you could call time-series designs "pre–pre–pre–pre–post–post–post–post" designs.

To illustrate the differences between the pretest–posttest design and the time-series design, suppose you are interested in seeing whether a professor's disclosures about her struggles to learn course material affect how students evaluate her. Let's start by examining how you would use a pretest–posttest design to find the effect of such disclosures.

With a pretest–posttest design, you would have a class evaluate the professor before she tells them about her struggles to learn course material. Then, you would have them rate her after she discloses her problems. If you observed a difference between pretest and posttest ratings, you would be tempted to say that the difference was due to the disclosure. However, the difference in ratings might really be due to history, maturation, testing, mortality, instrumentation, or regression. Because you have no idea of how much of an effect history, maturation, testing, mortality, and instrumentation may have had, you cannot tell if you had a treatment effect.

Estimating the Effects of Threats to Validity With a Time-Series Design

What if you extended the pretest–posttest design? That is, what if you had students rate the professor after every lecture for the entire term, even though the professor would not disclose her problems with learning material until the fifth week? Then, you would have a time-series design.

What do you gain by all these pretests? From plotting the average ratings for each lecture, you know how much of an effect maturation, testing, instrumentation, and mortality tend to have (see Table 14.7). In other words, when you observe changes from pretest to pretest, you know those changes are not due to the treatment. Instead, those differences must be due to maturation, testing, history, instrumentation, or mortality.

For example, suppose ratings steadily improve at a rate of .2 points per week during the 5-week, predisclosure period. If you then found an increase of .2 points from Week 5 (when the professor made the disclosures about her problems) to Week 6, you would not attribute that increase to the disclosures. Instead, you would view such a difference as being due to the effects of history, maturation, mortality, testing, or instrumentation. If, on the other hand, you found a much greater increase in ratings from Week 5 to Week 6 than you found between any other 2 consecutive weeks, you might conclude that the professor's disclosures about her struggles to learn course material improved her student evaluations (see Figure 14.4).

TABLE **14.7** How Pretest–Posttest Designs and Time-Series Designs Stack Up in Terms of Dealing With Campbell and Stanley's Threats to Internal Validity

Threat to Validity	Type of Design	
	Pretest–Posttest	Time-Series
Selection	Automatically eliminated.	Automatically eliminated.
Selection × Maturation Interactions	Automatically eliminated.	Automatically eliminated.
Mortality	Through logic and careful planning, this threat can be eliminated.	Through logic and careful planning, this threat can be eliminated.
Instrumentation	Through logic and careful planning, this threat can be eliminated.	Through logic and careful planning, this threat can be eliminated.
Regression!	Problem!	You should be able to determine whether regression is a plausible explanation for the difference between conditions.
Maturation	Problem	Often, you will be able to estimate the extent to which differences between conditions could be due to testing or maturation.
Testing	Problem!	You should be able to estimate the extent to which differences between conditions could be due to testing or maturation.
History	Problem!	Problem!

© Cengage Learning 2013

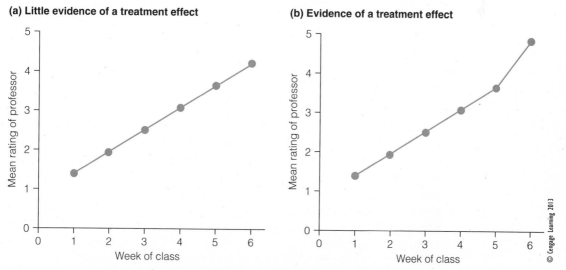

(a) Little evidence of a treatment effect

(b) Evidence of a treatment effect

© Cengage Learning 2013

FIGURE **14.4** Two very different patterns of results in a time-series design in which the treatment was introduced after the fifth week.

Problems in Estimating Effects of Nontreatment Factors

Unfortunately, that conclusion could be wrong. Your conclusion is valid only if you can correctly estimate the effects of history, maturation, mortality, testing, and instrumentation during the time that the treatment was administered. On the surface, it seems safe to assume that you can estimate the effects of those variables. After all, for the pretest period, you know what the effects of those variables were. Thus, you may feel safe assuming that the effects of those variables were the same during the treatment period as they were during the pretest period. But this assumption is correct only if the effects of history, maturation, mortality, instrumentation, and testing are relatively consistent over time. In other words, your conclusions about the treatment's effect could be wrong if there is a sudden change in any one of these nontreatment factors (see Table 14.8).

Sudden changes in these nontreatment factors are possible. As you will see, history and regression tend to produce sudden changes, and the effects of testing, instrumentation, mortality, and maturation are not always slow and consistent across time.

History. To see how history could produce a sudden change, imagine just some of the many specific events that could affect performance on the posttest. For instance, ratings of the professor might change as a result of students getting the midterm back, the professor becoming ill, the professor reading a book on teaching, and so on. Unlike the single-*n* design, the time-series design does not control all these history effects. Indeed, you could argue that the time-series design's lack of control over history, and thus its vulnerability to history, prevents it from reaching experimental-design status.

Although *history is the one threat to which the time-series is extremely vulnerable,* you can try to reduce its effects. One strategy is to have a very

---

TABLE **14.8** Threats to Time-Series Designs

- History: By far the most serious threat
- Regression (although you should be able to tell whether regression could be a problem)
- Any other inconsistent effect. Usually, the only inconsistent effects will be history and regression. Usually, maturation, mortality, testing, and instrumentation will have consistent effects that you can estimate. However, if their effect is inconsistent, it could imitate a treatment effect. Thus, the following are possible, but unlikely, threats to a time-series study's validity
  - Inconsistent maturation effects
  - Inconsistent testing effects
  - Inconsistent mortality effects (if you don't have any dropouts or if the number of dropouts is consistent throughout the study, you probably don't have a mortality problem)
  - Inconsistent instrumentation effects (if you do your study properly, you shouldn't have instrumentation effects)

short interval between testing sessions, thus giving history fewer opportunities to have an effect.

In addition to reducing the effects of history, you can also try to do a better job of estimating its effects. One key to estimating history's effects is to know the past by collecting extensive baseline data. Ideally, you would collect baseline data for several years to help you identify any patterns that might otherwise be mistaken for a treatment effect. For instance, your baseline would alert you to cyclical patterns in student evaluations, such as students being very positive toward the professor during the first 2 weeks of the term, more negative toward the professor after the midterm examination, and then becoming more favorably disposed toward the professor during the last week of the term. Consulting your baseline data would prevent you from mistaking these cyclical fluctuations for a treatment effect.

Regression. Like the effects of history, regression effects will not change steadily from week to week. After all, regression is due to chance measurement error, and chance measurement error will not change steadily and predictably from week to week. Although you cannot use a time-series design to measure regression's effect, you can use time-series designs to determine if regression is a likely explanation for your results. Specifically, you should suspect regression if

1. the ratings immediately before the treatment are extremely high or extremely low relative to the previous ratings, and
2. the posttreatment ratings, although very different from the most immediate pretreatment ratings, are not substantially different from earlier pretreatment ratings.

Inconsistent Effects From Threats That Are Often Consistent. The effects of history and regression are difficult to estimate because they are likely to be inconsistent. Although the effects of instrumentation, mortality, testing, and maturation are less likely to be inconsistent, when they are inconsistent, it causes problems.

Inconsistent Instrumentation Effects. If you administered the same rating scale in the same way for the first 5 weeks, your measurements from Weeks 1 through 5 would not be affected by instrumentation. As a result, your estimate for the amount of change to expect between Week 5 and Week 6 would not include any effect for instrumentation. However, suppose that you ran out of copies of the original rating scale during Week 6 and decided, while you were going to the trouble to run off more copies, that you would make some minor corrections to the form. Consequently, you handed out a refined version of your rating scale during Week 6—the same week the professor started telling her class about her struggles to learn course material. In that case, you might have an instrumentation effect that could not have been estimated based upon the previous weeks' data. Therefore, you might mistake an instrumentation effect for a treatment effect.

Inconsistent Mortality Effects. Similarly, if mortality does not follow a consistent pattern, you might mistake mortality's effects for treatment effects. For example, suppose that the last week to drop the course was the same week the professor started to tell the class about her problems. In that case, a disproportionate number of students who did not like the professor might drop out during that week. Consequently, the professor's ratings might improve because of attrition (mortality) rather than because of her disclosures.

Inconsistent Testing Effects. In the study we've been discussing, the effect of testing should be gradual and consistent. However, the effect of testing will not be consistent in every study. In some studies, for example, participants will, in a flash of insight, discover the rule behind the task, and, as soon as they discover the rule, their performance increases dramatically.

Inconsistent testing effects are not limited to situations in which participants are aware of having an insight. That is, practice does not always produce steady, continuous improvement. As you know from experience, after weeks of work with little to show for it, you may suddenly improve.

Inconsistent Maturation Effects. Similarly, maturation's effect may sometimes be discontinuous. For instance, suppose you measure young children every 3 months on a motor abilities test. Then, you expose them to an enriched environment and measure them again. Certainly, you will see a dramatic change, but is this change due to the treatment? Or, is it due to the children jumping to a more advanced developmental stage (for example, learning to walk)?

You cannot escape sudden, sporadic maturation by studying adults. Even in our teacher evaluation study, participants might mature at an inconsistent rate. That is, first-year students might grow up quickly after getting their first exams back, or students might suddenly develop insight into the professor's teaching style. If this sudden development occurred the same week the professor started to disclose her struggles to learn course material, maturation could masquerade as a treatment effect.

Eliminating, Rather Than Estimating, Threats to Internal Validity

In short, the time-series design can accurately estimate and thus rule out some threats to validity (see Table 14.9), but there are certain effects that it cannot accurately estimate. Therefore, when using a time-series design, do not focus so much on estimating the impact of the eight threats to validity that you don't try to eliminate those eight threats.

Try to eliminate the threat of instrumentation by using the same measuring instrument each time and administering it the same way. In our student evaluation study, we would give students the same rating scales and the same instructions each time.

Likewise, try to eliminate mortality. If you had students sign their rating sheets, you could eliminate mortality by analyzing data from only those students who had perfect attendance.

If you can't eliminate a threat, at least try to reduce its effects. Try to reduce the effects of both maturation and history by keeping the interval

TABLE **14.9** How the Time-Series Design Deals With Threats to Internal Validity

Threat	Approach
Selection	Automatically eliminated because testing and retesting the same participants.
Selection × *Maturation*	Automatically eliminated because testing and retesting the same participants.
Instrumentation	If effects are constant, effects can be estimated. In addition, try to use the same instrument in the same way every time.
Mortality	If effects are constant, effects can be estimated. In addition, if no participants drop out, mortality is not a problem.
Testing	If effects are constant, effects can be estimated.
Maturation	If effects are constant, effects can be estimated. In addition, study slowly maturing participants or make sure that time between the last pretest and the posttest is very brief.
Regression	Regression is unlikely if ratings prior to introducing the treatment were not extreme and did not differ greatly from previous ratings. Because regression capitalizes on random error, regression is less likely if you use a measure that is relatively free of random error: a reliable measure.
History	Try to collect extensive pretest data to predict history's effects. In addition, you may try to make sure that 1. time between the last pretest and the posttest is brief. 2. participants are isolated from outside events.

© Cengage Learning 2013

between pretest and posttest short. Minimize the likelihood of regression effects by choosing the time that you will administer the treatment well in advance—don't administer the treatment as an immediate reaction to extremely low ratings.

Variations on the Traditional Time-Series Design

Now that you are familiar with the basic logic behind the time-series design, you are ready to see how to extend that design. One simple way of extending a time-series design is to increase the number of pretest and posttest measurements you take. Increasing the number of pretest and posttest measurements you take has two advantages.

First, the more measurements you take, the better you should be at estimating the combined effects of maturation, history, mortality, testing, and instrumentation. Thus, you are less likely to mistake these effects for a treatment effect.

Second, the more measurements you have, the less likely it is that an unusual history, maturation, mortality, testing, or instrumentation effect would influence only the posttreatment measurement. To illustrate the advantages of having more measurements, suppose you measure student reactions on only the 5th, 6th, and 7th weeks. You administer the treatment between the 6th and 7th weeks. Would it be an unusual coincidence if history, maturation, mortality, or testing had more of an effect between the 6th and 7th weeks than between the 5th and 6th weeks? No—consequently, any of these threats to validity might easily imitate a treatment effect. However, what if you had students evaluate the teacher from Week 1 to Week 12? Then, it would be quite a coincidence for a threat to have an extraordinarily large effect between the 6th (the same week you gave the treatment) and 7th weeks, but not have such an effect between any of the other weeks.

Reversal Time-Series Designs. In addition to taking more measurements, you can extend your time-series design by administering and withdrawing the treatment. That is, you can imitate the single-*n* researcher's reversal design.

For example, you might test (pretest), administer the treatment, test again (posttest), withdraw the treatment, and test again. You might even withdraw and introduce the treatment several times.

To see the beauty of this reversal time-series design, imagine that you were able to get increases each time the professor tells her class about her struggles to learn course material, then decreases when she stops talking about her problems, followed by increases when the professor again tells her class about her studying woes. With that pattern of results, you would be confident that the disclosures made a difference.

Despite the elegance of the reversal design, ethical and construct validity problems may prevent you from using it. The ethical problems are the most serious: In some situations, you cannot ethically withdraw the treatment after you have administered it (e.g., psychotherapy, reinforcement for wearing seatbelts).

The construct validity problems can also be serious. Specifically, withdrawing and re-administering the treatment may alert participants to your hypothesis. Consequently, your results may be due to participants guessing the hypothesis and playing along.

To prevent participants from guessing the hypothesis or becoming resentful when you withdraw the treatment, use placebo treatments or multiple levels of the treatment. If you were to use this design for your student evaluations study, you might have a placebo condition in which the professor discloses innocuous facts about studying experiences. Alternatively, you might use several levels of disclosure ranging from innocuous to intimate.

Two-Group Time-Series Design. A final way of extending the time-series design is to collect time-series data on two groups. One group, the comparison group, would not get the treatment. The advantage of using a comparison group is that it allows you to rule out certain history effects. In your disclosure study, the comparison group might be another section of the same professor's class. If, after the treatment was administered, the ratings went down only in the treatment group, you could rule out general history effects (midterm blues, spring fever) as an explanation of the results.

However, you can't rule out every history effect because the two classes may have different histories. For example, the afternoon class may be subjected to an overheated classroom whereas the morning class is not.

The Nonequivalent Control-Group Design

You do not have to use a time-series design. For example, rather than using a two-group time-series design, you could simply (a) give one group the treatment, then (b) measure both groups. Such a study would be called a nonequivalent control-group design. Essentially, the **nonequivalent control-group design** is the simple experiment without random assignment.

Because of the nonequivalent control-group design's similarity to the simple experiment, it has many of the simple experiment's strengths. For example, because every participant is tested only once, the nonequivalent control-group design, like the simple experiment, is not vulnerable to maturation, testing, or instrumentation. Furthermore, because of the control group, the nonequivalent group design, like the simple experiment, can usually deal with the effects of history, maturation, and mortality.

The Nonequivalent Control-Group Design Is Extremely Vulnerable to Selection

Because this design does not use random assignment, the control and treatment groups are not equivalent. Indeed, to make it clear that the control group is not equivalent to the treatment group, some have argued that using the term *control group* to describe the no-treatment group is inappropriate and should be replaced with the term *contrast group* (Wilkinson & the Task Force on Statistical Inference, 1999). Because the no-treatment and treatment groups are not equivalent, selection is a serious threat in this design.

Why Matching Doesn't Make Groups Equivalent

To address the selection threat, investigators often attempt either to ensure that each participant in the control group is identical in several key respects to a participant in the treatment group or to ensure that groups have the same average scores on key variables. These key variables may be background variables (age, gender, IQ) that are expected to correlate with scores on the dependent measure or they may be actual scores on the dependent measure (pretest scores).

Although you might think that matching would succeed at making the nonequivalent control group equivalent to the treatment group, realize two important points about matched participants:

1. Matched participants or groups are matched only on a few variables rather than on every variable.
2. Matched participants or groups are not matched on characteristics directly, but on imperfect measures of those characteristics.

You Can't Match on Everything. Just because two groups are matched on a few variables, you shouldn't think that they are matched on all variables. They aren't. The unmatched variables may cause the two groups to score differently on the dependent measure (see Table 14.10).

For instance, suppose you decide to use a nonequivalent group design to test your hypothesis about the effect of self-disclosing problems with learning material. To make the two classes similar, you match the classes on IQ scores,

TABLE **14.10** How Two Nonequivalent Control-Group Designs Stack Up in Terms of Dealing With Threats to Internal Validity

Threat to Validity	Type of Nonequivalent Control Group	
	Unmatched	**Matched**
Selection	Big problem!	Problem
Selection × *Maturation*	Problem	Problem
Regression	Not a problem	Big problem!
Mortality	If participants do not drop out of your study, this will not be a problem.	If participants do not drop out of your study, this will not be a problem.
Instrumentation	If you use administer the same measure the same way each time, this will not be a problem	If you use administer the same measure the same way each time, this will not be a problem
Maturation	Automatically eliminated by the design	Automatically eliminated by the design
Testing	Automatically eliminated by the design	Automatically eliminated by the design
History	Automatically eliminated by the design	Automatically eliminated by the design

© Cengage Learning 2013

grade point averages, proportion of psychology majors, proportion of females and males, and proportion of sophomores, juniors, and seniors. However, you have not matched them in terms of interest in going on to graduate school, number of times they had taken classes from this professor before, and a few hundred other variables that might affect their ratings of the professor. These unmatched variables, rather than the treatment, may be responsible for the difference between your treatment and control groups.

Because investigators realize that they cannot match participants on every factor that may influence task performance, some investigators try to match participants on task performance (pretest scores). Yet, even when groups are matched on pretest scores, unmatched variables can cause the groups to score differently on the posttest. Just because two groups of students start out with the same enthusiasm for a course, you cannot be sure that they will end the term with the same enthusiasm. For example, one group may end the term with more enthusiasm because that group began the course with a clearer understanding of what the course would be like, what the tests would be like, and how much work was involved. Consequently, although both groups might rate the professor the same at first, the groups may differ after they get the first exam back. For instance, because the naïve group had misconceptions about what the professor's exams would be like, they may rate the professor more harshly than the experienced group.

Although this change in student attitudes toward the professor might appear to be a treatment effect, it is not. Instead, the difference between the two groups is due to the groups differing on variables that they were not matched on—and those differences causing the groups to grow apart.

Technically, there was a selection by maturation interaction. Because of interactions between selection and other variables, even matching on pretest scores does not free you from selection problems.

What can be done about interactions between selection and other variables? One approach is to assume that nature prefers simple, direct main effects to complex interactions. Thus, if an effect could be due to either a treatment main effect or an interaction between selection and maturation, assume that the effect is a simple treatment main effect. Be aware, of course, that your assumption could be wrong.

If you want to go beyond merely assuming that selection-maturation interactions are unlikely, you can make them less likely by making the groups similar on as many selection variables as possible. You can match the groups not only on pretest scores but also on other variables. With such extensive matching, there would be fewer variables on which the groups differed and, therefore, fewer selection variables to interact with maturation. Hence, you would reduce the chance of selection-maturation interactions occurring.

You have seen that one way to reduce interactions between selection variables and maturation is to reduce differences between groups that might contribute to selection. The other way to reduce interactions between selection and maturation is to reduce opportunities for maturation. After all, if neither group can mature, then you won't have a selection-maturation interaction. To reduce the potential for a selection by maturation interaction, you may decide to present the posttest as soon after the pretest as possible.

You Match on Measures of Variables—Not on Variables. As you have seen, failing to match on every relevant variable sets you up for selection-maturation interactions. Another problem with matching is that participants must be matched on observed scores, rather than on true scores.

Observed scores are not the same as true scores because observed scores are contaminated by measurement error. As a result of this measurement error, two groups might appear to be similar on certain variables, although they are actually different on those variables.

How can participants score the same on a measure of a variable, but actually be different in terms of that variable? To see how, suppose a researcher wanted to examine the effect of a drug on treating clinical depression. The researcher has received approval and patients' permissions to give the drug to the 10 individuals at her small psychiatric facility who have been diagnosed with severe depression. However, she realizes that if the participants improve after getting the drug, it proves nothing. Maybe the patients would get better anyway. She wants to have a comparison group that does not get the drug. After getting a phone call asking her to give a guest lecture at a nearby college, she gets an idea. She could use some college students as her comparison group. After testing hundreds of students, she obtains a group of 10 college students who score the same on the depression scale as her group of 10 individuals who are hospitalized for depression.

But are the two groups equal in terms of depression? Probably not. The college student participants' scores are extremely depressed relative to the average college student. The fact that the participants' scores are extremely different from the mean (average) sets up regression toward the mean.

Regression toward the mean occurs because extreme scores tend to have an extreme amount of random error. Thus, when the students are tested again, their scores will be less extreme because their scores will not be as dramatically swayed by random error. On the posttest, the college student participants will probably score more like average college students—less depressed. In this case, because of regression toward the mean, a drug that has no effect may appear to hurt recovery from depression.

How can you stop from mistaking such a regression effect for a treatment effect? One approach is to reduce regression. As we mentioned earlier, there are two ways to reduce the potential for regression effects. First, because regression takes advantage of random measurement error, you can reduce regression by using a measure that is relatively free of random measurement error: a reliable measure. Second, because extreme scores tend to be more influenced by random error than less extreme scores, don't select participants who have extreme pretest scores.

A trickier approach to combat regression is to obtain results that regression cannot account for. In our depression example, regression would tend to make it look like the college students had improved more than the individuals who were hospitalized for depression. However, if you found the opposite results—the individuals who were hospitalized for depression improved in mood more than college students—regression would not be an explanation for your results. Thus, one approach to eliminating regression is to get results exactly opposite from what regression would predict.

There is no way to guarantee that your treatment's effect will push scores in exactly the opposite direction of where regression would push scores. Furthermore, even if the treatment effect goes against the regression effect, regression effect may cancel or even overwhelm the treatment effect. That is, even though your treatment had a positive effect, the treatment group's scores—because of regression—may still decline. When regression and selection by maturation are both pushing scores in the opposite direction of the treatment's effect, they may overwhelm the effects of even moderately effective treatments.

To illustrate how regression and selection by maturation can hide a treatment's effect, consider research attempting to determine the effects of social programs. Sometimes researchers try to find the effect of a social program by matching a group of individuals who participate in the program with individuals who are not eligible. For example, researchers compared children who participated in Head Start with an upper-income group of children who had the same test scores. Unfortunately, this often meant selecting a group of upper-income children whose test scores were extremely low compared to their upper-income peers. Consequently, on retesting, these scores regressed back up toward the mean of upper-income children. Because of this regression toward the mean effect, scores in the no-treatment group increased more than scores in the Head Start group.

Not only was regression a problem, but there was also the potential for a selection by maturation interaction—especially for studies that looked for

TABLE **14.11** Problems with Trying to Make Groups Equivalent by Matching	
Problem	**Implication**
You cannot match on all variables.	Selection by maturation interactions possible.
You cannot match on true scores. Instead, you have to match on observed scores—and observed scores are affected by random error	Regression effects possible.

© Cengage Learning 2013

long-term effects of Head Start. Even if the groups started out the same, the upper-income group, because of superior health, nutrition, and schools, might mature academically at a faster rate than the disadvantaged group. Thus, not surprisingly, some early studies of Head Start that failed to take regression and selection by maturation into account made it look like Head Start harmed, rather than helped, children.

In conclusion, matching is not the perfect solution that it first appears to be (see Table 14.11). Therefore, the nonequivalent control-group design is a flawed way of establishing that a treatment has an effect.

Conclusions About Quasi-Experimental Designs

Unfortunately, all quasi-experimental designs are flawed methods of establishing that a treatment caused an effect. Although quasi-experiments ensure temporal precedence and assess covariation, quasi-experiments do not automatically rule out the effects of nontreatment factors. To compensate for the inability of their designs to automatically rule out the effects of nontreatment factors, quasi-experimenters use a variety of tactics.

Quasi-experimenters may combine two quasi-experimental designs, using one design to cover for another's weaknesses. For example, they may use a time-series design to rule out selection biases and then use a nonequivalent control-group design to rule out history effects.

Quasi-experimenters may also identify a specific threat to their study's internal validity and then take specific steps to minimize that threat (for a review, see Table 14.5). For instance, they may eliminate instrumentation biases by administering the same measure, the same way, every time.

Finally, they may rule out some threats by arguing that the particular threat is not a likely explanation for the effect. For example, they may argue that mortality was low and therefore not a threat or that pretest scores were not extreme and so regression was not a problem.

When arguing that nontreatment factors are unlikely explanations for their results, quasi-experimenters often cite the **law of parsimony**: the assumption that the explanation that is simplest, most straightforward, and makes the fewest assumptions is the most likely. Thus, the time-series researcher argues that the simplest assumption to make is that the effects of maturation, instrumentation, testing, and mortality are consistent over time. Therefore, a dramatic change after introducing the treatment should not be viewed as a complex, unexpected maturation effect, but as a simple, straightforward treatment effect.

Clearly, the quasi-experimenter's job is a difficult one, requiring much creativity and effort. But there are rewards. Quasi-experimenters can often study the effects of treatments that couldn't be studied with conventional experimental designs. For example, quasi-experimenters can study treatments that could not—or should not—be randomly assigned, such as the effects of disasters, new laws, new technology, and new social programs. Furthermore, because quasi-experimenters often study real-world treatments, their studies sometimes have more external validity than traditional experiments.

CONCLUDING REMARKS

Quasi-experiments and single-*n* designs are extremely useful—if you want to infer that a treatment causes an effect and you cannot use random assignment. If you want to infer causality and you can use random assignment, you should probably use one of the designs described in Chapters 10–13. If you do not want to infer causality, you should use one of the methods discussed in Chapters 7 and 8.

SUMMARY

1. To infer that a treatment causes an effect, you must show that changes in the amount of the treatment are accompanied by changes in participants' behavior (covariation), that changes in the treatment come before changes in the behavior (temporal precedence), and that nothing other than the treatment is responsible for the change in behavior (the change is not due to spuriousness).

2. By comparing treatment and nontreatment conditions, you can determine whether the suspected cause and the effect covary.

3. When you introduce the treatment, you make sure that the treatment comes before the change in behavior, thereby establishing temporal precedence.

4. Randomization is an effective way of making it unlikely that nontreatment factors may be responsible for the change in behavior.

5. Like randomized experiments, single-*n* designs introduce the treatment to ensure temporal precedence and compare conditions to assess covariation.

6. Single-*n* researchers try to identify the important, nontreatment variables, and then they try to stop those variables from varying within their study.

7. Single-*n* researchers prevent individual difference variables from varying within their study by limiting their study to examining a single participant. That is, differences between subjects (between-subjects variability) cannot make the treatment condition score higher than the control condition because the treatment condition subject and the control condition subject are the same individual.

8. Single-*n* researchers may keep many environmental variables constant by keeping the participant in a highly controlled environment.

9. The A–B–A reversal design and the multiple-baseline design are used by single-*n* researchers to rule out the effects of maturation and testing.

10. When it comes to construct validity, the single-*n* researcher and the within-subjects

researcher use very similar approaches. To prevent participants from figuring out the hypothesis, both researchers may use (a) few levels of the independent variable, (b) placebo treatments, and/or (c) gradual variations in the levels of the independent variable.

11. Unlike single-*n* researchers, quasi-experimenters cannot keep relevant nontreatment factors from varying.

12. Quasi-experimenters must explicitly rule out the eight threats to internal validity: history, maturation, testing, instrumentation, mortality (attrition), regression, selection, and selection by maturation interactions.

13. Instrumentation can be ruled out by using the same measure, the same way, every time.

14. You can rule out mortality (attrition) threat to your study's validity if you can prevent participants from dropping out of your study.

15. You can probably rule out regression if participants were not chosen on the basis of their extreme scores or if your measuring instrument is extremely reliable.

16. The time-series design is very similar to the A–B single-*n* design. The main differences are that the time-series design (a) studies more participants, (b) does not control the variables necessary to establish a stable baseline, and (c) doesn't isolate participants from history the way the single-*n* design does. Because of its lack of control over environmental variables, it is vulnerable to history effects.

17. The nonequivalent control-group design resembles the simple experiment. However, because participants are not randomly assigned to groups, selection is a serious problem in the nonequivalent control-group design.

18. Although quasi-experimental designs are not as good as experimental designs for inferring causality, they are more versatile.

KEY TERMS

A–B design *(p. 569)*
baseline *(p. 573)*
carryover effects *(p. 576)*
covariation *(p. 565)*
law of parsimony *(p. 601)*
maturation *(p. 573)*
multiple-baseline design
  (p. 578)

nonequivalent control-group
  design *(p. 597)*
pretest–posttest design
  (p. 588)
quasi-experiment *(p. 583)*
reversal design A–B–A design
  A–B–A reversal design
  (p. 575)

single-*n* design *(p. 569)*
spurious relationship *(p. 566)*
stable baseline *(p. 572)*
temporal precedence *(p. 566)*
testing *(p. 575)*
time-series design *(p. 590)*

EXERCISES

1. Suppose that the means for the treatment and no-treatment conditions are the same. If so, which requirement of establishing causality has not been met?

2. If the study does not manipulate the treatment, which requirement of establishing causality will be difficult to meet?

3. If participants are not randomly assigned to condition, which requirement for establishing causality will be almost impossible to meet?

4. Compare and contrast how single-*n* designs and randomized experiments account for nontreatment factors.

5. What arguments can you make for generalizing results from the single-*n* design?

6. How do the A–B design and the pretest–posttest design differ in terms of
   a. procedure?
   b. internal validity?

7. How does the single-*n* researcher's A–B–A design differ from the quasi-experimenter's reversal time-series design in terms of
   a. procedure?
   b. internal validity?

8. Design a quasi-experiment that looks at the effects of a course on simulating parenthood,

including an assignment that involves taking care of an egg, on changing the expectations of junior-high school students about parenting. What kind of design would you use? Why?

9. An ad depicts a student who has improved his grade-point average from 2.0 to 3.2 after a stint in the military. Consider Campbell and Stanley's "spurious eight." Is the military the only possible explanation for the improvement?

10. One study found that students who had been held back a grade did worse in school than students who had not been held back. Based on this evidence, some people concluded holding students back a grade harmed students.

a. Does this evidence prove that holding students back harms their performance? Why or why not?

b. If you were a researcher hired by the Department of Education to test the assertion that holding students back harms them, what design would you use? Why?

 ## WEB RESOURCES

1. Go to the Chapter 14 section of the book's student website and
   1. Look over the concept map of the key terms.
   2. Test yourself on the key terms.
   3. Take the Chapter 14 Practice Quiz.

2. Read the interactive story that reviews different threats to internal validity.

3. Consider an alternative to using quasi-experiments by reading "Web Appendix: Field Experiments."

CHAPTER 15

Putting It All Together:

Writing Research Proposals and Reports

It takes less time to do a thing right than it does to explain why you did it wrong.
—Henry Wadsworth Longfellow

If you fail to plan, you plan to fail.
—W. Clement Stone

CHAPTER OVERVIEW

Your research should be carefully planned before you test your first participant. Without such planning, you may fail to have a clear hypothesis, or you may fail to test your hypothesis properly. In short, poor planning leads to poor execution.

Poor execution can lead to unethical research. At best, it wastes participants' time; at worst, it harms participants. Therefore, the purpose of this chapter is to help you avoid unethical research by showing you how to plan and report the results of your study. If you follow our advice, your research should be humane, valid, and meaningful.

AIDS TO DEVELOPING YOUR IDEA

In this section, you will learn about two major research tools: the research journal and the research proposal. Many scientists regard the research journal and the research proposal as essential to the development and implementation of sound, ethical research.

The Research Journal

We recommend that you keep a **research journal**: a diary of your research ideas and your research experiences. Keeping a journal will help you in at least three ways: (1) you'll have a record of why and how you did what you did; (2) writing to yourself helps you think through decisions; and (3) a research journal can help you prepare your research proposal.

Because the journal is for your eyes only, it does not have to be neatly typed and free of grammatical errors. What is in the journal is much more important than how it is written.

What should you put in your journal? You should jot down every idea you have about your research project. At the beginning of the research process, when you are trying to develop a research hypothesis, use your research journal for brainstorming. Write down any research ideas that you think of and indicate what stimulated each idea. When you decide on a given idea, explain why you decided on that particular research idea. When reading related research, summarize and critique it in your journal. If you do quote any material, be sure to put that material in quotation marks. Otherwise, you will not be able to remember whether you have paraphrased or quoted that material when it comes time to write your paper. Whether you quote, paraphrase, summarize, or critique a source, write down the authors, year, title, and publisher for that source. This source information will come in handy when you write your research proposal. In short, whenever you have an insight, find a relevant piece of information, or make a design decision, record it in your journal.

To use the information in your journal, you will have to organize it. One key to effective organization is to write down only one idea per page. Another key is to rewrite or rearrange your entries every couple of days. Your goal in rearranging entries should be to put them in an order that makes sense to you. For example, your first section may deal with potential hypotheses, your second section may deal with ideas related to the Introduction section of your paper, and your third section may deal with how you would conduct the study.

The Research Proposal

Like the research journal, the purpose of the research proposal is to help you think through each step of your research project. In addition, the research proposal will let others, such as your professor, think through your research plan so that they can give advice that will improve your study. By writing the proposal, you will have the opportunity to try out ideas and explore alternatives without harming a single participant. In other words, the process of writing the proposal will help you make intelligent and ethical research decisions.

Although the research proposal builds on the research journal, it is much more formal than the journal. When you write the proposal, you will have to go through several drafts. The result of this writing and rewriting will be a proposal that is not only clear but also conforms in content, style, and organization to the guidelines given in the *Publication Manual of the American Psychological Association* (2010).

We want to emphasize that it is not enough to have good ideas: You must present them in a way that people will receive them. History is full of examples of people who had good ideas but got little credit because they expressed them poorly. Conversely, some people have become famous more for how well they expressed their ideas than for the originality of their ideas.

If you write using American Psychological Association (APA) style, you will have a better chance of expressing your ideas well. Think of APA style as a kind of language that makes it easier for professionals in the psychology field to communicate with one another.

If, on the other hand, you fail to write a research proposal that conforms to APA style, most professors will judge the content of your proposal more harshly. They will feel that if you cannot follow that style, you are incapable of doing good research.

To reiterate, following APA format is important. Indeed, one of the best known professors of research design cites learning APA style as one of the most important things students learn from his design class (Brewer, 1990).

As we have stressed, following APA style will help you communicate the content of your proposal. However, before you worry about how to communicate your content clearly, you need to have content: Style without substance is worthless. The substance of your proposal will be your statements regarding

1. why your general topic is important,
2. what your hypothesis is,
3. how your hypothesis is consistent with theory or past research,
4. how your study fits in with existing research,
5. how you define your variables,
6. who your participants will be,
7. what procedures you will follow,
8. how you will analyze your data, and
9. what implications you hope your results will have for theory, future research, or real life.

The research proposal's substance makes it the foundation for your study, and its substance and style makes it the foundation of the final research report. To be more specific, the Introduction and the Method sections you write for your research proposal should be highly polished drafts of the Introduction and Method sections of your final report, whereas other parts of the research proposal will serve as rough drafts of the abstract, results, and discussion sections of your final report.

WRITING THE RESEARCH PROPOSAL

Now that you know what a research proposal is, it is time for you to begin writing one. We will first show you how to write the Introduction.

General Strategies for Writing the Introduction

Like the opening scene in a television show, your **Introduction** should set the stage for what will come next while convincing the audience to stick around to the end. To grab your audience's attention, start your Introduction by showing that your general topic is interesting. For example, you might use a real-life example to illustrate how interesting the behavior, predictor, relationship, or phenomenon that you are studying is.

After showing that your general topic is important, show that your hypothesis makes sense, that testing your hypothesis makes sense, and that your way of testing that hypothesis makes sense. To show that your hypothesis makes sense, you will need to spell out the reasoning behind your hypothesis. In addition, you will probably need to cite research studies and theories that suggest that your hypothesis might be correct. To show that testing your hypothesis is important and that your way of testing your hypothesis is a good one, you will need to discuss and cite relevant research—and do so in a way that makes it look like your study is the logical next step in that research. In short, after reading the Introduction, your reader should know

1. why your research area is important,
2. what your hypothesis is,
3. why your hypothesis makes sense, and
4. why your study is the best way to test the hypothesis.

Establishing the Importance of Your Study

Before you can persuade people that your study is important and interesting, you must let them know exactly what concepts you are studying, and then explain why those concepts are important. To establish that your concepts are important, you will probably want to use one of the following three strategies:

1. *prevalence:* presenting statistical or other evidence of how often people encounter the basic principle or topic,
2. *relevance:* presenting an example that illustrates how the concept has important implications for real life or for testing a theory, or
3. *precedence:* demonstrating that the concept has captured the interest of other researchers.

Demonstrate the Concept's Prevalence. One strategy for showing that your study is important is to show that your general topic area is a common part of real life. Sometimes, authors boldly assert that the phenomenon they are studying is common. For example, authors may write "_____ is a part of everyday life" or "People are bombarded with _____."

Rather than asserting "Most people have experienced _____," you might document the prevalence of the concept by presenting statistical evidence. Thus, if you were studying widowhood, you might present statistics on the percentage of people who are widowed. In the absence of statistics, you could use quotations from influential people or organizations (e.g., the American Psychological Association) to stress the prevalence of your concept.

Demonstrate the Concept's Relevance to Real Life. Rather than emphasizing the concept's prevalence, you might emphasize its relevance. For example, you might stress the practical problems that might be solved by understanding the concept. Alternatively, you might demonstrate the problem's relevance by presenting a real-life example of your concept in action. Giving an example of the concept is a very good way to both define the concept and provide a vivid picture of its importance.

Demonstrate Historical Precedence. Finally, you might show that there is a historical precedence for your study. You could emphasize the great minds that have pondered the concept you will study, the number of people through the ages who have tried to understand the behavior, or the length of time that people have pondered the concept. Normally, you will also want to show that the research topic has—or should have—been important to both researchers and theorists.

Writing the Literature Review

One way of establishing historical precedence is to summarize research done on the topic. In addition to helping the reader understand your research question, citing research shows the reader that the field considers your general research area important. That is, if the field did not consider these concepts important, investigators would not be researching these areas, and their findings would not be published. Thus, it is not uncommon for Introductions to include statements such as, "The focus of research for the past 20 years ..." or "Historically, research has emphasized...."

However, even if you do not use the literature review to establish the importance of the general concepts, you will still want to write it to show how your particular study fits in with existing work. In other words, although you can select one of many ways to show that your general concepts are important, the only way to show that your particular research study is important is to write a literature review. Therefore, *all Introductions should contain a literature review*.

Goals of the Literature Review. Because the literature review is designed to sell your particular study, you need to do more than merely summarize previous work. You must also use the summary to set the stage for your study. You will do so by showing that your study (a) corrects a weakness in previous research or (b) builds on and extends previous research. In short, you need to make the reader feel that there is a need for your research.

Deciding Which Research to Review. We have addressed the goals of the literature review. You know why you should review the literature. Now, let's talk about what you will review. Instead of reviewing informal sources such as magazine articles and people's personal websites, you will review either printed or online versions of journal articles (see Web Appendix B to learn how to find articles to review). When citing research reported in scholarly journal articles, your focus will usually be on reviewing *recent* research. However, you may also review one or two older, classic works as well as recent research. Critiquing—rather than merely summarizing—the articles you cite will show that you have thought about what you have read. By analyzing the strengths and weaknesses of a number of articles, you will establish that you have done your homework.

Although critiquing these journal articles may establish you as a scholar, realize that your goal is not simply to establish your credibility. Instead, your primary goal is to show how your study follows from existing research.

You may feel that these two goals (establishing your expertise versus showing that your study follows from existing research) conflict. To establish your expertise, you may feel that you should cite all research ever done in the field. But to set up your research study, you want to cite and analyze only those studies that bear directly on your study. To help you resolve the apparent conflict between these two goals, we offer two tips.

First, realize that your main goal is to set up your study. Thus, you will be offering in-depth critiques of only those studies that directly apply to your study.

Second, realize that Introductions begin by talking about the general area and then focus on the specific research question. Thus, you should cite classic research that establishes the importance of your general topic. However, if those classic findings are only indirectly related to your work, you should probably cite them only in your first paragraph.

Not knowing what to include in a literature review is one of the two main problems students have in writing the literature review. The other main problem is how to write a literature review that is well organized.

If you are going to write an organized literature review, you must start organizing before you start writing. Begin by grouping together the studies that seem to have something in common. If you have summaries of all your studies on large index cards, you might find that you have the following four stacks of cards:

1. a stack that emphasizes the importance of the general concept,
2. a stack that deals with problems with previous research,
3. a stack that deals with reasons to believe that your hypothesis will not be supported, and
4. a stack that deals with reasons to believe that your hypothesis will be supported.

Alternatively, you may find that you have three piles:

1. a stack that deals with how to measure your outcome variable,
2. a stack that deals with studies that obtained a certain finding, and
3. a stack that deals with studies that obtained the opposite finding.

Regardless of the specific content of your stack, the fact that you have stacks shows that you have some way of organizing the studies. Now, you have to convert those organized stacks into an organized literature review. Your first step is to turn each stack into a paragraph.

To help convert these piles into meaningful paragraphs, write a sentence summarizing what all the cards have in common. Each stack's sentence could be the topic sentence for a paragraph, with the rest of the stack providing evidence and citations for the statements made in that sentence (Kuehn, 1989).

Once you have finished a draft of your literature review, read it aloud. After fixing problems you find as you read the literature review aloud, outline it. Then, rate your literature review on the following five-point scale:

1. Very few recent journal articles are cited.
2. Enough reports of recent research are cited, but the articles are not clearly summarized.
3. Enough articles are cited and articles are clearly summarized, but a reader might not understand either (a) why those articles are being cited or (b) why they are being cited in that order.
4. Enough recent articles are cited, and articles are *either* (a) integrated with other summaries or (b) critiqued.
5. Enough recent articles are cited and articles are *both* (a) integrated with other summaries and (b) critiqued.

Always keep in mind that your literature review should do more than evaluate other people's work: It should also show how that work led to your study. Your job is to convince the reader that, given the existing research in that area, your study is the next study that should be done. In other words, you want to help readers see how your study builds on existing research's strengths while avoiding existing work's weaknesses. So, as you comment on existing research, comment favorably on aspects of existing research that will be incorporated into your study (e.g., praise certain studies for using the measure that you will use) and comment negatively on aspects of existing research that will not be incorporated into your study (e.g., if you will use an experiment, criticize related studies for using nonexperimental designs). Your literature review should do such a good job of setting the stage for your study that just from reading your literature review, a clever reader could guess what your hypothesis is and how you plan to test it.

However, you won't make readers guess the rationale for your hypothesis and for your study. After summarizing the relevant research, spell out the reasoning that led to your hypothesis so clearly that your readers will know what your hypothesis is before you actually state it.

Stating Your Hypothesis

Even though your readers may have guessed your hypothesis, leave nothing to chance: *State your hypothesis!* To emphasize a point that can't be emphasized enough, state your hypothesis boldly and clearly so that readers can't miss it. Let them know what your study is about by writing, "The hypothesis of this study is...."

When you state your hypothesis, be sensitive to whether you'll be testing it with an experiment or with a correlational study. Because only experiments allow you to test cause–effect hypotheses, your hypothesis should include the word *causes* (or synonyms for *causes,* such as *affects*, *impacts*, *influences*, *leads to,* and *makes*) only if you plan to conduct an experiment. If you don't plan on directly manipulating your predictor variable, you have a correlational study and therefore can test only whether two or more variables are related. Thus, if you were surveying people about their lifestyles and moods, your hypothesis should not be "A sedentary lifestyle causes depression," but rather "A sedentary lifestyle is related to depression."

General Strategies for Writing an Introduction: Conclusions and Tips

We have given you some general advice about how to write an Introduction— an important section that will probably be between one-fourth and one-third the length of your final paper. You have seen the importance of clearly defining your concepts, critically summarizing research, carefully explaining the reasoning behind your hypothesis, and stating your hypothesis. Because summarizing research, explaining the reasoning behind hypotheses, and stating hypotheses are so important, you are probably not surprised to find that some Introductions have subheadings such as "Overview of Past Research," "Theoretical Background," and "Hypotheses." Although you do not need to include such subheadings, you should outline your Introduction, and your outline should include headings such as "Overview of Past Research" and "Hypotheses."

If you have followed our advice to this point, you should avoid a major problem that students have when they write the Introduction: disorganization. If your Introduction is disorganized, you will probably get a low grade because your professor may not understand why the work you cite is relevant, why your hypothesis makes sense, or why your way of testing the hypothesis makes sense. Thus, having a well-organized Introduction is important.

Organization, however, is not the only thing. You need to be careful when you write your Introduction: careful not to state an opinion as a fact, careful to back up your facts with citations, careful to cite both the author and the date of each works you cite, careful to use quotation marks when you quote, and careful to use a formal, professional, unbiased tone. To achieve this level of carefulness, check your Introduction against the "Introduction Checklist" in Appendix A.

Specific Strategies for Writing Introduction Sections for Different Types of Studies

We have pointed out that your Introduction should review the literature, be well-organized, be well-documented, look professional, and spell out why your study should be done. Although we have given you general advice about how to explain why your study should be done, you may feel that you need specific advice. Because the specific way you justify your study will depend on the kind of study you are doing, we will devote the next section to showing you how to justify six common types of studies:

1. exploratory,
2. direct replication,
3. systematic replication,
4. conceptual replication,
5. replication and extension, and
6. theory testing.

The Exploratory Study

In introducing an **exploratory study**—a study investigating a new area of research—you must take special care to justify your study, your hypothesis, and your procedures. You must compensate for the fact that your reader will not have any background knowledge about this new research area.

New Is Not Enough.   Because of the lack of research in the area, you will not be able to use the common strategy of showing that your topic is important by showing that it has inspired a lot of research. Although you will be able to show that your research area has been ignored, that will not be enough to justify your study: Many research areas (e.g., the psychology of tiddlywinks) have been ignored for good reasons. To justify your study, you must convince your readers that it is a tragedy that your research question has been overlooked. Make them believe this wrong must be righted to help psychology advance as a science.

One approach you can use to justify your exploratory study is to discuss hypothetical or real-life cases that could be solved or understood by answering your research question. For example, consider how Latané and Darley (1968) opened their pioneering work on helping behavior. Not only did they state that there was a lack of research on helping behavior, but they also bolstered their research justification by referring to the case of murder victim Kitty Genovese. Ms. Genovese was, according to a newspaper report,[1] brutally attacked for more than 30 minutes in the presence of more than 30 witnesses—none of whom intervened. Thus, Latané and Darley effectively convinced readers that understanding why people fail to help is an important research area.

To reiterate, your first step in justifying an exploratory study is to show that the area is important. Once you have convinced your readers that your area is important, you can further excite them by emphasizing that you are exploring new frontiers.

Spell Out Your Reasoning. In an exploratory study, as in all studies, you must spell out the rationale for your hypothesis. Because you are studying an unexplored dimension, you must give your readers the background to understand your predictions. Therefore, be extremely thorough in explaining the logic behind your predictions. Even if you think your predictions are just common sense, you need to spell them out because not everyone will immediately see how your predictions follow from common sense.

Beyond spelling out the common sense logic of your prediction, try to explain how your prediction is consistent with (a) theory and (b) research on related variables. For example, suppose you are interested in seeing how low-sensation seekers and high-sensation seekers differ in their reactions to stress. You might argue that your hypothesis is consistent with arousal theory—the theory that we all have an ideal level of arousal (Berlyne, 1971). That is, you might argue that high-sensation seekers like stress because it raises their arousal up to the optimal level, whereas low-sensation seekers hate stress because it raises their arousal beyond the optimal level.

In addition to—or instead of—using theory to support your hypothesis, you could use research on related concepts. Thus, in our example, you might start by arguing that introversion–extroversion and sensation-seeking are related concepts. Then, you might argue that because introversion and sensation-seeking are related, and because stress has different effects on introverts and extroverts, stress should also have different effects on low-sensation seekers vs. high-sensation seekers.

Defend Your Procedures. In addition to explaining your predictions, you may have to take special care in explaining your procedures. If you are studying variables that have never been studied before, you can't tell the reader that you are using familiar, well-accepted measures and manipulations. Instead, you may have to invent—and justify—your own measures and manipulations. Therefore, you will need to explain, either in the Introduction or in the Method section, why your manipulations and measures are probably valid.

---

[1] More than 30 years after the incident, some are now questioning the original newspaper account. For our purposes, however, the accuracy of the newspaper account is irrelevant.

The Direct Replication

Rather than doing a completely original exploratory study, you may decide to do the opposite. That is, you may decide to do a **direct (exact) replication**: a repetition of an original study. Before doing a direct replication, you must be very clear about why you are repeating the study. If you are not careful, the reader may think you performed the study before you realized that someone else had already done the study. Even if you do spell out why you repeated the study, some journal reviewers will find the fact that you did a direct replication a legitimate reason to reject the paper for publication (Fiske & Fogg, 1990). However, you can use a two-pronged strategy to persuade people that your study is worth doing.

Document the Original Study's Importance. First, to justify a direct replication, you should show that the original study was important. To do this, discuss its impact on psychology. To get some objective statistics about the number of times the study has been cited, you can use *Google Scholar* or the *Social Science Citation Index,* both of which are described in Web Appendix B.

Explain Why the Results Might Not Replicate. After establishing the study's importance, try to convince your readers that the study's results might not replicate. There are basically four arguments you can make in support of the idea that the findings won't replicate.

1. The findings appear to contradict other published work.
2. The original study's statistically significant results may be a Type 1 error (mistaking a coincidence for a reliable relationship; declaring a chance difference statistically significant).
3. The original study's null results (not obtaining significant results) may be a Type 2 error (a failure to find a real relationship).
4. People or times have changed so much from when the original study was performed that a replication would produce different results.

Perhaps the strongest argument you can make for replicating the study is to show that the findings appear to contradict other published work. The more you can make the case that other findings appear to contradict the findings of the study you wish to replicate, the stronger the case for a replication.

One reason that the original study may be inconsistent with other published work is that the statistically significant result in the original study could be the result of a Type 1 error. Thus, if you showed that the original study's results would not have been significant at the conventional $p = .05$ level or that the authors used an unconventional statistical technique that inflated their chances of making a Type 1 error, you would have a strong case for replicating the study.

But what if the original study reported nonsignificant (null) results? Then, you could argue that random error or poor execution of the study may have prevented it from finding an existing relationship. That is, you could argue that the null results were due to a Type 2 error. If the original study's findings seem to conflict with several other published papers that *did* find a significant relationship between those variables, you have a compelling rationale for replicating the study.

If you can't reasonably argue that the original results are due to either a Type 1 or Type 2 error, you still might be able to justify a direct replication on the grounds that the study would come out differently today. For example, you might want to replicate a conformity study because you believe that, as a result of cohort differences in parenting style, teenagers today are less conforming than teenagers were when the original study was conducted. Regardless of what approach you take, you must present a compelling rationale for any study that is merely a rerun of another study.

The Systematic Replication

Rather than repeating the study, you might conduct a **systematic replication**: a study that makes a minor modification of the original study. The systematic replication accomplishes everything the direct replication does and more. Therefore, every reason for doing a direct replication is also a reason for doing a systematic replication. In addition, you can justify a systematic replication by showing that modifying the procedures would improve the original study's power (ability to find relationships), construct validity, or external validity.

Improved Power. As we mentioned earlier, if you thought the original study's null results were due to Type 2 error, you could just redo the original study. However, if you just repeat the study, you may just repeat its Type 2 error. Therefore, instead of repeating the original study, you might make a minor change in procedure to improve its power to find relationships. For example, you might use more participants, more extreme levels of the predictor/ independent variable, or a more sensitive measure (e.g., replacing *yes/no* questions with *strongly agree, agree, neutral, disagree,* and *strongly disagree* questions) than the original study used.

Improved Construct Validity. You might also want to modify the original study if you thought that the original study's results were biased by demand characteristics (clues that suggest to the participant how the researcher wants the participant to behave). Thus, you might repeat the study using a double-blind procedure (making sure that neither the participant nor the person who has direct contact with the participant knows what type of treatment the participant has received) to reduce subject and researcher bias.

Improved External Validity. If you are replicating a study to improve external validity, you should explain why you suspect that the results may not generalize to different stimulus materials, levels of the treatment variable, or participants. For example, even if it seems obvious to you why a study done on rats might not apply to humans, spell out your reasons for suspecting that the results wouldn't generalize.

The Conceptual Replication

Most of the reasons for conducting a systematic replication are also relevant for introducing and justifying a **conceptual replication**: a study that is based on the original, but uses different methods to better assess the true relationships between the variables being studied. In addition to having the same advantages as the systematic replication, the conceptual replication has several other unique selling points, depending on how you changed the original study.

Using a Different Measure. Your conceptual replication might differ because you used a different way of measuring the dependent measure than the original authors did. In that case, you should show that your measure is more reliable, sensitive, or valid than the original measure. To make the case for your measure, you may want to cite other studies that used your measure. As in the legal arena, precedent carries weight in psychology: If someone else published a study using a given measure, the measure automatically gains some credibility.

Using a Different Manipulation. Instead of trying to use a different measure of a construct, you might want to use a different manipulation of a construct. For example, if one researcher induced stress in participants by suggesting that they would get painful electric shocks, you might decide to replicate the study, but induce stress by giving participants a very short period of time to do certain mathematical problems. If you use a different manipulation, you should start by defining the variable you are trying to manipulate. Next, you should discuss weaknesses of previous manipulations. Then, show how your manipulation avoids those weaknesses. Conclude by showing that your manipulation is consistent with definitions of the concept you are trying to manipulate.

Using a Different Design. If you are changing the original study's design, let your readers know why you are making the change. For example, suppose you believe that the original study failed to find a significant effect because it used a relatively low powered design: a between-subjects experiment that compared a treatment group to a no-treatment group. If you are repeating the study using a more powerful design—a within-subjects (repeated measures) experiment that compares each participant's response in the treatment condition to that participant's response in the no-treatment condition—tell your readers that you switched to a within-subjects design to boost your chances of finding a significant effect.

The Replication and Extension

Your study may go beyond a conceptual replication by looking at additional factors or measures. In that event, your Introduction would not only contain everything a conceptual replication would but also a rationale for the additional factors or measures.

Rationale for Additional Factors. For example, suppose the original author found that people loaf when working in groups. You might think of a situation (e.g., a group in which all members were good friends) in which social loafing wouldn't occur. Thus, you might include friendship as a factor in your design. Be sure to state (a) your reasons for including the factor and (b) your predictions regarding the factor.

Rationale for Additional Dependent Measures. Instead of adding a predictor/ independent variable to a study, you might add an outcome/dependent measure. Your purpose would be to discover how the treatment produces the effect. In other words, you are trying to show that a certain mental or physiological reaction is both (a) triggered by the treatment and (b) the mediating mechanism by which the treatment has its effect on behavior.

How would you go about finding out the invisible processes underlying an observable effect? In a social-loafing experiment, you might collect measures of participants' perceptions of others to uncover the mental processes responsible for social loafing (such as perceptions that their efforts are not being noticed). Or, you might monitor arousal levels in an attempt to discover the physiological reasons for social loafing (e.g., lower physiological arousal in a group setting). However, even if you found that working in a group reduced arousal or changed perceptions, you could not say that these changes, in turn, caused the loafing.

Not surprisingly, then, the tricky part about writing an Introduction to a *process* study is to persuade your readers that you really are going to be able to pin down the underlying causes of a phenomenon. You must do more than merely show that these processes occur before the phenomenon occurs, because these processes could be incidental side effects of the treatment. For instance, a fever may appear before you get ill—and may intensify as you get ill—but a fever doesn't cause you to be ill. It's a side effect of your illness. In the same way, a mental or physiological event may accompany a change in behavior but not be the cause of that behavioral change.

Critics usually will not accept evidence that the treatment had certain effects on a physiological or mental process as proof that the treatment works by altering that process. Instead, critics usually want more direct proof. For example, suppose you say that your treatment (A) has its effect on an outcome (C) through a mechanism (B). That is, you suggest that $A \rightarrow B \rightarrow C$. In that case, critics may not be satisfied by you showing that $A \rightarrow B$ and that $A \rightarrow C$. To convince them that $A \rightarrow B \rightarrow C$, they may insist that you to show that, by disrupting B (the mechanism by which A supposedly affects B), you disrupt the $A \rightarrow C$ relationship. That is, they may insist that you show that the treatment doesn't have its usual effect when you block the process through which the treatment supposedly operates (Sigall & Mills, 1998).

To show how clever some of these process-testing experiments can be, pretend that you and a friend are participants in the following experiment (Steele, Southwick, & Critchlow, 1981). Participants in your condition are asked to write an essay favoring a big tuition increase. According to dissonance theory, you will

1. feel unpleasant tension after writing this essay and
2. reduce this tension by being less opposed to tuition increases.

As dissonance theory would predict, you now are less opposed to tuition increases. But did you feel that unpleasant tension, and did you change your attitudes as a way to reduce that unpleasant tension? In other words, was dissonance the *mediating mechanism* for your attitude change?

To find out, let's look at participants in your friend's condition. Those participants also received the treatment. That is, like you, they wrote an essay favoring a big tuition increase. Unlike you, however, the experimenter set it up so that your friend and people in your friend's condition did not experience prolonged unpleasant tension after writing the essay. Specifically, the researcher set it up so that right after writing the essay, participants in your friend's group thought they had finished the dissonance study and were now participating in an unrelated study that involved judging alcoholic beverages. Actually, that "unrelated study" was a sneaky way of getting your friend and the other participants in that condition to drink alcohol without suspecting that the researchers were using alcohol to reduce any unpleasant arousal (dissonance) caused by writing an essay that disagreed with their attitudes.

Consistent with the researchers' predictions, the effect of preventing that group from feeling dissonance was to prevent attitude change: People in the alcohol condition did not change their views about tuition increases. This study provides strong evidence that having people write counter-attitudinal essays has its effect by creating unpleasant tension that people try to reduce. (For more about how to test hypotheses involving mediating variables, see pages 91–95.)

In summary, you can extend an existing study by adding measures or manipulations. Such extensions may provide insights into how a treatment has its effect. When proposing such a study, remember that you must (a) justify why you are adding the measure or manipulation and (b) explain your predictions regarding the additional measure or manipulation.

The Theory-Testing Study

A theory is a set of principles that explains existing research findings and can be used to make new predictions that can lead to new research findings. If you are testing a prediction from a theory, there's good news and bad news.

The good news is that you won't have to spend much effort justifying your study's importance. Scientists are interested in studies that provide information about the accuracy of their explanations (i.e., their theories). If your test supports your prediction, scientists have more confidence that the theory explains existing research findings, and they have seen yet another example that the theory can be used to make new predictions that lead to new research findings. If your test disconfirms your hypothesis, your study would raise doubts about the theory from which you derived the hypothesis. Indeed, because theories cannot be proven but can be disproven, your study could destroy a theory that has been supported by thousands of studies.

The bad news is that not everyone will agree that your predictions follow from the theory. To convince them, you must clearly spell out how your predictions follow from the theory. By being clear, everyone will follow your logic, and some may even agree with it.

Writing the Method Section

You have reviewed the literature, developed a hypothesis, decided how to measure your variables, and stated your reasons for testing your hypothesis. Your preliminary work, however, is still not done. You must now decide exactly what specific actions you will take. In other words, although you probably have decided on the general design (e.g., a simple experiment), your plan is not complete until each detail of your study has been thought through and written down.

In your journal, specify exactly what procedures you will follow. For example, what instructions will participants be given? Who will administer the treatment? Where? Will participants be run in groups or individually? How should the researcher interact with participants? Although your answers must be accountable to issues of validity, your paramount concern must always be ethics. You do not have the right to harm another. If you need help thinking through the ethical of practical issues, see Web Appendix D: Practical Tips for Conducting an Ethical Study.

Once you have thoroughly thought out each step of your study, you are ready to write the **Method section**[2] of your proposal. This is the "how" section—here you will explain exactly how you plan to conduct your study. However, keep in mind that—just like the Introduction—the Method section is written on two levels. As you will recall, at one level, the Introduction summarizes existing research; at another level, it sells the need for your study by pointing out deficiencies in existing research. Similarly, at one level, the Method section tells the reader what you are going to do; at another level, it sells the reader on the idea that what you plan to do is the correct thing.

To sell the reader on what you plan to do, tell your reader about the wise design choices you have made. Thus, in the Method section, point out that your measure is valid, that your manipulation is widely accepted, or that you are doing something a certain way to reduce demand characteristics, researcher biases, random error, or some other problem.

In short, selling the value of a research strategy is a never-ending job. If possible, you should sell your strategy in each of the Method section's subsections.

The Method must include two subsections: a participants section and a procedure section. However, it may include other sections, such as overview, design, apparatus, materials, manipulations, and dependent measures subsections.

---

[2] When writing your paper, please label this section "Method" rather than "Methods."

Participants

In the participants section, answer the question "Who (in a general sense) are these people—and how did/will they get into my study?" by describing your sample's characteristics. Specifically, state how many participants you plan to have, how many will be men, how many will be women, their ages, their probable ethnic composition, where your study will be conducted (e.g., Clarion, PA), and how you plan to obtain or recruit the participants. You should also indicate whether they will be tested individually or in groups. If they will be tested in groups, you should state the size of the groups. If you plan to exclude data from some participants, state the rule that you will use to exclude participants (such as excluding all participants who score above 16 on the Beck Depression Inventory). Finally, provide assurance that your participants will be treated ethically.

You may find it easier to write the Method section if you realize you have three goals. First, you want to help the reader judge to which people, places, and times your results can be generalized. Thus, the reader needs to know who (the age, gender, and ethnic composition of the sample) will be tested and where they will be tested (e.g., results from a study done in China may not apply to an Australian sample). Second, you want to assure the reader that your study will be ethical. Third, you want to be as clear and transparent as you can about the characteristics of your sample—without compromising your participants' confidentiality. Thus, you will be describing who (in a general sense) will be in your study, where they will be tested, when (year, and possibly month) they will be tested, how they will be recruited (and, if they will be divided into groups, how they will be assigned to those groups), how they will be treated (ethically), why participants will agree to be in the study, and what would cause you to keep someone out of the study (e.g, if the participant is under 18 or if the participant did not following directions).

Because the characteristics of all samples can be described in the same ways (e.g., all samples can be described in terms of gender, age, and racial composition) and because clarity is a key goal of the Participants section, the Participants section is written in a straightforward and somewhat mechanical fashion. In fact, it is so mechanical that you can often model yours after a participants section you find in an article or after the following sample participants section.

Participants

The participants will be 80 introductory psychology students (52 men and 28 women; 80% White, 10% Hispanic, 5% Black, 5% multiracial; $M_{age} = 19.12$) from Clarion University who will be given extra credit for their participation. Participants will be run individually and will be randomly assigned to experimental condition. Before this research is conducted, the Clarion University Institutional Review Board will give approval for this research.

Although it may help you to model your Participants section after someone else's, beware of two serious problems with copying or paraphrasing parts of articles that convey approximately what you want to say.[3]

---

[3] We thank an anonymous reviewer for noting these problems.

First, you may end up committing **plagiarism**: using someone else's words, thoughts, or work without giving proper credit. Plagiarism is considered a serious act of academic dishonesty. Indeed, at some institutions, students convicted of plagiarism are expelled. Furthermore, concerns about plagiarism are no longer limited to colleges and universities. The world economy is increasingly based on information. Thus, businesses and individuals are increasingly concerned about the theft of ideas (now called "intellectual property"). Therefore, if you quote someone's work, use quotation marks and cite the source, including the page number on which it appears in the publication, if available. You must also be careful to use the exact words and punctuation that are used in the original work, and indicate if you have omitted any part of it. In addition, if you paraphrase or in any sense borrow an idea from a source, you must also cite that source.

Second, you will rarely find a section that says exactly what you want to say. So, rarely copy things word for word. For example, do not copy our sample participants section verbatim. Instead, create one that best describes how participants will be recruited and assigned in your study.

Design or Design Summary (Optional)

Like the participants section, the design section is easy to write. Merely describe the design of your study. For an experiment, state the number of levels (values) of each independent (treatment) variable and whether the independent variable is a between-subjects variable (each participant gets only one level of the variable) or a within-subjects variable (each participant gets all the levels of the variable that are used in the experiment). Then, tell the reader what the dependent variable (measure) is. For example, you might write, "The design is a 2 (source expertise: nonexpert vs. expert) × 2 (information type: unimportant versus important) between-subjects design. The dependent measure is the number of items recalled."

Apparatus and Materials (Optional)

Apparatus refers to laboratory equipment—not computers you will use to type up instructions or copiers you will use to duplicate materials. If you are not using equipment to present stimuli or to collect responses from participants, you do not need an apparatus section.

You can describe specialized laboratory equipment in an apparatus section or in the procedure section. If you plan to use equipment made by a company, list the product's brand name and the model name and/or number. If you designed your equipment, briefly describe it. You need to give enough detail so that readers will have a general idea of what it looks like. If your apparatus is unusual, include a photo or diagram of it in the appendix.

If you are showing participants photographs, having them listen to a tape, or giving them a booklet of tests, you may need a materials section. If all your materials are measures, you may label your materials section "**Measures**"; if they are all tests, you may label your materials section "**Tests.**" If your tests and questionnaires are straightforward or well known, you may decide to embed your description of them in the procedure section rather than in a separate measures section.

If you used a test or questionnaire, provide at least one example of a typical item. This gives readers a feel for what the participants will see. In addition, if the measure has been published, reference the source of the measure. Finally, include a copy of your test or questionnaire in the appendix.

Procedure

As the name suggests, your procedure section will be a summary of what you actually are going to do. However, contrary to what the name suggests, the focus is on what happens to participants. Readers should be able to visualize what it would be like to be a participant in your study. Note how the sample paper (Appendix B) does a good job of showing what *happens from the participants' perspective.*

Like the authors of the sample paper, you can make it easy for readers to make a movie in their head of what happened to participants. Just use these two tactics:

1. Start with the first thing will happen to the participants, then discuss the second thing that will happen to participants, and continue in chronological order, so that the last part of the procedure deals with the last thing that happened to participants.
2. Keep the focus on participants by making the word "participants" the subject of most sentences. That is, most sentences should deal with what participants will do or see.

In addition to having trouble figuring out how to sequence and present information about procedures, beginning authors have trouble with what to put in and what to leave out of the procedure section. They know they are supposed to provide enough information so that a reader could replicate (repeat) the study, but they still ask, "How much detail should I include?" To help you answer this question, we offer five suggestions.

First, be sure to include enough information so that the reader will understand how you will operationalize your independent and dependent variables. To help the reader understand your independent variable manipulation, include key elements of instructions to participants, especially when the experimental group receives different instructions from the control group. To help the reader understand the dependent measure, introduce it in a simple and straightforward manner (e.g., "The dependent measure is …"). Then, if you will use a scale to measure it, give the name of the scale, the number of subscales, and sample question from each subscale (Cooper, 2011). After making it clear what you are trying to measure and how you are trying to measure it, make it clear that your way of measuring it makes sense. For example, you might cite the reliability of the measure and what steps you will take to reduce observer bias (e.g., training observers or making observers blind).

Second, include any methodological wrinkles that you believe are critical to the study's internal, external, or construct validity. For example, if you will use a placebo treatment or double-blind procedures to reduce bias, tell the reader.

Third, read the procedure sections of several related studies and mimic their style—but avoid plagiarism. Reading these sections will show you how to provide your readers with the right level of detail—enough so they could replicate the study, but not so much that they feel overwhelmed.

Fourth, leave out most of the "behind the scenes" details about events that participants don't see. Don't write, "Booklets will be made by cutting sheets of paper in half, typing them up on a computer, and then using a copier" and don't name the people who will take care of those details. Statements such as "Tom will randomly assign participants to groups" or "Our nice secretary will copy the booklets" do not belong in the Method section. Send those people a card, thank them when you win an Academy Award, or acknowledge their contribution in the Author Notes section of your paper—just keep their names out of the Method section.

Fifth, don't worry if your procedure section seems too brief. You can include your complete protocol (detailed list of what the researchers said and did) in the appendix of your proposal. Alternatively, if your procedure is very similar to a published study's, you can refer the reader who wants detailed information to that study.

Refining Your Method Section

Realize that your Method section should help any reader know what participants will do, let the critical reader know the strengths of your study, and let the highly motivated reader replicate your study (Cooper, 2011). To achieve this level of transparency, Method sections may eventually become videos. For now, they must be as clear as possible—especially their Procedure sections.

The first step to making your Procedure section clear is to make it well-organized. How do you make it well-organized? One important step is to organize it around what will happen to the participant. Describe the first thing the participant will experience, then the second thing, and so on. You can start from the moment they enter the lab ("The participants will be greeted with…") and end when they leave ("Participants will be debriefed and dismissed.").

Your first attempt to organize your Procedure section around what will happen to the participant may suffer from one of two problems. First, it may contain details that interrupt its flow. Second, it may lack important details.

If the problem is that the details interrupt the flow, remove those details from your Procedure section. But what do you do with those flow-destroying details once they are out of your Procedure section? Depending on how relevant those details are, you can remove them from your paper, move them to an Appendix, or move them to another part of the Method section.

Delete those details that nobody needs to know. For those details that only readers who will replicate your study need to know, refer the reader to another source of those details. For example, if the Method section of a published article describes those details, refer the reader to that article. If no published article contains those details, put those details in an appendix (e.g., Appendix A: Instructions to Participants) and refer to the reader to that appendix.

If the details that bog down your Procedure section are details that most readers need to know, include those details in the Method section in a subsection that comes before your Procedure section. That way, you can get those details out of the way before you describe how events unfolded for the participant. Thus, before the Procedure section, you might describe the Apparatus (in an Apparatus section), the measures (in a Measures section), the independent variable (in an Independent Variable section), or the steps you will take to improve your study's construct validity (in a "Training" or a "Methodological Improvements") subsection. When writing these subsections, you should have two goals: (1) to help the reader understand what you will do and (2) to help the reader understand how what you will do will improve accuracy or reduce bias.

After you think you have a well-organized Method section, see if you can outline it. If you can't make a satisfactory outline of your Method section, rewrite your Method section until you can outline it. Then, rewrite it again, this time focusing on trying to make each sentence as clear as possible. Then, compare your Method section to the "Method Checklist" in Appendix A.

Once your Method section passes all the checks in Appendix A's "Method Checklist," you probably have a good, clear recipe for your study. However, just as the proof of a good recipe is whether people can use it to make good tasting food, the proof a clear Method section is whether a friend who is unfamiliar with psychological research can read it and then tell you how the study will be conducted. So, find a friend to see how clear your recipe is—and what you can do to make it clearer.

Writing the Results Section

In a proposal, you may not have a results section. After all, the study hasn't been done, so there are no results to report. Thus, your professor may advise you to replace the results section with a "Design and Data Analysis" section or even eliminate the section entirely.

If you will have a results section or a data analysis section, your main goal is to show that you would know how to code and analyze participants' responses. As was the case with the Method section, your goal is not only to tell the reader what you are going to do, but also to sell the reader on the idea that you are doing the right thing. Thus, it is important to be clear about not only what analysis you are going to do, but why. Ideally, your proposal should answer four questions:

1. What data will be analyzed? That is, how will a participant's response be converted into a score?
2. What statistical test will be used on those scores?
3. Why can that statistical test be used? Thus, you might show that the data meet the assumptions of the statistical test or you might cite a text or article that supports the use of the test under these conditions.
4. Why should the analysis be done? Usually, you will remind the reader of the hypothesis you want to test. To emphasize the value of the analysis, you may want to describe what results of that analysis would support your hypothesis and what outcomes would not. You might even—with

your professor's permission—plug in imaginary outcomes of your study to give the reader a concrete example of how your proposed analyses will help test your hypotheses.

To illustrate how a results section might accomplish these goals, study the following sample results section:

Results

I will sum participants' responses to the two 5-point altruism items to come up with an altruism score for each participant that will range from 2 (*very low*) to 10 (*very high*). Those scores will be subjected to a 2×2 between-subjects analysis of variance.

I hypothesize that mood will affect altruism. If the results turn out as I predict, positive-mood participants will score significantly higher on my altruism scale than negative-mood participants. This significant main effect would indicate support for the hypothesis that mood influences altruism. Furthermore, I also predict that arousal will amplify mood's effect. Therefore, I expect a significant mood by arousal interaction. Analyses of simple main effects will show that negative-mood participants who are in the high-arousal condition will score lower on altruism than negative-mood participants in the low-arousal condition. In contrast, positive-mood participants in the high-arousal condition will score higher on altruism than positive-mood participants in the low-arousal condition.

As you can see, the results section shows the reader what data will be put into the analysis, what analyses will be done, and what results from the analyses will support the hypotheses.

Writing the Discussion Section

Once you have decided how you will analyze your data, you are ready to discuss how you will interpret them. By referring back to both the literature you discussed in the Introduction and the arguments you made there, you should be able to address two key questions:

1. What would be the implications for interpreting existing theory and research if your hypothesis is supported?
2. What would be the implications if the results don't support your hypothesis?

In addition to addressing these two key questions, the **Discussion** is the place to present the limitations of your study, to speculate about what research should be done to follow up on your study, and to discuss the implications of your study.

Writing this section is difficult because you do not know how the study will turn out. Probably the easiest thing to do is to imagine that your study turned out as you expected. In that case, your discussion can be primarily a rehash of the Introduction.

To be more specific, your Discussion should probably devote a paragraph to at least four of the following six points:

1. relating the predicted results to the hypothesis ("Consistent with my predictions, …")

2. relating the predicted results to previous research and theory discussed in the Introduction ("This study joins others in showing ..." or "The findings are consistent with _____ theory.")
3. discussing the limitations of the study ("However, because the current research is only correlational, one cannot say that the variables are causally related" or "The results may not generalize to noncollege students.")
4. discussing future research that would build on the present study ("Future research might consider testing the generality of this effect.")
5. discussing practical or theoretical implications of the research findings
6. summarizing the importance of remembering or building on the study's major findings ("To summarize, I found that the effectiveness of rewards depended on the participant's personality. This finding suggests that teachers should not use salient rewards on intrinsically motivated students. Furthermore, in light of these findings, learned industriousness theory must be revised. In short, this research takes a step toward better understanding creativity.")

Almost all authors devote the first paragraph of the discussion to the first of these six points (relating the results to the hypothesis). In fact, you will often see that the discussion section's first sentence has both the words *support* and *hypothesis* in it. Similarly, almost all authors devote the second paragraph to the second of these points: relating the predicted results to previous research and theory discussed in the Introduction. However, authors vary in how much they discuss the remaining four points.

Sometimes, authors use subheadings to signal which of the remaining four points they will highlight. Thus, if you browse discussion sections, you will see subheadings such as "Comparisons With Previous Research," "Limitations of the Current Study," "Suggestions for Future Research," "Implications," and "Concluding Remarks."

Do you need to have a separate paragraph for all six of these paragraphs? No. Usually, the content and organization of your discussion will be fine as long as you

1. outline it,
2. connect your Discussion to your Introduction, and
3. explain how your study will contribute to existing knowledge.

Putting on the Front and Back

You've written the Introduction, Method, Results, and Discussion sections.

Now it's time to return to the beginning of your proposal. Specifically, it's time to type the title page and the abstract.

Title and Title Page

The title is the first thing readers will see; therefore, it should be simple, direct, and informative. Ideally, your title should be a brief statement about the relationship between your predictor/independent and criterion/dependent variables. If you did not do an experiment, your title should not suggest that you determined the cause of some effect. Thus, if you did a survey or an observational study, your title should not include words that are synonyms for cause, such as "influences," "increases," "decreases," and "affects."

Avoid using a title that is too cute or obscure. If there is some catchy saying that you must include, use a colon and add a subtitle (e.g., "The Effect of Eating Sugar on Anxiety: A Bittersweet Dilemma").

The title should appear centered on a separate piece of paper. One double-spaced line below the title, center your name. One double-spaced line below your name, center your school's name.

Below your school's name, begin the Author Note with the centered heading: Author Note. In the first paragraph, put your name followed by a comma, then your school's name followed by a period. In the next paragraph, thank anyone who helped you with the study or with the paper. In the final paragraph, tell readers where to write you.

To write an Author Note, follow the example in the sample paper (Appendix B) and check your Author Note against the Author Note checklist on page 638.

For more information about typing and formatting the title page, see the title page checklist in Appendix A.

Abstract

Once readers have read your title, they will continue to the next section—the **Abstract**: a short, one-paragraph summary of your research proposal.

According to Jolley, Murray, and Keller (1992), most abstracts one sentence for each of the following topics

1. the general research topic (For example, "Love is a common topic in popular music.")
2. the number of participants and their treatment
3. how the dependent measure will be collected ("Participants will fill out the Reuben Love–Like Scale 10 min after receiving the treatment.")
4. the hypothesis ("The hypothesis is that listening to love ballads will raise scores on the Love–Like Scale.")
5. the main results—those that relate to your hypothesis (Include this in your final research report, not in the proposal.)
6. the implications of your results (a sentence or two at the end; this miniature version of your discussion section might read something like, "The findings call into question _____ theory's assumption that _____" or "The results suggest that future research needs to address whether _____ interacts with _____.")

Once you have a draft of your Abstract, compare it against the Abstract checklist in Appendix A. Then, revise it so that it meets the 12 key criteria listed in that checklist.

References

Now that you have the title and Abstract written, it's time to compile your Reference list. At this point, be sure to (a) include all your references and (b) put them in the right order.

To be sure that you have included all your references, start by comparing your notes and your sources to your paper to make sure that you have cited every source that you got an idea from or quoted from. Then, be sure that every source that you cited in your paper is listed in your references.

To put your sources in the right order, start by putting them in alphabetical order according to the main author's last name. Thus, a source written by Ambady would come before a source written by Zimbardo. If you have more than one work by the same author, put the oldest source first and the most recent last. Thus, if you cited a 2000, a 2006, and a 2012 article by Williams, you would list the 2000 article first, the 2006 article next, and the 2012 article last.

Once you have your references organized, you need to format them in APA style. The easiest way to do that is to use the reference section of the sample paper (Appendix B) as a model. Thus, if you need to write the reference for a journal article with two authors, follow the example of the sample paper's first reference.

One problem with using any model is that you may think you are doing what the model is doing when you aren't. To make sure that you are following the model in the sample paper, check your references page against the reference checklist in Appendix A. (For more specifics about how to reference, see this chapter's website.)

WRITING THE RESEARCH REPORT

If you wrote a research proposal, much of the work on your research report already has been done. Essentially, your research proposal was the first draft of your research report. The next few pages will help you convert that first draft into a polished, complete research report.

What Stays the Same or Changes Very Little

The title page and references from your proposal can be transferred to your research report without any changes. However, you will need to make minor changes to the Introduction, Method, and Abstract.

You need to revise your Introduction so that you describe what you did (past tense) rather than what you plan to do (future tense). For example, you no longer plan to test the hypothesis that X causes Y; instead, you tested the hypothesis that X caused Y.

Whereas you had to make only one change to the Introduction, you will need to make three changes in the Method section. First, you will need to change the Method section to reflect any changes you made in how you conducted the study. Usually, the procedures you initially proposed are not the ones you end up following. Sometimes, after reading your proposal, your professor will ask you to make some modifications. Sometimes an ethics committee may mandate some changes. Often, after testing out your procedures on a few participants, you will make some changes so that the actual study will run more smoothly.

Second, you probably will have to make some minor changes in the participants section. For example, prior to running the study, you can rarely anticipate the standard deviation and age of your participants, the exact ethnic composition of your participants, and the number of participants you will need to exclude for not following directions.

Third, you need to rewrite the Method section in the past tense. In the proposal, you told readers what you were going to do; in the report, you tell them what you did.

Like your Method section, your Abstract needs only minor modifications. Specifically, you need to add a sentence to describe the main results, a sentence to discuss the implications of those results, and you need to check that the participant and procedure information are still accurate.

Unlike the minor changes you made to the Abstract, Introduction, and Method, you will have to make major changes to the Results and Discussion sections before you can include them in your final report. Because these two sections change the most from proposal to final report, the rest of this chapter will be devoted to those two sections.

Writing the Results Section

There are two main purposes of the **Results section**: (a) to show the reader that you competently analyzed the data, and (b) to tell the reader what you found. To accomplish these goals, you will report from one to five kinds of results:

1. results describing the distribution of participants' scores,
2. results supporting the validity of your measure,
3. results of the manipulation check,
4. results relating to your hypothesis, and
5. other statistically significant results.

Results Describing the Distribution of Scores

At the beginning of your Results section, you might include a subsection that describes the distribution of scores on your dependent variable. Thus, you might give the mean and the standard deviation (or range) of scores.[4] For example, you might report, "The scores on the measure were normally distributed ($M = 75$, range $= 50 - 100$)."

Most authors do not include a section that describes the distribution of scores. If you do include such a section, it probably will be for one of the following five reasons:

1. to make a case that the sample is representative of some population by showing that the scores are very similar to the population's distribution of scores;
2. to argue that your data meet the assumptions of the statistical test you used (e.g., if you were doing a t test, you might show that the scores were normally distributed and that the different groups had similar variances);
3. to show that your data had to be transformed or that the data could not be analyzed by a certain statistical test because your data were not normally distributed;
4. to argue that there should be no problems due to ceiling effects, floor effects, or restriction of range because there was a wide range of scores and those scores were normally distributed; or

---

[4] If the data are *not* normally distributed, you may want to provide a graph of the raw scores.

5. to emphasize descriptive statistics that are of interest in their own right, as would be the case if reporting the percentage of the sample who had married before the age of 20.

Results Supporting the Measure's Validity

Like the section describing the distribution of scores, the section supporting the measure's validity often is omitted. If you are using an accepted, validated measure, you probably will omit this section. If you choose to include this section, you probably will stress the results that emphasize the measure's

1. test–retest reliability, indicating that the measure is not unduly influenced by random error, as shown by participants getting the same score from one day to the next (for more on test–retest reliability, see Chapter 5)
2. interobserver reliability, indicating that the measure is objectively scored, as shown by different observers giving the participants the same scores (for more on interobserver reliability, see Chapter 5)
3. internal consistency, indicating that the items of a test or subscale are all measuring the same thing, as shown by people who score high on a characteristic according to one question on the test also scoring high on that characteristic according to other questions on the test (for more on internal consistency, see Chapter 5)

Results of the Manipulation Check

If you used a manipulation check, you should put these findings near the beginning of the results section. Although these results usually will be statistically significant and unsurprising, it is important to demonstrate that you manipulated what you said you would manipulate. Reporting the outcome of your manipulation check is also a good lead into discussing results relating to your hypothesis: Once you have shown the reader that you manipulated the variable you planned to manipulate, the reader is ready to know whether that variable produced the effects you expected.

Results Relating to Your Hypothesis

Your results section does not have to describe the distribution of scores, provide evidence for the validity of the measure, or describe the results of a manipulation check. However, it must describe the results relating to the hypothesis. In writing the results section, your main goal should always be to make it very easy for your readers to know how the hypothesis did.

To make it easy for readers to know how the hypothesis fared, tell your readers what the hypothesis was and whether it was supported. Then, use descriptive statistics (usually averages like means) and the results of your statistical test to link the results to the hypothesis. For example, if your hypothesis was that people who own cats are more likely to hug their children, report what the data said about this hypothesis: "The hypothesis that people who own cats would be significantly more likely to hug their children was supported. Cat owners hugged their children on the average 4.6 ($SD = 1.6$) times a day per child, whereas people who did not own cats hugged their children on the average 2.3 ($SD = 1.5$) times a day, $F(1,64) = 18.2, p < .001$."

Other Significant Results

After reporting results relating to your hypothesis (whether or not the results were significant), you should report any other statistically significant results. Even if the results are unwanted and make no sense to you, significant results must be reported. Therefore you might report: "There was an unanticipated relationship between gender of the child and cat ownership. Parents of girls owned more cats ($M = 2.0$, $SD = 1.0$) than parents of boys ($M = 1.0$, $SD = 1.8$), $F(1,64) = 20.1$, $p < .001$."

Four Tips That Will Help You Write the Results Section

Although you now know what a Results section is, you may not believe that you know enough to write a decent one. To write a good Results section, you need to realize that its goal is to help the reader understand what you found. To help your reader understand what you found, we offer four tips: (1) start off simply, (2) explain what you are doing, (3) use means or other summary statistics to make the pattern of your results more concrete, and (4) focus on your hypothesis.

Tip 1: Start Off Simply.   Sometimes, beginning writers lose their audience at the very beginning of the Results section. To avoid that problem, start out simply and slowly. You might begin the section by just explaining what the scores meant. Your goal would be to give the reader some sense as to what a participant who had a low score did differently than a person getting a high score.

What if the meaning of the scores is too obvious? Or, what if you explained how scores were computed in the Method section? Then, you might start with a simple analysis. For example, you might discuss results relating to the degree to which the different coders coded the data similarly (e.g., "Raters agreed 98% of the time"). Or, you might discuss other results that should be predictable and easy to understand, such as the results of the manipulation check (e.g., "As predicted, the attractive ($M = 7.2$) pictures were rated as more attractive than the unattractive ($M = 2.1$) pictures, $t(28) = 81.2$, $p < .001$").

When discussing a set of related analyses, try the following two strategies. First, start off by discussing simple findings and then moving to more complex findings. For example, report relationships between two variables (e.g., main effects or simple correlations) before discussing how that relationship is moderated by a third variable (e.g., interactions or partial correlations). Second, discuss general findings before moving to more specific findings. Thus, you might first report that the treatment had an effect (e.g., by reporting the results of the overall F test) before talking about which particular groups differed from each other (e.g., by reporting the results of more specific follow-up tests).

Tip 2: Don't Report Results—Analyze Them.   Do not, however, merely report results. That is, do not, in effect, shove the results of the computer printout in the reader's face and say "Here, see if you can make sense of this!" Instead, follow our second tip, which is to give the reader your *analysis* of the results.

Your analysis will not include every statistic the computer generated. Instead, you will give the reader only those statistics that make a point.

In addition to giving the reader a statistic only if the statistic makes a point, you will tell the reader what the point is before the reader sees the statistic. The reader should not be left wondering, "Why is she giving me these numbers?" To help the reader, before presenting a statistic, you should introduce it by saying what you hoped to find out ("To test the hypothesis that ...") and what statistical test you used to find that out ("I used a between-subjects t test"). For example, if you are doing an analysis to see whether your manipulation check worked, you will let the reader know by writing a short paragraph like: "As a check on the attractiveness manipulation, I conducted a t test on participants' ratings of the pictures. As predicted, the attractive ($M = 7.2$) pictures were rated as more attractive than the unattractive ($M = 2.1$) pictures, $t(28) = 81.2$, $p < .001$." To help the reader even more, you might introduce that paragraph with the subheading: Attractiveness Manipulation Check.

Using subheadings like headlines to help the reader understand the purpose of the analysis is usually a good idea. However, even without subheadings, your results section should be readable. To make it more readable, have other people read your results section. If they do not understand it, rewrite until they can.

In rewriting your Results section, it may pay to forget about the numbers for awhile. Instead, just focus on (a) how you got the scores you put in the analysis, (b) what the analysis was does (e.g., "the independent t test is commonly used to determine whether the means of a treatment group and a control group are reliably different"), and (c) why you did the analysis (e.g., "To test the hypothesis that ..."). You might even defend your use of a particular test by pointing out (a) the assumptions of the test and (b) that your data appear to meet those assumptions. Thus, if you wrote a "Results" section for your research proposal, much of that material should be useful for your final report. After all, in the research proposal, you explained and justified how you will analyze your data; in the research report, you will explain and justify how you analyzed your data.

Once it is clear what analysis you did and why you did that analysis, then you can focus on telling the reader what you found. However, as you begin to tell the reader what you found, you don't need to give the reader any numbers. For example, you might start by saying whether the results supported the hypothesis.

Tip 3: Use Summary Statistics to Make the Findings More Concrete. In addition to telling the reader the general purpose of the analysis (e.g., "to test the hypothesis that ...") and the general outcome of the analysis (e.g., "In support of the hypothesis, the experimental group scored significantly higher than the control group."), present specific summary statistics that back up your general statements. By supplementing the general, abstract information (e.g., one group scored significantly higher than the other group) with specific, concrete information (the actual means), you go beyond giving the reader a general idea of what you found to letting the reader actually see what you found. So, make your Results section easier to understand by supplementing the results of statistical tests with the relevant means, frequencies, correlations, or other summary statistics.

Tip 4: Focus on the Hypothesis. The fourth, and most important tip, is to ask the question, "After reading the results section, will the reader know whether the results supported the hypothesis?" One way to determine whether you have achieved this goal is to ask a friend to read your results section. See whether your friend can answer these five questions:

1. What was the hypothesis?
2. Was the hypothesis supported?
3. What statistical test was used to find this out?
4. What were the results of that test (value of the statistic and the **probability value**, or p **value**: the chances of obtaining this pattern of results if only chance were at work)?
5. Did the averages (or some other summary statistic, such as percentages) for the different conditions help you understand whether the prediction was supported? If not, would a table or graph make things clearer?

By focusing on helping the reader understand whether the results supported the hypothesis, you will end up doing many of the things that we just suggested. You will leave out information that is irrelevant and distracting. You will include all information that helps the reader understand the results section such as what the scores represent, why the analysis is being done, and what the analysis shows.

Conclusions About Writing the Results Section

In short, you should try to make your Results section as clear and understandable as possible. If you focus your Results section on your hypothesis, have empathy for your reader, and use the "Results" checklist in Appendix A, you should be able to write an understandable and useful Results section.

Writing the Discussion Section

If the results matched your predictions, the Discussion section you wrote for your proposal might work as the Discussion section for the final report. However, there are two reasons why you will have to modify it. First, you will probably not get exactly the results you expected. Second, during the course of conducting the research or writing the paper, you probably will think of problems or implications that you did not think of when you wrote your proposal.

As you revise the Discussion section, realize that although you are making a case, you should argue like an impartial judge who has come to certain conclusions after carefully weighing all the evidence rather than like a crusading attorney who is trying to prove a point. In writing your discussion, be sure to take the following seven steps:

1. Briefly review the research hypothesis.
2. Briefly highlight the results as they relate to your hypothesis—without using numbers.

3. Interpret the results in light of the arguments made in your Introduction.
4. Acknowledge alternative explanations for your results, trying to dismiss these alternatives, if possible.
5. Discuss unexpected findings, and speculate on possible reasons for them.
6. Discuss, in general terms, future research. What would you do if you were to follow up on this research (Assume you had an unlimited budget)? Follow-up research might focus on improving the methodology of your study, exploring unexpected findings, trying to rule out alternative explanations for your findings, testing the generality of your findings, looking for practical implications of the findings, looking for variables that might have similar effects, or looking for mental or physiological factors that mediate the observed relationship.
7. Discuss the practical or theoretical implications of your findings.

Once you have written your Discussion section, you should have a first draft of your paper. However, you will need to write several drafts before you have a paper that meets APA's requirements for style and format. To meet APA standards, your paper must be clear and well organized. It must be free of grammatical errors, spelling errors, biased language, wordiness, and informal language. Fortunately, your computer's spelling and grammar checker can help you catch and fix spelling errors, typographical errors, and grammatical errors, as well as problems due to using sexist or overly informal language (for specific advice on how to use your computer to edit your paper, see the Chapter 15 section of this text's student website). To edit your paper so that it conforms to APA format, check your "next-to-final draft" against the checklist in Appendix A. In addition, *make sure that your paper matches the format of the model paper in Appendix B.*

CONCLUDING REMARKS

If you carefully followed the advice in this book, you should have just completed a carefully planned, meaningful, and ethical research project. Congratulations—and best wishes for your continued success as a researcher!

SUMMARY

1. The research journal and proposal will help you plan and conduct ethical and valid research.
2. The research proposal is more formal than the research journal and should conform to APA style.
3. In the Introduction of your proposal, you need to summarize and critique relevant research.
4. In the Introduction, begin by connecting the general research area to the reader's experience. Then, review the research that relates to your study and your hypothesis.

Realize that your review of the literature should set up (a) why your hypothesis makes sense, (b) why your hypothesis should be tested, and (c) why hypothesis should be tested the way you are going to test it. Finally, state your hypothesis.

5. Before writing the Method section, you should carefully plan out each step of your study.
6. Once you have planned out every detail of your study, you should formalize your plan in the Method, Results, and Discussion sections in your proposal.

7. The Method section is the "how" section in which you explain how you plan to conduct your study and why you are going to do it that way.

8. In the proposal's Results section, you will discuss how you plan to analyze your results.

9. In the Discussion section, you will explore the implications of your anticipated research findings for theory, future research, or real life.

10. Once you finish the body of the proposal, write the Abstract (a brief summary of the proposal), the title page, and the reference section. Much of your final report will be based on your proposal—provided you wrote a good proposal.

11. The title page and reference sections of your proposal can be transferred directly to your final report. After you change the appropriate parts of the Introduction to the past tense, it can be transferred to your final report. After you change the appropriate parts of the Method section to the past tense, it may also be transferred (with only minor modifications) to the final report.

12. Try to make the Results section as understandable as possible. Tell the reader what you are trying to find out by doing the analysis, and then explain what you actually did find out from doing the analysis.

13. In the Results section, be sure to stress whether the results supported or failed to support your hypothesis.

14. In the Discussion section, summarize the main findings of your study and relate these to the points you made in the Introduction.

15. Writing involves a great deal of rewriting.

16. Writing research proposals and reports involves writing very carefully. You must careful not to state an opinion as a fact, careful to back up your facts with citations, careful to cite both the author and the date of each work you use, careful to use quotation marks when you quote, careful to report all numbers accurately, and careful to use a formal, professional, and unbiased tone.

17. Do not plagiarize! Keep notes about what you read so that you can cite it. Realize that even if you didn't quote a source, you still have to cite it if you borrowed from it or got some ideas from it.

18. Whenever you write a research report, a lab report, or a research proposal, compare your paper to the checklist in Appendix A before you turn it in to your professor.

KEY TERMS

abstract *(p. 628)*
conceptual replication *(p. 617)*
direct (exact) replication
    (p. 615)
discussion *(p. 626)*

exploratory study
    (p. 613)
Introduction *(p. 608)*
Method section *(p. 620)*
plagiarism *(p. 622)*

probability value (*p* value)
    (p. 634)
research journal *(p. 606)*
results section *(p. 630)*
systematic replication *(p. 616)*

WEB RESOURCES

1. Go to the Chapter 15 section of the book's student website and
   a. Look over the concept map of the key terms.
   b. Test yourself on the key terms.
   c. Take the Chapter 15 Practice Quiz.

2. Get more tips on finding articles to cite in your paper by clicking on the "Literature Search" link.

Research Report and Proposal Checklist

TITLE PAGE

1. I have a separate title page.
2. Top of the page
   a. I put the following into the "header" of my document,
      1. "Running head:" followed, IN ALL CAPITAL LETTERS, by a two- to six-word phrase that, in fewer than 50 characters, describes my paper's topic.
      2. the number "1," indicating that it is page 1.
   b. When I printed out my title page, the phrase "Running head:" started on the left margin (about an inch from the left side of the page) and the number "1" was in the right hand corner (about an inch from the right side of the page).
   c. I did not capitalize the "h" in "Running head."
3. Middle (Top half) of page
   a. I centered the title.
   b. I capitalized the first letter of the title. I capitalized the first letter of the other words in the title that have four letters or more. I capitalized all words that had fewer than four letters—except for prepositions (e.g., "in," "of,"), articles (e.g., "a," "an," "the"), or conjunctions (e.g., "and," "but").
   c. I kept my title simple and to the point. I included the names of the key variables (in an experiment, the independent and dependent variables; in a correlational study, the predictor and criterion variables).
   d. If I did not do an experiment (e.g., I did a survey or an observational study), I did not use the word "cause" or any of its synonyms (e.g., makes, affects, leads to, influences, increases) in my title. Instead, I used words like "associated," "correlated," or "related."
   e. I did not say that a treatment "effects" the dependent variable. Instead, I said that the treatment "affects" the dependent variable.

  f. My name (first name, middle initial, and last name) is
   1. one double-spaced line below the title,
   2. centered, and
   3. not accompanied by the word "by."
  g. My school's name is
   1. one double-spaced line below my name and
   2. centered.
 4. Author Note
  a. On the bottom half of the page, I started my Author Note with the centered heading "Author Note."
  b. The rest of my Author Note is not centered.
  c. The first "paragraph" of my Author Note is indented but is not a sentence. Instead, it contains my name (including middle initial), a comma, my department (e.g., "Department of Psychology), a comma, and the name of my school.
  d. The second paragraph of my Author Note gives credit (and usually thanks) to people who helped me conduct my study, analyze my data, or write my paper. (If nobody helped you, omit this paragraph.)
  e. The final paragraph of my Author Note begins with "Address correspondence concerning this article to" followed by an appropriate mailing address. It ends with "E-mail:" followed by my e-mail address.

ABSTRACT

1. My Abstract is on its own separate page.
2. The running head (in all capital letters) is in the top left-hand corner.
3. I have the number "2," indicating that the Abstract is page 2, at the top right-hand corner.
4. The heading "Abstract" is centered at the top of page 2.
5. The text of my Abstract starts one double-spaced line below the heading ("Abstract").
6. My Abstract is a single, un-indented paragraph, and it contains fewer than 121 words.
7. My Abstract, like the rest of my paper, is double-spaced.
8. To keep my Abstract as brief as possible, I used digits rather than writing out numbers.
9. I avoided starting any of my sentences with a number.
10. I avoided using the first person (e.g., "I," "my," "our," or "we").
11. I included a brief summary of the following sections of my paper:
  a. the Introduction—my hypothesis (what I studied and why),
  b. the Participants section—who the participants were,
  c. the Procedure—what the participants did,
  d. the Results—whether the data supported the hypothesis, and
  e. the Discussion—the meaning of the results.
12. Starting one double-spaced line below the Abstract and indented five spaces, I have "*Keywords:*" (in italics) followed by terms that (a) relate to the main variables of my study and (b) are in PsycINFO's "Thesaurus of Psychological Terms."

CITING SOURCES

1. I gave credit where credit was due.
   a. I cited any source from which I got ideas—even if I did not quote that source. When I summarized or paraphrased from a source, I cited that source. To minimize the chances that I plagiarized, I did the following:
      i. If I had any paragraph without a citation in the Introduction or Discussion sections of my paper, I went back to my notes to make sure that I had not left out a citation.
      ii. If I had any doubt about whether to cite a source, I cited it.
   b. If I obtained information from a secondary source, I cited and referenced the secondary source.
2. I quoted appropriately.
   a. I listed the page number of the source from which I got the quote.
   b. I put quotation marks around quotes shorter than 40 words.
   c. For quotes of 40 or more words, I separated the quote from the rest of my paper by indenting the whole quotation five spaces from the left margin.
3. My citations are free of common content errors.
   a. When citing authors, I limited myself to stating authors' last names. I did not mention authors' first names, professional titles (e.g., "Dr."), or professional affiliations.
   b. When citing sources, I used parentheses. I did not use footnotes to cite sources.
   c. I did not mention any article titles in the text of my paper.
4. I followed the rules regarding parentheses.
   a. When I mentioned the authors in the sentence, I put only the date in parentheses: "Jolley and Mitchell (2012) argued that ...".
   b. If the authors' names are not part of the sentence, I put their names and the date in parentheses. I separated the last author's name from the date with a comma: "Some have argued that ... (Jolley & Mitchell, 2012)."
   c. If the multiple-author citation was part of the sentence, I used "and" to connect authors' last names; however, if the multiple-author citation was in parentheses, I used "&" to connect authors' last names.
   d. When I cited several articles within one set of parentheses, I did the following:
      i. I listed the articles in alphabetical order. I did not put them in order by date.
      ii. I separated the articles from each other with semicolons: "(Brickner, 1980; Jolley, 2009; Mitchell, 2010; Ostrom, 1965; Pusateri, 2005; Williams, 2012)."
5. I correctly cited multiple-author papers.
   a. If the paper has more than six authors, I listed only the first author's last name followed immediately (with no comma) by "et al." (e.g., Glick et al., 2012).
   b. If I discussed a paper with three to five authors, I mentioned all the authors' last names the first time I cited that paper.

      c. If I discussed a paper with three to five authors and had already cited the paper, I used the first author's last name, followed immediately (with no comma) by et al. (e.g., First et al., 2011).

      d. I checked all my citations that used the phrase "et al." to make sure that I had (a) correctly used such citations and (b) correctly punctuated such citations.

         i. When citing multiple-author papers, I did not overuse "et al." citations.

            1. I never used "et al." the first time I introduced a paper with fewer than six authors.

            2. I never used "et al." with a two-author paper.

        ii. I correctly punctuated my "et al." citations.

            1. I never put a period after "et" (e.g., "et al.").

            2. I never put a comma between the first author's last name and "et al" (e.g., "First et al.").

6. I cited the appropriate literature.

      a. It is clear what the articles cited have to do with my study.

      b. Most of my citations are to recent journal articles describing actual research studies. Few, if any, of my citations are to secondhand sources such as textbooks, magazines, and newspapers.

◼ INTRODUCTION

1. My Introduction begins on a separate page (page 3).
2. My Introduction is not titled "Introduction." Instead, the heading for my article is the article's title. Thus, my article's title is centered at the top of the first page of the Introduction.
3. To be sure that my Introduction was organized, I outlined it before writing it.
4. It is clear why my topic area is important.
5. It is clear why testing my hypothesis is important: I showed how my study builds on previous work or fills a gap in previous work.
6. It is clear why I believe that the hypothesis might be true. To make the logic behind my hypothesis clear, I explained relevant concepts and theories.
7. My hypothesis is clearly stated.
8. A reader should be able to foresee much of the rest of the paper (especially the essence of the Method section) after reading my Introduction.
9. I checked my Introduction against the "Citing Sources" checklist.
10. I checked my Introduction against the "General Rules" checklist.

◼ METHOD

1. I put the **boldfaced** and centered heading "**Method**"—not "Methods"— one double-spaced line below the last line of the Introduction.
2. I wrote the Method section of my research report in the past tense (The Method section of a research proposal would be written in the future tense).

3. I divided the Method section into at least two subsections (Participants and Procedure).
4. I put the boldfaced heading "**Participants**" flush against the left margin, and I put that heading one double-spaced line below the heading "Method." (I did not use the word "subjects" if I studied humans.)
5. I indented the text for the Participants section, and I began that text on the next double-spaced line after the "**Participants**" heading.
6. If I started a sentence with a number, I spelled out the number (for example, "Twenty undergraduates were participants"). Otherwise, if the number was 10 or more, I wrote it as a number. (So, I wrote: "Participants were 20 undergraduates," but I did not write: "20 undergraduates were participants.")
7. In my Participants section, I have been specific about
   a. how I selected or recruited my sample of participants or nonhuman animal subjects. (I did not say participants were randomly chosen unless I used specific procedures that guaranteed that participants were selected through some form of random sampling.)
   b. how participants were compensated (if they were compensated).
   c. major demographic characteristics, such as
      i. number of participants of each gender,
      ii. age of participants (average age and either the standard deviation or the range of ages), and
      iii. ethnic composition (e.g., 70% White, 10% Black, 10% Hispanic, 5% Asian, 5% multiracial).
   d. the number of participants who dropped out of the study,
   e. the number of participants whose data were not analyzed and the reasons for not analyzing those data, and
   f. how participants were assigned to condition.
8. I made it clear that participants were treated ethically.
9. I put the boldfaced heading "**Procedure**" (a) one double-spaced line after the last line of the previous subsection and (b) flush against the left margin.
10. I indented the text for the Procedure section five spaces, and I started it one double-spaced line below the "Procedure" heading.
11. If I used equipment that could usually only be obtained from a company that specialized in laboratory equipment—or if using different equipment would lead to different results—I identified the manufacturer and model name or number.
12. I used complete sentences and paragraphs. For example, I did not merely provide a bulleted list of my materials or an outline of my operational definitions.
13. I focused on what happened to participants and what participants did—and presented the information in order from the first thing that happened to the last thing that happened. Thus, many more of my sentences started with "the participants" than with "the experimenter."
14. My Procedure section includes information about procedures—but not about the participants' characteristics (information about the participants is in the Participants section).

15. My Procedure section is like a good, clear recipe in that someone reading my report could replicate (repeat) my study.
    a. It is clear how (under what conditions) I tested each participant.
    b. It is clear how I turned each participant's response into a score.
    c. It is clear what the operational definitions of my key variables are. Thus, in any study, the reader would, by reading my Method section (and any papers or Appendixes I referred to in that section), know how to measure the dependent variable. In addition, if I did an experiment, the reader would know how the different conditions differed from each other. Thus, if I did an experiment that had a control group, I clearly described what happened to that group—and especially how it different from the experimental group(s).
    d. If I used an apparatus that people might be unfamiliar with, I described that apparatus. If possible, I also included a diagram of that apparatus or a citation to a source that had more information about that apparatus.
    e. I had someone unfamiliar with my study read my Method section and then had that person play the role of experimenter while I played the role of participant.
16. I made a case for the study's validity.
    a. It is clear what I did to reduce the effects of researcher bias.
    b. It is clear how a control or comparison group ruled out an alternative explanation for a difference between groups.
    c. It is clear that the measure being used is reliable and valid.
       i. I cited evidence of the measure's reliability and validity.
       ii. I reported any data that I, as part of conducting the study, collected that related to the measure's validity. For example, if I had data related to the extent to which two different observers' scores agreed when scoring the same response, I reported those data.
    d. I gave my measure(s) and manipulation(s) names that are closely tied to my operational definitions (e.g., "Introversion Test Score") rather than using a general name that may not be valid (e.g., "shyness").
    e. I made it clear why my study is a good way to test my hypothesis.
17. Usage
    a. When referring to people, I used
       i. "who" instead of "that." Thus, I wrote "participants who responded" rather than "participants that responded."
       ii. "participants" instead of "subjects."
    b. I did not use the word "random" when I merely meant "haphazard," "arbitrarily," or "unsystematically" (e.g., I did not say "We used a random sample of people who were walking near our classroom" when I meant we observed anyone who happened to walk by).
    c. I did not use random sampling when I meant random assignment. (In experiments, participants are randomly assigned to condition, but participants are usually not a random sample of a population. In non-experimental studies, such as surveys and field observational studies, participants are not randomly assigned to condition, but may some-times be random samples of a population.)

d. I used the metric system. For example, instead of describing the length of a stimulus in inches, I described its length in centimeters; instead of describing the weight of a stimulus in pounds, I described its weight in kilograms.

18. I did not use the word "experiment" when I meant "study": All experiments are studies, but very few studies are experiments. For example, surveys, polls, and observational studies are not experiments.

19. I did not discuss
   a. my hypothesis.
   b. behind the scenes details that would not be necessary for replicating the study (e.g., "We used a Macintosh computer and Microsoft Word to make the coding sheets," "A team of us used the random numbers table to randomly assign participants to groups. We started in column 1....").

RESULTS

1. I centered and **boldfaced** the title "**Results**" one double-spaced line after the last line of the Method section. (I did not skip to a new page to begin the Results section.)

2. I wrote the Results section of my research report in the past tense.

3. I did not overwhelm the reader with a bunch of numbers. Instead, I gently introduced the numbers by
   a. letting the reader know what the scores meant: how behaviors were turned into scores and how a high score differed from a low score (e.g., "To compute each participant's accuracy score, ... Thus, the higher the score, the less accurate the participant was."), unless (a) I had already made the scoring clear in the Method section or (b) the scoring system was obvious.
   b. letting the reader know why I was going to discuss a certain analysis (e.g., "To test the hypothesis that ...") followed immediately by letting the reader know what that certain analysis was (e.g., "... I performed a t test").
   c. describing the results of that analysis by using both (a) summary statistics (usually averages) for the relevant groups (so that the reader could see which groups scored higher and would know more than whether the results were significant) and (b) p values (so that the reader could know whether these differences were significant).
   d. presenting the summary statistics (usually means and standard deviations, totals, or percentages) in the way that made it easiest for the reader to see the pattern in my data. Usually, this meant I presented these summary statistics in text (e.g., "The mean for the experimental group ($M = 4.6$, $SD = 1.1$) was significantly higher than the mean for the control group ($M = 3.9$, $SD = 1.3$"). However, I used a table or a graph if (a) I had more than four conditions or (b) my professor required one.

e. Using a comma to append the key details related to the statistical test to the end of the sentence in which I summarized the results of the test. For example, I wrote, "The mean for the experimental group ($M = 4.6$, $SD = 1.1$) was significantly higher than the mean for the control group ($M = 3.9$, $SD = 1.3$, $t (38) = 2.47$, $p = .018$" or "The results were statistically significant, $t (38) = 2.47$, $p = .018$."

f. Including the name or abbreviation of the statistic I calculated (e.g., "t"), the degrees of freedom for that statistic—in parentheses—e.g.,"(38)", an equals sign followed by the value of that statistic (e.g., "$= 2.47$"), and the p value resulting from the statistical test (e.g., "$p = .018$"). Thus, if I had done an F test with 1 and 24 degrees of freedom and obtained the statistically significant value of 4.84, after discussing the means, I added the following sentence: "The results were statistically significant, $F(1, 24) = 4.84$, $p = .038$." In other words, *except for the spacing,* my sentence would end with the statistical information presented in the form shown below:

Statistic	Degrees of freedom *(df)*	Numerical Value of the Test	Probability (of result being due to chance alone)
F	$(1, 46) =$	1.85,	$p = .18$
t	$(24) =$	2.0,	$p = .057$
χ^2	$(1) =$	4.1,	$p = .043$

For correlations, use the number of participants rather than the degrees of freedom. Thus, if you had 25 participants and a correlation of .5, you would write, "$r (25) = .5$, $p = 012$."

APA now prefers that you report the actual p value (e.g., $p = .04$) unless reporting that p is less than .001. Thus, if your p value was .0007, you would write $p < .001$. If you do not have access to a detailed t table, your professor may not require to you to provide the exact probability values. For example, if all you knew was the critical values for your test at the .05 level, you might be allowed to write "$p < .05$" (if the value of your statistic is greater than the critical value) or "$p > .05$" (if the value of your statistic is less than the critical value).

4. Because the purpose of writing up a study is to report what I found, I realize that misreporting what I found may lead to my failing the assignment—and to even more severe penalties. Therefore,

a. I double-checked to make sure that when I claimed to report summary statistics for a condition (e.g., the experimental group's mean), that those statistics belonged to that condition rather than to a different condition. For example, I did not say that the experimental group mean was 3.12 when that was the control group mean.

b. I double-checked to make sure that when I claimed to report a p value for a finding, I was reporting its p value rather than the statistic's value (e.g., its t value).

c. I triple-checked to make sure that the numbers I reported in my paper matched the numbers I had in my notes or in my computer print out.

5. I had someone unfamiliar with my research read this section and that person was able to tell what my hypothesis was and whether it was supported.

6. I italicized all the letters that represent statistical symbols and abbreviations for statistical variables (e.g., "p" as the abbreviation for probability value), except Greek letters (α, β).

Wrong: t (48) = 2.09, p = .042

Right: t (48) = 9.08, p = .042

7. If a result was not significant, I wrote that it was "not significant." I did not write that it was "insignificant" or "of no significance."
8. I remembered that the word "data" is plural for "datum." Thus, I wrote "data were" rather than "data was."
9. I used digits (e.g., 10) to express numbers rather than spelling out numbers (e.g., "ten")—unless I started a sentence with a number or if I used a phrase such as "two 4 × 4 designs."
10. I interpreted the Results only to the degree that I let the reader know what the statistical analyses said about my hypothesis. I did not discuss weaknesses in my study that might have affected the results—I left that for the Discussion.
11. I determined my professor's views about reporting analyses beyond that of significance tests. Thus, I knew whether my professor required me to follow the *Publication Manual of the American Psychological Association*'s suggestions that I
    a. include a confidence interval
    b. report an estimate of effect size, such as Cohen's d (e.g., "$F(2,46)$ = 3.85, p = .028, d = 0.27").
12. If I used tables, I made sure that
    a. each table added meaningful information beyond that which was presented in the text of my Results section.
    b. each table was referred to in the text of my paper (e.g., "As Table 1 indicates ...").
    c. each table's number corresponded to when I referred to it in text (e.g., the first table I mentioned in text was Table 1).
    d. I double-spaced everything in each table.
    e. I put each table on a separate page.
    f. I put all my tables located near the end of my report—after the References.
    g. My tables comply with the format illustrated by the following two tables.

TABLE 1

Pearson Product Moment Correlations for Self-Esteem

	Body Concept	
Group	**Attractiveness**	**Fitness**
Female	.65[*]	.50[**]
Male	.35[***]	.70[***]

[*]p < .05. [**]p < .01. [***]p < .001.

TABLE **2**

Analysis of Variance for Self-Esteem

Source	df	F
Exercise (E)	2	9.75**
Within-Group Error	57	(2.56)

Note: Value enclosed in parentheses represents mean square error.
**$p < .01$.

© Cengage Learning 2013

13. I referred to all graphs as figures ("Figure" 1, not "Graph" 1).
14. I gave each figure an informative heading.
15. I put each figure on a separate page.
16. I labeled both the x and y axes of each graph.

DISCUSSION

1. I centered and boldfaced the title "**Discussion**" and put it one double-spaced line after the last line of the Results section.
2. My first sentence describes whether the results support, fail to support, or refute my hypothesis, and my first paragraph summarizes my main findings.
3. I used the present tense when discussing my conclusions.
4. I compared my Discussion against the citing sources checklist.
5. I interpreted my results in the context of the theory, past research, and practical issues that I introduced in my Introduction. For example, I compared my results to what other researchers found.
6. I tried to explain results that were not predicted, and I admitted when my explanations were speculations.
7. I addressed alternative explanations for my results. I tried to rule out these alternative explanations, but when I could not, I admitted that I could not.
8. I pointed out the weaknesses and limitations of my study. I even sketched out future research that could be done to correct these weaknesses or overcome these limitations.
9. If I believed I could make a case for generalizing my results (I had a representative sample, the results were similar to what others had found, etc.), I made such a case.
10. I treated nonsignificant results as inconclusive.
    a. I did not use nonsignificant results as proof that a treatment had no effect.
    b. I did not use nonsignificant results that were almost significant as evidence that the treatment had an effect.

REFERENCES

1. I started my References section on a separate page. I indicated the start of that page by centering the word "References" at the top of the page. I did not boldface that heading.
2. Everything is double-spaced.

3. My references are listed in alphabetical order (according to the last name of the first author).

4. The first line of each reference is not indented. Instead, it is flush against the left margin.

5. When a reference took up more than one line, those additional lines were indented.

6. I started each reference with the authors' last names and initials, followed by the year of publication (in parentheses), and then a period.

7. I did not use the authors' first or middle names.

8. For all journal articles, I wrote down the volume number. (The volume number [e.g., 30] is different from the year [e.g., 2007]. The volume number can usually be found on the journal's cover, the journal's table of contents, and at the bottom of the first page of each article.)

9. For all recent journal articles, I wrote down the digital object identifier (DOI). The DOI is a long string of characters that usually starts with "10." If you look for "DOI:" on the article's first page, you will probably find the DOI. (For more on the DOI, go to this text's website.)

10. If my reference has a DOI, I ended the reference with "doi:" followed by the DOI (e.g., 10.117/0146167209347380).

11. Every reference—unless it ends with a URL or DOI—ends with a period.

12. If there is more than one author for a source, I separated authors' names with commas. (There is a comma after every author's name—except for the last author's name.)

13. I used italics correctly.
    a. I italicized the titles of all books.
    b. I italicized the titles of all journals.
    c. I italicized the volume numbers of every journal article cited.
    d. I put the titles of journal articles in normal, non-italicized type.

14. I correctly capitalized the names of books and titles of articles. That is, I capitalized only
    a. proper nouns (e.g., Asia, Skinner),
    b. the first word of titles of articles and books, and
    c. the first word following a colon in the title of an article or book.

15. I used abbreviations appropriately.
    a. When citing journal articles, I avoided both the word "pages" and the abbreviation "pp."
    b. When mentioning where a book was published, I abbreviated, rather than wrote out, the name of the state. I used the two-letter state abbreviations (e.g., PA for Pennsylvania) that the U.S. Postal Service uses. I remembered that these abbreviations are capitalized and do not contain periods.

16. All the references in this section are also cited in my paper. If a reference was not cited, I either added that citation to the body of my paper or I deleted the reference.

17. All the sources cited in my paper are also listed in this section except for
    a. "personal communication" citations,
    b. original works that I did not read but instead learned about through a secondary source, and
    c. classical works such as the Bible.

GENERAL RULES

1. I double-spaced everything.
2. I put the running head and the page number at the top of every page.
3. I started every paragraph by indenting five spaces with only two exceptions:
   a. I did not indent the Abstract.
   b. I did not indent Notes at the bottom of tables.
4. I did not use terms or labels that devalue, stereotype, exclude, or offend people belonging to groups other than my own (e.g., people older than I am, people with mental illnesses, people who have a different racial or ethnic background than I have). Instead, I used terms that were emotionally neutral or respectful.
5. I did not use sexist language (e.g., referring to males as "men" but referring to females as "girls," referring to the typical participant as "he" when most of your participants were women).
6. I spell-checked my document using my word processing function, and I manually double-checked for words that are spelled right, but used wrong (e.g., "their" for "there," "preformed" for "performed").
7. I used complete sentences (all of my sentences have subjects and verbs).
8. I was careful about not making statements that went beyond the evidence. Specifically, I
   a. reported what I observed (e.g., "the participant took 10 seconds before pressing the button") rather than what I inferred (e.g., "the participant hesitated and worried about potential embarrassment before pressing the button"),
   b. did not use the word "prove,"
   c. did not claim something was a fact when it was an opinion, and
   d. did not make cause–effect statements without having evidence from an experiment.
9. To help my paper sound professional, I
   a. outlined my paper to make sure it was organized;
   b. read my paper aloud;
   c. split up, shortened, or eliminated long sentences;
   d. eliminated unnecessary words and redundant sentences; and
   e. used a grammar checker to weed out wordy phrases, sexist language, and other unprofessional language.
10. My paper's appearance is professional. It
   a. is neatly typed,
   b. is free of typographical errors, and
   c. has at least 1-in. (2.54 cm) margins.
11. I did not hyphenate words at the end of a line.
12. I centered the following headings: Abstract, Method, Results, Discussion, and References. I capitalized only the first letters of words of these headings.
13. I centered and boldfaced the following headings: **Method, Results,** and **Discussion.**

14. I boldfaced major subheadings such as **Participants** and **Procedure**, and put them flush against the left margin (I did not indent them). I capitalized only the first letters of words of these subheadings.
15. I did not include anyone's first name or affiliation in my paper (except for putting my name and affiliation on the title page).
16. The order of the sections in my paper is as follows: title page, Abstract, Introduction, Method, Results, Discussion, References, Appendixes, tables, and figures.
17. I used white, 8.5 × 11 in (22 × 28 cm), 20-pound, bond, nonerasable paper.
18. Using the right words:
    a. When referring to people, I used "who" instead of "that." Thus, I wrote "participants who responded" rather than "participants that responded."
    b. I did not use "since" when I meant "because."
    c. I did not write "experiment" when I meant "study."
19. I typed my paper using
    a. 12-point type,
    b. Times New Roman font,
    c. black, easy-to-read print, and
    d. only one side of the paper.
20. I avoided using
    a. contractions,
    b. exclamation points,
    c. question marks,
    d. hyphens at the end of lines,
    e. headings at the bottom of a page,
    f. underlining, and
    g. one-sentence paragraphs.

Sample APA-Style Paper

Frank, M. G., & Gilovich, T. (1988). The dark side of self- and social perception: Black uniforms and aggression in professional sports. *Journal of Personality and Social Psychology, 54*(1), 74–85.

Running head: BLACK UNIFORMS AND AGGRESSION

1

The Dark Side of Self-Perception:

Black Uniforms and Aggression

Mark G. Frank and Thomas Gilovich

Cornell University

Author Note

Mark G. Frank, Department of Psychology, Cornell University; Thomas Gilovich, Department of Psychology, Cornell University.

We are grateful to Lauren Ostergren and Mark Schmuckler for their assistance in collecting our data and to Daryl Bem for commenting on an earlier version of the manuscript.

Address correspondence concerning this article to Mark Frank, Department of Communication, University of Buffalo, 359 Baldy Hall, Buffalo, NY 14260. E-mail: mfrank83@buffalo.edu

BLACK UNIFORMS AND AGGRESSION   2

Abstract

Black is viewed as the color of evil and death in virtually all cultures. With this association in mind, we were interested in whether a cue as subtle as the color of a person's clothing might have a significant impact on the wearer's behavior. To test this possibility, we performed a laboratory experiment to determine whether wearing a black uniform can increase a person's inclination to engage in aggressive behavior. We found that participants who wore black uniforms showed a marked increase in intended aggression relative to those wearing white uniforms. Our discussion focuses on the theoretical implications of these data for an understanding of the variable, or "situated," nature of the self.

Keywords: self, self-perception, aggressive behavior

The Dark Side of Self-Perception: Black Uniforms and Aggression

A convenient feature of the traditional American Western film was the ease with which the viewer could distinguish the good guys from the bad guys: The bad guys wore the black hats. Of course, film directors did not invent this connection between black and evil, but built upon an existing association that extends deep into American culture and language. Americans can be hurt by others by being "blacklisted," or "blackballed," or "blackmailed" (Williams, 1964). When the Chicago White Sox deliberately lost the 1919 World Series as part of a betting scheme, they became known as the Chicago Black Sox, and to this day the "dark" chapter in American sports history is known as the Black Sox Scandal. In a similar vein, Muhammad Ali has observed that Americans refer to white cake as "angel food cake" and dark cake as "devil's food cake."

These anecdotes concerning people's negative associations to the color black are consistent with the research literature on color meanings. In one representative experiment, groups of college students and seventh graders who were asked to make semantic differential rating of colors were found to associate black with evil, death, and badness (Williams & McMurty, 1970). Moreover, this association between black and evil is not strictly an American or Western phenomenon because college students in Germany, Denmark, Hong Kong, and India (Williams, Moreland, & Underwood, 1970) and Ndembu tribesmen in Central Africa (Turner, 1967) all report that the color black connoted evil and death. Thus, Adams and Osgood (1973) concluded that black is seen, in virtually all cultures, as the color of evil and death.

The intriguing question is whether these associations influence people's behavior in important ways. For example, does wearing black clothing lead the wearer to actually act more aggressively?

This possibility is suggested by studies on anonymity and "deindividuation" which show that a person's clothing can affect the amount of aggression he or she expresses. In one study, female participants in a "learning" experiment were asked to deliver shocks to another participant whenever she made a mistake. Under the pretense of minimizing individual identities, one half of the participants wore nurses' uniforms (a prosocial cue), and the other half wore outfits resembling Ku Klux Klan uniforms (an antisocial cue). As predicted, participants who wore nurses uniforms delivered less shock to the "learner" than did participants who wore the Ku Klux Klan uniforms, which demonstrates that the cues inherent in certain clothes can influence the wearer's aggressive behavior (Johnson & Downing, 1979).

Although such studies are suggestive, they involve rather contrived situations that raise troubling questions of experimental demand. Accordingly, we decided to seek parallel evidence for a link between clothing cues and aggressiveness by examining the effect of a much more subtle cue, the color of a person's uniform.

There are a couple of difficulties that confront any attempt to test whether wearing a black uniform tends to make a person more aggressive. First, any such test is fraught with the usual ethical problems involved in all research on human aggression. Second, because black is associated with violence, observers may be biased when judging the behavior of participants wearing black. The usual solution to these twin problems is to use some version of the bogus shock paradigm (Buss, 1961). However, we chose not to use this procedure because of the difficulty in finding participants who—given the publicity of Milgram's (1965, 1974) work—would not view the proceedings with extreme suspicion.

Our solution to these problems was to collect "behavioroid" data (Carlsmith, Ellsworth, & Aronson, 1976) in the form of the participants' intended aggressive behavior. Volunteers for an experiment on competition were led to believe that they would be vying against other participants in several competitive events. They were also led to believe that they could exercise some control over which events they were to participate in by selecting their 5 most preferred events from a list of 12. The 12 events varied in the amount of aggressiveness they called for, allowing us to use participants' choices as a measure of their readiness to engage in aggressive action. By means of a suitable cover story, we elicited participants' choices twice: once individually when wearing their usual clothes, and later as a team of 3 wearing black or white jerseys. We hypothesized that wearing black jerseys would induce participants to view themselves as more mean and aggressive and thus would produce more of a "group shift" toward aggressive choices by participants wearing black jerseys than by those wearing white (Drabman & Thomas, 1977; Jaffe, Shapir, & Yinon, 1981).

Method

Participants

The participants were 72 male students from Cornell University who were paid $3 for their participation. They were run in groups of 3, with the members of each group unacquainted with one another.

Procedure

As the participants reported for the experiment they were brought together in one room and led to believe that another group of participants was assembling in a different room. Participants were told

> You will be competing, as a team, on a series of five games against another group of 3 participants who are waiting in the next room. I matched the two teams for size as you came in, so the contests should be fair. This study is designed to mimic real-life competition as closely as possible … [and so] … we want you to choose the games you want to play.

Participants were then given a list of descriptions of 12 games and were asked to indicate, individually, which games they would like to play. They were asked to choose 5 of the

12 games and to rank order those 5. After reminding the participants not to discuss their choices with one another, the experimenter left the room, ostensibly to elicit the choices of the other team.

Upon his return, the experimenter collected the participants' individual choices and stated that "now I would like you to make a group decision as to which games you will play, because many times people's preferences are so divergent that we need to use a group choice to serve as a tie-breaker when deciding on which games to play." The experimenter further explained, "to make the experiment more like real-world competition and to build team cohesion, I would like you to put these uniforms on over your shirts. From now on you will be referred to as the black [white] team." The participants were then given black or white uniforms with silver duct-tape numerals (7, 8, and 11) on the backs.

The experimenter left the room to allow the participants to make their group choices. After 4 min, the experimenter returned and thoroughly debriefed the participants. All participants seemed surprised (and many disappointed) to learn that the experiment was over. The debriefing interview also made it clear that none of the participants had entertained the possibility that the color of the uniforms might have been the focus of the experiment.

Dependent Measure

The measure in this experiment was the level of aggressiveness involved in the games participants wanted to include in the competition. A group of 30 participants had earlier rated a set of descriptions of 20 games in terms of how much aggressiveness they involved. The 12 games that had received the most consistent ratings and that represented a wide spectrum of aggressiveness were then used as the stimulus set in this experiment. These 12 games were ranked in terms of these aggressiveness ratings and assigned point values consistent with their ranks, from the most aggressive (12, 11, and 10 points for "chicken fights," "dart gun duel," and "burnout," respectively) to the least aggressive (1, 2, and 3 points for "basket shooting," "block stacking," and "putting contest," respectively). Participants were asked to choose the

5 games that they wanted to include in the competition and to rank order their choices in terms of preference. To get an overall measure of the aggressiveness of each participant's preferences, we multiplied the point value of his first choice by 5, his second choice by 4, and so forth, and then added these five products. When comparing the choices made by the participants individually (without uniforms), we compared the average individual choices of the 3 participants with their group choice.

Results

The mean levels of aggressiveness in participants' individual and group choices are presented in Table 1. As expected, there was no difference in participants' individual choices across the two groups (Ms = 113.4 vs. 113.5) because they were not wearing different-colored uniforms at the time these choices were made. However, the participants who donned black uniforms subsequently chose more aggressive games (mean change in aggressiveness = 16.8), whereas those who put on white uniforms showed no such shift (mean change = 2.4). A 2 × 2 mixed ANOVA of participants' choices yielded a significant interaction between uniform color and individual-group choice $F(1,22)$ = 6.14, p = .021, d = 1.01, 95% CI [0.16, 1.86], indicating that the pattern of choices made by participants in black uniforms was different from that of those wearing white. Wearing black uniforms induced participants to seek out more aggressive activities, matched-pairs $t(11)$ = 3.21, p = .008; wearing white uniforms did not, matched-pairs $t(11)$ = 1.00, p = .338.

Discussion

The results of this experiment support the hypothesis that wearing a black uniform can increase a person's inclination to engage in aggressive behavior. Participants who wore black uniforms showed a marked increase in intended aggression relative to those wearing white uniforms.

It should be noted, however, that our demonstration involved only intended aggression. It did not involve actual aggression. It would have been interesting to have allowed our participants to compete against one another in their chosen activities and seen whether those in black jerseys performed more aggressively. We refrained from doing so because of ethical and methodological difficulties (i.e., the difficulty of objectively measuring aggression, especially given that observers tend to be biased toward viewing people wearing black uniforms as being more aggressive). Nevertheless, the results of this experiment make the important point that in a competitive setting at least, merely donning a black uniform can increase a person's willingness to seek out opportunities for aggression. If the wearing of a black uniform can have such an effect in the laboratory, there is every reason to believe that it would have even stronger effects on the playing field (or rink), where many forms of aggression are considered acceptable behavior.

One question raised by this research concerns the generality of the effect of uniform color on aggression. It is very unlikely that donning any black uniform in any situation would make a person more inclined to act aggressively. We do not believe, for example, that the black garments worn by Catholic clergymen or Hassidic Jews make them any more aggressive than their secular peers. Rather, it would seem to be the case that the semantic link between the color black and evil and aggressiveness would be particularly salient in domains that already possess overtones of competition, confrontation, and physical aggression.

Perhaps the most important question raised by this research concerns the exact mechanisms by which the color of a uniform might affect the behavior of the wearer. Our own explanation for this phenomenon centers upon the implicit demands on one's behavior generated by wearing a particular kind of uniform. To wear a certain uniform is to assume a particular identity, an identity that not only elicits a certain response from others but also compels a particular pattern of behavior from the wearer (Stone, 1962). Wearing an athletic uniform, for example, thrusts one into the role of athlete, and leads one to "try on" the image that such a role conveys. When the uniform is that of a football or hockey player, part of that image—and therefore part of what one "becomes"—involves toughness, aggressiveness, and "machismo." These elements are particularly salient when the color of one's uniform is black. Just as observers see those in black uniforms as tough, mean, and aggressive, so too does the person wearing that uniform (Bem, 1972). Having inferred such an identity, the person then remains true to the image by acting more aggressively in certain prescribed contexts.

More broadly construed, then, our results serve as a reminder of the flexible or "situated" nature of the self (Alexander & Knight, 1971; Goffman, 1959; Mead, 1934; Stone, 1962). Different situations, different roles, and even different uniforms can induce people to try on different identities. Around those who are socially subdued or shy, an individual may become a vivacious extrovert; around true socialites, that same individual may retreat into a more reserved role. Some of the identities that people try to adopt are unsuitable, and those identities are abandoned. Abandoning such identities reassures people that at their core lies a "true" self. To a surprising degree, however, the identities people are led to adopt do indeed fit, and people continue to play them out in the appropriate circumstances. Perhaps the best evidence for this claim is the existence of identity conflict, such as that experienced by college students who bring their roommates home to meet their parents. This is often a disconcerting experience for many students because they cannot figure out how they should behave or "who they should be"—with their parents they are one person and with their friends they are someone else entirely.

The present investigation demonstrates how a seemingly trivial environmental variable, the color of one's uniform, can induce such a shift in a person's identity. This is not to suggest, however, that in other contexts the direction of causality might not be reversed. The black uniforms worn by gangs like the Hell's Angels, for example, are no doubt deliberately chosen precisely because they convey the desired malevolent image. Thus, as in the world portrayed in the typical American Western, it may be that many inherently evil characters choose to wear black. However, the present investigation makes it clear that in certain contexts at least, some people become the bad guys because they wear black.

References

Adams, F. M., & Osgood, C. E. (1973). A cross-cultural study of the affective meanings of color. *Journal of Cross-Cultural Psychology, 4,* 135–156. doi:10.1177/002202217300400201

Alexander, C. N., & Knight, G. (1971). Situated identities and social psychological experimentation. *Sociometry, 34,* 65–82. doi:10.2307/2786351

Bem, D. J. (1972). Self-perception theory. In L. Berkowitz (Ed.), *Advances in experimental social psychology* (Vol. 6, pp. 1–62). New York, NY: Academic Press.

Buss, A. M. (1961). *The psychology of aggression.* New York, NY: Wiley.

Carlsmith, J. M., Ellsworth, P. C., & Aronson, E. (1976). *Methods of research in social psychology.* Reading, MA: Addison-Wesley.

Drabman, R. S., & Thomas, M. H. (1977). Children's imitation of aggressive and prosocial behavior when viewing alone and in pairs. *Journal of Communication, 27,* 199–205. doi:10.1111/j.1460-2466.1977.tb02148.x

Feshbach, S. (1955). The drive-reducing function of fantasy behaviour. *Journal of Abnormal and Social Psychology, 50,* 3–11. doi:10.1037/h0042214

Goffman, E. (1959). *The presentation of self in everyday life.* New York, NY: Doubleday.

Jaffe, Y., Shapir, N., & Yinon, Y. (1981). Aggression and its escalation. *Journal of Cross-Cultural Psychology, 12,* 21–36. doi:10.1177/0022022181121002

Johnson, R. D., & Downing, L. L. (1979). Deindividuation and valence of cues: Effects of prosocial and antisocial behavior. *Journal of Personality and Social Psychology, 37,* 1532–1538. doi:10.1037/0022-3514.37.9.1532

Mead, G. H. (1934). *Mind, self, and society.* Chicago, IL: University of Chicago Press.

Milgram, S. (1965). Some conditions of obedience and disobedience to authority. *Human Relations, 18,* 57–76. doi:10.1177/001872676501800105

BLACK UNIFORMS AND AGGRESSION   12

Milgram, S. (1974). *Obedience to authority*. New York, NY: Harper.

Murray, H. A. (1943). *Thematic Apperception Test manual*. Cambridge, MA: Harvard University Press.

Stone, G. P. (1962). Appearance and the self. In A. M. Rose (Ed.), *Human behavior and social process* (pp. 86–118). Boston, MA: Houghton Mifflin.

Turner, V. (1967). *The forest of symbols: Aspects of Ndembu ritual*. Ithaca, NY: Cornell University Press.

Williams, J. E. (1964). Connotations of color names among Negroes and Caucasians. *Perceptual and Motor Skills*, 18, 721–731.

Williams, J. E., & McMurty, C. A. (1970). Color connotations among Caucasian 7th graders and college students. *Perceptual and Motor Skills, 30*, 701–713.

Williams, J. E., Moreland, J. K., & Underwood, W. I. (1970). Connotations of color names in the U.S., Europe, and Asia. *Journal of Social Psychology, 82*, 3–14.

Table 1

Mean Level of Aggressiveness Contained in Participants' Chosen Activities as a Function of Uniform Condition

Uniform color	Mean individual choice (without uniforms)		Group choice (with uniforms)		Change in aggression	
	M	*SD*	*M*	*SD*	*M*	*SD*
White	113.4	23.9	115.8	25.4	2.4	8.5
Black	113.5	18.4	130.3	22.9	16.8	18.1

A Checklist for Evaluating a Study's Validity

QUESTIONS ABOUT CONSTRUCT VALIDITY (ARE THE RESEARCHERS MEASURING AND MANIPULATING THE VARIABLES THEY CLAIM TO BE?)

1. Was the manipulation valid (does it manipulate what it claims to manipulate)?
   a. Is the manipulation consistent with definitions of the construct that is allegedly being manipulated?
   b. If the treatment manipulation involved using a sample of specific stimuli (e.g., particular men's names and women's names) to represent a broad, general variable that has many members (e.g., all men's and all women's names), did the researcher use a good-enough sample of stimuli to make the case that the difference between conditions was due to differences in the underlying construct? For example, if "David" produced a different reaction than "Dana," that difference might be due to some factor other than gender ("David" is longer, more common, and more closely associated with the Bible's King David). Thus, we would be more confident saying that the effect was due to gender of the name if the researcher had obtained the same effect using several other pairs of names (e.g., "Larry" and "Mary"). Similarly, if the researchers used one male experimenter and one female experimenter and then talked about a gender of experimenter effect, the manipulation's effect may be due to some other difference between the experimenters besides gender.
   c. Did the researchers use a **manipulation check:** a question or set of questions designed to determine whether participants perceived the

manipulation in the way that the researcher intended? For example, the researcher might ask questions to see whether participants in the "good mood" condition rated themselves as being in a better mood than the participants in the "neutral mood" condition. (For more on manipulation checks, see Chapter 5.)

d. Are *more* or *better* control (comparison) groups needed? For example, if the researcher claims to be manipulating "violence of video game" by having participants play either a violent video game or a nonviolent video game, are both games equally interesting and equally challenging? If the games differ in respects unrelated to violence, the researcher should not claim that the manipulation is a violence manipulation. In short, the control condition[s] and the experimental [treatment] condition[s] should be identical except for those aspects directly related to the construct being manipulated. (For more on control groups, see Chapter 11.)

2. Is the measure **valid:** does it measure what it claims to measure?

a. Is it **reliable:** does it produce stable, consistent scores that are not strongly influenced by random error? Reliability (consistency) is a prerequisite for validity (accuracy). One index of reliability—called test–retest reliability—assesses whether participants score about the same when they are retested as when they were originally tested. If test–retest reliability is below .70, the measure is not very reliable. Indeed, many people are displeased with test–retest reliabilities below .80.

Even if the authors do not provide the measure's test–retest reliability—an index of the measure's overall resistance to random error—the authors may provide indexes of the measure's vulnerability to specific sources of random error. The specific index of reliability you would want would depend on what specific sources of unreliability concerned you the most. For example, if the measure involved making raters judge something, you should be concerned that the raters might not be reliable. Therefore, look for evidence that different raters judging the same thing made similar judgments. Percentage of times judges agreed, correlations between raters, and Cohen's kappa might all serve as evidence of observer agreement. If, on the other hand, participants are filling out a rating scale measure, you do not need to worry about scorers disagreeing with each other. Instead, you need to be worried about questions that disagree with each other (e.g., according to one question, the participant is outgoing; according to another question, the participant is shy). If the questions are measuring the same concept, their answers should agree with each other. In technical terminology, this within-the-test (*internal*) agreement (*consistency*) is called internal consistency. Therefore, to get at your concern that the questions may not be measuring the same concept, you would want some index of internal consistency (sometimes called internal reliability) such as inter-item correlations (which should be above .30) or Cronbach's alpha—often abbreviated as alpha, Cronbach's α, or just α—(which should be above .70).

b. If the score a participant gets depends on a scorer's judgment, is this judgment trustworthy? The author should provide some evidence that independent raters obtain similar scores (e.g., some measure of rater agreement such as percentage of times raters agree, correlations between raters, or Cohen's kappa). Furthermore, the researcher should have used scoring that was masked/blind. That is, scorers should not know what treatment the participant had received (if the study was an experiment) or the participant's gender and other characteristics (if the study was a correlational study).

c. Did research show that the measure correlated with other measures of that same construct? (You would expect the researcher's measure of outgoingness to correlate with other measures of outgoingness.)

d. Did research show that the measure was *un*correlated with measures of unrelated constructs? (You would expect their measure of outgoingness to be uncorrelated with agreeableness.)

e. Was the measure consistent with accepted definitions of the construct?

3. Could the researchers have biased the study's results?

a. Were researchers "blind"(also called "masked")—or did they know which participants were expected to score higher?

b. Did the lack of detailed and clearly spelled-out procedures make it easy for researchers to bias the results?

4. Could participants have figured out the hypothesis? If so, they might have tried to "help" the researcher get the "right" results.

a. Could participants have learned about the study from former participants?

b. Were participants experienced enough to figure out the hypothesis (for instance, senior psychology majors who had participated in several studies)?

c. Was the hypothesis a fairly easy one to figure out?

d. Did the research use a no-treatment (empty) control group—rather than a control group that got a fake (placebo) treatment? (If one group got a pill and one didn't, the participants getting the pill might expect their behavior to change, whereas participants not getting a pill would not expect their behavior to change.)

e. Did the researcher fail to make the study a double-blind study, thus allowing either the participants or the researcher to know which treatment the participants were receiving?

f. Was it obvious to participants what was being measured? For example, did participants fill out a self-report scale, such as "Rate your happiness on a 1–5 scale"?

g. Did the study lack **experimental (research) realism:** the ability to engage participants in the task? If participants do not take the task seriously, their responses probably should not be taken seriously: At best, the participants do not show any reaction to the manipulation; at worst, they show a false reaction—they fake the response they think will support the researcher's hypothesis.

h. Did the researchers fail to have an effective "cover story" that disguised the true purpose of the study? For example, rather than telling participants they were being given a cola to see its effect on arousal, it would be better to tell participants that they were drinking the cola as part of a taste test.

QUESTIONS ABOUT INTERNAL VALIDITY (CAN WE CONCLUDE THAT ONE FACTOR CAUSED AN EFFECT?)

1. Was an experimental design used? If not, the study probably does not have internal validity. To help determine whether a study is an experiment, realize that (a) all experiments involve manipulating (administering) a treatment and (b) all experiments are either (1) between-subjects experiments that compare individuals who were *randomly assigned* to receive a treatment to individuals who were given a different treatment (to see how to randomly assign participants, see Table 6 of Appendix F) or (2) within-subjects experiments that compare individuals when they were given a treatment with those same individuals when they were given a different treatment.

Most experiments are of the first type: between-subjects experiments that compare a group that was randomly assigned to receive the treatment with one or more groups that were randomly assigned to receive different treatment(s). (Such studies are sometimes called randomized controlled trials [RCTs].) Random assignment allows researchers to make a strong case that the difference between the actions of participants in the different conditions is due to the treatment manipulation rather than to nontreatment factors (for more on why random assignment helps establish internal validity, see Chapter 2 or Chapter 10).

Many studies that compare participants who received the treatment against those same participants when those participants had either not received the treatment or had received a different treatment are not within-subjects experiments. For such a study to be an experiment, the study must control for (a) participants naturally changing over time and (b) participants changing as a result of practice on the task. To show you that studies without such controls do not provide valid results, imagine that Dr. N. Ept does two studies. In the first study, he has participants eat, immediately gives them a vitamin pill, and then immediately has them eat again. He notes that participants eat less the second time and concludes that the vitamin pill decreases appetite (he and his design ignore the possibility that, pill or no pill, participants may not be as hungry after having just eaten). In the second study, he has participants play a video game, take a vitamin pill, and play the video game again. If participants score higher the second time they play the video game, Dr. N. Ept credits the pill (rather than practice). If participants score lower the second time they play the game, Dr. N. Ept blames the pill (rather than boredom or fatigue).

As you can see, when participants are compared with themselves, you must ask how the researchers were able to separate the effects of *when* participants received the treatment (e.g., receiving one treatment first and the other treatment second) from *what* treatment participants received (e.g., nonviolent video game vs. violent video game). Specifically, to do a version of Dr. N. Ept's study that had internal validity, the researchers must have used at least one of the following two techniques.

First, the researchers might randomize the order of treatments (e.g., a coin flip would determine whether the participant played the nonviolent or violent game first). Second, the researchers might randomly assign participants to two groups in a way that half the participants would play the nonviolent video game first, while the other half would play the violent video game first. The researchers, by making sure that half of the participants get the sequence violent game–nonviolent game and half get the sequence nonviolent game–violent game, have ensured that if participants tend to be more violent at the beginning of the study, this tendency will not affect the violent game condition more than the nonviolent condition. This technique of *balancing* out order effects by giving participants systematically different sequences is called *counterbalancing*.

Regardless of the type of experiment, ask what experimenters did to make it so they could say that the difference between treatment conditions was due to the treatment rather than to something else. Usually, experimenters will try to neutralize the effects of nontreatment factors in at least one of the following three ways:

1. Preventing the nontreatment factor from being a variable by keeping the nontreatment factor constant. Thus, to control for time of day, the researcher might test all participants at the same time of day.
2. Preventing the nontreatment factor from affecting one condition more than another by *counterbalancing*: systematically rotating it between conditions to balance out the effect of that variable. Thus, to control for time of day, the researcher might alternate testing sessions. For example, on the first day, the treatment group might be tested in the morning and the no-treatment group might be tested in the afternoon. On the next day, the situation would be reversed.
3. Using random assignment to randomize—and then statistically account for—nontreatment variables. Thus, to control for time of day, the researcher would randomly assign participants to condition. With random assignment, there would be no systematic difference between participants in terms of when they were tested. Instead, any differences in time of testing would be unsystematic differences. Consequently, if the difference between groups' scores was statistically significant, it is unlikely that the difference in scores is due solely to time of day—or to any other—unsystematic, chance difference.

2. If the study was a between-subjects experiment, did more participants drop out of the treatment group than out of the control group? If so, the groups' different dropout rates—not the groups' different treatments—may be responsible for the differences between the average scores of the groups.
3. If the study was an experiment, was there a reliable (statistically significant) difference between the scores in the different conditions?

If not, there is no evidence of an effect—and thus no point in talking about its cause.

4. If the study was not an experiment, the study probably does not have internal validity. Thus, if the researcher suggests a cause–effect conclusion, ask

a. **Could the researcher have cause and effect reversed?** In some nonexperimental research, what the researcher thinks is a cause may actually be an effect. For example, surveys show that people who watch more television tend to have lower self-esteem. If a researcher concluded that television-viewing caused low self-esteem, the researcher could be wrong. It might be that low self-esteem causes people to watch television (Moskalenko & Heine, 2003). Note that if the researchers measured participants on both variables several times (such designs are usually called either longitudinal designs or prospective designs), researchers may be able to determine which variable changed first.

b. **Could the researcher have ignored a third variable?** In some nonexperimental research, two variables may be statistically related because both are effects of some other variable. For example, both low self-esteem and television-viewing may be side effects of having few friends: People who have few friends may have low self-esteem and may watch a lot of television (Moskalenko & Heine, 2003). As we explain in Appendix E, some researchers who have nonexperimental data use statistical techniques such as partial correlations, multiple regression, and structural equation modeling to try to rule out third variables.

QUESTIONS ABOUT EXTERNAL VALIDITY (CAN THE RESULTS BE GENERALIZED TO OTHER PEOPLE, PLACES, AND TIMES?)

1. Do the study's conclusions describe what people do or think (e.g., "30% of Americans approve of the president.") or does the study focus on causes of behavior (e.g., "Negative ads cause drop in president's popularity.")? Usually, external validity is much more of a concern for studies that try to describe behavior than for studies that try to explain the causes of behavior.

2. Would results apply to the average person?
   a. Were participants human?
   b. Were participants distinct in any way?
   c. Were the participants too homogeneous? That is, were there certain types of individuals (women, minorities) who were not included in the study?
   d. Was the dropout rate high—or high among certain groups (e.g., were all the dropouts participants over 65 years old)? If so, the results apply only to those who stayed in the study.

e. Is there any specific reason to suspect that the results would not apply to a different group of participants?

f. If the researchers used a survey and tried to generalize their results to a wider group,

1. What was that larger group?

2. Did they have a large and random sample from that group?

3. Would the results generalize to different settings? Can you pinpoint a difference between the research setting and a real-life setting and give a specific reason why this difference would prevent the results from applying to real life?

4. Would the results generalize to different levels (amounts) of the treatment variable?

a. Was a wide range of treatment amounts tested?

b. Were realistic amounts of the treatment variable tested?

c. Were at least three different amounts of the treatment variable tested? If only two amounts are tested, it is extremely risky to generalize to untested levels.

QUESTIONS ABOUT POWER (HOW GOOD WAS THE STUDY AT FINDING DIFFERENCES?)

If the study failed to find a statistically significant difference between groups or conditions, ask the following six questions:

1. Were participants homogeneous (similar) enough so that differences between participants would not hide a treatment effect—or did between-subject differences mask the treatment effect? (To illustrate the impact of homogeneity, consider the following analogy. If all participants have the same singing range but one group is asked to sing a moderately high note whereas the other is not, you could easily hear the difference between the groups. If, however, some people had low voices and some had high voices, the group differences would be harder to detect.)

2. Were enough participants used? (In a sense, more participants means more voices, which makes differences between the groups easier to hear. Note that in our singing example, if we had 2 singers in each group and the singers had widely different ranges, we might have trouble hearing the difference [especially if the two lowest voices were randomly assigned to the group that was asked to sing high]. If, on the other hand, we had 100 participants in each group, the difference between groups would be easy to hear—regardless of whether singers' voices were homogeneous.)

3. Was the study sufficiently standardized? That is, did lack of consistency in how the study was conducted and lack of control over the testing environment create so much treatment-unrelated background noise that a treatment effect would not be heard?

4. Did conditions differ enough on the treatment/predictor variable? (To return to our singing analogy, if we had asked one group to sing very

high and others to sing very low, we would have easily detected a difference. If, however, we had asked one group to sing one note above their best note and the other group to sing their best note, we might not have detected a difference.)

5. Were the measures sensitive enough? Just as a sensitive bathroom scale can detect differences that less sensitive scales would miss, sensitive measures can detect differences an insensitive measure would miss. If our instrument to measure pitch was unreliable (the needle bounced around randomly, was not valid (it was affected by how loud instead of how high voices were), or did not provide a wide range of scores (it only registered "high" or "low" rather than B, C-sharp, etc.), our measure would be insensitive. Put another way, sensitive measures tend to be reliable, valid, and provide a range of scores (for more on sensitivity, see Chapter 6).

6. Could the failure to find a difference be due to a floor or ceiling effect?

   1. In a floor effect, a problem with the measure makes it so participants who are, in reality, extremely low on the variable do not score lower on the measure of the variable than people who are merely low on the variable. Because participants who actually differ from each other are not scoring differently from one another, the researcher may not find differences between conditions. Suspect a floor effect if everyone is scoring low on the measure.

   2. The ceiling effect is the reverse of the floor effect. Everyone is scoring so high on the measure that participants who are, in reality, very high on the variable can't score higher on the measure of the variable than participants who are somewhat high on that variable. Because participants are "maxing" out the scale, the researcher may not find differences between conditions. For example, if all the participants scored 100% on a memory test, the participants who have a memory for the information that is better than the average participant's are not able to show their better memory on this test. Thus, even if every participant in the treatment group had a better memory for the material than anyone in the no-treatment group, there would be no difference between the groups on the measure because both groups would average 100%. Suspect a ceiling effect if everyone is scoring high on the measure.

QUESTIONS ABOUT STATISTICAL ANALYSES

1. Do the data meet the assumptions of the statistical test? If the results were published in an APA or APS journal, you can assume that the data meet the assumptions of the test. If you are looking at an unpublished paper, however, you may need to ask questions. For example, if the researchers did an independent groups *t* test, was each participant's response independent—unaffected by how other participants responded? (To learn more about the independent groups *t* test, see Chapter 10 or Appendix E.)

2. Are the researchers running a high risk of making a **Type 1 error:** declaring a difference *statistically significant* (reliable) even though, in reality, the difference is not reliable? The purpose of statistical significance tests is to prevent us from mistaking a chance difference for a real one. However, bad luck or author recklessness sometimes defeats this safeguard. To determine whether they may be making a Type 1 error, ask the following three questions.

   a. Are they doing multiple statistical tests without correcting for the fact that their reported significance level is only valid if they are doing a single test? For example, if they use a .05 (5%) significance level, they are saying there is a less than a 5 in 100 chance of getting these results by chance alone. That's fair—if they did only one test. If, however, they did 100 tests, 5 tests could turn out significant by chance alone. In other words, it is one thing to do one test and have a 5% chance of getting a false positive; it is another thing to do 100 tests and be assured of false positives. Even worse, some authors, rather than telling you about the 95 tests that were not significant, will act like they only did the 5 tests that were significant. One clue that the authors are reporting only the tests that supported their position is if there are measures they mention in the Method section that are not discussed in the Results section. This practice of hiding failed analyses reminds us of the spam e-mailers who send half their list a prediction that a stock will go up and tell the other half that the stock will go down. If it goes up, they contact the first half of their list; if it goes down, they contact the second half. (They do not tell the group they re-contact about their wrong predictions.)

   b. Are they using unconventionally high significance levels (the higher the significance level, the higher the risk of a Type 1 error)? For example, if they are using a $p < .20$ level rather than the traditional $p < .05$ level, they are taking more than 4 times ($4 \times .05 = .20$) the risk of making a Type 1 error than most researchers take.

   c. Has the study been replicated? Replication, rather than statistical significance, is the best evidence that the findings are reliable. If you know of failures to replicate, or if you suspect that the studies that do not get significant results are not getting published, the significant results of the study may reflect a Type 1 error.

3. Did the authors represent differences between two means as real even though

   a. A statistical test had been performed and the differences were not statistically significant? For example, some researchers report nonsignificant results as "trends" or as "marginally significant."

   b. No statistical significance test had been performed that directly tested whether those two means were significantly different from each other?

4. Did the authors represent significant differences as being large without providing evidence of that claim? Statistical significance suggests the differences are reliable, not that they are big. To show how big a difference is, researchers must use effect size indexes (e.g., r, r^2, eta squared [η^2], omega-squared [ω^2], Cohen's d).

Practical Tips for Conducting an Ethical and Valid Study

For help on almost all the "nuts and bolts" of planning and conducting a study, go to www.cengage.com/psychology/mitchell or to www.jolley-mitchell.com.

APPENDIX E

Introduction to Statistics

For help on choosing, interpreting, or conducting statistical tests, go to www.cengage.com/psychology/mitchell or to www.jolley-mitchell.com.

Statistics and Random Numbers Tables

DIRECTIONS FOR USING TABLE 1

Use the left column (the column labeled *df*) to find the row labeled with the number of degrees of freedom (*df*) that your study had. For the simple experiment, that number equals the number of participants minus 2. Thus, if you had 32 participants, you would go down the *df* column until you reached the number 30. Then, unless you have a one-tailed test, read across that row until you find the entry in the column corresponding to your level of significance (e.g., if you were using a significance, probability, or alpha level of .05, you would stop at the entry in the .05 column). The number in that cell will be the critical value of *t* for your study. To be statistically significant, the absolute value of *t* that you obtain from your study must be greater than the value you found in the table. For example, suppose *df* = 30 and *p* < .05 (two-tailed test). In that case, to be statistically significant, the absolute value of the *t* you calculated must be greater than 2.042.

DIRECTIONS FOR USING TABLE 2

Use the column labeled *df* to find the row that has the same number of degrees of freedom that your study had. (To calculate your *df*, subtract one from the number of columns in your chi-square, then subtract one from the number of rows, and then multiply those results together. Thus, with a 2 × 2 chi-square, you would have 1 *df* [because (2 − 1) × (2 − 1) = 1 × 1 = 1], and with a 3 × 2 chi-square, you would have 2 *df* [because (3 − 1) × (2 − 1) = 2 × 1 = 2]). Then, unless you have a one-tailed test, go across the row until you find the entry in the column corresponding to your level of significance. The number in that cell will be the critical value of chi-square for your study. To be statistically significant, your chi-square value must be greater than the value you found in the table. For example, if *df* = 1 and your significance level is *p* < .05, then your chi-square value must be greater than 3.84146.

TABLE **1** Critical Values of *t*

Level of Significance for Two-Tailed *t* Test

df	p Levels			
	.10	.05	.02	.01
1	6.314	12.706	31.821	63.657
2	2.920	4.303	6.965	9.925
3	2.353	3.182	4.541	5.841
4	2.132	2.776	3.747	4.604
5	2.015	2.571	3.365	4.032
6	1.943	2.447	3.143	3.707
7	1.895	2.365	2.998	3.499
8	1.860	2.306	2.896	3.355
9	1.833	2.262	2.821	3.250
10	1.812	2.228	2.764	3.169
11	1.796	2.201	2.718	3.106
12	1.782	2.179	2.681	3.055
13	1.771	2.160	2.650	3.012
14	1.761	2.145	2.624	2.977
15	1.753	2.131	2.602	2.947
16	1.746	2.120	2.583	2.921
17	1.740	2.110	2.567	2.898
18	1.734	2.101	2.552	2.878
19	1.729	2.093	2.539	2.861
20	1.725	2.086	2.528	2.845
21	1.721	2.080	2.518	2.831
22	1.717	2.074	2.508	2.819
23	1.714	2.069	2.500	2.807
24	1.711	2.064	2.492	2.797
25	1.708	2.060	2.485	2.787
26	1.706	2.056	2.479	2.779
27	1.703	2.052	2.473	2.771
28	1.701	2.048	2.467	2.763
29	1.699	2.045	2.462	2.756
30	1.697	2.042	2.457	2.750
40	1.684	2.021	2.423	2.704
60	1.671	2.000	2.390	2.660
120	1.658	1.980	2.358	2.617
∞	1.645	1.960	2.326	2.576

TABLE **2** Critical Values for Chi-Square Tests

Level of Significance

df	p Levels			
	.10	**.05**	**.01**	**.001**
1	2.70554	3.84146	6.63490	10.828
2	4.60517	5.99147	9.21034	13.816
3	6.25139	7.81473	11.3449	16.266
4	7.77944	9.48773	13.2767	18.467
5	9.23635	11.0705	15.0863	20.515
6	10.6446	12.5916	18.5476	22.458
7	12.0170	14.0671	18.4753	24.322
8	13.3616	15.5073	20.0902	26.125
9	14.6837	16.9190	21.6660	27.877
10	15.9871	18.3070	23.2093	29.588
11	17.2750	19.6751	24.7250	31.264
12	18.5494	21.0261	26.2170	32.909
13	19.8119	22.3621	27.6883	34.528
14	21.0642	23.6848	29.1413	36.123
15	22.3072	24.9958	30.5779	37.697
16	23.5418	26.2962	31.9999	39.252
17	24.7690	27.5871	33.4087	40.790
18	25.9894	28.8693	34.8053	42.312
19	27.2036	30.1435	36.1908	43.820
20	28.4120	31.4104	37.5662	45.315
21	29.6151	32.6705	38.9321	46.797
22	30.8133	33.9244	40.2894	48.268
23	32.0069	35.1725	41.6384	49.728
24	33.1963	36.4151	42.9798	51.179
25	34.3816	37.6525	44.3141	52.620
26	35.5631	38.8852	45.6417	54.052
27	36.7412	40.1133	46.9630	55.476
28	37.9159	41.3372	48.2782	56.892
29	39.0875	42.5569	49.5879	58.302
30	40.2560	43.7729	50.8922	59.703
40	51.8050	55.7585	63.6907	73.402
50	63.1671	67.5048	76.1539	86.661
60	74.3970	79.0819	88.3794	99.607
70	85.5271	90.5312	100.425	112.317
80	96.5782	101.879	112.329	124.839
90	107.565	113.145	124.116	137.208
100	118.498	124.342	135.807	149.449

Source: This table is taken from Table 8 of the *Biometrika Tables for Statisticians* (Vol. 1, 3rd ed.) by E. S. Pearson and H. O. Hartley (Eds.), 1970, New York: Cambridge University Press. Used with the kind permission of the Biometrika trustees.

DIRECTIONS FOR USING TABLE 3

Find the column that matches your degrees of freedom for the effect (the first df) and then go down that column until you hit the row that matches the degrees of freedom for your error term (the second df). Thus, if you had 1 df for the effect and 23 for the error term, you would start at the column labeled "1" and go down until you reached the row labeled "23." There, you would find the critical value: 4.28. Thus, to be statistically significant at the $p < .05$ level, your obtained F would have to be greater than 4.28.

TABLE 3 Critical Values of F for $p < .05$

2nd df	1st df								
	1	2	3	4	5	6	7	8	9
1	161.4	199.5	215.7	224.6	230.2	234.0	236.8	238.9	240.5
2	18.51	19.00	19.16	19.25	19.30	19.33	19.35	19.37	19.38
3	10.13	9.55	9.28	9.12	9.01	8.94	8.89	8.85	8.81
4	7.71	6.94	6.59	6.39	6.26	6.16	6.09	6.04	6.00
5	6.61	5.79	5.41	5.19	5.05	4.95	4.88	4.82	4.77
6	5.99	5.14	4.76	4.53	4.39	4.28	4.21	4.15	4.10
7	5.59	4.74	4.35	4.12	3.97	3.87	3.79	3.73	3.68
8	5.32	4.46	4.07	3.84	3.69	3.58	3.50	3.44	3.39
9	5.12	4.26	3.86	3.63	3.48	3.37	3.29	3.23	3.18
10	4.96	4.10	3.71	3.48	3.33	3.22	3.14	3.07	3.02
11	4.84	3.98	3.59	3.36	3.20	3.09	3.01	2.95	2.90
12	4.75	3.89	3.49	3.26	3.11	3.00	2.91	2.85	2.80
13	4.67	3.81	3.41	3.18	3.03	2.92	2.83	2.77	2.71
14	4.60	3.74	3.34	3.11	2.96	2.85	2.76	2.70	2.65
15	4.54	3.68	3.29	3.06	2.90	2.79	2.71	2.64	2.59
16	4.49	3.63	3.24	3.01	2.85	2.74	2.66	2.59	2.54
17	4.45	3.59	3.20	2.96	2.81	2.70	2.61	2.55	2.49
18	4.41	3.55	3.16	2.93	2.77	2.66	2.58	2.51	2.46
19	4.38	3.52	3.13	2.90	2.74	2.63	2.54	2.48	2.42
20	4.35	3.49	3.10	2.87	2.71	2.60	2.51	2.45	2.39
21	4.32	3.47	3.07	2.84	2.68	2.57	2.49	2.42	2.37
22	4.30	3.44	3.05	2.82	2.66	2.55	2.46	2.40	2.34
23	4.28	3.42	3.03	2.80	2.64	2.53	2.44	2.37	2.32
24	4.26	3.40	3.01	2.78	2.62	2.51	2.42	2.36	2.30
25	4.24	3.39	2.99	2.76	2.60	2.49	2.40	2.34	2.28
26	4.23	3.37	2.98	2.74	2.59	2.47	2.39	2.32	2.27
27	4.21	3.35	2.96	2.73	2.57	2.46	2.37	2.31	2.25
28	4.20	3.34	2.95	2.71	2.56	2.45	2.36	2.29	2.24
29	4.18	3.33	2.93	2.70	2.55	2.43	2.35	2.28	2.22
30	4.17	3.32	2.92	2.69	2.53	2.42	2.33	2.27	2.21
40	4.08	3.23	2.84	2.61	2.45	2.34	2.25	2.18	2.12
60	4.00	3.15	2.76	2.53	2.37	2.25	2.17	2.10	2.04
120	3.92	3.07	2.68	2.45	2.29	2.17	2.09	2.02	1.96
∞	3.84	3.00	2.60	2.37	2.21	2.10	2.01	1.94	1.88

TABLE **3** Critical Value of *F* for *p* < .025

2nd *df*	1st *df*								
	1	**2**	**3**	**4**	**5**	**6**	**7**	**8**	**9**
1	647.8	799.5	864.2	899.6	921.8	937.1	948.2	956.7	963.3
2	38.51	39.00	39.17	39.25	39.30	39.33	39.36	39.37	39.39
3	17.44	16.04	15.44	15.10	14.88	14.73	14.62	14.54	14.47
4	12.22	10.65	9.98	9.60	9.36	9.20	9.07	8.98	8.9
5	10.01	8.43	7.76	7.39	7.15	6.98	6.85	6.76	6.68
6	8.81	7.26	6.60	6.23	5.99	5.82	5.70	5.6	5.52
7	8.07	6.54	5.89	5.52	5.29	5.12	4.99	4.9	4.82
8	7.57	6.06	5.42	5.05	4.82	4.65	4.53	4.43	4.36
9	7.21	5.71	5.08	4.72	4.48	4.32	4.20	4.10	4.03
10	6.94	5.46	4.83	4.47	4.24	4.07	3.95	3.85	3.78
11	6.72	5.26	4.63	4.28	4.04	3.88	3.76	3.66	3.59
12	6.55	5.10	4.47	4.12	3.89	3.73	3.61	3.51	3.44
13	6.41	4.97	4.35	4.00	3.77	3.60	3.48	3.39	3.31
14	6.30	4.86	4.24	3.89	3.66	3.5	3.38	3.29	3.21
15	6.20	4.77	4.15	3.8	3.58	3.41	3.29	3.20	3.12
16	6.12	4.69	4.08	3.73	3.50	3.34	3.22	3.12	3.05
17	6.04	4.62	4.01	3.66	3.44	3.28	3.16	3.06	2.98
18	5.98	4.56	3.95	3.61	3.38	3.22	3.10	3.01	2.93
19	5.92	4.51	3.90	3.56	3.33	3.17	3.05	2.96	2.88
20	5.87	4.46	3.86	3.51	3.29	3.13	3.01	2.91	2.84
21	5.83	4.42	3.82	3.48	3.25	3.09	2.97	2.87	2.80
22	5.79	4.38	3.78	3.44	3.22	3.05	2.93	2.84	2.76
23	5.75	4.35	3.75	3.41	3.18	3.02	2.90	2.81	2.73
24	5.72	4.32	3.72	3.38	3.15	2.99	2.87	2.78	2.7
25	5.69	4.29	3.69	3.35	3.13	2.97	2.85	2.75	2.68
26	5.66	4.27	3.67	3.33	3.10	2.94	2.82	2.73	2.65
27	5.63	4.24	3.65	3.31	3.08	2.92	2.80	2.71	2.63
28	5.61	4.22	3.63	3.29	3.06	2.9	2.78	2.69	2.61
29	5.59	4.2	3.61	3.27	3.04	2.88	2.76	2.67	2.59
30	5.57	4.18	3.59	3.25	3.03	2.87	2.75	2.65	2.57
40	5.42	4.05	3.46	3.13	2.90	2.74	2.62	2.53	2.45
60	5.29	3.93	3.34	3.01	2.79	2.63	2.51	2.41	2.33
120	5.15	3.80	3.23	2.89	2.67	2.52	2.39	2.30	2.22
∞	5.02	3.69	3.12	2.79	2.57	2.41	2.29	2.19	2.11

TABLE 3 Critical Values of *F* for *p* < .01

2nd *df*	1st *df*								
	1	2	3	4	5	6	7	8	9
1	4052	4999.5	5403	5625	5764	5859	5928	5982	6022
2	98.50	99.00	99.17	99.25	99.30	99.33	99.36	99.37	99.39
3	34.12	30.82	29.46	28.71	28.24	27.91	27.67	27.49	27.35
4	21.20	18.00	16.69	15.98	15.52	15.21	14.98	14.80	14.66
5	16.26	13.27	12.06	11.39	10.97	10.67	10.46	10.29	10.16
6	13.75	10.92	9.78	9.15	8.75	8.47	8.26	8.10	7.98
7	12.25	9.55	8.45	7.85	7.46	7.19	6.99	6.84	6.72
8	11.26	8.65	7.59	7.01	6.63	6.37	6.18	6.03	5.91
9	10.56	8.02	6.99	6.42	6.06	5.80	5.61	5.47	5.35
10	10.04	7.56	6.55	5.99	5.64	5.39	5.20	5.06	4.94
11	9.65	7.21	6.22	5.67	5.32	5.07	4.89	4.74	4.63
12	9.33	6.93	5.95	5.41	5.06	4.82	4.64	4.50	4.39
13	9.07	6.70	5.74	5.21	4.86	4.62	4.44	4.30	4.19
14	8.86	6.51	5.56	5.04	4.69	4.46	4.28	4.14	4.03
15	8.68	6.36	5.42	4.89	4.56	4.32	4.14	4.00	3.89
16	8.53	6.23	5.29	4.77	4.44	4.20	4.03	3.89	3.78
17	8.40	6.11	5.18	4.67	4.34	4.10	3.93	3.79	3.68
18	8.29	6.01	5.09	4.58	4.25	4.01	3.84	3.71	3.60
19	8.18	5.93	5.01	4.50	4.17	3.94	3.77	3.63	3.52
20	8.10	5.85	4.94	4.43	4.10	3.87	3.70	3.56	3.46
21	8.02	5.78	4.87	4.37	4.04	3.81	3.64	3.51	3.40
22	7.95	5.72	4.82	4.31	3.99	3.76	3.59	3.45	3.35
23	7.88	5.66	4.76	4.26	3.94	3.71	3.54	3.41	3.30
24	7.82	5.61	4.72	4.22	3.90	3.67	3.50	3.36	3.26
25	7.77	5.57	4.68	4.18	3.85	3.63	3.46	3.32	3.22
26	7.72	5.53	4.64	4.14	3.82	3.59	3.42	3.29	3.18
27	7.68	5.49	4.60	4.11	3.78	3.56	3.39	3.26	3.15
28	7.64	5.45	4.57	4.07	3.75	3.53	3.36	3.23	3.12
29	7.60	5.42	4.54	4.04	3.73	3.50	3.33	3.20	3.09
30	7.56	5.39	4.51	4.02	3.70	3.47	3.30	3.17	3.07
40	7.31	5.18	4.31	3.83	3.51	3.29	3.12	2.99	2.89
60	7.08	4.98	4.13	3.65	3.34	3.12	2.95	2.82	2.72
120	6.85	4.79	3.95	3.48	3.17	2.96	2.79	2.66	2.56
∞	6.63	4.61	3.78	3.32	3.02	2.80	2.64	2.51	2.41

Source: This table is abridged from Table 18 of the *Biometrika Tables for Statisticians* (Vol. 1, 3rd ed.) by E. S. Pearson and H. O. Hartley (Eds.), 1970, New York: Cambridge University Press. Used with the kind permission of the Biometrika trustees.

TABLE **4** Coefficients of Orthogonal Polynomials

Condition	3-Condition Case Trend		4-Condition Case Trend			5-Condition Case Trend			
	1	**2**	**1**	**2**	**3**	**1**	**2**	**3**	**4**
	(Lin)	**(Quad)**	**(Lin)**	**(Quad)**	**(Cubic)**	**(Lin)**	**(Quad)**	**(Cubic)**	
1	−1	1	−3	1	−1	−2	2	−1	1
2	0	−2	−1	−1	3	−1	−1	2	−4
3	1	1	1	−1	−3	0	−2	0	6
4			3	1	1	1	−1	−2	−4
5						2	2	1	1
Weighting Factor	2	6	20	4	20	10	14	10	70

Condition	6-Condition Case Trend					7-Condition Case Trend					
	1	**2**	**3**	**4**	**5**	**1**	**2**	**3**	**4**	**5**	**6**
	(Lin)	**(Quad)**	**(Cubic)**			**(Lin)**	**(Quad)**	**(Cubic)**			
1	−5	5	−5	1	−1	−3	5	−1	3	−1	1
2	−3	−1	7	−3	5	−2	0	1	−7	4	−6
3	−1	−4	4	2	−10	−1	−3	1	1	−5	15
4	1	−4	−4	2	10	0	−4	0	6	0	−20
5	3	−1	−7	−3	−5	1	−3	−1	1	5	15
6	5	5	5	1	1	2	0	−1	−7	−4	−6
7						3	5	1	3	1	1
Weighting Factor	70	84	180	28	252	28	84	6	154	84	924

Source: This table is adapted from Table VII of *Statistics* (pp. 662–664) by W. L. Hays, 1981, New York: Holt, Rinehart and Winston. Copyright © 1982 by Holt, Rinehart and Winston, Inc. Adapted by permission.

From Hays, Statistics 3/E, 3E. © 1981 Cengage Learning. This table is adapted from Table VII (pp. 662–664).

USING TABLE 4 TO COMPUTE TREND ANALYSES

Suppose you had the following significant effect for sugar on aggression.

	DF	*SS*	*MS*	*F*
Sugar Main Effect	2	126.95	63.47	6.35
Error Term	21	210.00	10.00	

© Cengage Learning 2013

How would you compute a trend analysis for this data? You would start by calculating an *F* ratio for the linear and quadratic effects so that you could complete the following ANOVA table.

	DF	SS	MS	F
Sugar Main Effect	2	126.95	63.47	6.35
Linear Trend	1			
Quadratic Trend	1			
Error Term	21	210.00	10.00	

© Cengage Learning 2013

Before you generate an *F* ratio for an effect, you must have a sum of squares for that effect. To compute the sum of squares for a trend, you must first get the sum of the scores for each group. In this case, you will need the sum (total) of the scores for these three groups: (a) the no-sugar group, (b) the 50 mg of sugar group, and (c) the 100 mg of sugar group. One way to get the sum of scores for a group is to add up (sum) all the scores for that group. Another way to get the sum of scores for a group is to multiply the group's average by the number of scores making up that average. For example, if one group's mean was 10, and there were 8 scores making up that mean, the sum for that condition would be 10 × 8, which is 80.

Once you have your sums, arrange these sums by placing the sum for the group getting the lowest level of the independent variable (e.g., no-sugar condition) first, the sum for the group getting the next highest amount of the treatment next, and so on. In our example, you would order your sums like the following:

Total Number of Violent Instances per Condition

Amount of sugar	Total number of violent instances
0 mg	10.0
50 mg	50.0
100 mg	12.0

© Cengage Learning 2013

Now that you have your sums in order, you are ready to consult the tables of orthogonal polynomials in Table 4. Because this example involves three conditions, you would look for the three-condition table. The table reads as follows:

Three-Condition Case

	Trend	
	Linear	Quadratic
Condition 1	−1	1
Condition 2	0	−2
Condition 3	1	1
Weighting Factor	2	6

© Cengage Learning 2013

To get the numerator for the sum of squares for the *linear* trend, multiply the sum for the first level of the independent variable by the first (Condition 1) value in the "*Linear*" column of the table (−1), the second sum by the second value in the "*Linear*" column of that table (0), and the third sum by the third value in the "*Linear*" column (+1). Next, get a sum by adding these three products together. Then, square that sum. So, for the sugar example we just described, you would do the following calculations:

$$[(-1 \times 10)+(0 \times 50)+(1 \times 12)]^2$$

which equals

$$(-10 + 0 + 12)^2$$

which equals

$$(2)^2$$

which equals

$$4$$

To get the denominator for the sum of squares, multiply the weighting factor for the linear trend (2) by the number of observations in each condition. Because there were *8* observations in each condition, the denominator would be 2 × 8, which is 16.

To get the sum of squares linear, divide the numerator by the denominator. In this case, the numerator (4) divided by the denominator (16) equals .25.

Once you have computed the sum of squares for the linear trend, the rest is easy. All you have to do is compute *F* ratio by dividing the mean square linear by the mean square error and then see if that result is significant.

Calculating the mean square linear involves dividing the sum of squares linear by the degrees of freedom linear. Because the degrees of freedom for any trend is always 1, you could divide your sum of squares (.25) by 1 and get .25. Or, you could simply remember that a trend's mean square is always the same as its sum of squares.

Getting the mean square error is also easy: Just find the mean square error in the printout (it is the same error term that was used to calculate the overall *F*). In this example, the MSE is 10.

So, to get the *F* value for this linear comparison, you would divide the mean square for the comparison (.25) by the mean square error used on the overall main effect (10.0). Thus, the *F* would be .25/10, or .025. Because the *F* is below 1.00, this result is not significant.

But how large would the *F* have had to be to be significant? That depends on how many trends you were analyzing. If you had decided to look only at the linear trend, the significant *F* at the .05 level would have to exceed the value in the *F* table for 1 degree of freedom (the *df* for any trend) and 21 degrees of freedom (the *df* for this study's error term). That value is 4.32.

If, however, you are going to analyze more than one trend, you must correct for the number of *F*s you are going to compute. The correction is simple: You divide the significance level you want (say .05) by the number of trends you will test. In this example, you are looking at two trends, so you are computing two Fs. Therefore, you should use the critical value of F for the significance level of .05/2, which is .025. So, rather than look in an F table listing the critical values of F for $p < .05$, you would look in an F table listing the critical values of F for $p < .025$. In this example, you would only declare a trend significant at the .05 level if the F for that trend exceeds the critical value for $F (1,21)$ at the .025 level: 5.83.

Obviously, the F for the linear component, $F(1,21) = .025$, falls far short of the critical value of 5.83. But what about the quadratic component? To determine whether the quadratic component is significant, you would follow the same steps as before. The only difference is that you would look at the "Quadratic" column of the table for the three-condition case instead of the "Linear" column.

Thus, you would first multiply each condition's treatment sums by the appropriate constants listed in the "Quadratic" column, add them together to get a sum, and square that sum. In other words,

$$[(1 \times 10) + (-2 \times 50) + (1 \times 12)]^2$$

which equals

$$[10 + (-100) + 12)]^2$$

which equals

$$(-78)^2$$

which equals

$$6084$$

Now that you have the numerator for your sum of squares (6084), you need to compute the denominator. As when you computed the denominator for the linear component's SS, you compute the denominator for the quadratic's SS by multiplying the number of observations in each condition (8) by the weighting factor. The difference is that whereas the weighting factor for the linear component was 2, the weighting factor for the quadratic component is, as the table tells us, 6. Thus, the denominator for the SS quadratic is 8 (the number of observations in each condition) × 6 (the weighting factor for the quadratic effect), which is 48.

To compute the SS quadratic, divide your numerator (6084) by your denominator (48). The result is 126.7 (because 6084/48 = 126.7).

Note that 126.7, your SS quadratic, is also your MS quadratic because the MS quadratic is always the same as the SS quadratic. (The reason the MS quadratic is always the same as the SS quadratic is that (a) the MS quadratic always equals the SS quadratic /df quadratic, (b) the df quadratic always equals 1, and (c) SS/1 always equals SS.)

To get the F for the quadratic trend, you would divide the MS quadratic (126.7) by MS error (10). Therefore, the F for the quadratic trend is 126.7/10 = 12.67. As before, the critical value for the comparison is the F value for the .025 significance level with 1 and 21 degrees of freedom is 5.83. Because our F of 12.67 exceeds the critical value of 5.83, we have a statistically significant quadratic trend.

By adding the results of these two trend analyses to the previous ANOVA results, we can now produce the following table:

	df	SS	MS	F
Sugar Main Effect	2	126.95	63.47	6.35*
Linear	1	0.25	0.25	0.02
Quadratic	1	126.70	126.70	12.67*
Error Term	21	210.00	10.00	

*Significant at .05 level.
© Cengage Learning 2013

From looking at the table, you see that if you add up the degrees of freedom for all the trends involved in the sugar main effect (1 + 1), you get the total df for the sugar main effect (2). More importantly, note that if you add up the sum of squares for the quadratic and linear trends (126.70 + .25), you get the sum of squares for the overall effect (126.95). This fact gives you a way to check your work. Specifically, if the total of the sums of squares for all the trends does not add up to the sum of squares for the overall effect, you have made a mistake.

USING TABLE 5 TO COMPUTE POST HOC TESTS

Suppose you do an experiment in which you compare the effects of three colors on mood. For example, one third of your participants are put in a blue room, one third are put in a green room, and one third are put in a yellow room. If your ANOVA tells you that the color has an effect, you still do not know *which* colors significantly differ from each other. To find out which conditions differ from each other, you can use post hoc tests, such as the Tukey test *after* an analysis of variance finds a significant main effect for a multilevel factor.

To see how you could use Table 5 to compute post hoc tests, suppose that an investigator studies 24 participants (8 in each group) to examine the effect of color (blue, green, or yellow) on mood. As you can see from the following table, the investigator's ANOVA table reveals a significant effect of color.

Source	Sum of squares	Degrees of freedom	Mean square	F
Color	64	2	32.0	4.0*
Error	168	21	8.0	

*Significant at .05 level.
© Cengage Learning 2013

The means for the three color conditions are

Blue	Green	Yellow
10.0	5.0	8.0

The question is, "Which conditions differ from one another?" Does yellow cause a different mood than green? Does blue cause a different mood than yellow? To find out, we need to do a post hoc test. For this example, we will do the Tukey test.

The formula for the Tukey test is

$$\frac{\text{Mean 1} - \text{Mean 2}}{\sqrt{(MSE \times 1/\text{number of observations per condition})}}$$

TABLE 5 Critical Values for the Tukey Test at the .05 Level of Significance

df error	Number of Means							
	2	**3**	**4**	**5**	**6**	**7**	**8**	**9**
10	3.15	3.88	4.33	4.65	4.91	5.12	5.30	5.46
11	3.11	3.82	4.26	4.57	4.82	5.03	5.20	5.35
12	3.08	3.77	4.20	4.51	4.75	4.95	5.12	5.27
13	3.06	3.73	4.15	4.45	4.69	4.88	5.05	5.19
14	3.03	3.70	4.11	4.41	4.64	4.83	4.99	5.13
15	3.01	3.67	4.08	4.37	4.59	4.78	4.94	5.08
16	3.00	3.65	4.05	4.33	4.56	4.74	4.90	5.03
17	2.98	3.63	4.02	4.30	4.52	4.70	4.86	4.99
18	2.97	3.61	4.00	4.28	4.49	4.67	4.82	4.96
19	2.96	3.59	3.98	4.25	4.47	4.65	4.79	4.92
20	2.95	3.58	3.96	4.23	4.45	4.62	4.77	4.90
21	2.95	3.57	3.95	4.22	4.43	4.60	4.75	4.88
30	2.89	3.49	3.85	4.10	4.30	4.46	4.60	4.72
40	2.86	3.44	3.79	4.04	4.23	4.39	4.52	4.63
60	2.83	3.40	3.74	3.98	4.16	4.31	4.44	4.55
120	2.80	3.36	3.68	3.92	4.10	4.24	4.36	4.47
∞	2.77	3.31	3.63	3.86	4.03	4.17	4.29	4.39

Because the mean square error is 8 (see original ANOVA table) and there are 8 participants in each group, the denominator in this example will always be

$$\sqrt{(8 \times 1/8)}$$

which equals

$$\sqrt{8/8}$$

which equals

$$\sqrt{1}$$

which equals

$$1$$

The numerator, because it is the difference between the means, will change, depending on what means you are comparing. If you are comparing blue mean and green mean, the numerator would be the blue mean (10) minus the green mean (5), which equals 5 (because $10 - 5 = 5$). So, to see whether the blue and green conditions differ significantly, you would do the following calculations:

$$\frac{10.0(\text{blue mean}) - 5.0(\text{green mean})}{\sqrt{(8 \times 1/8)}} = \frac{5.0}{\sqrt{1}} = \frac{5.0}{1.0} = 5.0$$

To find out whether 5.0 is significant, go to Table 5 and look at the column labeled "3" because you have three means (blue, green, yellow). Then, go down the column until you hit row 21 because you have 21 degrees of freedom in your error term (as you can see by looking at the original ANOVA table). The value in that table is 3.57. This is the critical value that you will use in all your comparisons. If your Tukey statistic for a pair of means is larger than this critical value, there is a significant difference between conditions. Because 5.0 is greater than 3.57, your result is significant at the .05 level.

But, do blue and yellow differ? To find out, compute the Tukey statistic using the blue mean (10) minus the yellow mean (8) as the numerator, as we have done below:

$$\frac{10.0 - 8.0}{\sqrt{(8 \times 1/8)}} = \frac{2.0}{\sqrt{1}} = \frac{2.0}{1.0} = 2.0$$

Because 2.0 is less than our critical value of 3.57, the difference between blue and yellow is not statistically significant at the .05 level.

Do yellow and green differ?

$$\frac{8.0 - 5.0}{\sqrt{(8 \times 1/8)}} = \frac{3.0}{\sqrt{1}} = \frac{3.0}{1.0} = 3.0$$

Because 3.0 is less than our critical value of 3.57, the difference between yellow and green is not statistically significant at the .05 level.

DIRECTIONS FOR USING TABLE 6

If you are doing an experiment, you can use Table 6 to randomly assign participants to treatment condition (to learn how, see the next page). If you are doing a survey, you can use Table 6 to generate a random sample (to learn how, see page 694).

TABLE 6 Table of Random Numbers

5	28	80	31	99	77	39	23	69	0	15	49	100	2	22	64	73	92	53
29	71	48	4	87	32	17	90	89	9	99	34	58	8	61	73	98	48	89
90	94	19	80	70	36	2	17	48	63	82	39	85	26	65	27	81	69	83
62	66	48	74	86	6	66	41	15	65	6	41	85	57	84	64	70	39	64
67	54	3	54	23	40	25	95	93	55	59	46	77	55	49	82	26	8	87
75	27	62	15	81	36	22	26	69	42	44	91	55	0	84	48	68	65	5
70	19	7	100	94	53	81	76	73	40	22	58	49	42	96	18	66	89	8
75	7	9	20	58	92	41	42	79	26	91	44	63	87	45	21	23	15	6
55	70	10	23	25	73	91	72	29	47	93	58	21	75	80	52	9	12	36
83	42	62	53	55	12	11	54	19	2	45	43	67	13	5	74	30	93	11
94	20	76	23	65	72	55	27	44	19	10	72	50	67	83	18	67	22	49
51	10	72	9	59	47	66	32	17	6	75	8	54	22	37	3	46	83	95
99	50	22	2	92	9	98	9	40	23	34	8	63	58	49	31	70	39	83
9	12	3	23	2	0	82	75	36	63	71	19	78	26	66	63	16	75	7
20	40	50	29	51	82	81	47	73	69	74	100	80	37	14	67	1	90	92
90	92	54	52	74	0	88	71	45	49	38	54	80	2	85	42	75	47	20
25	6	92	30	19	31	22	41	0	22	79	87	84	61	6	19	67	97	60
13	12	94	76	29	61	50	67	29	76	27	70	97	16	83	88	100	22	48
91	77	51	3	92	85	46	22	0	58	84	64	87	93	94	94	13	98	41
29	12	39	35	32	47	30	81	40	32	37	8	48	81	50	77	18	39	7
43	96	86	14	91	24	22	85	16	51	42	37	41	100	94	76	45	50	67
57	44	72	45	87	21	7	29	26	82	69	99	10	39	76	29	11	17	85
63	10	10	76	7	75	19	91	2	31	45	94	54	72	10	48	52	7	12
34	28	11	95	4	82	51	7	69	53	93	36	81	66	93	88	15	73	54

Source: This table is taken from the random numbers table in Appendix D of *Foundations of Behavioral Research*, 3rd ed. (pp. 642–643) by F. N. Kerlinger, 1986, New York: Holt, Rinehart and Winston. Copyright (c) 1986 by Holt, Rinehart and Winston. Reprinted by permission.

RANDOMLY ASSIGNING PARTICIPANTS TO GROUPS IN AN EXPERIMENT

STEP 1: Across the top of a piece of paper, write down your conditions. Under each condition, draw a line for each participant you will need. In this example, we had three conditions and needed 12 participants.

Group 1	Group 2	Group 3
_____	_____	_____
_____	_____	_____
_____	_____	_____
_____	_____	_____

STEP 2: Turn to Table 6. Roll a die to determine in which column in the table you will start.

STEP 3: Assign the first number in the column to the first space under Group 1, the second number to the second space, and so on. When you have filled the spaces for Group 1, put the next number under the first space under Group 2. Similarly, when you fill all the spaces under Group 2, place the next number in the first space under Group 3. Thus, if we had started in the second column, our sheet of paper would now look like this:

Group 1	Group 2	Group 3
28	54	70
71	27	42
94	19	20
66	7	10

STEP 4: Assign the first person who participates in your study to the condition with the lowest random number. The second participant will be in the condition with the second-lowest random number, and so on. Thus, in this example, your first participant would be in Group 2 and your second participant would be in Group 3. To be more specific,

Participant 1 (*7*) = Group 2

Participant 2 (*10*) = Group 3

Participant 3 (*19*) = Group 2

Participant 4 (*20*) = Group 3

Participant 5 (*27*) = Group 2

Participant 6 (*28*) = Group 1

Participant 7 (*42*) = Group 3

Participant 8 (*54*) = Group 2

Participant 9 (*66*) = Group 1

Participant 10 (*70*) = Group 3

Participant 11 (*71*) = Group 1

Participant 12 (*94*) = Group 1

USING TABLE 6 TO GET A RANDOM SAMPLE

STEP 1: Determine how large your sample will be.

STEP 2: Get a list of your population and put a line next to each individual's name.

STEP 3: Turn to Table 6. Roll a die to determine in which column in the table you will start.

STEP 4: Assign the first number in the column to the first name on your list, the second number to the second space, until you have assigned numbers to all your names.

STEP 5: Put your participants in order based on their random number. Thus, the individual with the lowest random number next to his or her name would be the first on the list, the individual with the second-lowest random number would be the second, and so on.

STEP 6: Go down the list to get your sample. If your sample size will be 50, pick the first 50 individuals on the list. If your sample size will be 100, pick the first 100 individuals on the list.

A–B design The simplest single-*n* design, consisting of measuring the participant's behavior at baseline (A) and then measuring the participant after the participant has received the treatment (B).

A–B–A reversal design See *reversal design.*

Abstract A short (fewer than 120 words), one-page summary of a research proposal or an article.

Alpha (α) If referring to a measure, see *Cronbach's alpha*; otherwise, see *probability value.*

Analysis of variance (ANOVA) A statistical test for analyzing data from experiments that is especially useful when the experiment has more than one independent variable or more than two levels of an independent variable.

Archival data Data from existing records and public archives.

Baseline A participant's behavior on the task before receiving the treatment. A measure of the dependent variable as it occurs without the experimental manipulation. Used as a standard of comparison in single-subject and small-*n* designs.

Between-groups variance (mean square treatment, mean square between) An index of the degree to which group means differ; an index of the combined effects of random error and treatment. This quantity is compared to the within-groups variance in ANOVA. It is the top half of the *F* ratio. If the treatment has no effect, the between-groups variance should be roughly the same as the within-groups variance. If the treatment has an effect, the between-groups variance should be larger than the within-groups variance.

Bias Systematic errors that can push the scores in a given direction. Bias may lead to "finding" the results that the researcher wanted.

Blind (also called *masked*) A strategy of making the participant or researcher unaware of which experimental condition the participant is in.

Blocked design A factorial design in which, to boost power, participants are first divided into groups (blocks) on a subject variable (e.g., low-IQ block and high-IQ block). Then, participants from each block are randomly assigned to an experimental condition. Ideally, a blocked design will be more powerful than a simple, between-subjects design.

Carryover (treatment carryover) effect When a treatment administered earlier in the experiment affects participants when those participants are receiving additional treatments. Carryover effects may make it hard to interpret the results of single-subject and within-subjects designs because they may make it hard to know whether the participant's change in behavior is a reaction to the treatment just administered or to a delayed reaction to a treatment administered some time ago.

Ceiling effect The effect of treatment(s) is underestimated because the dependent measure is not sensitive to psychological states above a certain level. The measure puts an artificially low ceiling on how high a participant may score.

Central limit theorem If numerous large samples (30 or more scores) from the same population are taken, and you plot the mean for each of these samples, your plot would resemble a normal curve—even if the population from which you took those samples was not normally distributed.

Chi square (χ^2) test A statistical test you can use to determine whether two or more variables are related. Best used when you have nominal data.

Coefficient of determination (r^2 or η^2) The square of the correlation coefficient; tells the degree to which knowing one variable helps to know another. This measure of effect size can range from 0 (knowing a participant's score on one variable tells you absolutely nothing about the participant's score on the second variable) to 1.00 (knowing a participant's score on one variable tells you the participant's exact score on the second variable). A coefficient of determination of .09 is considered medium, and a coefficient of determination of .25 is considered large.

Cohen's *d* A measure of effect size that tells you how different two groups are in terms of standard deviations. Traditionally, a Cohen's *d* of .2 is considered small, .5 is considered moderate, and .8 is considered large.

Conceptual replication A study that is based on the original study but uses different methods to assess the true relationships between the treatment and dependent variables better. In a conceptual replication, you might use a different manipulation or a different measure.

Confounding variables Variables, other than the independent variable, that may be responsible for the differences between your conditions. There are two types of confounding variables: ones that are manipulation irrelevant and ones that are the result of the manipulation. Confounding variables that are irrelevant to the treatment manipulation threaten internal validity. For example, the difference between groups may be due to one group being older than the other rather than to the treatment. Random assignment can control for the effects of those confounding variables. Confounding variables that are produced by the treatment manipulation hurt the construct validity of the study because even though we may know that the treatment manipulation had an effect, we don't know what it was about the treatment manipulation that had the effect. For example, we may know that an "exercise" manipulation increases happiness (internal validity), but not know whether the "exercise" manipulation worked because people exercised more, got more encouragement, had a more structured routine, practiced setting and achieving goals, or met new friends. In such a case, construct validity is questionable because it would be questionable to label the manipulation an "exercise" manipulation.

Construct A mental state such as love, intelligence, hunger, and aggression that cannot be directly observed or manipulated with our present technology.

Construct validity The degree to which a study, test, or manipulation measures and/or manipulates what the researcher claims it does. For example, a test claiming to measure aggressiveness would not have construct validity if what it actually measured was assertiveness.

Content analysis A method used to categorize a wide range of open-ended (unrestricted) responses. Content analysis schemes have been used to code the frequency of violence on certain television shows and are often used to code archival data.

Content validity The extent to which a measure represents a balanced and adequate sampling of relevant dimensions, knowledge, and skills. In many measures and tests, participants are asked a few questions from a large body of knowledge. A test has content validity if its content is a fair sample of the larger body of knowledge. Students hope that their psychology tests have content validity.

Control group Participants who are randomly assigned to *not* receive the experimental treatment. These participants are compared to the treatment group to determine whether the treatment had an effect.

Convenience sampling Including people in your sample simply because they are easy (convenient) to survey. It is hard to generalize the results accurately from a study that used convenience sampling.

Convergent validity Validity demonstrated by showing that the measure correlates with other measures of the construct.

Correlation coefficient A number that can vary from -1.00 to $+1.00$ and indicates the kind of relationship that exists between two variables (positive or negative as indicated by the sign of the correlation coefficient) and the strength of the relationship (indicated by the extent to which the coefficient differs from 0). Positive correlations indicate that the variables tend to go in the same direction (if a participant is low on one variable, the participant will tend to be low on the other). Negative correlations indicate that the variables tend to head in opposite directions (if a participant is low on one, the participant will tend to be high on the other).

Counterbalanced within-subjects design Design that gives participants the treatments in different sequences. These designs balance out routine order effects.

Covariation Changes in the treatment are accompanied by changes in the behavior. To establish causality, you must establish covariation.

Cronbach's alpha A measure of internal consistency. To be considered internally consistent, a measure's Cronbach's alpha should be at least above .70 (most researchers would like to see it above .80).

Crossover (disordinal) interaction When an independent variable has one kind of effect in the presence of one level of a second independent variable, but a different kind of effect in the presence of a different level of the second independent variable. Examples: Getting closer to people may increase their attraction to you if you have just complimented them, but may decrease their attraction to you if you have just insulted them. Called a *crossover interaction* because the lines in a graph will cross. Called *disordinal interaction* because it cannot be explained by having ordinal rather than interval data.

Debriefing Giving participants the details of a study at the end of their participation. Proper debriefing is one of the researcher's most serious obligations.

Degrees of freedom (*df*) An index of sample size. In the simple experiment, the *df* for your error term will always be two less than the number of participants.

Demand characteristics Characteristics of the study that suggest to the participant how the researcher might want the participant to behave.

Demographics Characteristics of a group, such as gender, age, social class.

Dependent groups *t* test A statistical test used with interval or ratio data to test differences between two conditions on a single dependent variable. Differs from the between-groups *t* test in that it is to be used only when you are getting two scores from each participant (within-subjects design) or when you are using a matched-pairs design.

Dependent variable (dependent measure) The factor that the experimenter predicts is affected by the independent variable; the participant's response that the experimenter is measuring.

Descriptive hypothesis A hypothesis about a group's characteristics or about the correlations between variables; a hypothesis that does not involve a cause–effect statement.

Dichotomous questions Questions that allow only two responses (usually "yes" or "no").

Direct (exact) replication Repeating a study as exactly as possible, usually to determine whether or not the same results will be obtained. Direct replications are useful for establishing that the findings of the original study are reliable.

Discriminant validity When a measure does not correlate highly with a measure of a different construct. Example: A violence measure might have a degree of discriminant validity if it does not correlate with the measures of assertiveness, social desirability, and independence.

Discussion The part of the article, immediately following the results section, that discusses the research findings and the study in a broader context and suggests research projects that could be done to follow up on the study.

Disordinal interaction See *crossover (disordinal) interaction.*

Double-barreled question A statement that contains more than one question. Responses to a double-barreled question are difficult to interpret. For example, if someone responds, "No," to the question "Are you hungry and thirsty?" we do not know whether he is hungry, but not thirsty; not hungry, but thirsty; or neither hungry nor thirsty.

Double-blind technique A strategy for improving construct validity that involves making sure that neither the participants nor the people who have direct contact with the participants know what type of treatment the participants have received.

Empty control group A group that does not get any kind of treatment. The group gets nothing, not even a placebo. Usually, because of participant and experimenter biases that may result from such a group, you will want to avoid using an empty control group.

Environmental manipulation A manipulation that involves changing the participant's environment rather than giving the participant different instructions.

Eta squared (η^2) An estimate of effect size that ranges from 0 to 1 and is comparable to *r*-squared.

Ethical Conforming to the American Psychological Association's principles of what is morally correct behavior. To learn more about these guidelines and standards, see Appendix D.

Ex post facto research When a researcher goes back, after the research has been completed, looking to test hypotheses that were not formulated prior to the beginning of the study. The researcher is trying to take advantage of hindsight. Often an attempt to salvage something out of a study that did not turn out as planned.

Experiment A study that allows researchers to disentangle treatment effects from natural differences between groups, usually by randomly assigning participants to treatment group. In medicine, such studies may be called controlled clinical trials or randomized clinical trials.

Experimental design A design in which (a) a treatment manipulation is administered and (b) that manipulation is the only variable that systematically varies between treatment conditions.

Experimental group Participants who are randomly assigned to receive the treatment.

Experimental hypothesis A prediction that the treatment will cause an effect.

Experimental (research) realism When a study engages the participant so much that the participant is not merely playing a role (helpful participant, good person).

Experimenter bias Experimenters being more attentive to participants in the treatment group or giving different nonverbal cues to treatment group participants than to other participants are examples of experimenter bias. When experimenter bias is present, differences between groups' results may be due to the experimenter treating the two groups differently rather than to the treatment.

Exploratory study A study investigating (exploring) a new area of research. Unlike replications, an exploratory study does not follow directly from an existing study.

External validity The degree to which the results of a study can be generalized to other participants, settings, and times.

Extraneous factor Factor other than the treatment. If we cannot control or account for extraneous variables, we can't conclude that the treatment had an effect. That is, we will not have internal validity.

F ratio Analysis of variance (ANOVA) yields an F ratio for each main effect and interaction. In between-subjects experiments, the F ratio is a ratio of between-groups variance to within-groups variance. If the treatment has no effect, F will tend to be close to 1.0.

Face validity The extent to which a measure looks, on the face of it, to be valid. Face validity has nothing to do with actual, scientific validity. That is, a test could have face validity and not real validity or could have real validity, but not face validity. However, for practical/political reasons, you may decide to consider face validity when comparing measures.

Factor analysis A statistical technique designed to explain the variability in several questions in terms of a smaller number of underlying hypothetical factors.

Factorial experiment An experiment that examines two or more independent variables (factors) at a time.

Fatigue effect Decreased participant performance on a task due to participants being tired or less enthusiastic as a study continues. In a within-subjects design, this decrease in performance might be incorrectly attributed to a treatment.

File drawer problem A situation in which the research not affected by Type 1 errors languishes in researchers' file cabinets, whereas the Type 1 errors are published.

Fixed-alternative question Item on a test or questionnaire in which a person must choose an answer from among a few specified alternatives. Multiple-choice, true–false, and rating-scale questions are all fixed-alternative questions.

Floor effect The effects of treatment(s) are underestimated because the dependent measure artificially restricts how low scores can be.

Frequency distribution A graph on which the frequencies of the scores are plotted. Thus, the highest point on the graph will be over the most commonly occurring score. Often, frequency distributions will look like the normal curve.

Functional relationship The shape of a relationship. Depending on the functional relationship between the independent and dependent variable, a graph of the relationship might look like a straight line or might look like a U, an S, or some other shape.

Hawthorne effect When members of the treatment group change their behavior not because of the treatment itself, but because they are getting special treatment.

History Events in the environment—other than the treatment—that have changed. Differences between conditions that may seem to be due to the treatment may really be due to history.

Hypothesis A testable prediction about the relationship between two or more variables.

Hypothesis-guessing When participants alter their behavior to conform to their guess as to what the research hypothesis is. Hypothesis-guessing can be a serious threat to construct validity, especially if participants guess correctly.

Hypothesis testing The use of inferential statistics to determine if the relationship found between two or more variables in a particular sample holds true in the population.

Hypothetical construct See *construct*.

Illusory correlation When there is in fact no relationship (a zero correlation) between two variables, but people perceive that the variables are related.

Independence Factors are independent when they are not causally or correlationally linked. Independence is a key assumption of most statistical tests. In the simple experiment, observations must be independent. That is, what one participant does should have no influence on what another participant does and what happens to one participant should not influence what happens to another participant. Individually assigning participants to the treatment or no-treatment condition and individually testing each participant are ways to achieve independence.

Independent random assignment Randomly determining for each individual participant which condition he will be in. For example, you might flip a coin for each participant to determine to what group he will be assigned.

Independent variable The variable being manipulated by the experimenter.

Participants are assigned to a level of independent variable by independent random assignment.

Inferential statistics Procedures for determining the reliability and generalizability of a particular research finding.

Informed consent If participants agree to take part in a study after they have been comprehensively told what is going to happen to them, you have their informed consent.

Institutional review board (IRB) A committee of at least five members—one of whom must be a nonscientist—that reviews proposed research in an effort to protect research participants.

Instructional manipulation Manipulating the treatment by giving written or oral instructions.

Instrumentation bias The way participants were measured changed from pretest to posttest. In instrumentation bias, the actual measuring instrument changes or the way it is administered changes. Sometimes people may think they have a treatment effect when they really have an instrumentation effect.

Interaction An interaction occurs when a relationship between two variables (e.g., X and Y) is affected by (is moderated by, depends on) the amount of a third variable (Z). You are probably most familiar with interactions involving drugs (e.g., two drugs may both be helpful but the combination of the two drugs is harmful or a drug is helpful, except for people with certain conditions). If you need to know how much of one variable participants have received to say what the effect of another variable is, you have an interaction between those two variables. If you graph the results from an experiment that has two or more independent variables, and the lines you draw between your points are not parallel, you may have an interaction. See also *moderator variable*.

Internal consistency The degree to which each question on a scale correlates with the other questions. Internal consistency is high if answers to each item correlate highly with answers to all other items.

Internal validity The degree to which a study establishes that a factor causes a difference in behavior. If a study lacks

internal validity, the researcher may falsely believe that a factor causes an effect when it really doesn't.

Interobserver (judge) agreement The percentage of times the raters agree.

Interobserver reliability An index of the degree to which different raters give the same behavior similar ratings.

Interval scale data Data that give you numbers that can be meaningfully ordered along a scale (from lowest to highest) and in which equal numerical intervals represent equal psychological intervals. That is, the difference between scoring a "2" and a "1" and the difference between scoring a "7" and a "6" are the same not only in terms of scores (both are a difference of 1), but also in terms of the actual psychological characteristic being measured. Interval scale measures allow us to compare participants in terms of how much of a quality participants have—and in terms of how much more of a quality one group may have than another.

Interview A survey in which the researcher orally asks questions.

Interviewer bias When an interviewer influences a participant's responses. For example, the interviewer might—consciously or unconsciously—verbally or nonverbally reward the participant for giving responses that support the research hypothesis.

Introduction The part of the article that occurs right after the abstract. In the introduction, the authors tell you what their hypothesis is, why their hypothesis makes sense, how their study fits in with previous research, and why their study was worth doing.

IRB See *Institutional Review Board*.

Known-groups technique A way of making the case for your measure's convergent validity that involves seeing whether groups known to differ on the characteristic you are trying to measure also differ on your measure (e.g., ministers should differ from atheists on an alleged measure of religiosity).

Laboratory observation A technique of observing participants in a laboratory setting.

Law of parsimony The assumption that the explanation that is simplest, most straightforward, and makes the fewest assumptions is the most likely.

Leading question Question structured to lead respondents to the answer the researcher wants (such as, "You like this book, don't you?").

Levels of an independent variable When the treatment variable is given in different kinds or amounts, these different values are called *levels*. In the simple experiment, you only have two levels of the independent variable.

Likert-type item Item that typically asks participants whether they strongly agree, agree, are neutral, disagree, or strongly disagree with a certain statement. These items are assumed to yield interval data.

Linear relationship A relationship between an independent and dependent variable that is graphically represented by a straight line.

Loose-protocol effect Variations in procedure because the written procedures (the protocol) is not detailed enough. These variations in procedure may result in researcher bias.

Main effect See *overall main effect.*

Manipulation check A question or set of questions designed to determine whether participants perceived the manipulation in the way that the researcher intended.

Matched-pairs design An experimental design in which the participants are paired off by matching them on some variable assumed to be correlated with the dependent variable. Then, for each matched pair, one member is randomly assigned to one treatment condition, and the other gets the other treatment condition. This design usually has more power than a simple, between-groups experiment.

Matching Choosing your groups so that they are similar (they match) on certain characteristics. Matching reduces, but does not eliminate, the threat of selection bias.

Maturation Changes in participants due to natural growth or development. A researcher may think that the treatment had an effect when the difference in behavior is really due to maturation.

Mean An average calculated by adding up all the scores and then dividing by the number of scores.

Median If you arrange all the scores from lowest to highest, the middle score will be the median.

Median split The procedure of dividing participants into two groups ("highs" and "lows") based on whether they score above or below the median.

Mediating variable Variables inside the individual (such as thoughts, feelings, or physiological responses) that come between a stimulus and a response. In other words, the stimulus has its effect because it causes changes in mediating variables, which, in turn, cause changes in behavior.

Method section The part of the article immediately following the introduction. Whereas the introduction explains *why* the study was done, the method section describes *what* was done. For example, it will tell you what design was used, what the researchers said to the participants, what measures and equipment were used, how many participants were studied, and how participants were selected. The method section could also be viewed as a "how we did it" section. The method section is usually subdivided into at least two subsections: participants and procedure.

Mixed design An experimental design that has at least one within-subjects factor and one between-subjects factor.

Mode The score that occurred most often; the most frequent score. For example, 2 is the mode of the following data set: 2, 2, 2, 6, 10, 50.

Moderator variable Variable that can intensify, weaken, or reverse the effects of another variable. For example, the effect of wearing perfume may be moderated by gender: If you are a woman, wearing perfume may make you more liked; if you are a man, wearing perfume may make you less liked.

Mortality (attrition) Participants dropping out of a study before the study is completed. Sometimes, differences between conditions may be due to participants dropping out of the study rather than the treatment.

Multiple-baseline design A single-subject or small-*n* design in which different behaviors receive baseline periods of varying lengths prior to the introduction of the treatment variable. Often, the goal is to show that the behavior being rewarded changes, whereas the other behaviors stay the same until they too are reinforced.

Multiple regression A statistical technique that can take data from several predictors and an outcome variable to create a formula that weights the predictors in such a way as to make the best possible estimates of the outcome variable given those predictors. In linear multiple regression, this equation is for the straight line that best predicts the outcome data. Often, with multiple regression, you not only are able to predict your outcome variable with accuracy but you are also able to tell which predictors are most important for making accurate predictions. For more information on multiple regression, see Appendix E.

Naturalistic observation A technique of observing events as they occur in their natural setting.

Negative correlation An inverse relationship between two variables (such as number of suicide attempts and happiness).

95% confidence interval A range in which the parameter you are estimating (usually the population mean) falls 95% of the time.

Nominal–dichotomous item A question that presents participants with only two—usually very different—options (e.g., "Are you for or against animal research?"). Such questions are often yes/no questions and often ask the participant to classify herself or himself into one of two different categories.

Nominal-scale numbers Numbers that do not represent different amounts of a characteristic but instead represent different kinds of characteristics (qualities, types, or categories); numbers that substitute for names.

Nonequivalent control-group design A quasi-experimental design that, like a simple experiment, has a treatment group and a no-treatment comparison group. However, unlike the simple experiment, random assignment does not determine which participants get the treatment and which do not.

Nonreactive measure Measurement that is taken without changing the participant's behavior; also referred to as *unobtrusive measure.*

Nonresponse bias The problem caused by the refusal of people who were in your sample to participate in your study. Nonresponse bias is one of the most serious threats to a survey design's external validity.

Nonsignificant results See *null results*.

Normal curve A bell-shaped, symmetrical frequency distribution that has its center at the mean.

Normal distribution If the way the scores are distributed follows the normal curve, scores are said to be normally distributed. For example, a population is said to be normally distributed if 68% of the scores are within 1 standard deviation of the mean, 95% are within 2 standard deviations of the mean, and 99% of the scores are within 3 standard deviations of the mean. Many statistical tests, including the *t* test, assume that sample means are normally distributed.

Null hypothesis The hypothesis that there is no relationship between two or more variables. The null hypothesis can be disproven, but it cannot be proven.

Null results (nonsignificant results) Results that fail to disconfirm the null hypothesis; results that fail to provide convincing evidence that the factors are related. Null results are inconclusive because the failure to find a relationship could be due to your design lacking the power to find the relationship. In other words, many null results are Type 2 errors.

Observer bias Bias created by the observer seeing what he or she wants or expects to see.

Open-ended question Question that does not ask participants to choose between the responses provided by the researcher (e.g., choosing "a," "b," or "c" on a multiple-choice question or choosing a number between 1 and 5 on a rating scale measure) but instead asks the participant to generate a response. Essay and fill-in-the-blank questions are open-ended questions.

Operational definition A publicly observable way to measure or manipulate a variable; a "recipe" for how you are going to measure or manipulate your factors.

Order The place in a sequence (first, second, third, etc.) when a treatment occurs.

Order (trial) effects A big problem with within-subjects designs. The order in which the participant receives a

treatment (first, second, etc.) will affect how participants behave.

Ordinal scale numbers Numbers that can be meaningfully ordered from lowest to highest. Ranks (e.g., class rank, order in which participants finished a task) are ordinal scale numbers.

Overall main effect The overall or average effect of an independent variable.

p < .05 level A traditional significance level; if the variables are unrelated, results significant at this level would occur less than 5 times out of 100. Traditionally, results that are significant at the *p* < .05 level are considered statistically reliable and thus replicable.

Parameter estimation The use of inferential statistics to estimate certain characteristics of the population (parameters) from a sample of that population.

Parameters Measurements describing populations; often inferred from statistics, which are measurements describing a sample.

Parsimony See *law of parsimony*.

Participant bias Participants trying to behave in a way that they believe will support the researcher's hypothesis.

Participant observation An observation procedure in which the observer participates with those being observed. The observer becomes "one of them."

Placebo treatment A fake treatment that we know has no effect, except through the power of suggestion. It allows experimenters to see if the treatment has an effect beyond that of suggestion. For example, in medical experiments, participants who are given placebos (pills that do not contain a drug) may be compared to participants who are given pills that contain the new drug.

Plagiarism Using someone else's words, thoughts, or work without giving proper credit.

Population The entire group that you are interested in. You can estimate the characteristics of a population by taking large random samples from that population.

Positive correlation A relationship between two variables in which the two

variables tend to vary together—when one increases, the other tends to increase. (For example, height and weight have a positive correlation: The taller one is, the more one tends to weigh; the shorter one is, the less one tends to weigh.)

Post hoc test Usually refers to a statistical test that has been performed after an ANOVA has obtained a significant effect for a factor. Because the ANOVA says only that at least two of the groups differ from one another, post hoc tests are performed to find out which groups differ from one another.

Post hoc trend analysis A type of post hoc test designed to determine whether a linear or curvilinear relationship is statistically significant (reliable).

Power The ability to find statistically significant differences when differences truly exist; the ability to avoid making Type 2 errors.

Practice effect The change in a score on a test (usually a gain) resulting from previous practice with the test. In a within-subjects design, this improvement might be incorrectly attributed to participants having received a treatment.

Pretest–posttest design A before–after design in which each participant is given the pretest, administered the treatment, then given the posttest.

Probability value (p value) The chances of obtaining a certain pattern of results if there really is no relationship between the variables.

Proportionate stratified random sampling Technique ensuring that the sample is similar to the population in certain respects (for instance, percentage of men and women) and then randomly sampling from these groups (strata) and having all the advantages of random sampling but with even greater accuracy.

Psychological Abstracts A useful resource that contains abstracts from a wide variety of journals. The *Abstracts* can be searched by year of publication, topic of article, or author. For more about the *Abstracts*, see Web Appendix B.

Psychological construct See *construct*.

PsycINFO The computerized version of *Psychological Abstracts*.

Quadratic relationship A relationship on a graph shaped like a "U" or an upside down "U."

Quasi-experiment A study that resembles an experiment except that random assignment played no role in determining which participants got which level of treatment. Usually, quasi-experiments have less internal validity than experiments.

Questionnaire A written survey instrument.

Quota sampling Technique ensuring that you get the desired number of (meet your quotas for) certain types of people (certain age groups, minorities, etc.). This method does not involve random sampling and usually gives you a less representative sample than random sampling would. However, it may be an improvement over convenience sampling.

Random assignment See *independent random assignment*.

Random-digit dialing Finding participants for telephone interviews by taking the area code and the 3-digit prefixes that you are interested in and then adding random digits to the end to create 10-digit phone numbers. You may use this technique when (a) you cannot afford to buy a list of phone numbers and then randomly select numbers from that list or (b) you want to contact people with unlisted numbers.

Random error Variations in scores due to unsystematic, chance factors.

Randomized controlled trials (RCTs) Laboratory experiments in which participants are randomly assigned to one of two (or more) groups. These studies have impressive internal validity, especially relative to correlational studies.

Random sampling A sample that has been randomly selected from a population. If you randomly select enough participants, those participants will usually be fairly representative of the entire population. That is, your random sample will reflect its population. Often, random sampling is used to maximize a study's external validity. Note that random sampling—unlike random assignment—does not promote internal validity.

Randomized within-subjects design As in all within-subjects designs, all participants receive more than one level or type of treatment. However, to make sure that not every participant receives the series of treatments in the same sequence, the researcher randomly determines which treatment comes first, which comes second, and so on. In other words, participants all get the same treatments, but they receive different sequences of treatments.

Ratio scale data The highest form of measurement. With ratio scale numbers, the difference between any two consecutive numbers is the same (see *interval scale*). But in addition to having interval scale properties, in ratio scale measurement, a zero score means the total absence of a quality. (Thus, Fahrenheit is not a ratio scale measure of temperature because 0 degrees Fahrenheit does not mean there is no temperature.) If you have ratio scale numbers, you can meaningfully form ratios between scores. If IQ scores were ratio (they are not; very few measurements in psychology are), you could say that someone with a 60 IQ was twice as smart as someone with a 30 IQ (a ratio of 2 to 1). Furthermore, you could say that someone with a 0 IQ had absolutely no intelligence whatsoever.

Regression (toward the mean) The tendency for scores that are extremely unusual to revert back to more normal levels on the retest. If participants are chosen because their scores were extreme, these extreme scores may be loaded with extreme amounts of random measurement error. On retesting, participants are bound to get more normal scores as random measurement error abates to more normal levels. This regression effect could be mistaken for a treatment effect.

Reliability A general term, often referring to the degree to which a participant would get the same score if retested (test–retest reliability). Reliability can, however, refer to the degree to which scores are free from random error. A measure can be reliable, but not valid. However, a measure cannot be valid if it is not also reliable.

Repeated-measures design See *within-subjects design*.

Replicable Repeatable. A researcher should be able to repeat another researcher's study and obtain the same pattern of results.

Replicate Repeat, or duplicate, an original study.

Research journal A relatively informal notebook in which you jot down your research ideas and observations. The research journal can be a useful resource when it comes time to write the research proposal. Note: Despite the fact that they sound similar, the term "research journal" is not similar to the term "scientific journal." The term "scientific journal" is used to distinguish journals from magazines. In contrast to magazines, scientific journals tend (1) not to have ads for popular products, (2) not to have full-page color pictures, (3) to have articles that follow APA format (having abstract, introduction, method, results, discussion, and reference sections), and (4) to have articles that have been peer-reviewed.

Researcher effect Ideally, you hope that the results from a study would be the same no matter who was conducting it. However, it is possible that the results may be affected by the researcher. If the researcher is affecting the results, there is a researcher effect.

Research Ethics Board See *Institutional Research Board*.

Researcher-expectancy effect When a researcher's expectations affect the results. This is a type of researcher bias.

Response set Habitual way of responding on a test or survey that is independent of a particular test item (for instance, a participant might always check "agree" no matter what the statement is).

Restriction of range To observe a sizable correlation between two variables, both must be allowed to vary widely (if one variable does not vary, the variables cannot vary together). Occasionally, investigators fail to find a relationship between variables because they study only one or both variables over a highly restricted range. Example: comparing NFL offensive linemen and saying that weight has nothing to do with playing offensive line in the NFL on the basis of your finding that great offensive tackles do not weigh much more than poor offensive tackles. Problem: You compared only people who ranged in weight from 315 to 330 pounds.

Results section The part of an article, immediately following the method section, that reports statistical results and relates those results to the hypotheses. From reading this section, you should know whether the results supported the hypotheses.

Retrospective self-report Participants telling you what they said, did, or believed in the past. In addition to problems with ordinary self-report (response sets, giving the answer that a leading question suggests, etc.), retrospective self-report is vulnerable to memory biases. Thus, retrospective self-reports should *not* be accepted at face value.

Reversal design (A–B–A design, A–B–A reversal design) A single-subject or small-*n* design in which baseline measurements are made of the target behavior (A), then an experimental treatment is given (B), and the target behavior is measured again (A). The A–B–A design makes a more convincing case for the treatment's effect than the A–B design.

Scatterplot A graph made by plotting the scores of individuals on two variables (e.g., each participant's height and weight). By looking at this graph, you should get an idea of what kind of relationship (positive, negative, zero) exists between the two variables.

Selection (or selection bias) Apparent treatment effects being due to comparing groups that differed even before the treatment was administered (comparing apples with oranges).

Selection by maturation interaction When treatment and no-treatment groups, although similar at one point, would have grown apart (developed differently) even if no treatment had been administered.

Self-administered questionnaire A questionnaire filled out in the absence of an investigator.

Semistructured interview An interview constructed around a core of standard questions; however, the interviewer may expand on any question in order to explore a given response in greater depth.

Sensitive, sensitivity The degree to which a measure is capable of distinguishing between participants who differ on a variable (e.g., have different amounts of a construct or who do more of a certain behavior).

Sensitization After getting several different treatments and performing the dependent variable task several times, participants may realize (become sensitive to) what the hypothesis is. Sensitization is a problem in within-subjects designs.

Sequence effect When participants who receive one sequence of treatments score differently (i.e., significantly lower or higher) than those participants who receive the same treatments in a different sequence.

Significance level See *probability value.*

Simple experiment A study in which participants are independently and randomly assigned to one of two groups, usually to either a treatment group or to a no-treatment group. It is the easiest way to establish that a treatment causes an effect.

Simple main effect The effects of one independent variable at a specific level of a second independent variable. The simple main effect could have been obtained merely by doing a simple experiment.

Single blind To reduce either subject biases or researcher biases, you might use a single-blind experiment in which either the participant (if you are most concerned about subject bias) or the person running participants (if you are more concerned about researcher bias) is unaware of who is receiving what level of the treatment. If you are concerned about both subject and researcher bias, then you should probably use a double-blind study.

Single-*n* designs See *single-subject design.*

Single-subject design Design that tries to establish causality by studying a single participant and arguing that the covariation between treatment and changes in behavior could not be due to anything other than the treatment. A key to this approach is to prevent factors other than the treatment from varying. Single-*n* designs are common in operant conditioning and psychophysical research. See also *A–B design, A–B–A reversal design, multiple-baseline design.*

Social desirability bias A bias resulting from participants giving responses that make them look good rather than giving honest responses.

Spurious When the covariation observed between two variables is not due to the variables influencing each other, but is because both are being influenced by some third variable. For example, the relationship between ice cream sales and assaults in New York is spurious—not because it does not exist (it does!)—but because ice cream does not cause assaults, and assaults do not cause ice cream sales. Instead, high temperatures probably cause both increased assaults and ice cream sales. Beware of spuriousness whenever you look at research that does not use an experimental design.

Stable baseline When the participant's behavior, prior to receiving the treatment, is consistent. Single-*n* experimenters try to establish a stable baseline.

Standard deviation A measure of the extent to which individual scores deviate from the population mean. The more scores vary from each other, the larger the standard deviation will tend to be. If, on the other hand, all the scores are the same as the mean, the standard deviation would be zero.

Standard error of the difference An index of the degree to which random sampling error may cause two sample means representing the same populations to differ. In the simple experiment, if we are to find a treatment effect, the difference between our experimental-group mean and control-group mean will usually be at least twice as big as the standard error of the difference. To find out the exact ratio between our observed difference and the standard error of the difference, we conduct a *t* test.

Standard error of the mean An index of the degree to which random error may cause the sample mean to be an inaccurate estimate of the population mean. The standard error will be small when the standard deviation is small, and the sample mean is based on many scores.

Standardization Treating each participant in the same (standard) way. Standardization can reduce both bias and random error.

Statistical regression See *regression (toward the mean).*

Statistical significance When a statistical test says that the relationship we have observed is probably not due to chance alone, we say that the results are statistically significant. In other words, because the relationship is probably not due to chance, we conclude that there probably is a real relationship between our variables.

Stimulus set The particular stimulus materials that are shown to two or more groups of participants. Researchers may use more than one stimulus set in a study so that they can see whether the treatment effect replicates across different stimulus sets. In those cases, stimulus sets would be a replication factor.

Stooge Confederate who pretends to be a participant but is actually a researcher's assistant. The use of stooges raises ethical questions.

Stratified sampling See *proportionate stratified sampling*.

Straw theory An oversimplified version of an existing theory. Opponents of a theory may present and attack a straw version of that theory but claim they have attacked the theory itself.

Structured interview An interview in which all respondents are asked a standard list of questions in a standard order.

Subject bias (subject effects) When the participants bias the results by guessing the hypothesis and playing along or by giving the socially correct response.

Summated score When you have several Likert-type questions that all tap the same dimension (such as attitude toward democracy), you can add up each participant's responses to those questions to get an overall, total (summated) score.

Survey A non-experimental design useful for describing how people think, feel, or behave. The key is to design a valid questionnaire, test, or interview and administer it to a representative sample of the group you are interested in.

Systematic replication A study that varies from the original study only in some minor aspect. For example, a systematic replication may use more participants, more standardized procedures, more levels of the independent variable, or a more realistic setting than the original study.

t **test** The most common way of analyzing data from a simple experiment. It involves computing a ratio between two things: (1) the difference between your group means and (2) the standard error of the difference (an index of the degree to which group means could differ by chance alone). If the difference you observe is more than three times bigger than the difference that could be expected by chance, then your results are probably statistically significant. We can only say "probably" because the exact ratio that you need for statistical significance depends on your level of significance and on how many participants you have.

Temporal precedence When the causal factor comes before the change in behavior. Because the cause must come before the effect, researchers trying to establish causality must establish that the factor alleged to be the cause was introduced before the behavior changed.

Test–retest reliability A way of assessing the amount of random error in a measure by administering the measure to participants at two different times and then correlating their results. If the measure is free of random error, scores on the retest should be highly correlated with scores on the original test.

Testing effect When participants score differently on the posttest as a result of what they learned from taking the pretest. Occasionally, people may think the participants' behavior changed because of the treatment when it really changed due to experience with the test.

Theory A set of principles that explain existing research findings and that can be used to make new predictions can lead to new research findings.

Time-series design A quasi-experimental design in which a series of observations are taken from a group of participants before and after they receive treatment. Because it uses many times of measurement, it is an improvement over the pretest–posttest design. However, it is still extremely vulnerable to history effects.

Trend analysis See *post hoc trend analysis*.

Type 1 error Rejecting the null hypothesis when it is in fact true. In other words, declaring a difference statistically significant when the difference is really due to chance.

Type 2 error Failure to reject the null hypothesis when it is in fact false. In other words, failing to find a relationship between your variables when there really is a relationship between them.

Unobtrusive measurement Recording a particular behavior without the participant knowing you are measuring that behavior. Unobtrusive measurement reduces subject biases such as social desirability bias and obeying demand characteristics.

Unstructured interview When the interviewer has no standard set of questions that he or she asks each participant—a virtually worthless approach for collecting scientifically valid data.

Valid Usually, a reference to whether a conclusion or claim is justified. A measure is considered valid when it measures what it claims to measure. See also *construct validity*, *internal validity*, and *external validity*.

Variability between group means See *between-groups variance*.

Within-groups variance (mean square within, mean square error, error variance) An estimate of the amount of random error in your data. The bottom half of the *F* ratio in a between-subjects analysis of variance.

Within-subjects design (repeated-measures design) An experimental design in which each participant is tested under more than one level of the independent variable. The sequence in which the participants receive the treatments is usually randomly determined. See also *randomized within-subjects design* and *counterbalanced within-subjects designs*.

Zero correlation When there doesn't appear to be a linear relationship between two variables. For practical purposes, any correlation between −.10 and +.10 may be considered so small as to be nonexistent.

Abelson, R. P. (1995). *Statistics as principled argument*. Mahwah, NJ: Erlbaum.

Abelson, R. P. (1997). On the surprising longevity of flogged horses: Why there is a case for the significance test. *Psychological Science, 8,* 12–15.

Ai, A. L., Park, C. L., Huang, B., Rodgers, W., & Tice, T. N. (2007). Psychosocial mediation of religious coping styles: A study of short-term psychological distress following cardiac surgery. *Personality and Social Psychology Bulletin, 33,* 867–882. doi:10.1177/0146167207301008

Ainsworth, M. D. S., & Bell, S. M. (1970). Attachment, exploration, and separation: Illustrated by the behavior of one-year-olds in a strange situation. *Child Development, 41,* 49–67.

Alexander, C. N., Langer, E. J., Newman, R. I., Chandler, H. M., & Davies, J. L. (1989). Transcendental meditation, mindfulness, and longevity: An experimental study with the elderly. *Journal of Personality and Social Psychology, 57,* 950–964.

Ambady, N., & Rosenthal, R. (1993). Half a minute: Predicting teacher evaluations from thin slices of nonverbal behavior and physical attractiveness. *Journal of Personality and Social Psychology, 64,* 431–444.

American Psychological Association. (1982). *Ethical principles in the conduct of research with human behavior*. Washington, DC: Author.

American Psychological Association. (1996a). *Guidelines for ethical conduct in the care and use of animals*. Washington, DC: Author.

American Psychological Association. (1996b). *Task force on statistical inference initial report*. Washington, DC: Author.

American Psychological Association. (2010). *Publication manual of the American Psychological Association* (6th ed.). Washington, DC: Author.

American Psychological Association. (2002). Ethical principles of psychologists and code of conduct. *American Psychologist, 57,* 1597–1611.

Anastasi, A. (1982). *Psychological testing* (5th ed.). New York, NY: Macmillan.

Anderson, C. A., & Bushman, B. J. (1997). External validity of "trivial" experiments: The case of laboratory aggression. *Review of General Psychology, 1,* 19–41.

Anderson, C. A., & Bushman, B. J. (2002, June/July). Media violence and the American public revisited. *American Psychologist, 57,* 448–450.

Anderson, C. A., Carnagey, N. L., & Eubanks, J. (2003). Exposure to violent media: The effects of songs with violent lyrics on aggressive thoughts and feelings. *Journal of Personality and Social Psychology, 84,* 960–971.

Anderson, C. A., Lindsay, J. J., & Bushman, B. J. (1999). Research in the psychological laboratory: Truth or triviality? *Current Directions in Psychological Science, 8,* 3–9.

Antill, J. K. (1983). Sex role complementarity versus similarity in married couples. *Journal of Personality and Social Psychology, 45,* 145–155.

Ariely, D. (2008). *Predictably irrational: The hidden forces that shape our decisions*. New York, NY: Harper.

Ariely, D., & Loewenstein, G. (2006). The heat of the moment: The effect of sexual arousal on decision making. *Journal of Behavioral Decision Making, 19,* 87–98.

Arkes, H. R. (2003). Psychology in Washington: The nonuse of psychological research at two federal agencies. *Psychological Science, 14,* 1–6.

Aronson, E. (1990). Applying social psychology to desegregation and energy conservation. *Personality and Social Psychology Bulletin, 16,* 118–131.

Aronson, E., & Carlsmith, J. M. (1968). Experimentation in social psychology. In G. Lindzey & E. Aronson (Eds.), *Handbook of social psychology* (2nd ed., pp. 1–79). Reading, MA: Addison-Wesley.

Asch, S. E. (1946). Forming impressions of personality. *Journal of Abnormal and Social Psychology, 41,* 258–290.

Asch, S. E. (1955). Opinions and social pressure. *Scientific American, 193,* 31–35.

Ayllon, T., & Azrin, N. H. (1968). *The token economy: A motivational system for therapy and rehabilitation*. New York, NY: Appleton-Century-Crofts.

Banaji, M. R., & Crowder, R. G. (1989). The bankruptcy of everyday memory. *American Psychologist, 44,* 1185–1193.

Banaji, M. R., & Crowder, R. G. (1991). Some everyday thoughts on ecologically valid methods. *American Psychologist, 46,* 78–79.

Baron, R. M., & Kenny, D. A. (1986). The moderator-mediator distinction in social psychological research: Conceptual, strategic, and statistical considerations. *Journal of Personality and Social Psychology, 51,* 1173–1182.

Basson, R., McInnes, R., Smith, M., Hodgson, G., & Koppiker, N. (2002). Efficacy and safety of sildenafil citrate in women with sexual dysfunction associated with female sexual arousal disorder. *Journal of Women's Health and Gender Based Medicine, 11,* 331–333.

Batson, C. D., Kobrynowicz, D., Dinnerstein, J. L., Kampf, H. C., & Wilson, A. D. (1997). In a very different voice: Unmasking moral hypocrisy. *Journal of Personality and Social Psychology, 77,* 525–537.

Baumeister, R. F., DeWall, C. N., Ciarocco, N. J., & Twnege, J. M. (2005). Social exclusion impairs self-regulation. *Journal of Personality and Social Psychology, 88,* 589–604.

Begley, S. (2007, May 7). Just say no—to bad science. *Newsweek,* p. 57.

Begley, S. (2007, June 18). Get shrunk at your own risk. *Newsweek,* p. 49.

Beilock, S. L., & Carr, T. H. (2005). When high-powered people fail: Working memory and choking under pressure in math. *Psychological Science, 16,* 101–105.

Beins, B. C. (1993). Using the Barnum effect to teach about ethics and deception in research. *Teaching of Psychology, 20,* 30–35.

Benjamin, L. T., Jr., & Baker, D. B. (2004). *From séance to science.* Belmont, CA: Wadsworth/Thomson.

Berkowitz, L. (1981, June). How guns control us. *Psychology Today,* 11–12.

Berlyne, D. E. (1971). *Conflict, arousal, and curiosity.* New York, NY: McGraw-Hill.

Bernhardt, P. C., Dabbs, J. M., Jr., Fielden, J. A., & Lutter, C. D. (1998). Testosterone changes during vicarious experiences of winning and losing among fans at sporting events. *Physiology and Behavior, 65,* 59–62.

Bernieri, F. J. (1991). Interpersonal sensitivity in teaching interactions. *Personality and Social Psychology Bulletin, 17,* 98–103.

Berscheid, E., Dion, K., Walster, E., & Walster, G. W. (1971). Physical attractiveness and dating choice: A test of the matching hypothesis. *Journal of Experimental Social Psychology, 7,* 173–189.

Blough, D. S. (1957). Effect of lysergic acid diethylamide on absolute visual threshold in the pigeon. *Science, 126,* 304–305.

Blumberg, S. J., & Luke, J. V. (2008). Wireless substitution: Early release of estimates from the National Health Interview Survey, July–December 2007. National Center for Health Statistics. Retrieved May 13, 2008 from: http://www.cdc.gov/nchs/nhis.htm

Boese, A. (2007). *Elephants on acid and other bizarre experiments.* New York, NY: Harcourt.

Brackett, M.A. & Mayer, J.D. (2003). Convergent, discriminant, and incremental validity of competing measures of emotional intelligence. *Personality and Social Psychology Bulletin, 29,* 1147–1158.

Brady, J. V. (1958). Ulcers in executive monkeys. *Scientific American, 199,* 95–100.

Brennan, C. (2005, March 24). E-mail surveys may be first step in effort to cripple progress of Title IX. *USA Today,* p. D5.

Brescoll, V., & LaFrance, M. (2004). The correlates and consequences of newspaper reports of research on sex differences. *Psychological Science, 15,* 515–520.

Brewer, C. L. (1990). *Teaching research methods: Three decades of pleasure and pain.* Presentation at the 98th Annual Convention of the American Psychological Association, Boston, MA.

Briggs, C. S. (2001). *Supermaximum security prisons and institutional violence: An impact assessment.* Unpublished master's thesis. Southern Illinois University, Carbondale, Illinois.

Broad, W. J., & Wade, N. (1982). Science's faulty fraud detectors. *Psychology Today, 16,* 50–57.

Brown, J. D. (1991). Staying fit and staying well: Physical fitness as a moderator of life stress. *Journal of Personality and Social Psychology, 60,* 555–561.

Buchanan, M. (2007). *The social atom.* New York, NY: Bloomsbury.

Burger, J. (2007). Replicating Milgram. *APS Observer, 20*(12), 15–17.

Burke, J. (1978). *Connections.* Boston, MA: Little, Brown.

Burke, J. (1985). *The day the university changed.* Boston, MA: Little, Brown.

Bushman, B. J., & Anderson, C. A. (2001, June/July). Media violence and the American public: Scientific facts versus media misinformation. *American Psychologist, 56,* 477–489.

Buss, D. M. (1994). *The evolution of desire.* New York, NY: Basic Books.

Byrne, B. M. (2004, July). *A beginner's guide to structural equation modeling: Basic concepts and applications.* Workshop presented at the meeting of the American Psychological Association, Honolulu, HI.

Byrne, D. (1961). Interpersonal attraction and attitude similarity. *Journal of Abnormal and Social Psychology, 62,* 713–715.

Byrne, D. (1971). *The attraction paradigm.* New York, NY: Academic Press.

Cacioppo, J. T., Hawkley, L. C., & Berntson, G. G. (2004). The anatomy of loneliness. In J. B. Ruscher & E. Y. Hammer (Eds.), *Current directions in social psychology* (pp. 171–177). Upper Saddle River, NJ: Pearson.

Campbell, D. T., & Stanley, J. C. (1963). *Experimental and quasi-experimental designs for research.* Chicago, IL: Rand McNally.

Carbaugh, B. T., Schein, M. W., & Hale, E. B. (1962). Effects of morphological variations of chicken models on sexual responses of cocks. *Animal behaviour, 10,* 235–238.

Carrere, S., & Gottman, J. (1999). Predicting divorce among newlyweds from the first three minutes of marital conflict discussion. *Family Process, 38,* 292–301.

Carroll, R. T. (2003). *The skeptic's dictionary.* Hoboken, NJ: Wiley.

Centers for Disease Control (n.d.), *U.S. Public Health Service Syphilis Study* at Tuskegee. Retrieved June, 12, 2008 from http://www.cdc.gov/tuskegee/timeline.htm

Chapman, L. J., & Chapman, P. J. (1967). Genesis of popular but erroneous psychodiagnostic observations. *Journal of Abnormal Psychology, 72*, 193–204.

Cialdini, R. B. (2005, April). Don't throw in the towel: Use social influence research. *APS Observer, 18*, 33–34.

Clark, H. H. (1973). The languageas-fixed-effect fallacy: A critique of language statistics in psychological research. *Journal of Verbal Learning and Verbal Behavior, 12*, 335–359.

Clark, R. D., III, & Hatfield, E. (2003). Love in the afternoon. *Psychological Inquiry, 14*, 227–231.

Cohen, J. (1976). Random means random. *Journal of Verbal Learning and Verbal Behavior, 15*, 261–262.

Cohen, J. (1990). Things I have learned (so far). *American Psychologist, 45*, 1304–1312.

Cohen, J. (1994). The earth is round (*p* < .05). *American Psychologist, 49*, 997–1003.

Cohen, J., & Cohen, P. (1983). *Applied multiple regression/correlation analysis for the behavioral science.* Hillsdale, NJ: Erlbaum.

Coile, D. C., & Miller, N. E. (1984). How radical animal activists try to mislead humane people. *American Psychologist, 39*, 700–701.

Coleman, E. B. (1979). Generalization effects vs. random effects. *Journal of Verbal Learning and Verbal Behavior, 18*, 243–256.

Coles, C. D. (1993). Saying "goodbye" to the "crack baby." *Neurotoxicology and Teratology, 15*, 290–292.

Condry, J., & Condry, S. (1976). Sex differences: A study of the eye of the beholder. *Child Development, 47*, 812–819.

Cook, T. D., & Campbell, D. T. (1979). *Quasi-experimentation: Design and analysis for field settings.* Chicago, IL: Rand McNally.

Cooper, H. (2011). *Reporting research in psychology: How to meet journal article reporting standards.* Washington, DC: American Psychological Association.

Cronbach, L. J. (1957). The two disciplines of scientific psychology. *American Psychologist, 12*, 671–684.

Cumming, G. (2008). Replication and *p* intervals: *p* values predict the future only vaguely, but confidence intervals do much better. *Perspectives on Psychological Science, 3*(4), 286–300. doi:10.1111/j.1745-6924.2008.00079.x

Cumming, G., & Finch, S. (2005). Inference by eye: Confidence intervals and how to read pictures of data. *American Psychologist, 60*, 170–180.

Custer, S. (1985). *The impact of backward masking.* Presented at the Thirteenth Annual Western Pennsylvania Undergraduate Psychology Conference in Clarion, Pennsylvania.

Custers, R., & Aarts, H. (2010, July 2). The unconscious will: How the pursuit of goals operates outside of conscious awareness. *Science, 329*, 47. DOI:10.1126/science.1188595.

Dabbs, J. M. Jr., & Dabbs, M. G. (2000) *Heroes, rogues, and lovers: Testosterone and behavior.* New York, NY: McGraw-Hill.

Danner, D., Snowden, D., & Friesen, W. (2001). Positive emotions in early life and longevity: Findings from the nun study. *Journal of Personality and Social Psychology, 80*, 804–813.

Darley, J., & Batson, D. (1973). From Jerusalem to Jericho: A study of situational and dispositional variables on helping behavior. *Journal of Personality and Social Psychology, 27*, 100–119.

Darley, J., & Latané, B. (1968). Bystander intervention in emergencies: Diffusion of responsibility. *Journal of Personality and Social Psychology, 8*, 377–383.

Davis, D., Shaver, P. R., & Vernon, M. L. (2004). Attachment style and subjective motivations for sex. *Personality and Social Psychology Bulletin, 30*, 1076–1090.

Dawes, R. M. (1994). *A house of cards: Psychology and psychotherapy built on myth.* New York, NY: Free Press.

Dawkins, R. (1998). *Unweaving the rainbow: Science, delusion, and the appetite for wonder.* Boston, MA: Houghton Mifflin.

De Leeuw, E. (1992). *Data quality in mail, telephone, and face-to-face surveys.* Amsterdam: TT Publications.

Dermer, M. L., & Hoch, T. A. (1999). Improving descriptions of single subject experiments in research texts written for undergraduates. *The Psychological Record, 49*, 49–66.

DeWall, C. N., & Baumeister, R. F. (2007). From terror to joy: Automatic tuning to positive affective information following mortality salience. *Psychological Science, 18*, 984–990.

Diener, E., & Diener, C. (1996) Most people are happy. *Psychological Science, 3*, 181–185.

Diener, E., & Seligman, M. E. (2002). Very happy people. *Psychological Science, 13*, 81–84.

Diener, E., & Seligman, M. (2004). Beyond money: Toward an economy of well-being. *Psychological Science in the Public Interest, 5*(1) (November), 1–31.

Dillman, D. A. (1978). *Mail and telephone surveys: The total design method.* New York, NY: Wiley.

Dillman, D. A. (2000). *Mail and internet surveys: The tailored design method.* New York, NY: Wiley.

Dillon, K. (1990). Generating research ideas; or, that's Salada tea…. *High School Psychology Teacher, 21*, 6–7.

Doherty, P. (2002). In M. Thomas (Ed.), *The right words at the right time* (p. 89). New York, NY: Atria Books.

Duffield, L. F. (2007, November/December). What is common about common sense? *Skeptical Inquirer*, pp. 62–63.

Edwards, T. (1990, July 23). Marketing grads told to take a reality check. *AMA News*, p. 9.

Ehrenberg, R. G., Brewer, D. J., Gamoran, A., & Williams, J. D. (2001). Class size and student achievement. *Psychological Science in the Public Interest, 2*, 1–30.

Elias, M. (2005, March 7). Study: A happy marriage can help mend physical wounds. *USA Today*, p. D7.

Elicker, J., Englund, M., & Sroufe, L. A. (1992). Predicting peer competence and peer relationships in childhood from early parent-child relationships. In R. D. Parke & G. W. Ladd (Eds.), *Family-peer relationships: Modes of linkage.* Hillsdale, NJ: Erlbaum.

Emmons, R. A., & McCullough, M. E. (2003). Counting blessings versus burdens: An experimental investigation of gratitude and subjective well-being in daily life. *Journal of Personality and Social Psychology, 84*, 377–389.

Estes, W. K. (1997). Significance testing in psychological research: Some persisting issues. *Psychological Science, 8*, 18–20.

Faurie, C., & Raymond, M. (2005). Handedness, homicide, and negative frequency-dependent selection. Proceedings of the Royal Society of London B.

Festinger, L., & Carlsmith, J. M. (1959). Cognitive consequences of forced compliance. *Journal of Abnormal and Social Psychology, 58*, 203–210.

Field, T. (1993). The therapeutic effects of touch. In G. C. Brannigan, & M. R. Merrens (Eds.), *The undaunted psychologist: Adventures in research*. New York, NY: McGraw-Hill.

Finkel, E. J., & Eastwick, P. W. (2008). Speed-dating. *Current Directions in Psychological Science, 17*, 193–197.

Fisher, R. A. (1938). Presidential Address to the First Indian Statistical Congress.

Fiske, D. W., & Fogg, L. (1990). But the reviewers are making different criticisms of my paper! Diversity and uniqueness in reviewer comments. *American Psychologist, 45*, 591–598.

Fitzsimons, G. M., & Kay, A. C. (2004). Language and interpersonal cognition: Causal effects of variations in pronoun usage on perceptions of closeness. *Personality and Social Psychology Bulletin, 30*, 547–557.

Forer, B. R. (1949). The fallacy of personal validation: A classroom demonstration of gullibility. *Journal of Abnormal and Social Psychology, 44*, 118–123.

Forsyth, D. (2004, Fall). IRBism: Prejudice against Institutional Review Boards. *Dialogue*, pp. 14–15, 20.

Forsyth, D. (2008, Spring). Defining deception as the "waiver of an element." *Dialogue*, p. 7.

Frank, M. G., & Gilovich, T. (1988). The dark side of self-and other perceptions: Black uniforms and aggression in professional sports. *Journal of Personality and Social Psychology, 54*, 74–85.

Franz, C. E., McClelland, D. C., & Weinberger, J. (1991). Childhood antecedents of conventional social accomplishment in midlife adults: A 36-year prospective study. *Journal of Personality and Social Psychology, 60*, 586–595.

Frederick, D. A., & Hazelton, M. G. (2007). Why is muscularity sexy? Tests of the fitness indicator hypothesis. *Personality and Social Psychology Bulletin, 33*, 1167–1183.

Frederickson, B. L., Roberts, T., Noll, S. M., Quinn, D. M., & Twenge, J. M. (1998). That swimsuit becomes you: Sex differences in self-objectification, restrained eating, and math performance. *Journal of Personality and Social Psychology, 75*, 269–284.

Frederickson, N. (1986). Toward a broader conception of human intelligence. *American Psychologist, 41*, 445–452.

Frijda, N. H. (1988). The laws of emotion. *American Psychologist, 43*, 349–358.

Gailliot, M. T., & Baumeister, R. F. (2007). The physiology of willpower: Linking blood glucose to self control. *Personality and Psychology Review, 11*, 303–327.

Garner, D. M., & Garfinkel, P. E. (1979). The eating attitudes test: An index of the symptoms of anorexia nervosa. *Psychological Medicine, 9*, 273–279.

Gazzaniga, M. S., & Heatherton, T. F. (2006). *Psychological science* (2nd ed.). New York, NY: Norton.

Gernsbacher, M. A. (2007, May). The value of undergraduate training in psychological science. *APS Observer*, p. 5, 13.

Gesn, P. R., & Ickes, W. (1999). The development meaning contexts of empathic accuracy: Channel and sequence effects. *Journal of Personality and Social Psychology, 77*, 746–761.

Gilovich, T., Vallone, R., & Tversky, A. (1985). The hot hand in basketball: On the misperception of random sequences. *Cognitive Psychology, 17*, 295–314.

Gladue, B. A., & Delaney, H. J. (1990). Gender differences in perception of attractiveness of men and women in bars. *Personality and Social Psychology Bulletin, 16*, 378–391.

Gladwell, M. (1996, July 8). Conquering the coma. *New Yorker*.

Gladwell, M. (2007, November 12). Dangerous minds: Criminal profiling made easy. *New Yorker*.

Glass, D. C., & Singer, J. E. (1972). *Urban stress: Experiments on noise and social stressors*. New York, NY: Academic Press.

Glick, P., Gottesman, D., & Jolton, J. (1989). The fault is not in the stars: Susceptibility of skeptics and believers in astrology to the Barnum effect. *Personality and Social Psychology Bulletin, 15*, 559–571.

Goldman, B. A., & Mitchell, D. F. (1990). *Directory of unpublished experimental mental measures* (Vol. 5). Dubuque, IA: Wm. C. Brown.

Goldman, B. A., & Mitchell, D. F. (2007). *Directory of unpublished experimental mental measures* (Vol. 9). Washington, DC: American psychological Association.

Gosling, S. D., Ko, S. J., Mannarelli, T., & Morris, M. E. (2002). A room with a cue: Personality judgments based on offices and bedrooms. *Journal of Personality and Social Psychology, 82*, 379–398.

Gosling, S. D., Vazire, S., Srivastava, S., & John, O. P. (2004). Should we trust web-based studies? A comparative analysis of six preconceptions about Internet questionnaires. *American Psychologist, 59*, 93–104.

Gottman, J. M. (1993). *What predicts divorce? The relationship between marital processes and marital outcomes*. Hillsdale, NJ: Erlbaum.

Greenberger, E., & Steinberg, L. (1986). *When teenagers work: The psychological and social costs of adolescent employment*. New York, NY: Basic Books.

Greenwald, A. G. (1975). Significance, nonsignificance, an interpretation of an ESP experiment. *Journal of Experimental Social Psychology, 11*, 180–191.

Greenwald, A. G. (1976). Withinsubjects designs: To use or not to use? *Psychological Bulletin, 83*, 314–320.

Greenwald, A. G., Gonzalez, R., Harris, R. J., & Guthrie, D. (1996). Effect sizes and p values: What should be reported and what should be replicated. *Psychophysiology, 33*, 175–183.

Greenwald, A. G., McGhee, D. E., & Schwartz (1998). Measuring individual differences in implicit cognition: The implicit association test. *Journal of Personality and Social Psychology, 74,* 1464–1480.

Groopman, J. (2004, January 26). The grief industry. *New Yorker,* p. 30.

Groves, R. M., & Kahn, R. L. (1979). *Surveys by telephone: A national comparison with personal interviews.* New York, NY: Academic Press.

Haack, S. (2004, July/August). Defending science—within reason: The critical common-sensist manifesto. *Skeptical Inquirer,* pp. 28–34.

Hadaway, C. K., Marler, P. L., & Chaves, M. (1993). What the polls don't show: A closer look at U.S. church attendance. *American Sociological Review, 58,* 741–752.

Hagemann, N., Strauss, B., & Leising, J. (2008. When the referee sees red. *Psychological Science, 19,* 769–771. doi:10.1111/j.1467-9280.2008. 02154.x

Hagen, R. L. (1998). A further look at wrong reasons to abandon statistical testing. *American Psychologist, 53,* 801–803.

Hagemann, N., Strauss, B., & Leising, J. (2008). When the referee sees red. *Psychological Science, 19,* 769–771.

Haidt, J. (2006). *The happiness hypothesis.* New York, NY: Basic books.

Harker, L., & Keltner, D. (2001). Expressions of positive emotion in women's college yearbook pictures and their relationship to personality and life outcomes across adulthood. *Journal of Personality and Social Psychology, 80,* 112–124.

Harlow, H. (1958). The nature of love. *American Psychologist, 13,* 673–685.

Harris, R. J. (1997). Significance tests have their place. *Psychological Science, 8,* 8–11.

Haselton, M. G., Buss, D. M., Oubaid, V., & Angleitner, A. (2005). Sex, lies, and strategic interference: The psychology of deception between the sexes. *Personality and Social Psychology Bulletin, 31,* 3–23.

Hays, W. L. (1981). *Statistics* (3rd ed.). New York, NY: Holt, Rinehart and Winston.

Hebl, M. R., King, E. B., & Lin, J. (2004). The swimsuit becomes us all: Ethnicity, gender, and vulnerability to self-objectification. *Personality and Social Psychology Bulletin, 30,* 1322–1331.

Hebl, M., & Mannix, L. (2003). The weight of obesity in evaluating others: A mere proximity effect. *Personality and Social Psychological Bulletin, 29,* 28–38.

Hedges, L. (1987). How hard is hard science, how soft is soft science? *American Psychologist, 42,* 443–455.

Helliwell, J. F. (2003). How's life? Combining individual and national variables to explain subjective wellbeing. *Economic Modeling, 20,* 331–360.

Henderson, M. A. (1985). *How con games work.* Secaucus, NJ: Citadel Press.

High-handed professor's comments called hot error. (1985, August). *USA Today,* p. C2.

Hill, C. A. (1991). Seeking emotional support: The influence of affiliative need and partner warmth. *Journal of Personality and Social Psychology, 60,* 112–121.

Hilton, J. L., & Fein, S. (1989). The role of typical diagnosticity on stereotype-based social judgments. *Journal of Personality and Social Psychology, 57,* 201–211.

Hilton, J. L., & von Hippel, W. (1990). The role of consistency in the judgment of stereotype-relevant behaviors. *Personality and Social Psychology Bulletin, 16,* 723–727.

Hirsh, J. B., DeYoung, C. G., Xu, X., & Peterson, J. B. (2010). Compassionate liberals and polite conservatives: Associations of agreeableness with political ideology and moral values. *Personality and Social Psychology Bulletin, 36,* 655–664.

Holmes, D. S., & Will, M. J. (1985). Expression of interpersonal aggression by angered and nonangered persons with Type A and Type B behavior patterns. *Journal of Personality and Social Psychology, 48,* 723–727.

Honomichl, J. (1990, August 6). Answering machines threaten survey research. *Marketing News,* p. 11.

Huberty, C. J., & Morris, J. D. (1989) Multivariate analysis versus multiple univariate analyses. *Psychological Bulletin, 105,* 302–308.

Ickes, W. (2003). *Everyday mind reading: Understanding what other people think and feel.* Amherst, NY: Prometheus Books.

Ickes, W., & Barnes, R. D. (1978). Boys and girls together and alienated: On enacting stereotyped sex roles in mixed-sex dyads. *Journal of Personality and Social Psychology, 36,* 669–683.

Ickes, W., Robertson, E., Tooke, W., & Teng, G. (1986). Naturalistic social cognition: Methodology, assessment, and validation. *Journal of Personality and Social Psychology, 51,* 66–82.

Injury quiets Rams. (1984, August 6). *USA Today,* p. C7.

Iyengar, S. S., & Lepper, M. R. (2000). When choice is demotivating: Can one desire too much of a good thing? *Journal of Personality and Social Psychology, 79,* 995–1006.

Jackson, J. M., & Padgett, V. (1982). With a little help from my friend: Social loafing and the Lennon-McCartney songs. *Personality and Social Psychology Bulletin, 8,* 672–677.

Jackson, J. M., & Williams, K. D. (1985). Social loafing on difficult tasks: Working collectively can improve performance. *Journal of Personality and Social Psychology, 49,* 937–942.

Jolley, J. M., Murray, J. D., & Keller, P. A. (1992). *How to write psychology papers: A student's survival guide for psychology and related fields.* Sarasota, FL: Professional Resource Exchange.

Jordan, C. H., & Zanna, M. P. (1999). How to read a journal article in social psychology. In R. F. Baumeister (Ed.), *The self in social psychology* (pp. 461–470). Philadelphia, PA: Psychology Press.

Kasser, T., & Ryan, R. M. (1993). The dark side of the American dream: Correlates of financial success as a central life aspiration. *Journal of Personality and Social Psychology, 65,* 410–422.

Kay, A. C., Jimenez, M. C., & Jost, J. T. (2002). Sour grapes, sweet lemons, and the rationalization of the status quo. *Personality and Social Psychology Bulletin, 28,* 1300–1312.

Keith, K. (2004, October). Presentation at the Teaching of Psychology Society Conference: "Finding Out: Best Practices in Teaching Research Methods and Statistics in Psychology," Atlanta, GA.

Kennedy, T. (1977, August 3). Meeting of the U.S. Senate Select Committee on Intelligence, and Subcommittee on Health and Scientific Research of the Committee on Human Resources. Retrieved June 12, 2008 from http://www.druglibrary.org/schaffer/history/e1950/mkultra/Hearing01.htm

Kenny, D. A., & Smith, E. R. (1980). A note on the analysis of designs in which subjects receive each stimulus only once. *Journal of Experimental Social Psychology*, 16, 497–507.

Keyser, D. J., & Sweetland, R. C. (Eds.). (1984). *Test critiques*. Kansas City, MO: Test Corporation of America.

Kiesler, C. A. (1982). Public and professional myths about mental hospitalization: An empirical reassessment of policy-related beliefs. *American Psychologist*, 37, 1323–1339.

Kiesler, C. A., & Sibulkin, A. E. (1987). *Mental hospitalization: Myths and facts about a national crisis*. Beverly Hills, CA: Sage.

Kimble, G. A. (1990). A search for principles in principles of psychology. *Psychological Science*, 1, 151–155.

Kincher, J. (1992). *The first honest book about lies*. Minneapolis, MN: Free Spirit.

Kirsch, I., Moore, T. J., Scoboria, A., & Nicholls, S. S. (2002). The emperor's new clothes: An analysis of antidepressant medication data submitted to the U.S. Food and Drug Administration. *Prevention & Treatment*, 5, Retrieved July 23, 2002, from http://journals.apa.org/prevention/volume5/toc-jul15-02.htm

Kitty, A. (2005). *Don't believe it: How lies become news*. New York, NY: The Disinformation Company.

Klarreich, E. (2004, December 4). Take a chance: Scientists put randomness to work. *Science News*, pp. 362–364.

Kline, R. B. (1998). *Principles and practice of structural equation modeling*. New York, NY: Guilford Press.

Klinesmith, J., Kasser, T., & McAndrew, F. T. (2006). Guns, testosterone, and aggression: An experimental test of a mediational hypothesis. *Psychological Science*, 17, 568–571.

Koens, F., Cate, O. T., & Custers, J. F. (2003). Context-dependent memory in a meaningful environment for medical education: In the classroom

and at the bedside. *Advances in Health Sciences Education*, 8(2), 155–163.

Kohlberg, L. (1981). *The meaning and measurement of moral development*. Worcester, MA: Clark University Press.

Kohn, A. (1988). You know what they say: Are proverbs nuggets of truth or fool's gold? *Psychology Today*, 22(4), 36–41.

Kolata, G. (2003). *Ultimate fitness: The quest for truth about exercise and health*. New York, NY: Picador.

Kolata, G. (2007). *Rethinking thin: The new science of weight loss—and the myths and realities of dieting*. New York, NY: Farrar, Straus and Giroux.

Kosko, B. (2002, June 7). Scientific jargon goes over heads of judges, jurors. *The Atlanta Journal-Constitution*, p. A23.

Kramer, J. J., & Conoley, J. C. (Eds.). (1992). *The eleventh mental measurements yearbook*. Lincoln, NE: Buros Institute of Mental Measurements.

Krosnick, J. A., & Schuman, H. (1988). Attitude intensity, importance, and certainty and susceptibility to response effects. *Journal of Personality and Social Psychology*, 54, 940–952.

Kuehn, S. A. (1989). *Prospectus handbook for Comm 352*. Unpublished manuscript. Landau, M. J., Solomon, S., Greenberg, J., Cohen, F., Pyszczynski, T., Arndt, J., Miller, C. H., Ogilivie, D. M., & Cook, A. (2004). Deliver us from evil: The effects of mortality salience and reminders of 9/11 on support for President George W. Bush. *Personality and Social Psychology Bulletin*, 30, 1136–1150.

Kunstman, J., & Maner, J. K. (2011). Sexual overperception: Power, mating goals, and biases in social judgment. *Journal of Personality and Social Psychology*, 100, 282–294. doi:10.1037/a0021135

Langer, E. J., Blank, A., & Chanowitz, B. (1978). The mindlessness of ostensibly thoughtful action: The role of "placebic" information in interpersonal interaction. *Journal of Personality and Social Psychology*, 36, 635–642.

Langer, E. J., & Rodin, J. T. (1976). The effects of choice and enhanced personal responsibility for the aged: A field experiment in an

institutional setting. *Journal of Personality and Social Psychology*, 34, 909–917.

Lardner, R. (1994, Feb. 21). Common nonsense: The age of reason (1974–1994). *The Nation*, 258, 232–234.

Larrick, R. P., Timmerman, T. A., Carton. A. M. , & Abrevaya, J. (2011). Temper, temperature, and temptation: Heat-related retaliation in baseball. *Psychological Science*, 22, 423–428. doi:10.1177/0956797611399292

Lassiter, G. D., Diamond, S. S., Schmidt, H. C., & Elek, J. K. (2007). Evaluating videotaped confessions: Expertise provides no defense against the cameraperspective effect. *Psychological Science*, 18, 224–225.

Latané, B., & Darley, J. M. (1968). Group inhibition of bystander intervention in emergencies. *Journal of Personality and Social Psychology*, 10, 215–221.

Latané, B., & Darley, J. M. (1970). *The unresponsive bystander: Why doesn't he help?* New York, NY: Appleton-Century-Crofts.

Latané, B., Williams, K., & Harkins, S. (1979). Many hands make light the work: The causes and consequences of social loafing. *Journal of Personality and Social Psychology*, 37, 822–832.

Lavrakas, P. J., Shuttles, C. D., Steeh, C., & Fienberg, H. (2007). The state of surveying cell phone numbers in the United States: 2007 and beyond. *Public Opinion Quarterly*, 71(5), 840–854. doi:10.1093/poq/ nfm054

Lawson, T. J. (1999). Assessing critical thinking as a learning outcome for psychology majors. *Teaching of Psychology*, 26, 207–209.

Lee, L., Frederick, S., & Ariely, D. (2006). Try it, you'll like it: The influence of expectation, consumption, and revelation on preferences for beer. *Psychological Science*, 17, 1054–1058.

Lehman, D. R., Lempert, R. O., & Nisbett, R. E. (1988). The effects of graduate training on reasoning: Formal discipline and thinking about everyday-life events. *American Psychologist*, 43, 431–442.

Lesko, W. A. (2009). *Readings in social psychology: General, classic, and contemporary selections* (7th ed.). New York, NY: Pearson.

Levitt, S. D., & Dubner, S. J. (2005). *Freakonomics: A rogue economist explores the hidden side of everything.* New York, NY: William Morrow.

Levy, D. A. (2010). *Tools of critical thinking: Metathoughts for psychology* (2nd ed.). Long Grove, IL: Waveland Press.

Levy, S. (2005, January 31). Does your iPod play favorites? *Newsweek,* p. 10.

Levy-Leboyer, C. (1988). Success and failure in applying psychology. *American Psychologist, 43,* 779–785.

Lewis, D., & Greene, J. (1982). *Thinking better.* New York, NY: Rawson, Wade.

Liberman, V., Samuels, S. M., & Ross, L. (2004). The name of the game: Predictive power of reputations versus situational labels in determining prisoner's dilemma game moves. *Personality and Social Psychology Bulletin, 30,* 1175–1185.

Lilienfeld, S. O. (2007). Psychological treatments that cause harm. *Perspectives on Psychological Science, 2,* 53–70.

Lilienfeld, S. O., Lynn, S. J., Namy, L. L., & Woolf, N. J. (2009). *Psychology: From inquiry to understanding.* New York, NY: Pearson.

Lin, H. Y. (2004, August). Responses on anonymous and computer administered survey: A good way to reduce social desirability effect. Poster session presented at the annual meeting of the American Psychological Association, Honolulu, HI.

Lippa, R. A. (2006). Is high sex drive associated with increased sexual attraction to both sexes? It depends on whether you are male or female. *Psychological Science, 17,* 46–52.

Loftus, E. F. (1975). Leading questions and the eyewitness report. *Cognitive Psychology, 7,* 560–572.

Luo, S., & Klohnen, C. C. (2005). Assortative mating and marital quality in newlyweds: A couplecentered approach. *Journal of Personality and Social Psychology, 88,* 304–326.

MacCallum, R. C., Zhang, S., Preacher, K. J., & Rucker, D. D. (2002). On the practice of dichotomization of quantitative variables. *Psychological Methods, 7*(1), 19–40.

Madey, S. F., Simo, M., Dillworth, D., Kemper, D., Toczynski, A., & Perella, A. (1996). They do get more attractive a closing time, but only when you are not in a relationship. *Basic and Applied Social Psychology, 18,* 387–393.

Maslow, A. H. (1970). Cited in S. Cunningham, Humanists celebrate gains, goals. *APA Monitor, 16,* 16.

Mayer, J. D., Salovey, P., & Caruso, D. (2002). *Mayer-Salovey-Caruso Emotional Intelligence Test (MSCEIT), Version 2.0.* Toronto, Canada: Multi-Health Systems.

McDougal, Y. B. (2007, July/August). Psychic events workshop fails APA curriculum requirement. *Skeptical Inquirer,* p. 9.

McNeil, J. M. & Fleeson, W. (2006). The causal effects of extraversion on positive affect and neuroticism on negative affect: Manipulating state extraversion and state neuroticism in an experimental approach. *Journal of Research in Personality, 40,* 529–550.

McNulty, J. K., & Karney, B. R. (2004). Positive expectations in the early years of marriage: Should couples expect the best or brace for the worst? *Journal of Personality and Social Psychology, 86,* 729–745.

Medvec, V. H., Madey, S. F., & Gilovich, T. (1995). When less is more: Counterfactual thinking and satisfaction among Olympic medalists. *Journal of Personality and Social Psychology, 69,* 603–610.

Mehl, M. R., Vazire, S., Ramirez-Esparza, N., Slatcher, R. B., & Pennebaker, J. W. (2007, July 6). Are women really more talkative than men? *Science, 317,* p. 82. doi:10.1126/science.1139940

Middlemist, R. D., Knowles, E. S., & Matter, C. F. (1976). Personal space invasions in the lavatory: Suggestive evidence for arousal. *Journal of Personality and Social Psychology, 33,* 541–546.

Milgram, S. (1974). *Obedience to authority: An experimental view.* New York, NY: Harper and Row.

Milgram, L., Bickman, L., & Berkowitz, L. (1969). Note on the drawing power of crowds of different size. *Journal of Personality and Social Psychology, 13,* 79–82.

Miller, E. T., Neal, D. J., Roberts, L. J., Baer, J. S., Cressler, S. O., Metrik, J., & Marlatt, G. A. (2002).

Test–retest reliability of alcohol measures: Is there a difference between Internet-based assessment and traditional methods? *Psychology of Addictive Behaviors, 16,* 56–63.

Miller, R. L., Wozniak, W. J., Rust, M. R., Miller, B. R., & Slezak, J. (1996). Counterattitudinal advocacy as a means of enhancing instructional effectiveness: How to teach students what they do not want to know. *Teaching of Psychology, 23,* 215–219.

Mitchell, J. V. (Ed.). (1983). *Tests in print III: An index to tests, test reviews, and the literature on specific tests.* Lincoln, NE: Buros Institute of Mental Measurements.

Mlodinow, L. (2008). The drunkard's walk: How randomness rules our lives. New York, NY: Pantheon.

Moerman, D. E. (2002). "The Loaves and the Fishes": A Comment on "The emperor's new drugs: An analysis of antidepressant medication data submitted to the U.S. Food and Drug Administration." *Prevention& Treatment, 5.* Retrieved July 23, 2002 from http://journals.apa.org/prevention/volume5/toc-jul15-02.htm

Mook, D. G. (1983). In defense of external invalidity. *American Psychologist, 38,* 379–387.

Morris, J. D. (1986). MTV in the classroom. *Chronicle of Higher Education, 32,* 25–26.

Moskalenko, S., & Heine, S. J. (2003). Watching your troubles away: Television viewing as a stimulus for subjective self-awareness. *Personality and Social Psychology Bulletin, 29,* 76–85.

Myers, D. G. (1999). *Social psychology* (6th ed.). New York, NY: McGraw-Hill.

Myers, D. G. (2002a). *Social psychology* (7th ed.). New York, NY: McGraw-Hill.

Myers, D. G. (2002b). *Intuition: Its powers and perils.* New Haven, CT: Yale University Press.

Myers, D. G. (2004). Exploring social psychology (4th ed.). New York, NY: McGraw-Hill.

Mynatt, C. R., & Doherty, M. E. (1999). *Understanding human behavior.* Needham Heights, MA: Allyn & Bacon.

Nairne, J. S., Thompson, S. R., & Pandeirada, J. N. (2007). Adaptive memory: Survival processing enhances retention. *Journal of Experimental Psychology: Learning, Memory, and Cognition, 33,* 263–273.

Neisser, U. (1984). *Ecological movement in cognitive psychology.* Invited address at the 92nd Annual Convention of the American Psychological Association in Toronto, Canada.

Nickell, J. (2005, May/June). Alleged Reagan astrologer dies. *Skeptical Inquirer, 29,* 8–9.

Nida, S. A., & Koon, J. (1983). They get better looking at closing time around here, too. *Psychological Reports, 15,* 258–264.

Nisbett, R. E., & Ross, L. (1980). *Human inference: Strategies and shortcomings of social judgment.* Englewood Cliffs, NJ: Prentice-Hall.

Nisbett, R. E., & Wilson, T. D. (1977). Telling more than we can know: Verbal reports on mental processes. *Psychological Review, 84,* 231–259.

Nosek, B. A., Bamaji, M. R., & Greenwald, A. G. (2002). E-Research: Ethics, security, design, and control in psychological research on the Internet. *Journal of Social Issues, 58,* 161–176.

Novella, S. (2007, November/ December). The anti-vaccination movement. *Skeptical Inquirer,* pp. 25–31.

Orne, M. (1962). On the social psychology of the psychological experiment: With particular reference to demand characteristics and their implications. *American Psychologist, 17,* 776–783.

Oishi, S., Diener, E., & Lucas, R.E. (2007). The optimum level of wellbeing: Can people be too happy? *Perspectives on Psychological Science, 2,* 346–360.

Omarzu, J. (2004, October). *Two birds, one stone: Asking students to create departmental surveys.* Poster presented at the conference Best Practices in Teaching Research Methods and Statistics, Atlanta, GA.

Padgett, V. R., & Jorgenson, D. O. (1982). Superstition and economic threat: Germany 1918–1940. *Personality and Social Psychology Bulletin, 8*(4), 736–741. doi:10.1177/0146167282084021

Painter, K. (2008, February 4). Alternative therapy: Healing or hooey? *USA Today,* p. 8D.

Park, R. L. (2000). *Voodoo science: The road from foolishness to fraud.* New York, NY: Oxford University Press.

Pennebaker, J. W., Dyer, M. A., Caulkins, R. S., Litowitz, D. L., Ackerman, P. L., Anderson, D. B., & McGraw, K. M. (1979). Don't the girls get prettier at closing time: A country and western application to psychology. *Personality and Social Psychology Bulletin, 5,* 122–125.

Peterson, M. (2008). *Our daily meds.* New York, NY: Farrar, Straus, and Giroux.

Pettijohn, T. F., & Jungeberg, B. J. (2004). *Playboy* playmate curves: Changes in facial and body feature preferences across social and economic conditions. *Personality and Social Psychology Bulletin, 30,* 1186–1197.

Pfungst, O. (1911). *Clever Hans.* New York, NY: Henry Holt.

Phillips, D. P. (1979). Suicide, motor vehicle fatalities, and the mass media: Evidence toward a theory of suggestion. *American Journal of Sociology, 84,* 1150–1174.

Pliner, P., Chaiken, S., & Flett, G. L. (1990). Gender differences in concern with body weight and physical appearance over the life span. *Personality and Social Psychology Bulletin, 16,* 263–273.

Plomin, R. (1993). Nature and nurture: Perspective and prospective. In R. Plomin, R. McClearn, & G. E. McClearn (Eds.), *Nature, nurture, and psychology.* Washington, DC: American Psychological Association.

Porter, T. M. (1997). *Trust in numbers: The pursuit of objectivity in science and public life.* Princeton, NJ: Princeton University Press.

Prentice, D. A., & Miller, D. T. (1992). When small effects are impressive. *Psychological Bulletin, 112,* 160–164.

Pronin, E., & Wegner, D. M. (2006). Manic thinking: Independent effects of thought speed and thought content on mood. *Psychological Science, 17,* 807–813.

Pronin, E., Wegner, D. M., McCarthy, K., & Rodriguez, S. (2006). Everyday magical powers: The role of apparent mental causation in the overestimation of personal influence. *Journal of Personality and Social Psychology, 91,* 218–231.

Provine, R. F. (2004). Laughing, tickling, and the evolution of speech and self. *Current Directions in Psychological Science, 13,* 215–218.

Radford, B. (2007, November/ December). Interview with Roy Richard Ginker. *Skeptical Inquirer,* pp. 36–38.

Ralof, J. (1998, September 19). The science of museums. *Science News,* 184–186.

Ranieri, D. J., & Zeiss, A. M. (1984). Induction of a depressed mood: A test of opponent-process theory. *Journal of Personality and Social Psychology, 47,* 1413–1422.

Revelle, W. (2007). Experimental approaches to the study of personality. In R. Robins, C. Fraley, & R. Krueger (Eds.), *Handbook of personality research methods* (pp. 37–61). New York, NY: Guilford.

Redelmeier, D. A., & Tibshirani, R. J. (1997). Association between cellular-telephone calls and motor vehicle collisions. *The New England Journal of Medicine, 336,* 453–458.

Reifman, A. S., Larick, R. P., & Fein, S. (1991). Temper and temperature on the diamond: The heataggression relationship in Major League Baseball. *Personality and Social Psychology Bulletin, 17,* 580–585.

Reis, H. T., & Stiller, J. (1992). Publication trends in JPSP: A threedecade review. *Personality and Social Psychology Bulletin, 18,* 465–472.

Richter, M. L., & Seay, M. B. (1987). ANOVA designs with subjects and stimuli and random effects: Applications to prototype effects on recognition memory. *Journal of Personality and Social Psychology, 53,* 470–480.

Rietzschel, E. F., De Dreu, C. K., & Nijstad, B. A. (2007). Personal need for structure and creative performance: The moderating influence of fear of invalidity. *Personality and Social Psychology Bulletin, 33,* 855–863.

Risen, J. L., & Gilovich, T. (2007). Another look at why people are reluctant to exchange lottery tickets. *Journal of Personality and Social Psychology, 93,* 12–22.

Robins, C. J. (1988). Attributions and depression: Why is the literature so inconsistent? *Journal of Personality and Social Psychology, 54,* 880–889.

Robinson, D. N. (Speaker). (1997). *The great ideas of psychology, part 1, minds possessed: Witchery and the search for explanations.* (Cassette Recording). Washington, DC: The Teaching Company.

Robinson, M. D., Vargas, P. T., Tamir, M., & Solberg, E. C. (2004). Using and being used by categories: The case of negative evaluations and daily well-being. *Psychological Science, 15,* 515–520.

Roediger, H. L., III, & Karpicke, J. D. (2006). Test-enhanced learning: Taking memory tests improves long-term retention. *Psychological Science, 17,* 249–255.

Roethlisberger, F. J., & Dickson, W. J. (1939). *Management and the worker.* Cambridge, MA: Harvard University Press.

Rogers, C. R. (1985). Cited in S. Cunningham, Humanists celebrate gains, goals. *APA Monitor, 16,* 16.

Rosa, L. R., Rosa, E. R., Sarner, L. S., & Barrett, S. B. (1998). Close look at therapeutic touch. *Journal of the American Medical Association, 279,* 1005–1010.

Rosenthal, R. (1966). *Experimenter effects in behavioral research.* New York, NY: Appleton-Century-Crofts.

Rosenthal, R. (1992, Fall). Computing contrasts: On sharpening psychological science. *Dialogue,* p. 3.

Rosenzweig, P. (2007). The halo effect … and eight other business delusions that deceive managers. New York, NY: Free Press.

Rowland, I. (2005). *The full facts book of cold reading* (4th ed.). Kent, England: Author.

Rubin, Z. (1970). Measurement of romantic love. *Journal of Personality and Social Psychology, 16,* 265–273.

Ruchlis, H., & Oddo, S. (1990). *Clear thinking: A practical introduction.* Buffalo, NY: Prometheus.

Rucker, D. D., & Petty, R. E. (2003). Effects of accusations on the accuser: The moderating role of accuser culpability. *Personality and Social Psychology Bulletin, 29,* 1259–1271.

Sagan, C. (1993). *Broca's brain.* New York, NY: Ballantine.

Sagaspe, P., Taillard, J., Chaumet, G., Moore, N., Bioulac, B., & Philip, P. (2007). Aging and nocturnal driving: Better with coffee or a nap? A randomized study. *Sleep, 30,* 1808–1813.

Sargent, M. J. (2004). Less thought, more punishment: Need for cognition predicts support for punitive responses to crime. *Personality and Social Psychology Bulletin, 30,* 1485–1493.

Saucier, D. A., & Miller, C. T. (2003). The persuasiveness of racial arguments as a subtle measure of racism. *Personality and Social Psychology Bulletin, 29,* 1303–1315.

Scarr, S. (1997). Rules of evidence: A larger context for the statistical debate. *Psychological Science, 8,* 16–17.

Schachter, D. L., Gilbert, D. T., & Wegner, D. M. (2009). *Psychology.* New York, NY: Worth.

Schachter, S. (1959). *The psychology of affiliation.* Stanford, CA: Stanford University Press.

Schmit, J. (2005, May 31). A winded FDA races to keep up with drug ads that go too far. *USA Today,* 1A, p. A4.

Schultz, D., & Schultz, S. E. (2006). *Psychology and work today: An introduction to industrial and organizational psychology* (9th ed.). Upper Saddle River, NJ: Pearson Prentice Hall.

Schwarz, N. (1999). Self-Reports: How the questions shape the answers. *American Psychologist, 54,* 93–105.

Schwarz, N., & Oyserman, D. (2001). Asking questions about behavior: Cognition, communication, and questionnaire construction. *American Journal of Evaluation, 22*(2), 127–160.

Scott, A. J. (2007, May/June). Danger! Scientific inquiry hazard. *Skeptical Inquirer, 31,* 40–45.

Seligman, C., & Sorrentino, R. M. (2002, Fall). The control agenda in Canada's governance of ethical review of human research. *Dialogue,* pp. 22–24.

Seligman, M. E. P. (1975). *Helplessness: On depression, development, and death.* San Francisco, CA: Freeman.

Seligman, M. E. P. (1990). *Learned optimism: How to change your mind and your life.* New York, NY: Pocket.

Seligman, M. E. P. (2002). *Authentic happiness.* New York, NY: Free Press.

Semon, T. (1990, April 16). Beware of bedazzling number mongers. *Marketing News,* p. 13.

Shedler, J., & Block, J. (1990). Adolescent drug use and psychological health: A longitudinal inquiry. *American Psychologist, 45,* 612–637.

Shefrin, H. M., & Statman, M. (1986). How not to make money in the stock market. *Psychology Today, 20,* 52–57.

Sherman, L. W., & Berk, R. A. (1984). The specific deterrent effects of arrest for domestic assault. *American Sociology Review, 49,* 262–272.

Sherman, S. J. (1980). On the selferasing nature of errors of prediction. *Journal of Personality and Social Psychology, 39,* 211–221.

Shermer, M. (2002, September). Smart people believe weird things. *Scientific American Magazine.* Retrieved January 9, 2008 from http://www. sciam.com/article.cfm? id=0002F4E6-8CF7-1D49- 90FB809EC5880000

Shorter, E. (1997). *A history of psychiatry.* New York, NY: Wiley.

Shrout, P. E. (1997). Should significance tests be banned? Introduction to a special section exploring the pros and cons. *Psychological Science, 8,* 1–2.

Sigall, H., & Mills, J. (1998). Measures of independent variables and mediators are useful in social psychology experiments: But are they necessary? *Personality and Social Psychology Review, 2,* 218–226.

Skinner, B. F. (1948). "Superstition" in the pigeon. *Journal of Experimental Psychology, 38,* 168–172.

Slovic, P. S., & Fischoff, B. (1977). On the psychology of experimental surprises. *Journal of Experimental Psychology: Human Perception and Performance, 3,* 455–471.

Snyder, M. (1984). When belief creates reality. In L. Berkowitz (Ed.), *Advances in experimental social psychology,* Vol. 18. New York, NY: Academic Press.

Specter, M. (2009). *Denialism: How irrational thinking harms the planet and threatens our lives.* New York, NY: Penguin Press.

Spencer, S. J., Zanna, M. P., & Fong, G. T. (2005). Establishing a causal chain: Why experiments are often more effective than mediational analyses in examining psychological processes. *Journal of Personality and Social Psychology, 89,* 845–851.

Stanovich, K. E. (2009). *How to think straight about psychology* (9th ed.). Glenview, IL: Scott, Foresman.

Steele, C. M., Southwick, L. L., & Critchlow, B. (1981). Dissonance and alcohol: Drinking your troubles away. *Journal of Personality and Social Psychology, 41,* 831–846.

Steinberg, L., & Dornbusch, S. M. (1991). Negative correlates of parttime employment during adolescence: Replication and elaboration. *Developmental Psychology, 27,* 304–313.

Stern, P. C. (1993). A second environmental science: Humanenvironmental interactions. *Science, 260,* 1997–1999.

Sternberg, R. J. (1986). *Intelligence applied: Understanding and increasing your intellectual skills.* New York, NY: Harcourt Brace Jovanovich.

Sternberg, R. J. (1994, Spring). Love is a story. *The General Psychologist, 30,* 1–11.

Stirman, S. W., & Pennebaker, J. W. (2001). Word use in the poetry of suicidal and non-suicidal poets. *Psychosomatic Medicine, 63,* 517–522.

Stone, J., Aronson, E., Crain, A. L., Winslow, M. P., & Fried, C. (1994). Introducing hypocrisy as a means of encouraging young adults to use condoms. *Personality and Social Psychology Bulletin, 20,* 116–128.

Stone, P. J., Dunphy, D. C., Smith, M. S., & Ogilvie, D. M. (1966). *The general inquirer: A computer approach to content analysis.* Cambridge, MA: MIT Press.

Strayer, D. L., & Drews, F. A. (2008). Cell-phone-induced driver distraction. *Current Directions in Psychological Science, 17,* 128–131.

Strayer, D. L., Drews, F. A., & Crouch, D. J. (2006). Comparing the cell-phone driver and the drunk driver. *Human Factors, 48,* 381–391.

Strayer, D. L., & Johnston, W. A. (2001). Driven to distraction: Dualtask studies of simulated driving and conversing on a cellular phone. *Psychological Science, 12,* 462–466.

Surowiecki, J. (2004). *The wisdom of crowds.* New York, NY: Doubleday.

Swann, W. B., Jr., & Rentfrow, P. J. (2001). Blirtatiousness: Cognitive, behavioral, and physiological consequences of rapid responding. *Journal of Personality and Social Psychology, 81,* 1160–1175.

Swets, J. A., & Bjork, R. A. (1990). Enhancing human performance: An evaluation of "new age" techniques considered by the U.S. Army. *Psychological Science, 1,* 85–96.

Tashiro, T., & Mortensen, L. (2006). Translational research: How social psychology can improve psychotherapy. *American Psychologist, 61,* 959–966.

Tavris, C., & Aronson, E. (2007). *Mistakes were made (but not by me).* New York, NY: Harcourt.

Tetlock, P.E. (2005). *Expert political judgment. How Good Is it? How can we know?* Princeton, NJ: Princeton University.

Thomas, G., Fletcher, G., & Lange, C. (1997). On-line empathic accuracy in marital interaction. *Journal of Personality and Social Psychology, 72,* 939–850.

Thomas, M. (Ed.). (2002). *The right words at the right time.* New York, NY: Atria Books.

Trzesniewski, K. H., Donnellan, M. B., & Lucas, R. E. (Eds.). (2011). *Secondary data analysis: An introduction for psychologists.* Washington, DC: American Psychological Association.

Tversky, B. (1973). Encoding processes in recognition and recall. *Cognitive Psychology, 5,* 275–287.

Twenge, J. M. (2002). The age of anxiety? The birth cohort change in anxiety and neuroticism, 1952–1993. *Journal of Personality and Social Psychology, 79,* 1007–1021.

Vohs, K. D., Mead, N. L., & Goode, M. R. (2008). Merely activating the concept of money changes personal and interpersonal behavior. *Current Directions in Psychological Science, 17,* 208–212.

Wadman, M. (2005, June 9). One in three scientists confesses to having sinned. *Nature, 435,* p. 718.

Wansink, B., van Ittersum, K., & Painter. J. E. (2005). How descriptive food names bias sensory perceptions in restaurants," *Food Quality and Preference, 16,* 393–400.

Ward, W. C., & Jenkins, H. M. (1965). The display of information and the judgment of contingency. *Canadian Journal of Psychology, 19,* 231–241.

Wedell, D. H., & Parducci, A. (1988). The category effect in social judgment: Experimental ratings of happiness. *Journal of Personality and Social Psychology, 55,* 341–356.

Wenger, D. M., Fuller, V. A., & Sparrow, B. (2003). Clever hands: Uncontrolled intelligence in facilitated communication. *Journal of Personality and Social Psychology, 85,* 5–19.

Weiten, W. (1992). *Psychology: Themes and variations.* Pacific Grove, CA: Brooks/Cole.

Whyte, J. (2005). *Crimes against logic.* New York, NY: McGraw-Hill.

Wickens, T. D., & Keppel, G. (1983). On the choice of design and of test statistic in the analysis of experiments with sampled materials. *Journal of Verbal Learning and Verbal Behavior, 22,* 296–309.

Wike, E. L., & Church, J. D. (1976). Comments on Clark's "The language-as-fixed-effect fallacy." *Journal of Verbal Learning and Verbal Behavior, 15,* 249–255.

Wilkinson, L., & The Task Force on Statistical Inference. (1999). Statistical methods in psychology journals: Guidelines and explanations. *American Psychologist, 54,* 594–604.

Williams, K. D., Nida, S. A., Baca, L. D., & Latané, B. (1989). Social loafing and swimming: Effects of identifiability on individual and relay performance of intercollegiate swimmers. *Basic and Applied Social Psychology, 10,* 73–81.

Williams, K. D., & Sommer, K. L. (1997). Social ostracism by one's coworkers: Does rejection lead to loafing or compensation? *Personality and Social Psychology Bulletin, 23,* 693–706.

Williams, R. L., & Long, J. D. (1983). *Toward a self-managed lifestyle* (3rd ed.). Boston, MA: Houghton-Mifflin.

Wilson, M., & Daly, M. (1985). Competitiveness, risk taking, and violence: The young male syndrome. *Ethology and Sociobiology, 6,* 59–73.

Wilson, T. D. (2002). *Strangers to ourselves: Discovering the adaptive unconscious*. Cambridge, MA: The Belknap Press of Harvard University Press.

Wilson, T. D., & Schooler, J. W. (1991). Thinking too much: Introspection can reduce the quality of preferences and decisions. *Journal of Personality and Social Psychology, 60*, 181–192.

Wohlford, P. (1970). Initiation of cigarette smoking: Is it related to parental smoking behavior? *Journal of Consulting and Clinical Psychology, 34*, 148–151.

Woods, N. S., Eyler, F. D., Conlon, M., Behnke, M., & Wobie, K. (1998). Pygmalion in the cradle: Observer bias against cocaine-exposed infants. *Developmental and Behavioral Pediatrics, 19*, 283–285.

Zajonc, R. B. (1965). Social facilitation. *Science, 149*, 269–274.

Zajonc, R. B. (1968). The attitudinal effects of mere exposure. *Journal of Personality and Social Psychology, 9*, 1–27.

Zajonc, R. B., & Sales, S. M. (1966). Social facilitation of dominant and subordinate responses. *Journal of Experimental Social Psychology, 2*, 160–168.

Zebrowitz, L. A., Montepare, J. M., & Lee, H. K. (1993). They don't all look alike: Individual impressions of other racial groups. *Journal of Personality and Social Psychology, 65*, 85–101.

Zimbardo, P., Haney, C., Banks, W. C., & Jaffe, D. (1975). The psychology of imprisonment: Privation, power, and pathology. In D. Rosenhan & P. London (Eds.), *Theory and research in abnormal psychology* (pp. 272–287). New York, NY: Holt, Rinehart and Winston.

Zuckerman, M. (1993). Out of sensory deprivation and into sensation seeking: A personal and scientific journey. In G. C. Brannigan & M. R. Merrens (Eds.), *The undaunted psychologist: Adventures in research*. New York, NY: McGraw-Hill.

INDEX